NFPA® 921

Guide for

Fire and Explosion Investigations

2011 Edition

This edition of NFPA 921, *Guide for Fire and Explosion Investigations,* was prepared by the Technical Committee on Fire Investigations. It was issued by the Standards Council on December 14, 2010, with an effective date of January 3, 2011, and supersedes all previous editions.

This edition of NFPA 921 was approved as an American National Standard on January 3, 2011.

Origin and Development of NFPA 921

NFPA 921, *Guide for Fire and Explosion Investigations,* was developed by the Technical Committee on Fire Investigations to assist in improving the fire investigation process and the quality of information on fires resulting from the investigative process. The guide is intended for use by both public sector employees who have statutory responsibility for fire investigation and private sector persons conducting investigations for insurance companies or litigation purposes. The goal of the committee is to provide guidance to investigators that is based on accepted scientific principles or scientific research.

The first edition of the document, issued by NFPA in 1992, focused largely on the determination of origin and cause of fires and explosions involving structures. The 1995 edition of the document included revised chapters on the collection and handling of physical evidence, safety, and explosions. NFPA 907M, *Manual for the Determination of Electrical Fire Causes,* was withdrawn as an individual document and was integrated with revisions into this document as a separate chapter. Elements of NFPA 907M that relate to other chapters of this document were relocated appropriately. New chapters dealing with the investigation of motor vehicle fires, management of major investigations, incendiary fires, and appliances were added.

The 1998 edition of the document included a new chapter on fuel gas systems in buildings and the impact of fuel gases on fire and explosion investigations. The chapter on electricity and fire was rewritten to improve organization, clarify terminology, and add references. In the chapter on fire patterns, several sections were revised. Other revisions were made in the chapter on physical evidence on the subject of preservation of the fire scene and of physical evidence. The edition also included new text regarding ignitible liquid detection canine/handler teams.

The 2001 edition of this document included new chapters on building systems, fire-related human behavior, failure analysis and analytical tools, fire and explosion deaths and injuries, and wildfire investigations. An updated chapter on motor vehicle fires was written. The document was organized to group chapters into subjects that made it more usable.

The 2004 edition of this document included a revision of the document to comply with the new *Manual of Style for NFPA Technical Committee Documents,* and a new chapter titled, "Analyzing the Incident for Cause and Responsibility," a rewrite of the chapter on Legal Considerations, and a revision of the chapter on Recording the Scene.

The 2008 edition of this document included rewrites of Chapter 5, Basic Fire Science; Chapter 6, Fire Patterns; Chapter 17, Origin Determination; Chapter 25, Motor Vehicle Fires; and Chapter 27, Management of Complex Investigations. A new Chapter 28, Marine Fire Investigations, was added to the document.

The seventh edition of this document includes important changes to Chapter 4, which has a new section on Report Review Procedures. A revision of Chapter 12, Safety, includes chemical and contamination exposure to the fire investigator. Chapter 18, Fire Cause Determination, was completely revised to be like Chapter 17, following the scientific method. This is also where the committee discusses that it is improper to base a hypothesis on the absence of any supportive evidence, also know as negative corpus. Chapter 21 on Explosions was completely revised with significantly more illustrations. Chapter 23 on Fire Deaths and Injuries was completely rewritten. Chapter 25 on Motor Vehicles was substantially expanded in sections addressing recreational vehicles and agricultural equipment. Chapter 26 was completely reorganized and rewritten with more photographs and illustrations.

The NFPA Technical Committee on Fire Investigations would like to dedicate the 2011 edition of NFPA 921 to the memory of our dear friend, colleague, and long-time staff liaison Frank Florence (1943–2010). Frank passed away on July 27, 2010, after a brief illness. Frank served with the Salt Lake City Fire Department for 31 years before retiring as Fire Chief. After retiring from the fire service, Frank joined NFPA in the Public Fire Protection Division. Since the 2001 edition of NFPA 921, Frank has served as our Staff Liaison. Frank was an invaluable resource to our committee and a strong supporter of the fire investigation profession. He will be dearly missed. Our thoughts and prayers continue to go out to his wife, Diane, and his sons, Robert and Randy.

Technical Committee on Fire Investigations

Charles R. Watson, *Chair*
SEA, Ltd., GA [SE]

Russell K. Chandler, *Secretary*
Virginia Department of Fire Programs, VA [E]

Vytenis Babrauskas, Fire Science and Technology Inc., WA [SE]
Michael Beasley, London Fire Brigade, United Kingdom [U]
Craig L. Beyler, Hughes Associates, Inc., MD [SE]
Steve Campolo, Leviton Manufacturing Company, Inc., NY [U]
Rep. National Electrical Manufacturers Association
Joseph Carey, Robinson & Cole LLP, CT [C]
Michael G. Chionchio, Delaware State Fire Marshal's Office, DE [E]
Rep. International Fire Marshals Association
Daniel L. Churchward, Kodiak Enterprises, Inc., IN [SE]
John Comery, U.S. Bureau of Alcohol, Tobacco, Firearms & Explosives, OR [E]
Philip E. Crombie, Jr., Travelers Insurance Company, CT [I]
Richard L. P. Custer, Arup Fire, MA [SE]
Richard A. Dyer, Kansas City Fire Department, MO [E]
Rep. International Association of Fire Chiefs
David S. Evinger, Robins, Kaplan, Miller & Ciresi L.L.P., MN [C]
James M. Finneran, ElectroTek Consultants, Inc., IN [SE]
Terry-Dawn Hewitt, McKenna Hewitt, CO [C]
Ronald L. Hopkins, Eastern Kentucky University, KY [U]
Rep. NFPA Fire Service Section
Thomas W. Horton, Jr., South Carolina Farm Bureau Insurance Company, SC [I]
David J. Icove, University of Tennessee, TN [U]
Patrick M. Kennedy, John A. Kennedy & Associates, Inc., FL [U]
Rep. National Association of Fire Investigators
Michael E. Knowlton, State of New York, NY [E]
John J. Lentini, Scientific Fire Analysis, LLC, FL [SE]
Rep. ASTM E30-Forensic Sciences
Jeffrey D. Long, Salt Lake City Fire Department, NM [U]
Daniel Madrzykowski, U.S. National Institute of Standards & Technology, MD [RT]
Ronald E. Orlando, General Motors Corporation, MI [M]
Rep. Society of Automotive Engineers
Edward S. Paulk, State of Alabama, AL [E]
Rep. National Association of State Fire Marshals
Richard J. Roby, Combustion Science & Engineering, Inc., MD [SE]
Joseph J. Sesniak, Forensic Fire Consultants, Ltd., AZ [U]
Rep. International Association of Arson Investigators, Inc.
Stuart A. Sklar, Fabian, Sklar and King, P.C., MI [C]
David M. Smith, Associated Fire Consultants, AZ [M]
Rep. International Fire Service Training Association
Michael E. Weyler, Weyler Engineering, PC, VA [SE]

Alternates

John G. Atherton, Burgoyne Incorporated, GA [U]
(Alt. to M. Beasley)
Robert P. Bailey Virginia Department of Fire Programs, VA [E]
(Alt. to R. K. Chandler)
Quentin A. Baker, Baker Engineering & Risk Consultants, Inc., TX [SE]
(Alt. to M. E. Weyler)
Robert D. Banta, Banta Technical Services LLC, MI [M]
(Alt. to R. E. Orlando)
Randall E. Bills, SEA, Ltd., OH [SE]
(Alt. to C. R. Watson)
Douglas Carpenter, Combustion Science & Engineering, Inc., MD [SE]
(Alt. to R. J. Roby)
Wayne Chapdelaine, Metro-Rural Fire Forensics, Canada [M]
(Alt. to D. M. Smith)
Michael DiMascio, Arup Fire, MA [SE]
(Alt. to R. L. P. Custer)
John J. Golder, U.S. Bureau of Alcohol, Tobacco, Firearms & Explosives, NC [E]
(Alt. to J. Comery)
Daniel T. Gottuk, Hughes Associates, Inc., MD [SE]
(Alt. to C. L. Beyler)
John H. Kane, Robinson & Cole LLP, CT [C]
(Alt. to J. Carey)
Kathryn C. Kennedy, John A. Kennedy & Associates, Inc., FL [U]
(Alt. to P. M. Kennedy)
Patrick A. King, Fabian, Sklar and King, P.C., MI [C]
(Alt. to S. A. Sklar)
Hal C. Lyson, Fire Cause Analysis, ND [C]
(Alt. to D. S. Evinger)
Wayne J. McKenna, McKenna Hewitt, CO [C]
(Alt. to T. D. Hewitt)
Rodney J. Pevytoe, Wisconsin Department of Justice, WI [U]
(Alt. to J. J. Sesniak)
Willard F. Preston, III, Goldfein & Joseph, PC, DE [E]
(Alt. to M. G. Chionchio)
James H. Shanley, Jr., Travelers Insurance Company, CT [I]
(Alt. to P. E. Crombie, Jr.)
Dennis W. Smith, Kodiak Enterprises, Inc., IN [SE]
(Alt. to D. L. Churchward)
Russell M. Whitney, Salt Lake City Fire Department, UT [U]
(Alt. to J. D. Long)

Committee Scope: This Committee shall have primary responsibility for documents relating to techniques to be used in investigating fires, and equipment and facilities designed to assist or be used in developing or verifying data needed by fire investigators in the determination of the origin and development of hostile fires.

Contents

NFPA 921

Guide for

Fire and Explosion Investigations

2011 Edition

IMPORTANT NOTE: This NFPA document is made available for use subject to important notices and legal disclaimers. These notices and disclaimers appear in all publications containing this document and may be found under the heading "Important Notices and Disclaimers Concerning NFPA Documents." They can also be obtained on request from NFPA or viewed at www.nfpa.org/disclaimers.

NOTICE: An asterisk (*) following the number or letter designating a paragraph indicates that explanatory material on the paragraph can be found in Annex A.

Changes other than editorial are indicated by a vertical rule beside the paragraph, table, or figure in which the change occurred. These rules are included as an aid to the user in identifying changes from the previous edition. Where one or more complete paragraphs have been deleted, the deletion is indicated by a bullet (•) between the paragraphs that remain.

A reference in brackets [ ] following a section or paragraph indicates material that has been extracted from another NFPA document. As an aid to the user, the complete title and edition of the source documents for extracts in advisory sections of this document are given in Chapter 2 and those for extracts in the informational sections are given in Annex C. Editorial changes to extracted material consist of revising references to an appropriate division in this document or the inclusion of the document number with the division number when the reference is to the original document. Requests for interpretations or revisions of extracted text should be sent to the technical committee responsible for the source document.

Information on referenced publications can be found in Chapter 2 and Annexes A and C.

Chapter 1 Administration

1.1 Scope. This document is designed to assist individuals who are charged with the responsibility of investigating and analyzing fire and explosion incidents and rendering opinions as to the origin, cause, responsibility, or prevention of such incidents, and the damage and injuries which arise from such incidents.

1.2 Purpose.

1.2.1 The purpose of this document is to establish guidelines and recommendations for the safe and systematic investigation or analysis of fire and explosion incidents. Fire investigation or analysis and the accurate listing of causes are fundamental to the protection of lives and property from the threat of hostile fire or explosions. It is through an efficient and accurate determination of the cause and responsibility that future fire incidents can be avoided. This document has been developed as a model for the advancement and practice of fire and explosion investigation, fire science, technology, and methodology.

1.2.2 Proper determination of fire origin and cause is also essential for the meaningful compilation of fire statistics. Accurate statistics form part of the basis of fire prevention codes, standards, and training.

1.3 Application. This document is designed to produce a systematic, working framework or outline by which effective fire and explosion investigation and origin and cause analysis can be accomplished. It contains specific procedures to assist in the investigation of fires and explosions. These procedures represent the judgment developed from the NFPA consensus process system that if followed can improve the probability of reaching sound conclusions. Deviations from these procedures, however, are not necessarily wrong or inferior but need to be justified.

1.3.1 The reader should note that frequently the phrase *fire investigation* is used in this document when the context indicates that the relevant text refers to the investigation of both fires and explosions.

1.3.2 As every fire and explosion incident is in some way unique and different from any other, this document is not designed to encompass all the necessary components of a complete investigation or analysis of any one case. The scientific method, however, should be applied in every instance.

1.3.3 Not every portion of this document may be applicable to every fire or explosion incident. It is up to investigators (depending on their responsibility, as well as the purpose and scope of their investigation) to apply the appropriate recommended procedures in this guide to a particular incident.

1.3.4 In addition, it is recognized that the extent of the fire investigator's assignment, time and resource limitations, or existing policies may limit the degree to which the recommendations or techniques in this document will be applied in a given investigation.

1.3.5 This document is not intended as a comprehensive scientific or engineering text. Although many scientific and engineering concepts are presented within the text, the user is cautioned that these concepts are presented at an elementary level and additional technical sources, training, and education may often need to be utilized in an investigation.

1.4* Units of Measure. Metric units of measurement in this guide are in accordance with the modernized metric system known as the International System of Units (SI). The unit of liter is outside of but recognized by SI and is commonly used in international fire protection. These units are listed in Table 1.4.

Table 1.4 SI Units and Equivalent U.S. Customary Units

SI	U.S.
2.54 cm	1 in.
0.3048 m	1 ft
0.09290 m^2	1 ft^2
28.32 L	1 ft^3
0.02832 m^3	1 ft^3
3.785 L	1 U.S. gal
0.4536 kg	1 lb
28.35 g	1 oz (weight)
0.3048 m/sec	1 ft/sec
16.02 kg/m^3	1 lb/ft^3
0.06308 L/sec	1 gpm
Pressure exerted by 760 mm of mercury of standard density at 0°C, 14.7 $lb/in.^2$ (101.3 kPa).	1 atmosphere
1.055 kW	1 Btu/sec
1055 J	1 Btu
0.949 Btu/sec	1 kW
248.8 Pa = 0.036 psi	1 in. w.c.
1 atmosphere	27.7 in. w.c.

Chapter 2 Referenced Publications

2.1 General. The documents or portions thereof listed in this chapter are referenced within this guide and shall be considered part of the requirements of this document.

2.2 NFPA Publications. National Fire Protection Association, 1 Batterymarch Park, Quincy, MA 02169-7471.

NFPA 30, *Flammable and Combustible Liquids Code,* 2008 edition.

NFPA 33, *Standard for Spray Application Using Flammable or Combustible Materials,* 2011 edition.

NFPA 45, *Standard on Fire Protection for Laboratories Using Chemicals,* 2011 edition.

NFPA 54, *National Fuel Gas Code,* 2009 edition.

NFPA 58, *Liquefied Petroleum Gas Code,* 2011 edition.

NFPA 68, *Standard on Explosion Protection by Deflagration Venting,* 2007 edition.

NFPA 70®, *National Electrical Code*®, 2011 edition.

FPA 72®, *National Fire Alarm and Signaling Code,* 2010 edition.

NFPA 77, *Recommended Practice on Static Electricity,* 2007 edition.

NFPA 120, *Standard for Fire Prevention and Control in Coal Mines,* 2010 edition.

NFPA 170, *Standard for Fire Safety and Emergency Symbols,* 2009 edition.

NFPA 220, *Standard on Types of Building Construction,* 2009 edition.

NFPA 302, *Fire Protection Standard for Pleasure and Commercial Motor Craft,* 2010 edition.

NFPA 303, *Fire Protection Standard for Marinas and Boatyards,* 2011 edition.

NFPA 400, *Hazardous Materials Code,* 2010 edition.

NFPA 501, *Standard on Manufactured Housing,* 2010 edition.

NFPA 555, *Guide on Methods for Evaluating Potential for Room Flashover,* 2009 edition.

NFPA 654, *Standard for the Prevention of Fire and Dust Explosions from the Manufacturing, Processing, and Handling of Combustible Particulate Solids,* 2006 edition.

NFPA 1192, *Standard on Recreational Vehicles,* 2011 edition.

NFPA 1194, *Standard for Recreational Vehicle Parks and Campgrounds,* 2011 edition.

NFPA 1144, *Standard for Reducing Structure Ignition Hazards from Wildland Fire,* 2008 edition.

NFPA 1403, *Standard on Live Fire Training Evolutions,* 2007 edition.

NFPA 1500, *Standard on Fire Department Occupational Safety and Health Program,* 2007 edition.

NFPA 1971, *Standard on Protective Ensembles for Structural Fire Fighting and Proximity Fire Fighting,* 2007 edition.

NFPA 1977, *Standard on Protective Clothing and Equipment for Wildland Fire Fighting,* 2011 edition.

Fire Protection Handbook, 15th (1981), 16th (1986), 17th (1991), 18th (1997), and 19th (2003) edition.

Fire Protection Guide to Hazardous Material, 12th edition, 1997. edition.

National Fuel Gas Code Handbook, 2002 edition.

The SFPE Engineering Guide to Human Behavior in Fire, 2003 edition.

The SFPE Handbook of Fire Protection Engineering, Society of Fire Protection Engineers, Quincy, MA, 2000 edition.

SPP 51 Flash Point Index of Trade Name Liquids, 1978 edition.

2.3 Other Publications.

2.3.1 ABYC Publications. American Boat and Yacht Council, 613 Third Street, Suite 10, Annapolis, MD 21403.

ABYC A-3, *Galley Stoves,* 2007.

ABYC A-7, *Boat Heating Systems,* 2006.

ABYC A-26, *LPG and CNG Fueled Appliances,* 2006.

ABYC A-30, *Cooking Appliances with Integral LPG Cylinders,* 2006.

ABYC H-24.13, *Gasoline Fuel Systems,* 2005.

ABYC H-32, *Ventilation of Boats Using Diesel Fuel,* 2007.

ABYC P-1, *Installation of Exhaust Systems,* 2002.

2.3.2 ANSI Publications. American National Standards Institute, Inc., 25 West 43rd Street, 4th Floor, New York, NY 10036.

ANSI Z129.1, *Precautionary Labeling of Hazardous Industrial Chemicals,* 2000.

ANSI Z400.1, *Material Safety Data Sheets — Preparation,* 1998.

ANSI Z535.1, *Safety Color Code,* 1998.

ANSI Z535.2, *Environmental and Facility Safety Signs,* 1998.

ANSI Z535.3, *Criteria for Safety Symbols,* 1998.

ANSI Z535.4, *Product Safety Signs and Labels,* 1998.

ANSI Z535.5, *Accident Prevention Tags,* 1998.

2.3.3 API Publications. American Petroleum Institute, 1220 L Street, NW, Washington, DC 20005-4070.

API/RP 2003, *Protection Against Ignitions Arising Out of Static, Lightning, and Stray Currents,* 1991.

API 2216, *Ignition Risk of Hydrocarbon Vapors by Hot Surfaces in the Open Air,* 2003.

2.3.4 ASME Publications. American Society of Mechanical Engineers, Three Park Avenue, New York, NY 10016-5990.

Boiler and Pressure Vessel Code.

2.3.5 ASTM Publications. ASTM International, 100 Barr Harbor Drive, P.O. Box C700, West Conshohocken, PA 19428-2959.

ASTM D 56, *Standard Test Method for Flash Point by Tag Closed Tester,* 2002.

ASTM D 86, *Standard Test Method for Distillation of Petroleum,* 2007.

ASTM D 92, *Standard Test Method for Flash and Fire Points by Cleveland Open Cup,* 2002.

ASTM D 93, *Standard Test Method for Flash Point by Pensky-Martens Closed Cup Tester,* 2002.

ASTM D 1230, *Standard Test Method for Flammability of Apparel Textiles,* 2001.

ASTM D 1265, *Standard Practice for Sampling Liquefied Petroleum (LP) Gases (Manual Method),* 2002.

ASTM D 1310, *Standard Test Method for Flash Point and Fire Point of Liquids by Tag Open-Cup Apparatus,* 2001.

ASTM D 1929, *Standard Test Method for Determining Ignition Temperature of Plastics,* 2001.

ASTM D 2859, *Standard Test Method for Flammability of Finished Textile Floor Covering Materials,* 1993.

ASTM D 2887, *Standard Test Method for Boiling Range Distribution of Petroleum Fractions by Gas Chromatography*, 2006.

ASTM D 3065, *Standard Test Methods for Flammability of Aerosol Products*, 2001.

ASTM D 3828, *Standard Test Methods for Flash Point by Small Scale Closed Tester*, 2002.

ASTM D 4809, *Standard Test Method for Heat of Combustion of Liquid Hydrocarbon Fuels by Bomb Calorimeter (Precision Method)*, 2000.

ASTM D 5305, *Standard Test Method for Determination of Ethyl Mercaptan in LP-Gas Vapor*, 1997.

ASTM E 84, *Standard Test Method for Surface Burning Characteristics of Building Materials*, 2003.

ASTM E 108, *Standard Test Method for Fire Tests of Roof Coverings*, 2000.

ASTM E 119, *Standard Methods of Tests of Fire Endurance of Building Construction and Materials*, 2000.

ASTM E 603, *Standard Guide for Room Fire Experiments*, 2001.

ASTM E 648, *Standard Test Method for Critical Radiant Flux of Floor-Covering Systems Using a Radiant Heat Energy Source*, 2000.

ASTM E 659, *Standard Test Method for Autoignition Temperature of Liquid Chemicals*, 2000.

ASTM E 681, *Standard Test Method for Concentration Limits of Flammability of Chemicals*, 2001.

ASTM E 800, *Standard Guide for Measurement of Gases Present or Generated During Fires*, 2001.

ASTM E 860, *Standard Practice for Examining and Testing Items That Are or May Become Involved in Litigation*, 1982.

ASTM E 906, *Standard Test Method for Heat and Visible Smoke Release Rates for Materials and Products*, 1999.

ASTM E 1188, *Standard Practice for Collection and Preservation of Information and Physical Items by a Technical Investigator*, 1995.

ASTM E 1226, *Test Method for Pressure and Rate of Pressure Rise for Combustible Dusts*, 2000.

ASTM E 1352, *Standard Test Method for Cigarette Ignition Resistance of Mock-up Upholstered Furniture Assemblies*, 2002.

ASTM E 1353, *Standard Test Methods for Cigarette Ignition Resistance of Components of Upholstered Furniture*, 2002.

ASTM E 1354, *Standard Test Method for Heat and Visible Smoke Release Rates for Materials and Products Using an Oxygen Consumption Calorimeter*, 2003.

ASTM E 1387, *Standard Test Method for Ignitible Liquid Residues in Extracts from Fire Debris Samples by Gas Chromatography*, 2001.

ASTM E 1459, *Standard Guide for Physical Evidence Labeling and Related Documentation*, 1998.

ASTM E 1618, *Standard Guide for Ignitible Liquid Residues in Extracts from Fire Debris Samples by Gas Chromatography–Mass Spectrometry*, 2001.

2.3.6 FMC Publications. FM Global, 1301 Atwood Avenue, Johnston, RI 02919.

FMC Product Safety Sign and Label System Manual, 1985.

2.3.7 Military Standards Publications. SAE, 1620 I Street, NW, Suite 210, Washington, DC 20006.

MIL-Std-202F, *Test Method for Electronic and Electrical Components*.

2.3.8 SAE International Publications. SAE International, 400 Commonwealth Drive, Warrendale, PA 15096-0001.

SAE J2578, *Recommended Practice for General Fuel Cell Vehicle Safety*, 2009.

2.3.9 UL Publications. Underwriters Laboratories Inc., 333 Pfingsten Road, Northbrook, IL 60062-2096.

ANSI/UL 263, *Standard for Safety Fire Tests of Building Construction and Materials*, 2003.

ANSI/UL 969, *Standard for Marking and Labeling Systems*, 1995, revised 2008.

UL 1500, *Standard for Safety Ignition Protection Test for Marine Products*, 1997, revised 2007.

2.3.10 USFA Publication. U.S. Fire Administration, 16825 S. Seton Avenue, Emmitsburg, MD 21727.

"Minimum Standards on Structural Fire Fighting Protective Clothing and Equipment," 1992.

2.3.11 U.S. Government Publications. U.S. Government Printing Office, Washington, DC 20402.

"Consumer Safety Act" (15 USC, Sections 2051–2084, and Title 16, Code of Federal Regulations, Part 1000).

"Federal Food, Drug and Cosmetic Act" [15 USC, Section 321 (m), and Title 21, Code of Federal Regulations, Part 600].

"Flammable Fabrics Act" (15 USC, Sections 1191–1204 and Title 16, Code of Federal Regulations, Parts 1615, 1616, and 1630–1632.

Hazardous Substances Act (15 USC, Section 1261 et seq., and Title 16, Code of Federal Regulations, Part 1500).

OSHA Regulations (Title 29, Code of Federal Regulations, Part 1910).

Title 24, Code of Federal Regulations, Part 3280, "Manufactured Home Construction and Safety Standards (HUD Standard.)"

Title 29, Code of Federal Regulations, Part 1910, "Federal Hazards Communication Standard."

Title 33, Code of Federal Regulations, Part 173, "Vessel Numbering and Casualty and Accident Reporting."

Title 33, Code of Federal Regulations, Part 181, "Manufacturer Requirements."

Title 33, Code of Federal Regulations, Part 183, "Boats and Associated Equipment."

Title 46, Code of Federal Regulations, Chapter 1, subchapter C, "Shipping."

Title 49, Code of Federal Regulations, Part 129.625, "Fire Related Human Behavior," U.S. Fire Administration, 257, 1994.

Title 49, Code of Federal Regulations, Part 173, "General Requirements for Shipments and Packagings."

Title 49, Code of Federal Regulations, Part 178, "Shipping Container Specifications."

Title 49, Code of Federal Regulations, Part 192, "Transportation of Natural and Other Gases by Pipeline Minimum Safety Standards."

Title 49, Code of Federal Regulations, Part 568, Vehicles Manufactured in Two or More Stages."

United States Federal Rules of Evidence as amended through 2002.

U.S. Senate Committee on Government Operations, *Chart of the Organization of Federal Executive Departments and Agencies.*

2.3.12 Other Publications. Babrauskas, V. *Ignition Handbook.* Issaquah, WA: Fire Science and Technology, Inc., 2003.

Baumeister, T., E. A. Avallone, and T. Baumeister III. *Mark's Standard Handbook for Mechanical Engineers,* 10th edition. New York, NY: McGraw-Hill, 1996.

Beyler, C. "Flammability Limits of Premixed and Diffusion Flames." In *SFPE Handbook of Fire Protection Engineering,* ed. P. DiNenno. Quincy, MA: National Fire Protection Association, 2002.

Braisie, N., and N. Simpson. "Guide for Estimating Damage," *Explosion Loss Prevention,* 1968.

Bull, J. P., and J. C. Lawrence. "Thermal Conditions to Produce Skin Burns," *Fire and Materials* 3(2) (1979): 100–05.

Bustin, W. M., and W. G. Duket. *Electrostatic Hazards in the Petroleum Industry.* London, UK: Research Studio Press, July 1983.

Cole, L. *The Investigation of Motor Vehicle Fires: A Guide for Law Enforcement, Fire Department and Insurance Personnel,* 3rd ed. Lincoln, NE: Lee Books, 1992.

Coltharp, D. R. "Blast Response Tests of Reinforced Concrete Box Structures," Department of Defense, 1983.

Crowl, D. A., and J. F. Louvar. Chemical Process Safety, 2nd ed. Englewood Cliffs, NJ: Prentice Hall, 2001.

Derkson, W. L., T. I. Monohan, and G. P. deLhery. "The Temperature Associated with Radiant Energy Skin Burns," *Temperature — Its Measurement and Control in Science and Industry* 3(3) (1963): 171–75.

Douglas, J. E., A. W. Burgess, and R. K. Ressler. *Crime Classification Manual.* New York, NY: Lexington Books, 1992.

Drysdale, D. *An Introduction to Fire Dynamics.* Chichester, UK: John Wiley and Sons, 1999.

Drysdale, D. "Fire Dynamics," *ISFI Proceedings, International Symposium on Fire Investigation Science and Technology.* Sarasota, FL: National Association of Fire Investigators, 2006.

Fang, J. B., and J. N. Breese, *Fire Development in Basement Rooms.* Gaithersburg, MD: National Institute of Standards and Technology, 1980.

Garner, B. A., and H. C. Black. *Black's Law Dictionary,* 7th ed. Saint Paul, MN: West Publishing Company, 1999.

Gieck, K., and R. Gieck. Engineering Formulas. New York, NY: McGraw-Hill, 1997.

Gottuck & White, *Liquid Fuel Fires SFPE Handbook of Fire Protection Engineering,* NFPA, 2002.

Grant, G., and D. Drysdale. "Numerical Modeling of Early Flame Spread in Warehouse Fires," *Fire Safety Journal* 24(3) (1995): 247–78.

Guide to Plastics (Plastics Handbook). New York, NY: McGraw-Hill, 1989.

Kennedy & Shanley, Report on the USFA Program for the Study of Fire Pattern, Interflam '96 Proceedings.

Hagglund, B., and S. Persson, *An Experimental Study of the Radiation from Wood Flames.* FOA Report C 4589-D6(A3). Stockholm, Sweden: Forsvarerts Forskningsanstalt, 1976.

Hilado, C. J. *Flammability Handbook for Plastics,* 4th ed. Lancaster, PA: Technomic Publishing, 1990.

Krasny, J. *Cigarette Ignition of Soft Furnishings — A Literature Review With Commentary.* Washington, DC: Center for Fire Research, National Bureau of Standards, June 1987.

Kransny, J., W. Parker, and V. Babrauskas. *Fire Behavior of Upholstered Furniture and Mattresses.* Park Ridge, NJ: Noyes Publications, 2001.

LaPointe, N., C. Adams, and J. Washington. "Autoignition of Gasoline on Hot Surfaces," *Fire and Arson Investigator,* 2005.

Lattimer, B. "Heat Fluxes from Fires to Surfaces," in *SFPE Handbook of Fire Protection Engineering,* ed. P. DiNenno. Quincy, MA: National Fire Protection Association, 2002.

Lawson, J. *An Evaluation of Fire Properties of Generic Gypsum Board Products* (NBSIR 77-1265). Washington, DC: NIST, Center for Fire Research, 1977.

Lee, B. T. *Heat Release Rate Characteristics of Some Combustible Fuel Sources in Nuclear Power Plants.*

Lees, F. *Loss Prevention in the Process Industries.* Boston, MA: Butterworth-Heinemann, 1996.

Lide, D. R., ed. *Handbook of Chemistry and Physics,* 71st ed. Boca Raton, FL: CRC Press, 1990–1991.

McGrattan, K., A. Hamins, and D. Stroup. *Sprinkler, Smoke and Heat Vent, Draft Curtain Interaction: Large Scale Experiments and Model Development.* Technical Report NISTIR 6196-1. Gaithersburg, MD: National Institute of Standards and Technology, 1998.

McRae, T. G., H. C. Goldwire, W. J. Hogan, and D. L. Morgan. "Effects of Large-Scale LNG/Water RPT Explosions," Department of Energy, 1984.

Merriam-Webster's Collegiate Dictionary, 11th edition, Merriam-Webster, Inc., Springfield, MA, 2003.

National Propane Gas Association Bulletin T133. *Purging LP-Gas Containers.* Washington, DC: NPGA, 1989.

Orloff, L., J. deRis, and G. Markstein. "Upward Turbulent Fire Spread and Burning of Fuel Surface," *Fifteenth Symposium (International) on Combustion.* Pittsburgh, PA: The Combustion Institute, 1994, pp. 183–92.

Quintiere, J. "Surface Flame Spread." In *SFPE Handbook of Fire Protection Engineering,* ed. P. DiNenno. Quincy, MA: National Fire Protection Association, 2002.

Saito, K., J. G. Quintiere, and F. A. Williams. "Upward Turbulent Flame Spread," *Fire Safety Science.* International Association for Fire Safety Science, 1986. *Proceedings, 1st International Symposium.* C. E. Grant and P. J. Pagni, eds. New York, NY: Hemisphere Publishing Corp., pp. 75–86.

Snyder, E. Health Hazard Evaluation Report 2004–0368–3030, Bureau of Alcohol, Tobacco, Firearms and Explosives, Austin, TX, January 2007.

Society of Fire Protection Engineers. *SFPE Handbook of Fire Protection Engineering,* ed. P. DiNenno. Quincy, MA: National Fire Protection Association, 2002.

Stoll, A., and L. C. Greene. "Relationship Between Pain and Tissue Damage Due to Thermal Radiation," *Journal of Applied Physiology* 14 (1959): 373–83.

Stoll, A., and M. A. Chianta. "Method and Rating System for Evaluation of Thermal Protection," *Aerospace Medicine* 40 (1969): 1232–38.

Thomas, P. "The Growth of Fire-Ignition to Full Involvement." In *Combustion Fundamentals of Fire,* ed. G. Cox. London, UK: Academic Press, 1995.

Wood, P. G. *Fire Research Note #953.* Borehamwood, UK: Building Research Establishment, 1973.

Wu, P., L. Orloff, and A. Tewarson. "Assessment of Material Flammability with the FG Propagation Model and Laboratory Test Methods," Thirteenth Joint Panel Meeting of the UJNR Panel on Fire Research and Safety, Gaithersburg, MD, 1996.

2.4 References for Extracts in Advisory Sections.

NFPA 53, *Recommended Practice on Materials, Equipment, and Systems Used in Oxygen-Enriched Atmospheres,* 2011 edition.

NFPA 68, *Standard on Explosion Protection by Deflagration Venting,* 2007 edition.

NFPA 70®, National Electrical Code®, 2011 edition.

NFPA 72®, National Fire Alarm and Signaling Code, 2010 edition.

NFPA 318, *Standard for the Protection of Semiconductor Fabrication Facilities,* 2009 edition.

NFPA 654, *Standard for the Prevention of Fire and Dust Explosions from the Manufacturing, Processing, and Handling of Combustible Particulate Solids,* 2006 edition.

Chapter 3 Definitions

3.1 General. The definitions contained in this chapter shall apply to the terms used in this guide. Where terms are not defined in this chapter or within another chapter, they shall be defined using their ordinarily accepted meanings within the context in which they are used. *Merriam-Webster's Collegiate Dictionary,* 11th edition, shall be the source for the ordinarily accepted meaning.

3.2 NFPA Official Definitions.

3.2.1* Approved. Acceptable to the authority having jurisdiction.

3.2.2* Code. A standard that is an extensive compilation of provisions covering broad subject matter or that is suitable for adoption into law independently of other codes and standards.

3.2.3* Guide. A document that is advisory or informative in nature and that contains only nonmandatory provisions. A guide may contain mandatory statements such as when a guide can be used, but the document as a whole is not suitable for adoption into law.

3.2.4* Recommended Practice. A document that is similar in content and structure to a code or standard but that contains only nonmandatory provisions using the word "should" to indicate recommendations in the body of the text.

3.2.5* Standard. A document, the main text of which contains only mandatory provisions using the word "shall" to indicate requirements and which is in a form generally suitable for mandatory reference by another standard or code or for adoption into law. Nonmandatory provisions shall be located in an appendix or annex, footnote, or fine-print note and are not to be considered a part of the requirements of a standard.

3.3 General Definitions.

3.3.1* Absolute Temperature. A temperature measured in Kelvins (K) or Rankines (R).

3.3.2 Accelerant. A fuel or oxidizer, often an ignitible liquid, used to initiate a fire or increase the rate of growth or spread of fire.

3.3.3 Accident. An unplanned event that interrupts an activity and sometimes causes injury or damage or a chance occurrence arising from unknown causes; an unexpected happening due to carelessness, ignorance, and the like.

3.3.4 Ambient. Someone's or something's surroundings, especially as they pertain to the local environment; for example, ambient air and ambient temperature.

3.3.5 Ampacity. The current, in amperes, that a conductor can carry continuously under the conditions of use without exceeding its temperature rating. [**70,** Article 100]

3.3.6 Ampere. The unit of electric current that is equivalent to a flow of one coulomb per second; one coulomb is defined as 6.24×10^{18} electrons.

3.3.7 Arc. A high-temperature luminous electric discharge across a gap or through a medium such as charred insulation.

3.3.8 Arcing Through Char. Arcing associated with a matrix of charred material (e.g., charred conductor insulation) that acts as a semiconductive medium.

3.3.9 Area of Origin. A structure, part of a structure, or general geographic location within a fire scene, in which the "*point of origin*" of a fire or explosion is reasonably believed to be located. *(See also 3.3.127, Point of Origin.)*

3.3.10 Arrow Pattern. A fire pattern displayed on the cross-section of a burned wooden structural member.

3.3.11* Arson. The crime of maliciously and intentionally, or recklessly, starting a fire or causing an explosion.

3.3.12 Autoignition. Initiation of combustion by heat but without a spark or flame.

3.3.13 Autoignition Temperature. The lowest temperature at which a combustible material ignites in air without a spark or flame.

3.3.14 Backdraft. A deflagration resulting from the sudden introduction of air into a confined space containing oxygen-deficient products of incomplete combustion.

3.3.15 Bead. A rounded globule of re-solidified metal at the end of the remains of an electrical conductor that was caused by arcing and is characterized by a sharp line of demarcation between the melted and unmelted conductor surfaces.

3.3.16 Blast Pressure Front. The expanding leading edge of an explosion reaction that separates a major difference in pressure between normal ambient pressure ahead of the front and potentially damaging high pressure at and behind the front.

3.3.17 BLEVE. Boiling liquid expanding vapor explosion.

3.3.18 Bonding. The permanent joining of metallic parts to form an electrically conductive path that ensures electrical continuity and the capacity to conduct safely any current likely to be imposed.

3.3.19 British Thermal Unit (Btu). The quantity of heat required to raise the temperature of one pound of water 1°F at the pressure of 1 atmosphere and temperature of 60°F; a British thermal unit is equal to 1055 joules, 1.055 kilojoules, and 252.15 calories.

3.3.20 Burning Rate. See 3.3.94, Heat Release Rate (HRR).

3.3.21 Calorie. The amount of heat necessary to raise 1 gram of water 1°C at the pressure of 1 atmosphere and temperature of 15°C; a calorie is 4.184 joules, and there are 252.15 calories in a British thermal unit (Btu).

3.3.22 Cause. The circumstances, conditions, or agencies that brought about or resulted in the fire or explosion incident, damage to property resulting from the fire or explosion incident, or bodily injury or loss of life resulting from the fire or explosion incident.

3.3.23 Ceiling Jet. A relatively thin layer of flowing hot gases that develops under a horizontal surface (e.g., ceiling) as a result of plume impingement and the flowing gas being forced to move horizontally.

3.3.24 Ceiling Layer. A buoyant layer of hot gases and smoke produced by a fire in a compartment.

3.3.25 Char. Carbonaceous material that has been burned or pyrolyzed and has a blackened appearance.

3.3.26 Char Blisters. Convex segments of carbonized material separated by cracks or crevasses that form on the surface of char, forming on materials such as wood as the result of pyrolysis or burning.

3.3.27 Clean Burn. A fire pattern on surfaces where soot has been burned away.

3.3.28* Combustible. Capable of undergoing combustion.

3.3.29* Combustible Gas Indicator. An instrument that samples air and indicates whether there are ignitible vapors or gases present.

3.3.30 Combustible Liquid. Any liquid that has a closed-cup flash point at or above 37.8°C (100°F). *(See also 3.3.74, Flammable Liquid.)*

3.3.31 Combustion. A chemical process of oxidation that occurs at a rate fast enough to produce heat and usually light in the form of either a glow or flame.

3.3.32 Combustion Products. The heat, gases, volatilized liquids and solids, particulate matter, and ash generated by combustion.

3.3.33 Competent Ignition Source. An ignition source that has sufficient energy and is capable of transferring that energy to the fuel long enough to raise the fuel to its ignition temperature. (See 18.4.2.)

3.3.34 Conduction. Heat transfer to another body or within a body by direct contact.

3.3.35 Convection. Heat transfer by circulation within a medium such as a gas or a liquid.

3.3.36 Creep. The tendency of a material to move or deform permanently to relieve stresses.

3.3.37 Current. A flow of electric charge.

3.3.38 Deductive Reasoning. The process by which conclusions are drawn by logical inference from given premises.

3.3.39 Deflagration. Propagation of a combustion zone at a velocity that is less than the speed of sound in the unreacted medium. [**68,** 2007]

3.3.40 Density. The weight of a substance per unit volume, usually specified at standard temperature and pressure. The density of water is approximately 1 gram per cubic centimeter. The density of air is approximately 1.275 grams per cubic meter.

3.3.41 Detection. (1) Sensing the existence of a fire, especially by a detector from one or more products of the fire, such as smoke, heat, infrared radiation, and the like. (2) The act or process of discovering and locating a fire.

3.3.42 Detonation. Propagation of a combustion zone at a velocity that is greater than the speed of sound in the unreacted medium. [**68,** 2007]

3.3.43 Diffuse Fuel. A gas, vapor, dust, particulate, aerosol, mist, fog, or hybrid mixture of these, suspended in the atmosphere, which is capable of being ignited and propagating a flame front.

3.3.44 Diffusion Flame. A flame in which fuel and air mix or diffuse together at the region of combustion.

3.3.45 Drop Down. The spread of fire by the dropping or falling of burning materials. Synonymous with "fall down."

3.3.46 Effective Fire Temperatures. Temperatures reached in fires that produce physical effects that can be related to specific temperature ranges.

3.3.47 Electric Spark. A small, incandescent particle created by some arcs.

3.3.48 Entrainment. The process of air or gases being drawn into a fire, plume, or jet.

3.3.49 Explosion. The sudden conversion of potential energy (chemical or mechanical) into kinetic energy with the production and release of gases under pressure, or the release of gas under pressure. These high-pressure gases then do mechanical work such as moving, changing, or shattering nearby materials.

3.3.50 Explosive. Any chemical compound, mixture, or device that functions by explosion.

3.3.51 Explosive Material. Any material that can act as fuel for an explosion.

3.3.52 Exposed Surface. The side of a structural assembly or object that is directly exposed to the fire.

3.3.53 Extinguish. To cause to cease burning.

3.3.54 Failure. Distortion, breakage, deterioration, or other fault in an item, component, system, assembly, or structure that results in unsatisfactory performance of the function for which it was designed.

3.3.55 Failure Analysis. A logical, systematic examination of an item, component, assembly, or structure and its place and function within a system, conducted in order to identify and analyze the probability, causes, and consequences of potential and real failures.

3.3.56 Fall Down. See 3.3.45, Drop Down.

3.3.57 Finish Rating. The time in minutes, determined under specific laboratory conditions, at which the stud or joist in contact with the exposed protective membrane in a protected combustible assembly reaches an average temperature rise of 121°C (250°F) or an individual temperature rise of 163°C (325°F) as measured behind the protective membrane nearest the fire on the plane of the wood.

3.3.58 Fire. A rapid oxidation process, which is a chemical reaction resulting in the evolution of light and heat in varying intensities.

3.3.59 Fire Analysis. The process of determining the origin, cause, development, responsibility, and, when required, a failure analysis of a fire or explosion.

3.3.60 Fire Cause. The circumstances, conditions, or agencies that bring together a fuel, ignition source, and oxidizer (such as air or oxygen) resulting in a fire or a combustion explosion.

3.3.61* Fire Dynamics. The detailed study of how chemistry, fire science, and the engineering disciplines of fluid mechanics and heat transfer interact to influence fire behavior.

3.3.62 Fire Investigation. The process of determining the origin, cause, and development of a fire or explosion.

3.3.63 Fire Hazard. Any situation, process, material, or condition that can cause a fire or explosion or that can provide a ready fuel supply to augment the spread or intensity of a fire or explosion, all of which pose a threat to life or property.

3.3.64 Fire Patterns. The visible or measurable physical changes, or identifiable shapes, formed by a fire effect or group of fire effects.

3.3.65 Fire Propagation. See 3.3.68, Fire Spread.

3.3.66 Fire Scene Reconstruction. The process of recreating the physical scene during fire scene analysis investigation or through the removal of debris and the placement of contents or structural elements in their pre-fire positions.

3.3.67* Fire Science. The body of knowledge concerning the study of fire and related subjects (such as combustion, flame, products of combustion, heat release, heat transfer, fire and explosion chemistry, fire and explosion dynamics, thermodynamics, kinetics, fluid mechanics, fire safety) and their interaction with people, structures, and the environment.

3.3.68 Fire Spread. The movement of fire from one place to another.

3.3.69 Flame. A body or stream of gaseous material involved in the combustion process and emitting radiant energy at specific wavelength bands determined by the combustion chemistry of the fuel. In most cases, some portion of the emitted radiant energy is visible to the human eye. [**72**, 2007]

3.3.70 Flame Front. The flaming leading edge of a propagating combustion reaction zone.

3.3.71 Flameover. The condition where unburned fuel (pyrolysate) from the originating fire has accumulated in the ceiling layer to a sufficient concentration (i.e., at or above the lower flammable limit) that it ignites and burns; can occur without ignition of, or prior to, the ignition of other fuels separate from the origin.

3.3.72 Flammable. Capable of burning with a flame.

3.3.73 Flammable Limit. The upper or lower concentration limit at a specified temperature and pressure of a flammable gas or a vapor of an ignitible liquid and air, expressed as a percentage of fuel by volume that can be ignited.

3.3.74 Flammable Liquid. A liquid that has a closed-cup flash point that is below 37.8°C (100°F) and a maximum vapor pressure of 2068 mm Hg (40 psia) at 37.8°C (100°F). *(See also 3.3.30, Combustible Liquid.)*

3.3.75 Flammable Range. The range of concentrations between the lower and upper flammable limits. [**68**, 2007]

3.3.76 Flash Fire. A fire that spreads by means of a flame front rapidly through a diffuse fuel, such as dust, gas, or the vapors of an ignitible liquid, without the production of damaging pressure.

3.3.77 Flash Point of a Liquid. The lowest temperature of a liquid, as determined by specific laboratory tests, at which the liquid gives off vapors at a sufficient rate to support a momentary flame across its surface.

3.3.78 Flashover. A transition phase in the development of a compartment fire in which surfaces exposed to thermal radiation reach ignition temperature more or less simultaneously and fire spreads rapidly throughout the space, resulting in full room involvement or total involvement of the compartment or enclosed space.

3.3.79 Forensic (Forensic Science). The application of science to answer questions of interest to the legal system.

3.3.80 Fuel. A material that will maintain combustion under specified environmental conditions. [**53**, 2004]

3.3.81 Fuel Gas. Natural gas, manufactured gas, LP-Gas, and similar gases commonly used for commercial or residential purposes such as heating, cooling, or cooking.

3.3.82 Fuel Load. The total quantity of combustible contents of a building, space, or fire area, including interior finish and trim, expressed in heat units or the equivalent weight in wood.

3.3.83 Fuel-Controlled Fire. A fire in which the heat release rate and growth rate are controlled by the characteristics of the fuel, such as quantity and geometry, and in which adequate air for combustion is available.

3.3.84 Full Room Involvement. Condition in a compartment fire in which the entire volume is involved in fire.

3.3.85 Gas. The physical state of a substance that has no shape or volume of its own and will expand to take the shape and volume of the container or enclosure it occupies.

3.3.86 Glowing Combustion. Luminous burning of solid material without a visible flame.

3.3.87 Ground. A conducting connection, whether intentional or accidental, between an electrical circuit or equipment and earth or to some conducting body that serves in place of the earth.

3.3.88 Ground Fault. An unintended current that flows outside the normal circuit path, such as (a) through the equipment grounding conductor; (b) through conductive material in contact with lower potential (such as earth), other than the electrical system ground (metal water or plumbing pipes, etc.); or (c) through a combination of these ground return paths.

3.3.89 Hazard. Any arrangement of materials that presents the potential for harm.

3.3.90* Heat. A form of energy characterized by vibration of molecules and capable of initiating and supporting chemical changes and changes of state.

3.3.91 Heat and Flame Vector. An arrow used in a fire scene drawing to show the direction of heat, smoke, or flame flow.

3.3.92 Heat Flux. The measure of the rate of heat transfer to a surface, expressed in kilowatts/m^2, kilojoules/$m^2 \cdot$ sec, or Btu/$ft^2 \cdot$ sec.

3.3.93* Heat of Ignition. The heat energy that brings about ignition.

3.3.94* Heat Release Rate (HRR). The rate at which heat energy is generated by burning.

3.3.95 High Explosive. A material that is capable of sustaining a reaction front that moves through the unreacted material at a speed equal to or greater than that of sound in that medium [typically 1000 m/sec (3000 ft/sec)]; a material capable of sustaining a detonation. *(See also 3.3.43, Detonation.)*

3.3.96 High-Order Damage. A rapid pressure rise or high-force explosion characterized by a shattering effect on the confining structure or container and long missile distances.

3.3.97 Hypergolic Material. Any substance that will spontaneously ignite or explode upon exposure to an oxidizer.

3.3.98 Ignitible Liquid. Any liquid or the liquid phase of any material that is capable of fueling a fire, including a flammable liquid, combustible liquid, or any other material that can be liquefied and burned.

3.3.99 Ignition. The process of initiating self-sustained combustion.

3.3.100 Ignition Energy. The quantity of heat energy that should be absorbed by a substance to ignite and burn.

3.3.101* Ignition Temperature. Minimum temperature a substance should attain in order to ignite under specific test conditions.

3.3.102 Ignition Time. The time between the application of an ignition source to a material and the onset of self-sustained combustion.

3.3.103 Incendiary Fire Cause. A classification of the cause of a fire that is intentionally ignited under circumstances in which the person igniting the fire knows the fire should not be ignited.

3.3.104 Inductive Reasoning. The process by which a person starts from a particular experience and proceeds to generalizations. The process by which hypotheses are developed based upon observable or known facts and the training, experience, knowledge, and expertise of the observer.

3.3.105 Interested Party. Any person, entity, or organization, including their representatives, with statutory obligations or whose legal rights or interests may be affected by the investigation of a specific incident.

3.3.106 Investigation Site. For the purpose of Chapter 27, the terms "site" and "scene" will be jointly referred to as the "investigation site," unless the particular context requires the use of one or the other word.

3.3.107 Investigative Team. A group of individuals working on behalf of an interested party to conduct an investigation into the incident.

3.3.108 Isochar. A line on a diagram connecting points of equal char depth.

3.3.109 Joule. The preferred SI unit of heat, energy, or work. A joule is the heat produced when one ampere is passed through a resistance of one ohm for one second, or it is the work required to move a distance of one meter against a force of one newton. There are 4.184 joules in a calorie, and 1055 joules in a British thermal unit (Btu). A watt is a joule/second. *[See also 3.3.19, British Thermal Unit (Btu), and 3.3.21, Calorie.]*

3.3.110 Kilowatt. A measurement of energy release rate.

3.3.111 Kindling Temperature. See 3.3.101, Ignition Temperature.

3.3.112 Layering. The systematic process of removing debris from the top down and observing the relative location of artifacts at the fire scene.

3.3.113 Low Explosive. An explosive that has a reaction velocity of less than 1000 m/sec (3000 ft/sec).

3.3.114 Low-Order Damage. A slow rate of pressure rise or low-force explosion characterized by a pushing or dislodging effect on the confining structure or container and by short missile distances.

3.3.115 Material First Ignited. The fuel that is first set on fire by the heat of ignition; to be meaningful, both a type of material and a form of material should be identified.

3.3.116* Noncombustible Material. A material that, in the form in which it is used and under the condition anticipated, will not ignite, burn, support combustion, or release flammable vapors when subjected to fire or heat.

3.3.117 Nonflammable. (1) Not readily capable of burning with a flame. (2) Not liable to ignite and burn when exposed to flame. Its antonym is *flammable.*

3.3.118 Ohm. The SI unit of electrical impedance or, in the direct current case, electrical resistance.

3.3.119 Origin. The general location where a fire or explosion began. *(See 3.3.127, Point of Origin, or 3.3.9, Area of Origin.)*

3.3.120 Overcurrent. Any current in excess of the rated current of equipment or the ampacity of a conductor; it may result from an overload *(see 3.3.122),* short circuit, or ground fault.

3.3.121 Overhaul. A fire fighting term involving the process of final extinguishment after the main body of the fire has been knocked down. All traces of fire must be extinguished at this time.

3.3.122* Overload. Operation of equipment in excess of normal, full-load rating or of a conductor in excess of rated ampacity that when it persists for a sufficient length of time would cause damage or dangerous overheating. An overload current is usually but might not always be confined to the normal intended conductive paths provided by conductors and other electrical components of an electrical circuit. Operation of the equipment or wiring under current flow conditions leading to temperatures in excess of the temperature rating of the equipment or wiring.

3.3.123 Oxygen Deficiency. Insufficiency of oxygen to support combustion. *(See also 3.3.181, Ventilation-Controlled Fire.)*

3.3.124 Piloted Ignition Temperature. See 3.3.101, Ignition Temperature.

3.3.125* Plastic. Any of a wide range of natural or synthetic organic materials of high molecular weight that can be formed by pressure, heat, extrusion, and other methods into desired shapes.

3.3.126 Plume. The column of hot gases, flames, and smoke rising above a fire; also called *convection column, thermal updraft,* or *thermal column.*

3.3.127 Point of Origin. The exact physical location within the area of origin where a heat source and the fuel interact, resulting in a fire or explosion.

3.3.128 Premixed Flame. A flame for which the fuel and oxidizer are mixed prior to combustion, as in a laboratory Bunsen burner or a gas cooking range; propagation of the flame is governed by the interaction between flow rate, transport processes, and chemical reaction.

3.3.129 Preservation. Application or use of measures to prevent damage, change or alteration, or deterioration.

3.3.130 Products of Combustion. See 3.3.32, Combustion Products.

3.3.131 Protocol. A description of the specific procedures and methodologies by which a task or tasks are to be accomplished.

3.3.132 Proximate Cause. The cause that directly produces the effect without the intervention of any other cause.

3.3.133 Pyrolysate. Product of decomposition through heat; a product of a chemical change caused by heating.

3.3.134 Pyrolysis. A process in which material is decomposed, or broken down, into simpler molecular compounds by the effects of heat alone; pyrolysis often precedes combustion.

3.3.135 Pyrophoric Material. Any substance that spontaneously ignites upon exposure to atmospheric oxygen.

3.3.136 Radiant Heat. Heat energy carried by electromagnetic waves that are longer than light waves and shorter than radio waves; radiant heat (electromagnetic radiation) increases the sensible temperature of any substance capable of absorbing the radiation, especially solid and opaque objects.

3.3.137 Radiation. Heat transfer by way of electromagnetic energy.

3.3.138 Rate of Heat Release. See 3.3.94, Heat Release Rate (HRR).

3.3.139 Rekindle. A return to flaming combustion after apparent but incomplete extinguishment.

3.3.140 Responsibility. The accountability of a person or other entity for the event or sequence of events that caused the fire or explosion, spread of the fire, bodily injuries, loss of life, or property damage.

3.3.141 Risk. The degree of peril; the possible harm that might occur that is represented by the statistical probability or quantitative estimate of the frequency or severity of injury or loss.

3.3.142 Rollover. See 3.3.71, Flameover.

3.3.143 Scene. The general physical location of a fire or explosion incident (geographic area, structure or portion of a structure, vehicle, boat, piece of equipment, etc.) designated as important to the investigation because it may contain physical damage or debris, evidence, victims, or incident-related hazards.

3.3.144 Scientific Method. The systematic pursuit of knowledge involving the recognition and formulation of a problem, the collection of data through observation and experiment, and the formulation and testing of a hypothesis.

3.3.145 Seat of Explosion. A craterlike indentation created at the point of origin of some explosions.

3.3.146 Seated Explosion. An explosion with a highly localized point of origin, such as a crater.

3.3.147 Secondary Explosion. Any subsequent explosion resulting from an initial explosion.

3.3.148 Self-Heating. The result of exothermic reactions, occurring spontaneously in some materials under certain conditions, whereby heat is generated at a rate sufficient to raise the temperature of the material.

3.3.149 Self-Ignition. Ignition resulting from self-heating, synonymous with *spontaneous ignition.*

3.3.150 Self-Ignition Temperature. The minimum temperature at which the self-heating properties of a material lead to ignition.

3.3.151 Short Circuit. An abnormal connection of low resistance between normal circuit conductors where the resistance is normally much greater; this is an overcurrent situation but it is not an overload.

3.3.152 Site. The general physical location of the incident, including the scene and the surrounding area deemed significant to the process of the investigation and support areas.

3.3.153 Smoke. The airborne solid and liquid particulates and gases evolved when a material undergoes pyrolysis or combustion, together with the quantity of air that is entrained or otherwise mixed into the mass. [**318**, 2006]

3.3.154 Smoke Condensate. The condensed residue of suspended vapors and liquid products of incomplete combustion.

3.3.155 Smoke Explosion. See 3.3.14, Backdraft.

3.3.156 Smoldering. Combustion without flame, usually with incandescence and smoke.

3.3.157 Soot. Black particles of carbon produced in a flame.

3.3.158 Spalling. Chipping or pitting of concrete or masonry surfaces.

3.3.159 Spark. A moving particle of solid material that emits radiant energy due either to its temperature or the process of combustion on its surface. [**654,** 2006]

3.3.160 Specific Gravity (air) (vapor density). The ratio of the average molecular weight of a gas or vapor to the average molecular weight of air.

3.3.161 Specific Gravity (of a liquid or solid). The ratio of the mass of a given volume of a substance to the mass of an equal volume of water at a temperature of 4°C.

3.3.162 Spoliation. Loss, destruction, or material alteration of an object or document that is evidence or potential evidence in a legal proceeding by one who has the responsibility for its preservation.

3.3.163* Spontaneous Heating. Process whereby a material increases in temperature without drawing heat from its surroundings.

3.3.164 Spontaneous Ignition. Initiation of combustion of a material by an internal chemical or biological reaction that has produced sufficient heat to ignite the material.

3.3.165 Suppression. The sum of all the work done to extinguish a fire, beginning at the time of its discovery.

3.3.166 Target Fuel. A fuel that is subject to ignition by thermal radiation such as from a flame or a hot gas layer.

3.3.167* Temperature. The degree of sensible heat of a body as measured by a thermometer or similar instrument.

3.3.168 Thermal Column. See 3.3.126, Plume.

3.3.169* Thermal Expansion. The increase in length, volume, or surface area of a body with rise in temperature.

3.3.170 Thermal Inertia. The properties of a material that characterize its rate of surface temperature rise when exposed to heat; related to the product of the material's thermal conductivity (k), its density (ρ), and its heat capacity (c).

3.3.171 Thermoplastic. Plastic materials that soften and melt under exposure to heat and can reach a flowable state.

3.3.172 Thermoset Plastics. Plastic materials that are hardened into a permanent shape in the manufacturing process and are not commonly subject to softening when heated; typically form char in a fire.

3.3.173 Time Line. Graphic representation of the events in a fire incident displayed in chronological order.

3.3.174 Total Burn. A fire scene where a fire continued to burn until most combustibles were consumed and the fire self extinguished due to a lack of fuel or was extinguished when the fuel load was reduced by burning and there was sufficient suppression agent application to extinguish the fire.

3.3.175 Understanding or Agreement. A written or oral consensus between the interested parties concerning the management of the investigations.

3.3.176 Upper Layer. See 3.3.24, Ceiling Layer.

3.3.177 Vapor. The gas phase of a substance, particularly of those that are normally liquids or solids at ordinary temperatures. *(See also 3.3.85, Gas.)*

3.3.178 Vapor Density. See 3.3.160, Specific Gravity (air) (vapor density).

3.3.179 Vent. An opening for the passage of, or dissipation of, fluids, such as gases, fumes, smoke, and the like.

3.3.180 Ventilation. Circulation of air in any space by natural wind or convection or by fans blowing air into or exhausting air out of a building; a fire-fighting operation of removing smoke and heat from the structure by opening windows and doors or making holes in the roof.

3.3.181 Ventilation-Controlled Fire. A fire in which the heat release rate or growth is controlled by the amount of air available to the fire.

3.3.182 Venting. The escape of smoke and heat through openings in a building.

3.3.183 Volt (V). The unit of electrical pressure (electromotive force) represented by the symbol "E"; the difference in potential required to make a current of one ampere flow through a resistance of one ohm.

3.3.184 Watt (W). Unit of power, or rate of work, equal to one joule per second, or the rate of work represented by a current of one ampere under the potential of one volt.

3.3.185 Work Plans. An outline of the tasks to be completed as part of the investigation including the order or timeline for completion. See Chapter 14, Planning the Investigation.

Chapter 4 Basic Methodology

4.1* Nature of Fire Investigations. A fire or explosion investigation is a complex endeavor involving skill, technology, knowledge, and science. The compilation of factual data, as well as an analysis of those facts, should be accomplished objectively, truthfully, and without expectation bias, preconception, or prejudice. The basic methodology of the fire investigation should rely on the use of a systematic approach and attention to all relevant details. The use of a systematic approach often will uncover new factual data for analysis, which may require previous conclusions to be reevaluated. With few exceptions, the proper methodology for a fire or explosion investigation is to first determine and establish the origin(s), then investigate the cause: circumstances, conditions, or agencies that brought the ignition source, fuel, and oxidant together.

4.2 Systematic Approach. The systematic approach recommended is that of the scientific method, which is used in the physical sciences. This method provides for the organizational and analytical process desirable and necessary in a successful fire investigation.

4.3 Relating Fire Investigation to the Scientific Method. The scientific method *(see Figure 4.3)* is a principle of inquiry that forms a basis for legitimate scientific and engineering processes, including fire incident investigation. It is applied using the following steps outlined in 4.3.2 through 4.3.9.

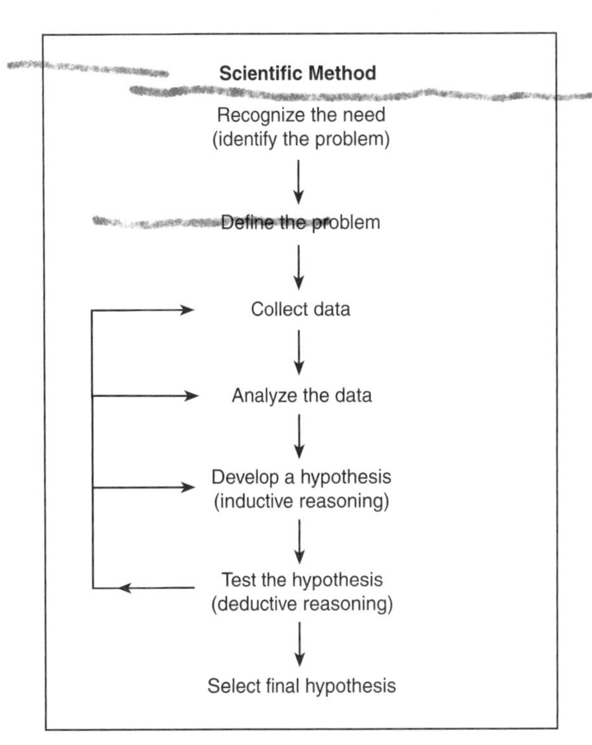

FIGURE 4.3 Use of the Scientific Method.

4.3.1 Recognize the Need. First, one should determine that a problem exists. In this case, a fire or explosion has occurred and the cause should be determined and listed so that future, similar incidents can be prevented.

4.3.2 Define the Problem. Having determined that a problem exists, the investigator or analyst should define the manner in which the problem can be solved. In this case, a proper origin and cause investigation should be conducted. This is done by an examination of the scene and by a combination of other data collection methods, such as the review of previously conducted investigations of the incident, the interviewing of witnesses or other knowledgeable persons, and the results of scientific testing.

4.3.3 Collect Data. Facts about the fire incident are now collected by observation, experiment, or other direct data-gathering means. The data collected is called empirical data because it is based on observation or experience and is capable of being verified or known to be true.

4.3.4* Analyze the Data. The scientific method requires that all data collected be analyzed. This is an essential step that must take place before the formation of the final hypothesis. The identification, gathering, and cataloging of data does not equate to data analysis. Analysis of the data is based on the knowledge, training, experience, and expertise of the individual doing the analysis. If the investigator lacks expertise to properly attribute meaning to a piece of data, then assistance should be sought. Understanding the meaning of the data will enable the investigator to form hypotheses based on the evidence, rather than on speculation.

4.3.5* Develop a Hypothesis (Inductive Reasoning). Based on the data analysis, the investigator produces a hypothesis, or hypotheses, to explain the phenomena, whether it be the nature of fire patterns, fire spread, identification of the origin, the ignition sequence, the fire cause, or the causes of damage or responsibility for the fire or explosion incident. This process is referred to as inductive reasoning. These hypotheses should be based solely on the empirical data that the investigator has collected through observation and then developed into explanations for the event, which are based upon the investigator's knowledge, training, experience, and expertise.

4.3.6* Test the Hypothesis (Deductive Reasoning). The investigator does not have a valid hypothesis unless it can stand the test of careful and serious challenge. Testing of the hypothesis is done by the principle of deductive reasoning, in which the investigator compares his or her hypothesis to all the known facts as well as the body of scientific knowledge associated with the phenomena relevant to the specific incident. A hypothesis can be tested either physically by conducting experiments or analytically by applying scientific principles in "thought experiments." When relying on experiments or research of others, the investigator must ensure that the conditions and circumstances are sufficiently similar. When the investigator relies on previously conducted research, references to the research relied upon should be noted. If the hypothesis cannot be supported, it should be discarded and alternate hypotheses should be developed and tested. This may include the collection of new data or the reanalysis of existing data. The testing process needs to be continued until all feasible hypotheses have been tested and one is determined to be uniquely consistent with the facts, and with the principles of science. If no hypothesis can withstand an examination by deductive reasoning, the issue should be considered undetermined.

4.3.6.1* Any hypothesis that is incapable of being tested is an invalid hypothesis. A hypothesis developed based on the absence of data is an example of a hypothesis that is incapable of being tested. The inability to refute a hypothesis does not mean that the hypothesis is true.

4.3.7 Avoid Presumption. Until data have been collected, no specific hypothesis can be reasonably formed or tested. All investigations of fire and explosion incidents should be approached by the investigator without presumption as to origin, ignition sequence, cause, fire spread, or responsibility for the incident until the use of scientific method has yielded testable hypotheses, which cannot be disproved by rigorous testing.

4.3.8 Expectation Bias. Expectation bias is a well-established phenomenon that occurs in scientific analysis when investigator(s) reach a premature conclusion without having examined or considered all of the relevant data. Instead of collecting and examining all of the data in a logical and unbiased manner to reach a scientifically reliable conclusion, the investigator(s) uses the premature determination to dictate investigative processes, analyses, and, ultimately, conclusions, in a way that is not scientifically valid. The introduction of expectation bias into the investigation results in the use of only that data that supports this previously formed conclusion and often results in the misinterpretation and/or the discarding of data that does not support the original opinion. Investigators are strongly cautioned to avoid expectation bias through proper use of the scientific method.

4.3.9* Confirmation Bias. Different hypotheses may be compatible with the same data. When using the scientific method, testing of hypotheses should be designed to disprove the hypothesis. Confirmation bias occurs when the investigator instead tries to prove the hypothesis. This can result in failure to consider alternate hypotheses. A hypothesis can be said to be valid only when rigorous testing has failed to disprove the hypothesis.

4.4 Basic Method of a Fire Investigation. Using the scientific method in most fire or explosion incidents should involve the steps shown in 4.4.1 through 4.4.6.

4.4.1 Receiving the Assignment. The investigator should be notified of the incident, told what his or her role will be, and told what he or she is to accomplish. For example, the investigator should know if he or she is expected to determine the origin, cause, and responsibility; produce a written or oral report; prepare for criminal or civil litigation; make suggestions for code enforcement, code promulgation, or changes; make suggestions to manufacturers, industry associations, or government agency action; or determine some other results.

4.4.2 Preparing for the Investigation. The investigator should marshal his or her forces and resources and plan the conduct of the investigation. Preplanning at this stage can greatly increase the efficiency and therefore the chances for success of the overall investigation. Estimating what tools, equipment, and personnel (both laborers and experts) will be needed can make the initial scene investigation, as well as subsequent investigative examinations and analyses, go more smoothly and be more productive.

4.4.3 Conducting the Investigation.

4.4.3.1 It is during this stage of the investigation that an examination of the incident fire or explosion scene is conducted. The fundamental purpose of conducting an examination of any incident scene is to collect all of the available data and document the incident scene. The investigator should conduct an examination of the scene if it is available and collect data necessary to the analysis.

4.4.3.2 The actual investigation may include different steps and procedures, which will be determined by the purpose of the assignment. These steps and procedures are described in detail elsewhere in the document. A fire or explosion investigation may include all or some of the following tasks: a scene inspection or review of previous scene documentation done by others; scene documentation through photography and diagramming; evidence recognition, documentation, and preservation; witness interviews; review and analysis of the investigations of others; and identification and collection of data from other appropriate sources.

4.4.3.3 In any incident scene investigation, it is necessary for at least one individual/organization to conduct an examination of the incident scene for the purpose of data collection and documentation. While it is preferable that all subsequent investigators have the opportunity to conduct an independent examination of the incident scene, in practice, not every scene is available at the time of the assignment. The use of previously collected data from a properly documented scene can be used successfully in an analysis of the incident to reach valid conclusions through the appropriate use of the scientific method. Thus, the reliance on previously collected data and scene documentation should not be inherently considered a limitation in the ability to successfully investigate the incident.

4.4.3.4 The goal of all investigators is to arrive at accurate determinations related to the origin, cause, fire spread, and responsibility for the incident. Improper scene documentation can impair the opportunity of other interested parties to obtain the same evidentiary value from the data. This potential impairment underscores the importance of performing comprehensive scene documentation and data collection.

4.4.4 Collecting and Preserving Evidence. Valuable physical evidence should be recognized, documented, properly collected, and preserved for further testing and evaluation or courtroom presentation.

4.4.5 Analyzing the Incident. All collected and available data should be analyzed using the principles of the scientific method. Depending on the nature and scope of one's assignment, hypotheses should be developed and tested explaining the origin, ignition sequence, fire spread, fire cause or causes of damage or casualties, or responsibility for the incident.

4.4.6 Conclusions. Conclusions, which are final hypotheses, are drawn as a result of testing the hypotheses. Conclusions should be drawn according to the principles expressed in this guide and reported appropriately.

4.5 Level of Certainty. The level of certainty describes how strongly someone holds an opinion (conclusion). Someone may hold any opinion to a higher or lower level of certainty. That level is determined by assessing the investigator's confidence in the data, in the analysis of that data, and testing of hypotheses formed. That level of certainty may determine the practical application of the opinion, especially in legal proceedings.

4.5.1 The investigator should know the level of certainty that is required for providing expert opinions. Two levels of certainty commonly used are probable and possible:

(1) Probable. This level of certainty corresponds to being more likely true than not. At this level of certainty, the likelihood of the hypothesis being true is greater than 50 percent.
(2) Possible. At this level of certainty, the hypothesis can be demonstrated to be feasible but cannot be declared probable. If two or more hypotheses are equally likely, then the level of certainty must be "possible."

4.5.2 If the level of certainty of an opinion is merely "suspected," the opinion does not qualify as an expert opinion. If the level of certainty is only "possible," the opinion should be specifically expressed as "possible." Only when the level of certainty is considered "probable" should an opinion be expressed with reasonable certainty.

4.5.3 Expert Opinions. Many courts have set a threshold of certainty for the investigator to be able to render opinions in court, such as "proven to an acceptable level of certainty," "a reasonable degree of scientific and engineering certainty," or "reasonable degree of certainty within my profession." While these terms of art may be important for the specific jurisdiction or court in which they apply, defining these terms in those contexts is beyond the scope of this document.

4.6 Review Procedure. A review of a fire investigator's work product (e.g., reports, documentation, notes, diagrams, photos, etc.) by other persons may be helpful, but there are certain limitations. This section describes the types of reviews and their appropriate uses and limitations.

4.6.1 Administrative Review. An administrative review is one typically carried out within an organization to ensure that the investigator's work product meets the organization's quality assurance requirements. An administrative reviewer will determine whether all of the steps outlined in an organization's procedure manual, or required by agency policy, have been followed and whether all of the appropriate documentation is present in the file, and may check for typographical or grammatical errors.

4.6.1.1 Limitations of Administrative Reviews. An administrative reviewer may not necessarily possess all of the knowledge, skills, and abilities of the investigator or of a technical reviewer. As such, the administrative reviewer may not be able to provide a substantive critique of the investigator's work product.

4.6.2 Technical Review. A technical review can have multiple facets. If a technical reviewer has been asked to critique all aspects of the investigator's work product, then the technical reviewer should be qualified and familiar with all aspects of proper fire investigation and should, at a minimum, have access to all of the documentation available to the investigator whose work is being reviewed. If a technical reviewer has been asked to critique only specific aspects of the investigator's work product, then the technical reviewer should be qualified and familiar with those specific aspects and, at a minimum, have access to all documentation relevant to those aspects. A technical review can serve as an additional test of the various aspects of the investigator's work product.

4.6.2.1 Limitations of Technical Reviews. While a technical review may add significant value to an investigation, technical reviewers may be perceived as having an interest in the outcome of the review. Confirmation bias (attempting to confirm a hypothesis rather than attempting to disprove it) is a subset of expectation bias *(see 4.3.8)*. This kind of bias can be introduced in the context of working relationships or friendships. Investigators who are asked to review a colleague's findings should strive to maintain a level of professional detachment.

4.6.3 Peer Review. Peer review is a formal procedure generally employed in prepublication review of scientific or technical documents and screening of grant applications by research-sponsoring agencies. Peer review carries with it connotations of both independence and objectivity. Peer reviewers should not have any interest in the outcome of the review. The author does not select the reviewers, and reviews are often conducted anonymously. As such, the term "peer review" should not be applied to reviews of an investigator's work by coworkers, supervisors, or investigators from agencies conducting investigations of the same incident. Such reviews are more appropriately characterized as "technical reviews," as described above.

4.6.3.1 The methodologies used and the fire science relied on by an investigator are subject to peer review. For example, NFPA 921 is a peer-reviewed document describing the methodologies and science associated with proper fire and explosion investigations.

4.6.3.2 Limitations of Peer Reviews. Peer reviewers should have the expertise to detect logic flaws and inappropriate applications of methodology or scientific principles, but because they generally have no basis to question an investigator's data, they are unlikely to be able to detect factual errors or incorrectly reported data. Conclusions based on incorrect data are likely to be incorrect themselves. Because of these limitations, a proper technical review will provide the best means to adequately assess the validity of the investigation's results.

4.7 Reporting Procedure. The reporting procedure may take many written or oral forms, depending on the specific responsibility of the investigator. Pertinent information should be reported in a proper form and forum to help prevent recurrence.

Chapter 5 Basic Fire Science

5.1 Introduction.

5.1.1* General. The fire investigator should have a basic understanding of ignition and combustion principles and should be able to use them to help in interpretation of evidence at the fire scene and in the development of conclusions regarding the origin and causes of the fire. The body of knowledge associated with combustion and fire easily fills several textbooks. The discussion presented in this chapter should be considered introductory. The user of this guide is urged to consult the reference material listed in Annex A and Annex C for additional details.

5.1.2 Fire Tetrahedron. The combustion reaction can be characterized by four components: the fuel, the oxidizing agent, the heat, and the uninhibited chemical chain reaction. These four components have been classically symbolized by a four-sided solid geometric form called a tetrahedron *(see Figure 5.1.2)*. Fires can be prevented or suppressed by controlling or removing one or more of the sides of the tetrahedron.

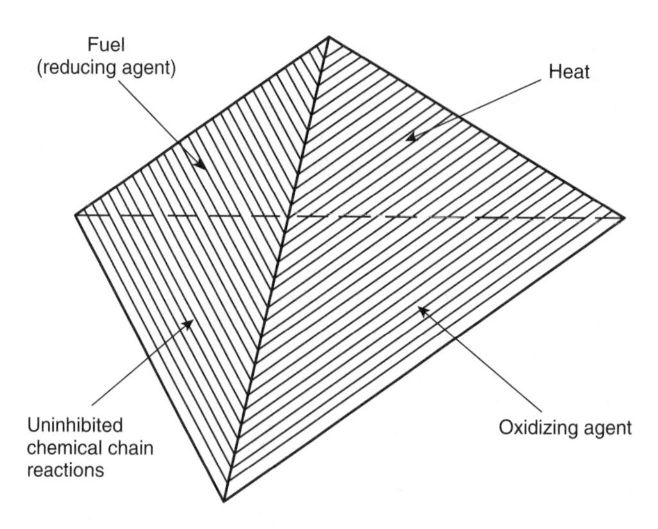

FIGURE 5.1.2 Fire Tetrahedron.

5.1.2.1 Fuel. A fuel is any substance that can undergo combustion. The majority of fuels encountered are organic, which simply means that they are carbon-based and may contain other elements such as hydrogen, oxygen, and nitrogen in varying ratios. Examples of organic fuels include wood, plastics, gasoline, alcohol, and natural gas. Inorganic fuels contain no carbon and include combustible metals, such as magnesium or sodium. All matter can exist in one of three states: solid, liquid, or gas. The state of a given material depends on the temperature and pressure and can change as conditions vary. If cold enough, carbon dioxide, for example, can exist as a solid (dry ice). The normal state of a material is that which exists at NTP (normal temperature and pressure) conditions: 20°C (68°F) temperature, and a pressure of 101.6 kPa (14.7 psi), or 1 atmosphere at sea level.

5.1.2.1.1 Combustion of liquid fuels and most solid fuels takes place above the fuel surface in a region of vapors created by heating the fuel surface. The heat can come from the ambient conditions, from the presence of an ignition source, or from exposure to an existing fire. The application of heat causes vapors or pyrolysis products to be released into the atmosphere, where they can burn if in the proper mixture with an oxidizer and if a competent ignition source is present or if the fuel's autoignition temperature is reached. Ignition is discussed in Section 5.3.

5.1.2.1.2 Gaseous fuels do not require vaporization or pyrolysis before combustion can occur. Only the proper mixture with an oxidizer and an ignition source are needed.

5.1.2.1.3 For the purposes of the following discussion, the term *fuel* is used to describe vapors and gases rather than solids.

5.1.2.2 Oxidizing Agent. In most fire situations, the oxidizing agent is the oxygen in the earth's atmosphere. Fire can occur in the absence of atmospheric oxygen, when fuels are mixed with chemical oxidizers. Many chemical oxidizers contain readily released oxygen. Ammonium nitrate fertilizer (NH_4NO_3), potassium nitrate (KNO_3), and hydrogen peroxide (H_2O_2) are examples.

5.1.2.2.1 Certain gases can form flammable mixtures in atmospheres other than air or oxygen. One example is a mixture of hydrogen and chlorine gas.

5.1.2.2.2 Every fuel–air mixture has an optimum ratio at which point the combustion will be most efficient. This ratio occurs at or near the mixture known by chemists as the stoichiometric ratio. When the amount of air is in balance with the amount of fuel (i.e., after burning there is neither unused fuel nor unused air), the burning is referred to as stoichiometric. This condition rarely occurs in fires except in certain types of gas fires. *(See 21.8.2.1.)*

5.1.2.3 Heat. The heat component of the tetrahedron represents heat energy above the minimum level necessary to release fuel vapors and cause ignition. Heat is commonly defined in terms of intensity or heating rate (kilowatts) or as the total heat energy received over time (kilojoules). In a fire, heat produces fuel vapors, causes ignition, and promotes fire growth and flame spread by maintaining a continuous cycle of fuel production and ignition.

5.1.2.4 Uninhibited Chemical Chain Reaction. Combustion is a complex set of chemical reactions that results in the rapid oxidation of a fuel, producing heat, light, and a variety of chemical by-products. Slow oxidation, such as rust or the yellowing of newspaper, produces heat so slowly that combustion does not occur. Self-sustained combustion occurs when sufficient excess heat from the exothermic reaction radiates back to the fuel to produce vapors and cause ignition in the absence of the original ignition source. For a detailed discussion of ignition, see Section 5.7.

5.2* Fire Chemistry.

5.2.1 General. Fire chemistry is the study of chemical processes that occur in fires, including changes of state, decomposition, and combustion.

5.2.2 Phase Changes and Thermal Decomposition. The response of fuels to heat is quite varied. Figure 5.2.2 illustrates the wide range of processes that can occur.

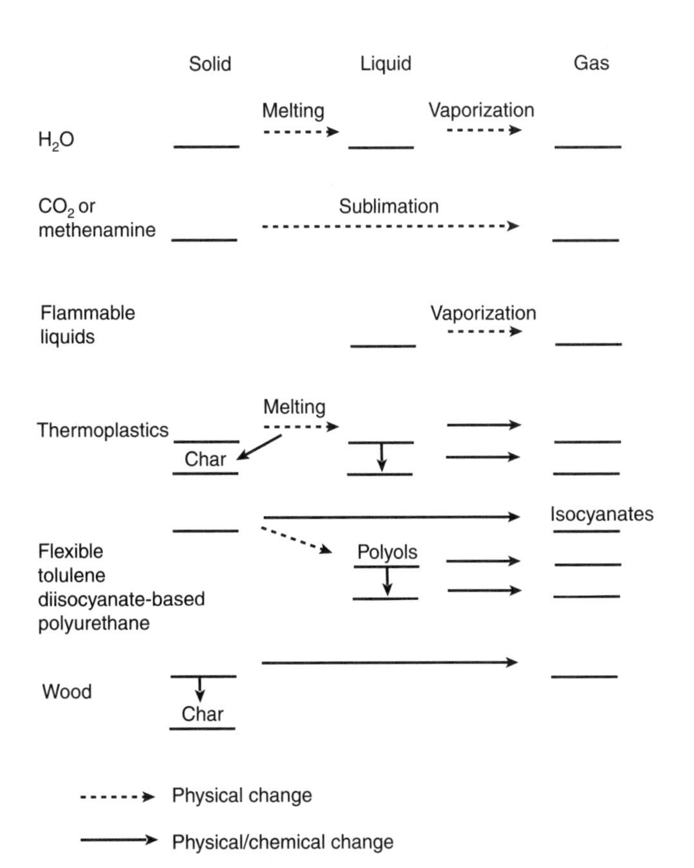

FIGURE 5.2.2 Physical and Chemical Changes During Thermal Decomposition. [*Adopted from Beyler and Hirschler(2002).*]

5.2.2.1 Phase changes most relevant in fire are melting and vaporization. In melting, the material changes from a solid to a liquid with no change in the chemical structure of the material (e.g., melting of candle wax). In vaporization, the material changes from a liquid to a vapor with no change in chemical structure of the material (e.g., evaporation of molten candle wax on the wick to form the vapor that burns in the candle flame). Phase changes are reversible events, that is, upon cooling, vapors will return to the liquid state and liquids will solidify.

5.2.2.2 Thermal decomposition involves irreversible changes in the chemical structure of a material due to the effects of heat (pyrolysis). Thermal decomposition of a solid or liquid most often results in the production of gases. Wood decomposes to create char and vapors, some of which are flammable. Under vigorous heating, flexible polyurethane decomposes to form a liquid and flammable gases or vapors. At more moderate heating conditions, flexible polyurethane decomposes to a char and flammable gases or vapors.

5.2.3 Combustion. The combustion reactions can be characterized by the fire tetrahedron *(see 5.1.2)* and may occur with the fuel and oxidizing agent already mixed (premixed burning) or with the fuel and oxidizing agent initially separate (diffusion burning). Both premixed and diffusion flames are important in fire.

5.2.3.1 Premixed burning occurs when fuel vapors mix with air in the absence of an ignition source and the fuel–air mixture is subsequently ignited. Examples of premixed fuel and air include a natural gas release into the environment and evaporation of gasoline. Upon application of an ignition source to the fuel–air mixture, a premixed flame quickly propagates through the volume of fuel–air. Premixed flame spread can proceed as a deflagration (subsonic combustion) or as a detonation (supersonic combustion). Deflagration velocities normally range from cm/sec to m/sec, though velocities into the hundreds of m/sec are possible. Detonation velocities are normally in the thousands of m/sec. Premixed flame propagation in a confined volume is normally termed a combustion explosion.

5.2.3.2 In premixed flames not all mixtures of fuel and oxidizer can burn. The lowest and highest concentrations of fuel in a specified oxidant are known as the lower and upper flammability limits, also known as the lower and upper explosive limits (LEL and UEL). For example, the lower and upper flammable limits of methane are 5 percent and 15 percent, respectively, in air at ordinary temperatures. At concentrations below 5 percent and above 15 percent methane, methane will not burn in air at ordinary temperatures. The difference between the lower and upper limits is called the flammable or explosive range. The extent (width) of the flammable or explosive range of a material, as well as its LEL and UEL, are among the properties that describe the fire hazard of a material. For example, the flammable range of hydrogen is 71 percent (4 percent to 75 percent). When considering the fire hazard of ignitible gases and vapors, the lower the LEL, the higher the UEL, and the wider the flammable or explosive range, the greater the fire hazard of the material.

5.2.3.3 Diffusion flame burning is the ordinary sustained burning mode in most fires. Fuel vapors and oxidizer are separate and combustion occurs in the region where they come together. A diffusion flame is typified by a candle flame in which the luminous flame zone exists where the air and the fuel vapors meet.

5.2.3.4* Diffusion flames can only occur for certain concentrations of the mixture components. The lowest oxygen concentration in nitrogen is termed the limiting oxygen index (LOI). For most fuel vapors, the LOI is in the range of 10 percent to 14 percent by volume at ordinary temperatures (Beyler 2002). Similarly, the fuel gas stream can be diluted with nitrogen or other inert gas to the extent where burning is no longer possible. For example, methane diluted with nitrogen to below 14 percent methane will not burn with air at normal temperatures.

5.2.3.5 Transitions from premixed burning to diffusion flame burning are common during the ignition of liquid and solid fuels. For instance, if an ignition source is applied to a pan of gasoline, the ignition source ignites gasoline vapors mixed with air above the pan. These vapors are quickly consumed and the burning of fuel vapors from the pan of gasoline occurs as a diffusion flame.

5.3* Products of Combustion.

5.3.1 The chemical products of combustion can vary widely, depending on the fuels involved and the amount of air available. Complete combustion of hydrocarbon fuels containing only hydrogen and carbon will produce carbon dioxide and water. Materials containing nitrogen, such as silk, wool, and polyurethane foam, can produce nitrogen oxides or hydrogen cyanide as combustion products under some combustion conditions. Literally hundreds of compounds have been identified as products of incomplete combustion of wood.

5.3.2 When less air is available for combustion, as in ventilation-controlled fires, the production of carbon monoxide increases as does the production of soot and unburned fuels.

5.3.3 Combustion products exist in all three states of matter: solid, liquid, and gas. Solid material makes up the ash and soot products that represent the visible "smoke." Many of the other products of incomplete combustion exist as vapors or as extremely small tarry droplets or aerosols. These vapors and droplets often condense on surfaces that are cooler than the smoke, resulting in smoke patterns that can be used to help determine the origin and spread of the fire. Such surfaces include walls, ceilings, and glass. Because the condensation of residue results from temperature differences between the smoke body and the affected surface, the presence of a deposit is evidence that smoke did engulf the surface, but the lack of deposit or the presence of a sharp line of demarcation is not evidence of the limits of smoke involvement.

5.3.4 Soot and tarry products often accumulate more heavily on ceramic-tiled surfaces than on other surrounding surfaces due to the heat conduction properties of ceramic tile. Those surfaces that remain the coolest the longest tend to collect the most condensate.

5.3.5 Some fuels, such as alcohol or natural gas, burn very cleanly, while others, such as fuel oil or styrene, will produce large amounts of sooty smoke even when the fire is fuel controlled.

5.3.6 Smoke is generally considered to be the collection of the solid, liquid, and gaseous products of incomplete combustion.

5.3.7 Smoke color is not necessarily an indicator of what is burning. While wood smoke from a well-ventilated or fuel-controlled wood fire is light-colored or gray, the same fuel under low-oxygen conditions, or ventilation-controlled conditions in a post-flashover fire, can be quite dark or black. Black smoke also can be produced by the burning of other materials, including most plastics and ignitible liquids.

5.3.8 The action of fire fighting can also have an effect on the color of the smoke being produced. The application of water can produce large volumes of condensing vapor that will appear white or gray when mixed with black smoke from the fire. This result is often noted by witnesses at the fire scene and has been misinterpreted to indicate a change of fuel being burned.

5.3.9 Smoke production rates are generally less in the early phase of a fire but increase greatly with the onset of flashover, if flashover occurs.

5.4* Fluid Flows.

5.4.1 General. Flows can be generated by mechanical forces (like fans) or by buoyant forces generated by temperature differences. In most instances, buoyant flows are most significant in fires. Important buoyant flows in fire include fire plumes above burning objects, ceiling jet flows when plume gases strike the ceiling and move along the ceiling, and the flow of hot gases out of a door or window (vent flows).

5.4.2 Buoyant Flows. Buoyant flows occur because hot gases are less dense than cold gases. This causes the hot gases to rise, just as a hot air balloon rises.

5.4.3 Fire Plumes. The primary engine for flows is the creation of hot gases by the fire itself. The hot gases created by the fire rise above the fire source as a fire plume. As the hot gases rise, they mix with or entrain the surrounding air so that the flow of gases in the plume increases with height above the fire and at the same time the temperature of the plume is reduced by the entrainment of air. It is the entrainment of air into the plume that causes the plume to increase in diameter as it rises.

5.4.4 Ceiling Jets. When a fire plume reaches the ceiling of a room, the gases turn to move laterally along the ceiling as a ceiling jet. The ceiling jet flows along the ceiling until the flow encounters a vertical obstruction such as a wall. The hot ceiling jet is generally responsible for the operation of ceiling-mounted detectors or sprinklers.

5.4.5 Vent Flows. The buoyancy of gases in a compartment fire causes flow into and out of a compartment through vents. In a compartment fire with a single vent opening, hot gases flow out through the upper portion of the opening, and fresh air enters in the lower portions of the opening.

5.5* Heat Transfer.

5.5.1 General. Heat transfer is classically defined as the transport of heat energy from one point to another caused by a temperature difference between those points. The heat transfer rate per unit area (also known as heat flux) is normally expressed in kW/m^2. The transfer of heat is a major factor in fires and has an effect on ignition, growth, spread, decay (reduction in energy output), and extinction. Heat transfer is also responsible for much of the physical evidence used by investigators who attempt to establish a fire's origin and cause.

5.5.1.1 It is important to distinguish between heat and temperature. Temperature is a measure that expresses the degree of molecular activity of a material compared to a reference point, such as the freezing point of water. Heat is the energy that is needed to change the temperature of an object. When heat energy is transferred to an object, the temperature increases. When heat is transferred away from an object, the temperature decreases.

5.5.1.2 Unless work is done on the system by outside forces, heat is naturally transferred from a higher temperature mass to a lower temperature mass. Heat transfer is measured in terms of energy flow per unit of time (kilowatts). The greater the temperature difference between the objects, the more energy transferred per unit of time and the higher the heat transfer rate. Temperature can be compared to the pressure in a fire hose and heat or energy transfer to the water flow in gallons per minute.

5.5.1.3 Heat transfer is accomplished by three mechanisms: conduction, convection, and radiation. All three mechanisms play a role in fire, and an understanding of each is necessary in the investigation of a fire.

5.5.2 Conduction. Conduction is the form of heat transfer that takes place within solids when one portion of an object is heated. Energy is transferred from the heated area to the unheated area at a rate dependent on the difference in temperature and the thermal conductivity *(k)* of the material. The thermal conductivity *(k)* of a material is a measure of the amount of heat that will flow across a unit area with a temperature gradient of 1 degree per unit of length (W/m-K, Btu/hr-ft-°F). The heat capacity (specific heat) of a material is a measure of the amount of heat necessary to raise the temperature of a unit mass 1 degree, under specified conditions (J/kg-K, Btu/lb-°F).

5.5.2.1 If thermal conductivity *(k)* is high, the rate of heat transfer through the material is high. Metals have high thermal conductivities *(k)*, while plastics and glass have low thermal conductivity *(k)* values. High-density materials conduct heat faster than low-density materials. Therefore, low-density materials make good insulators. Materials with a high heat capacity *(c)* require more energy to raise the temperature than materials with low heat capacity values.

5.5.2.2* When one portion of a solid is exposed to a high temperature and another portion of that solid is at a lower temperature, then heat energy will be transferred into and through the solid from the higher to the lower temperature areas. Initially, the heat energy moving through the solid will raise the temperature at all interior points to some level of temperature between the extreme high and extreme low. When the temperatures at all interior points have stopped increasing, the temperature and heat transfer within the solid is said to be in a steady state thermal condition. During steady state heat transfer, a condition that is rare in most fire scenarios, thermal conductivity *(k)* is the dominant heat transfer property. Figure 5.5.2.2 shows the steady state maximum surface temperature achievable as a function of the incident radiant flux. While achieving these steady state temperatures might take an unrealistic time period, the plot is illustrative of the maximum possible surface temperature for a given incident radiant heat flux.

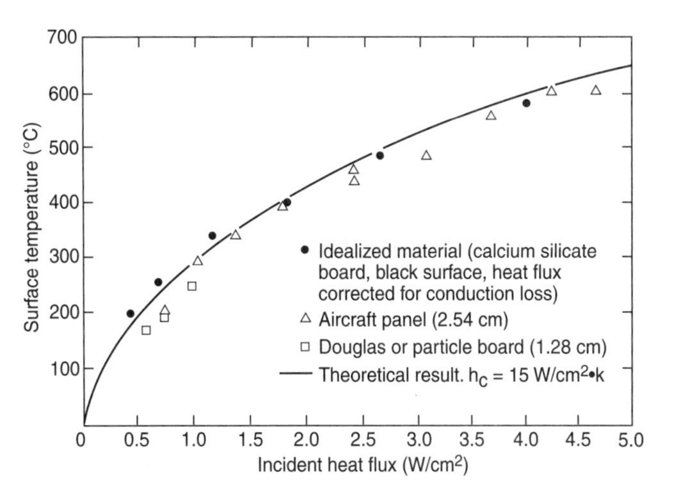

FIGURE 5.5.2.2 Maximum Surface Temperature Achievable from a Steady State Radiant Heat Flux in an Environment at Normal Ambient Temperature [20°C (68°F)].

5.5.2.3 Thermal Inertia. During transient heating, a more common condition, the result is changing rates of heat transfer and temperature. During this period, all three properties — thermal conductivity *(k)*, density (ρ), and heat capacity *(c)* — play a role. Multiplied together as a mathematical product, these properties are called the *thermal inertia*, *k*ρ*c*, of a material. The thermal inertia of a material is a measure of how easily the surface temperature of the material will increase when heat flows into the material. Low-density materials like polyurethane foam have a low thermal inertia and the surface temperature will increase quickly upon exposure to a heat flux. Conversely, metals have a high thermal inertia due to their high thermal conductivity and high density. As such, when exposed to a flame, the surface temperature of a metal object increases relatively slowly compared to the surface temperature of a plastic or wood object. Table 5.5.2.3 provides data for some common materials at room temperature. Thermal properties are generally a function of temperature.

5.5.2.4 The influence of the thermal inertia on the surface temperature of a thick material occurs principally during the time the surface temperature is increasing. Eventually, as the material reaches a steady temperature, the effects of density (ρ) and heat capacity *(c)* become insignificant relative to thermal conductivity. Therefore, thermal inertia of a material is most important at the initiation and early stages of a fire.

5.5.2.5 The effect of conduction of heat into a material is an important aspect of ignition. Thermal inertia determines how fast the surface temperature will rise. The lower the thermal inertia of the material, the faster the surface temperature will rise.

5.5.2.6 Conduction is also a mechanism of fire spread. Heat conducted through a wall or along a pipe or beam can cause ignition of combustibles in contact with the heated object. Thermally thin materials are those materials that are physically thin or have a very high thermal conductivity. The full thickness of the material is at approximately the same temperature during heating. The rate of temperature rise is dependent on the thermal mass of the material, which is the mass per unit area multiplied by the heat capacity of the material. Subjected to the same heat source, a thin curtain will heat more rapidly than a thick drapery. This effect has a direct impact on ignitibility and flame spread.

5.5.3 Convection. Convection is the transfer of heat energy by the movement of heated liquids or gases from the source of heat to a cooler part of the environment. In most cases, convection will be present in any environment where there are temperature differences, although in a few cases a stably-stratified condition may be found that does not cause fluid movement.

5.5.3.1 Heat is transferred by convection to a solid when hot gases pass over cooler surfaces. The rate of heat absorbed by the solid is a function of the temperature difference between the hot gas and the surface, the thermal inertia of the material being heated, the surface area exposed to the hot gas, and the velocity of the hot gas. The higher the velocity of the gas, the greater the rate of convective heat transfer. Because a flame itself is a hot gas, flame contact involves heat transfer by convection.

5.5.3.2 In the early part of a fire, convection plays a major role in heating the surfaces exposed to gases heated by the fire. As the room temperature rises, convection continues, but

Table 5.5.2.3 Thermal Properties of Selected Materials

Material	Thermal Conductivity (k) (W/(m K))	Density (ρ) (kg/m^3)	Heat Capacity (c) (J/(kg-K))	Thermal Inertia ($k\rho c$) (W^2 · s/k^2 m^4)
Copper	387	8940	380	1.31E+09
Concrete	0.8–1.4	1900–2300	880	1.34E+06–2.83E+06
Gypsum plaster	0.48	1440	840	5.81E+05
Oak	0.17	800	2380	3.24E+05
Pine (yellow)	0.14	640	2850	2.55E+05
Polyethylene	0.35	940	1900	6.25E+05
Polystyrene (rigid)	0.11	1100	1200	1.45E+05
Polyvinylchloride	0.16	1400	1050	2.35E+05
Polyurethane*	0.034	20	1400	9.52E+02

*Typical values and properties vary with temperature. *Source:* Drysdale (1999).

the role of radiation increases rapidly and becomes the dominant heat transfer mechanism.

5.5.3.3 Convection heat transfer occurs by two mechanisms, natural and forced convection. In forced convection, the velocity of the gas flowing over the material is externally imposed (e.g., by a fan). In natural convection, the velocity of the gas flowing over the material is the result of buoyancy-induced flows associated with the temperature difference between the surface and the gas. Heat transfer from a hot surface in a quiescent environment is by natural convection. The hot gas plume above the hot surface results from the high temperature of the hot surface relative to the environment. Conversely, when that plume reaches a sprinkler at the ceiling, heat transfer is by forced convection. The flow of hot gases over the sprinkler is externally imposed by the hot gas plume or ceiling jet and is not created by the sprinkler itself.

5.5.4 Radiation. Radiation is the transfer of heat energy from a hot surface or gas, the radiator, to a cooler material, the target, by electromagnetic waves without the need of an intervening medium. For example, the heat energy from the sun is radiated to earth through the vacuum of space. Radiant energy can be transferred only by line of sight and will be reduced or blocked by intervening materials. Intervening materials do not necessarily block all radiant heat. For example, radiant heat is reduced by about 50 percent by some glazing materials. Radiators and targets are not limited to solids but can be liquids and gases, as well. For example, the smoke and hot gases that collect at ceiling level in a compartment fire are the source of radiant heat that may lead to ignition of materials.

5.5.4.1 The rate of heat transfer from a radiating material is proportional to that material's absolute temperature raised to the fourth power. For example, doubling the absolute temperature of a radiating material will result in a 16-fold increase in radiation from that material. Figure 5.5.4.1 illustrates this relation. Since all materials emit radiant energy proportional to the fourth power of their absolute temperatures, then the net heat radiation between two materials separated in space is proportional to the difference in the fourth powers of the absolute temperatures. Absolute temperatures are measured in Kelvins (°C + 273).

5.5.4.2 The rate of radiant heat transfer is also strongly affected by the distance between the radiator and the target. As the distance increases, the amount of energy falling on a unit of area falls off in a manner that is related to both the size of the radiating source and the distance to the target. For example, when the distance between the radiator and the target doubles, the amount of net radiant heat transfer may not change significantly or may drop to as little as one fourth of its original value, depending on the size of the radiator relative to the distance involved. Table 5.5.4.2 provides general information on the effects of radiant heat fluxes.

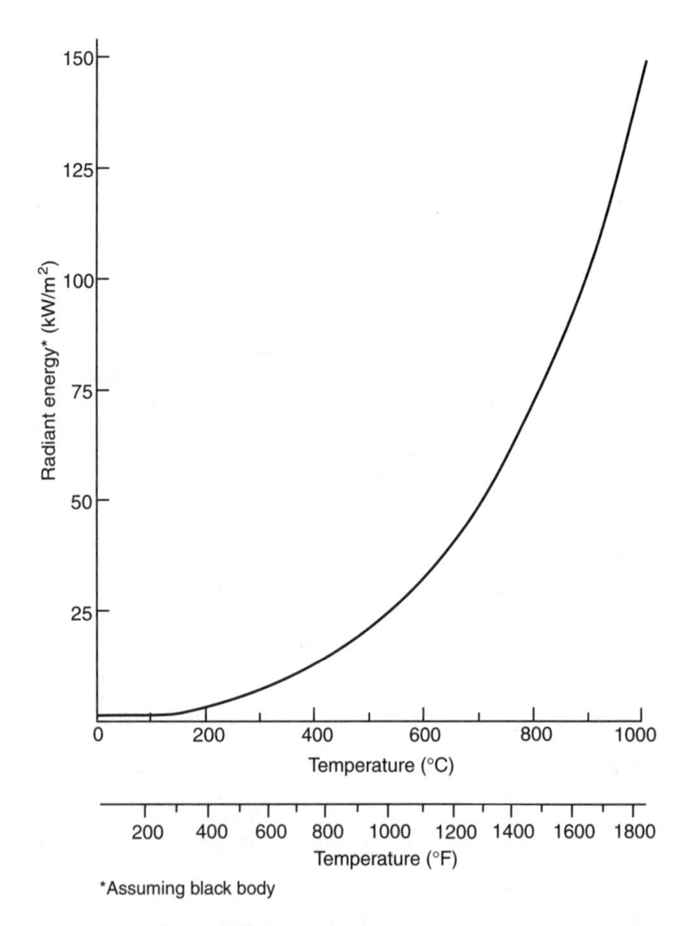

FIGURE 5.5.4.1 Relation of Radiation to Temperature.

5.6* Fuel Load, Fuel Packages, and Properties of Flames. This section deals with the combustion properties of fuels and collections of fuels.

Table 5.5.4.2 Effect of Radiant Heat Flux

Approximate Radiant Heat Flux (kW/m^2)	Comment or Observed Effect
170	Maximum heat flux as currently measured in a postflashover fire compartment.
80	Heat flux for protective clothing Thermal Protective Performance (TPP) Test.[a]
52	Fiberboard ignites spontaneously after 5 seconds.[b]
29	Wood ignites spontaneously after prolonged exposure.[b]
20	Heat flux on a residential family room floor at the beginning of flashover.[c]
20	Human skin experiences pain with a 2-second exposure and blisters in 4 seconds with second-degree burn injury.[d]
15	Human skin experiences pain with a 3-second exposure and blisters in 6 seconds with second-degree burn injury.[d]
12.5	Wood volatiles ignite with extended exposure[e] and piloted ignition.
10	Human skin experiences pain with a 5-second exposure and blisters in 10 seconds with second-degree burn injury.[d]
5	Human skin experiences pain with a 13-second exposure and blisters in 29 seconds with second-degree burn injury.[d]
2.5	Human skin experiences pain with a 33-second exposure and blisters in 79 seconds with second-degree burn injury.[d]
2.5	Common thermal radiation exposure while fire fighting.[f] This energy level may cause burn injuries with prolonged exposure.
1.0	Nominal solar constant on a clear summer day.[g]

Note: The unit kW/m^2 defines the amount of heat energy or flux that strikes a known surface area of an object. The unit kW represents 1000 watts of energy and the unit m^2 represents the surface area of a square measuring 1 m long and 1 m wide. For example, 1.4 kW/m^2 represents 1.4 multiplied by 1000 and equals 1400 watts of energy. This surface area may be that of the human skin or any other material.

[a]From NFPA 1971, *Standard on Protective Ensembles for Structural Fire Fighting and Proximity Fire Fighting.*

[b]From Lawson, "Fire and Atomic Bomb."

[c]From Fang and Breese, "Fire Development in Residential Basement Rooms."

[d]From Society of Fire Protection Engineering Guide: "Predicting 1st and 2nd Degree Skin Burns from Thermal Radiation," March 2000.

[e]From Lawson and Simms, "The Ignition of Wood by Radiation," pp. 288–292.

[f]From U.S. Fire Administration, "Minimum Standards on Structural Fire Fighting Protective Clothing and Equipment," 1997.

[g]*SFPE Handbook of Fire Protection Engineering,* 2nd edition. NFPA, Quincy, MA

5.6.1 Fuel Load.

5.6.1.1 The term *fuel load* is used to describe the amount of fuel present, usually within a compartment. For instance, a room that is filled with shelving units containing records stored in cardboard boxes is said to be a high fuel load compartment. It is commonly expressed in terms of wood-fuel equivalent mass (kg or lb) or the potential combustion energy (MJ) associated with that fuel mass.

5.6.1.2 The potential combustion energy is determined by multiplying the mass of fuel by the heat of combustion of the fuels. Heats of combustion typically range from 10 to 45 MJ/kg. While the total fuel load for a compartment is a measure of the total heat available if all the fuel burns, it does not determine how fast the fire will develop once the fire starts. Fuel load can be used in conjunction with the size of vent openings to estimate the duration of fully developed burning in a compartment.

5.6.1.3 The term *fuel load density* is the potential combustion energy output per unit floor area [MJ/m^2 (Btu/ft^2)] or the mass of fuel per unit floor area [kg/m^2 (lb/ft^2)]. Fuel load densities are most often associated with particular occupancies or used as a means to characterize the fire load characteristics of the room contents. The fuel load of a compartment is determined by multiplying the fuel load density by the compartment floor area.

5.6.2 Fuel Items and Fuel Package.

5.6.2.1 A fuel item is any article that is capable of burning. A fuel package is a collection or array of fuel items in close proximity with one another such that flames can spread throughout the array of fuel items. Single fuel item fuel packages are possible when the fuel item is located away from other fuel items. A chair that is located away from other fuels is an example of a single item fuel package. Fuel packages are generally identifiable by the separation of the array of fuel items from other fuel items. Typical fuel packages include the following:

(1) A group of abutting office workstations separated from other fuel arrays by aisles
(2) A collection of pieces of living room furniture in close proximity to one another, separated from other fuel arrays by space
(3) A double row rack in a warehouse, separated from other shelves by aisles
(4) A forklift truck with a pallet of goods located away from other combustibles

5.6.2.2 Fire spread from one fuel package to another is generally by radiative ignition of the target fuel package.

5.6.3 Heat Release Rate.

5.6.3.1 Total fuel load in the room has no bearing on the rate of growth of a given fire in its preflashover phase. During this period of development, the rate of fire growth is determined by the heat release rate (HRR) from burning of individual fuel arrays. The HRR describes how the available energy is released. This quantity characterizes the power or energy release rate [watts(Joules/sec) or kilowatts] of a fire and is a quantitative measure of the size of the fire. A generalized HRR curve can be characterized by an initial growth stage, a period of steady state burning, and a decay stage as shown in Figure 5.6.3.1. The largest value of the HRR measured is defined as the peak heat release rate. Representative peak HRRs for a number of fuel items are listed in Table 5.6.3.1. These values should only be considered as representative values for comparison purposes. Fuel items with the same function (e.g., sofas) can have significantly different HRRs. The actual peak heat release rate for a particular fuel item is best determined by test. The heat release rate during the growth phase generally increases as a result of increasing flame spread rates over the fuel package. The peak or steady period of heat release is characterized by full involvement of the surface of the package in flames. The decay phase reflects the reduction in remaining fuel and fuel area available to burn.

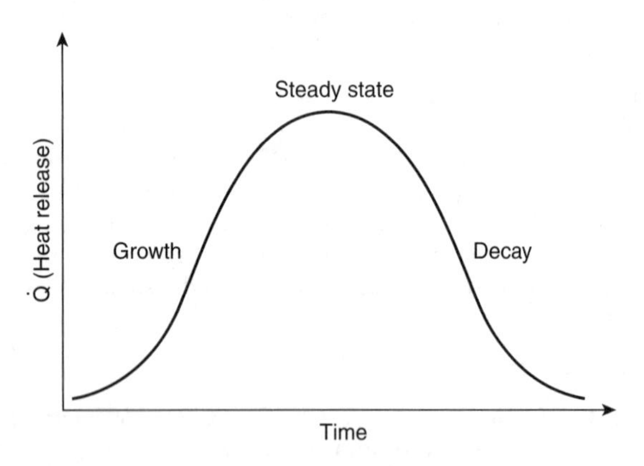

FIGURE 5.6.3.1 Idealized Heat Release Rate Curve.

5.6.3.2 In a compartment fire, as additional items ignite, their individual HRRs combine to obtain an HRR for the compartment. Tests for measuring the HRR of fuel items or packages are usually performed in the "open," where radiant effects of a compartment are not present. However, when a fuel package is exposed to thermal radiant energy, such as from the hot upper layer of a room, this can increase the HRR for that fuel package. In general, the HRR is controlled by the incident heat flux, the chemical and physical properties of the fuel, and the surface area of the fuel, but this does not mean that simple formulas are available to compute the HRR from known variables.

5.6.4 Properties of Flames. The objective of this section is to provide information about the relationship between heat release rate and visible fire size, about the temperatures and velocities achieved within the visible flame, and about heat fluxes from fires to adjacent surfaces.

5.6.4.1 Color of Flame. The color of flame is not necessarily an accurate indicator of what is burning, or of the temperature of the flame.

5.6.4.2 The visible size of a flame is normally expressed as the flame height and the fire dimensions (length and width diameter of the involved fuel package). Observing a fire over time reveals that the height of the flame fluctuates over time. The following three visual measures of flame height are often employed:

(1) Continuous flame height — the height over which flames are visible at all instances
(2) Average flame height — the height over which flames are visible 50 percent of the time
(3) Flame tip height — the greatest height over which flames are visible at any time

Table 5.6.3.1 Representative Peak Heat Release Rates (Unconfined Burning)

Fuel	Weight		Peak HRR (kW)
	kg	lb	
Wastebasket, small	0.7–1.4	1.5–3	4–50
Trash bags, 42 L (11 gal) with mixed plastic and paper trash	1.1–3.4	2½–7½	140–350
Cotton mattress	11.8–13.2	26–29	40–970
TV sets	31.3–32.7	69–72	120 to over 1500
Plastic trash bags/paper trash	1.2–14.1	2.6–31	120–350
PVC waiting room chair, metal frame	15.4	34	270
Cotton easy chair	17.7–31.8	39–70	290–370
Gasoline/kerosene in 0.185 m^2 (2 ft^2) pool	19	—	400
Christmas trees, dry	6–20	13–44	3000–5000
Polyurethane mattress	3.2–14.1	7–31	810–2630
Polyurethane easy chair	12.2–27.7	27–61	1350–1990
Polyurethane sofa	51.3	113	3120
Wardrobe, wood construction	70–121	154–267	1900–6400

Sources: Values are from the following publications:

Babrauskas, V., "Heat Release Rates," in *SFPE Handbook of Fire Protection Engineering,* 3rd ed., National Fire Protection Assn., Quincy MA (2002).

Babrauskas, V. and Krasny, J. (1985) *Fire Behavior of Upholstered Furniture,* NBS Monograph 173 Fire Behavior of Upholstered Furniture, National Bureau of Standards, Gaithersburg MD.

Lee, B.T. (1985), *Heat Release Rate Characteristics of Some Combustible Fuel Sources in Nuclear Power Plants,* NBSIR 85-3195, National Bureau of Standards, Gaithersburg MD.

NFPA 72, National Fire Alarm and Signaling Code, 2010 edition, Annex B.

5.6.4.3 The following flame height definitions define the three regions of a fire:

(1) Continuously flaming region (lower portion of visible flame)
(2) Intermittently flaming region (upper portion of the visible flame)
(3) Plume region (above the visible flame)

5.6.4.4 These heights are best determined from frame-by-frame analysis of a videotape of the fire. Casually observed flame height determinations tend to be most consistent with the flame tip height, as our eyes seem to focus on the tip of the flame. The most widely reported flame height in the fire science literature is the average flame height.

5.6.4.5 Flame Height. Figure 5.6.4.5 shows the flame height of a circular fire source with heat release rates per unit area of 250, 500, and 1000 kW/m^2. Flame heights were calculated from the widely used Heskestad correlation. The figure illustrates that flame height is not strictly a function of the heat release rate. In addition, it is clear from the figure that even for a given heat release rate per unit area, small variations in observed flame height yield much larger variations in estimated heat release rate.

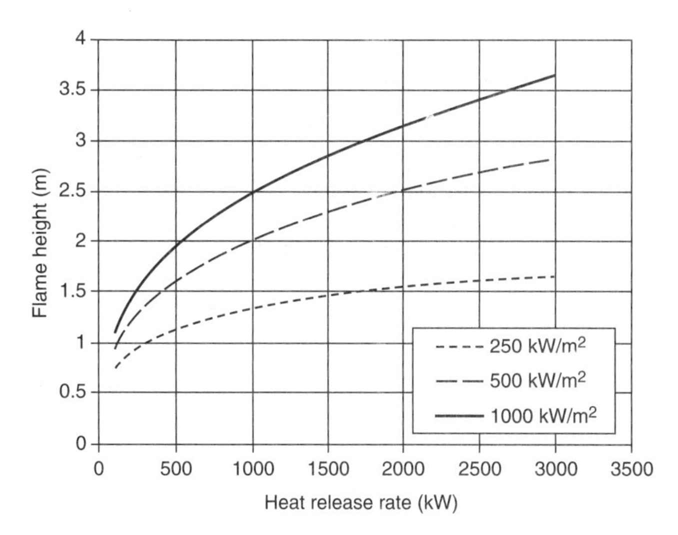

FIGURE 5.6.4.5 Average Flame Height (50 Percent Intermittency) as a Function of Heat Release Rate for a Range of Heat Release Rates per Unit Area of Fuel Package.

5.6.4.6* Fuel Package Location.

5.6.4.6.1 Air Entrainment. When a burning fuel package is positioned away from a wall, air is free to flow into the plume from all directions and mix with the fuel gases. If the fuel package is placed against a wall or in a corner (formed by the intersection of two walls), air entrainment into the plume can be restricted, creating an imbalance in the airflow. As a result of the imbalance in airflow, the flame and thermal plumes will bend toward the restricting surface(s).

5.6.4.6.2 Flame and Plume Attachment. In cases where the flame or thermal plume bends sufficiently to become attached to the wall(s), the air entrainment is reduced. The fuel package must be sufficiently close to the wall(s) to cause the flame or thermal plume to attach to the wall(s) in order for the effects of restricted air entrainment to occur. The extent of the bending of the flame toward and the attachment to the wall(s) is dependent on the geometry of the fuel and the position of the fuel package relative to the wall(s).

5.6.4.6.3 Effect of Reduced Air Entrainment. A decrease in air entrainment has an effect on plume and upper layer temperatures as well as on the height of the flame.

5.6.4.6.3.1 Plume and Upper Layer Temperatures. A reduction in ambient air being entrained into the thermal plume lessens the amount of mixing of cooler ambient air with the thermal plume, resulting in less dilution and higher temperatures. Since the plume transports thermal energy to the upper layer, an increase in temperature in the plume will also produce an increase in the upper layer temperature.

5.6.4.6.3.2 Flame Height. For diffusion flames, the mixing of fuel vapor and air controls the location where flaming combustion occurs; thus, the flame height at any given time represents the vertical distance (i.e., the mixing length) over which the fuel and air must be transported to complete the combustion process. Therefore, a reduction in air entrainment can result in greater flame heights, since the fuel vapor must be transported over a longer mixing length in order to completely mix with the reduced amount of air.

5.6.4.6.4 Effect of Walls. If the fuel package is positioned adjacent to one wall in a manner sufficient to reduce the air entrainment, there will be an increase in the absolute temperature of the upper layer when compared with the same fire positioned away from the wall. In contrast, experimental results have shown no significant increase in flame length for fire against a wall. Figure 5.6.4.6.4(a) and Figure 5.6.4.6.4(b) provide an example of this finding for a fire away from and against a wall.

5.6.4.6.5 Effect of Corners. When the same fuel package is placed in a corner sufficient to further reduce the air entrainment, there will also be an increase in the absolute temperature of the upper layer when compared with the same fire positioned away from corners. Similarly, a significant increase in the flame height is observed when the flames are attached

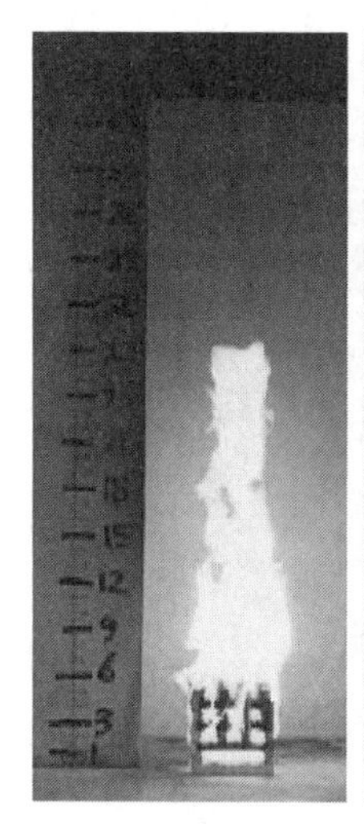
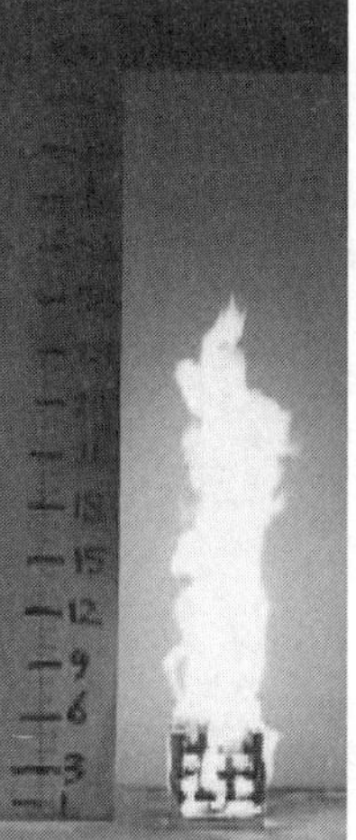

FIGURE 5.6.4.6.4(a) Average Flame Heights for Replicate Wood Crib Fires in the Open. The range of measured heat release rates and estimated average flame heights were 24 kW to 26 kW and 27 in. to 30 inches, respectively.

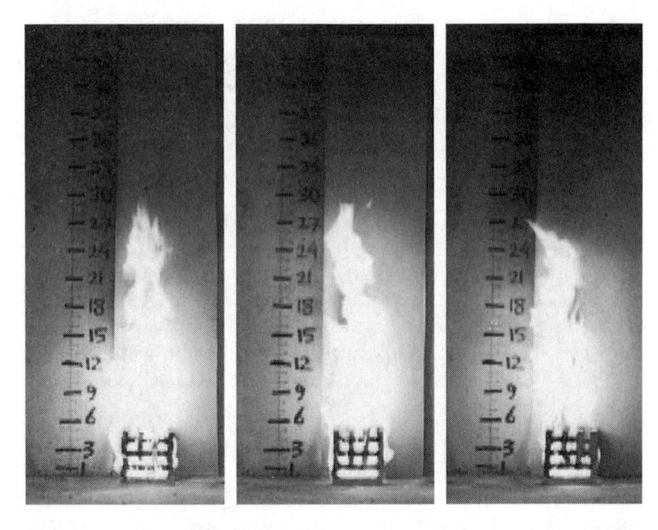

FIGURE 5.6.4.6.4(b) Average Flame Heights for Replicate Wood Crib Fires Against the Wall. The range of measured heat release rates and estimated average flame heights were 21 kW to 25 kW and 27 in. to 20 in., respectively.

to the walls in a corner configuration. Figure 5.6.4.6.5 provides an example of the increase in flame height for a fire in a corner configuration.

5.6.4.6.6 Analysis of Wall Effects. The possible effect of the location of wall(s) relative to the fire should be considered in the analysis of the fire and/or the interpretation of damage patterns produced by the fire.

5.6.4.6.7 Outdoor Fires. It should be noted that similar effects to those described above for indoor fires will also be observed for outdoor fires.

5.6.4.7* Flames that have flame heights in excess of the ceiling height result in flame extensions along the ceiling. If the free flame height is much greater than the ceiling height, the flame extension generally results in longer flames than would exist in the absence of a ceiling *(see Figure 5.6.4.7)*. The total length of a flame becomes longer ($H + h_r$) when cut off by a ceiling, compared to its free height (h_f).

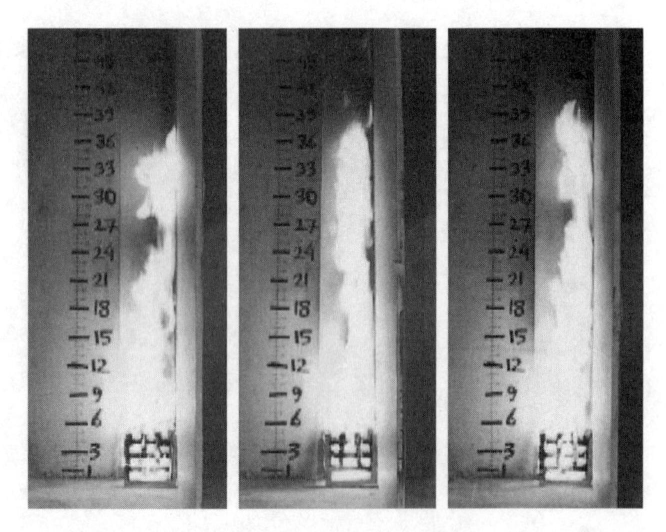

FIGURE 5.6.4.6.5 Average Flame Heights for Replicate Wood Crib Fires in a Corner Configuration. The range of measured heat release rates and estimated average flame heights were 25 kW to 26 kW and 37 in. to 40 in., respectively.

5.6.4.8 Factors such as ceiling height and distance from the plume can have significant effects on the response time of fire protection devices, such as heat and smoke detectors and automatic sprinklers. For a given device and fire size (as determined by HRR), the response time of the device will increase with higher ceilings and with increasing distance from the plume. Stated another way, the higher the ceiling or the farther away the device, the larger the heat output from the fire will be at the time the device responds. These factors should be considered when attempting to understand why a fire appears to be larger than expected at the time of alarm or sprinkler operation.

5.6.5 Thermal Structure of a Flame.

5.6.5.1 Continuous Flaming Region. Maximum time-averaged flame temperatures at a height occur at the centerline of the fire. In the continuously flaming region, centerline temperatures are approximately constant around 1000°C (1832°F). As indicated by data in Table 5.6.5.1, there is little variation in this temperature with the fuel. Methanol flames have higher temperatures due to the low radiant output of the flame, while sootier, more radiative flames are somewhat lower in temperature. In very large pool fires, the sootier flames can reach temperatures of 1200°C (2192°F) because radiative losses are relatively smaller. Flame temperatures for accelerants are not higher than for ordinary fuels, like wood or plastics.

5.6.5.2 Intermittent Flame Region. Centerline time-averaged temperatures in the intermittent region fall from about 1000°C (1832°F) at the continuous flame region to about 300°C (572°F) at the plume region. The time-averaged temperature at the average flame height (50 percent intermittency) is about 500°C (932°F).

5.6.5.3 Plume Region. Centerline time-averaged temperatures in the plume region fall from about 300°C (572°F) at the intermittent flame region to ambient temperatures well above the visible flame.

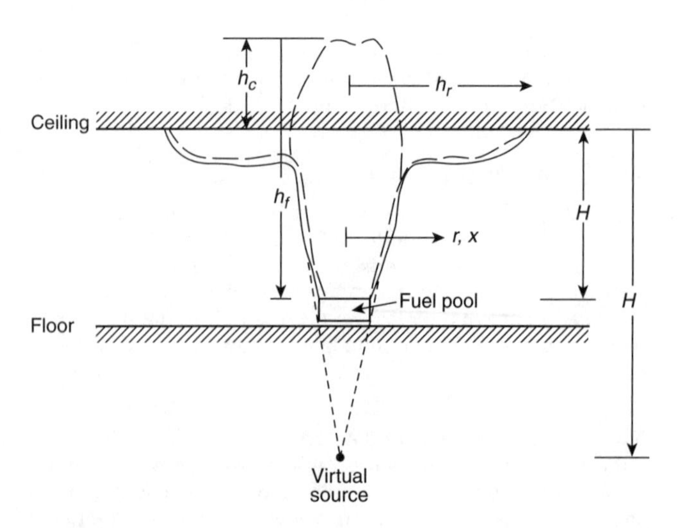

FIGURE 5.6.4.7 Representation of Theoretical Flame Heights in a Room with a Ceiling.

Table 5.6.5.1 Maximum Time-Averaged Flame Temperatures Measured on the Centerline of Fires Involving a Range of Fuels

Source	Temperature °C	Temperature °F
Flames		
Benzene[a]	920	1690
Gasoline[a]	1026	1879
JP-4[b]	927	1700
Kerosene[a]	990	1814
Methanol[a]	1200	2190
Wood[c]	1027	1880

[a]From Drysdale, *An Introduction to Fire Dynamics.*

[b]From Hagglund, B., Persson, L. E. (1976), Heat Radiation From Petroleum Fires, National Defence Research Inst., Stockholm, Sweden, FOA Report C20126-D6(A3).

[c]From Hagglund, B., Persson, L. E (1974), Experimental Study of the Radiation From Wood Flames, National Defence Research Inst., Stockholm, Sweden, FOA Report C4589-D6(A3).

5.6.6 Heat Fluxes from Flames. The thermal impact of a flame on nearby materials (combustibles or noncombustibles) and surfaces is measured in terms of the heat flux history to those surfaces. For example, the thermal decomposition and ignition of combustibles and the calcination of gypsum are governed by the incident heat flux history. As such, fire spread and fire-generated patterns are directly governed by the distribution of heat flux from flames to adjacent surfaces.

5.6.6.1 Heat Fluxes from Flames to Contacted Surfaces.

5.6.6.1.1 Walls. Figure 5.6.6.1.1(a) shows the distribution of heat flux from a fire in a room corner to the wall surface. The fire source had a heat release rate of 300 kW and the flames reached the ceiling. Figure 5.6.6.1.1(b) shows the same condition, but in the absence of a ceiling. The shapes of the heat flux contours are clearly different, with the ceiling case showing a more pronounced V-pattern nature than the no ceiling case.

5.6.6.1.2 Ceilings. Figure 5.6.6.1.2 shows the heat flux contours on a ceiling. The maximum heat flux occurs at the area of flame impact, and the fluxes are reduced with increasing distance from the impact area.

5.6.6.2* Heat Fluxes from Flames to Remote Surfaces. Heat fluxes from flames to a remote surface decrease rapidly with distance. Figure 5.6.6.2(a) shows the maximum heat flux as a function of distance from a chair or couch fire to targets at a number of heights. fire. Figure 5.6.6.2(b) shows the heat flux histories for a number of target distances with the target 0.4 m above the floor. Figure 5.6.6.2(c) shows the heat flux as a function of L/D, the distance of the ground level target from a pool fire at the center of a circular pool fire divided by the pool diameter (e.g., $L/D = 0.5$ is the edge of the pool). Both these figures illustrate that heat fluxes are markedly reduced at target distances comparable to the fire diameter.

5.7* Ignition. Forms and mechanisms of ignition vary with the form of the material (gas, liquid, solid), the chemical properties of the material, and the form and intensity of heating. Classifications of ignition include smoldering vs. flaming ignition, and piloted vs. autoignition. Piloted ignition occurs when an external ignition source acts to ignite flammable vapors. Pilot sources include small flames, sparks, and hot surfaces. The following is a general introduction.

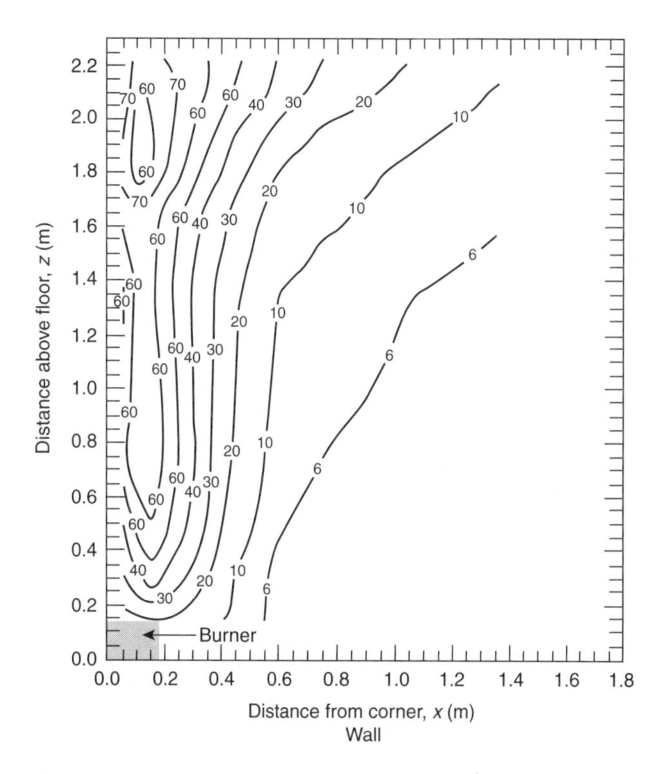

FIGURE 5.6.6.1.1(a) Wall Heat Flux (kW/m²) Contours from a 300 kW Fire in a Corner Configuration. A ceiling is present at 2.3 m above the floor and the fuel burned was propane. [*Adopted from Lattimer (2002).*]

FIGURE 5.6.6.1.1(b) Wall Heat Flux (kW/m²) Contours from a 300 kW Fire in a Corner Configuration. No ceiling is present and the fuel burned was propane. [*Adopted from Lattimer (2002).*]

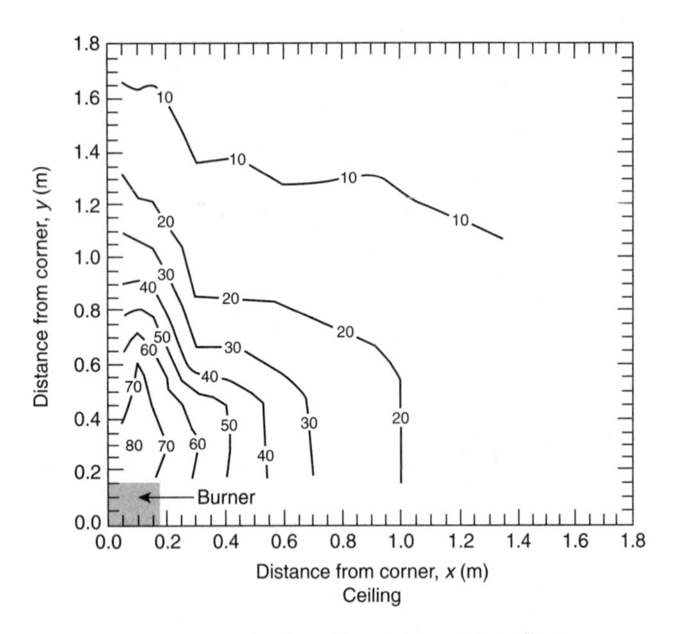

FIGURE 5.6.6.1.2 Ceiling Heat Flux (kW/m²) Contours from a 300 kW Fire in a Corner Configuration. A Ceiling is Present at 2.3 m Above the Floor and the Fuel Burned was Propane. [*Adopted from Lattimer (2002).*]

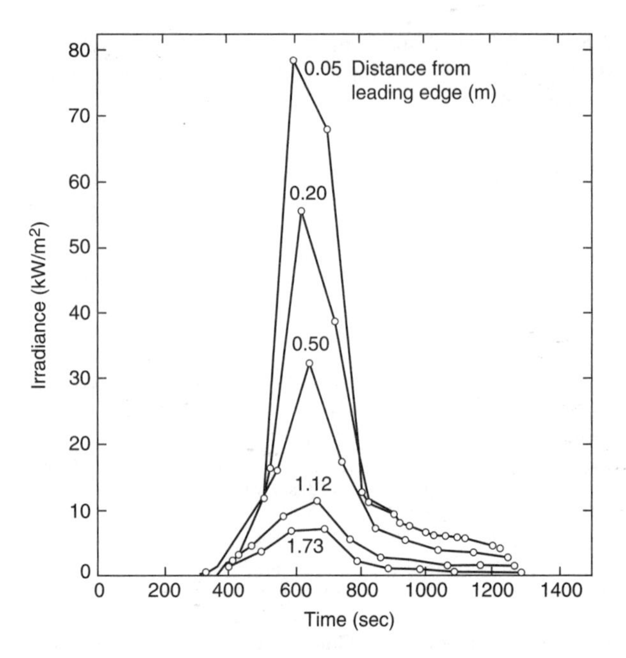

FIGURE 5.6.6.2(b) Radiant Heat Flux Histories to Targets at a Height of 0.41 m Facing a Wicker Couch Fire. [*Adopted from Krasny, Parker, and Babrauskas(2001).*]

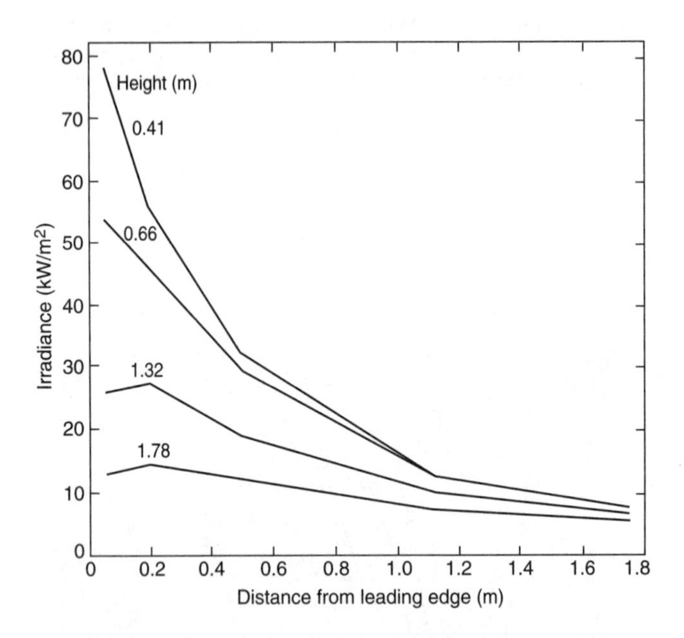

FIGURE 5.6.6.2(a) Maximum Radiant Heat Flux to Targets Facing a Wicker Couch Fire. [*Adopted from Krasny, Parker, and Babrauskas(2001).*]

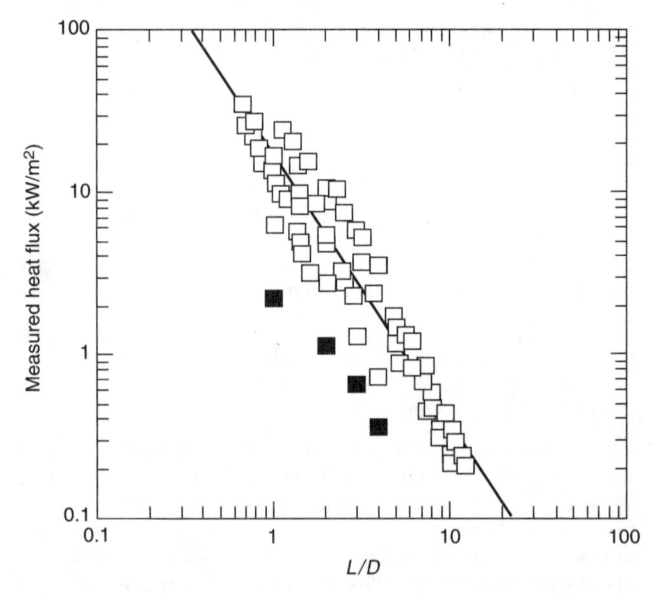

FIGURE 5.6.6.2(c) Radiant Heat Flux to a Target at Ground Level Facing Pool Fires. Data from 1 m to 30 m diameter pool fire tests are shown; data from larger diameter pool fire tests do not follow the correlation and are omitted here.

5.7.1 General. In order for most materials to be ignited, they generally must be in a gaseous or vapor state. A few materials may burn directly in a solid state or glowing form of combustion, including some forms of carbon (such as charcoal) and magnesium. Gases or vapors from ordinary fuels must be present in the atmosphere in sufficient quantity to form a flammable mixture. Liquids with flash points below ambient temperature do not require additional heat to produce a flammable mixture. The temperature of the fuel vapors produced must then be raised to their ignition temperature. The time and energy required for ignition to occur is a function of the energy of the ignition source, the thermal inertia (k, ρ, c) of the fuel, the minimum ignition energy, and the geometry of the fuel. If the fuel is to increase in temperature, the rate of heat transfer to the fuel must be greater than the sum of the conduction losses, convection losses, radiation losses, energy associated with phase changes (such as the heat of vaporization), and energy associated with chemical changes (such as pyrolysis). In some cases, chemical changes in the fuel during heating may also produce heat prior to combustion (exothermic reaction). If the fuel is to reach its ignition temperature, the heat source itself must have a temperature higher than the fuel's ignition temperature. Spontaneous ignition is an exception.

5.7.1.1 Table 5.7.1.1 shows the temperature of selected ignition sources. A few materials, such as cigarettes, upholstered furniture, sawdust, and cellulosic insulation, are permeable and readily allow air infiltration. These materials can undergo solid phase combustion, known as smoldering. This is a flameless form of combustion whose principal heat source is char oxidation. Smoldering produces more toxic compounds than flaming combustion per unit mass burned, and it provides a chance for flaming combustion from a heat source too weak to produce flame directly.

Table 5.7.1.1 Reported Burning and Sparking Temperatures of Selected Ignition Sources

Source	Temperature	
	°C	°F
Flames		
Benzene[a]	920	1690
Gasoline[a]	1026	1879
JP-4[b]	927	1700
Kerosene[a]	990	1814
Methanol[a]	1200	2190
Wood[c]	1027	1880
Embers[d]		
Cigarette (puffing)	830–910	1520–1670
Cigarette (free burn)	500–700	930–1300
Mechanical sparks[e]		
Steel tool	1400	2550
Copper–nickel alloy	300	570

[a]From Drysdale, *An Introduction to Fire Dynamics.*
[b]From Hagglund, B., Persson, L. E. (1976), Heat Radiation From Petroleum Fires, National Defence Research Inst., Stockholm, Sweden, FOA Report C20126-D6(A3).
[c]From Hagglund, B., Persson, L. E (1974), Experimental Study of the Radiation From Wood Flames, National Defence Research Inst., Stockholm, Sweden, FOA Report C4589-D6(A3).
[d]From Krasny, J. (1987) *Cigarette Ignition of Soft Furnishings — A Literature Review with Commentary,* NBSIR 87-3509; National Bureau of Standards, Gaithersburg MD.
[e]From NFPA *Fire Protection Handbook,* 15th ed., Section 4, p. 167.

5.7.2 Ignition of Flammable Gases.

5.7.2.1 Flammable gases can only be ignited by a spark or pilot flame over specific ranges of gas concentration. These limits are normally expressed as the lower flammable/explosive limit (LFL/LEL), the lowest concentration by volume of flammable gas in air that will support flame propagation, and the upper flammable/explosive limit (UFL/UEL), the highest concentration of flammable gas in air that will support flame propagation. These limit concentrations fluctuate with temperature and pressure changes, and with changes in oxygen concentration.

5.7.2.2 In the absence of a spark or pilot flame, a flammable gas–air mixture can autoignite if the temperature of the mixture is sufficiently high. The lowest temperature at which a flammable gas–air mixture can be ignited without a pilot is termed the autoignition temperature (AIT). The AIT is strongly dependent upon the size and geometry of the gas volume and the flammable gas concentration. Typically, large volumes and stoichiometric flammable gas–air mixtures favor ignition at lower temperatures. Because the AIT is dependent upon the conditions, a handbook AIT determined using standard test methods is primarily of value in comparing different gases. Comparisons of different gases must be made in the same apparatus and conditions to be meaningful. Open clouds of flammable gas–air mixtures can ignite on hot surfaces, with ignition occurring at lower temperatures for larger hot surface areas.

5.7.3 Ignition of Liquids.

5.7.3.1 Flashpoint. The ignition of a liquid in a flashpoint test occurs when a sufficient vapor concentration is generated above the liquid surface to allow ignition of the flammable vapors above the liquid surface by a pilot source. The flammable vapor concentration at the surface must reach the lower flammability limit *(see 5.7.2, Ignition of Flammable Gases).* The liquid temperature above which an ignitible concentration of flammable vapors is generated is known as the flash point. At the flash point temperature, the vapors above the liquid can be ignited, but typically sustained burning of the liquid does not occur. The liquid must be heated to a slightly higher temperature, known as the fire point, at which burning of the vaporizing liquid fuel can be sustained as a pool fire. For some liquids, the flash point and fire point temperatures are the same.

5.7.3.2 Liquids at bulk temperatures below the fire point temperature cannot be ignited by a pilot flame or spark. However, liquids can be heated locally to achieve ignition and the fire can then spread to involve the pool. Local heating mechanisms can include flame impingement on the liquid surface or burning of the pooled liquid at a wick formed by material wetted by the liquid. Local application and ignition of a liquid above its flash point to a liquid below its flash point is another method that can cause ignition of a liquid that is otherwise below its flash point temperature.

5.7.3.3 Atomized liquids or mists (those having a high surface area to mass ratio) can be more easily ignited than the same liquid in the bulk form. In the case of sprays or mists, piloted ignition can occur at temperatures below the published flash point of the bulk liquid, and even very high flashpoint liquids (several hundred degrees °C) have been shown to be ignitible when in the form of a spray.

5.7.3.4 Some liquids are capable of being oxidized in the liquid phase. Most often this only leads to ignition when the liquid is supported on a porous substrate (e.g.,. linseed oil on rags) This subject is treated in the solid fuels section on self-heating below. However, in some industrial situations contact between two liquid phases can result in an exothermic reaction (not necessarily oxidation) sufficient to cause an explosion.

5.7.3.5 Autoignition of a liquid can occur if the flammable vapors produced above the liquid surface are sufficiently hot so as to support gas phase autoignition as discussed above in the Ignition of Gases section. AITs for a given liquid vary with the scale and configuration, as they do for gases. Quantitative AIT determinations in the same apparatus are useful for comparing the behavior of different liquids.

5.7.4 Ignition of Solids. There are three forms of ignition that occur with solid fuels: smoldering ignition or, more generally, initiation of solid phase burning; piloted flaming ignition; and flaming autoignition.

5.7.4.1 Smoldering Ignition and Initiation of Solid Phase Burning.

5.7.4.1.1 General. Smoldering is a solid phase burning process, which normally includes a thermal decomposition step to create a char, followed by solid phase burning of the char produced.

5.7.4.1.1.1 The thermal decomposition process, often called pyrolysis, may be a purely thermal process or may involve interaction with oxygen. When oxygen is known to be involved, this is often referred to as oxidative pyrolysis. The initial thermal decomposition process is normally endothermic [i.e., it requires or uses energy rather than producing heat or energy (which would be exothermic)].

5.7.4.1.1.2 While some virgin materials are capable of solid phase oxidation (e.g., carbon or magnesium), most materials that smolder must be pyrolyzed to form a carbonaceous char, which subsequently oxidizes in the solid phase. The most common class of materials that smolder in this manner includes wood, paper, and other lignocellulosic products.

5.7.4.1.1.3 Materials that are neither capable of solid phase burning as a virgin fuel, nor capable of being pyrolyzed to form a char that can burn cannot smolder. As such, most thermoplastic materials are not capable of smoldering. Some thermosetting polymers (e.g., polyurethane foam), often decompose to form a liquid product when vigorously heated, but do form a char under more modest heating conditions.

5.7.4.1.1.4 The term *smoldering* is sometimes inappropriately used to describe a nonflaming response of a solid fuel to an external heat flux. Solid fuels, such as thermoplastics, when subjected to a sufficient heat flux, will degrade, gasify, and release vapors. There usually is little or no oxidation involved in this gasification process, and thus it is endothermic. This process is pyrolysis, and not smoldering. Smoldering must involve a solid phase exothermic process (i.e., it must be self-sustained).

5.7.4.1.1.5 Spontaneous combustion due to self-heating is a special form of smoldering ignition that does not involve an external heating process. An exothermic reaction within the material is the source of the energy that leads to ignition and burning. The key concept in ignition by self-heating is the ability of the material to dissipate the heat generated by the internal exothermic reactions. If the heat generated by the reaction cannot be dissipated to the surroundings, the material will rise in temperature to an extent that the reaction rates accelerate (i.e., runaway), and a smolder front is formed. Key variables in self-heating include the ambient temperature, the pile size, and the reaction kinetics of the exothermic process. As the ambient temperature rises, the baseline reaction rate increases, and as the pile size increases, the ability to dissipate heat to the surroundings decreases. Both high ambient temperatures and large pile sizes favor self-heating processes. See the following section for more detailed information concerning self-heating in piles.

5.7.4.1.1.6 While self-heating is most often associated with ignition processes in piles due to the inability of the material to dissipate the heat from internal exothermic reactions, all smoldering ignition mechanisms can be understood in the context of the fundamentals of self-heating theory. Smolder initiation by radiative heating, smolder initiation by contact with a hot surface, smoldering ignition by contact with hot objects (e.g., contact with hot welding slag, burning embers, or a cigarette), smoldering ignition of layers or other accumulations of dust in a dryer are all governed by the fundamental laws of self-heating theory. If heat from oxidizing the material cannot be adequately dissipated, a thermal runaway, resulting in smoldering, will occur.

5.7.4.1.1.7 Because all smoldering ignition mechanisms are governed by self-heating laws, there is no generally or widely applicable "standard" ignition temperature that can be assigned. For a specific pile size of a specific material, there is a critical ambient temperature (CAT) above which ignition is expected to occur. For a wood surface heated by radiation, there is a specific surface temperature above which smoldering will occur. However, these ignition temperatures are only applicable to the conditions under which they were experimentally determined. They are not generally or broadly applicable.

5.7.4.1.2 Self-Heating and Self-Ignition.

5.7.4.1.2.1 Self-heating is a process whereby a material undergoes a chemical reaction and increases in temperature solely due to exothermic reactions between the material (normally a solid) and the surrounding atmosphere (normally air).

5.7.4.1.2.2 Most organic materials and metals capable of reacting with oxygen will oxidize at some critical temperature with the evolution of heat. The evolution of heat is not restricted to oxidation reactions, but can also be due to various other chemical reactions, for example, polymerization, where liquids react to form solids. Generally, self-heating and spontaneous combustion (self-ignition) are commonly encountered in organic materials, such as animal and vegetable fats and oils, because these materials contain polyunsaturated fatty acids. Such fatty acids react with oxygen to generate heat. Unsaturated molecules contain carbon-to-carbon double bonds, which are reactive.

5.7.4.1.2.3 Self-heating and spontaneous combustion (self-ignition) of oils containing mostly saturated hydrocarbons, such as motor oil or lubricating oil, occur only under elevated temperature conditions (e.g., an oil-soaked rag wrapped around a steam pipe) or in very large piles at lower temperatures. Saturated hydrocarbons contain carbon-to-carbon single bonds, which are far less reactive than unsaturated oils. Unlike highly unsaturated oils such as linseed oil, consumer quantities of motor oil or lubricating oil on rags are not expected to self-heat to ignition.

5.7.4.1.2.4 Certain inorganic materials, such as metal powders, may undergo rapid oxidation, self-heating, and self-ignition to form metal oxides, given appropriate conditions.

5.7.4.1.3 Mechanism of Self-Heating to Ignition. Spontaneous combustion requires certain steps for it to occur. First, the material must be capable of self-heating and must be subjected to conditions where self-heating is elicited. Next, the self-heating must proceed to thermal runaway (i.e., the heat generated exceeds the heat losses to the environment). Thermal runaway, in theory, means a temperature rise so large that stable conditions can no longer exist. In practice, it means that the material will undergo an internal temperature rise (often at or near its middle) on the order of several hundred degrees Celsius. Next, thermal runaway must result in self-sustained smoldering. The opposite of this is a condition where the material chars locally, but fails to establish a propagating smolder front.

5.7.4.1.3.1 Thermal runaway is an instability that occurs when heat generation exceeds heat loss within the material. It is a contest between exothermic chemistry and heat loss to the surroundings. Heat generation has its best chance of winning at the most insulated parts of the fuel package, that is, at the center, and this is usually where the highest temperatures are found. How well-insulated the interior of the fuel package is depends on the distance to the boundary and the temperature there. During self-heating, the center temperature is typically higher than the surrounding temperatures.

5.7.4.1.3.2 Self-heating to ignition requires a porous, permeable, and oxidizable material — the material must have all three properties. When smoldering, the fuel must char without melting, otherwise the porous and permeable qualities will be lost and self-heating will be inhibited. The solid may initially serve primarily as an inert substrate, as in the case of linseed-oiled rags, or the substrate also may act as the fuel. The most common self-heating substrates are organic solids derived from plant materials, such as cotton fabrics, wood and wood products, agricultural products, and coal. Self-heating may occur when the surroundings are at ordinary ambient conditions, for example, a pile of linseed-oiled rags, or it may require an elevated temperature.

5.7.4.1.3.3 The tendency to self-heat is dependent on the size and shape of the fuel package and its surrounding conditions. This tendency is not exclusively a material property. Therefore, evaluation of a material's self-ignition potential is incomplete (except for the elimination of non-self-heating fuel packages) without considering the particulars of the material's size, shape, and surroundings. For a given volume, low surface area shapes, such as spheres or cubes, promote self-heating more than high surface area shapes, such as thin sheets. The small external surface area reduces heat loss and the outer parts of the fuel package insulate the interior, promoting the rise in the interior temperature from self-heating. For example, linseed-oiled rags in a pile are more likely to self-heat than the same rags on a clothesline or laid flat. Figure 5.7.4.1.3.3 is a graphical representation of the conditions required for spontaneous ignition to occur.

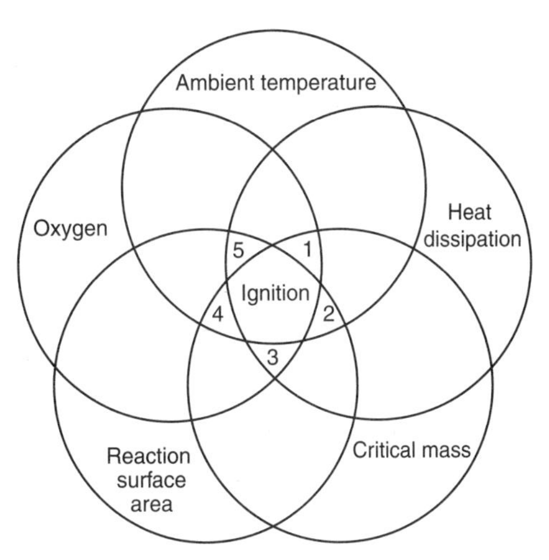

FIGURE 5.7.4.1.3.3 Conditions Required for Spontaneous Ignition to Occur in Materials Capable of Self-Heating.

5.7.4.1.3.4 The fuel package's initial temperature may be the crucial factor in whether ignition occurs. This is sometimes encountered in industrial drying of plant-derived materials such as wood products and agricultural products, and drying of oiled fabrics in clothes dryers. If fuel packages are assembled into sufficiently large symmetrical shapes while too hot, they may proceed to thermal runaway and ignition. If the material is dried to moisture content lower than equilibrium, part of the internal heat generation results from the latent heat of moisture absorption — the opposite of evaporative cooling.

5.7.4.1.3.5 Self-heating and the resulting smoldering within a pile may not be noticeable until the smoldering front reaches the surface. First visual indications of self-heating may be a wet spot on the surface of the pile resulting from condensation of water or from other products of the reactions. Redeposition of smoke within the pile often leads to little or no observable smoke above or around the pile. Musty odors may first be noticeable where piles are in enclosed areas.

5.7.4.1.3.6 Enclosure of the fuel in a sealed container or envelope may arrest self-heating because the enclosure eliminates one of the necessary conditions for self-heating-permeability, which allows oxygen diffusion into the solid. Without a supply of oxygen, oxidation and heat generation are inhibited unless the oxidizer is present within the material. For example, linseed-oiled rags in a closed paint can might not self-heat significantly before consuming the oxygen in the can. The use of containers or vapor barriers has been successfully employed to mitigate self-heating hazards, though depending on the amount of oxidizer present in the container and the physical properties and integrity of the container or barrier, self-heating to ignition may occur. Investigation of the particulars of the material storage container is required to assess the potential for self-heating within the container.

5.7.4.1.3.7 The minimum surroundings or exposure temperature necessary for ignition via self-heating is generally lower than the minimum ignition temperature for the same material without self-heating. For example, linseed oil on a cotton substrate can ignite when the surroundings are at ordinary ambient temperatures (20°C or 68°F), yet in the pure liquid form its flash point is reported as 222°C (428°F) and its AIT is 343°C (585°F).

5.7.4.1.3.8 Haystacks and other large packages of biomass that are assembled at ambient temperature may begin self-heating with biological activity. If the fuel's moisture content is appropriate, this biologically driven self-heating may be supplanted by oxidation, and thermal runaway followed by ignition may result.

5.7.4.1.3.9* Wood Ignition. Wood, like many other cellulosic materials, is subject to self-heating when exposed to elevated temperatures below its ignition temperature. However, the temperatures at which self-heating of wood will occur is not an intrinsic property of the material. Rather, it is dependent on factors such as the nature of the heat exposure, pile size, and geometry. For short-term heating (less than one day), wood requires a minimum temperature of approximately 250°C (482°F) to ignite, although this value rises as the heat flux increases. For wood subjected to long-term, low-temperature heating, exothermicity due to self-heating is increasingly important. Factors such as the nature of the heat exposure, the size of the wood specimen, and the geometry of the specimen play a deciding role. The scientific community has not reached consensus concerning the self-heating ignition of wood subjected to long-term heating.

5.7.4.1.3.10* Charcoal Briquettes. Charcoal briquettes have been suspected of self-heating to ignition even when packaged in household-sized bags [<9 kg (<20 lb)] at ordinary ambient temperatures. Of particular suspicion is the effect of wetting briquettes with water followed by evaporation. However, rigorous laboratory testing has shown that common-size bags of briquettes do not even approach self-ignition, even when placed in unusually high ambient temperatures, such as in a closed automobile in the sun. Charcoal briquettes have no greater tendency for self-heating than other common materials such as sawdust or hay, even after wetting. Spontaneous combustion of bagged charcoal briquettes in commercially available sizes is not possible under any normal ambient conditions. The effect of contaminants upon self-heating of charcoal briquettes has not been studied.

5.7.4.1.3.11* Contaminants. Contaminants in wood or other cellulosic materials are known to have an effect on thermal decomposition and oxidation. Contaminants that have been identified include fatty acids, vegetable oils, iron, iron oxide, iron sulphides, cobalt, copper, magnesium, lead carbonate, potassium carbonate, lead acetate, sodium acetate, and vanadium pentoxide. Quantitative work to deduce the effect of such contaminants on ignition behavior remains to be done.

5.7.4.1.4* Common Materials Subject to Self-Heating.

5.7.4.1.4.1 The most commonly encountered forms of self-heating and self-ignition by fire investigators are the following:

(1) Polymerization of fatty acids (animal fats, cooking oils, and drying oils) in cellulose materials (wood, cloth, and paper)
(2) Oxidation of carbonaceous materials (coal and charcoal)
(3) Biologically induced oxidation (hay bales and compost)
(4) Heat-induced oxidation of lingocellulose materials (usually wood fiber and cloth)
(5) Polymerization reactions (plastics, rubbers, adhesives, and paint overspray particles) *(See* NFPA 33, *Standard for Spray Application Using Flammable or Combustible Materials.)*

5.7.4.1.4.2 Because of the many possible combinations of these controlling or influencing factors, it is difficult to predict when the material will self-heat. Annex A of the NFPA *Fire Protection Handbook,* 19th edition includes a list of materials suseptable to spontaneous ignition. Omission of any material does not necessarily indicate that it is not subject to self-heating.

5.7.4.1.5 Oxidizer Fires. An oxidizing agent is a chemical substance that, while not necessarily combustible by itself, can rapidly increase the rate of burning of other substances, or result in spontaneous combustion when combined with other substances. Many of these oxidizers are only found in industrial situations, but swimming pool sanitizers, such as calcium hypochlorite and the salts of dichloroisocyanuric or trichloroisocyanuric acid (stabilized chlorine) will undergo spontaneous heating or spontaneous combustion when contaminated with certain organic materials, particularly hypergolic substances, or with other oxidizers. Once an oxidizer fire starts, a decomposition reaction may take place within the mass of oxidizer, which, while not technically combustion, evolves large quantities of heat and light. These reactions may be quite violent. For a list of oxidizing substances, and the regulations governing their classification and storage requirements, see NFPA 400, *Hazardous Materials Code.*

5.7.4.1.6 Pyrophoric Materials. Certain elements, particularly white phosphorus, sodium, potassium, and some finely divided metals, such as zirconium, spontaneously ignite when exposed to air. Materials that undergo spontaneous combustion upon exposure to air are known as *pyrophoric.*

5.7.4.1.7 Transition to Flaming Combustion.

5.7.4.1.7.1 Smoldering can transition to flaming if the smoldering creates sufficient flammable vapors for piloted flaming ignition. This normally occurs when the smoldering becomes vigorous as a result of enhanced airflow to the smolder region. This can occur as a result of the spread of smoldering to generate additional airflows, the creation of a hole or channel by smoldering that then acts as a chimney, or by external imposition of an airflow. When a flammable concentration of vapors is developed, the glowing char can act as the ignition source for ignition of the vapors.

5.7.4.1.7.2* The time required from smolder initiation to transition to flaming is not predictable. Transitions to flaming in upholstered furniture have been observed in times ranging from 20 minutes to many hours. Times for transitions to flaming in large piles can be measured in days or months. Because transition to flaming is governed by changes in airflow and the creation of holes or channels, the time to transition to flaming combustion, if it happens at all, appears to be largely random.

5.7.4.1.7.3 When flaming combustion is initiated by a smoldering source such as a cigarette, or by self-heating, the process leading up to the appearance of the first flame may be quite slow. Once flaming combustion begins, however, the development of the fire may be faster than if the original ignition source were a flame, due to preheating of the fuel.

5.7.4.2 Piloted Flaming Ignition of Solid Fuels. For solid fuels to burn with a flame, the substance must either be melted and vaporized (e.g., thermoplastics) or pyrolyzed into gases or vapors (e.g., wood or thermoset plastics). In both cases, heat must be supplied to the fuel to generate the flammable vapors. In piloted ignition, these flammable vapors are ignited by a pilot source, in the form of a small flame, a spark, an ember, or a hot surface.

5.7.4.2.1 While the concept of a piloted ignition temperature for solids is an engineering approximation rather than a reproducibly measurable property, the principle is sufficiently robust to allow general application of experimentally determined ignition temperatures. The range of ignition temperatures for solids ranges from about 270°C to 450°C (518°F to 842°F). Ignition temperatures for non-fire retardant plastics tend to range from 270°C to 360°C (518°F to 680°F), wood based products tend to range from 330°C to 375°C (626°F to 707°F). Ignition temperatures above 400°C (752°F) are generally observed only in inherently fire-retardant materials or materials significantly loaded with fire retardants.

5.7.4.2.2 Associated with the ignition temperature concept is the concept of a minimum radiant flux for piloted ignition. Based upon a thick, well-insulated rear (unexposed) surface, the ignition temperature, along with simple steady state heat transfer concepts, can be applied to deduce the minimum radiant heat flux that can cause ignition. For most materials, the critical radiant heat flux is in the range of 10 to 15 kW/m^2 with some materials exhibiting higher values. This is in contrast with minimum heat fluxes for smoldering ignition of 7 to 8 kW/m^2.

5.7.4.2.3 The experimentally determined value of the minimum heat flux depends upon the duration of the test, with shorter test periods resulting in higher minimum heat flux values. Piloted ignitions have been observed up to about one hour of radiant heating. Most minimum heat flux testing uses a test duration of 15 to 20 minutes. This could increase the minimum heat flux from 10 to 12 kW/m^2.

5.7.4.2.4 As discussed in the Conduction Heat Transfer section, the transient surface temperature rise in a thick material during exposure to a specific heat flux is governed by the thermal inertia of the material. Because thermal inertias vary widely and ignition temperatures range narrowly, variations in material behavior result largely from variations in the thermal inertia. Generally, conductivity and thermal inertia are proportional to the material density. High-density materials of the same generic type (woods, plastics) conduct energy away from the area of the ignition source more rapidly than low-density materials, which act as insulators and allow the energy to remain at the surface. For example, given the same ignition source, oak takes longer to ignite than a soft pine, and low-density foam plastic ignites more quickly than high-density plastic. It is relatively easy to ignite a pile of thin pine shavings, while ignition of a one-pound solid block of wood is more difficult.

5.7.4.2.5 The thickness of the material also markedly affects the ignitibility of a sample. For instance, a piece of paper or wood shaving is much easier to ignite than a thick block of wood. As the thickness increases, the time to ignition increases until thermally thick behavior dominates and increasing the thickness has little additional impact. Thin materials are also easier to ignite because two-sided heating at an edge is also possible. It is easier to ignite the edge of a piece of paper than the center of the sheet.

5.7.4.3 Flaming Autoignition of Solids. Where no pilot sources are available, ignition of solids relies upon autoignition of the flammable gases generated by heating. While autoignition temperatures are not well defined in that they are highly dependent upon environmental conditions, it is common to find reported AITs in the 400°C to 600°C (752°F to 1112°F) range. Drysdale (1999) reports two temperatures for wood to autoignite. These are heating by radiation, 600°C (1112°F), and heating by convection, 490°C (914°F). For autoignition to occur as a result of radiative heat transfer, the volatiles released from the surface need to be hot enough to produce a flammable mixture above its autoignition temperature when it mixes with unheated air. With convective heating, on the other hand, the air is already at a high temperature and the volatiles need not be as hot.

5.8* Flame Spread.

5.8.1 General. The growth of a fire normally includes the spread of flame over involved fuel surfaces. The rapidity of the fire growth depends upon the fuel properties and the orientation of the fuel surfaces. Broadly speaking, flame spread can be classified as concurrent flame spread or counterflow flame spread. These terms relate the direction of flame spread compared to the direction of gas flow. Examples of these types of flame spread are shown in Figure 5.8.1.

5.8.1.1 Counterflow Flame Spread. Counterflow flame spread, also known as opposed flow flame spread, occurs where the flame spread direction is counter to or opposed to the gas flow. Notable examples of this are lateral spread on a horizontal surface (Figure 5.8.1) or downward flame spread on a vertical surface. Counterflow flame spread is generally slow as a result of the limited ability of the flame to heat the fuel ahead of the flame front.

5.8.1.2 Concurrent Flame Spread. Concurrent flame spread, also known as wind-aided flame spread, occurs where the flame spread direction is the same as the gas flow or wind direction. Notable examples of concurrent flow flame spread include upward flame spread on a wall. Concurrent flame spread is generally quite rapid as a result of the direct contact of the flame with the fuel ahead of the flame front.

5.8.1.3 Fire Spread on Sloped Surfaces. Figure 5.8.1.3 illustrates the counterflow and concurrent flow on surfaces at varying slopes. Fires on sloped surfaces, such as sloping combustible walls, ramps, or stairs, display the effects of concurrent flow fire spreads *(see Figure 5.8.1.3)*. Most such ramp or stair constructions in structures are sloped at 30 degrees to 50 degrees from the horizontal. The sloped fire spread is a combined effect of the preheating of the combustible surface above the flame by conductive, convective, and radiant heat transfer mechanisms and the exponentially increased radiant effect on the surface above the flame by the concurrent air entrainment from the side facing the slope "bending" the flame down toward the surface.

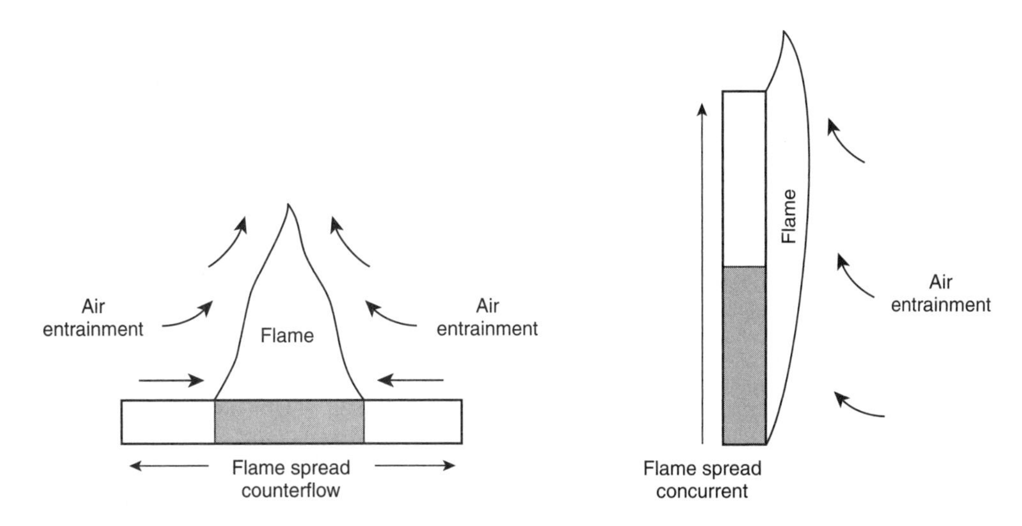

FIGURE 5.8.1 Examples of Counterflow and Concurrent Flame Spread. [*Adopted from Beyler and DiNenno (1994).*]

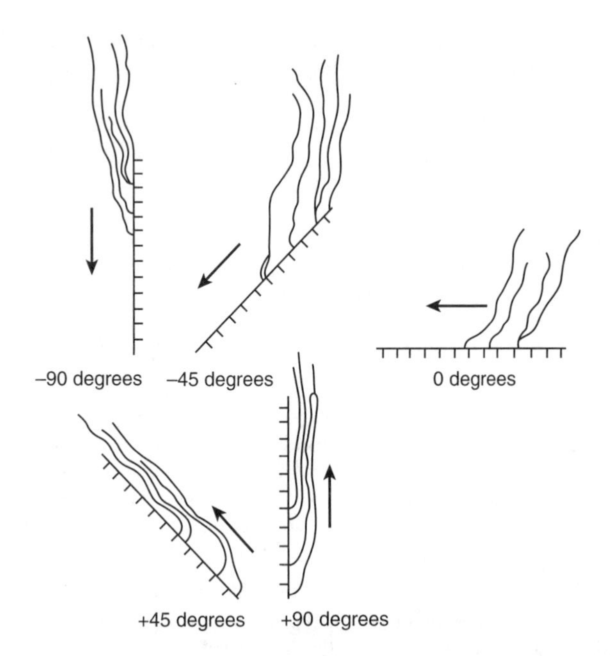

FIGURE 5.8.1.3 Interaction Between the Flame and the Fuel for Flame Spread at Different Angles of Inclination. Counterflow: −90 degrees, −45 degrees, 0 degrees; concurrent flow: +45 degrees +90 degrees.

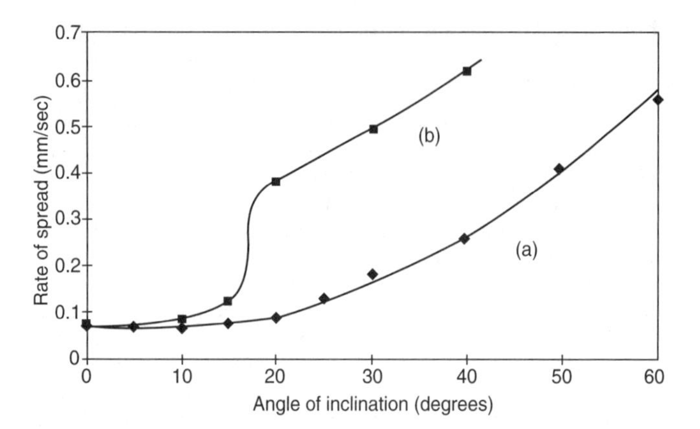

FIGURE 5.8.1.3.1 Variation of the Rate of Upward Flame Spread Over 60 mm Wide Slabs of PMMA at Different Angles. [*From "Fire Dynamics," ISFI (2006).*]

5.8.1.3.1* Fire Spread in an Inclined Trench. The 1987 King's Cross Underground ("tube") (subway) Station escalator fire in London drew attention to a previously unrecognized mechanism (at least for fires in buildings), the "Inclined Trench." An "Inclined Trench" is a sloped combustible surface that is laterally bounded by vertical side walls. The best example is an enclosed stairway, with walls on either side. Upward flame spread (concurrent flow) on vertical surfaces is rapid, but this can be enhanced by the local geometry. If there are two inward facing combustible surfaces, such as in a corner or within the side walls of an enclosed combustible stairway, upward spread is enhanced by cross-radiation (feedback radiation) between the burning surfaces. Cross-radiation between the surfaces can greatly enhance the rate of spread and the rate of burning. A relatively simple change to the way in which air could gain access to the flames has an enormous effect on the rate of heat transfer from the flame to the surface, which can result in extremely rapid upward flame spread. It is important to understand the mechanism in order to appreciate its behavior. For upward spread, the physical configurations around the spreading flame can have a dramatic effect. The fire at the King's Cross Underground Station in London in November 1987 involved a wooden escalator inclined at an angle of 30 degrees. A fire became established across the full width of the escalator and flames spread up the escalator "trench" also at an angle of 30 degrees, rather than rising vertically. This was caused by the confinement provided by the sides of the escalator. The rate of upward spread was totally unexpected and was probably equal to the rate that would have been experienced if the escalator had been vertical. The effect is illustrated in Figure 5.8.1.3.1, which shows how the rate of (upward) flame spread on the slabs of polymethylmethacrylate (PMMA) changes as the angle of the slab is increased from 0 degrees (horizontal) to 60 degrees. Without sidewalls, the rate of spread remained unchanged until about 20 degrees, and thereafter increased slowly. However, with sidewalls in place (mimicking the escalator side walls), the rate of spread increased dramatically at slopes above 15 degrees.

5.8.2 Flame Spread on Liquids.

5.8.2.1 Flame spread on liquid fuels depends upon the liquid temperature relative to the flash point. Below the flash point, flame spread is via liquid flow, and above the flash point, the flame spread is via gas phase spread mechanisms.

5.8.2.2 Liquid Phase Flame Spread. Most liquid phase flame spread is counterflow flame spread, though downwind flame spread in a pool fire or upward flame spread on fuel cascading down a wall are examples of concurrent flame spread. Counterflow flame spread on liquids is aided by surface tension–driven liquid flows within the pool, which accelerate flame heated fuel ahead of the flame front. For very thin fuel layer thicknesses, the surface tension–driven flows are retarded. Several investigators have observed that liquid phase flame spread does not occur at fuel thicknesses of less than 2 mm. Free spills of fuel on flat, horizontal surfaces are typically about 1 mm deep, though deeper spills are expected on realistically leveled floor surfaces. Liquid-driven flame spread rates are generally in the 1 to 10 cm/sec range. See Figure 5.8.2.2 for data on JP-8, a military fuel similar to commercial jet fuel (Jet A).

5.8.2.3 Gas Phase Flame Spread. When the liquid is above its flash point, flammable concentrations of fuel vapors exist near the fuel surface. Flame spread occurs through the premixed fuel vapor/air layer at speeds typical of premixed flames, 1 to 2 m/sec *[see Figure 5.8.2.2 for data on JP-8, a military fuel similar to commercial jet fuel (Jet A).]*

5.8.3 Flame Spread on Solids.

5.8.3.1 Flame spread on solid fuels can be thought of in terms of a continuous ignition process. At each location, the fuel needs to be heated to conditions where it can ignite and burn. As such, all the factors that affect ignition of a solid fuel also affect flame spread rates. Flame spread rates on solids depend upon the flame spread mechanism (concurrent vs. counterflow) as well as the thickness and thermal properties of the fuel.

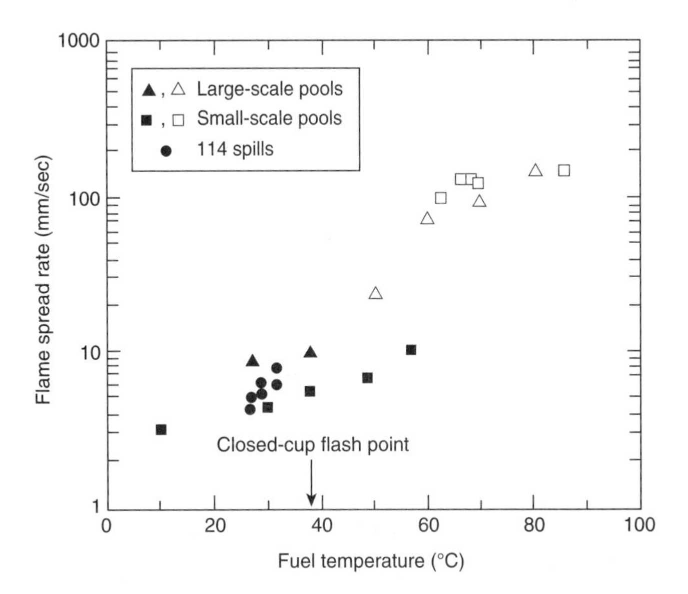

FIGURE 5.8.2.2 Counterflow Flame Spread Rates Over 5 mm (0.2 in.) Deep JP-8 as a Function of Fuel Temperature. JP-8 is a military fuel very similar to Jet B used in commercial jet aircraft. Large-scale data are from tests on a pool 5 ft by 40 ft while the small-scale data are from tests on a pool 8 in. by 5 ft. The transition to gas phase flame spread moves toward the flash point with increasing scale. [*Adopted from Gottuk and White (2002).*]

5.8.3.2 Counterflow Flame Spread on Thin Fuels. Counterflow flame spread on thin fuels normally occurs for downward flame spread. The flame attaches to the surface of the fuel on both sides of the sheet and the active burning region of the fuel is usually fairly short. Flame spread down a match stick or flame spread down a piece of paper are typical examples of this type of behavior. Flame spread rates are generally in the range of 0.2 to 2 mm/sec with the highest rates occurring for the thinnest fuel.

5.8.3.3 Concurrent Flame Spread on Thin Fuels. Concurrent flame spread on thin fuels normally occurs for upward flame spread. The flame attaches to the surface of the fuel on both sides of the sheet. Because the flame spread rate is faster than counterflow flame spread, the active burning region of the fuel is longer. Flame spread up a curtain or drapery or flame spread up a piece of paper are typical examples of concurrent flame spread on thin fuels. Flame spread rates are generally in the range of tens of cm/sec with the highest rates occurring for the thinnest fuel. No general trends of flame spread rate with fuel thickness are possible because thinner fuels ignite faster, but also burn out faster (yielding shorter flame lengths).

5.8.3.4* Counterflow Flame Spread on Thick Fuels. Counterflow flame spread on thick fuels normally occurs for downward flame spread on a wall or horizontal spread on an upward facing horizontal surface (see Figure 5.8.3.4). Heat transfer from the flame foot (see Figure 5.8.3.4 insert), is via both the gas phase and the solid phase. Heating rates are limited by the very small region of heat transfer, normally a heating length measured in millimeters. Heat is also lost into the thick material. Unenhanced flame spread rates are generally on the order of 0.1 mm/sec. Polyurethane foam spread rates of 2 to 4 mm/sec result from the very low density. Many thick materials cannot support counterflow flame spread without external heating. A common example of this is thick wood. With preheating to 100–200°C, most wood products will support counterflow flame spread. Sources of such preheating include radiation from hot compartment gases or from a nearby flame. Flue space geometries are well suited to provide mutual radiative support for surface burning and flame spread. With external heating, flame spread rates on thick solid fuels can approach the gas phase flame spread rates of liquid fuels.

5.8.3.5* Concurrent Flame Spread on Thick Fuels. Concurrent flame spread on thick fuels normally occurs for upward flame spread on walls or on the underside of combustible ceilings. With a thick fuel, the flame length that heats the material ahead of the burning area is continuously increasing. This allows for the possibility of unlimited acceleratory flame spread rates. Measurement of flame spread rate on a 1.6 m PMMA walls yields flame spread rates up to 6 mm/sec and measurements on a 3.5 m wall yield rates up to 10 mm/sec [Orloff et. al. (1974), Wu et. al. (1996)]. Similar measurements on a 1.3 m tall corrugated cardboard wall yielded flame spread rates up to 6 to 9 mm/sec [Grant and Drysdale (1995)]. Faster flame spread rates are observed in the flue spaces of rack storage of commodities in cardboard boxes. From an igniter at floor level at time zero, flames can involve fuel at a height of 2 m in 30 seconds and involve cardboard at the 5 m height in 42 seconds [McGrattan et. al. (1998).] Increasing flame spread rates do not occur for all materials and all exposing flames. If the fire exposing a wall is not sufficiently

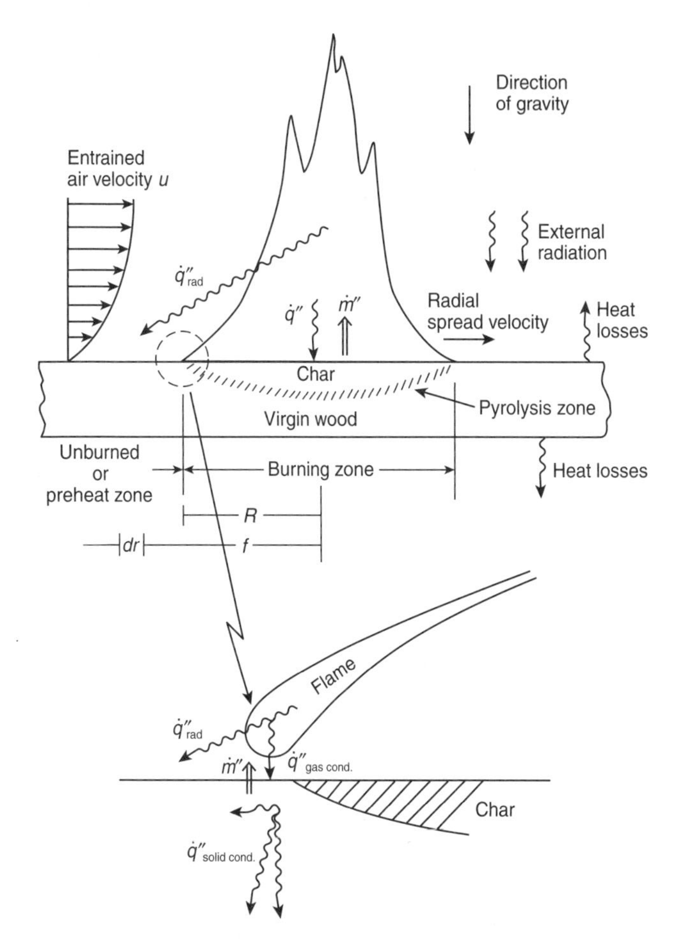

FIGURE 5.8.3.4 Flame Spread on a Horizontal Surface of Wood. [*Adopted from Atreya (1984).*]

large, upward flame spread may not occur at all or may grow to a limiting height before spread ceases [Saito et al. (1986)].

5.8.3.6 Role of Melting and Dripping in Flame Spread. Flames may spread differently on different portions of the same fuel package, for example, a flame may spread slowly across the horizontal surface of a chair cushion with spread increasing rapidly as the back of the chair becomes involved. If a material melts, it is likely to drip and pool on horizontal surfaces. The dripping of liquefied polyurethane foam from furniture can form a pool fire under that furniture. The ignition of the underside of the furniture by the pool fire increases the rate of burning of the furniture and results in rapid flame spread. The melting and dripping of material can also result in the removal of fuel from the spreading flame front. This can significantly hinder the ability of the flame front to propagate across the surface of the material.

5.8.3.7 Role of External Heating on Flame Spread. External heating of a fuel surface by a nearby flame or by radiation from upper layer hot gases building up in a compartment can significantly enhance flame spread rates.

5.9* Fire Spread in a Compartment.

5.9.1 The heat release rate of a fire is generally a function of the amount and type of fuel that is involved at any given time and the ventilation conditions. An increase in the heat release rate can occur through flame spread on an individual fuel package, through the ignition of additional fuels, or by changes in ventilation. As the fire grows in size, it increases the potential for fire spread to other compartments or areas within the building. Flame spread is the movement of flames on an individual fuel package (e.g., sofa or combustible wall), and fire spread is described as the ignition of additional fuel packages that can spread the fire throughout a compartment or building.

5.9.2 Fire Spread. Fire spread, as opposed to flame spread, involves the ignition of more remote fuel packages. The fuel packages can be located within the same or in adjacent compartments. The fire can spread either by direct flame impingement or by remote ignition of adjacent fuel packages.

5.9.2.1 Fire Spread by Flame Impingement. As a fire grows, flame spread can be aided by the flow created by hot gases rising within the compartment. For example, the flames from a pool fire in the center of a compartment can be deflected by the flow of ambient air into the compartment through a doorway. The deflection of these flames from one fuel package can impinge on a second fuel package located adjacent to the first. In addition, due to lack of symmetry of airflow into a fire plume, flames from fuel packages located against a wall or in a corner can attach to vertical surfaces. If these surfaces are combustible, they can be ignited through direct flame contact.

5.9.2.2 Fire Spread by Remote Ignition. Remote ignition can occur through the three modes of heat transfer. A transfer of heat through conduction that results in ignition can occur through ceilings and walls. A common example would be the ignition of wood studs behind noncombustible masonry walls or partitions due to the heat from a compartment fire being conducted through the masonry to the combustible structural members.

5.9.2.2.1 Thermal radiant heat transfer that results in ignition can occur from flame radiation or hot gas layer radiation to remote fuel packages. The dominant method of spreading fire from one remote location to another remote location is through radiation. The hot smoke layer will transmit thermal radiation to other fuel packages, which may ignite. If the initial fire in a compartment grows large enough and involves enough fuel, the radiative output from the fire can become large enough to heat surfaces of other remote fuel packages up to the point when autoignition occurs and flame spread then begins on the remote fuel. This effect is observed at the point when a compartment transitions to flashover. Some of the factors that can affect this phenomenon are the size of the fire, the amount of energy radiated, the geometry between the two fuel objects (i.e., facing each other or at an angle, one large object exposing a small object), and the distance between the two objects.

5.9.2.2.2 Fire spread can also occur as a result of "drop down." This occurs when a flaming material drops down and ignites combustibles that it falls on or near. When a flame is burning on an elevated structure or material within a compartment, the possibility exists that before the fuel is completely consumed, the elements will lose integrity and cause the fuel to fall to other areas within the compartment or to a lower level or floor of a building. If this occurs, the still-burning fuel could be next to combustible material that has yet to ignite. Drop down also occurs with burning thermoplastic materials, curtains, draperies, or other thin fuel items.

5.10 Compartment Fire Development. The following is a list of references that pertain to compartment fire development:

(1) Custer, R. (2003), "Dynamics of Compartment Fire Growth," NFPA Fire Protection Handbook, 19th ed., Section 2.4.

(2) Thomas, P. (1995), "The Growth of Fire-Ignition to Full Involvement," *Combustion Fundamentals of Fire*, ed. Cox, G., Academic Press, London.

(3) Quintiere, J. (2002), "Compartment Fire Modeling," *SFPE Handbook of Fire Protection Engineering*, ed. DiNenno, P., National Fire Protection Association, Quincy, MA.

(4) Walton, D., Thomas, P. (2002), "Estimating Temperatures in Compartment Fires," *SFPE Handbook of Fire Protection Engineering*, ed. DiNenno, P., National Fire Protection Association, Quincy, MA.

(5) Cooper, L. (2002), "Compartment Fire-Generated Environment and Smoke Filling," *SFPE Handbook of Fire Protection Engineering*, ed. DiNenno, P., National Fire Protection Association, Quincy, MA.

5.10.1 General.

5.10.1.1 The rate and pattern of fire development depend on a complex relationship between the burning fuel and the surrounding environment.

5.10.1.2 In a compartment fire, the collection of heat at the top of the room can raise the temperature at the ceiling and produce a large body of high-temperature smoke and hot gases. The radiation from this upper portion of the space can significantly enhance the rate of fire spread and heat release from a burning item. This can result in heat release rates significantly greater than are observed in open burning.

5.10.1.3 Compartments can also restrict the heat release rate by limiting the inflow of air to support combustion. This section summarizes the development of a fire in a compartment.

5.10.1.4 The rate of fire growth as determined by witness statements is highly subjective. Many times witnesses are reporting the fire growth from time of discovery, which cannot be directly correlated to ignition time. The rate of fire growth is dependent on many factors besides fuel load, including fuel configuration, compartment size, compartment properties, ventilation, ignition source, and first fuel ignited. Eyewitnesses reporting a rapid rate of fire growth should not be construed as data supporting an incendiary fire cause.

5.10.2 Compartment Fire Phenomena.

5.10.2.1 When a fire plume reaches the ceiling of a compartment, the flow of smoke and hot gases and the growth of the fire will be affected. Figure 5.10.2.1 depicts a room with a door opening. There are multiple fuel items in the room; one is the item first ignited, and the others are "target" fuels. The fire plume strikes the ceiling and the flow is diverted to flow under the ceiling as a ceiling jet. Ceiling jet gases flow in all directions until the gases strike the walls of the compartment. As the ceiling jet flow reaches the walls and can no longer spread horizontally, the gases turn downward and begin the creation of a layer of hot gases below the ceiling.

5.10.2.2 Smoke Filling. The fire acts as a "pump," adding mass (hot gases) and energy to the upper layer. This continued addition of hot gases to the layer causes the upper layer to grow in depth. The depth of the upper layer of hot gases will continue until the hot upper gas layer reaches the base of the fire or until the layer fills down to the top of an opening.

5.10.2.3 When the hot smoky layer reaches the top of the door opening, as illustrated in Figure 5.10.2.3, it will begin to flow out of the compartment. The descent of the hot gas layer will be stopped when the flow of hot gases out of the opening is equal to the rate of hot gas addition to the layer by the plume.

5.10.2.4 If the fire grows in size, the bottom of the ceiling layer, the layer interface, will continue to descend, the temperature of the hot smoke and gases will increase, and radiant heat from the layer will begin to heat the unignited target fuel, as shown in Figure 5.10.2.4. A well-defined flow pattern will be established at the opening, with the hot combustion products flowing out the top and cool air flowing into the compartment under the smoke layer.

5.10.2.5 In the early stage of burning, there is sufficient air to burn all of the materials being pyrolyzed. This is referred to as *fuel-controlled burning.* As the burning progresses, the availability of air may continue and the fire may have sufficient oxygen even as it grows. Normally, this would be a compartment that had a large door or window opening. In such cases, the gases collected at the upper portion of the room, while hot, will contain significant oxygen and relatively small amounts of unburned fuel.

5.10.2.6 As the fire continues to grow, the ceiling layer gas temperature and the intensity of the radiation on the exposed combustible contents in the room increase. Figure 5.10.2.6

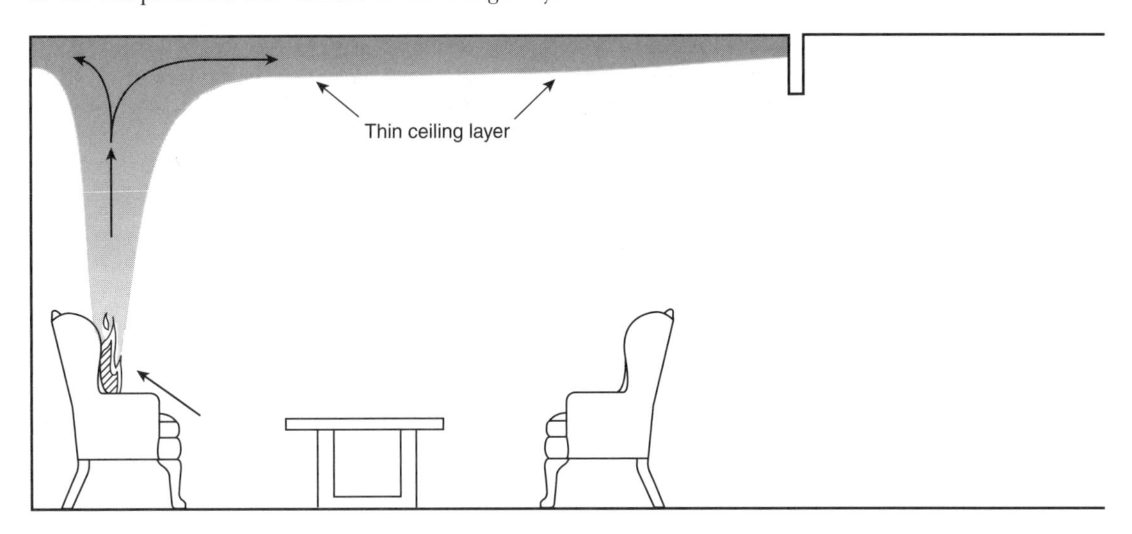

FIGURE 5.10.2.1 Early Compartment Fire Development.

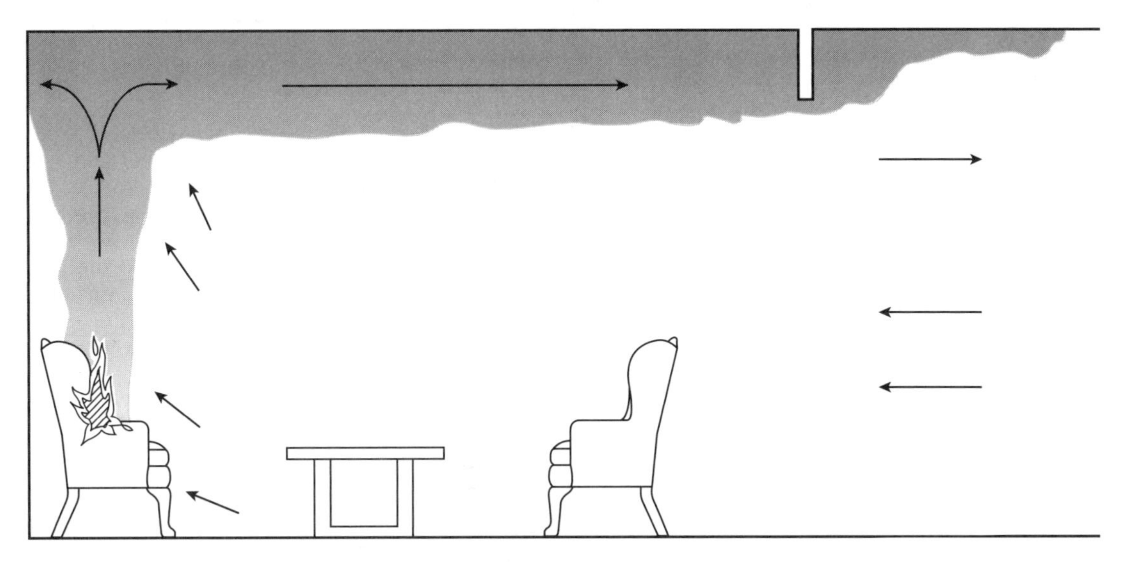

FIGURE 5.10.2.3 Upper Layer Development in Compartment Fire.

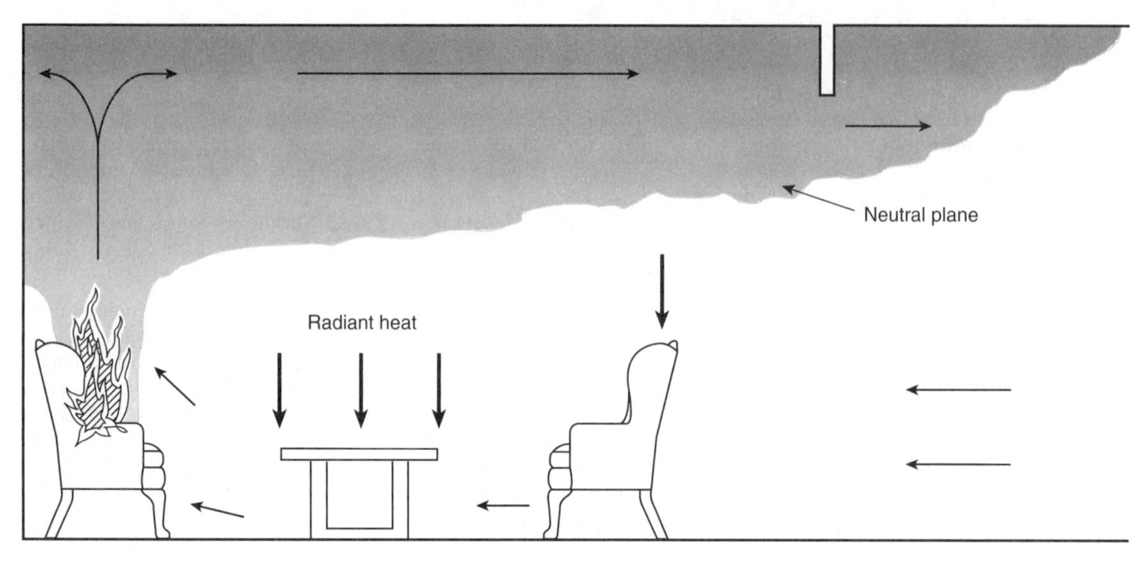

FIGURE 5.10.2.4 Preflashover Conditions in Compartment Fire.

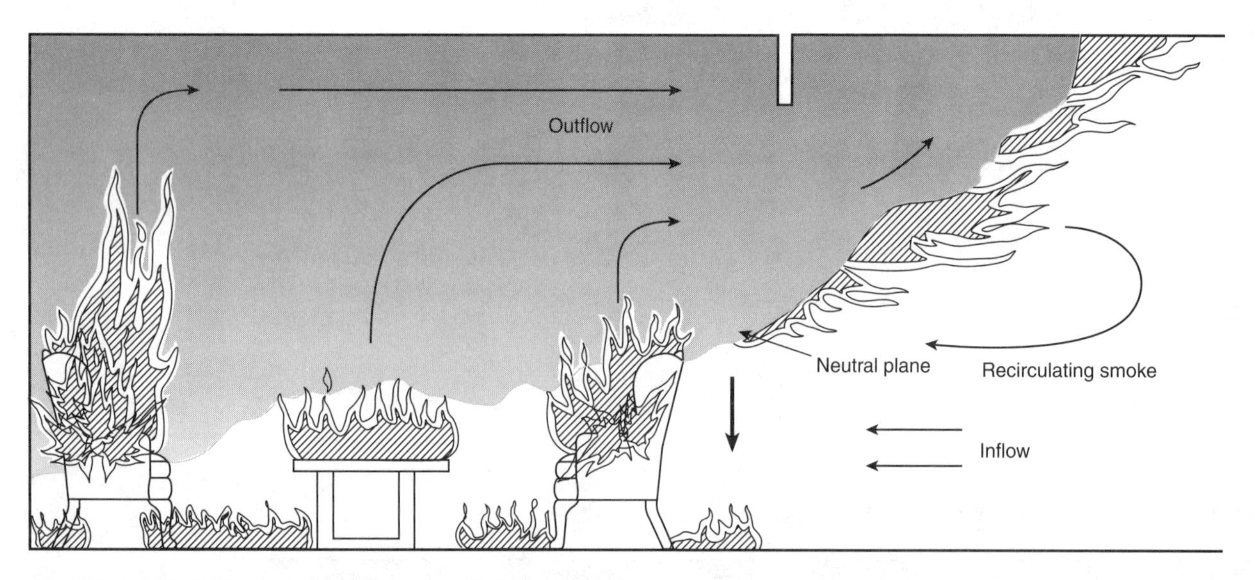

FIGURE 5.10.2.6 Flashover Conditions in Compartment Fire.

shows the evolution in the relative importance of convection and radiation heat transfer. As the fire develops, both convective and radiative heat fluxes increase, but radiation comes to dominate the overall heat transfer. The surface temperature of these combustible contents rises, and pyrolysis gases are produced. When the upper layer temperature reaches approximately 590°C (1100°F), the combustible contents ignite, involving all of the combustible surfaces exposed to upper layer radiation. This phenomenon, known as flashover, is illustrated in Figure 5.10.2.6. The terms *flameover* and *rollover* are often used to describe the condition where flames propagate through or across the ceiling layer only and do not involve the surfaces of target fuels. Flameover or rollover generally precede flashover but may not always result in flashover. As the fire develops, the relative significance of radiation heat transfer comes to dominate over convection heat transfer.

5.10.2.7 If the air flow into the compartment is not sufficient to burn all of the combustibles being pyrolyzed by the fire, the fire will shift from fuel controlled (where the heat release rate of the fire depends on the amount of fuel involved) to ventilation controlled (where all the fuel is on fire, and the heat release rate is controlled by the amount of oxygen available). In a ventilation-controlled fire, the hot gas layer will contain high levels of unburned pyrolysis products and carbon monoxide. See Figure 5.10.2.7.

5.10.2.8 During postflashover burning, the position of bottom of the ceiling layer and the existence and size of flaming on target fuels within the layer can vary between the conditions shown in Figure 5.10.2.6 and Figure 5.10.2.7. While the burning of floors or floor coverings is common, such burning may not always extend under target fuels or other shielding surfaces. This fully developed fire stage is typically *ventilation-controlled burning*.

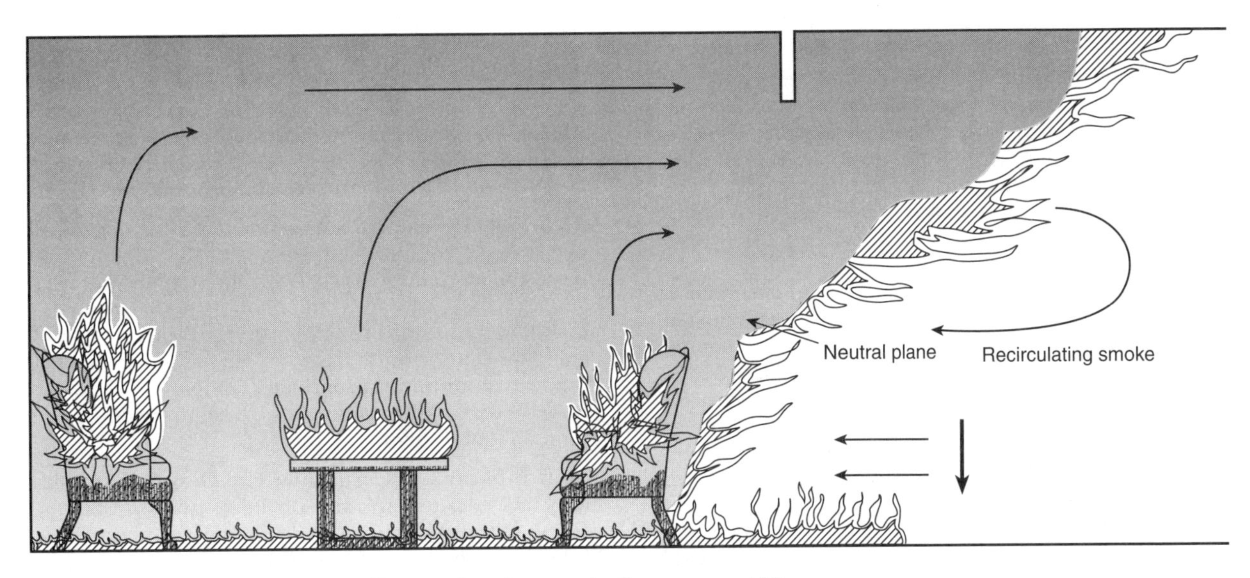

FIGURE 5.10.2.7 Postflashover or Full Room Involvement in Compartment Fire.

5.10.3 Compartment Vent Flows.

5.10.3.1 Airflows required to support a compartment fire can be provided by mechanical ventilation (HVAC flows) or by natural ventilation via openings (vents). Except where natural ventilation openings are extremely limited, natural ventilation generally dominates over mechanical ventilation.

5.10.3.2 Natural ventilation via an existing open door or window occurs due to buoyancy associated with the fire and the hot gas layer in the fire compartment. The hot gases, having a much lower density than ambient air, flow out of the top of the opening and the air that replaces the lost gases flows into the lower portion of the opening. The height at which the flow changes direction is known as the "neutral plane." This demarcation is often visible in the fire patterns on the door frame. An upward sweeping region of thermal damage on an open door records the flow of hot gases.

5.10.3.3 Single Opening Flows.

5.10.3.3.1 When the upper hot layer interface is near the top of the opening, the outflow of gases through the opening occurs at and above the height of the interface. Under these conditions the layer interface and the neutral plane are at the same height.

5.10.3.3.2 When the hot layer interface moves down near the base of the opening, the neutral plane is no longer at the interface height. When the hot upper gas layer within the room extends below the bottom of the opening, the neutral plane is normally at ⅓ to ½ of the opening height.

5.10.3.3.3 For a single vent opening, the air flow into the compartment is proportional to the ventilation factor, $A\sqrt{H}$, where A is the opening area and H is the opening height (i.e., from the bottom of the opening to the top of the opening. The $\sqrt{H}$ dependence arises from the increased buoyant pressures that are achieved with a greater height of the opening. The maximum heat release rate supportable by the air flow (stoichiometric assumption that the fuel will be burned completely based on the available air) into the compartment with a single vent is given by the following:

$$\dot{Q}_{stoich} = 1500\ A_o\sqrt{H_o} \qquad \textbf{[kW, m]}$$

where:

$\dot{Q}_{stoich}$ = maximum heat release rate based on air flow (kW)

A_o = area of opening (m^2)

H_o = height of opening (m)

5.10.3.3.4 A door opening 0.9 m (3 ft) wide and 2.1 m (7 ft) tall can support a heat release rate of about 4000 kW or 4 MW. If we block the lower 1 meter of the doorway to make it a window, the height of the opening is reduced to 1.1 meter and the heat release rate that the window opening can support is only 1600 kW or 1.6 MW.

5.10.3.4 Multiple Opening Flows.

5.10.3.4.1 A fire compartment has only a single neutral plane height, even if it has multiple openings at different elevations. Above the neutral plane height, flows are out of the compartment. Below the neutral plane, flows are into the compartment.

5.10.3.4.2 Because some openings may exist only above the neutral plane, such vents will be fully outflow vents. Other vents may be entirely below the neutral plane and act solely as inflow vents. Such inflow vents will of course not have any external plume smoke or heat damage.

5.10.3.4.3 Changes in the neutral plane height will occur as a result of opening additional vents during a fire. For instance, a window may show two directional flow during a portion of the fire. Later, that vent may transition to solely an inflow vent when the fire breaches the ceiling of the compartment. At that time, the plume from the window would disappear and smoke from the ceiling vent would issue into an upper floor. If the ceiling is the roof, a plume would appear at the roof.

5.10.3.5 Vent Openings.

5.10.3.5.1 Some compartments have existing vent openings (e.g., open windows and/or doors). Closed windows become vent openings when the glass fails. While glass cracks when the glass reaches 60–100°C, glass does not generally fall out until flashover, except when flames directly impinge on the glass. Vents can also be created by the destruction of doors or barriers.

5.10.4 Flashover.

5.10.4.1 Flashover represents a transition from a condition where the fire is dominated by burning of the first item ignited (and nearby items subject to direct ignition) to a condition where the fire is dominated by burning of all items in the compartment. This transition is generally characterized as the transition from "a fire in a room" to "a room on fire." It is important for investigators to be aware of the fact that flashover is a triggering condition, not a close-ended event. The postflashover condition is called *full room involvement*. The onset of flashover occurs when the hot gas layer imposes radiant energy levels (flux) on unignited fuels sufficient to ignite them. A heat flux from the hot gas layer of approximately 20 kW/m^2 at floor level is generally considered sufficient to cause flashover. At this heat flux, crumpled newsprint will ignite in seconds. Flux levels in full room involvement are considerably higher than at flashover. Heat fluxes at a floor level of 170 kW/m^2 have been recorded.

5.10.4.2 Once flashover conditions have been reached, full room involvement will follow in the majority of fires unless the fuel is exhausted, the fire is oxygen deprived, or the fire is extinguished. In full room involvement, the hot layer can be at floor level, but tests and actual fires have shown that the hot layer is not always at floor level. Full room involvement may be achieved by fire growth that does not involve flashover.

5.10.4.3 Not all compartments can achieve full room involvement. Poorly ventilated compartments or very large compartments may be fully consumed without ever involving all the fuel packages at the same time. Large warehouse fires often spread and burn as a flame front from one end to the other, rather than burning over the entire area of the warehouse at once. Very elongated compartments with a vent at one end tend also to burn from the vent end to the back of the compartment over time, even if the initial ignition is not near the vent.

5.10.4.4 Ventilation Opening. The minimum size (in kilowatts) of a fire that can cause a flashover in a given room is a function of the ventilation provided through an opening. This function is known as the *ventilation factor* and is calculated as the area of the opening (A_o) times the square root of the height of the opening (H_o).

5.10.4.5* An approximation of the minimum heat release rate required for flashover for a compartment with a single opening can be found from the following relationship:

$$\dot{Q}_{fo} = 750\ A_o\sqrt{H_o}$$

where:

$\dot{Q}_{fo}$ = heat release rate for flashover (kW)

A_o = area of opening (m^2)

H_o = height of opening (m)

5.10.4.5.1 For a door opening 0.9 m (3 ft) wide and 2.1 m (7 ft) high, a heat release rate of about 2000 kW or 2 MW would be needed to cause flashover. The above relationship is time-independent, but studies show that a higher heat release rate value is needed to cause flashover if the peak heat release rate is of short duration.

5.10.4.6 Flashover times of 3 to 5 minutes are not unusual in residential room fire tests and even shorter times to flashover have been observed in nonaccelerated room fires.

5.10.5 Fully Developed Compartment Fires. In a fully developed fire, the compartment door typically becomes a restriction to the amount of air available for combustion inside the compartment, and significant amounts of the pyrolysis products will burn outside the compartment. Flameover or rollover generally occurs prior to flashover but may not always result in flashover conditions throughout a compartment, particularly where there is a large volume or high ceiling involved, or there is limited fuel present.

5.10.6 Effects of Enclosures on Fire Growth. For a fire in a given fuel package, the size of the ventilation opening, the volume of the enclosure, the ceiling height, and the location of the fire with respect to the walls and corners will affect the overall fire growth rate in the enclosure.

5.10.6.1 Room Volume and Ceiling Height. Development of a ceiling layer of sufficient temperature to cause radiant ignition of exposed combustible fuels is necessary for flashover. High ceilings or large compartment volumes will delay this buildup of temperature and therefore delay or possibly prevent flashover from occurring. The distance between the bottom of the hot layer and the combustible fuel is also a factor, but of less importance.

5.10.6.2* Location of the Fire in the Compartment. When a burning fuel package is away from a wall, air is free to flow into the plume from all directions and mix with the fuel gases. This brings air for combustion into the flame zone and cools the upper part of the plume by entrainment. (*See 5.5.2.*) If the fuel package or the fire plume is against a wall (not in a corner), a given fire size will lead to a 30 percent greater hot layer absolute temperature than the same fire size away from the wall. When the same fuel package is placed in a corner, a given fire size will lead to a 70 percent greater hot layer absolute temperature than the same fire size away from walls or corners.

5.11 Fire Spread Between Compartments. Fire spread between compartments can occur via a connecting opening or through the barrier (wall or ceiling) connecting two compartments.

5.11.1 Fire Spread via Openings. Fire spread via openings can occur by several mechanisms as follows:

(1) Direct contact of a flame from the burning compartment opening with fuels in the target compartment
(2) Radiant ignition of fuels in the target room due to radiation from the burning room opening, the flame at the opening, and from the hot gas layer formed by flows into the target compartment from the burning compartment
(3) Ignition of fuels in the target compartment by embers transported from the burning compartment to the target compartment via the opening

5.11.2 Fire Spread via Barriers. Fire spread via barriers can occur by several mechanisms as follows:

(1) Conduction of heat through the barrier from the burning compartment to fuels adjacent to the barrier in the target compartment.
(2) Physical penetration of the barrier by the fire to create an opening between the burning compartment and the target

compartment (actual spread occurs via opening mechanisms noted above). The penetration of the barrier can be caused by degradation of the wall materials (e.g., crumbling gypsum board) or the creation of cracks in the barrier.

(3) Structural collapse of the barrier due to fire effects on the barrier or on structural members whose deformation causes damage to the barrier.

5.12 Paths of Smoke Spread in Buildings. The flow of gases across an opening is the result of differences in pressure. Therefore, smoke can flow through doors, windows, and other openings. Because no compartment is hermetically sealed, there are leakage areas where smoke can flow between compartments. Plenum spaces above false ceilings are also a significant path for smoke travel. Smoke movement by temperature difference can result from the heat from the fire. In tall buildings, smoke at a distance from the fire may be the same temperature as the ambient air but may be moving due to the "building stack effect." Stack effect pressures result from differences between the temperatures inside and outside of the building. Pressure from HVAC systems can also transport smoke from one compartment to another.

Chapter 6 Fire Patterns

6.1 Introduction.

6.1.1 The major objective of any fire scene examination is to collect data as required by the scientific method (*see 4.3.3*). Such data include the patterns produced by the fire. A fire pattern is the visible or measurable physical changes or identifiable shapes formed by a fire effect or group of fire effects. Fire effects are the observable or measurable changes in or on a material as a result of exposure to the fire. The collection of fire scene data requires the recognition and identification of fire effects and fire patterns. The data can also be used for fire pattern analysis (i.e., the process of interpreting fire patterns to determine how the patterns were created). This data and analysis can be used to test hypotheses as to the origin of the fire as discussed in Chapter 17. The purpose of the discussion in this chapter is to aid the investigator in the recognition and identification of fire effects and fire patterns as well as the interpretation of patterns through fire pattern analysis.

6.2 Fire Effects.

6.2.1 To identify fire patterns, the investigator must recognize the changes that have occurred in materials due to fire. These changes are referred to as fire effects, which are the observable or measurable changes in or on a material as the result of a fire.

6.2.2 Temperature Estimation Using Fire Effects. If the investigator knows the approximate temperature required to produce an effect, such as melting, color change, or deformation a material, an estimate can be made of the temperature to which the material was raised. This knowledge may assist in evaluating the intensity and duration of the heating, the extent of heat flow, or the relative rates of heat release from fuels.

6.2.2.1 When using materials such as glass, plastics, and white pot metals for estimating temperature, the investigator is cautioned that there is a wide variety of material properties for these generic materials. The best method for utilizing such materials as temperature indicators is to take a sample of the material and have its properties ascertained by a competent laboratory, materials scientist, or metallurgist.

6.2.2.2* Wood and gasoline burn at essentially the same flame temperature. The turbulent diffusion flame temperatures of all hydrocarbon fuels (plastics and ignitible liquids) and cellulosic fuels are approximately the same, although the fuels release heat at different rates. Burning metals and highly exothermic chemical reactions can produce temperatures significantly higher than those created by hydrocarbon- or cellulosic-fueled fires.

6.2.2.3 Heat transfer is responsible for much of the physical evidence used by fire investigators. The temperature achieved by a material at a specific location within a structure depends on the amount of heat energy transferred to the material. As discussed in Section 5.5, heat energy is transferred by three modes: conduction, convection, and radiation. All three modes can contribute to a change in the temperature of a specific material in a fire. The temperature achieved will depend on the individual contribution from each mode of heat transfer. The individual contribution associated with each mode is dependant on the variables discussed in Section 5.5.

6.2.2.4 Identifiable temperatures achieved in structural fires rarely remain above 1040°C (1900°F) for long periods of time. These identifiable temperatures are sometimes called *effective fire temperatures*, because they reflect physical effects that can be defined by specific temperature ranges. The investigator can use the analysis of the melted materials to assist in establishing the minimum temperatures present in specific areas.

6.2.3 Mass Loss of Material.

6.2.3.1 Fires convert fuel and oxygen into combustion products, heat, and light. This process results in mass loss of the fuel (consumption of the material). During a fire, combustible and noncombustible materials may also lose mass due to evaporation, calcination, or sublimation.

6.2.3.2 The mass loss of a material consumed in a fire may be determined by comparing fire-damaged materials to exemplar materials. The edges or surfaces of remaining material may be used to estimate the size and shape of the material before the fire. Materials existing prior to a fire may be determined from duplicate or similar materials not consumed by the fire or from drawings, plans, photographs, or interviews with individuals familiar with conditions prior to the fire.

6.2.3.3 The mass loss of material is often used as an indication of the duration and intensity of the fire. While this may be valid in many instances, it is not valid in all cases. The mass loss rate results from a complex combination of factors involving material properties and fire conditions.

6.2.3.4 The rate of mass loss normally changes throughout the course of a fire. The rate of mass loss is generally dependent on the heat flux to the material surface, fire growth rate, and rate of heat release of the material itself. As the fire grows in size and intensity, the rate of mass loss increases.

6.2.4 Char.

6.2.4.1 Introduction. Charred material is likely to be found in nearly all structural fires. When exposed to elevated temperatures, wood undergoes pyrolysis, a chemical decomposition that drives off gases, water vapor, and various pyrolysis products as smoke. The solid residue that remains is mainly carbon. Char shrinks as it forms, and develops cracks and blisters.

6.2.4.2 Surface Effect of Char. Many surfaces are decomposed in the heat of a fire. The binder in paint will char and darken the color of the painted surface. Wallpaper and the paper surface of gypsum wallboard will char when heated. Vinyl and other plastic surfaces on walls, floors, tables, or counters also will discolor, melt, or char. Wood surfaces will char, but, because of the greater prevalence of wood char, it is treated in further detail in 6.2.4.5. The degree of discoloration and charring can be compared to adjacent areas to find the areas of greatest burning.

6.2.4.3 Appearance of Char. In the past, the appearance of the char and cracks had been given meaning by the fire investigation community beyond what has been substantiated by controlled testing. The presence of large shiny blisters (alligator char) is not evidence that a liquid accelerant was present during the fire, or that a fire spread rapidly or burned with greater intensity. These types of blisters can be found in many different types of fires. There is no justification for the inference that the appearance of large, curved blisters is an indicator of an accelerated fire. Figure 6.2.4.3, showing boards exposed to the same fire, illustrates the variability of char blister.

FIGURE 6.2.4.3 Variability of Char Blister.

6.2.4.3.1 It is sometimes claimed that the surface appearance of the char, such as dullness, shininess, colors, or appearance under ultraviolet light sources, has some relation to the use of a hydrocarbon accelerant or the rate of fire growth. There is no scientific evidence that such a correlation exists, and the investigator is advised not to claim indications of accelerant or a rapid fire growth rate on the basis of the appearance of the char.

6.2.4.4* Rate of Wood Charring. The correlation of 2.54 cm (1 in.) in 45 minutes for the rate of charring of wood is based on ventilation-limited burning. Fires may burn with more or less intensity during the course of an uncontrolled fire than in a controlled laboratory fire. Laboratory char rates from exposure to heat from one side vary from 1 cm (0.4 in.) per hour to 25.4 cm (10 in.) per hour. Care needs to be exercised in solely using depth of char measurements to determine the duration of burning. A more in-depth discussion of the appropriate use of depth of char measurements appears in Chapter 17.

6.2.4.4.1 The rate of charring of wood varies widely depending upon variables, including the following:

(1) Rate and duration of heating
(2) Ventilation effects
(3) Surface area-to-mass ratio
(4) Direction, orientation, and size of wood grain
(5) Species of wood (pine, oak, fir, etc.)
(6) Wood density
(7) Moisture content
(8) Nature of surface coating
(9) Oxygen concentration of the hot gases
(10) Velocity of the impinging gases
(11) Gaps/cracks/crevices and edge effect of materials

6.2.4.4.2 The rate of charring and burning of wood in general has no relation to its age once the wood has been dried. Wood tends to gain or lose moisture according to the ambient temperature and humidity. Thus, old, dry wood is no more combustible than new kiln-dried wood if they have both been exposed to the same atmospheric conditions.

6.2.4.4.3 The investigator is cautioned that no specific time of burning can be determined based solely on depth of char.

6.2.4.5 Depth of Char. Analysis of the depth of charring is more reliable for evaluating fire spread, rather than for the establishment of specific burn times or intensity of heat from adjacent burning materials. The relative depth of char from point to point is the key to appropriate use of charring, locating the places where the damage was most severe due to exposure, ventilation, or fuel placement. The investigator may then deduce the direction of fire spread, with decreasing char depths being farther away from the heat source.

6.2.4.6 Nature of Char. Overall, the use of the nature of char to make determinations about fuels involved in a fire should be done with careful consideration of all the variables that can affect the speed and severity of burning.

6.2.5* Spalling. Spalling is characterized by the loss of surface material resulting in cracking, breaking, and chipping or in the formation of craters on concrete, masonry, rock, or brick.

6.2.5.1 Fire-Related Spalling. Fire-related spalling is the breakdown in surface tensile strength of material caused by changes in temperature resulting in mechanical forces within the material. These forces are believed to result from one or more of the following *(see Figure 6.2.5.1)*:

(1) Moisture present in uncured or "green" concrete
(2) Differential expansion between reinforcing rods or steel mesh and the surrounding concrete
(3) Differential expansion between the concrete mix and the aggregate (most common with silicon aggregates)
(4) Differential expansion between the fire-exposed surface and the interior of the slab

6.2.5.1.1* A mechanism of spalling is the expansion or contraction of the surface while the rest of the mass expands or contracts at a different rate; one example is the rapid cooling of a heated material by water.

6.2.5.1.2* Spalled areas may appear lighter in color than adjacent areas. This lightening can be caused by exposure of clean subsurface material. Adjacent areas may also tend to be darkened by smoke deposition.

6.2.5.1.3* Another factor in the spalling of concrete is the loading and stress in the material at the time of the fire. Because these high-stress or high-load areas may not be related to the fire location, spalling of concrete on the underside of ceilings or beams may not be directly over the origin of the fire. *(See Figure 6.2.5.1.3.)*

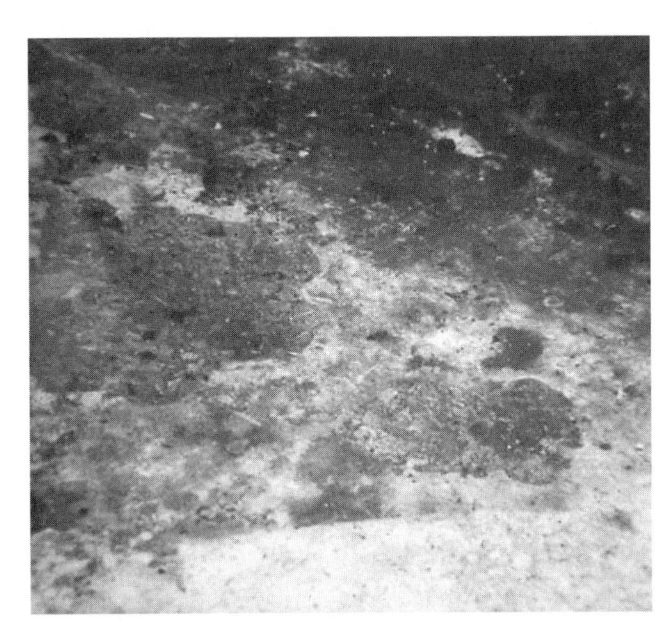

FIGURE 6.2.5.1 Spalled Concrete Floor.

FIGURE 6.2.5.1.3 Spalling on Ceiling.

6.2.5.2 The presence or absence of spalling at a fire scene should not, in and of itself, be construed as an indicator of the presence or absence of liquid fuel accelerant. The presence of ignitible liquids will not normally cause spalling beneath the surface of the liquid. Rapid and intense heat development from an ignitible liquid fire may cause spalling on adjacent surfaces, or a resultant fire may cause spalling on the surface after the ignitible liquid burns away.

6.2.5.3 Non-Fire-Related Spalling. Spalling of concrete or masonry surfaces may be caused by many factors, including heat, freezing, chemicals, abrasions, mechanical movement, shock, force, or fatigue. Spalling may be more readily induced in poorly formulated or finished surfaces. Because spalling can occur from sources other than fires, the investigator should determine whether spalling was present prior to the fire.

6.2.6* Oxidation. Oxidation is the basic chemical process associated with combustion. Oxidation of some non-combustible materials can produce lines of demarcation and fire patterns of use to fire investigators. For these purposes, oxidation may be defined as a combination of oxygen with substances such as metals, rock, or soil that is brought about by high temperatures. Deposition of smoke aerosols containing acidic components may lead to the oxidation of material surfaces and discernible fire patterns. Surfaces may also be oxidized due to deposition of fire suppression agents such as dry or wet chemicals. *(See 6.2.10.)*

6.2.6.1 The effects of oxidation include change of color and change of texture. The higher the temperature and the longer the time of exposure, the more pronounced the effects of oxidation will be. The extent of post-fire oxidation will be a function of the ambient humidity and exposure time.

6.2.6.2 With mild heating, bare galvanized steel may acquire a dull whitish surface due to oxidation of the zinc coating. This oxidation may also eliminate the corrosion protection that the zinc provided. If the unprotected steel is wet for some time, it will rust, which is another form of oxidation. Thus, there can be a pattern of rust compared to non-rusted galvanized steel.

6.2.6.3 When uncoated iron or steel is oxidized in a fire, the surface first acquires a blue-gray dullness. At elevated temperatures, iron may also combine with oxygen to form black oxides. Oxidation can produce thick layers of oxide that can flake off. After the fire, if the metal has been wet, the usual rust-colored oxide may appear.

6.2.6.4* Heavily oxidized steel may exhibit a visual appearance similar to melting. It is frequently not possible to determine by visual observation alone whether the steel has melted. A metallurgical examination of a polished, etched cross-section of the steel is necessary to make a determination of melted steel.

6.2.6.5 On stainless steel surfaces, mild oxidation can result in color fringes, and severe oxidation will produce a dull gray color.

6.2.6.6 Copper forms a dark red or black oxide when exposed to heat. The color is not significant. What is significant is that the oxidation can form a line of demarcation. The thickness of the oxide depends upon the duration and intensity of the heat exposure. The more it is heated, the greater the oxidation.

6.2.6.7 Rocks and soil, when heated to very high temperatures, will often change colors that may range from yellowish to red.

6.2.7* Color Changes. Color changes are a source of information pertaining to the exposure of materials to various temperatures. The above sections have already addressed color changes in a number of specific materials. This section addresses generic color changes applicable to many other materials. Materials have a certain color due to the absorption, reflection, or transmission of light.

6.2.7.1 Color is a subjective quality unless quantitatively measured. People perceive and describe color differently. The intensity, color, and angle of the light source affect the viewer's interpretation of the color of the object. The surface characteristics of the material impact the viewer's color perception. For example, a dark blue car under some lighting conditions may appear black.

6.2.7.2 Color changes in general can be brought about by many non-fire factors. When first examining perceived color change evidence, the investigator should consider pre- and post-fire factors as varied as sun or chemical exposures. These exposures may cause dyes and color additives to undergo chemical changes that alter their original color.

6.2.7.3 Material deposited on a translucent surface, such as glass, may exhibit a different color than the same material deposited on an opaque surface. This effect can be observed directly by holding a film negative against a wall, where it may look dark or even black, but the same negative held over a light source will then be observed as lighter and having a visible image.

6.2.7.4 Fabric dyes may be subject to color changes after exposure to a fire. Fabrics may show variations of color from the burned area to a completely unburned area. While the color change is generally related to the heat exposure, without a detailed understanding of the dye breakdown behavior it is difficult to quantify the observation.

6.2.8 Melting of Materials.

6.2.8.1 General. The melting of a material is a physical change caused by exposure to heat. The border between the melted and non-melted portions of a fusible material can produce lines of heat and temperature demarcation that the investigator can use to define fire patterns.

6.2.8.2 Many solid materials soften or melt at elevated temperatures ranging from a little over room temperature to thousands of degrees. A specific melting temperature or range is characteristic for each material. *(See Table 6.2.8.2.)*

6.2.8.3 Melting temperatures of common metals range from as low as 170°C (338°F) for solder to as high as 1460°C (2660°F) for steel. When the metals or their residues are found in fire debris, some inferences concerning the temperatures in the fire can be drawn.

6.2.8.4 Thermoplastics soften and melt over a range of relatively low temperatures, from around 75°C (167°F) to near 400°C (750°F). Thus, the melting of plastics can give information on temperatures, but mainly where there have been hot gases and little or no flame in that immediate area. *(See Figure 6.2.8.4.)*

6.2.8.5 Glass softens over a range of temperatures. Nevertheless, glass can give useful information on temperatures during a fire.

6.2.8.6* Alloying of Metals. Alloying should be considered when analyzing post-fire metal specimens. The melting of certain metals may not always be caused by fire temperatures higher than the metals' stated melting point; it may be caused by alloying. Alloying refers to the mixing of, generally, two or more metals in which one or more of the metals is in a liquefied state, resulting in an alloy. Metals such as copper and iron (steel) can be affected by alloying with lower melting point metals such as aluminum, zinc, and lead. *(See Table 6.2.8.2.)*

6.2.8.6.1* During a fire, a metal with a relatively low melting point can soften or liquefy and contact other metals with melting temperatures that exceed the temperatures achieved. If a lower-melting-temperature metal, such as zinc, contacts the surface of a higher-melting-temperature metal, such as copper, the two metals can combine to create a zinc-copper alloy (a brass) with an alloy melting temperature lower than copper. In such instances, it is often possible to see the yellow-colored brass.

6.2.8.6.2* The resultant alloy will have a melting point that is less than the higher melting point component in the mixture. In some cases, the alloy can have a melting temperature less than either metal component. *(See Figure 6.2.8.6.2.)*

6.2.8.6.3* When metals with high melting temperatures are found to have melted due to alloying, it is not an indication that accelerants or unusually high temperatures were present in the fire.

6.2.9* Thermal Expansion and Deformation of Materials. Many materials change shape temporarily or permanently during fires. Nearly all materials expand when heated. That expansion can affect the integrity of solid structures when they are made from different materials. If one material expands more than another material in a structure, the difference in expansion can cause the structure to fail. Deformation is the change in shape characteristics of an object separate from the other changing characteristics defined elsewhere in this chapter. Deformation can result from a variety of causes ranging from thermal effects to chemical and mechanical effects. In order to make determinations about heat flow based upon deformation, the investigator should determine that the deformation occurred as a result of the fire and is not due to some other cause of deformation.

Table 6.2.8.2 Approximate Melting Temperatures of Common Materials

Material	Melting Temperatures	
	°C	°F
Aluminum (alloys)[a]	566–650	1050–1200
Aluminum[b]	660	1220
Brass (red)[a]	996	1825
Brass (yellow)[a]	932	1710
Bronze (aluminum)[a]	982	1800
Cast iron (gray)[b]	1350–1400	2460–2550
Cast iron (white)[b]	1050–1100	1920–2010
Chromium[b]	1845	3350
Copper[b]	1082	1981
Fire brick (insulating)[b]	1638–1650	2980–3000
Glass[b]	593–1427	1100–2600
Gold[b]	1063	1945
Iron[b]	1540	2802
Lead[b]	327	621
Magnesium (AZ31B alloy)[a]	627	1160
Nickel[b]	1455	2651
Paraffin[b]	54	129
Plastics (thermo)		
ABS[d]	88–125	190–257
Acrylic[d]	90–105	194–221
Nylon[d]	176–265	349–509
Polyethylene[d]	122–135	251–275
Polystyrene[d]	120–160	248–320
Polyvinylchloride[d]	75–105	167–221
Platinum[b]	1773	3224
Porcelain[b]	1550	2820
Pot metal[e]	300–400	562–752
Quartz (SiO_2)[b]	1682–1700	3060–3090
Silver[b]	960	1760
Solder (tin)[b]	135–177	275–350
Steel (carbon)[a]	1516	2760
Steel (stainless)[a]	1427	2600
Tin[b]	232	449
Wax (paraffin)[c]	49–75	120–167
White pot metal[e]	300–400	562–752
Zinc[b]	375	707

[a]From Lide, ed., *Handbook of Chemistry and Physics.*
[b]From Baumeister, Avallone, and Baumeister III, *Mark's Standard Handbook for Mechanical Engineers.*
[c]From NFPA *Fire Protection Guide to Hazardous Materials.*
[d]From McGraw-Hill, *Plastics Handbook.*
[e]From Gieck and Gieck, *Engineering Formulas.*

FIGURE 6.2.8.4 Melted Plastic Lighting Fixture, Indicating Heating from Right to Left.

FIGURE 6.2.9.1 Steel I-Beam Girders Deformed by Heating Under Load.

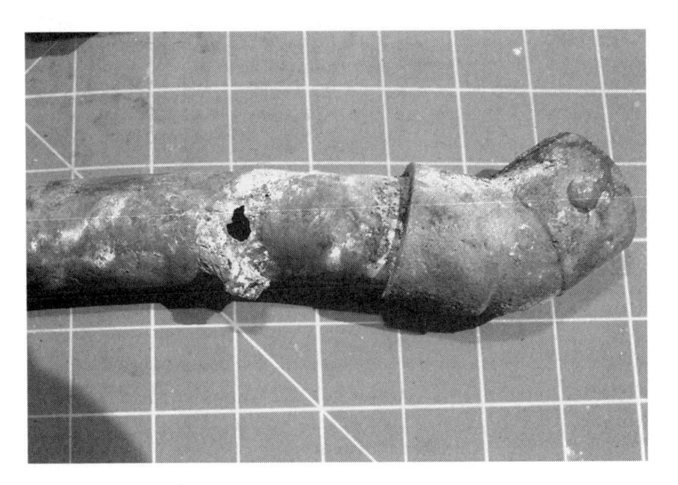

FIGURE 6.2.8.6.2 Hole in Copper Gas Line Caused by Alloying When Molten Aluminum [Melting Temperature ~649°C (~1200°F)] Dripped onto the Copper Pipe [Melting Temperature ~1083°C (~1980°F)]

FIGURE 6.2.9.2 Damage to an Outside Brick Wall Caused by Thermal Expansion of an I-Beam in the Basement.

6.2.9.1 Bending and buckling (deformation) of steel beams and columns occurs when the steel temperature exceeds approximately 538°C (1000°F). At elevated temperatures, steel exhibits a progressive loss of strength. When there is a greater fire exposure, the load required to cause deformation is reduced. Deformation is not the result of melting. A deformed element is not one that has melted during the fire, and therefore the occurrence of such deformation does not indicate that the material was heated above its melting temperature. On the contrary, a deformed as opposed to melted item indicates that the material's temperature did not exceed its melting point. Thermal expansion can also be a factor in the bending of the beam, if the ends of the beam are restrained. *(See Figure 6.2.9.1.)*

6.2.9.2 Metal Construction Elements. Studs, beams, columns, and the construction components that are made of high-melting-point metal, such as steel, can be distorted by heating. The higher the coefficient of thermal expansion of the metal, the more prone it is to heat distortion. The amount and location of distortion in a particular metal construction can indicate which areas were heated to higher temperatures or for longer times. In some cases, elongation of beams can result in damage to walls, as shown in Figure 6.2.9.2. This photo demonstrates that the beam inside the basement during the fire was heated above normal ambient temperatures, which led to expansion of the beam. The increased length of the beam pushed out the bricks, causing the wall damage. After the fire, when the beam had cooled, it may have returned to its approximate pre-fire length, but the structural damage to the wall remained as observable evidence of the beam expansion.

6.2.9.3 Piping systems and, specifically, fittings on piping systems may undergo deformation during a fire. These deformations can often be seen as one-way deformations where a fitting, even after complete cooling, does not return to its original shape and dimensions. For example, post-fire, a threaded elbow may be loose on the pipe to which it was originally secured. Due to the compressive and tensile forces of the connection and the heating and cooling exposure to which the connection was exposed, the elbow, which was a tight connection pre-fire, may be loose post-fire. This looseness is caused by the failure of the elbow to return to pre-fire dimensions even after complete cooling. Consideration should be given to the various materials used as sealants in pipe joining.

6.2.9.4 Plastered surfaces are also subject to thermal expansion. Locally heated portions of plaster walls and ceilings may expand and separate from their support lath. In addition to plaster separations from lath, joint compound (sometimes referred to as mud or spackle), joint tape, and patches on gypsum wallboard may fall off.

6.2.10* Deposition of Smoke on Surfaces. Smoke contains particulates, liquid aerosols, and gases. These particulates and liquid aerosols are in motion and may adhere upon collision with a surface. They may also settle out of the smoke over time. Carbon-based fuels produce particles that are predominantly carbon (soot). Petroleum products and most plastics are generally strong soot producers. When flames touch walls and ceilings, particulates and aerosols will commonly be deposited. Smoke deposits can collect on surfaces by settling and deposition.

6.2.10.1 Smoke deposits can collect on cooler surfaces of a building or its contents, often on upper parts of walls in rooms adjacent to the fire. Smoke condensates can be wet and sticky, thin or thick, or dried and resinous. Smoke, especially from smoldering fires, tends to condense on walls, windows, and other cooler surfaces.

6.2.10.2 It should be noted that the color and texture of smoke deposits do not indicate the nature of the fuel or its heat release rate. Chemical analysis of the smoke deposit may indicate the nature of the fuel. For example, smoke from candles may contain paraffin wax, and cigarette smoke may contain nicotine.

6.2.10.3* Enhanced Soot Deposition (Acoustic Soot Agglomeration) on Smoke Alarms. In many cases, the nature of soot deposition on certain surfaces of typical single- or multiple-station smoke alarms can show that the smoke alarm sounded or did not sound during a fire. Enhanced soot deposition (acoustic soot agglomeration) is a phenomenon whereby the soot particulate in smoke forms identifiable patterns on such surfaces of the smoke alarm as the internal and external surfaces of the smoke alarm cover near the edges of the "horn" (sound) outlet(s), the edges of and "horn" sound outlet(s) of the interior "horn" enclosures if present, and surfaces of the "horn" disks themselves. *[See Figure 6.2.10.3(a) through Figure 6.2.10.3(d).]*

6.2.10.3.1 Scene investigators should be cognizant of the importance of smoke alarms that may bear physical evidence of alarm activation and consider more detailed documentation, examination, and collection of such evidence.

6.2.10.3.2 Enhanced soot deposition acoustic agglomeration evidence can be delicate and easily disturbed or wiped away by careless handling or evidence packaging of the smoke alarm(s) in question. Care should be taken not to disturb any suspected soot deposits.

6.2.10.3.3 Evidence of enhanced soot deposition acoustic agglomeration on smoke alarms can be subtle and sometimes difficult to identify. Examination may require microscopic magnification.

FIGURE 6.2.10.3(a) An Unpowered (Non-Functioning) Smoke Alarm After Exposure to a Sooty Atmosphere.

FIGURE 6.2.10.3(b) Close-Up of the External "Horn" (Sound) Outlet of the Unpowered Smoke Alarm Displayed in Figure 6.2.10.3(a) After Exposure to a Sooty Atmosphere, Showing No Enhanced Soot Deposition.

FIGURE 6.2.10.3(c) A Duplicate Powered (Functioning) Smoke Alarm After Exposure to the Same Sooty Atmosphere as the Smoke Alarm in Figure 6.2.10.3(a) and Figure 6.2.10.3(b), Displaying Typical Enhanced Soot Deposition.

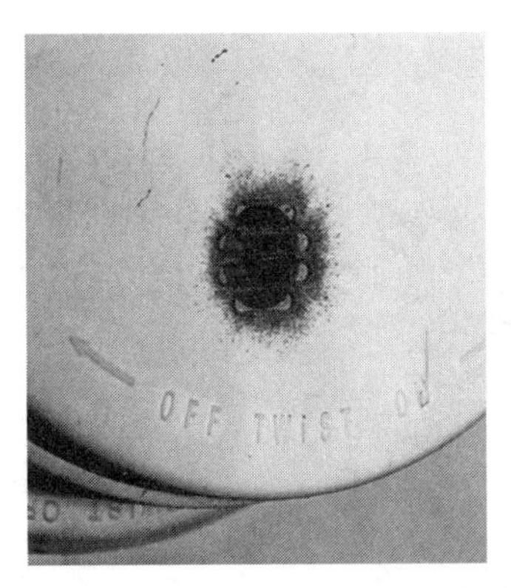

FIGURE 6.2.10.3(d) Close-Up of the External "Horn" (Sound) Outlet of the Powered Smoke Alarm Displayed in Figure 6.2.10.3(c) After Exposure to a Sooty Atmosphere, Showing Enhanced Soot Deposition.

6.2.10.3.4 Smoke alarms should may be taken into evidence after being photographed in place and should not be altered by applying power, removing or inserting batteries, or pushing the test button. Alarms still on the wall or ceiling should be secured intact with mounting hardware, electrical boxes, and wired connections. Removing a section of wall material with the alarm may be needed to preserve the condition of the alarm and all electrical power connections.

6.2.10.3.5 Investigators should keep in mind that acoustic smoke agglomeration deposits are persistent. The presence of acoustic smoke agglomeration deposits may not necessarily indicate when the agglomeration occurred, without additional data.

6.2.11 Clean Burn. Clean burn is a phenomenon that appears on noncombustible surfaces when the soot and smoke condensate that would normally be found adhering to the surface is burned off. This produces a clean area adjacent to areas darkened by products of combustion, as shown in Figure 6.2.11. Clean burn is produced most commonly by direct flame contact or intense radiated heat. Smoke deposits on surfaces are subject to oxidation. The dark char of the paper surface of gypsum wallboard, soot deposits, and paint can be oxidized by continued flame exposure. The carbon will be oxidized to gases and disappear from the surface.

6.2.11.1 Although they can be indicative of intense heating in an area, clean burn areas by themselves do not necessarily indicate areas of origin. The lines of demarcation between the clean-burned and darkened areas may be used by the investigator to determine direction of fire spread or differences in intensity or time of burning.

6.2.11.2 The investigator should be careful not to confuse the clean burn area with spalling. Clean burn does not show the loss of surface material that is a characteristic of spalling.

6.2.12* Calcination.

6.2.12.1 General. Calcination is used by fire investigators to describe numerous chemical and physical changes that occur in gypsum wallboard surfaces during a fire. Calcination of gypsum wallboard involves driving the free and chemically bound water out of the gypsum as well as other chemical and physical changes to the gypsum component itself. Calcination involves a chemical change of the gypsum to another mineral, anhydrite. Calcined gypsum wallboard is less dense than non-calcined wallboard. The deeper the calcination into the wallboard the greater the total amount of heat exposure (heat flux and duration).

FIGURE 6.2.11 Clean Burn on Wall Surface.

6.2.12.1.1 Gypsum wallboard has a predictable response to heat. First the paper surface will char and might also burn off. The gypsum on the side exposed to fire changes color from pyrolysis of the organic binder and destiffener in it. With further heating, the color change may extend all the way through, and the paper surface on the backside will char. The face exposed to fire will become whiter as the surface carbon is burned away (clean burn). When the entire thickness of wallboard has turned whitish, there will be no paper left on either face, and the gypsum will be chemically dehydrated and converted to a more crumbly, less dense solid. Such wallboard might stay on a vertical wall but will frequently drop off of an overhead surface, particularly if it has absorbed significant quantities of extinguishment water or post-fire precipitation. Fire-rated gypsum wallboard contains mineral fibers or vermiculite particles embedded in the gypsum to preserve the strength of the wallboard during fire exposure. The fibers add strength to the wallboard even after it has been thoroughly calcined.

6.2.12.1.2 Color changes other than shades of gray may occur after gypsum wall surfaces are exposed to heat. The color itself has no significance to the fire investigator. However, the difference between colors may show lines of demarcation.

6.2.12.1.3* The relationship between the calcined and noncalcined areas on gypsum wallboard can also display visible lines of demarcation on the surface. Significant mass loss and corresponding decrease in density occur within the calcined portion of the gypsum wallboard during the calcination process. Depth of calcination measurements can be plotted to display patterns not visible on the surface. See 17.4.4.

6.2.12.2 General Indications of Calcination. The calcination of gypsum board is an indicator demonstrating the heat exposure sustained by the material. The areas of greatest heat exposure may be indicated by both visual appearance and the depth of calcination. The relative differences in color and depth of calcination from point to point may be used as an indicator to establish the areas of greater or lesser heat exposure due to all fire condition variables, such as area of origin, ventilation, and fuel load.

6.2.13* Window Glass. Many texts have related fire growth history or fuels present to the type of cracking and deposits that resulted on window glass. There are several variables that affect the condition of glass after fire, which include the type and thickness of glass, rate of heating, degree of insulation to the edges of the glass provided by the glazing method, degree of restraint provided by the window frame, history of the flame contact, and cooling history.

6.2.13.1 Breaking of Glass. If a pane of glass is mounted in a frame that protects the edges of the glass from radiated heat of fire, a temperature difference occurs between the unprotected portion of the glass and the protected edge. Experimental research estimates that a temperature difference of about 70°C (126°F) between the center of the pane of glass and the protected edge can cause cracks that start at the edge of the glass. The cracks appear as smooth, undulating lines that can spread and join together. Depending on the degree of cracking, the glass may or may not fall out of its frame.

6.2.13.1.1 If a pane of glass has no edge protection from radiated heat of fire, the glass will break at a higher temperature difference. Also, experimental research suggests that fewer cracks are formed, and the pane is more likely to stay whole.

6.2.13.1.2 Glass that has received an impact will exhibit a characteristic "cobweb" pattern. The cracks will be in straight lines and numerous. The glass may have been broken before, during or after the fire.

6.2.13.1.3 If flame contacts one side of a glass pane while the unexposed side is relatively cool, a stress can develop between the two faces and the glass can fracture between the faces.

6.2.13.1.4 Crazing is a term used to describe a complicated pattern of short cracks in glass. These cracks may be straight or crescent-shaped and may or may not extend through the thickness of the glass. Crazing has been claimed to be the result of very rapid heating of one side of the glass while the other side remains cool. Despite widespread publication of this claim, there is no scientific basis for it. In fact, published research has shown that crazing cannot be caused by rapid heating, but can only be caused by rapid cooling. Regardless of how rapidly it was heated, hot glass will reproducibly craze when sprayed with water. *(See Figure 6.2.13.1.4.)*

FIGURE 6.2.13.1.4 Crazed Window Glass.

6.2.13.1.5 Occasionally with small-size panes, differential expansion between the exposed and unexposed faces may result in the pane popping out of its frame.

6.2.13.1.6 The pressures developed by fires in buildings generally are not sufficient either to break glass windows or to force them from their frames. Pressures required to break ordinary window glass are in the order of 2.07 kPa to 6.90 kPa (0.3 psi to 1.0 psi), while pressures from fire are in the order of 0.014 kPa to 0.028 kPa (0.002 psi to 0.004 psi). If an overpressure has occurred — such as a deflagration, backdraft, or detonation — glass fragments from a window broken by the pressure will be found some distance from the window. For example, an overpressure of 10.3 kPa (1.5 psi) can cause fragments to travel as far as 30.3 m (100 ft).

6.2.13.1.7 The investigator is urged to be careful not to make conclusions from glass-breaking morphology alone. Both crazing and long, smooth, undulating cracks have been found in adjacent panes.

6.2.13.2 Tempered Glass.

6.2.13.2.1 Tempered glass, whether broken when heated by fire impact or when exploded, will break into many small cube-shaped pieces. Such glass fragments should not be confused with crazed glass. Tempered glass fragments are more uniformly shaped than the complicated pattern of short cracks of crazing.

6.2.13.2.2 Tempered glass is commonly found in applications where safety from breakage is a factor, such as in shower stalls, patio doors, TV screens, motor vehicles, and in commercial and other public buildings.

6.2.13.3 Staining of Glass.

6.2.13.3.1 Glass fragments that are free of soot or condensates have likely been subjected to rapid heating, failure early in the fire, fracture prior to the fire, or flame contact. The proximity of the glass to a heat source and ventilation are factors that can affect the degree of staining.

6.2.13.3.2 The presence of a thick, oily soot on glass, including hydrocarbon residues, has been interpreted as positive proof of the presence or use of liquid accelerant. Such staining can also result from the incomplete combustion of other fuels such as wood and plastics and should not be interpreted as having come from an accelerant.

6.2.14* Collapsed Furniture Springs. The collapse of furniture springs may provide the investigator with clues concerning the direction, duration, or intensity of the fire. However, the collapse of the springs cannot be used to indicate exposure to a specific type of heat or ignition source, such as smoldering ignition or the presence of an ignitible liquid. The results of laboratory testing indicate that the annealing of springs, and the associated loss of tension (tensile strength), is a function of the application of heat. These tests reveal that short-term heating at high temperatures and long-term heating at moderate temperatures over 400°C (750°F) can result in the loss of tensile strength and in the collapse of the springs. Tests also reveal that the presence of a load or weight on the springs while they are being heated increases the loss of tension.

6.2.14.1 The value of analyzing the furniture springs is in comparing the differences in the springs to other areas of the mattress, cushion, or frame. Comparative analysis of the springs can assist the investigator in developing hypotheses concerning the relative exposure to a particular heat source. For example, if at one end of the cushion or mattress the springs have lost their strength, and at the other end they have not, then hypotheses may be developed concerning the location of the heat source. The hypotheses should take into consideration other circumstances, effects (such as ventilation), and evidence at the scene concerning duration or intensity of the fire, area of origin, direction of heat travel, or relative proximity of the heat source. The investigator should also consider that bedding, pillows, and cushions may shield the springs, or provide an additional fuel load. The portion with the loss of spring strength may indicate more exposure to heat than those areas without the loss of strength. The investigator should also consider the condition of the springs prior to the fire.

6.2.15 Distorted Lightbulbs. Incandescent lightbulbs can sometimes show the direction of heat impingement. As the side of the bulb facing the source of heating is heated and softened, the gases inside a bulb of greater than 25 W can begin to expand and bubble out the softened glass. This has been traditionally, albeit misleadingly, called a *pulled* lightbulb, though the action is really a response to internal pressure rather than a pulling. The bulged or pulled portion of the bulb will be in the direction of the source of the heating, as shown in Figure 6.2.15.

6.2.15.1 Because they contain a vacuum, bulbs of 25 watts or less can be pulled inward on the side in the direction of the source of heating.

6.2.15.2 Often these light bulbs will survive fire extinguishment efforts and can be used by the investigator to show the direction of fire travel. In evaluating a distorted light bulb, the investigator should be careful to ascertain that the bulb has not been turned in its socket or that the socket itself has not turned as a result of coming loose during or after the fire.

6.2.16 Rainbow Effect. Oily substances, which do not mix with water, float and create interference patterns on the surface of water. This results in a "rainbow" or "sheen" appearance. Such rainbow effects are common at fire scenes. Although ignitible liquids will create a rainbow effect, the observation of a rainbow effect should not be interpreted as an indication of the presence of ignitible liquids unless confirmed by a laboratory analysis. Building materials, such as asphalt, plastics, and wood produce oily substances upon pyrolysis that can produce rainbow effects.

FIGURE 6.2.15 A Typical "Pulled" Bulb Showing That the Heating Was from the Right Side.

6.2.17* Victim Injuries. A body will exhibit a material response to exposure to heat and fire. The skin, fat, muscle, and bones will develop a sequential response to heat exposure. Even heavily damaged bodies can be analyzed for body position, orientation to heat sources, and differential exposure and protection that should be correlated with burn patterns of the body and scene.

6.2.17.1 Skin can change color or physical shape, and it can burn. The color changes can vary from reddening to the black of char. Skin can tighten, shrink, and pull apart. Splits in the skin, as a result of exposure to fire, are superficial and distinguishable from traumatic penetrating injuries that deform and bulge along the wound tract. Skin can blister from either pre-mortem or post-mortem exposure.

6.2.17.2 Body fat can melt and burn as a liquid fuel. The burning of body fat typically requires the presence of a porous wick-like material such as cellulose fabric, wood, carpet, or other absorbent/carbonized materials.

6.2.17.3 Muscle can change shape, char, and burn. Heat causes dehydration and shortening of tendons and muscles. Bulkier flexor muscles, such as the biceps in the arms and quadriceps in the legs, shorten and contract, causing a body position known as the pugilistic posture. Shorter muscles of the torso cause arching of the neck and back. The pugilistic posture is a common postmortem response of muscle to heat and is not indicative of a behavioral response to events prior to or during the fire. Deviations from the pugilistic posture should be correlated with the scene to determine if circumstances such as fallen debris, entrapment, or body position (e.g., motor vehicle accidents pinning the body with dashboard or steering wheel) prevented the pugilistic response. Other considerations could be criminal attempts to restrain the body (arms behind the back, ligatures, preexisting traumatic injury, dismemberment, etc.), where the circumstances prevented or altered the expected pugilistic position. Tissues of the body are a fuel load and can continue to burn after surrounding materials self-extinguish.

6.2.17.4 In a fire, bones can change color, change composition, char, and fragment. The color change within bones is related to pyrolysis and is not an indicator of temperatures encountered in a fire. Calcination of the bone can occur when the organic components are burned. The small bones of the extremities, such as feet, hands, fingers, and toes, can appear to be consumed. However, these small bones may have fallen off or fragmented when the surrounding tissues were consumed. As a result, the debris around and under the body should be sifted in an effort to retrieve the small bones and fragments.

6.2.17.5 The skull can exhibit a fragmented, or fractured, appearance regardless of the presence or absence of preexisting traumatic injury. This fragmentary appearance can be caused by numerous actions: the burning off of organic material that renders the bone brittle, trauma, impact of a fire fighter's hose stream on the body, impacting debris, vertical falling of the body through furniture, flooring, or spatial levels, or post-fire movement. Prior to movement, the head should be stabilized and/or a protective bag or wrapping placed around it to minimize further fragment loss. Additional cranial fragments found around the head or body should be collected, as they can retain evidence of traumatic injury from gunshot wounds, blunt-force trauma, or sharp-force trauma. Forensic evaluations of these remains are necessary to determine the cause of death. All cranial fragments and teeth should be collected.

6.2.17.6 The body is evidence and should be examined within the original scene context, if practicable. Unlike other materials in the fire, a body is unique in that during the exposure to the fire, the victim can have purposely changed locations, or positions, prior to death. The investigator should carefully document the location, orientation, and condition of the body. The relationship of the victim to other objects or victims should be documented. The area around the victim should be documented as to significant fuels or collapsed material, which could have caused prolonged burning, protection from the fire, or impact damage to the body.

6.2.17.7 Autopsy reports and photographs provide useful information regarding burn damage. If possible, the fire investigator should attend the autopsy, as certain fire effects that could be significant to the fire investigator may not be significant to the official who examines the body to determine cause and manner of death. The autopsy can provide an opportunity for better examination and documentation of the fire effects to the body. The investigator can correlate the autopsy findings with burn patterns at the scene.

6.2.17.8 Individual victim variables such as age, weight, and health can affect how the body burns and what may survive after burning. Infants and children have developing bones and extra bones that later join to form a mature adult bone. Developing juvenile bones are less dense and may be more fragile and susceptible to damage than adult bones. Mature adult bones are denser and have a higher resistance to fragmenting during the fire. Elderly individuals lose bone density with age (osteoporosis), and their bones are more easily fragmented from heat and recovery. Obese individuals possess higher amounts of body fat than thin or emaciated bodies, thereby contributing more fuel for burning of the body.

6.2.17.9 Victims who survive the fire, but suffer injuries, should also be documented as soon as possible. The nature of their actions within the fire, their clothing, and their injuries should be documented. Interviews and photographing of injuries and clothing can provide immediate documentation. Medical records may be difficult to acquire at a later date.

6.3 Fire Patterns. A fire pattern is the visible or measurable physical changes or identifiable shapes formed by a fire effect or group of fire effects.

6.3.1 Introduction. Fire effects are the underlying data that are used by the investigator to identify fire patterns. The circumstances of every fire are different from every other fire because of the differences in the structures, fuel loads, ignition factors, airflow, ventilation, and many other variables. This discussion, therefore, cannot cover every possible variation in fire patterns and how they come about. The basic principles are covered here, and the investigator should apply them to the particular fire incident under investigation.

6.3.1.1 Dynamics of Pattern Production. The recognition, identification, and proper analysis of fire patterns depend on an understanding of the dynamics of fire development and heat and flame spread. This recognition, identification, and proper analysis require an understanding of the way that conduction, convection, and radiation produce the fire effects and the nature of flame, heat, and smoke movement within a structure. *(See Chapter 5.)*

6.3.1.2 Lines or Areas of Demarcation. Lines or areas of demarcation are the borders defining the differences in certain heat and smoke effects of the fire on various materials. They appear between the affected area and adjacent, less-affected areas.

6.3.1.2.1 The production of lines and areas of demarcation depends on a combination of variables: the material itself, the rate of heat release of the fire, fire suppression activities, temperature of the heat source, ventilation, and the amount of time that the material is exposed to the heat. For example, a wooden wall may display the same heat exposure patterns from exposure to a low-temperature heat source for a long period of time as to a high-temperature heat source for a shorter period of time. The investigator should keep this concept in mind while analyzing the nature of fire patterns.

6.3.1.2.2 The patterns seen by an investigator can represent much of the history of the fire. Each time another fuel package is ignited or the ventilation to the fire changes, the rate of energy production and heat distribution will change. Any burning item can produce a plume and thus a fire pattern. Determining which pattern was produced at the point of origin by the first material ignited usually becomes more difficult as the size and duration of the fire increases.

6.3.2 Causes of Fire Patterns. There are three basic causes of fire patterns: heat, deposition, and consumption. These causes of patterns are defined largely by the fire dynamics discussed in Section 5.5. A systematic analysis of fire patterns can be used to lead back to the heat source that produced them. Some patterns may be interpreted as defining fire intensity (heat/fuel) or spread (movement). See Section 6.4.

6.3.2.1* Plume-Generated Patterns. Most fire patterns are generated directly by fire plumes, which are three-dimensional. Fire patterns represent demarcation lines of fire effects upon materials created by the three-dimensional (conical) shape of the fire plume being cut (truncated) by an intervening two-dimensional surface such as a ceiling or a wall. When the plume intersects with surfaces it creates effects that are interpreted as patterns (conical sections). The rate of heat release of the burning fuel has a profound effect on the shape of the fire patterns produced. These fire patterns include the following:

(1) V patterns
(2) Inverted cone patterns
(3) Hourglass patterns
(4) U-shaped patterns
(5) Pointer and arrow patterns
(6) Circular-shaped patterns

6.3.2.1.1 As the buoyant column of flames, hot gases, and smoke rising above a fire in the plume are cooled by air entrainment, the plume temperatures approach that of the surrounding air (decreased temperatures with increasing height in the plume). Therefore, the production of fire patterns is most prominent when the surface displaying the patterns has been exposed to plume temperatures near or above its minimum pyrolysis temperature. The presence of a physical barrier, such as a ceiling, will contribute to the lateral extension of the plume boundary in a ceiling jet.

6.3.2.1.2 When no ceiling exists over a fire, and the fire is far from walls, the hot gases and smoke of the unconfined plume continue to rise vertically until they ultimately cool to the ambient air temperature. At that point, the smoke and hot gases will stratify and diffuse in the air. Such conditions exist for an unconfined fire outdoors. The same conditions can exist in a building fire at the very early stages of a fire, when the fire has a low heat release rate, when the plume is small, or if the fire is in a very large-volume space with a high ceiling such as an atrium.

6.3.2.1.3 The plume width varies with the size of the base of the fire and will increase over time as the fire spreads. A narrow base pattern will develop from a small surface area fire, and a wide base pattern will develop from a fire with a large surface area. *(See Figure 6.3.2.1.3.)*

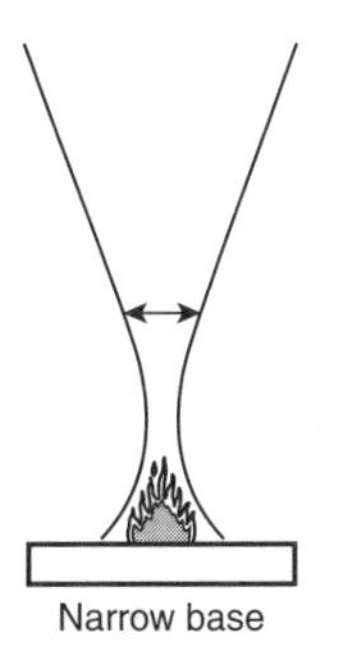

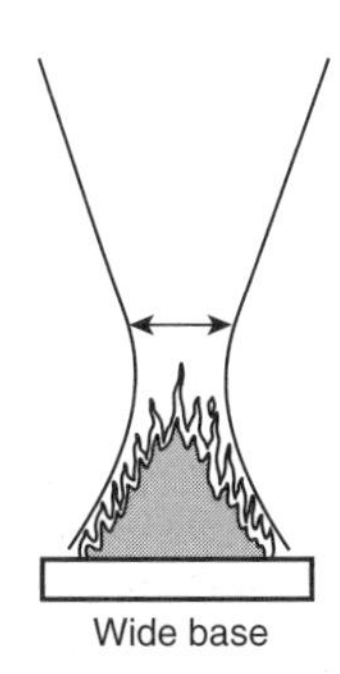

FIGURE 6.3.2.1.3 Effects of Fire Base on Fire Pattern Width.

6.3.2.1.4 An incipient stage fire may produce a fire pattern that has the appearance of an inverted cone. As the heat release rate and flame height increase, this inverted cone pattern may evolve into a subsequent pattern that is more columnar in appearance. Likewise, the growing fire can cause the columnar pattern to evolve into conical patterns such as a V pattern, U pattern, or hourglass pattern. The first patterns will be observable only if the fire goes out, whether from suppression, lack of oxygen, or fuel depletion. For this reason, observation of patterns gives the investigator insight into the fire development. It must also be understood that the lack of an observable inverted cone, hourglass, or columnar pattern after the fire does not mean that one was not present earlier in the fire's growth. If the fire achieves flashover and full room involvement, the patterns formed early in the growth of the fire are often changed by the intense convective and radiant heat transfer.

6.3.2.2 Ventilation-Generated Patterns. Ventilation of fires and hot gases through windows, doors, or other openings in a structure greatly increases the velocity of the flow over combustible materials. In addition, well-ventilated fires burn with higher heat release rates that can increase the rate of char and spall concrete or deform metal components. Areas of great damage are indicators of a high heat release rate, ventilation effects, or long exposure. Such areas, however, are not always the point of fire origin. For example, fire could spread from slow-burning fuels to rapid-burning fuels, with the latter producing most of the fire damage.

6.3.2.2.1 Airflow over coals or embers can raise temperatures, and more heat is transferred as the velocity of the hot gas increases. These phenomena can generate enough heat to spall concrete, melt metals, or burn holes through floors. If a building burns extensively and collapses, embers in debris can produce holes in floors. Once a hole is made, air can flow through the hole, and the burning rate can increase. Careful interpretation of these patterns should be exercised, because they may be mistaken for patterns originating from ignitible liquids.

6.3.2.2.2 When a door is closed on a fire-involved compartment, hot gases (being lighter) can escape through the space at the top of the closed door, resulting in charring. Cool air may enter the compartment at the bottom of the door, as in Figure 6.3.2.2.2(a). In a fully developed room fire where the hot gases extend to the floor, the hot gases may escape under the door and cause charring under the door and possibly through the threshold, as in Figure 6.3.2.2.2(b). Charring can also occur if glowing debris falls against the door either on the inside or the outside, as in Figure 6.3.2.2.2(c). Ignitible liquids burning under wood doors may cause charring of the doors.

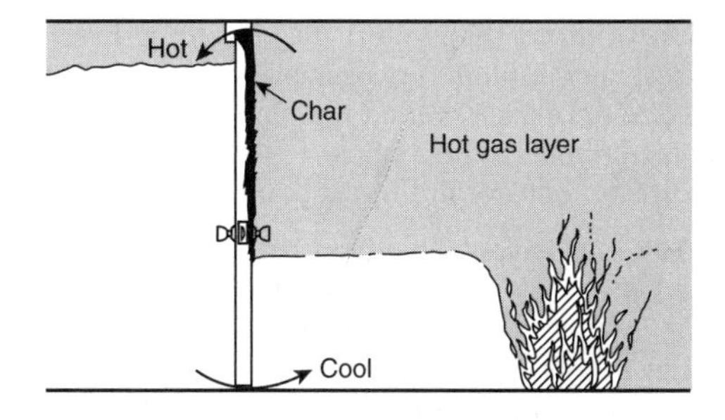

FIGURE 6.3.2.2.2(a) Airflow Around Door.

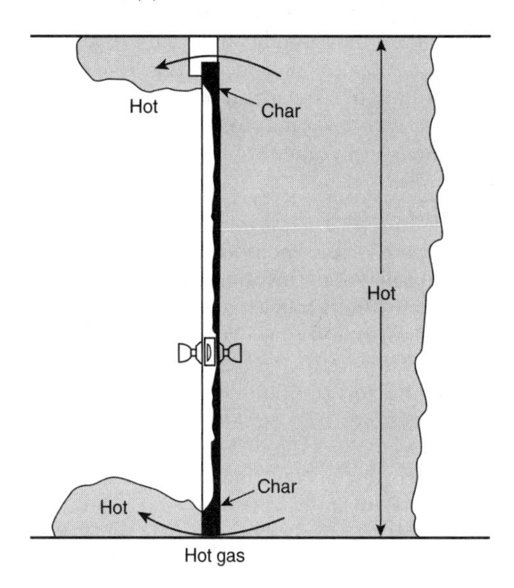

FIGURE 6.3.2.2.2(b) Hot Gases Under Door.

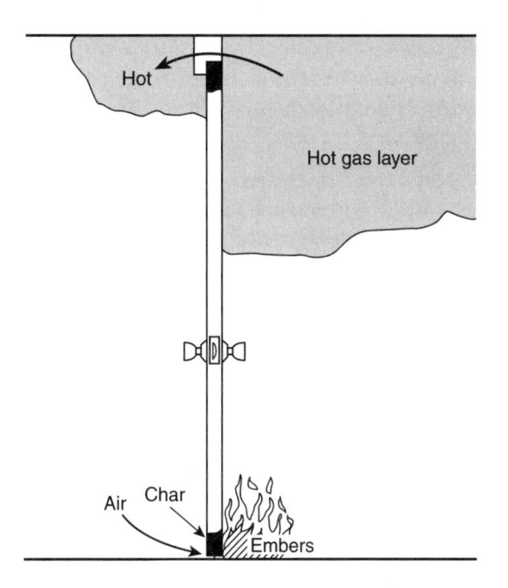

FIGURE 6.3.2.2.2(c) Glowing Embers at Base of Door.

6.3.2.2.3* Effects of Room Ventilation on Pattern Magnitude and Location.

6.3.2.2.3.1 The ventilation of the room has a significant effect on the growth and heat release rate of a fire, and for this reason greatly affects pattern formation.

6.3.2.2.3.2 Besides affecting the fire intensity, ventilation can affect the location, shape, and magnitude of fire patterns. Where fresh air ventilation is available to a fire through open windows and doors, it is common to find locally heavy damage effects on combustible items close to the ventilation opening. These patterns may not indicate a point of origin.

6.3.2.3 Hot Gas Layer–Generated Patterns. The radiant flux from the hot gas layer can produce damage to the upper surfaces of contents and floor covering materials. This process commonly begins as the environment within the room approaches flashover conditions. Similar damage to floor surfaces from radiant heat frequently occurs in adjacent spaces immediately outside rooms that are fully involved in fire. Damage to hallway floors and porches are examples. Protected surfaces may not exhibit any damage. At this time in the fire development, a line of demarcation representing the lower extent of the hot gas layer may form on vertical surfaces. The degree of damage generally will be uniform except where there is drop down, where there is burning of isolated items that are easily ignited, or where there are protected areas.

6.3.2.4 Full Room Involvement–Generated Patterns. If a fire progresses to full room involvement *(see 5.10.2.1 through 5.10.2.8)*, damage found at low levels in the room down to and including the floor can be more extensive due to the effects of radiant flux and the convected heat from the descending hot gas layer and the contribution of an increasing number of burning fuel packages. The radiant heat flux has the greatest impact on surfaces with a direct "view" of the hot gas layer. As the hot gas layer descends to floor level, damage will significantly increase. Damage can include charring of the undersides of furniture, burning of carpet and floor coverings under furniture and in corners, burning of baseboards, and burning on the undersides of doors. Full room involvement can result in holes burned through carpet and floor coverings. The effects of protected areas and floor clutter on low burn patterns should be considered *(see 6.3.3.2.8)*. Although the degree of damage will increase with time, the extreme conditions of the full room involvement can produce major damage in a few minutes, depending on ventilation and fuels present.

6.3.2.5 Suppression-Generated Patterns. Water or other agents used for fire suppression are capable of producing or altering patterns. Hose streams are capable of altering the spread of the fire and creating fire damage in places where the fire would not move in the absence of the hose stream. Additionally, fire department ventilation operations can influence fire patterns. Some fire departments use positive pressure ventilation (PPV) fans that can create patterns that may be difficult to interpret, particularly if the investigator is unaware of PPV use. The history of suppression-generated patterns can only be understood through communication with the responding fire suppression personnel.

6.3.3 Locations of Patterns. Fire patterns may be found on any surface that has been exposed to the effects of the fire or its by-products. These surfaces include interior surfaces, external surfaces and structural members, and outside exposures surrounding the fire scene. Interior surfaces commonly include walls, floors, ceilings, doors, windows, furnishings, appliances, machinery, equipment, other contents, personal property, confined spaces, attics, closets, and the insides of walls. Exterior surfaces commonly include walls, eaves, roofs, doors, windows, gutters and downspouts, utilities (e.g., meters, service drops), porches, and decks. Outside exposures commonly include outbuildings, adjacent structures, trees and vegetation, utilities (e.g., poles, lines, meters, fuel storage tanks, and transformers), vehicles, and other objects. Patterns can also be used to determine the height at which burning may have begun within the structure.

6.3.3.1 Walls and Ceilings. Fire patterns are often found on walls and ceilings. As the hot gas zone and the flame zone of the fire plume encounter these obstructions, patterns are produced that investigators may use to trace a fire's origin. *(See Sections 5.5 and 5.6.)*

6.3.3.1.1 Walls. Patterns on walls may appear as lines of demarcation on the surfaces of the walls or may be manifested as deeper burning. Once the surface coverings of a wall are destroyed by burning, the underlying construction can also display various patterns. These patterns are most commonly V patterns, U patterns, hourglass patterns, and spalling. Surfaces behind wall coverings, even when the covering is still in place, can sometimes also display patterns.

6.3.3.1.2 Ceilings. The investigator should examine patterns that occur on ceilings or the underside of such horizontal surfaces as tabletops or shelves. The buoyant nature of fire gases concentrates the heat energy at horizontal surfaces above the heat source. Therefore, the patterns that are created on the underside of such horizontal surfaces can indicate the locations of heat sources. Although areas immediately over the source of heat and flame will generally experience heating before the other areas to which the fire spreads, circumstances can occur where fuel at the origin burns out quickly, but the resulting fire spreads to an area where a larger supply of fuel can ignite and burn for a longer period of time. This process can cause more damage to the ceiling in that area than in the area immediately over the origin.

6.3.3.1.2.1 These horizontal patterns are roughly circular. Portions of circular patterns are often found where walls meet ceilings or shelves and at the edges of tabletops and shelves. The investigator should determine the approximate center of the circular pattern and investigate below this center point for a heat source.

6.3.3.1.2.2 Fire damage can be found inside walls and ceilings as a result of heat transfer through surfaces. It is possible for the heat of a fire to be conducted through a wall or ceiling surface and to ignite wooden structural members within the wall or ceiling.

6.3.3.1.2.3 The ability of the surface to withstand the passage of heat over time is called its finish rating. The finish rating of a surface material only represents the performance of the material in a specific laboratory test (e.g., as shown in ANSI/UL 263, *Standard for Safety Fire Tests of Building Construction and Materials*) and not necessarily the actual performance of the material in a real fire. Knowledge of the concept can be of value to an investigator's overall fire spread analysis.

6.3.3.1.2.4 This heat transfer process can be observed by the charring of the wooden structural element covered by the protective membrane, shown in Figure 6.3.3.1.2.4.

6.3.3.2 Floors. The investigators should examine patterns that occur on floor coverings and floors. The transition through flashover to full room involvement is associated with a radiant heat flux that exceeds approximately 20 kW/m^2 (2 W/cm^2) at floor level, a typical value for the radiant ignition of common combustible materials. Post-flashover or full room involvement conditions can typically produce fluxes in excess of 170 kW/m^2 and may create, modify, or obliterate patterns.

6.3.3.2.1 Since 1970, carpeting and rugs manufactured or imported to be sold in the United States have been resistant to ignition or fire spread. Typically, cigarettes or matches dropped on carpets will not set them on fire. ASTM D 2859, *Standard Test Method for Flammability of Finished Textile Floor Covering Materials* (Methenamine Pill Test), describes the test used to measure the ignition characteristics of carpeting from a small ignition source. Carpeting and rugs passing the pill test will have very limited ability to spread flame or char in a horizontal direction when exposed to small ignition sources such as a cigarette or match.

6.3.3.2.2* Fire will not spread across a room on the surface of these carpets or rugs without the input of additional energy, such as from a fire external to the carpet or fuel burning on the carpet, in which case the fire spread on the carpet will terminate at a point where the radiant energy from the exposing fire is less than the minimum needed to support flame spread on the carpet (critical radiant flux). Carpet can be expected to ignite and burn when exposed to flashover conditions because the radiant heat flux that produces flashover exceeds the carpet's critical radiant flux.

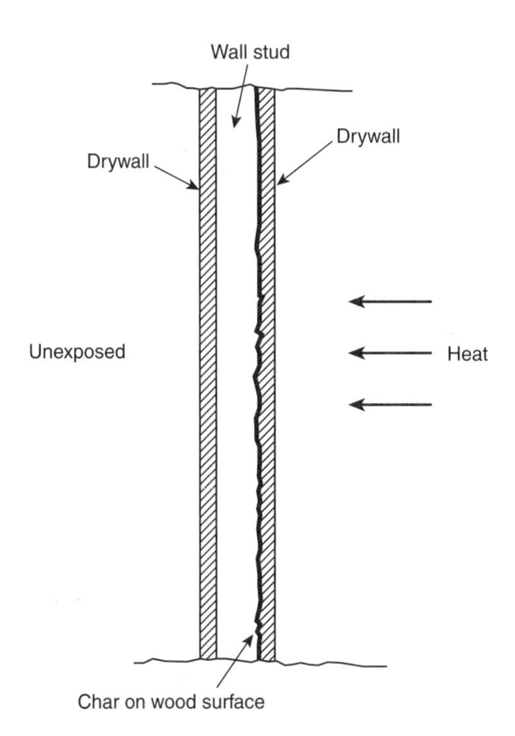

FIGURE 6.3.3.1.2.4 Charring of Wooden Structural Elements by Heat Conduction Through Wall Surface Material.

6.3.3.2.3 Burning between seams or cracks of floorboards or around door thresholds, sills, and baseboards may or may not indicate the presence of an ignitible liquid. Standard tests involving flooring materials such as ASTM E 648, *Standard Test Method for Critical Radiant Flux of Floor-Covering Systems Using a Radiant Heat Energy Source*, regularly produce burning between seams or cracks of floorboard assemblies from radiant heating alone. The knowledge of the pre-fire condition of floorboards, sills, and baseboards can assist in this assessment.

6.3.3.2.4 Full room involvement can also produce burning of floors or around door thresholds, sills, and baseboards due to radiation, the presence of hot combustible fire gases, or air sources (ventilation) provided by the gaps in construction. These gaps can provide sufficient air for combustion of, on, or near floors *(see 6.3.2.2)*.

6.3.3.2.5 Holes in floors may be caused by glowing combustion, radiation, or an ignitible liquid. The surface below a liquid remains cool (or at least below the boiling point of the liquid) until the liquid is consumed. Holes in the floor from burning ignitible liquids may result when the ignitible liquid has soaked into the floor or accumulated below the floor level. Evidence other than the hole or its shape is necessary to confirm the cause of a given pattern.

6.3.3.2.6 Fire-damaged vinyl floor tiles often exhibit curled tile edges, exposing the floor beneath. The curling of tile edges can frequently be seen in non-fire situations and is due to natural shrinkage and loss of plasticizer. In a fire, the radiation from a hot gas layer will produce the same patterns. These patterns can also be caused by ignitible liquids, although confirmation of the presence of ignitible liquids requires laboratory analysis.

6.3.3.2.7 The collection of samples and laboratory verification of the presence or absence of ignitible liquid residues may assist the investigator in developing hypotheses and drawing conclusions concerning the development of floor patterns.

6.3.3.2.8 Unburned areas present after a fire can reveal the location of items that protected the floor or floor covering from radiation damage or smoke staining.

6.3.3.2.9 Outside Surfaces. External surfaces of structures can display fire patterns. In addition to the regular patterns, both vertical and horizontal external surfaces can display burn-through. All other variables being equal, these burn-through areas can identify areas of intense or long-duration burning.

6.3.3.2.10 Drop Down (Fall Down). Burning debris can fall and burn upward, creating a new pattern from this heat source. This occurrence is known as *drop down.* Drop down can ignite other combustible materials, producing low burn patterns.

6.3.4 Location of Objects. Certain types of patterns can be used to locate the positions of objects as they were during a fire.

6.3.4.1 Heat Shadowing. Heat shadowing results from an object blocking the travel of radiant heat from its source to a target material on which the pattern is produced. The object blocking the travel of the heat energy may be a solid, liquid, or gas, combustible or noncombustible. Any object that absorbs or reflects the heat energy may cause the production of a pattern on the material it protects. *(See Figure 6.3.4.1.)*

6.3.4.1.1 Heat shadowing can change, mask, or inhibit the production of identifiable lines of demarcation that may have appeared on that material. Patterns produced by the heat shadowing, may, however, assist the fire investigator in the process of reconstruction during origin determination.

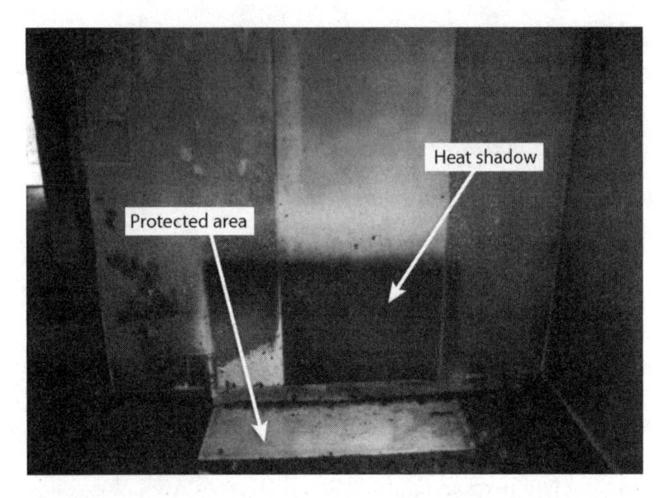

FIGURE 6.3.4.1 Heat Shadow and Protected Areas (USFA Fire Pattern Project).

6.3.4.2 Protected Areas. Closely related in appearance to the resulting pattern of heat shadowing is a protected area. A protected area results from an object preventing the products of combustion from depositing on the material that the object protects, or prevents the protected material from burning. The object may be a solid or liquid, combustible or noncombustible. Any object that prevents the deposition of the products of combustion, or prevents the burning of the material, may produce a protected area. Figure 6.3.4.2 provides an example.

6.3.5 Penetrations of Horizontal Surfaces. Penetration of horizontal surfaces, from above or below, can be caused by radiant heat, direct flame impingement, or localized smoldering with or without the effects of ventilation.

6.3.5.1 Penetrations in a downward direction are often considered unusual because the more natural direction of heat movement is upward because of the buoyancy of heated gases. In fully involved compartments, however, hot gases may be forced through small, pre-existing openings in a floor, resulting in a penetration. Penetrations may also arise as the result of intense burning under furniture items such as polyurethane mattresses, couches, or chairs. Flaming or smoldering under collapsed floors or roofs can also cause floor penetrations.

6.3.5.2 Whether a hole burned into a horizontal surface was created from above or below may be identified by an examination of the sloping sides of the hole. Sides that slope downward from above toward the hole are indicators that the fire was from above. Sides that are wider at the bottom and slope upward toward the center of the hole indicate that the fire was from below. During the course of the fire it is possible for both upward and downward burning to occur through a hole. The investigator should keep in mind that only the last burning direction through the hole may be evident. *(See Figure 6.3.5.2.)*

6.3.5.3 Structural elements, such as studs or joists, can influence the patterns created by fire penetrating up or down or laterally through a building surface. For example, a fire that moves upward through a floor may exhibit patterns significantly influenced by the joists, as opposed to a fire that moves downward through the same floor. The investigator should keep in mind that only the last burning direction through the surface may be evident.

6.3.6 Depth of Char Patterns with Fuel Gases. Flash fires involving fuel gases can produce widely distributed, even charring. However, in areas of pocketed fuel gas, deeper charring

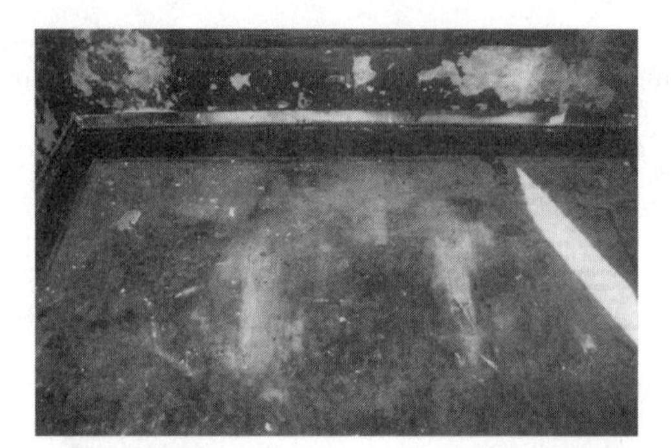

FIGURE 6.3.4.2 Photograph on Top, Showing Protected Area; Photograph at Bottom, Showing How the Chair Was Positioned During the Fire.

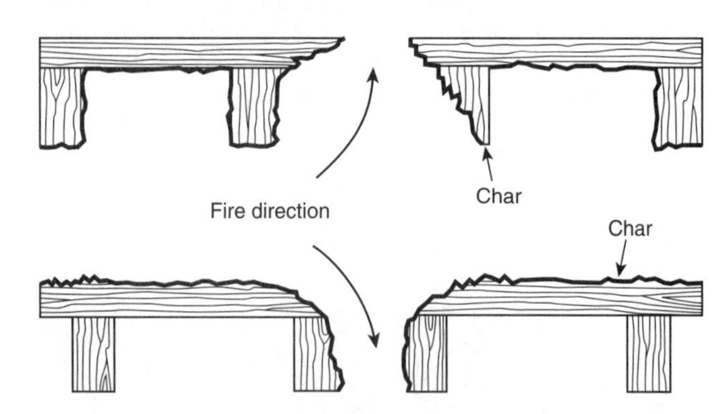

FIGURE 6.3.5.2 Burn Pattern with Fire from Above and Below.

can occur. In close proximity to the point of continuing gas leakage, deeper charring may exist, as burning may continue there after the original quantity of fugitive gas is consumed. This charring may be highly localized because of the pressurized gas jets that can exist at the immediate point of leakage and may assist the investigator in locating the leak.

6.3.7 Pattern Geometry. Various patterns having distinctive geometry or shape are created by the effects of fire and smoke exposure on building materials and contents. In order to identify them for discussion and analysis, they have been described in the field by terms that are indicative of their shapes. While these terms generally do not relate to the manner in which the pattern was formed, the descriptive nature of the terminology makes the patterns easy to recognize. The discussion that follows will refer to patterns by common names and provide some information about how they were formed and how they can be interpreted. Additional information can be found in 6.3.2. Because the interpretation of all possible fire patterns cannot be traced directly to scientific research, the user of this guide is cautioned that alternative interpretations of a given pattern are possible. In addition, patterns other than those described may be observed.

6.3.7.1 V Patterns on Vertical Surfaces. The V-shaped pattern is created by flames, convective or radiated heat from hot fire gases, or smoke within the fire plume. *(See 6.3.2.1.)* The V pattern often appears as lines of demarcation (*see 6.3.1.2*) defining the borders of the fire effects as shown in Figure 6.3.7.1(a) and Figure 6.3.7.1(b).

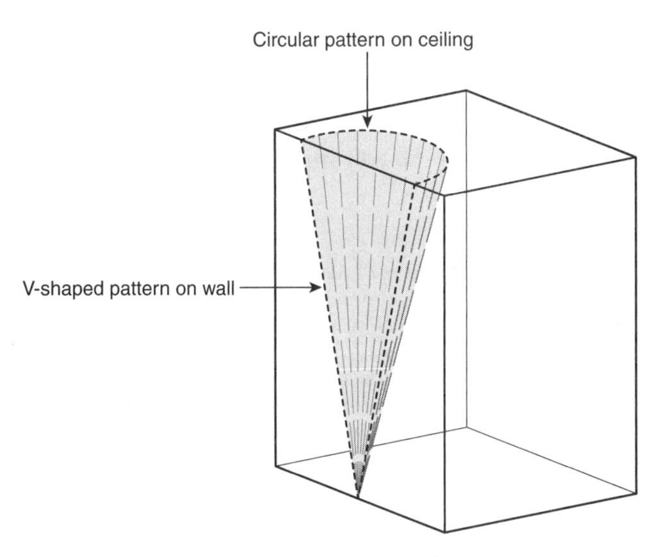

FIGURE 6.3.7.1(a) Idealized Formation of V Pattern and Circular Pattern.

FIGURE 6.3.7.1(b) V Pattern Showing Wall and Wood Stud Damage.

6.3.7.1.1 The angle of the V-shaped pattern is dependent on several variables *(see 6.3.2.1)*, including the following:

(1) Heat release rate (HRR)
(2) Geometry of the fuel
(3) Effects of ventilation
(4) Combustibility of the surface on which the pattern appears
(5) Presence of horizontal surfaces such as ceilings, shelves, table tops, or the overhanging construction on the exterior of a building *(See 6.3.2.1.)*

6.3.7.1.2 The angle of the borders of the V pattern does not indicate the speed of fire growth or rate of heat release of the fuel alone; that is, a wide V does not indicate a slowly growing ("slow") fire and a narrow V does not indicate a rapidly growing ("fast") fire.

6.3.7.2 Inverted Cone (Triangular) Patterns. Inverted cones are commonly caused by the vertical flame plumes not reaching the ceiling. The characteristic two-dimensional shape is triangular with the base at the bottom. *[See Figure 6.3.7.2(a) and Figure 6.3.7.2(b).]*

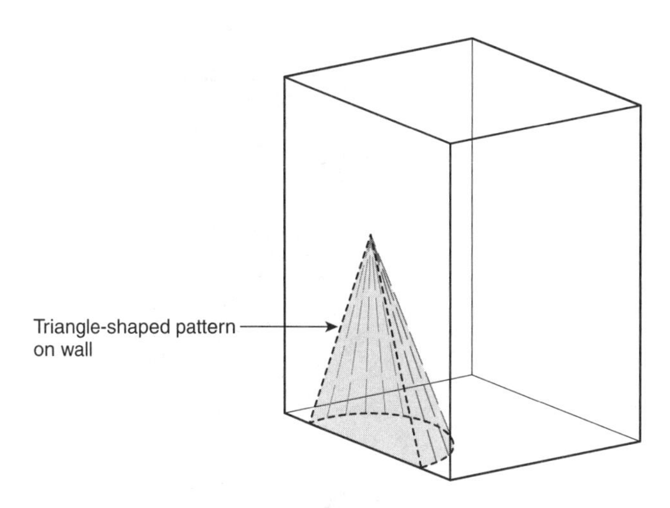

FIGURE 6.3.7.2(a) Idealized Formation of an Inverted Cone Pattern.

FIGURE 6.3.7.2(b) Inverted Cone Pattern Produced by Burning a Small Pile of Newspapers.

6.3.7.2.1 Interpretation of Inverted Cone Patterns. Inverted cone patterns are manifestations of relatively short-lived or low HRR fires that do not fully evolve into floor-to-ceiling flame plumes or have flame plumes that are not vertically restricted

6.3.7.2.2 Inverted cone patterns have been interpreted as proof of ignitible liquid fires, but any fuel source (leaking fuel gas, Class A fuels, etc.) that produces flame zones that do not become vertically restricted by a horizontal surface, such as a ceiling or furniture, can produce inverted cone patterns.

6.3.7.2.3 Inverted Cone Patterns with Natural Gas. The burning of leaking natural gas tends to produce inverted cone patterns, especially if the leakage occurs from below floor level and escapes above at the intersection of the floor and a wall, as in Figure 6.3.7.2.3. The subsequent burning often does not reach the ceiling and is manifested by a characteristic triangular inverted cone pattern shape.

FIGURE 6.3.7.2.3 Inverted Cone Pattern Fueled by a Natural Gas Leak Below the Floor Level.

6.3.7.3 Hourglass Patterns. The plume is a hot gas zone shaped like a V with a flame zone at its base. The flame zone is shaped like an inverted V. When the hot gas zone intersects a vertical surface, the typical V pattern is formed. If the fire itself is very close to or in contact with the vertical surface, the resulting pattern will show the effects of both the hot gas zone and the flame zone together as a large V above an inverted V. The inverted V is generally smaller and may exhibit more intense burning or clean burn. The overall pattern that results is called an hourglass.

6.3.7.4 U-Shaped Patterns. U patterns are similar to the more sharply angled V patterns but display gently curved lines of demarcation and curved rather than angled lower vertices. *(See Figure 6.3.7.4.)* The lowest lines of demarcation of the U patterns are generally higher than the lowest lines of demarcation of corresponding V patterns that are closer to the heat source.

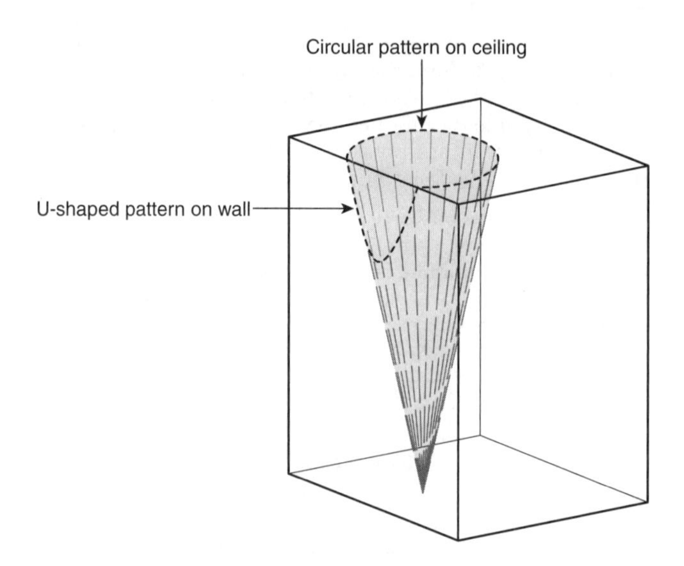

FIGURE 6.3.7.4 Idealized Formation of U-Shaped Pattern.

6.3.7.5 Truncated Cone Patterns. Truncated cone patterns, also called truncated plumes, are three-dimensional fire patterns displayed on both horizontal and vertical surfaces. *(See Figure 6.3.7.5(a) and Figure 6.3.7.5(b).)* It is the intersection or truncating of the natural cone-shaped or hourglass-shaped plume by these vertical and horizontal surfaces that causes the patterns to be displayed. Many fire patterns, such as V patterns, U patterns, circular patterns, and "pointer or arrow" patterns, are related directly to the three-dimensional cone of heat created by the fire.

6.3.7.5.1 Due to air entrainment, the width of the plume cone increases with increasing height. When the fire plume encounters an obstruction to its vertical movement, such as the ceiling of a room, the hot gases move horizontally. Thermal damage to a ceiling will generally extend beyond the circular area attributed to a "truncated cone" due to this horizontal movement. The truncated cone pattern combines two-dimensional patterns such as V-shaped patterns on vertical surfaces, with circular patterns displayed on horizontal surfaces. The combination of more than one two-dimensional pattern on perpendicular, vertical, and horizontal surfaces reveals the plume's three-dimensional shape.

6.3.7.6 Pointer and Arrow Patterns. These fire patterns may be on a series of combustible elements such as wooden wall studs whose surface sheathing has been destroyed by fire. The direction of fire spread along a wall can often be identified and traced back toward its source by an examination of the relative heights and burned-away shapes of the wall studs left standing after a fire. In general, shorter and more severely charred studs will be closer to a source of heat than taller studs. The heights of the remaining studs increase as distance from a source of fire increases. The difference in height and severity of charring may be observed and documented, as shown in Figure 6.3.7.6.

6.3.7.6.1 The shape of the studs' cross-section will tend to produce "arrows" pointing back toward the general area of the source of heat. This is caused by the burning off of the sharp angles of the edges of the studs on the sides toward the heat source that produces them, as shown in Figure 6.3.7.6.1.

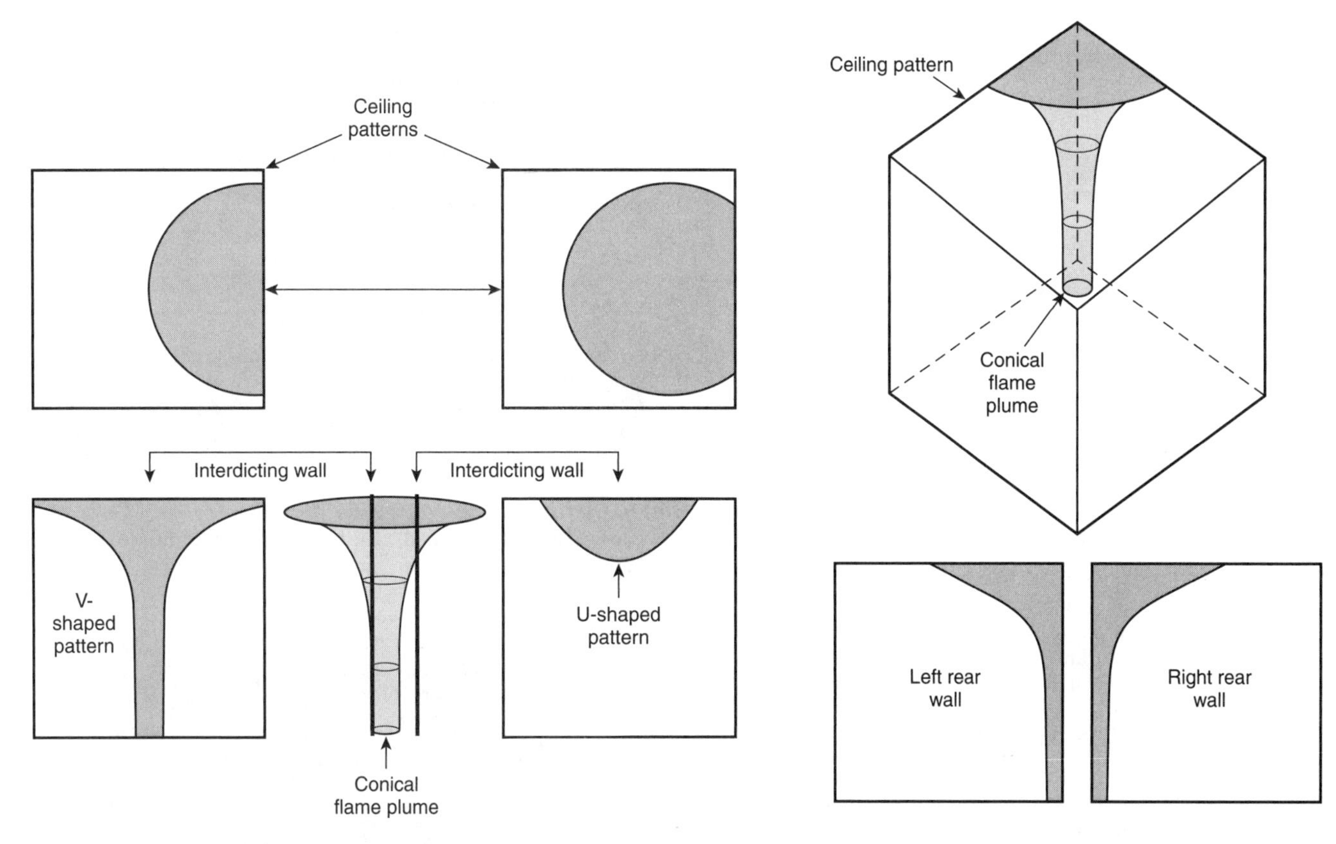

FIGURE 6.3.7.5(a) Idealized Truncated Cone Pattern Formation.

FIGURE 6.3.7.5(b) Truncated Cone Pattern Displayed on Perpendicular Walls.

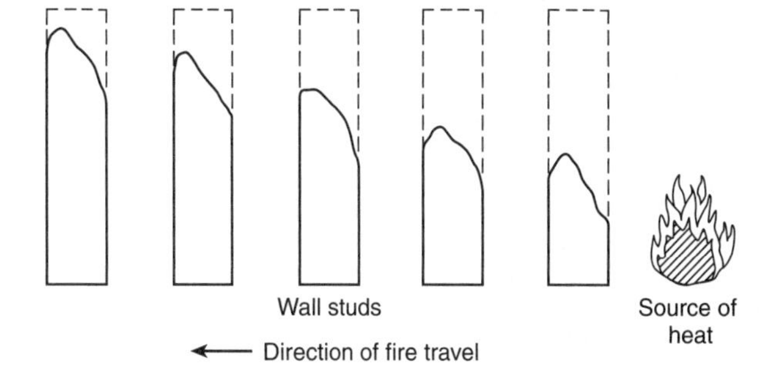

FIGURE 6.3.7.6 Wood Wall Studs Showing Decreasing Damage as Distance from Fire Increases.

6.3.7.6.2 More severe charring can be expected on the side of the stud closest to the heat source.

6.3.7.7 Circular-Shaped Patterns. Patterns on the underside of horizontal surfaces, such as ceilings, tabletops, and shelves, can appear in roughly circular shapes. The farther the heat source is from the wall, the more circular the patterns may appear. Portions of circular patterns can appear on the underside of surfaces that partially block the heated gases or fire plumes. This appearance can occur when the edge of the surface receiving the pattern does not extend far enough to show the entire circular pattern or when the edge of the surface is adjacent to a wall. Within the circular pattern, the center may show more heat degradation, such as deeper charring. By locating the center of the circular pattern, the investigator may find a valuable clue to the source of greatest heating, immediately below.

6.3.7.8 Irregular Patterns. Irregular, curved, or "pool-shaped" patterns on floors and floor coverings should not be identified as resulting from ignitible liquids on the basis of visual appearance alone. In cases of full room involvement, patterns similar in appearance to ignitible liquid burn patterns can be produced when no ignitible liquid is present.

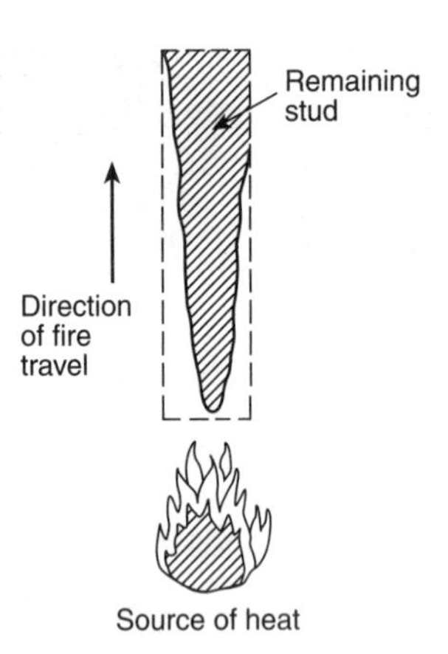

FIGURE 6.3.7.6.1 Cross-Section of Wood Wall Stud Pointing Toward the Heat Source.

6.3.7.8.1 The lines of demarcation between the damaged and undamaged areas of irregular patterns range from sharp edges to smooth gradations, depending on the properties of the material and the intensity of heat exposure. Denser materials like oak flooring will generally show sharper lines of demarcation than polymer (e.g., nylon) carpet. The absence of a carpet pad often leads to sharper lines.

6.3.7.8.2 Irregular patterns are common in situations of post-flashover conditions, long extinguishing times, or building collapse. These patterns may result from the effects of hot gases, flaming and smoldering debris, melted plastics, or ignitible liquids. If the presence of ignitible liquids is suspected, supporting evidence in the form of a laboratory analysis should be sought. It should be noted that many plastic materials release hydrocarbon fumes when they pyrolyze or burn. These fumes may have an odor similar to that of petroleum products and can be detected by combustible gas indicators when no ignitible liquid accelerant has been used. A "positive" reading should prompt further investigation and the collection of samples for more detailed chemical analysis. It should be noted that pyrolysis products, including hydrocarbons, can be detected in laboratory analysis of fire debris in the absence of the use of accelerants. It can be helpful for the laboratory, when analyzing carpet debris, to burn a portion of the comparison sample and run a gas chromatographic-mass spectrometric analysis on both samples. By comparing the results of the burned and unburned comparison samples with those from the fire debris sample, it may be possible to determine whether or not hydrocarbon residues in the debris sample were products of pyrolysis or residue of an accelerant. In any situation where the presence of ignitible liquids is suggested, the effects of flashover, airflow, hot gases, melted plastic, and building collapse should be considered.

6.3.7.8.3 When overall fire damage is limited and small, or isolated irregular patterns are found, further examination should be conducted for supporting evidence of ignitible liquids. *[See Figure 6.3.7.8.3(a) and Figure 6.3.7.8.3(b).]* Even in these cases, radiant heating may cause the production of patterns on some surfaces that can be misinterpreted as liquid burn patterns. *[See Figure 6.3.7.8.3(c).]*

FIGURE 6.3.7.8.3(a) Irregular Burn Patterns on a Floor of a Room Burned in a Test Fire in Which No Ignitible Liquids Were Used.

FIGURE 6.3.7.8.3(b) Irregularly Shaped Pattern on Carpet Resulting from Poured Ignitible Liquid; Burned Match Can be Seen at Lower Left.

FIGURE 6.3.7.8.3(c) "Pool-Shaped" Burn Pattern Produced by a Cardboard Box Burning on an Oak Parquet Floor.

6.3.7.8.4 Pooled ignitible liquids that soak into flooring or floor covering materials as well as melted plastic can produce irregular patterns. These patterns can also be produced by localized heating or fallen fire debris. *[See Figure 6.3.7.8.4(a) and Figure 6.3.7.8.4(b).]*

6.3.7.8.5 The term *pour pattern* implies that a liquid has been poured or otherwise distributed, and therefore, is demonstrative of an intentional act. Because fire patterns resulting from burning ignitible liquids are not visually unique, the use of the

FIGURE 6.3.7.8.4(a) Non-Accelerated Test Burns Demonstrating Melting, Dripping, Pooling, and Burning of Melted Polyurethane Foam Mattress.

FIGURE 6.3.7.8.4(b) Non-Accelerated Test Burns Demonstrating Melting, Dripping, Pooling, and Burning of Melted Upholstered Chair Padding.

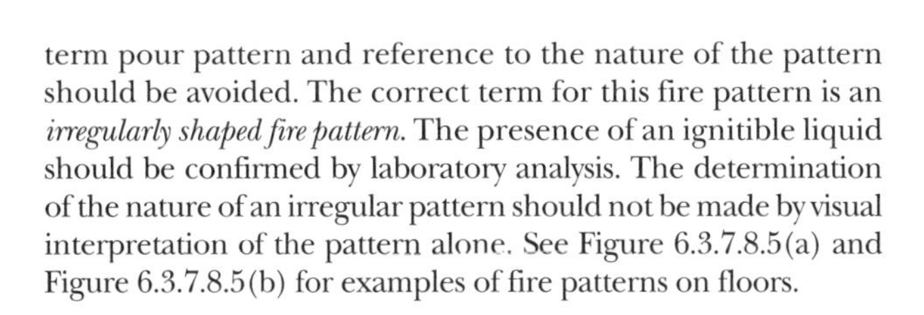

term pour pattern and reference to the nature of the pattern should be avoided. The correct term for this fire pattern is an *irregularly shaped fire pattern.* The presence of an ignitible liquid should be confirmed by laboratory analysis. The determination of the nature of an irregular pattern should not be made by visual interpretation of the pattern alone. See Figure 6.3.7.8.5(a) and Figure 6.3.7.8.5(b) for examples of fire patterns on floors.

6.3.7.8.6 Liquids Versus Melted Solids. Many plastic materials will burn. Thermoplastics react to heating by first liquefying, and then, when they burn as liquids, they produce irregularly shaped or circular patterns. When found in unexpected places, such patterns can be erroneously identified as ignitible liquid patterns and associated with an incendiary fire cause. The investigator should be careful to identify properly the fuel sources for any irregularly shaped or circular patterns.

FIGURE 6.3.7.8.5(a) Fire Patterns on Floor Resulting from Fully Developed (Post-Flashover) Fire in Full Scale Test Burn of Residential Structure. Floor Was Carpeted and Room Had Typical Residential Furnishings; No Ignitible Liquids Were Present.

FIGURE 6.3.7.8.5(b) Fire Patterns on Linoleum Floor Resulting from Fully Developed (Post-Flashover) Fire in Full-Scale Test Burn of Residential Structure; No Ignitible Liquids Were Present.

6.3.7.9 Doughnut-Shaped Patterns. A doughnut-shaped pattern, where an irregularly shaped burn area surrounds a less burned area, may result from an ignitible liquid. When a liquid causes this pattern, shown in Figure 6.3.7.9(a), it is due to the effects of the liquid cooling the center of the pool as it burns, while flames at the perimeter of the doughnut produce charring of the floor or floor covering. When this condition is found, further examination should be conducted for supporting evidence of ignitible liquids, especially on the interior of the pattern. See Figure 6.3.7.9(b).

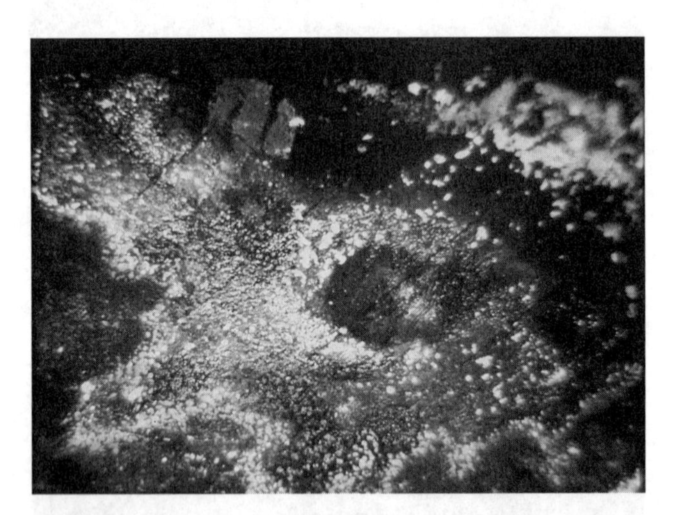

FIGURE 6.3.7.9(a) Doughnut-Shaped Fire Pattern on a Carpeted Floor.

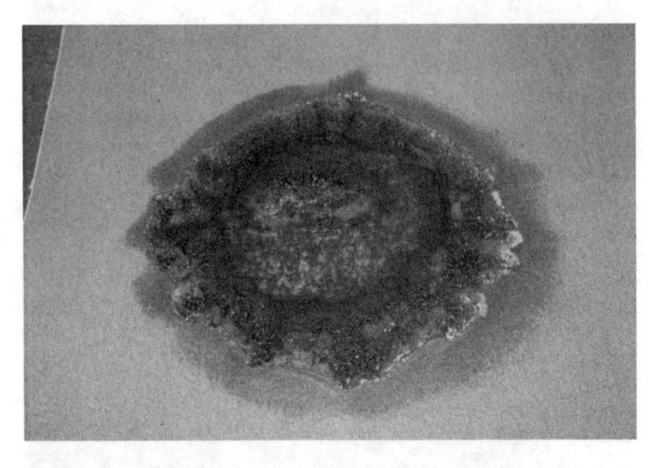

FIGURE 6.3.7.9(b) Doughnut-Shaped Fire Pattern on a Carpeted Floor Burn Test.

6.3.7.10 Linear Patterns. Patterns that have overall linear or elongated shapes can be called *linear patterns.* Linear patterns usually appear on horizontal surfaces.

6.3.7.10.1 Trailers. In many incendiary fires, when fuels are intentionally distributed or "trailed" from one area to another, the elongated patterns may be visible. Such fire patterns, known as "trailers," can be found along floors to connect separate fire sets, or up stairways as shown in Figure 6.3.7.10.1. Fuels used for trailers may be ignitible liquids, solids, or combinations of these.

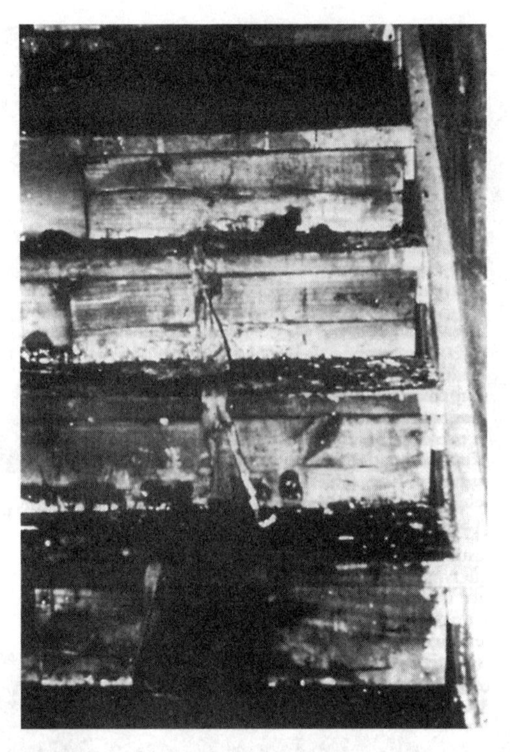

FIGURE 6.3.7.10.1 Trailer Running Up a Stairway.

6.3.7.10.2 Protected Floor Areas. Often when the floor area is cleared of debris to examine damage, long, wide, straight patterns will be found, showing areas of extensive heat damage bounded on each side by undamaged or less damaged areas. These patterns often have been interpreted to be "trailers." While this is possible, the presence of furniture, stock, counters, or storage may result in these linear patterns. These patterns may also result from wear on floors and the floor covering due to high traffic. Irregularly shaped objects on the floor, such as clothing or bedding, may also provide protection and produce patterns that may be inaccurately interpreted.

6.3.7.10.3 Fuel Gas Jets. Jets of ignited fuel gases, such as LP-Gas or natural gas, can produce linear patterns or lines of demarcation, particularly on noncombustible surfaces.

6.3.7.11 Area Patterns. Some patterns may appear to cover entire rooms or large areas without any readily identifiable source. These patterns are often formed when the fuels that create them are above the lower flammable limit and widely dispersed before ignition, or when the movement of the fire through the areas is very rapid, as in a flash fire.

6.3.7.11.1 Flashover and Full Room Involvement. In the course of a flashover transition, fire spreads rapidly to all exposed combustible materials as the fire progresses to full room involvement. *(See 5.10.2.6.)* This process can produce relatively uniform depths of char or calcination. If the fire is terminated before full room involvement, relatively uniform burning can be evident on vertical surfaces above the bottom of the hot layer. When the fire has progressed to full room involvement, the area pattern may be uneven and may extend to the floor. The uniformity described in this section may not be consistent throughout the room or space. Some exposed surfaces may exhibit little or no damage due to the ventilation effects or the locations of furnishings or fixtures that may prevent charring, darkening, or discoloration of wall and ceiling surfaces.

6.3.7.11.2 Flash Fires. The ignition of gases or vapors of liquids does not necessarily always cause explosions. Whether an explosion occurs depends on the location and concentration of diffuse fuels and on the geometry, venting, and strength of the confining structure.

6.3.7.11.2.1 If the diffuse fuels are near the lower flammable or lower explosive limit and there is no explosion, the fuels may burn as a flash fire, and there may be little or no subsequent burning. In the instance where the first fuel to be ignited is a diffuse fuel–air mixture, the area of greatest destruction may not, and generally does not, coincide with the area where the heat source ignites the mixture. The greatest destruction will occur where the flash fire from the burning mixture encounters a secondary fuel load that is capable of being ignited by the momentary intense temperature in the flame front. Likewise, once secondary ignition occurs, the dynamics of the fire spread will be dictated by the compartment and fuel geometry and the relative heat release rates of these secondary fuels. The relatively short duration of the burning may have little impact on the flashover in the compartment as compared to the burning of the secondary fuels. Therefore, origin determination of such a flash fire can be supported by accurate witness observations and the analysis of the potential ignition sources in the areas where the vapor or gas could have existed. When the analysis of fire patterns is the only means of determining the origin, the investigator should be aware that the resultant ignition of secondary fuels and compartment flashover could have altered or obliterated the subtle patterns created by the flash fire.

6.3.7.11.2.2 The difficulty in detecting patterns caused by flash fires is the result of the total consumption of available fuel without significantly raising the temperatures of other combustibles. In this case, the fire patterns may be superficial and difficult to trace to any specific point of ignition as in Figure 6.3.7.11.2.2. In addition, separate areas of burning from pocket fuel gas may exist and further confuse the tracing of fire spread.

FIGURE 6.3.7.11.2.2 Blistering of Varnish on Door and Slight Scorching of Draperies, the Only Indications of the Natural Gas Flash Fire.

6.3.7.12 Saddle Burns. *Saddle burns* are distinctive U- or saddle-shaped patterns that are sometimes found on the top edges of floor joists. They are caused by fire burning downward through the floor above the affected joist. Saddle burns display deep charring, and the fire patterns are highly localized and gently curved. They also may be created by radiant heat from a burning material in close proximity to the floor, including materials that may melt and burn on the floor (e.g., polyurethane foam). Ventilation caused by floor openings may also contribute to the development of these patterns, shown in Figure 6.3.7.12.

FIGURE 6.3.7.12 Saddle Burn in a Floor Joist.

6.4 Fire Pattern Analysis. Fire pattern analysis is the process of identifying and interpreting fire patterns to determine how the patterns were created and their significance.

6.4.1 Types of Fire Patterns. There are two basic types of fire patterns: movement patterns and intensity patterns. These types of patterns are defined by the fire dynamics discussed in Section 5.10. Often a systematic use of more than one type of fire pattern at a fire scene can be used in combination to lead back to the heat source that produced them. Some patterns may display aspects defining both movement and intensity (heat/fuel).

6.4.1.1 Fire Spread (Movement) Patterns. Flame, heat, and smoke produce patterns as a result of fire growth and fire spread. Movement patterns are produced by the growth, spread, and flow of products of combustion away from an initial heat source. If accurately identified and analyzed, these patterns can be traced back to the origin of the heat source that produced them.

6.4.1.2 Heat (Intensity) Patterns. Flames and hot gases produce patterns as a result of the response of materials to heat exposure. The various heat effects on materials can produce lines of demarcation. These lines of demarcation may be helpful to the investigator in determining the characteristics and quantities of fuel materials, as well as the direction of fire spread.

6.4.1.3 Combination of Patterns. Fire patterns may exhibit a combination of effects. The investigator should be aware of the influence each type of pattern may have on the other and the sequence of their production. Failure to consider these factors may lead the investigator to erroneous conclusions regarding fire dynamics.

Chapter 7 Building Systems

7.1* Introduction. Understanding the reaction of buildings and building assemblies to fire is of prime importance to the fire investigator. Development, spread, and control of a fire within a structure often depends on the type of construction, the ability of structural elements to remain intact, and the interface of fire protection and other building systems. Interior layout, occupant circulation patterns, interior finish materials, and building services can be important factors in the start, development, and spread of the fire. This chapter will assist the investigator to specifically track and document building systems as related to the fire.

7.1.1 It should be noted that this chapter only highlights general building information. Included in the reference section are a number of related texts that will provide the investigator with the opportunity to obtain greater detail and understanding on building construction and building systems. More detailed information can be found in the 18th edition of the NFPA *Fire Protection Handbook.*

7.1.2 In addition to building design and construction elements, there are important fuel-oriented considerations for the fire investigator. For example, during the preflashover, growth stage of a fire the heat release rate of a fuel package has a significant influence on the rate of fire growth. *(See Section 5.5.)*

7.2 Features of Design, Construction, and Structural Elements in Evaluating Fire Development.

7.2.1 General. The architectural design of a building has a significant influence on its fire safety capabilities. Interior layout, circulation patterns, interior finish materials, and building services are all important factors in fire safety. How the building design affects manual suppression of fires is another important consideration.

7.2.1.1 The way a fire develops and spreads can be influenced by the building design in the way that the structure is planned, shaped, built, and by the materials chosen. The nature of the occupancy or purpose to which the structure is used can also affect the way it burns. The investigator must evaluate the fire development and spread in light of the knowledge that how the building is formed can influence these factors.

7.2.1.2 Changes in occupancy types may create a hazard to fire-fighting efforts and may have an effect on the development of the fire. As an example, there can be an ordinary retail business that is then converted to a paint store that is deemed to be a hazardous occupancy. The increased fuel load will most probably affect fire intensity and spread, and the original design may be insufficient to withstand the fire.

7.2.2 Building Design.

7.2.2.1 General. Fire spread and development within a building is largely the effect of radiant and convective heating. In compartment fires, much of the fire spread is also a function of the state of confinement of heated upper gas layers. For a given fuel package, room size, room lining material, shape, ceiling height, and the placement and areas of doors and windows can profoundly affect the formation of ceiling jets, radiation feedback, the production and confinement of upper gas layers, ventilation, flameover, and the time to flashover of a compartment fire. All of these factors influence how a fire develops.

7.2.2.1.1 Compartmentation is a primary fire protection concept. Keeping fire confined in its room of origin and minimizing smoke movement to other areas of a building have long been goals of fire protection engineering designers and fire code organizations. The design of fire-resistive constructions, fire-stopped pipe chases and utility openings in fire walls, and construction techniques that minimize smoke and flame movement can aid in effective compartmentation. Designs that are less fire safe have just the opposite effect.

7.2.2.1.2 Extreme architectural designs such as atriums; large enclosed areas like stadiums or tunnels; and glass or unusual structural, wall, ceiling, roof, or finish materials also pose interesting considerations for the fire investigator, especially in the analysis of the way these features have affected fire growth and spread.

7.2.2.1.3 Most of these aspects are initially under the control of the building architect or systems designers. Small changes in the specifications for a structure can have profound effects on the overall fire safety of the building. When necessary and possible, the fire investigator should review design plans and fire code requirements for that structure. Modifications to structural and nonstructural areas of the building may change the fire-resistive capability of the building. For example, existing ceilings that have added drop-down ceilings create a void and may have significant impact on the fire and smoke travel.

7.2.2.2 Building Loads. The effects of undesigned loads, such as added dead and live loads, wind, water, and impact loads, may change the structural integrity of the building. Dead loads are the weight of materials that are part of a building, such as the structural components, roof coverings, and mechanical equipment. Live loads are the weight of temporary loads that need to be designed into the weight-carrying capacity of the structure, such as furniture, furnishings, equipment, machinery, snow, and rain water. Snow on the roof is an example of a live load; an additional layer of roofing is an example of a dead load. The function of a building's structure is to resist forces. As long as these forces remain in balance, the building will stand, but when the balance is lost the building may collapse. Building loads may become unbalanced when a building is subjected to fire and the structural components of the structure are damaged.

7.2.2.3 Room Size.

7.2.2.3.1 For a given fuel package, that is, heat release rate, the room's volume, ceiling height, size of the ventilation opening, and location of the fire will affect the rate of fire growth in the room. The speed of development of a hot gas upper layer, and the spread of a ceiling jet from the fire plume are among the important factors that determine if and when the room will flash over. Flashover, in turn, has a great effect on the spread of fire out of the room of origin.

7.2.2.3.2 The ignition and burning of a fuel load in the room produce heat, flame, and hot gases at a given rate. The area and volume of the room affect the time to flashover: the smaller the area and volume of the room, the sooner the room may flash over and the sooner the fire may spread outside the room, provided all other variables remain constant. Extremely large rooms may never have a sufficient heat energy transfer to cause flashover.

7.2.2.4 Compartmentation.

7.2.2.4.1 The common mode of fire spread in a compartmented building is through open doors, unenclosed stairways and shafts, unprotected penetrations of fire barriers, and non-fire-stopped combustible concealed spaces. Even in buildings of combustible construction, the common gypsum wallboard or plaster on lath protecting wood stud walls or wood joist floors provides a significant amount of resistance to a fully developed fire. When such barriers are properly constructed and maintained and have protected openings, they normally will contain fires of maximum expected severity in light-hazard occupancies. Even a properly designed, constructed, and maintained barrier will not reliably protect against fire spread indefinitely. Fire can also spread horizontally and vertically beyond the room or area of origin and through compartments or spaces that do not contain combustibles. Combustible surfaces on ceilings and walls of rooms, stairways, and corridors, which in and of themselves may not be capable of transmitting fire, will be heated and produce pyrolysis products. These products add to those of the main fire and increase the intensity and length of flames. Fire spread rarely occurs by heat transfer through floor/ceiling assemblies. Fire spread through floor/ceiling assemblies may occur in the later stages of fire development or through breaches of these assemblies.

7.2.2.4.2 The investigator will want to analyze the reasons that compartmentation of the fire failed or did not occur and which aspects of the design of the building may have been responsible for this failure.

7.2.2.5 Concealed and Interstitial Spaces. Concealed and other interstitial spaces can be found in most buildings. These spaces can create increased rates of fire spread and prolonged fire duration. Both of these factors aggravate the damage expected to be encountered.

7.2.2.5.1 Interstitial spaces in a high-rise building are generally associated with the space between the building frame and interior walls and the exterior facade, and with spaces between ceilings and the bottom face to the floor or deck above. These spaces may not have fire stops, the lack of which aids in the vertical spread of fire. Those spaces provided with fire stops should be examined to determine the type and effectiveness of the installation.

7.2.2.5.2 Fire investigators should consider the impact of concealed spaces when they conduct a fire investigation. Failure to consider the effects of fire travel through concealed spaces may lead to misreading the fire patterns. Care must be taken when examining areas such as attics, roofs, and lowered ceilings in rooms that can conceal fire and smoke until the fire is out of control.

7.2.2.6 Planned Design as Compared to "As-Built" Condition. The investigator should be aware that building specifications, plans, and schematic drawings, prepared before construction, are not always the "as-built" condition. After permit issuance or on-site inspections, the actual as-built condition may not always have met the approved design. If necessary or possible, the investigator should verify the original approved drawings, the actual as-built condition, and the current building condition. This verification can be accomplished by requesting the original building plans from the local building department or the original architect, by an examination of the fire scene, or if this is not possible due to fire damage or the unavailability of the fire scene, by examination of similar houses or buildings built by the same contractor at the same time, or by witness interviews.

7.2.2.6.1 When the investigator is comparing the original plans, the as-built plans, and the current construction, careful attention should be given to the location of current walls and the current electrical wiring construction, as these are often changed without required permits.

7.2.3 Materials. The nature of the materials selected and used in a building design can have a substantial effect on the fire development and spread. The nature of material is important from both its physical and chemical properties. How easily the material ignites and burns, resists heating, resists heat-related physical or chemical changes, conducts heat, and gives off toxic by-products are important to an overall evaluation of the design of the structure.

7.2.3.1 Ignitibility. How easily a specific material may be ignited, its minimum ignition temperature, minimum ignition energy, and a time–temperature relationship for ignition are basic considerations when the use of the material in a building design is evaluated.

7.2.3.2 Flammability. Once a material is ignited, either in flaming or smoldering combustion, how it burns and transmits its heat energy is also a consideration for the fire investigator. Such factors as heat of combustion, average and peak heat release rate, and perhaps even mass loss rate, can be important considerations in its overall fire safety and suitability for use. The entrainment of air has an important role in the way a fire develops upon the material.

7.2.3.3 Thermal Inertia. The thermal inertia of a material (specific heat × density × thermal conductivity) is a key factor in considering the material's reaction to heating and ease of ignition. These factors will need evaluation if the investigator is making determinations about the material's suitability for use or its role in the transition to flashover of a compartment in which the material is a liner.

7.2.3.4 Thermal Conductivity. Good conduction of heat from the surface of the fuel to its interior keeps the surface temperature lower than if it has poor conduction. Conduction impacts the change in temperature of the fuel. Conduction can be the means of transferring heat to the unexposed face of a material, such as a steel partition.

7.2.3.5 Toxicity. Though not directly related to the development and spread of a fire, the toxicity of the products of combustion of a material are a very important consideration in the overall fire safety of a design. Materials that give off large quantities of poisonous or debilitating gases or products of combustion can incapacitate or kill fire victims long before any heat or flames reach them. Toxicity is an important issue for fire investigators involved in evaluating how the design and condition of a building, building materials, and contents affected the occupants. In most fire situations, carbon monoxide is the dominant toxic species, which is particularly true of fire products produced in a flashed-over space.

7.2.3.6 Physical State and Heat Resistance. At what temperature the material under scrutiny changes in phase from solid to liquid or liquid to gas may be a factor in evaluating its fire performance. In general, liquids require less energy to ignite than solids, and gases require still less energy than liquids.

7.2.3.6.1 Characteristics of plastics, such as whether they are thermoplastics (which transform from solids to liquids and then to ignitible gases) or thermoset plastics (which pyrolyze directly to ignitible gases), may affect whether they are selected as a structural or surface material.

7.2.3.6.2 Materials that tend to melt and liquefy during the course of a fire may be more likely to cause fall-down damage or ignitible liquid fire spread. The choice of such materials in the design of a structure could become important considerations to the fire investigator.

7.2.3.7 Orientation, Position, and Placement.

7.2.3.7.1 Many materials burn differently, depending upon their orientation, position, or placement within a building. Generally, materials burn more rapidly when they are in a vertical rather than horizontal position. For example, carpeting that is designed and tested to be used in a horizontal position, in full contact with a horizontal flat surface, may burn at a rate well below the maximum standard set by code. When the same carpet material is mounted vertically as a wall covering or curtain, it is likely to exhibit a fire performance that is worse than would have been expected from the fire test results in the horizontal orientation. An adhesive may have an effect on the burning rate of the carpet.

7.2.3.7.2 Flame spread indexes, commonly used in codes and standards to quantify the flammability of a material, are usually based on testing in the ASTM E 84, *Standard Test Method for Surface Burning Characteristics of Building Materials*, often called the Steiner Tunnel Test. The Steiner Tunnel Test burns a sample of the material in a horizontal orientation, suspended on the ceiling of the 24-foot long Steiner Tunnel. Many of the materials tested in the Steiner Tunnel are not designed or intended to be applied in building designs as wall or ceiling coverings. The actual flame spread of the material as used in construction is often different and might bear no real relationship to its ASTM E 84 flame spread index classification. For similar materials, ASTM E 84 usually will rank order them in terms of flame spread index. This generalization breaks down if the tested material rapidly falls from the ceiling, as occurs with foam thermoplastic materials like polystyrene or thin film materials. Therefore, in practice, the flame spread index results from the ASTM E 84 test become invalid if the material cannot stay in place during the test.

7.2.4 Occupancy. When considering how the building elements affected the way in which a fire developed and spread, the investigator should consider whether the occupancy was acceptable for the design and condition of the building. A change in the occupancy of a building can produce much greater fire loads, ventilation effects, total heats of combustion, and heat release rates than originally expected. For example, a warehouse that was originally designed to store automotive engine parts will have a totally different reaction to a fire if the occupancy is changed to the high-rack storage of large quantities of ignitible liquids. The original design may have been adequate for the first fuel load, but inadequate for the subsequent fuel load with its increased hazard.

7.2.5 Computer Fire Model Survey of Building Component Variations. In analyzing the effects of building design upon the development, spread, and ultimate damage from a fire, the use of computer fire models can be very helpful. Through the use of models, the investigator can view the various effects of a number of design variables. By modeling differing building design components, the investigator can see how the changes in a component can change the computed development and growth of the fire.

7.2.6 Explosion Damage.

7.2.6.1 The amount and nature of damage to a building from an explosion is also affected by the design of the structure. The stronger the construction of the exterior or interior confining walls, the more a building can withstand the effects of a low-pressure or slow rate-of-pressure-rise explosion. Conversely, the more brisant or demolishing damage will result from a high-pressure or rapid rate-of-pressure-rise explosion. The shape of the explosion-confining room can also have an effect on the resulting damage. *(See 21.5.3, 21.5.5, and 21.14.3.1 on explosions for more information.)*

7.2.6.2 In a low order explosion, the more windows, doors, or other available vents within the confining structure, the less structural damage will be sustained.

7.3 Types of Construction.

7.3.1 General.

7.3.1.1 The following discussion concerning the types of construction is based on the methods of construction and materials rather than the descriptions used in classification systems of the model building codes. When necessary, the fire investigator should obtain the building construction classifications and descriptions that are a part of the particular building code that is enforced in the jurisdiction in which the fire occurred and should use them as a part of the scene documentation. For further detail, the investigator is directed to the NFPA *Fire Protection Handbook.*

7.3.1.2 The investigator should document the types of construction by looking at the main structural elements. Documentation may include main structural components, breaches, structural changes, or other factors that may influence structural integrity or fire spread.

7.3.2 Wood Frame. Wood frame construction is often associated with residential construction and contemporary lightweight commercial construction. Buildings with wood structural members and a masonry veneer exterior are considered wood frame. Lightweight wood frame construction is usually used in buildings of limited size. Floor joists in such construction are normally spaced 406 mm (16 in.) on center, and the vertical supports are often nominal two by four or nominal two by six wall-bearing studs, again spaced 406 mm (16 in.) on center. Wood frame construction has little fire resistance because flames and hot gases can penetrate into the spaces between the joists or the studs, allowing fire spread outside of the area of origin. *(See 6.3.3.)* Wood frame construction is classified as Type V construction, as defined in NFPA 220, *Standard on Types of Building Construction.* Wood frame construction can be sheathed with a fire-resistive membrane (e.g., gypsum board, lath and plaster, mineral tiles) to provide up to 2-hour fire resistance when tested in accordance with ASTM E 119, *Standard Methods of Tests of Fire Endurance of Building Construction and Materials.* Such high fire resistances in frame construction are unusual but may be encountered in special occupancies such as one- or two-story nursing homes.

7.3.2.1 Platform Frame Construction.

7.3.2.1.1 Platform frame construction is the most common construction method currently used for residential and lightweight commercial construction. In this method of construction, separate platforms or floors are developed as the structure is built. The foundation wall is built; joists are placed on the foundation wall; then a subfloor is placed. The walls for the first floor are then constructed, with the ceiling joists placed on the walls. The rafter, ridgepole, or truss construction methods are used for the roof assembly. An important fire concern other than the fact that combustible materials are used in construction is that there are concealed spaces in soffits and other areas for fire to spread without detection.

7.3.2.1.2 Platform construction inherently provides fire barriers to vertical fire travel as a result of the configuration of the stud channels. However, these barriers in wood frame construction are combustible and may be breached over the course of the fire, allowing the fire to spread to other spaces. Vertical fire, spread may also occur in platform construction through utility paths, such as electrical, plumbing, and HVAC. Openings for utilities in wall stud spaces may allow easy passage of the fire from floor to floor.

7.3.2.2 Balloon Frame.

7.3.2.2.1 In this type of construction, the studs go from the foundation wall to the roofline. The floor joists are attached to the walls by the use of a ribbon board, which creates an open stud channel between floors, including the basement and attic. This type of construction is typical in many homes built prior to 1940.

7.3.2.2.2 Almost all building codes have for many years required fire stopping of all vertical channels in balloon frame construction. Where fire stopping is present, buildings of balloon frame construction respond to fire similarly to buildings of platform frame construction. Fire stopping can be in the form of wood boards or by filling of the void space with noncombustible materials, historically with brick or dirt, and more recently with insulation. Where such fire stops were not installed or later removed (typically to install a utility such as wiring, HVAC, or other services), balloon frame construction provides unobstructed vertical channels, in concealed spaces behind interior finish, for rapid undetected vertical fire spread. Rapid fire spread and horizontal extension is further enhanced by the open connections of the floor joists to the vertical channels. Fire can spread upward to other floors or attic spaces and horizontally through floor spaces. Balloon frame construction will also allow fall down from above to ignite lower levels. Fire originating on lower levels can extend into the open vertical channels and may break out in one or more floors above where the fire originated. There can be more extensive burning at the upper level than where the fire originated. This result may be recognized by the attic fire that consumes the top of the structure while the fire actually originated at some lower level.

7.3.2.3 Plank and Beam.

7.3.2.3.1 In plank and beam framing, a few large members replace the many small wood members used in typical wood framing; that is, large dimension beams more widely spaced replace the standard floor and/or roof framing of smaller dimensioned members. The decking for floors and roofs is planking in minimum thickness as opposed to plywood sheeting. Instead of bearing partitions supporting the floor or roof joist or rafter systems, the beams are supported by posts. There is an identifiable skeleton of larger timbers that are visible. Generally, there is only a limited amount of concealed spaces to allow a fire to spread. This method of construction is often thought of as the ancestor to modern high-rise construction, as the major load-bearing portion of the structure is the frame and the rest is filler. The exterior veneer finish is of no structural value. Most planks will be tongue and groove, which will slow the progression of the fire.

7.3.2.3.2 This type of construction provides for larger spans of unsupported finish material than does framed construction. This property may result in failure of structural sections with large frame members still standing. Interior finishes in these constructions often have large areas of exposed, combustible construction surface that may allow flame spread over its surface.

7.3.2.4 Post and Frame. Post and frame construction is similar to plank and beam construction in that the structure utilizes larger elements, and the frame included is provided to attach the exterior finish. An example of this construction is a barn, with the major support coming from the posts, and the frame providing a network for the exterior finish to be applied.

7.3.2.5 Heavy Timber. Heavy timber is a construction type in which structural members, that is, columns, beams, arches, floors, and roofs, are basically of unprotected wood, solid or laminated, with large cross-sectional areas [200 mm or 150 mm (8 in. or 6 in.) in the smallest dimension, depending on reference]. No concealed spaces are permitted in the floors and roofs or other structural members, with minor exceptions. Floor assemblies are frequently large joists and matched lumber flooring [50 mm (2 in.) thick tongue and grooved, usually end matched].

7.3.2.5.1 When the term *heavy timber* is used in building codes and insurance classifications to describe a type of construction, it includes the requirement that all bearing walls, exterior or interior, be masonry or other 2-hour-rated noncombustible materials. *(See 7.3.4.)* Many buildings have heavy timber elements in combination with other materials such as smaller dimension wood and unprotected steel.

7.3.2.5.2 Contemporary log homes use specially milled logs for the exterior walls and for many of the structural elements. The remainder of the construction is usually nominal two-by-four wood frame construction. Open spans and spaces and large areas of combustible interior finish are common to this type of construction. Due to the interior finish, wood frame components, and open spaces, fire spread may be rapid. The rapid spread and failure frequently appears in conflict with the timber walls and structural elements that often remain standing.

7.3.2.6 Alternative Residential Construction. While wood frame site-built is traditionally associated with residential construction, there are other forms and materials being utilized.

7.3.2.6.1 Manufactured Homes (Mobile Homes). A manufactured home is a structure that is transportable in one or more sections and that, in the traveling mode, is 2.4 m (8 ft) or more in width and 12.2 m (40 ft) or more in length or, when erected on site, is 29.7 m^2 (320 ft^2) or more. This structure is built on a permanent chassis (frame) and designed to be used as a dwelling with or without a permanent foundation when connected to the required utilities. *(See* NFPA 501, *Standard on Manufactured Housing.)* In the U.S., since June 15, 1976, a manufactured home must be designed and constructed in accordance with 24 CFR 3280, "Manufactured Home Construction and Safety Standards (HUD Standard)."

(A) Manufactured homes consist of four major components or subassemblies: chassis, floor system, wall system, and roof system. The chassis is the structural base of the manufactured home, receiving all vertical loads from the wall, roof, and floor, and transferring them to stability devices that may be piers or footings or to a foundation. The chassis generally consists of two longitudinal steel beams, braced by steel cross members. Steel outriggers cantilevered from the outsides of the main beams bring the width of the chassis to the approximate overall width of the superstructure. The floor system consists of its framing members, with sheet decking glued and nailed to the joists, fiberglass insulation blankets installed between the joists, and a vapor barrier sealing the bottom of the floor. Ductwork and piping are often installed longitudinally within the floor system. The floor finish is generally carpeting, resilient flooring, linoleum, or tile.

(B) In newer HUD Standard homes, exterior siding is metal, vinyl, or wood on wood studs, and interior surfaces of exterior walls are most often gypsum wallboard. In older, pre–HUD Standard homes, walls are typically wood studs with aluminum exterior siding, and combustible interior wall surfaces are usually wood paneling.

(C) The roof system in HUD Standard homes consists of either the framed wood roof rafter and ceiling joist system or a wood truss system. Roof decking is generally oriented strand board or plywood attached to the top of the roof rafters or trusses. Finished roofing is often composition shingles. In newer HUD Standard units, gypsum wallboard may be attached directly to the bottom of the ceiling joists or to the bottom chords of the trusses. Blown rock wool or cellulose insulation or insulation blankets provide the roof insulation. In older, pre–HUD Standard homes, exterior roofing is often galvanized steel or aluminum. Interior ceiling surfaces may be combustible material or gypsum wallboard.

(D) Steel tie plates reinforce connections between wall and floor systems. Diagonal steel strapping binds the floors and roof into a complete unit.

(E) Older units that consist of metal exteriors and interiors of wood paneling may experience fires of greater intensity and rapidity than fires in site-built single family structures. The short burn-through time of the walls and ceiling results in quick involvement of the stud walls and the roof supports and decking. These units tend to have smaller rooms that may result in greater fuel load per unit volume than generally exist in other housing. The exterior metal shell results in increased radiation heat feedback after it is exposed to an interior fire. Metal roofing nominally prevents auto vertical ventilation that results in greater fire involvement.

(F) In newer homes, the use of gypsum wallboard on walls and ceilings, reduced flame spread ratings of materials around heating and cooking equipment, and mandatory smoke detectors, where maintained and operable, tends to result in fire incidents similar to those seen in traditional site-built homes of wood frame construction. In older, pre–HUD Standard units, combustible interior finish ignition of combustible materials adjacent to heating and cooking equipment and lack of smoke detectors are among identified fire problems.

7.3.2.6.2 Modular Homes. A modular home is constructed in a factory and placed on a site-built foundation, all or in part, in accordance with a standard adopted, administered, and enforced by the regulatory agency, or under reciprocal agreement with the regulatory agency, for conventional site-built dwellings.

7.3.2.6.3 Steel Frame Residential Construction. Many builders today are adopting steel framing for residential building. As cold-formed steel construction is becoming more prevalent in residential building, the model building codes are addressing the structural and fire safety characteristics of steel framing. Steel framing has many similarities to conventional wood framing construction. Steel framing methods are available for site-built (balloon or platform), panelized, and pre-engineered systems. Steel, like masonry construction, is noncombustible; however, steel framing can lose its structural capacity under severe exposure to heat. Tests have demonstrated that exposed steel beams and joists that may exist in unfinished spaces may fail in periods as short as 3 minutes during flashover fire conditions.

7.3.2.7 Manufactured Wood Structural Elements. Laminated timbers will behave similarly to heavy timbers until the heat of the fire begins to affect the structural stability adversely. If failure occurs, the investigator should document the overall dimensions of the beam as well as the dimensions of the glued pieces. Laminated beams are like heavy timber because their mass will remain and support loads longer than dimension lumber and unprotected steel beams. Laminated beams are generally designed for interior use only. The effects of weather may decrease the load-bearing capabilities of the beam and should be considered if the beam has been exposed to water or other similar conditions.

7.3.2.7.1 Wood "I beams" are constructed with small dimension or engineered lumber, as the top and bottom chord, with oriented strand board or plywood as the web of the beam. Newer floor joist assemblies can be made totally of laminated top and bottom chords with "chip" plywood. These members are generally thinner than the floor joists and typical structural members they replace. As a result, burn-through of the web and resulting failure can occur more quickly than is generally predicted with the use of dimensional lumber. Also increasing the rate of web burn-through is the use of fabricated lumber, such as plywood and oriented strand board, which may have adhesive failure, causing delamination and disintegration. The failure can cause early collapse of floor/ceiling assemblies. Breaches in the web for utilities may allow for fire spread through the spaces and result in earlier failure. Unlike wood trusses, wood I beams will confine fire to the joist space for a period of time.

7.3.2.7.2 Wood trusses are similar to trusses of other materials in their general design and construction. The truss members are often fastened using nail or gusset plates. The gussets can lead to earlier failure than burn-through of the members. This failure occurs because the metal gussets conduct heat rapidly into the wood, causing charring, and because the actual fastening penetrating tines are short. The charring causes the wood to "release" the gusset, leading to collapse of the truss. Failure of one truss will induce loads on adjacent trusses that may lead to a rapid collapse.

7.3.3 Ordinary Construction.

7.3.3.1 The difference between ordinary and frame construction lies mostly with the construction of the exterior walls. In frame construction, the load-bearing components of the walls are wood. In ordinary construction, the exterior walls are masonry or other noncombustible materials. The interior partitions, floor, and roof framing are wood assemblies and, in general, utilize either the platform or braced framing methods. Ordinary construction is classified as Type III construction as defined in NFPA 220, *Standard on Types of Building Construction.*

7.3.3.2 There are a number of factors that affect fire spread in this type of construction, including combustible materials and open vertical shafts. In addition to these items, there may be many other factors that can influence fire spread, including multiple ceilings, utility penetrations, structural failure, and premature collapse.

7.3.4 Mill Construction. Mill construction is a type of heavy timber construction where there are only beams and no girders so that the span of the floor is one bay. This creates small bays in the building but produces a strong resistance to ignition and an extended ability to maintain its load and to resist burn-through during fire. Semi-mill construction is similar but produces larger bays through the utilization of both beams and girders. Semi-mill buildings have a strong resistance to burn-through and have the ability to maintain their load-carrying capability during fire conditions, though the capability is usually considered a little less than that of full mill construction.

7.3.5 Noncombustible Construction.

7.3.5.1 General.

7.3.5.1.1 Noncombustible construction is principally used in commercial, industrial, storage, and high-rise buildings. The major structural components are noncombustible. The major feature of interest in noncombustible construction is that the structure itself will not add fuel to the fire and will not directly contribute to fire spread. Noncombustible construction may or may not be fire-resistive construction, although all construction has some inherent fire resistance.

7.3.5.1.2 Brittle materials, such as brick, stone, cast iron, and unreinforced concrete, are strong in compression but weak in tension and shear. Columns and walls, but not beams, can be constructed of these materials. Ductile materials such as steel will deform before failure during fire conditions. If this is in the elastic range of the member, it will resume its previous shape with no loss of strength after the load is removed. If it is in its plastic range, the member will be permanently deformed, but may continue to bear the load. In either event, elongation or deformation can produce building collapse or damage.

7.3.5.2 Metal Construction. Exposed steel beams and joists typically exist in unfinished spaces. These exposed elements have been shown by test to fail in as little as 3 minutes in a typical flashed-over fire exposure.

7.3.5.2.1 The structural elements used in noncombustible construction are primarily steel, masonry, and concrete. Wrought-iron elements can be encountered in older buildings, and copper and alloys such as brass and bronze are used primarily in decorative rather than load-bearing applications. Aluminum is rarely encountered as a structural element, although it is used in curtain wall construction and as siding in both combustible and noncombustible construction. Aluminum will melt at a temperature well below those encountered in fires. An energized wire may come into contact with conductive materials, causing a fault. This conduction may result in what appears to be a secondary point of origin. Concrete and masonry will generally absorb more heat than steel because these materials require more mass to obtain the necessary strength relative to steel. Concrete and masonry are good thermal insulators, so they do not heat up quickly and do not transfer heat quickly into or through them. Steel is a good conductor of heat, so it will absorb heat and transfer heat much faster than masonry or concrete.

7.3.5.2.2 Steel will lose its ability to carry a load at much lower temperatures than will concrete or masonry and will fail well below temperatures encountered in a fire. Steel structural elements can distort, buckle, or collapse as a result of fire exposure. Wrought iron will withstand higher temperatures than steel but even wrought iron columns can distort when exposed to building fires. The susceptibility of a steel structural element to damage in a fire depends on the intensity and duration of the fire, the size of the steel element, and the load carried by the steel element.

7.3.5.2.3 Although all construction assemblies have some inherent fire resistance, fire-rated assemblies are types that have been tested under specific procedures established for hourly fire ratings. Fire resistance ratings may refer to a structural system's ability to support a load during a fire or to prevent the spread of a fire. Fire resistance ratings are determined on the basis of a specific test and will not necessarily indicate how long a system will perform in any actual fire.

7.3.5.3 Concrete or Masonry Construction. Other major construction materials include concrete and masonry. These materials have an inherent resistance to effects from fire due to their mass, high density, and low thermal conductivity.

7.3.5.3.1 Concrete and masonry are found in many forms and applications. Masonry assemblies and concrete have high strength in compression and relatively low strength in tension. Consequently, both need reinforcement for tensile strength.

7.3.5.3.2 Fireground failures in these types of materials generally relate to reinforcement failure and failure in the connection between components. Reinforcement failure can result from heat transfer through the concrete or masonry, or from surface spacing and exposure of the reinforcement to the fire temperatures. Connections generally are made of steel, and their failures occur at temperatures well within the range found in structure fires.

7.4 Construction Assemblies.

7.4.1 General. *Assemblies,* as used in this chapter, can be described as a collection of components, such as structural elements, to form a wall, floor/ceiling, or other. Components may be assemblies such as doors that form a part of a larger complete unit. The way assemblies react in a fire often influences how the fire grows and spreads, as well as how they maintain their structural integrity during the fire. Assemblies are often interdependent, and failure of one may contribute to failure of another.

7.4.1.1 Assemblies may or may not be fire resistance–rated; however, most assemblies will provide some resistance to fire. It also should be noted that most assemblies are not rated for smoke penetration with the exception of smoke control dampers. Scene documentation should include what was there for later comparison with applicable code requirements.

7.4.1.2 An assembly without a rating means that we do not know what the standard fire tests would indicate, given the failure time under the test fire conditions.

7.4.1.3 Assemblies are rated for a specific fire test criterion and under test conditions. The actual fire conditions found in the structure may be more severe and may cause the assembly to fail at a time that is actually less than the hourly rating assigned as a result of the fire test. Failure of an assembly in and of itself is not an indicator of the cause of the fire; however, it is appropriate to determine the circumstances associated with the failure of the assembly.

7.4.1.4 When assemblies are evaluated after a fire, consideration should be given to any local deficiency of a component in the overall assembly, such as a hole in the wall, missing tiles in a ceiling, or even a door blocked open.

7.4.2 Floor/Ceiling/Roof Assemblies. Floor/ceiling assemblies fail in a number of ways, including collapse, deflection, distortion, heat transmission, or penetration of fire, allowing vertical fire spread. Failure depends on a number of factors, including the type of structural elements; the protection of the elements; and span, load, and beam spacing. Rated floor assemblies are tested for fire exposure from below and not from above. There are limited experimental data for fires burning downward, which can occur in a number of mechanisms such as hot layer radiation or drop down. Live loads and water weight can contribute to the collapse of floors and ceilings.

7.4.2.1 Penetrations are regularly found in floor/ceiling assemblies. Penetrations are often used to provide access for utilities, HVAC systems, plumbing, computer data and communication, and other functions. Penetrations in fire-rated floor/ceiling assemblies are required to be sealed to maintain the rating. Unsealed penetrations facilitate the passage of fire and smoke through the floor/ceiling assembly.

7.4.2.2 Roof assemblies affect structural stability during the fire rather than the resistance to the spread of the fire. Roofs can have a major impact on the fire dynamics if the roof fails in a fire.

7.4.3 Walls. Walls perform a number of fire safety–related functions, the most obvious of which is compartmentation, which tends to limit fire spread. Compartment walls are constructed to various standards, ranging from nonrated partitions to self-supporting parapeted fire walls.

7.4.3.1 Walls may or may not be fire rated and rated walls may or may not be load bearing. Also, load-bearing walls may be fire rated even though their function is not to stop the spread of fire. The fire wall will not be effective if it is not continuous through the ceiling and attic section of the structure.

7.4.3.2 A fire wall is a wall separating buildings or subdividing a building to prevent the spread of fire and having a fire resistance rating and structural stability.

7.4.3.3 A fire barrier wall is a wall, other than a fire wall, having a fire resistance rating. Fire walls and fire barrier walls do not need to meet the same requirements as smoke barriers.

7.4.3.4 Smoke barriers are continuous membranes, either vertical or horizontal, such as a wall, floor, or ceiling assembly, designed and constructed to restrict the movement of smoke. A smoke barrier might or might not have a fire resistance rating. Such barriers might have protected openings.

7.4.3.5 Penetrations are regularly found in wall assemblies. The penetrations often are used to provide access for doors, utilities, HVAC systems, plumbing, computer data and communication, and other functions. Penetrations in fire-rated wall assemblies are required to be sealed to maintain the rating. Unsealed penetrations facilitate the passage of fire and smoke through the wall assembly, allowing the fire to spread horizontally.

7.4.3.6 A fire barrier wall is not required to be constructed of noncombustible materials. Fire barrier walls constructed of combustible materials include the use of wood studs with Type X gypsum board on the exterior surfaces. Where the structure has a load-bearing party wall assembly, combustible materials can again be used. In this instance, there are two separate stud walls built; the exterior finish is gypsum board; between the two stud walls plywood is attached; and there is an air space between the walls. Most requirements for a fire barrier wall will have Type X on both sides of the wall to make it fire resistive.

7.4.3.7 There are a number of other walls found in structures. While these walls have not been subjected to fire tests in order to be rated, they will still provide some resistance to the spread of fire within a building.

7.4.4 Doors. Doors may be a key factor in the spread of fire. Doors can be made of a variety of materials and be fire rated or non–fire rated. It should be noted that if there is a door opening in a fire-rated wall or partition, it would be required to be provided with an appropriate fire-rated door, installed as an entire assembly. Fire-rated door assemblies are required to include rated frames, hinges, closures, latching devices, and if provided (and allowed), glazing. Fire doors may be of wood, steel, or steel with an insulated core of wood or mineral material. While some doors have negligible insulating value, others may have a heat transmission rating of 121°C, 232°C, and 343°C (250°F, 450°F, and 650°F). This means the doors will limit temperature rise on the unexposed side to that respective value when exposed to the standard time–temperature for 30 minutes. This insulating value aids egress, particularly in stairwells in multistory buildings, and provides some protection against autoignition of combustibles near the opening's unexposed side. In addition to the rating of the door, to be effective in limiting the spread of fire from one compartment to another, the door must be closed. The position of doors can change during and after a fire for a variety of reasons, including automatic closure systems, personnel movement, and fire suppression activities.

7.4.5 Concealed Spaces. Spaces that are generally inaccessible or limited-access areas of a structure such as interstitial space above a ceiling, below a floor, or between walls. Attics, accessible or not, may also be considered a concealed space. Concealed spaces provide a hidden path for fire to grow or spread without being identified early in the event. By the time fire moves out of the concealed space, it often has already spread extensively throughout the structure. Fires in concealed spaces are difficult to extinguish. Concealed spaces are found in almost all types of construction and may have built-in fire protection features such as sprinklers, barriers, and automatic detection. The presence, performance, or absence of these protective features may have a dramatic effect on progression of the fire. For those concealed spaces identified as noncombustible, all components, materials, or equipment used in the construction of the concealed space must be of noncombustible or fire-resistive assemblies, or must have been provided with listed fire-protective coating. Concealed spaces normally classified as noncombustible may still contain some combustible materials such as fire-retardant-treated lumber, communications and power wiring cable, and plastic pipe. Fires can still start and spread in concealed spaces that are classified as noncombustible.

Chapter 8 Electricity and Fire

8.1* Introduction. This chapter discusses the analysis of electrical systems and equipment. The primary emphasis is on buildings with 120/240-volt, single-phase electrical systems. These voltages are typical in residential and commercial buildings. This chapter also discusses the basic principles of physics that relate to electricity and fire.

8.1.1 Prior to beginning an analysis of a specific electrical item, it is assumed that the person responsible for determining the cause of the fire will have already defined the area or point of origin. Electrical equipment should be considered as an ignition source equally with all other possible sources and not as either a first or last choice. The presence of electrical wiring or equipment at or near the origin of a fire does not necessarily mean that the fire was caused by electrical energy. Often the fire may destroy insulation or cause changes in the appearance of conductors or equipment that can lead to false assumptions. Careful evaluation is warranted.

8.1.2 Electrical conductors and equipment that are used appropriately and protected by properly sized and operating fuses or circuit breakers do not normally present a fire hazard. However, the conductors and equipment can provide ignition sources if easily ignitible materials are present where they have been improperly installed or used. A condition in the electrical wiring that does not conform to the *NFPA 70, National Electrical Code,* might or might not be related to the cause of a fire.

8.2 Basic Electricity.

8.2.1 General. The purpose of this section is to present basic electrical terms and concepts briefly and simply in order to develop a working understanding of them.

8.2.2 Comparing Electricity to Hydraulics. Water flowing through a pipe is familiar to everyone. This phenomenon has some similarities to electrical current flowing in an electrical system. Because of these similarities, a limited comparison between a hydraulic system and an electrical system can be used to understand an electrical system.

8.2.2.1 Elements of Hydraulic and Electrical Systems. Table 8.2.2.1(a) compares selected hydraulic system hardware with analogous electrical system hardware. Table 8.2.2.1(b) compares selected hydraulic quantities and hydraulic units with analogous electrical quantities and electrical units.

Table 8.2.2.1(a) Hydraulic System Hardware and Analogous Electrical System Hardware

Hydraulic System Hardware	Electrical System Hardware
Constant-pressure pump	DC voltage source such as a battery
Pipe	Wire
Water turbine	DC motor
Differential pressure meter	DC voltmeter
Flow meter	DC ammeter
Shutoff valve	Switch

8.2.2.2 Comparing Hydraulic Pressure to Voltage.

8.2.2.2.1 In a hydraulic system, a pump creates a pressure differential that can force liquid through pipes and through hydraulic components. Common pressure gauges are connected to one point in a hydraulic system. This could cause the misconception that pressure is not a differential phenomenon. In reality, all pressures are differential. Common pressure gauges register the difference between pressure at the measured point and atmospheric pressure. It is important to understand that it is a differential pressure from one end of a pipe (or component) to the other end, not simply the pressure at one end, that forces liquid through a pipe (or component). This differential can be determined by computing the difference between the readings on two common pressure gauges or by using a differential pressure gauge which is equipped with two pressure sensing ports. Hydraulic pressure is often measured in kilopascals (kPa) or pounds per square inch (psi).

Table 8.2.2.1(b) Hydraulic Quantities and Units and Analogous Electrical Quantities and Units

Hydraulic Quantities	Hydraulic Units	Electrical Quantities	Electrical Units
Pressure	Pounds per square inch (psi)	Potential (voltage)	Volts
Differential pressure	Pounds per square inch (psi)	Potential difference (voltage)	Volts
Pressure loss	Pounds per square inch (psi)	Voltage drop (voltage)	Volts
Flow rate	Gallons per minute (gpm)	Current = flow rate of charge	Amperes = coulombs per second
Friction/ resistance to flow	Pounds per square inch (psi)	Resistance	Ohms
Pipe diameter (inside)	Inches	Electric wire diameter	AWG
Water	Gallons	Charge	Coulombs

Note: For SI conversion, 1 psi = 6.89 kPa.

8.2.2.2.2 In an electrical system, a battery or a dc generator creates a potential difference or voltage that can force charge through a wire and through electrical components. Common voltmeters record the potential difference between two points. All voltages are differential. The unit of measurement of potential difference or voltage is usually volts.

8.2.2.3 Comparing Water Flow to Current. In the hydraulic system, it is water that flows in a useful way. In the electrical system, it is charge that flows in a useful way. Charge is an accumulation of electrons just as water is an accumulation of water molecules. Charge is measured in coulombs, and water can be measured in gallons. The rate of charge flow is called electrical current, and current is measured in amperes. An electrical flow rate of one coulomb per second is equivalent to one ampere. The amount of current in amperes (A) can be measured with an ammeter. The flow rate of water flow can be expressed in gallons per minute (gpm) or liters per minute (lpm), and it can be measured with a flow meter. Electric current can be either direct current (dc), such as supplied by a battery, or alternating current (ac), such as supplied by the electric utility companies or inverters that convert dc to ac.

8.2.2.4 Direct Current and Alternating Current. Direct current flows in only one direction, as in a circulating water system, while alternating current flows back and forth with a specific frequency. For many of the applications encountered in this text, it is useful to visualize ac circuits as if they were dc circuits. In the United States, utilities provide electricity at a frequency of 60 hertz, or 60 cycles per second. Transformers, motors, capacitors, and other circuit elements that are not mostly resistive, and some electronic devices cannot be satisfactorily analyzed or fully described using dc techniques. In addition, three-phase circuits cannot be analyzed or described with dc circuit techniques.

8.2.2.5 Comparing Water Pipes to Conductors. The water pipe provides the pathway for the water to flow. In the electrical system, conductors such as wires provide the pathway for the current.

8.2.2.6 Comparing Closed Hydraulic Systems to Electrical Circuits. In a closed circulating hydraulic system (as opposed to a fire hose delivery system where water is discharged out of the end), water flows in a loop, returning to the pump, where it is circulated again through the loop. When the valve is closed, the flow stops everywhere in the system. When the valve is opened, the flow resumes. An electrical system must be a closed system, in that the current must flow in a loop or in a completed circuit. When the switch is turned on, the circuit is completed and the current flows. When the switch is turned off, the circuit is opened and current flow stops everywhere in the circuit. Refer to Figure 8.2.2.6.

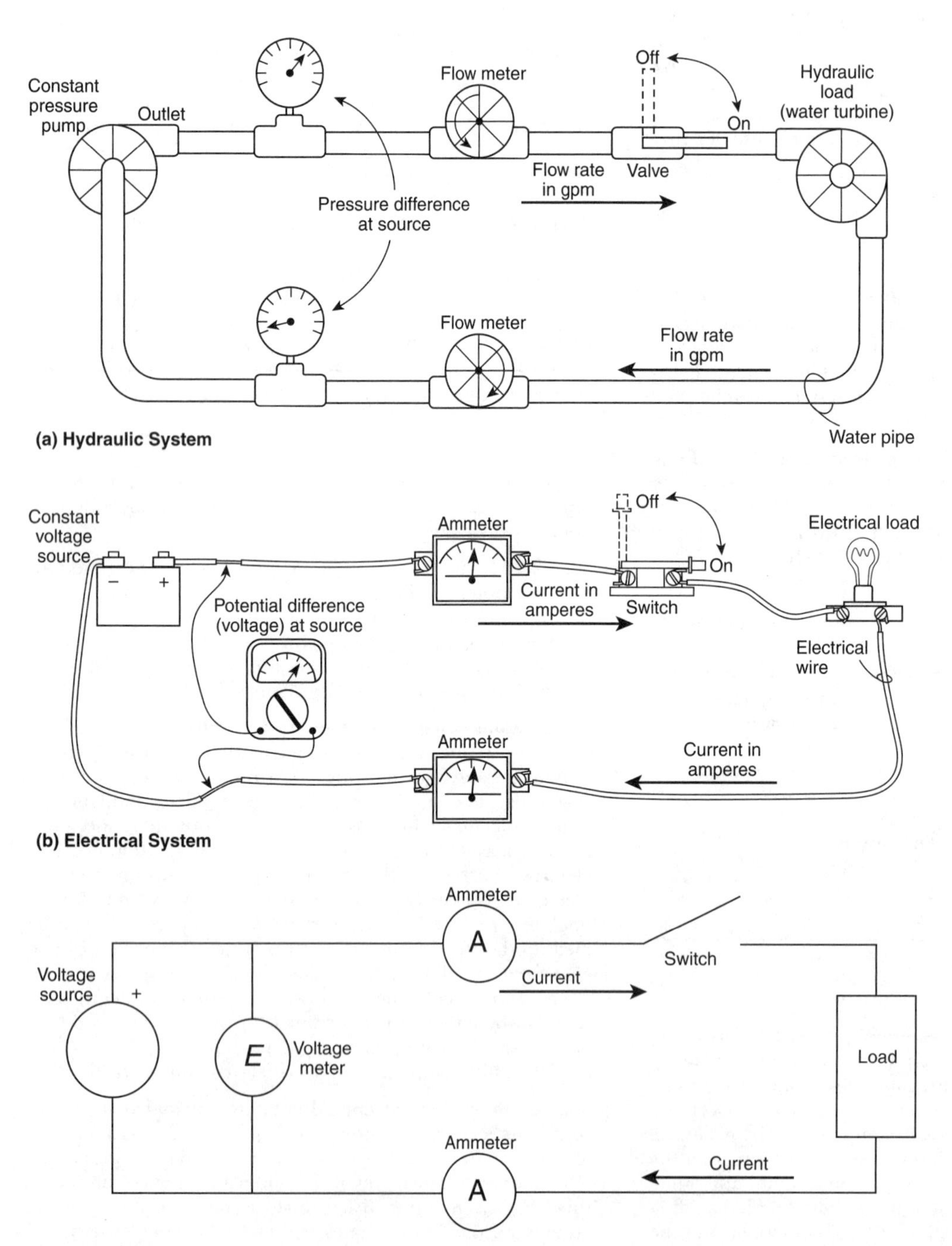

FIGURE 8.2.2.6 A Hydraulic Circuit, an Analogous Electrical Circuit, and a Schematic Representation of the Same Electrical Circuit.

8.2.2.7 Comparing Hydraulic Friction Loss to Electrical Resistance.

8.2.2.7.1 Friction losses in pipes result in pressure drops. Electrical friction (i.e., resistance) in conductors and other parts also results in electrical pressure drops or voltage drops. To express resistance as a voltage drop, Ohm's law must be used. *(See 8.2.5.)*

8.2.2.7.2 When electricity flows through a conducting material, such as a conductor, a pipe, or any piece of metal, heat is generated. The amount of heat depends on the resistance of the material through which the current is flowing and the amount of current. Some electrical equipment, such as a heating unit, is designed with appropriate resistance to convert electricity to heat.

8.2.2.8 Comparing Pipe Size to Wire Gauge. The flow of water in a pipe at a given pressure drop is controlled by the pipe size. A larger pipe will allow more volume per minute of water to flow than will a smaller pipe at a given pressure drop. Similarly, larger conductors allow more current to flow than do smaller conductors. Conductor sizes are given American Wire Gauge (AWG) numbers. The larger the number, the smaller the conductor diameter. Small conductors, such as 22 AWG, are used in telephone and other signal circuits where small currents are involved. Larger conductors, such as 14, 12, and 10 AWG, are used in residential circuits. The larger the diameter (and hence the larger the cross-sectional area) of the conductor, the lower the AWG number and the less the resistance of the conductor. This means that a 12 AWG copper conductor can safely carry a larger current than a smaller 14 AWG copper conductor. *(See Figure 8.2.2.8.)*

Solid Copper Wire

	Diameter	Resistance in ohms per 1000 ft (305 m) at 158°F (70°C)
14 AWG	0.064 in. (1.63 mm)	3.1
12 AWG	0.081 in. (2.06 mm)	2.0
10 AWG	0.102 in. (2.60 mm)	1.2

FIGURE 8.2.2.8 Conductors. American Wire Gauge (AWG) Sizes, Diameters of Cross Sections, and Resistance of Conductors Commonly Found in Building Wiring.

8.2.3 Ampacity. The ampacity of a conductor is the current in amperes a conductor can carry continuously under the conditions of use without exceeding its temperature rating. This depends on the ambient temperature the conductor is operating in as well as other factors, such as whether the conductor is in conduit with other conductors carrying similar current, alone, or in free air, and so forth. For example, Table 310.16 of *NFPA 70, National Electrical Code*, lists the ampacity of 8 AWG copper conductor with TW insulation (moisture-resistant thermoplastic) as 40 amperes. This rating is based on an ambient temperature of 30°C (86°F) and on being installed in a conduit or raceway in free air containing no more than three conductors. Any changes — such as more conductors in a raceway, higher ambient temperature, or insulation around the conduit — that reduce the loss of heat to the environment will decrease the ampacity. This same size conductor is rated at 50 amperes with THWN insulation (moisture- and heat-resistant thermoplastic); the THWN insulation has a temperature rating of 75°C (167°F) compared to 60°C (140°F) for the TW insulation. The temperature rating of the insulation is the maximum temperature at any location along its length that the conductor can withstand for a prolonged period of time without serious degradation.

8.2.3.1 The ampacity values for a conductor depend on the heating of the conductor caused by the electric current, the ambient temperature that the conductor is operating in, the temperature rating of the insulation, and the amount of heat dissipated from the conductor to the surroundings. Current passing through an aluminum conductor generates more heat than the same current passing through a copper conductor of the same diameter; the ampacity of an aluminum conductor is less than that for the same size copper conductor. Also, the ampacity of a conductor is reduced when it is operated at an elevated temperature or when it is covered with a material that provides thermal insulation. Conversely, the actual ampacity of a single conductor in open air or in a conduit will be higher than that given in the tables. The actual as-used ampacity may be an important consideration in evaluating the cause of electrical faulting.

8.2.3.2 A safety factor is included in ampacity values. Simply demonstrating that ampacity has been exceeded does not mean the fire had an electrical ignition source.

8.2.4 Conductivity of Conductors. Some conductor materials conduct current with less resistance than do other materials. Silver conducts better than copper. Copper conducts better than aluminum. Aluminum conducts better than steel. This means that a 12 AWG copper conductor will have less resistance than the same size 12 AWG aluminum conductor. There will be less heat generated in a copper conductor than in an aluminum conductor for the same current and AWG size.

8.2.5 Ohm's Law. The following discussion can be applied accurately only to dc (direct current) circuits. Similar but somewhat more complex equations are required in the analysis of ac (Alternating Current) circuits. Ohm's law states that the voltage *(see Figure 8.2.5)* in a circuit is equal to the current multiplied by the resistance, or

$$E = I \times R$$

where:
E = voltage
I = amperage
R = resistance

8.2.5.1 Voltage *(E)* is measured in volts, current *(I)* is measured in amperes, and resistance *(R)* is measured in ohms.

8.2.5.2 Using this simple law, the voltage drop can be found if the current and resistance are known. Rearranging the terms, we can solve for current if voltage and resistance are known:

$$\text{current} = \frac{\text{voltage}}{\text{resistance}} \text{ or amperes} = \frac{\text{volts}}{\text{ohms}}$$

8.2.5.3 Also, resistance can be found if the current and voltage are known:

$$\text{resistance} = \frac{\text{voltage}}{\text{current}} \text{ or ohms} = \frac{\text{volts}}{\text{amperes}}$$

8.2.5.4 A voltmeter and an ammeter can be used to determine the resistance. If the resistance and the voltage can be measured, the amperage can be calculated.

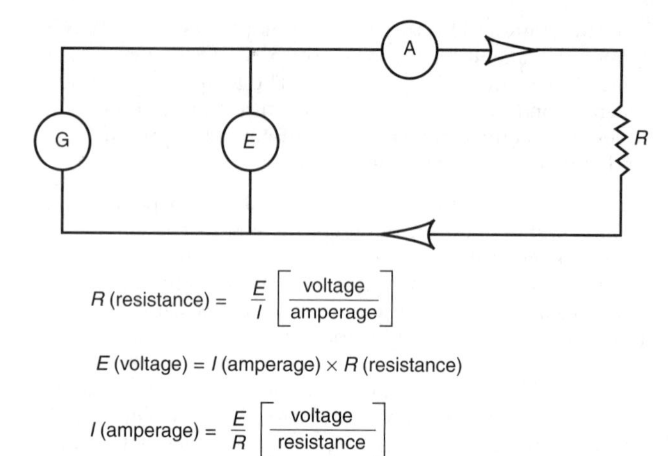

FIGURE 8.2.5 Ohm's Law in a Simple Circuit.

8.2.6 Electrical Power. When electrons are moved (electrical current) through a resistance, electrical energy is spent. This energy may appear in a variety of ways, such as light in a lamp or heating of a conductor.

8.2.6.1 The rate at which energy is used is called power. The amount of power is expressed in watts (W). A 100 W lightbulb produces more light and heat than a 60 W lightbulb. *(See Figure 8.2.6.1.)*

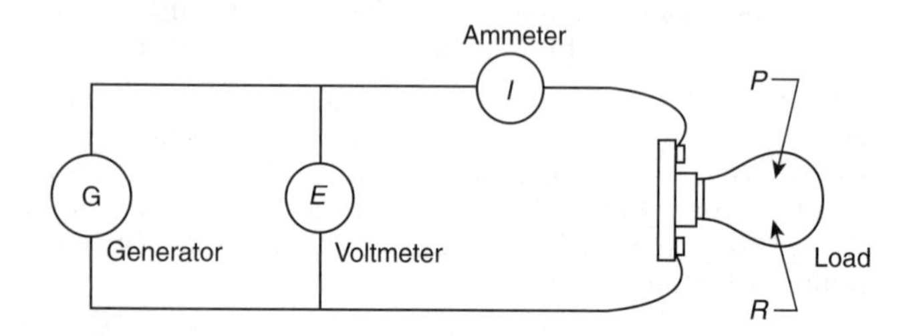

FIGURE 8.2.6.1 The Power in Watts *(P)* Consumed by a Lightbulb, a Product of the Current *(I)* Squared and the Resistance *(R)* of the Lightbulb.

8.2.6.2 Energy may be expressed in many different ways. For electrical applications, energy is usually measured in watt-seconds or watt-hours. A watt-second is equal to 1 joule, and a watt-hour is equal to 3600 joules (3.413 Btu).

8.2.7 Ohm's Law Wheel. Power in electrical systems *(P)* is measured in watts. Resistive appliances such as a hair dryer or lightbulb are rated in watts. Power is computed as shown in the Ohm's law wheel, in Figure 8.2.7. The relationships among power, current, voltage, and resistance are important to fire investigators because of the need to find out how many amperes were drawn in a specific case. See Figure 8.2.7 for a summary of these relationships. If, for example, several appliances were found plugged into one extension cord or many appliances were plugged into several receptacles on the same circuit, the investigator could calculate the current draw to find whether the ampacity of the conductor was exceeded.

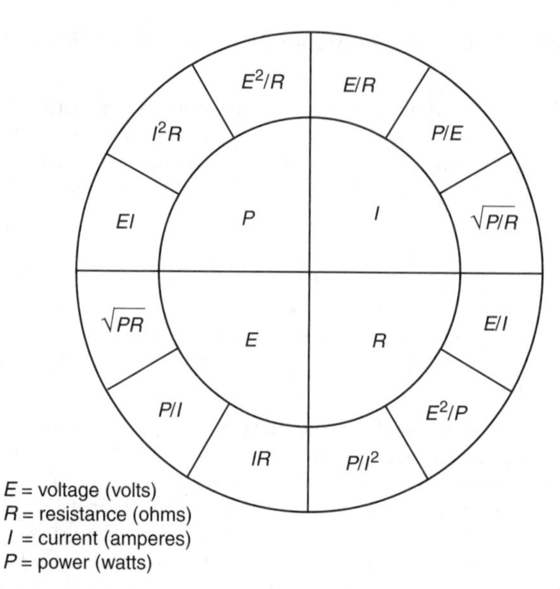

FIGURE 8.2.7 Ohm's Law Wheel for Resistive Circuits.

8.2.7.1 The calculations given in the following example give only approximate results, as they have been simplified to avoid the more complex calculations required for ac circuits. For example, a hair dryer designed to operate on 120 volts draws 1500 watts:

$$\text{current } (I) = \frac{\text{power } (P)}{\text{voltage } (E)} = \frac{1500 \text{ watts}}{120 \text{ volts}} = 12.5 \text{ amperes}$$

$$\text{resistance } (R) = \frac{\text{voltage}^2 \ (E^2)}{\text{power } (P)} = \frac{120^2 \text{ volts}}{1500 \text{ watts}} = 9.6 \text{ ohms}$$

8.2.7.2 To check results, do the following computation:

$$\text{volts } (E) = I \times R = 12.5 \times 9.6 = 120 \text{ V}$$

$$\text{watts} = (I)^2 \times R = (12.5)^2 \times 9.6 = 1500 \text{ W}$$

8.2.8 Applying Ohm's Law. The following example will show how to find the total amperes, assuming the heater and circuit protection are turned on and are carrying current. A portable electric heater and cooking pot are plugged into a 18 AWG extension cord. The heater is rated at 1500 W and the cooking pot is 900 W. The previous relationships showed that current equaled power divided by voltage.

$$\text{amperes } (I) = \frac{\text{watts } (P)}{\text{volts } (E)} \text{ or } \frac{1500}{120}$$
$$= 12.5 \text{ A for the heater}$$

$$\text{amperes } (I) = \frac{\text{watts } (P)}{\text{volts } (E)} \text{ or } \frac{900}{120}$$
$$= 7.5 \text{ A for the pot}$$

8.2.8.1 The total amperage of a circuit is the sum of the amperage of each device that is plugged into the circuit. The total amperage for a circuit consisting of three receptacles is the total amperage of all devices plugged into these receptacles. Similarly, the total amperage on an extension cord is the sum of the amperage of each device plugged into the extension cord.

8.2.8.2 In the example illustrated in Figure 8.2.8.2, the calculated amperages were 12.5 A and 7.5 A, so the total amperage of that extension cord when both appliances were operating was 12.5 A + 7.5 A = 20.0 A. Tables of allowable ampacities [from *NFPA 70, National Electrical Code,* Table 400.5(a)] show that the maximum current should be 10 A in the 18 AWG extension cord. Therefore, the cord was carrying an overcurrent. The question to be determined is whether this created an overload. Did the overcurrent last long enough to cause dangerous overheating? In a situation such as shown in Figure 8.2.8.2, where it appears an overload existed, it is necessary to show that these conditions will create enough temperature rise to cause ignition. An overload is not absolute proof of a fire cause. If an overload occurred, this cord could be considered as a possible ignition source, particularly if the heat was confined or trapped, such as under a rug or between a mattress and box spring, preventing dissipation.

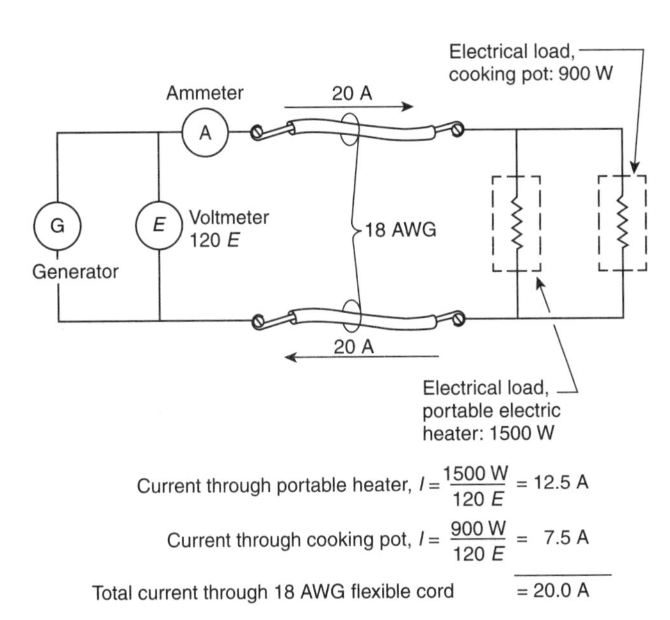

FIGURE 8.2.8.2 Total Current Calculation.

8.2.8.3 A similar situation exists when a short circuit occurs by conductor-to-conductor contact. This is, by definition, a connection of comparatively low resistance. As seen by Ohm's law, when the resistance goes down, the current goes up. Although a short circuit does cause a large current flow, the circuit overcurrent protection devices normally prevent this current from flowing long enough to cause overheating of the conductors.

8.3 Building Electrical Systems.

8.3.1* General. This section provides a description of the electrical service into and through a building. It is intended to assist an investigator in recognizing the various devices and in knowing generally what their functions are. The main emphasis is on the common 120/240 V, single-phase service with limited information on three-phase and higher voltage service. This section does not provide detailed information on codes. That information should come from the appropriate documents.

8.3.2 Electrical Service.

8.3.2.1 Single-Phase Service. Most residences and small commercial buildings receive electricity from a transformer, which is a device that lowers or raises voltages to desired levels. This electricity is delivered through three conductors, either overhead from a pole or underground. The two insulated conductors, called the "hot legs" or "phases," have their alternating currents flowing in opposite directions (reversing 120 times per second for 60-cycle power) so that the current goes back and forth at the same instant but in opposite directions (180° out of phase). This alternating current is called "single phase." The third conductor is grounded to serve as the neutral conductor, and it may be uninsulated. The voltage between either of the hot conductors and the grounded conductor is 120 V, as shown in Figure 8.3.2.1(a). The voltage between the two hot conductors is 240 V. The incoming conductors are large multistranded cables intended to carry large currents safely. As illustrated in Figure 8.3.2.1(b), they all may hang separately, or the two hot conductors may be wrapped around the neutral in a configuration called a triplex drop.

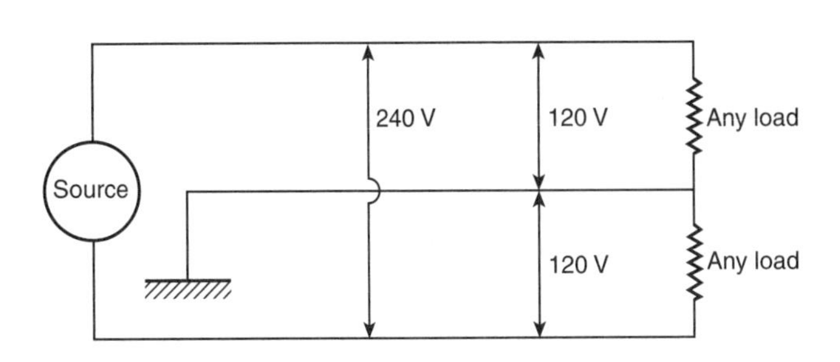

FIGURE 8.3.2.1(a) Relation of Voltages in 120/240 V Service.

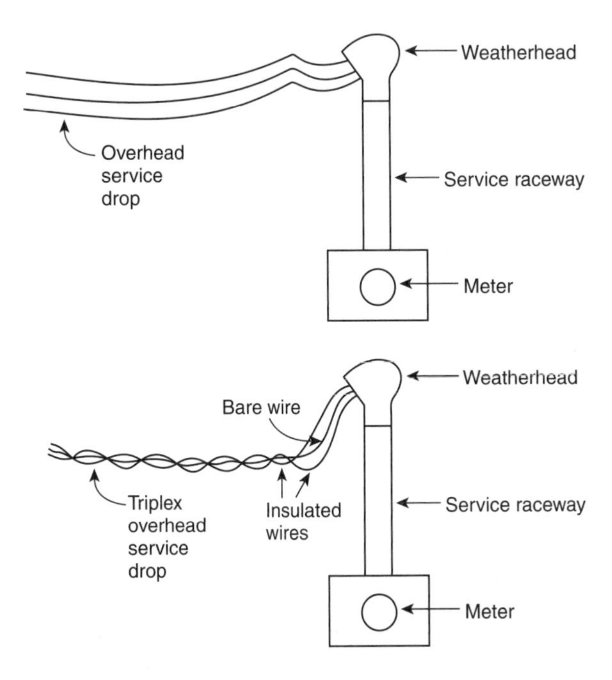

FIGURE 8.3.2.1(b) Overhead Service.

8.3.2.1.1 If the cables come from a transformer on a pole, they are called a "service drop." If they come from a transformer in or on the ground, they will be buried and are called a "service lateral." *(See Figure 8.3.2.1.1.)*

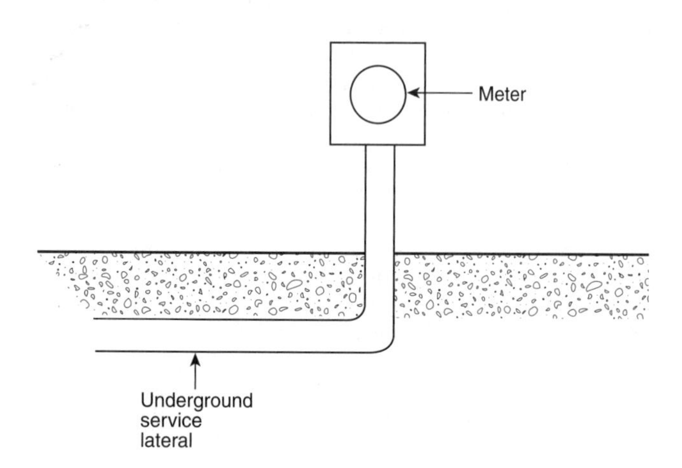

FIGURE 8.3.2.1.1 Underground Service.

8.3.2.1.2 The terms *hot, neutral,* and *ground* will usually be used in this document for installed conductors. The proper terms for them are *ungrounded, grounded,* and *grounding,* respectively.

8.3.2.2 Three-Phase Service. Industrial and large commercial buildings, large multifamily dwellings, and other large buildings normally are supplied with three-phase electrical service. Three-phase service consists of three alternating currents that go back and forth at different instants (out of phase with one another). There will be three current-carrying conductors and usually a fourth, which is the neutral and is at ground potential. The voltage between current-carrying conductors is typically 480 V, 240 V, or 208 V. The voltage between the conductors and ground depends on the wiring arrangement and may be 277 V, 208 V, or 120 V. The 480/277 V four-conductor system is a common service for large commercial and industrial buildings. Modern lighting systems in these buildings commonly operate at 277 V. In very large buildings, there might be more than one electrical service entrance. In some industrial buildings, the service entrance voltage may be very high (e.g., 4000 V). Transformers within buildings then reduce the voltage for utilization, including 120 V for lights and receptacles.

8.3.3 Meter and Base. The cables of a service drop go into a weatherhead, which is designed to keep water from entering the system, and then down a service raceway to a meter base. A watt-hour meter plugs into the meter base and connects the service cables so that electricity can flow into the structure. In newer structures, the meter base is normally mounted on the outside. Cables go from the meter base to the service equipment in the structure, as shown in Figure 8.3.3. In larger facilities, the entry cables may be connected directly to the service equipment without passing through the meter. In that case, the meter is operated from current transformers that surround each entry cable and sense current flow.

8.3.4 Significance. The service entry can be significant in fire investigations because damage to the insulation on the conductors can result in sustained high-power faulting by either short circuits or ground faults that can ignite most combustibles. Between the utility transformer and the main protection in the structure, there is usually no overcurrent protection of the cables, and faulting may begin and continue. Once there are fault currents, either causing a fire or resulting from a fire, continued faulting can damage all or part of a service entrance.

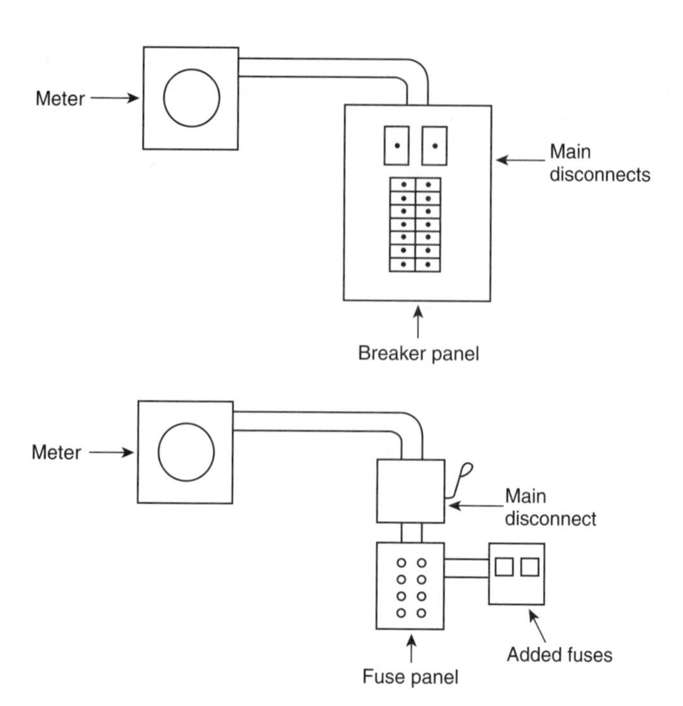

FIGURE 8.3.3 Service Entrance and Service Equipment.

8.4 Service Equipment. The cables from the meter base go to the service equipment, which consists of a main switch and fuses or circuit breakers. *(See Figure 8.3.3.)* The service equipment must be located close to where the cables enter the structure. The service equipment has three functions: to provide means for turning off power to the entire electrical system, to provide protection against electrical malfunctions, and to divide the power distribution into several branch circuits. Either a main switch or the main circuit breaker is the primary disconnect that can shut off all electricity to the building. From the cabinet of fuses or circuit breakers, electricity is distributed through branch circuits to the rest of the building.

8.5 Grounding.

8.5.1 General. All electrical installations must be grounded at the service equipment. Grounding is a means of making a solid electrical connection between the electrical system and the earth. Grounding is accomplished by bonding the breaker or fuse panel to a metallic cold water pipe if the pipe extends at least 3.0 m (10 ft) into soil outside. In the absence of a suitable metallic cold water pipe, a grounding electrode must be used. The grounding electrode may be a galvanized steel rod or pipe or a copper rod of at least 2.4 m (8 ft) in length driven into soil to a level of permanent moisture.

8.5.1.1 In all installations, the service equipment must be bonded to the metallic cold water piping or a grounding electrode. Bonding is the connecting of items of equipment by good conductors to keep the equipment bodies at the same voltage, which is essentially zero if bonded to ground. Bonding of the service equipment to ground is accomplished by a copper or aluminum conductor from the grounding block in the fuse or breaker cabinet to a clamp that is securely fixed to the metallic cold water pipe or grounding electrode. An example is shown in Figure 8.5.1.1. The purpose of grounding an electrical system is to make sure that any housings or exposed metal objects in the system or connected to it cannot become electrically charged. If an ungrounded conductor (the hot conductor) contacts a grounded object, the resulting surge of ground-fault current will open the protection.

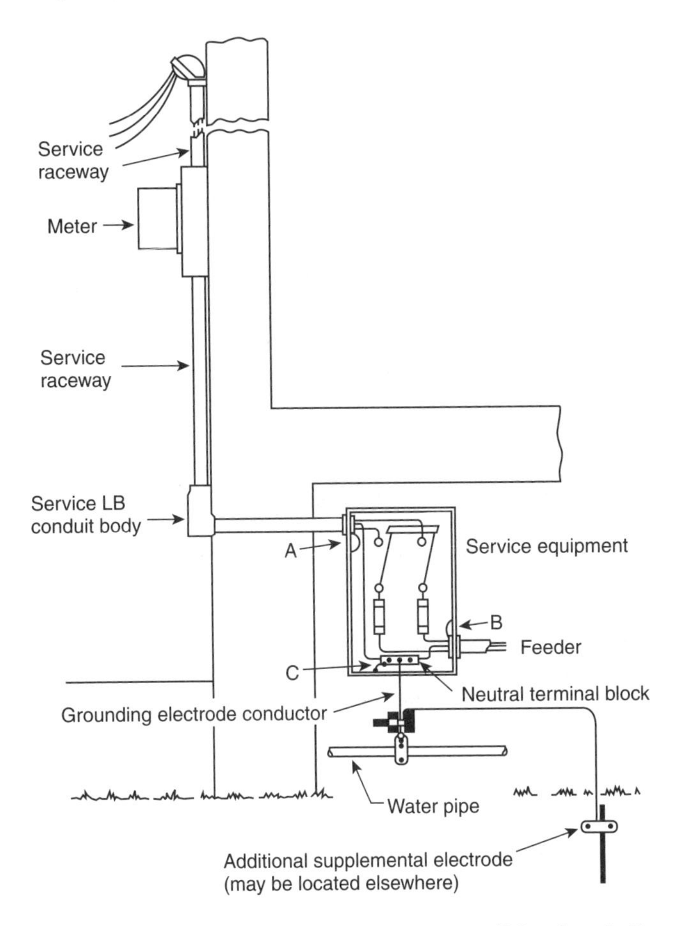

FIGURE 8.5.1.1 Grounding at a Typical Small Service. A, B, and C are bonding connections that provide a path to ground.

8.5.1.2 All parts of the system must be grounded, including cabinets, raceways, fittings, junction and outlet boxes, switches, receptacles, and any conductive objects attached to or plugged into the system. That is usually accomplished with a grounding conductor that accompanies the circuit conductors, although grounding can be accomplished through metallic conduit. Flexible metallic conduit may be used for grounding only if its length does not exceed 6 ft (1.8 m).

8.5.2 Floating Neutral (Open Neutral). An electrical installation with an open neutral conductor will not have a fixed point of zero voltage between the two legs. There will still be 240 V between the two legs, but instead of the voltages of the two legs being fixed at 120 V to neutral each, they may vary to some other values that add up to 240 V. *(See Figure 8.5.2.)* All line to neutral circuits will be affected. The actual voltages in the legs will depend on the loads on the two legs at any particular time. For example, the voltages might b[illegible] in Figure 8.5.2. The higher voltage can overhea[illegible] some equipment, and the lower voltage can da[illegible] electronic equipment. Occupants would have seen [illegible] cent lights that were too bright or too dim or applianc[illegible] overheated or malfunctioned in some way. A floating ne[illegible] condition is not dependent on proper grounding of the se[illegible] vice. Removing the grounding electrode connection does not cause an open neutral. Only a break in the neutral conductor can cause a floating neutral condition to occur.

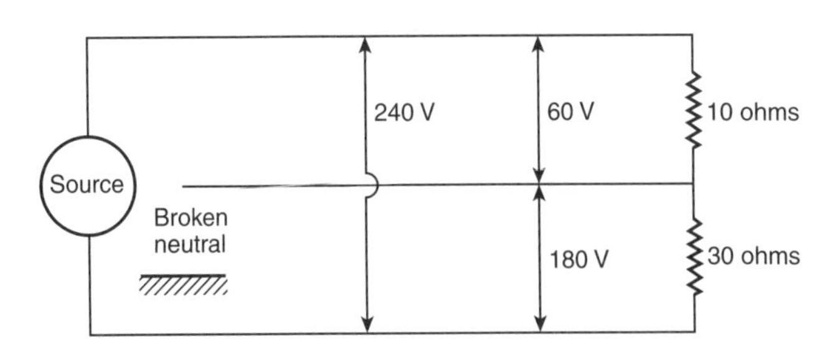

FIGURE 8.5.2 An Example of the Relation of Voltages in 120/240 V with an Open Neutral.

8.6 Overcurrent Protection.

8.6.1 General. Fuses and circuit breakers provide protection against electrical short circuits, ground faults, and load currents that might be damaging (i.e., overloads). In general, such an overcurrent device must be installed where each ungrounded (hot) branch conductor is connected to the power supply, and the device must function automatically.

8.6.1.1 Overcurrent devices are attached to bus bars in cabinets that are mounted in or on a wall. Examples are shown in Figure 8.6.1.1(a), Figure 8.6.1.1(b), Figure 8.6.1.1(c), and Figure 8.6.1.1(d).

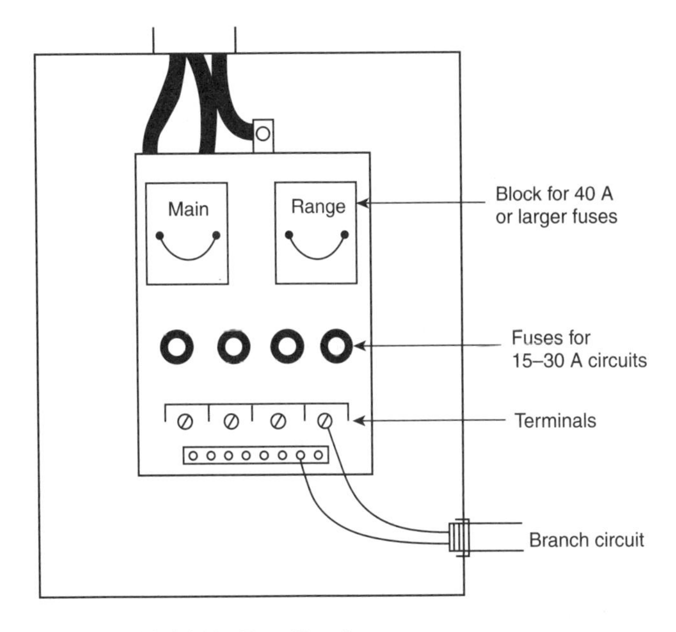

FIGURE 8.6.1.1(a) Fuse Panel.

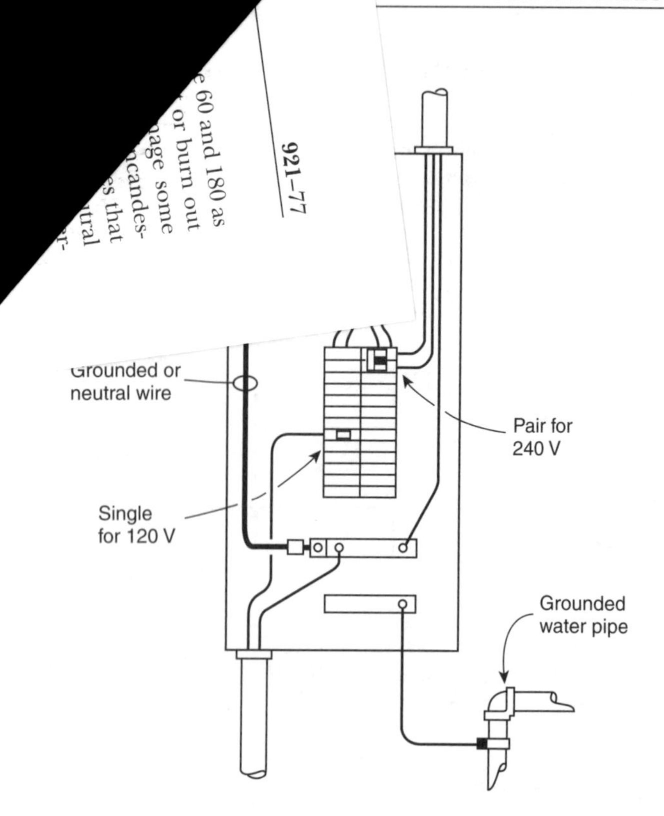

FIGURE 8.6.1.1(b) Common Arrangement for a Circuit Breaker Panel.

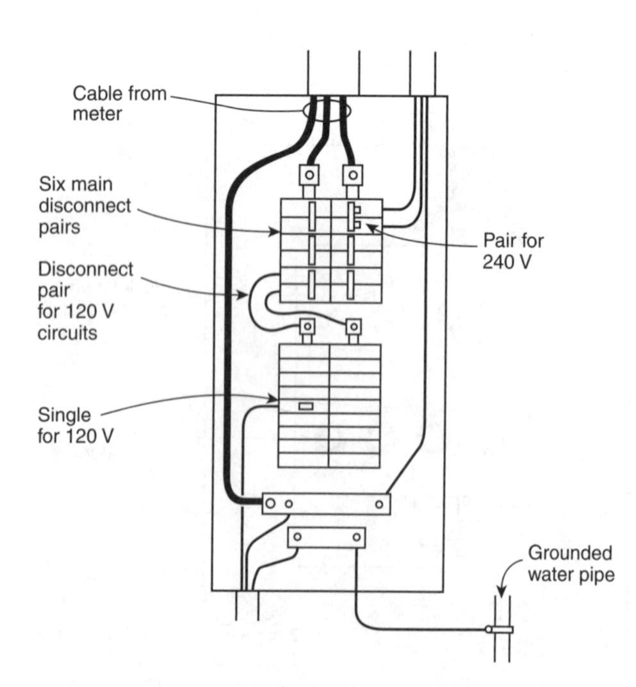

FIGURE 8.6.1.1(c) Common Arrangement for a Split-Bus Circuit Breaker Panel.

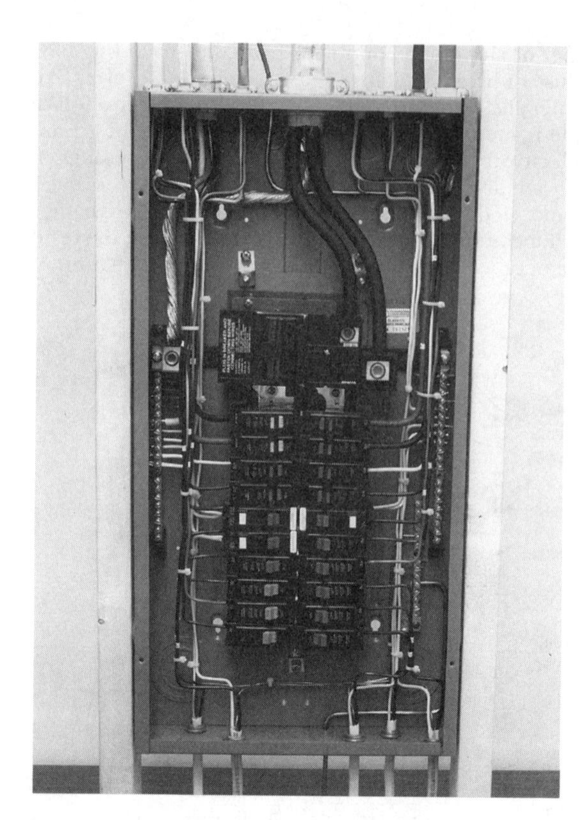

FIGURE 8.6.1.1(d) Circuit Breaker Panel.

8.6.1.2 Protective devices have two current ratings, the regular current rating and the interrupting current rating. The regular rating is the level of current above which the device will open, such as 15 A, 20 A, or 50 A. The interrupting rating is the level of current that the device can safely interrupt. A common value for circuit breakers is 10,000 A.

8.6.2 Fuses.

8.6.2.1 Operations. Fuses are basically nonmechanical devices with a fusible element in a small enclosure. The fusible element is made of a metal conductor or strip with enough resistance so that it will heat to melting at a selected level of current. Fuses have essentially no mechanical action; they operate only on the electrical and physical properties of the fuse element. Some fuses may contain a spring to help the separation of the fuse element on melting. Dual element fuses contain one element that operates most effectively with overloads and the other element that operates most effectively with short circuits. Ordinary fuses are single use, but some large fuses have replaceable elements.

8.6.2.1.1 There are two types of fuses: the plug type that screws into a base and the cartridge type that fits into a holder. They are shown in Figure 8.6.2.1.1(a), Figure 8.6.2.1.1(b), and Figure 8.6.2.1.1(c). Fuses are not resettable.

8.6.2.1.2 Fuses are mounted in a panelboard consisting of bus bars, connecting lugs, fuse holders, and supporting structures. Residential installations will usually be a combination of plug fuses for circuits of 30 A or less and cartridge fuses in removable holders for fuses with regular ratings greater than 30 A. The interrupting ratings on non-time-delay fuses are in the order of 100,000 A.

FIGURE 8.6.2.1.1(a) A Typical, Edison-Based Nonrenewable Fuse, Single Element, for Replacement Purposes Only.

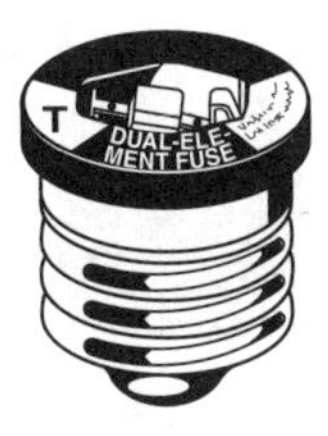

FIGURE 8.6.2.1.1(b) Another Edison-Based Nonrenewable Fuse, Dual Element, for Replacement Purposes Only.

FIGURE 8.6.2.1.1(c) A Type S Nonrenewable Fuse and Adapter. The time-lag type of fuse is acceptable but not required.

8.6.2.2 Plug Fuses. For circuits intended for 30 A or less, plug-type fuses have been used. The fuses have Edison bases so that all ampacities will fit in the same base. Thirty-ampere fuses could be put in where only 15 A fuses had been intended. Because of that overfusing and the ease with which the fuses could be bypassed (e.g., with a penny), they are not allowed in new installations. Such fuses are still available for replacement of burned-out fuses in existing installations.

8.6.2.3 Type S Fuses.

8.6.2.3.1 In an effort to minimize improper fusing, Type S fuses were developed. They are designed to make tampering or bypassing more difficult. They screw into adapters that fit into Edison bases. After an adapter has been properly installed, it cannot be removed without damaging the fuse base. The adapter prevents a larger-rated fuse from being used with a lower-rated circuit and makes bypassing the fuse more difficult.

8.6.2.3.2 *NFPA 70, National Electrical Code*, requires that fuseholders for plug fuses of 30 A or less shall not be used unless they are designed to use this Type S fuse or are made to accept a Type S fuse through use of an adapter.

8.6.2.4 Time-Delay Fuses. Whether a fuse is Type S or has an Edison base, the time-delay type of fuse permits overcurrents of short duration, such as starting currents for motors, without opening the circuit. While these momentary surges can be up to six times greater than the normal running c harmless because they last only a short time. possible to use time-delay fuses in sizes small eno better protection than a type without time delay. would have to be oversized to allow for such surges. event of short circuits or high-current ground faults, how the time-delay type will operate and open the circuit as rapi as the non-time-delay type. Time-delay fuses can be designed with dual elements or by modification of the fusing element.

8.6.2.5 Cartridge Fuses. For circuits intended for greater than 30 A, cartridge fuses are used. As shown in Figure 8.6.2.5(a) and Figure 8.6.2.5(b), they consist of a cylinder containing the fusing element and either caps or blades on each end to make electrical contact in its holder. Cartridge fuses may be made for either fast action or time delay. They also come in single-use or replaceable-element types. Cartridge fuses may be found in fuse panels of residential installations for high current loads, such as water heaters and ranges, and at the main disconnect. Large fuses of 100 A rating or greater are more common in commercial or industrial installations.

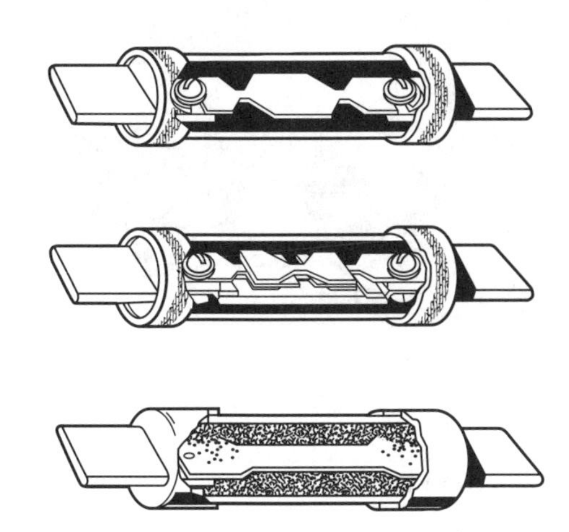

FIGURE 8.6.2.5(a) Three Types of Cartridge Fuses: (top), Ordinary Drop-Out Link Renewable Fuse; (center), Super-Lag Renewable Fuse; and (bottom) One-Time Fuse.

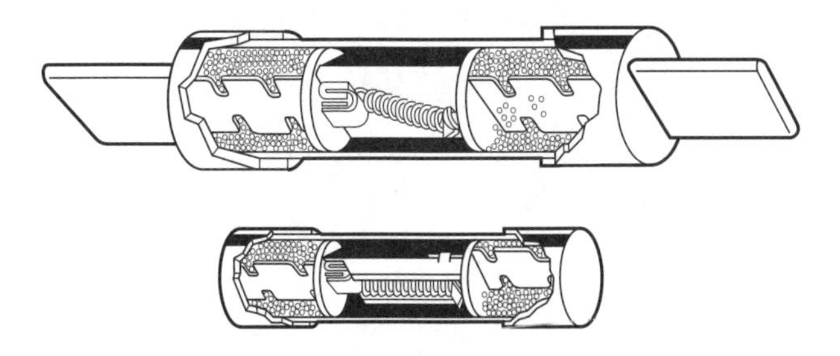

FIGURE 8.6.2.5(b) Dual-Element Cartridge Fuses, Blade and Ferrule Types.

8.6.3 Circuit Breakers.

8.6.3.1 Operations. A circuit breaker is a switch that opens either automatically with overcurrent or manually by pushing a handle. The current rating of the breaker is usually, but not always, given on the face of the handle. Breakers are designed so that the internal workings will trip with excessive current even if the handle is held in the ON position. The ON and

er on the handle or on the
ure 8.6.3.1(b).] The tripped
reakers, however some cir-
tion. *[See Figure 8.6.3.1(c).]*
be manually placed in the
he panel. However, if the
it breaker can usually be
)FF position, and then to
g rating for circuit break-

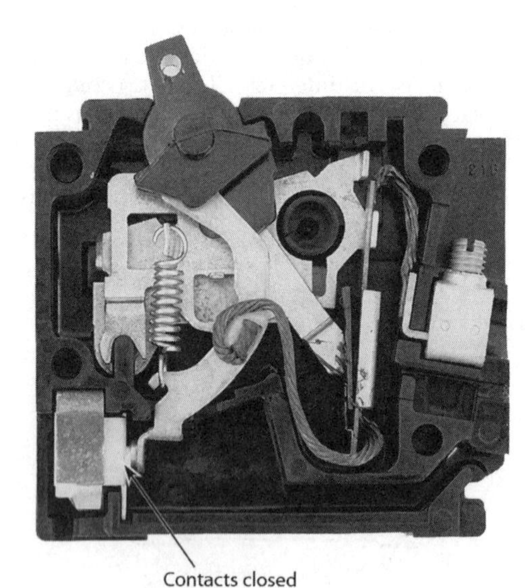

FIGURE 8.6.3.1(a) A 15 A Residential-Type Circuit Breaker in the Closed (ON) Position.

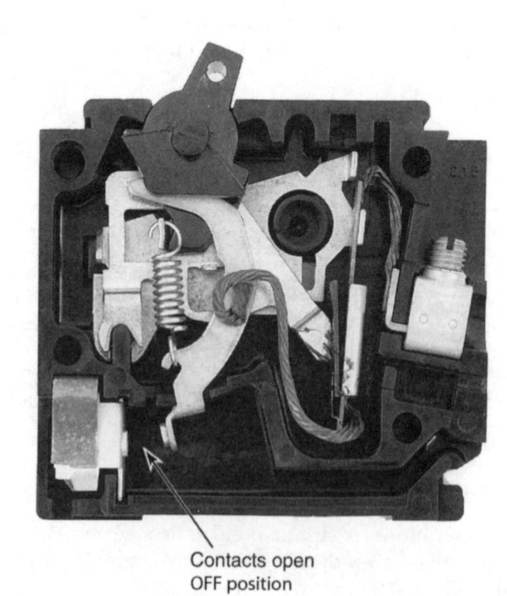

FIGURE 8.6.3.1(b) A 15 A Residential-Type Circuit Breaker in the Open (OFF) Position.

FIGURE 8.6.3.1(c) A 15 A Residential-Type Circuit Breaker in the Open (Tripped) Position.

8.6.3.1.1 Most residential circuit breakers are of the thermal-magnetic type. The thermal element, usually a bimetal, provides protection for moderate levels of overcurrent. The magnetic element provides protection for short circuits and for low-resistance ground faults, during which the fault currents are very high. Circuit breakers are mechanical devices that require movement of their components for operation. It is possible for them to fail to open, especially if they have not been operated either manually or by overcurrent in a long time and especially if they have been in a corrosive atmosphere.

8.6.3.1.2 The bodies of circuit breakers are usually made of molded phenolic plastic, which does not melt and does not sustain burning but which can be destroyed by fire impingement. Circuit breakers on a panelboard are directly connected to bus bars that are fed from the main disconnect. A cover plate over the rows of breakers exposes only the tops of the breakers so that no energized parts of the panel or wiring are exposed.

8.6.3.2 Main Breakers.

8.6.3.2.1 In the three-wire system used in most modern homes, the main disconnect in a breaker panel is a pair of circuit breakers of ampacity large enough to carry the entire current draw of the installation, commonly 100 A to 200 A in residences. The handles of the two breakers (one on each leg) are fastened together or are molded as one unit so that only one motion is needed to turn off both legs. Also, if one leg has a fault that trips the one breaker, the fastener will pull off the other breaker. Three-phase service uses three main breakers in a single body or with the handles fastened together and three bus bars to feed the breakers. Older homes and small homes such as summer cottages may have a two-wire system with only a single circuit breaker.

8.6.3.2.2 There are many split-bus panelboards in use. *[See Figure 8.6.1.1(c).]* They usually have six 2-pole breakers or pairs of breakers fastened together to make 240 V circuits. All of them must be off to cut off all power to the installation. One of the breaker pairs serves as a main for the lower bus bars that feed the 120 V circuits. Split-bus panelboards are not allowed in new installations.

8.6.3.3 Branch Circuit Breakers.

8.6.3.3.1 The circuit breakers for individual branch circuits are rated for the maximum intended current draw (ampacity). Circuits of 120 V will be fed from a single breaker, whereas circuits of 240 V will be fed from a double pole breaker or a pair of breakers of equal ampacity with the handles fastened together. General lighting and receptacle circuits will be 15 A or 20 A. Large appliances, such as ranges and water heaters, will have 30 A, 40 A, or 50 A breakers. Some small permanent appliances might have dedicated circuits with 15 A or 20 A breakers.

8.6.3.3.2 Three-phase service uses three bus bars to feed the breakers. Motors and other equipment that use three-phase power will be fed by three branch circuit breakers of equal ampacity with the handles fastened together.

8.6.3.4 Ground Fault Circuit Interrupter (GFCI). In newer installations, a GFCI is required for specific circuits such as those serving bathrooms, kitchens, and outside receptacles. Such interrupters often have a button labeled "push to test." This breaker houses a GFCI. It trips with a slight ground fault of about 5 milliamperes to give better protection for persons against electric shock at any level of amperage in the circuit. In addition, the breaker operates with overcurrents as an ordinary circuit breaker. The GFCI circuits are intended for bathrooms, patios, kitchens, or other locations where a person might be electrically grounded while near or using electrical appliances.

8.6.3.5 Arc Fault Circuit Interrupter (AFCI). AFCIs are designed to protect against fires caused by arcing faults in home electrical wiring. The AFCI circuitry continuously monitors current flow. AFCIs use special circuitry to discriminate between normal and unwanted arcing conditions. Once an unwanted arcing condition is detected, the control circuitry in the AFCI opens the internal contacts, thus de-energizing the circuit and reducing the potential for a fire to occur. An AFCI should not trip during normal arcing conditions, which can occur when a switch is opened or a plug is pulled from a receptacle. *NFPA 70, National Electrical Code,* requires that bedroom circuits be protected by AFCI circuit breakers.

8.6.4 Circuit Breaker Panels.

8.6.4.1 Circuit breaker panels often use plastic materials for insulation between energized parts and between the conductors, the metal enclosure, and the cover of the panel. The wires and cables and the molded circuit breaker enclosures are also typically insulated with plastic materials. These plastic materials may melt or decompose when exposed to heat. The heat causing them to decompose may be generated by sources inside the panel, or by external heat sources such as fire exposure. If the plastic insulation decomposes, deteriorates, or melts, energized conductors, or an energized conductor, may touch a grounded surface, producing arcing faults or overcurrent situations. Arcing may melt holes in or through the circuit breaker enclosure, sever or melt wires, and destroy portions of the components inside the panel. The arcing may be extensive due to the lack of overcurrent protection and the high available fault currents present at the circuit breaker panel. The presence of arc damage inside a circuit breaker panel does not by itself indicate that the panel was the source of ignition of the fire.

8.6.4.2 A circuit breaker panel after fire exposure is often fragile and should be recovered and secured with care if it is determined that additional examination of the panel or its contents is necessary. If possible, the panel should not be disassembled at the fire scene, but secured and analyzed under laboratory conditions. All small parts and debris should be preserved with the panel. The circuit breaker handles should not be operated or moved until the appropriate examination occurs. Supply and circuit wires should be examined for arc damage external to the circuit breaker panel if it is determined that additional examination of the panel or its contents is necessary.

8.6.4.3 Analysis of arc damage in a circuit breaker panel should include if possible the examination of the connections inside the panel, including those between the supply or service cable, the connections between the circuit breakers and the bus bars, and the connections between the branch circuit wires and the circuit breakers. The origin of any arcing fault should be determined if possible. Typically, the location of the initial arcing fault can be located and identified as to being on the load or line side of the circuit breakers. Locating the initial arcing fault may assist in determining the cause of the arcing. Examination of circuit conductors supplied by the circuit breaker panel can determine if downstream arcing had occurred outside of the panel. Comparison of arcing locations on the external circuit conductors to the arcing events inside the circuit breaker panel may reveal the sequence of arcing, thus indicating whether the arcing inside the panel was the source of ignition or was the result of heat impingement on the building's electrical system by the fire.

8.7 Branch Circuits. The individual circuits that feed lighting, receptacles, and various fixed appliances are the branch circuits. Each branch circuit should have its own overcurrent protection. The circuit consists of an ungrounded conductor (hot conductor) attached to a protective device and a grounded conductor (neutral conductor) attached to the grounding block in the cabinet. Both of those conductors carry the current that is being used in the circuit. In addition, there should be a grounding conductor (i.e., the ground). The grounding conductor normally does not carry any current but is there to allow fault current to go to ground and thereby open the protection. Some installations might have the grounding through metallic conduit, and some very old installations might not have a grounding conductor at all. The lack of a separate means of grounding has no effect on the operation of devices powered by that circuit.

8.7.1 Conductors. Conductors in electrical installations usually consist of copper or aluminum because they are economical and good conductors of electricity.

8.7.2 Sizes of Conductors. The sizes of conductors are measured in the American Wire Gauge (AWG). The larger the AWG number, the smaller the conductor. The branch circuit conductors for lighting and small appliances are usually solid copper, 14 AWG for 15 A circuits and 12 AWG for 20 A circuits. Circuits of larger ampacity will have larger conductors such as 10 or 8 AWG, as listed in Table 8.7.2. Conductors of 6 AWG or larger size will be multistranded to give adequate flexibility.

8.7.2.1 Aluminum branch wiring has been used and might be found in some installations. Because of problems with heating at the connections, aluminum conductors are not used in branch circuits without approved connectors, although aluminum cables such as 3/0 and 4/0 cables are used for service drops and service entry.

Table 8.7.2 Ampacity and Use of Branch Circuits

Wire Size (AWG)			
Copper	Copper-Clad Aluminum and Aluminum	Ampacity (A)	Use
14	12	15	Branch circuit conductors supplying other than kitchen
12	10	20–25	Small-appliance circuit conductors supplying outlets in kitchen for refrigerators, toasters, electric frying pans, coffee makers, and similar appliances
10	8	30	Large appliances such as ranges and dryers
8	6	40	
6	4	55	

8.7.2.2 The size of the conductor used in a circuit is chosen so as to carry the circuit current safely, with consideration given to such factors as the type of wire insulation and whether the wires are bundled. A circuit breaker of the proper size is then selected to protect that wire. The conductor should not be smaller than the allowed size but may be larger. The basic reason for regulating the allowed size is to prevent heating of the conductor enough to damage its insulation. Because conductors have some resistance, heat will be generated as current passes through them. Small conductors have more resistance than large conductors and so heat more. The *NFPA 70, National Electrical Code*, tables show how much current is allowed in various size conductors with various kinds of insulation.

8.7.3 Copper Conductors.

8.7.3.1 The chemical element copper is used in a pure form to make conductors. The copper is heated and drawn through progressively smaller dies to squeeze it down to the desired size. There is no identifiable crystal structure in such copper. Impurities or alloying elements would make the copper less conductive to electricity. Pure copper melts at 1082°C (1980°F).

8.7.3.2 Copper conductors oxidize in fires when the insulation has been lost. The surface usually becomes blackened with cupric oxide. For some conductors in a chemically reducing condition, such as glowing char before cooling, the surface may appear either to be bare of oxide or to be coated with a reddish cuprous oxide.

8.7.4* Aluminum Conductors.

8.7.4.1 Pure Aluminum. The chemical element aluminum is used in a pure form to make conductors. Pure aluminum melts at 660°C (1220°F). A skin of aluminum oxide forms on the surface, but the oxide does not mix with the metallic aluminum. Therefore, the melting temperature is not reduced, and the aluminum tends to melt through the whole cross section at one time instead of leaving an unmelted core as copper does. Melted aluminum can flow through the skin of oxide and have odd shapes when it solidifies. These shapes include pointed drips, and round and teardrop-shaped globules. Aluminum has a lower conductivity than does copper. Thus, for the same ampacity of a circuit, an aluminum conductor must be two AWG sizes larger than a copper conductor. For example, 10 AWG aluminum is equivalent in ampacity to 12 AWG copper.

8.7.4.2 Copper-Clad Aluminum. Copper-clad aluminum conductors have been used but are not common. Because they are aluminum conductors with just a skin of copper, their melting characteristics are essentially the same as those of aluminum conductors.

8.7.5 Insulation.

8.7.5.1 General. Conductors are insulated to prevent current from taking unwanted paths and to protect against dangerous voltages in places that would be hazardous to people. Insulation could be made of almost any material that can be applied readily to conductors, does not conduct electricity, and retains its properties for a long time even at elevated temperatures. For a summary of the types of insulation in use, see Table 310.13 of *NFPA 70, National Electrical Code*. Air serves as an insulator when bare conductors and energized parts are kept separated. At high voltage, air contamination by dust, pollution, or products of combustion can break down the insulating effects of air, resulting in arcs.

8.7.5.1.1 The type of insulation on individual conductors is marked in a code, along with the temperature rating, the manufacturer, and other information. Nonmetallic sheathed cable has the identifications printed on the sheath. The coding for the insulation material is given in Table 310.13 of *NFPA 70, National Electrical Code*.

8.7.5.1.2 Insulation on individual conductors is made in a variety of colors, some of which indicate specific uses. A grounding conductor must be bare, covered with green (or green with a yellow stripe) insulation, or marked where accessible in accordance with *NFPA 70, National Electrical Code*. A grounded conductor (neutral) may be white or light gray. An ungrounded conductor (hot) may be any color except green, white, or gray. In 120 V circuits it is commonly black. In 240 V circuits with nonmetallic cable, the two hot legs are commonly black and red. Where individual conductors are pulled through the conduit, the colors might vary more widely, especially if more than one circuit is in the conduit.

8.7.5.2 Polyvinyl Chloride (PVC). PVC is a commonly used thermoplastic insulating material for wiring. PVC must be blended with plasticizers to make it soft. Pigments and other modifiers may also be added. PVC, on aging, can slowly lose the plasticizers and become hard and brittle. In a fire, PVC may char and give off hydrogen chloride, a corrosive gas. Hydrogen chloride may combine with moisture to form hydrochloric acid. Hydrogen chloride or hydrochloric acid, if confined, may produce localized corrosion of metals. This corrosion can occur inside electrical enclosures.

8.7.5.3 Rubber. Rubber was the most common insulating material until approximately the 1950s. Rubber insulation contains pigments and various modifiers and antioxidants. In time it may become oxidized and brittle, especially if it was hot for long periods. Embrittled rubber has little strength and can be broken off the conductor if it is bent or scraped. Rubber insulation chars when exposed to fire or very high temperatures and leaves an ash when the rubber is burned away.

8.7.5.4 Other Materials. Polyethylene and other closely related polyolefins are used as insulation, more commonly on large cables than on insulation for residential circuits. Nylon jackets are put around other insulating materials (usually

PVC) to increase the thermal stability of the insulation. Silicone and fluorinated polyolefin (e.g., Teflon®) insulations are used on conductors that are expected to be installed where elevated temperatures will persist, particularly in appliances.

8.8 Outlets and Devices.

8.8.1 Switches. Switches are installed to turn the current on or off in parts of circuits that supply installed lights and equipment. Sometimes one or more receptacles are fed from a switch so that a table lamp can be turned on or off at the lamp or from the switch. The hot (black) conductor goes to both terminals of the switch while the neutral (white) conductor goes on to the light or device being controlled. The switch should always be put in the run of the black conductor for safety, although the switch will perform properly if put in the run of the white conductor. The switches may have screw, push-in, or slot terminals.

8.8.2 Receptacles. Receptacles for 15 A and 20 A circuits, illustrated in Figure 8.8.2(a) and Figure 8.8.2(b), are usually duplex. Receptacles for large appliances (30 A or more) are single. Receptacles now must be polarized and of the grounding type, although there are still many nongrounding and nonpolarized receptacles in older installations. The grounding type has a third hole that allows any appliance with a grounding prong in its plug to ground that appliance. In polarized receptacles, the neutral slot is longer than the hot slot. A two-prong plug with a wide neutral prong (polarized plug) can be inserted into the receptacle only with the wide prong in the wide slot and not in the reverse way. All grounding receptacles and plugs are inherently polarized.

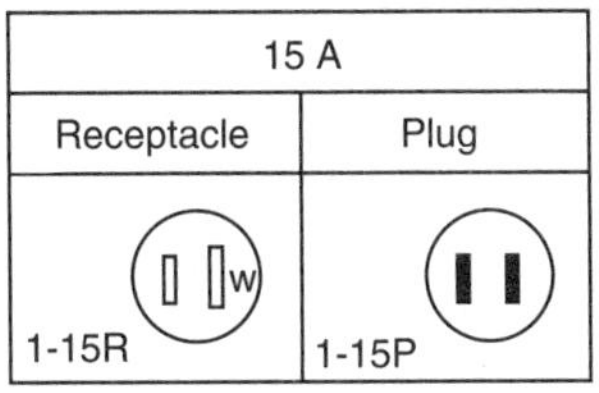

15 A	
Receptacle	Plug
W 1-15R	1-15P

FIGURE 8.8.2(a) Nongrounding-Type Receptacle.

15 A		20 A	
Receptacle	Plug	Receptacle	Plug
G W 5-15R	G W 5-15P	G W 5-20R	G W 5-20P

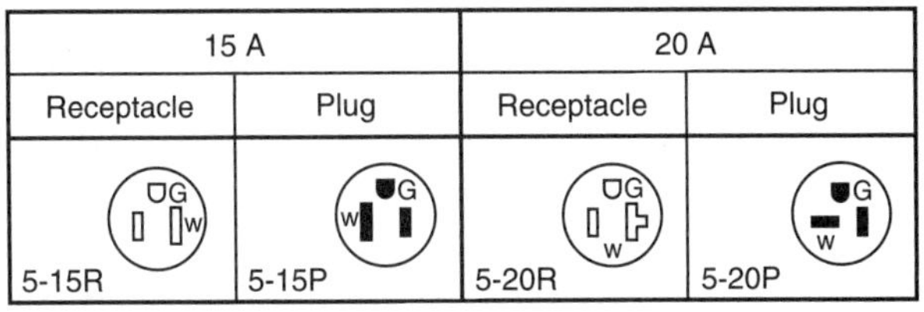

FIGURE 8.8.2(b) Grounding-Type Receptacle.

8.8.2.1 In bathrooms or other areas where personal safety is a concern, receptacles may have a built-in GFCI. *(See 8.6.3.4.)*

8.8.2.2 Depending on the receptacle style, hot and neutral wires are secured to the receptacle by inserting the bared ends into holes or slots (push-in terminals), by securing the bared ends under screw terminals, or by tightening screw-driven clamps over the bared ends. Some receptacles have provisions for both push-in attachment and screw attachment. The hot conductor (usually black or red but may be any color except green, white, gray, or bare) should be connected to the brass or darker-colored screw or to the terminal marked "hot." The neutral conductor (white insulation) should be connected to the silver or lighter-colored screw or to the terminal marked "neutral" or "white." On grounding-type receptacles, the bare wire (or bared end of the green wire) is usually secured under a green-colored screw. Some outlets having push-in terminals for the hot and neutral conductors also have push-in terminals for the ground conductor. This push-in terminal is likely to be marked "green," "bare," or "ground." On receptacles in which the wires are put in slots, the ground wire is also put in a slot.

8.8.3 Other Outlets, Devices, or Equipment. Permanent lighting fixtures are attached to electrical boxes in the wall or ceiling as appropriate with a wall switch or dimmer in its individual part of the circuit. Thermostats may be mounted in walls to control permanently installed heating units.

8.8.3.1 In commercial and industrial installations, much of the electrically powered equipment is permanently connected to the basic wiring. Because of the large current draws, much of the equipment may be switched on and off by contactors rather than directly by switches.

8.8.3.2 In installations where explosive atmospheres might occur, explosionproof outlets and fixtures should be used. The outlet boxes, fittings, and attached devices are designed so that even if explosive concentrations of gases get into the system, an internal ignition will not let a flame front out to ignite the surrounding atmosphere.

8.9 Ignition by Electrical Energy.

8.9.1 General. For ignition to be from an electrical source, the following must occur:

(1) The electrical wiring, equipment, or component must have been energized from a building's wiring, an emergency system, a battery, or some other source.
(2) Sufficient heat and temperature to ignite a close combustible material must have been produced by electrical energy at the point of origin by the electrical source.

8.9.1.1 Ignition by electrical energy involves generating both a sufficiently high temperature and heat (i.e., competent ignition source) by passage of electrical current to ignite material that is close. Sufficient heat and temperature may be generated by a wide variety of means, such as short-circuit and ground-fault parting arcs, excessive current through wiring or equipment, resistance heating, or by ordinary sources such as lightbulbs, heaters, and cooking equipment. The requirement for ignition is that the temperature of the electrical source be maintained long enough to bring the adjacent fuel up to its ignition temperature, with air present to allow combustion.

8.9.1.2 The presence of sufficient energy for ignition does not assure ignition. Distribution of energy and heat loss factors need to be considered. For example, an electric blanket spread out on a bed can continuously dissipate 180 W safely. If that same blanket is wadded up, the heating will be concentrated in a smaller space. Most of the heat will be held in by the outer layers of the blanket, which will lead to higher internal temperatures and possibly ignition. In contrast to the 180 W used by a typical electric blanket, just a few watts used by a small flashlight bulb will cause the filament to glow white hot, indicating temperatures in excess of 2204°C (4000°F).

8.9.1.3 In considering the possibility of electrical ignition, the temperature and duration of the heating must be great enough to ignite the initial fuels. The type and geometry of the fuel must be evaluated to be sure that the heat was sufficient to generate combustible vapors and for the heat source still to be hot enough to ignite those vapors. If the suspect electrical component is not a competent ignition source, other causes should be investigated.

8.9.1.4 Before a fire can properly be determined to have been caused by electricity, the source of electrical heat must be identified. The heat generated must be sufficient to cause ignition of the first ignited fuel. A path or method of heat transfer between the heat source and the first ignited fuel must be identified.

8.9.2 Resistance Heating.

8.9.2.1 General.

8.9.2.1.1 Whenever electric current flows through a conductive material, heat will be produced. See 8.2.2.7.2 for the relationships of current, voltage, resistance, and power (i.e., heating). With proper design and compliance with the codes, wiring systems and devices will have resistances low enough that current-carrying parts and connections should not overheat. Some specific parts, such as lamp filaments and heating elements, are designed to become very hot. However, when properly designed and manufactured and when used according to directions, those hot parts should not cause fires.

8.9.2.1.2 The use of copper or aluminum conductors of sufficient size in wiring systems (e.g., 12 AWG for up to 20 A for copper) will keep the resistance low. What little heat is generated should be readily dissipated to the air around the conductor under normal conditions. When conductors are thermally insulated and operating at rated currents, enough energy may be available to cause a fault or ignition.

8.9.2.2 Heat-Producing Devices. Common heat-producing devices can cause fires when misused or when certain malfunctions occur during proper use. Examples include combustibles placed too close to incandescent lamps or to heaters or coffee makers, and deep-fat fryers whose temperature controls fail or are bypassed. *(See Section 24.6.)*

8.9.2.3 Poor Connections. When a circuit has a poor connection such as a loose screw at a terminal, increased resistance causes increased heating at the contact, which promotes formation of an oxide interface. The oxide conducts current and keeps the circuit functional, but the resistance of the oxide at that point is significantly greater than in the metals. A spot of heating develops at that oxide interface that can become hot enough to glow. If combustible materials are close enough to the hot spot, they can be ignited. Generally, the connection will be in a box or appliance, and the probability of ignition is greatly reduced. The wattage of well-developed heating connections in wiring can be up to 30–40 W with currents of 15–20 A. Heating connections of lower wattage have also been noted at currents as low as about 1 A.

8.9.3 Overcurrent and Overload. Overcurrent is the condition in which more current flows in a conductor than is allowed by accepted safety standards. The magnitude and duration of the overcurrent determine whether there is a possible ignition source. For example, an overcurrent at 25 A in a 14 AWG copper conductor should pose no fire danger except in circumstances that do not allow dissipation of the heat, such as when thermally insulated or when bundled in cable applications. A large overload of 120 A in a 14 AWG conductor, for example, would cause the conductor to glow red hot and could ignite adjacent combustibles.

8.9.3.1 Large overcurrents that persist (i.e., overload) can bring a conductor up to its melting temperature. There is a brief parting arc as the conductor melts in two. The melting opens the circuit and stops further heating.

8.9.3.2 In order to get a large overcurrent, either there must be a fault that bypasses the normal loads (i.e., short circuit) or far too many loads must be put on the circuit. To have a sustained overcurrent (i.e., overload), the protection (i.e., fuses or circuit breakers) must fail to open or must have been defeated or been rendered ineffective by the circuit design or installation. Ignition by overload is rare in circuits that have the proper size conductors throughout the circuit, because most of the time the protection opens and stops further heating before ignition conditions are obtained. When there is a reduction in the conductor size between the load and the circuit protection, such as an extension cord, the smaller size conductor may be heated beyond its temperature rating. This overheating can occur without activating the overcurrent protection. For an example, see 8.2.8.3.

8.9.4 Arcs.

8.9.4.1 General. An arc is a high-temperature luminous electric discharge across a gap or through a medium such as charred insulation. Temperatures within the arc are in the range of several thousand degrees, depending on circumstances, including current, voltage drop, and metal involved. For an arc to jump even the smallest gap in air spontaneously, there must be a voltage difference of at least 350 V. In the 120/240 V systems being considered here, arcs do not form spontaneously under normal circumstances. *(See Section 8.12.)* In spite of the very high temperatures in an arc path, arcs may not be competent ignition sources for many fuels. In most cases, the arcing is so brief and localized that solid fuels such as wood structural members cannot be ignited. Fuels with high surface-area-to-mass ratio, such as cotton batting, tissue paper, and combustible gases and vapors, may be ignited when in contact with the arc.

8.9.4.2 High-Voltage Arcs.

8.9.4.2.1 High voltages can get into a 120/240 V system through accidental contact between the distribution system of the power company and the system on the premises. Whether there is a momentary discharge or a sustained high voltage, an arc may occur in a device for which the separation of conductive parts is safe at 240 V but not at many thousands of volts. If easily ignitible materials are present along the arc path, a fire can be started.

8.9.4.2.2 Lightning can send extremely high voltage surges into an electrical installation. Because the voltages and currents from lightning strikes are so high, arcs can jump at many places, cause mechanical damage, and ignite many kinds of combustibles. *(See 8.12.8.)*

8.9.4.3 Static Electricity. Static electricity is a stationary charge that builds up on some objects. Walking across a carpet in a dry atmosphere will produce a static charge that can produce an arc when discharged. Other kinds of motion can cause a buildup of charge, including the pulling off of clothing, operation of conveyor belts, and the flowing of liquids. *(See Section 8.12.)*

8.9.4.4 Parting Arcs. A parting arc is a brief discharge that occurs as an energized electrical path is opened while current is flowing, such as by turning off a switch or pulling a plug. The arc usually is not seen in a switch but might be seen when a plug is pulled while current is flowing. Motors with brushes may produce a nearly continuous display of arcing between the brushes and the commutator. At 120/240 V ac, a parting arc is not sustained and will quickly be quenched. Ordinary parting arcs in electrical systems are usually so brief and of low enough energy that only combustible gases, vapors, and dusts can be ignited.

8.9.4.4.1 In arc welding, the rod must first be touched to the workpiece to start current flowing. Then the rod is withdrawn a small distance to create a parting arc. If the gap does not become too great, the arc will be sustained. A welding arc involves enough power to ignite nearly any combustible material. However, the sustained arc during welding requires specific design characteristics in the power supply that are not present in most parting arc situations in 120/240 V wiring systems.

8.9.4.4.2 Another kind of parting arc occurs when there is a direct short circuit or ground fault. The surge of current melts the metals at the point of contact and causes a brief parting arc as a gap develops between the metal pieces. The arc quenches immediately but can throw particles of melted metal (i.e., sparks) around. *(See 8.9.5.)*

8.9.4.5* Arcing Across a Carbonized Path. Arcing between two conductors separated by a solid insulator can become possible if the insulator becomes carbonized. The two primary means by which carbonization is created is by flow of electric current or by thermal means not involving electricity. If carbonization is due to flow of electric current, the phenomenon is commonly called arc tracking. Nonelectrical means of creating carbonization usually involve some kind of heating; this can be due to heat-producing devices, or it can be due to fire itself.

8.9.4.5.1 Arc Tracking. Arcs may occur on surfaces of nonconductive materials if they become contaminated with salts, conductive dusts, or liquids. It is thought that small leakage currents created through such contamination cause degradation of the base material leading to the arc discharge, charring or igniting combustible materials around the arc. Arc tracking can be a problem not only at high voltages, but also in 120/240 V ac systems. PVC insulation is susceptible to arc tracking, and recent studies indicate that it can be susceptible to a unique form of failure: self-induced wetting. When PVC insulation containing the commonly used filler calcium carbonate has been heated to 110°C or higher, chemical degradation reactions occur that subsequently cause moisture from the air to be hydrophilically deposited onto the surface, potentially initiating a process of arc tracking failure.

8.9.4.5.2 Electrical current will flow through water or moisture only when that water or moisture contains contaminants such as dirt, dusts, salts, or mineral deposits. This stray current may promote electrochemical changes that can lead to electrical arcing. Most of the time the stray currents through a contaminated wet path cause enough warming that the path will dry. Then little or no current flows and the heating stops. If the moisture is continuously replenished so that the currents are sustained, deposits of metals or corrosion products can form along the electrical pathway. That effect is more pronounced in direct current situations. More energetic arcing between deposits might cause a fire under the right conditions. More study is needed to more clearly define the conditions needed for causing a fire.

8.9.5 Sparks. Sparks are luminous particles that can be formed when an arc melts metal and spatters the particles away from the point of arcing. The term *spark* has commonly been used for a high voltage discharge as with a spark plug in an engine. For purposes of electrical fire investigation, the term *spark* is reserved for particles thrown out by arcs, whereas an arc is a luminous electrical discharge across a gap.

8.9.5.1 Short circuits and high-current ground faults, such as when the ungrounded conductor (i.e., hot conductor) touches the neutral or a ground, produce violent events. Because there may be very little resistance in the short circuit, the fault current may be many hundreds or even thousands of amperes. The energy that is dissipated at the point of contact is sufficient to melt the metals involved, thereby creating a gap and a visible arc and throwing sparks. Protective devices in most cases will open (i.e., turn off the circuit) in a fraction of a second and prevent repetition of the event.

8.9.5.2 When just copper and steel are involved in arcing, the spatters of melted metal begin to cool immediately as they fly through the air. When aluminum is involved in faulting, the particles may actually burn as they fly and continue to be extremely hot until they burn out or are quenched by landing on some material. Burning aluminum sparks, therefore, may have a greater ability to ignite fine fuels than do sparks of copper or steel. However, sparks from arcs in branch circuits are inefficient ignition sources and can ignite only fine fuels when conditions are favorable. In addition to the temperature, the size of the particles is important for the total heat content of the particles and the ability to ignite fuels. For example, sparks spattered from a welding arc can ignite many kinds of fuels because of the relatively large size of the particles and the total heat content. Arcing in entry cables can produce more and larger sparks than can arcing in branch circuits.

8.9.6 High-Resistance Faults. Depending on the nature of the fault and the extent of the fire damage, evidence of a high resistance fault may be difficult to find after a fire. Examples of high resistance faults are an energized conductor coming into contact with a poorly grounded object, or a poor plug blade-to-receptacle connection. See 8.10.4 for examples of evidence of a high resistance fault that may be found after a fire.

8.10 Interpreting Damage to Electrical Systems.

8.10.1 General. Abnormal electrical activity will usually produce characteristic damage that may be recognized after a fire. Evidence of this electrical activity may be useful in locating the area of origin. The damage may occur on conductors, contacts, terminals, conduits, or other components. However, many kinds of damage can occur from nonelectrical events. This section will give guidelines for deciding whether observed damage was caused by electrical energy and whether it was the cause of the fire or a result of the fire. These guidelines are not absolute, and many times the physical evidence will be ambiguous and will not allow a definite conclusion.

8.10.2* Short-Circuit and Ground-Fault Parting Arcs. Whenever an energized conductor contacts a grounded conductor or a metal object that is grounded with nearly zero resistance in the circuit, there will be a surge of current in the circuit and melting at the point of contact. This event may be caused by heat-softened insulation due to a fire, or by failure to protect wire insulation properly where the wire passes over a sharp metal edge or penetrates a metal box. The high current flow produces heat that can melt the metals at the points of contact of the objects involved, thereby producing a gap and the parting arc. A solid copper conductor typically appears as though it had been notched with a round file, as shown in Figure 8.10.2. The notch may or may not sever the conductor. The conductor will break easily at the notch upon handling. The surface of the notch can be seen by microscopic examination to have been melted. Sometimes, there can be a projection of porous copper in the notch.

FIGURE 8.10.2 A Solid Copper Conductor Notched by a Short Circuit.

8.10.2.1 The parting arc melts the metal only at the point of initial contact. The adjacent surfaces will be unmelted unless fire or some other event causes subsequent melting. In the event of subsequent melting, it may be difficult to identify the site of the initial short circuit or ground fault. If the conductors were insulated prior to the faulting and the fault is suspected as the cause of the fire, it will be necessary to determine how the insulation failed or was removed and how the conductors came in contact with each other. If the conductor or other metal object involved in the short circuit or ground fault was bare of insulation at the time of the faulting, there may be spatter of metal onto the otherwise unmelted adjacent surfaces.

8.10.2.2 Stranded conductors, such as for lamp and appliance cords, appear to display effects from short circuits and ground faults that are less consistent than those in solid conductors. A stranded conductor may exhibit a notch with only some of the strands severed, or all of the strands may be severed with strands fused together or individual strands melted. *(See Figure 8.10.2.2.)*

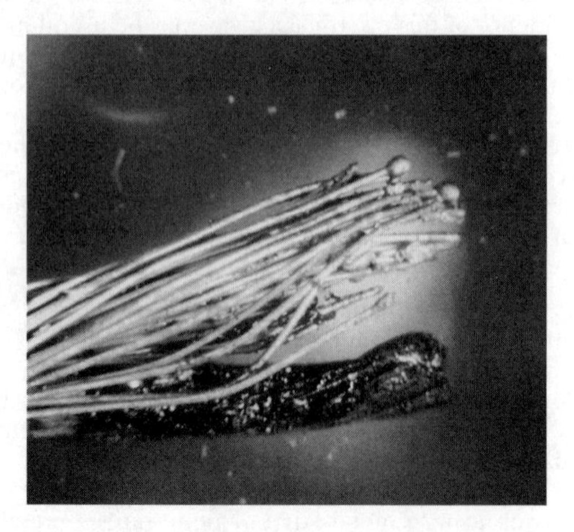

FIGURE 8.10.2.2 Stranded Copper Lamp Cord That Was Severed by a Short Circuit.

8.10.3* Arcing Through a Carbonized Path Due to Thermal Means (Arcing Through Char). Insulation on conductors, when exposed to direct flame or radiant heat, may be charred before being melted. That char is conductive enough to allow sporadic arcing through the char. That arcing can leave surface melting at spots or can melt through the conductor, depending on the duration and repetition of the arcing. There often will be multiple points of arcing. Several inches of conductor can be destroyed, either by melting or severing of several small segments.

8.10.3.1 When conductors are subject to highly localized heating, such as from arcing through char, the ends of individual conductors may be severed. When severed, they will have beads on the end, as shown in Figure 8.10.3.1(a). The bead may weld two conductors together, as shown in Figure 8.10.3.1(b). If the conductors are in conduit, holes may be melted in the conduit. Beads can be differentiated from globules, which are created by nonlocalized heating such as overload or fire melting. Beads are characterized by the distinct and identifiable line of demarcation between the melted bead and the adjacent unmelted portion of the conductor. Figure 8.10.3.1(c), Figure 8.10.3.1(d), Figure 8.10.3.1(e), and Figure 8.10.3.1(f) show examples.

FIGURE 8.10.3.1(a) Copper Conductors Severed by Arcing Through the Charred Insulation.

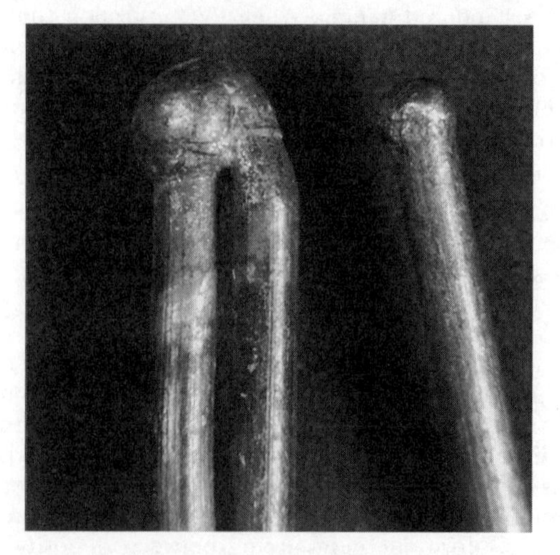

FIGURE 8.10.3.1(b) Copper Conductors Severed by Arcing Through the Charred Insulation with a Large Bead Welding the Two Conductors Together.

FIGURE 8.10.3.1(c) Stranded Copper Conductors Severed by Arcing through Charred Insulation with the Strands Terminated in Beads.

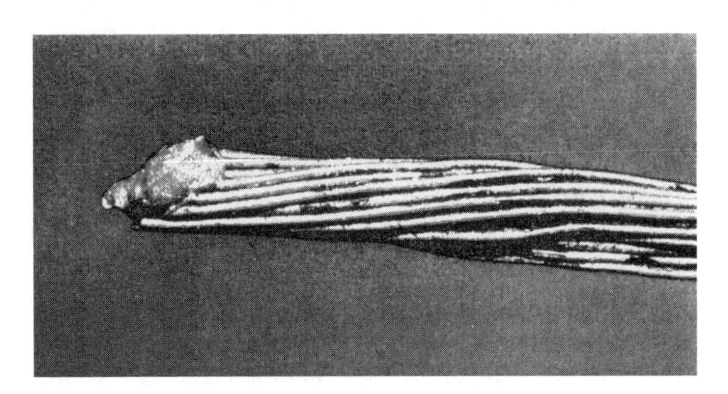

FIGURE 8.10.3.1(d) Arc Damage to 18 AWG Cord by Arcing through Charred Insulation.

8.10.3.2 The conductors downstream from the power source and the point where the conductors are severed become de-energized. Those conductors will likely remain in the debris with part or all of their insulation destroyed. The upstream remains of the conductors between the point of arc-severing and the power supply may remain energized if the overcurrent protection does not function. Those conductors can sustain further arcing through the char. In a situation with multiple arc-severing on the same circuit, arc-severing farthest from the power supply occurred first. It is necessary to find as much of the conductors as possible to determine the location of the first arcing through char. This will indicate the first point on the circuit to be compromised by the fire and may be useful in determining the area of origin. In branch circuits, holes extending for several inches may be seen in the conduit or in metal panels to which the conductor arced.

8.10.3.3 If the fault occurs in service entrance conductors, several feet of conductor may be partly melted or destroyed by repeated arcing because the overcurrent protection is on the primary side of the transformer. An elongated hole or series of holes extending several feet may be seen in the conduit.

FIGURE 8.10.3.1(e) Spot Arc Damage to 14 AWG Conductor Caused by Arcing Through Charred Insulator (Lab Test).

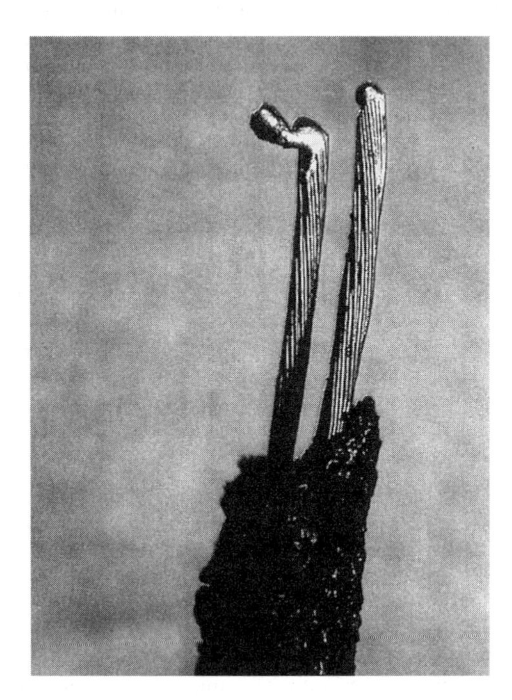

FIGURE 8.10.3.1(f) Arc Damage to 18 AWG Cord by Arcing Through Charred Insulation (Lab Test).

8.10.3.4* Arcing Involving Uninsulated Conductors. Some conductors, such as bus bars, are not insulated over their entire surface, but are held away from other conductors or metal panels by the use of plastic or ceramic insulators, and are close and parallel to each other or to the grounded enclosure. These conductors can be found as bus bars in circuit breaker panels or in commercial or industrial electrical panels and switchgear. These bus bars are often designed to carry hundreds or thousands of amperes. If an arcing fault occurs between bus bars or between a bus bar and a grounded panel, large amounts of melting and thermal damage can be created. Overcurrent protection devices are designed to allow large currents to flow through the conductors. Should an arc develop between the uninsulated bus bars or between the uninsulated bus bar and the grounded enclosure, the arc will travel down the bus bars away from the power source. This may result in more arc damage where the arc reaches the end of the bus bars rather than where the arc was initiated.

8.10.4* Overheating Connections. Connection points are the most likely place for overheating to occur on a circuit. The most likely cause of the overheating will be a loose connection or the presence of resistive oxides at the point of connection. Metals at an overheating connection will be more severely oxidized than similar metals with equivalent exposure to the fire. For example, an overheated connection on a duplex receptacle will be more severely damaged than the other connections on that receptacle. The conductor and terminal parts may have pitted surfaces or may have sustained a loss of mass where poor contact has been made. This loss of mass can appear as missing metal or tapering of the conductor. These effects are more likely to survive the fire when copper conductors are connected to steel terminals. Where brass or aluminum are involved at the connection, the metals are more likely to be melted than pitted. This melting can occur either from resistance heating or from the fire. Pitting also can be caused by alloying. *(See 8.10.6.3.)* Overheating at a connection can result in the thermal damage and charring of materials adjacent to the connection. Heat can be transferred along conductors attached to the overheated connection, resulting in charring or loss of the conductor's insulation. The charring or loss of plastic insulation may allow arcing to occur. Such arc damage may survive the fire.

8.10.5* Overload. Overcurrents that are large enough and persist long enough to cause damage or create a danger of fire are called overloads. Under any circumstance, suspected overloads require that the circuit protection be examined. The most likely place for an overload to occur is on an extension cord. Overloads are unlikely to occur on wiring circuits with proper overcurrent protection.

8.10.5.1 Overloads cause internal heating of the conductor. This heating occurs along the entire length of the overloaded portion of the circuit and may cause sleeving. Sleeving is the softening and sagging of thermoplastic conductor insulation due to heating of the conductor. If the overload is severe, the conductor may become hot enough to ignite fuels in contact with it as the insulation melts off. Severe overloads may melt the conductor. If the conductor melts in two, the circuit is opened and heating immediately stops. The other places where melting had started may become frozen as offsets. This effect has been noted in copper, aluminum, and Nichrome® conductors. *(See Figure 8.10.5.1.)* The finding of distinct offsets is an indication of a large overload. Evidence of overcurrent melting of conductors is not proof of ignition by that means.

8.10.5.2 Overload in service entrance cables is more common than in branch circuits but is usually a result of fire. Faulting in entrance cables produces sparking and melting only at the point of faulting unless the conductors maintain continuous contact to allow the sustained massive overloads needed to melt long sections of the cables.

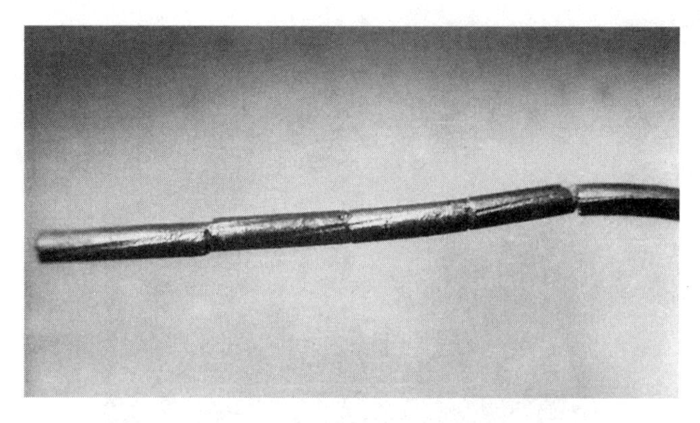

FIGURE 8.10.5.1 Aluminum Conductor Severed by Overcurrent Showing Offsets.

8.10.6 Effects Not Caused by Electricity. Conductors may be damaged before or during a fire by other than electrical means and often these effects are distinguishable from electrical activity.

8.10.6.1 Conductor Surface Colors. When the insulation is damaged and removed from copper conductors by any means, heat will cause dark red to black oxidation on the conductor surface. Green or blue colors may form when some acids are present. The most common acid comes from the decomposition of PVC. These various colors are of no value in determining cause because they are nearly always results of the fire condition.

8.10.6.2 Melting by Fire. When exposed to fire or glowing embers, copper conductors may melt. At first, there is blistering and distortion of the surface, as shown in Figure 8.10.6.2(a). The striations created on the surface of the conductor during manufacture become obliterated. The next stage is some flow of copper on the surface with some hanging drops forming. Further melting may allow flow with thin areas (i.e., necking and drops), as shown in Figure 8.10.6.2(b). In that circumstance, the surface of the conductor tends to become smooth. The resolidified copper forms globules. Globules caused by exposure to fire are irregular in shape and size. They are often tapered and may be pointed. There is no distinct line of demarcation between melted and unmelted surfaces.

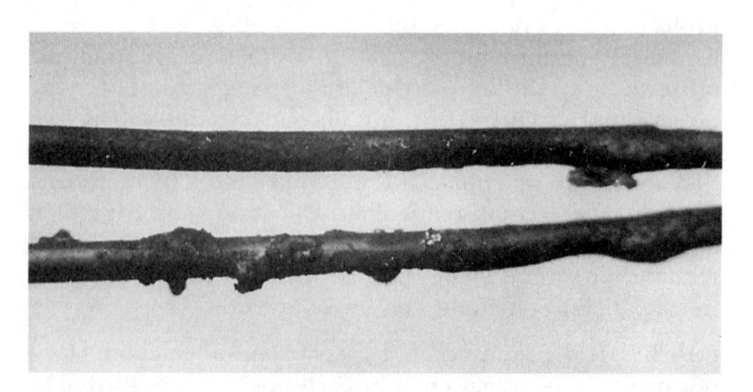

FIGURE 8.10.6.2(a) Copper Conductors Fire-Heated to the Melting Temperature, Showing Regions of Flow of Copper, Blistering, and Surface Distortion.

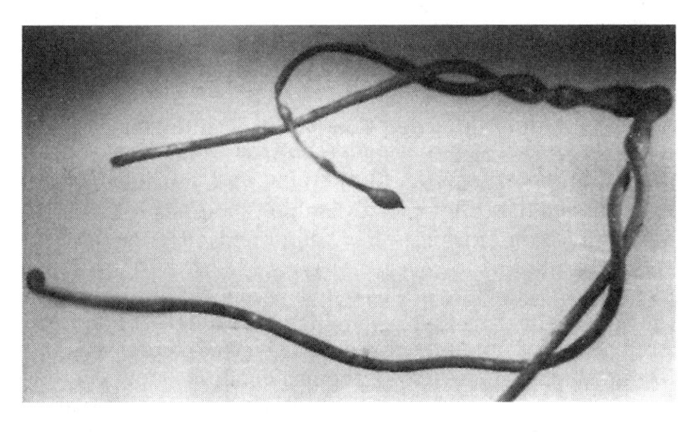

FIGURE 8.10.6.2(b) Fire-Heated Copper Conductors, Showing Globules.

8.10.6.2.1 Stranded conductors that just reach melting temperatures become stiffened. Further heating can let copper flow among the strands so that the conductor becomes solid with an irregular surface that can show where the individual strands were, as shown in Figure 8.10.6.2.1(a). Continued heating can cause the flowing, thinning, and globule formation typical of solid conductors. Magnification is needed to see some of these effects. Large-gauge stranded conductors that melt in fires can have the strands fused together by flowing metal or the strands may be thinned and stay separated. In some cases, individual strands may display a bead-like globule even though the damage to the conductor was from melting. Figure 8.10.6.2.1(b) and Figure 8.10.6.2.1(c) show some examples.

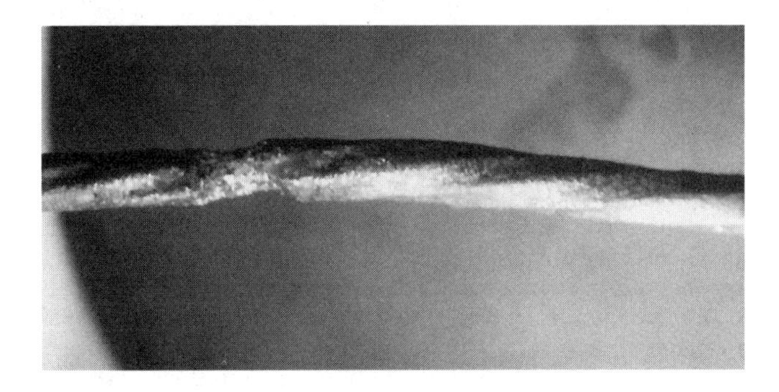

FIGURE 8.10.6.2.1(a) Stranded Copper Conductor in Which Melting by Fire Caused the Strands to Be Fused Together.

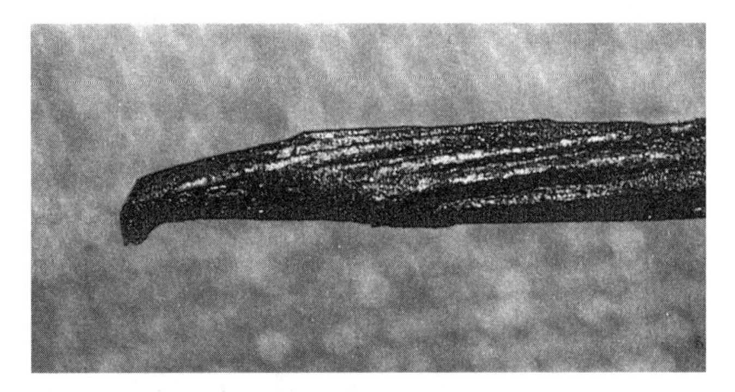

FIGURE 8.10.6.2.1(b) Fire Melting of Stranded Copper Wire.

FIGURE 8.10.6.2.1(c) Another Example of Fire Melting of Stranded Copper Wire.

8.10.6.2.2 Aluminum conductors melt and resolidify into irregular shapes that are usually of no value for interpreting cause, as shown in Figure 8.10.6.2.2. Because of the relatively low melting temperature, aluminum conductors can be expected to melt in almost any fire and rarely aid in finding the cause.

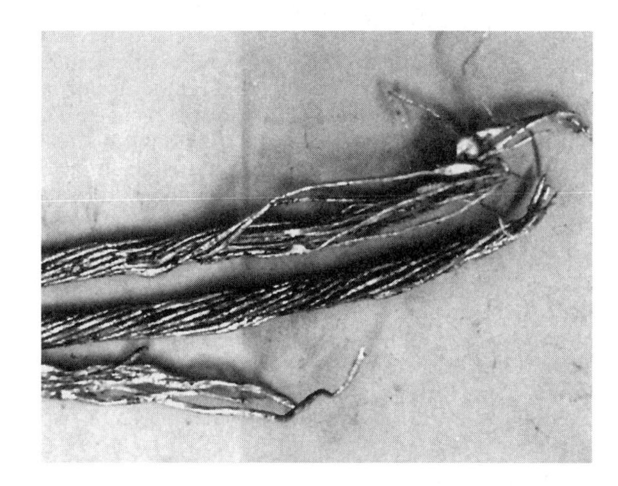

FIGURE 8.10.6.2.2 Aluminum Cables That Were Melted by Fire, Showing Thinned Areas, Bulbous Areas, and Pointed Ends.

8.10.6.3* Alloying. Metals such as aluminum and zinc can form alloys when melted in the presence of other metals. If aluminum drips onto a bare copper conductor during a fire and cools, the aluminum will be just lightly stuck to the copper. If that spot is further heated by fire, the aluminum can penetrate the oxide interface and form an alloy with the copper that melts at a lower temperature than does either pure metal. After the fire, an aluminum alloy spot may appear as a rough gray area on the surface, or it may be a shiny silvery area. The copper–aluminum alloy is brittle, and the conductor may readily break if it is bent at the spot of alloying. If the melted alloy drips off the conductor during the fire, there will be a pit that is lined with alloy. The presence of alloys can be confirmed by chemical analysis.

8.10.6.3.1 Aluminum conductors that melt from fire heating at a terminal may cause alloying and pitting of the terminal pieces. There is no clear way of visually distinguishing alloying from the effects of an overheating connection. Zinc forms a brass alloy readily with copper. It is yellowish in color and not as brittle as the aluminum alloy.

8.10.6.3.2 Copper and silver may also form alloys. This can occur at temperatures below their melting point. The alloys may be seen on contacts, electrical switches, thermostats, thermal protectors, contactors, relays, and similar items.

8.10.6.3.3 Copper conductors, terminated in connections, or terminals containing solder may have areas of alloying, globules, rounded ends, or pitting after a fire. These effects are caused by the interaction between the copper and the solder.

8.10.6.4* Mechanical Gouges. Gouges and dents that are formed in a conductor by mechanical means can usually be distinguished from arcing marks by microscopic examination. Mechanical gouges will usually show scratch marks from whatever caused the gouge. Dents will show deformation of the conductors beneath the dents. Dents or gouges will not show the fused surfaces caused by electrical energy.

8.11 Identification of Arc Melting of Electrical Conductors. Melted electrical conductors can be examined to determine if the damage is evidence of electrical arcing or melting by fire.

8.11.1 Melting Caused by Electrical Arcing.

8.11.1.1 Electrical arcing produces very high temperatures and localized heating in the path of the arc, which typically melts electrical conductors at the locations where the arc makes contact with them. Because the arc itself is normally small in area and short in duration, the arc damage is localized, with a sharp line of demarcation between the melted and unmelted portions of the conductor. Magnification may be necessary to detect the demarcation between the melted and unmelted regions on a conductor.

8.11.1.2 The result of arc damage can be notches in the sides of the conductors *[see Figure 8.10.2 and Figure 8.10.3.1(e)]*, or rounded or irregular-shaped beading on the end of a severed conductor *[see Figure 8.10.3.1(a), Figure 8.10.3.1(b), Figure 8.10.3.1(c), Figure 8.10.3.1(d), and Figure 8.10.3.1(f)]*. Arcing often produces sparks that are sprayed from the arc location and that may collect on nearby areas of the conductor.

8.11.2 Melting Caused by Fire. In contrast to melting caused by an arc, when conductors are melted by fire, the damage is spread over a larger area without a distinct line of demarcation between the melted and unmelted regions *(see 8.10.6.2)*. Conductors melted by fire may exhibit irregular or rounded globules, or smooth or rough tapered ends.

8.11.3 Considerations and Cautions.

8.11.3.1 Laboratory experiments, combined with the knowledge of basic chemical, physical, and electrical sciences, indicate that some prior beliefs are incorrect or are correct only under limited circumstances.

8.11.3.2 The investigator is cautioned to review the underlying scientific studies and research to determine what limitations or conditions are associated with experiments and their conclusions. Failure to take into account such underlying limitations can lead the investigator to erroneous conclusions.

8.11.4 Undersized Conductors. Undersized conductors, such as a 14 AWG conductor in a 20 A circuit, are sometimes thought to overheat and cause fires. There is a large safety factor in the allowed ampacities. Although the current in a 14 AWG conductor is supposed to be limited to 15 A, the extra heating from increasing the current to 20 A would not necessarily indicate a fire cause. The higher operating temperature would deteriorate the insulation faster but would not melt it or cause it to fall off and bare the conductor without some additional factors to generate or retain heat. The presence of undersized conductors or overfused protection is not proof of a fire cause. *(See 8.2.8.)*

8.11.5 Nicked or Stretched Conductors. Conductors that are reduced in cross-section by being nicked or gouged are sometimes thought to heat excessively at the nick. Calculations and experiments have shown that the additional heating is negligible under conductors tested at their rated currents. Notched (gouged) conductors that are carrying excessive currents may fail with a parting arc at the notch (gouge). Also, it is sometimes thought that pulling conductors through conduit can stretch them like taffy and reduce the cross-section to a size too small for the ampacity of the protection. Copper conductors do not stretch that much without breaking at the weakest point. Whatever stretching can occur before the range of plastic deformation is exceeded would not cause either a significant reduction in cross-section or excessive resistance heating.

8.11.6 Collecting Evidence.

8.11.6.1 Damage to electrical conductors should be treated as potential evidence. The damaged portion of the relevant conductors should be documented at the fire scene before being disturbed. The documentation should include the location of the damage and whether the wiring was from a branch circuit or from an electrical device.

8.11.6.2 If the damaged conductor is to be cut from its circuit, the cut should not be made in the damaged portion. Instead, it should be cut far enough away from the damaged area to include a section of unmelted or undamaged conductor. The conductors should not be cleaned, because the surface material is evidence that may be needed for future analysis and evaluation. The evidence should be preserved and packaged so as to protect it from mechanical abrasion, accidental fracture of the wires, or other damage. Different pieces of evidence should be packaged separately.

8.11.7 Deteriorated Insulation. When thermoplastic insulation deteriorates with age and heating, it tends to become brittle and will crack if bent. Those cracks do not allow leakage current unless conductive solutions get into the cracks. Rubber insulation does deteriorate more easily than thermoplastic insulation, and loses more mechanical strength. Thus, rubber-insulated lamp or appliance cords that are subject to being moved can become hazardous because of embrittled insulation breaking off. However, simple cracking of rubber insulation, as with thermoplastic insulation, does not allow leakage of current unless conductive solutions get into the cracks.

8.11.8* Overdriven or Misdriven Staple.

8.11.8.1 Staples driven too hard over nonmetallic cable have been thought to cause heating or some kind of faulting. The suppositions range from induced currents because of the staple being too close to the conductors to actually cutting through the insulation and touching the conductors. A properly installed cable staple with a flattened top cannot be driven through the insulation. If the staple is bent over, the edge of it can be driven through the insulation to contact the conductors. In that case, a short circuit or a ground fault would occur. That event should be evident after a fire by bent points of the staple and by melt spots on the staple or on the conductors unless these items were obliterated by the ensuing fire. A short circuit should cause the circuit overcurrent protection to operate and prevent any further damage. There would not be any continued heating at the contact, and the brief parting arc would not ignite the insulation on the conductor or the wood to which it was stapled.

8.11.8.2 If a staple is misdriven so that one leg of the staple penetrates the insulation and contacts both an energized conductor and a grounded conductor, then a short circuit or ground fault will result. If the staple or a nail severs the energized conductor without contacting a grounded conductor, or a staple or nail severs the neutral conductor without contacting the energized conductor, heating connections may be formed at the copper to steel contacts.

8.11.8.3 A misdriven staple or nail can apply a point of pressure against conductor insulation, or slightly pierce it without making contact with the conductor. The insulation on each conductor may creep and give way to a point of pressure. This can, over time, result in a high resistance contact between conductors and/or the staple or nail. This type of high resistance contact can result in a fire months or years after the damage has been inflicted. If the nail or staple that caused the damage can be positively identified by finding evidence of a transfer of metal between the wire and the staple or nail, this may support a hypothesis that a high resistance electrical path has been created. It should be noted that conductor metal transfer to a nail or staple may be the result of an advancing fire. For this source of heat to be an ignition source, easily ignitible fuels must be in close proximity to the staple. Wood studs or other structural members are not considered easily ignitible fuels.

8.11.9 Short Circuit. A short circuit (i.e., low resistance and high current) in wiring on a branch circuit has been thought to ignite insulation on the conductors and to allow fire to propagate. Normally, the quick flash of a parting arc prior to operation of the circuit protection cannot heat insulation enough to generate ignitible fumes, even though the temperature of the core of the arc may be several thousand degrees. If the overcurrent protection is defeated or defective, then a short circuit may become an overload and, as such, may become an ignition source.

8.11.10 Beaded Conductor. A bead on the end of a conductor in and of itself does not indicate the cause of the fire.

8.12 Static Electricity.

8.12.1 Introduction to Static Electricity.

8.12.1.1 Static electricity is the electrical charging of materials through physical contact and separation and the various effects that result from the positive and negative electrical charges formed by this process. Static electricity is accomplished by the transfer of electrons (negatively charged) between bodies, one giving up electrons and becoming positively charged, and the other gaining electrons and becoming oppositely, but equally, negatively charged.

8.12.1.2 Common sources of static electricity include the following:

(1) Pulverized materials passing through chutes or pneumatic conveyors
(2) Steam, air, or gas flowing from any opening in a pipe or hose, when the stream is wet or when the air or gas stream contains particulate matter
(3) Nonconductive power or conveyor belts in motion
(4) Moving vehicles
(5) Nonconductive liquids flowing through pipes or splashing, pouring, or falling
(6) Movement of clothing layers against each other or contact of footwear with floors and floor coverings while walking
(7) Thunderstorms that produce violent air currents and temperature differences that move water, dust, and ice crystals, creating lightning
(8) Motions of all sorts that involve changes in relative position of contacting surfaces, usually of dissimilar liquids or solids

8.12.2 Generation of Static Electricity.

8.12.2.1 General. The generation of static electricity cannot be prevented absolutely, but this generation is of little consequence because the development of electrical charges may not in itself be a potential fire or explosion hazard. For there to be an ignition, there must be a discharge or sudden recombination of the separated positive and negative charges in the form of an electric arc in an ignitible atmosphere. When an electrical charge is present on the surface of a nonconducting body, where it is trapped or prevented from escaping, it is called static electricity. An electric charge on a conducting body that is in contact only with nonconductors is also prevented from escaping and is therefore nonmobile or *static*. In either case, the body is said to be *charged*. The charge may be either positive (+) or negative (–).

8.12.2.2* Ignitible Liquids. Static is generated when liquids move in contact with other materials. This phenomenon commonly occurs in operations such as flowing through pipes, and in mixing, pouring, pumping, spraying, filtering, or agitating. Under certain conditions, particularly with liquid hydrocarbons, static may accumulate in the liquid. If the accumulation of charge is sufficient, a static arc may occur. If the arc occurs in the presence of a flammable vapor–air mixture, an ignition may result.

8.12.2.2.1 Filtering with some types of clay or microfilters substantially increases the ability to generate static charges. Tests and experience indicate that some filters of this type have the ability to generate charges 100 to 200 times higher than achieved without such filters.

8.12.2.2.2 The electrical conductivity of a liquid determines its ability to accumulate and hold a change. The lower the conductivity, the greater the ability of the liquid to create and hold a charge. Common liquids that have low conductivity and therefore represent a hazardous static potential are given in Table 8.12.2.2.2. For comparison, distilled water has a conductivity of 100,000,000 pico-siemen.

Table 8.12.2.2.2 Common Liquids That Have Low Conductivity

Typical Conductivity Product	Conductance per Meter [picosiemen (pS)[a]]
Highly purified hydrocarbons[ab]	0.01
Light distillates[ab]	0.01 to 10
Commercial jet fuel[bc]	0.2 to 50
Kerosene[bc]	1 to 50
Leaded gasoline[bc]	Above 50
Fuel with antistatic additives[bc]	50 to 300
Black oils[ab]	1000 to 100,000

[a]A Picosiemen (pS) is the reciprocal of an ohm. In fact, siemens are a measure of conductance, which is the reciprocal of resistance, measured in ohms. In other words, siemens (not picosiemens) are the reciprocal of ohms.
[b]API RP 2003, *Protection Against Ignitions Arising Out of Static, Lightning, and Stray Currents.*
[c]Bustin and Duket, *Electrostatic Hazards in Petroleum Industry.*

8.12.2.3 Charges on the Surface of a Liquid. If an electrically charged liquid is poured, pumped, or otherwise transferred into a tank or container, the unit charges of similar polarity within the liquid will be repelled from each other toward the outer surfaces of the liquid, including not only the surfaces in contact with the container walls but also the top surface adjacent to the air or vapor space, if any. It is the latter charge, often called the *surface charge,* that is of most concern in many situations. In most cases, the container is of metal, and hence electrically conductive.

8.12.2.3.1 Even if the tank shell is grounded, the time for the charge to dissipate, known as relaxation time, may be as little as a few seconds up to several minutes. This relaxation time is dependent on the conductivity of the liquid and the rate and manner that the liquid is being introduced into the tank, therefore, the rate at which the electrostatic charge is being accumulated.

8.12.2.3.2 If the electrical potential difference between any part of the liquid surface and the metal tank shell should become high enough, the air above the liquid may become ionized and an arc may discharge to the shell. However, an arc to the tank shell is less likely than an arc to some projection or to a conductive object lowered into the tank. These projections or objects are known as spark (i.e., arc) promoters. No bonding or grounding of the tank or container can remove this internal charge.

8.12.2.3.3 If the tank or container is ungrounded, the charge can also be transmitted to the exterior of the tank and can arc to any grounded object brought into proximity to the now-charged tank external surface.

8.12.2.4* Switch Loading. *Switch loading* is a term used to describe a product being loaded into a tank or compartment that previously held a product of different vapor pressure and flash point. Switch loading can result in an ignition when a low vapor pressure/higher flash point product, such as fuel oil, is put into a cargo tank containing a flammable vapor from a previous cargo, such as gasoline. Discharge of the static normally developed during the filling can ignite the vapor–air mixture remaining from the low flash point liquid.

8.12.2.5 Spraying Operations. High-pressure spraying of ignitible liquids, such as in spray painting, can produce significant static electric charges on the surfaces being sprayed and the ungrounded spraying nozzle or gun.

8.12.2.5.1 If the material being sprayed can create an ignitible atmosphere, such as with paints utilizing flammable solvents, a static discharge can ignite the fuel–air mixture.

8.12.2.5.2 In general, high-pressure airless spraying apparatus have a higher possibility for creating dangerous accumulations of static than low-pressure compressed air sprayers.

8.12.2.6 Gases. When flowing gas is contaminated with metallic oxides or scale particles, with dust, or with liquid droplets or spray, static electric accumulations may result. A stream of particle-containing gas directed against a conductive object will charge the object unless the object is grounded or bonded to the liquid discharge pipe. If the accumulation of charge is sufficient, a static arc may occur. If the arc occurs in the presence of an ignitible atmosphere, an ignition may result.

8.12.2.7 Dusts and Fibers. Generation of a static charge can happen during handling and processing of dusts and fibers in industry. Dust dislodged from a surface or created by the pouring or agitation of dust-producing material, such as grain or pulverized material, can result in the accumulation of a static charge on any insulated conductive body with which it comes in contact. The minimum electrical energy required to ignite a dust cloud is typically in the range of 10 to 100 mJ. Thus, many dusts can ignite with less energy than might be expended by a static arc from machinery or the human body.

8.12.2.8 tatic Electric Discharge from the Human Body. Numerous incidents have resulted from static electric discharges from people. The human body can accumulate an electric charge. The potential is greater in dry atmospheres than in humid atmospheres *(see Table 8.12.2.8).* If a person is insulated from ground, that person can accumulate a significant charge by walking on an insulating surface, by touching a charged object, by brushing surfaces while wearing nonconductive clothing, or by momentarily touching a grounded object in the presence of charges in the environment. During normal activity, the potential of the human body can reach 10 kV to 15 kV, and the energy of a possible arc can reach 20 mJ to 30 mJ. By comparing these values to the minimum ignition energies (MIEs) of gases or vapors, the hazard is readily apparent.

Table 8.12.2.8 Electrostatic Voltages Resulting from Triboelectric Charging at Two Levels of Relative Humidity

	Electrostatic Voltages (kV)	
Situation	**RH 10–20%**	**RH 65–90%**
Walking across carpet	35	1.5
Walking over vinyl floor	12	0.25
Working at bench	6	0.1
Vinyl envelopes for work instructions	7	0.6
Poly bag picked up from bench	20	1.2
Work chair padded with polyurethane foam	18	1.5

8.12.2.9 Clothing. Outer garments can build up considerable static charges when layers of clothing are separated, moved away from the body, or removed entirely, particularly when of dissimilar fabrics. For some materials (particularly synthetic polymers) and/or low humidity conditions, an electrostatic charge may be accumulated. The use of synthetic fabrics and the removal of outer garments in ignitible atmospheres can become an ignition source.

8.12.3* Incendive Arc. An arc that has enough energy to ignite an ignitible mixture is said to be incendive. A nonincendive arc does not possess the energy required to cause ignition even if the arc occurs within an ignitible mixture. An ignitible mixture is commonly a gas, vapor of an ignitible liquid, or dust.

8.12.3.1 When the stored energy is high enough, and the gap between two bodies is small enough, the stored energy is released, producing an arc. The energy so stored and released by the arc is related to the capacitance of the charged body and the voltage in accordance with the following formula:

$$E_s = \frac{CV^2}{2}$$

where:
E_s = energy (J)
C = capacitance (F)
V = voltage (V)

8.12.3.2 Static arc energy is typically reported in thousandths of a joule (millijoules, or mJ).

8.12.3.3 Arcs Between Conductors. Arcs from ungrounded charged conductors, including the human body, are responsible for most fires and explosions ignited by static electricity. Arcs are typically intense capacitive discharges that occur in the gap between two charged conducting bodies, usually metal. The energy of an arc discharge is highly concentrated in space and in time.

8.12.3.3.1 The ability of an arc to produce ignition is governed largely by its energy, which will be some fraction of the total energy stored in the system.

8.12.3.3.2 To be capable of causing ignition, the energy released in the discharge must be at least equal to the minimum ignition energy (MIE) of the ignitible mixture. Other factors, such as the shape of the charged electrodes and the form of discharge, influence conditions for the static electric discharge and its likelihood of causing ignition.

8.12.3.4 Discharges Between Conductors and Insulators. Arcs often occur between conductors and insulators. Examples of such occurrences include situations in which plastic parts and structures, insulating films and webs, liquids, and particulate material are handled. The charging of these materials can result in surface discharges and sparks, depending on the accumulated charge and the shape of nearby conductive surfaces. The variable (both in magnitude and polarity) charge density observed on insulating surfaces is the effect of these discharges spreading over a limited part of the insulating surface.

8.12.4* Ignition Energy. The ability of an arc to produce ignition is governed largely by its energy and the minimum ignition energy of the exposed fuel. The energy of the static arc will necessarily be some fraction of its total stored energy. Some of the total stored energy will be expended in heating the electrodes. With flat plane electrodes, the minimum arc voltage to jump a gap (0.01 mm) is 350 V. Increased gap widths require proportionately larger voltages; for example, 1 mm requires approximately 4500 V.

8.12.4.1 Though as little as 350 V are required to arc across a small gap, it has been shown by practical and experimental experience that, because of heat loss to the electrodes, arcs arising from electrical potential differences of at least 1500 V are required to be incendive.

8.12.4.2 Dusts and fibers require a discharge energy of 10 to 100 times greater than gases and vapors for arc ignitions of optimum mixtures with air.

8.12.5 Controlling Accumulations of Static Electricity. A static charge can be removed or can dissipate naturally. A static charge cannot persist except on a body that is electrically insulated from its surroundings unless it is regenerated more rapidly than it is being removed.

8.12.5.1 Humidification. Many commonly encountered materials that are not usually considered to be electrical conductors — such as paper, fabrics, carpet, clothing, and cellulosic and other dusts — contain certain amounts of moisture in equilibrium with the surrounding atmosphere. The electrical conductivity of these materials is increased in proportion to the moisture content of the material, which depends on the relative humidity of the surrounding atmosphere.

8.12.5.1.1 Under conditions of high relative humidity, 50 percent or higher, these materials and the atmosphere will reach equilibrium and contain enough moisture to make the conductivity adequate to prevent significant static electricity accumulations. With low relative humidities of approximately 30 percent or less, these materials dry out and become good insulators, so static accumulations are more likely.

8.12.5.1.2 Materials such as plastic or rubber dusts or machine drive belts, which do not appreciably absorb water vapor, can remain insulating surfaces and accumulate static charges even though the relative humidity approaches 100 percent.

8.12.5.1.3 The conductivity of the air itself is not appreciably increased by humidity.

8.12.5.2 Bonding and Grounding. Bonding is the process of electrically connecting two or more conductive objects. Grounding is the process of electrically connecting one or more conductive objects to ground potential and is a specific form of bonding.

8.12.5.2.1 A conductive object may also be grounded by being bonded to another conductive object that is already at ground potential. Some objects, such as underground metal pipe or large metal tanks resting on the earth, may be inherently grounded by their contact with the earth.

8.12.5.2.2 Bonding minimizes electrical potential differences between objects. Grounding minimizes potential differences between objects and the earth. Examples of these techniques include metal-to-metal contact between fixed objects and pickup brushes between moving objects and earth.

8.12.5.2.3 Investigators should not take the conditions of bonding or grounding for granted just by the appearance or contact of the objects in question. Specific electrical testing should be done to confirm the bonding or grounding conditions. Many factors, such as corrosion and earth settlement or shifting, can greatly affect the original state of the ground path.

8.12.5.2.4 If static arcing is suspected as an ignition source, examination and testing of the bonding, grounding, or other conductive paths should be made by qualified personnel using the criteria in NFPA 77, *Recommended Practice on Static Electricity*.

8.12.6 Conditions Necessary for Static Arc Ignition. In order for a static discharge to be a source of ignition, five conditions must be fulfilled:

(1) There must be an effective means of static charge generation.
(2) There must be a means of accumulating and maintaining a charge of sufficient electrical potential.
(3) There must be a static electric discharge arc of sufficient energy. *(See Section 18.3.)*
(4) There must be a fuel source in the appropriate mixture with a minimum ignition energy less than the energy of the static electric arc. *(See Section 18.4.)*
(5) The static arc and fuel source must occur together in the same place and at the same time.

8.12.7 Investigating Static Electric Ignitions. Often the investigation of possible static electric ignitions depends on the discovery and analysis of circumstantial evidence and the elimination of other ignition sources, rather than on direct physical evidence.

8.12.7.1 In investigating static electricity as a possible ignition source, the investigator should identify whether or not the five conditions necessary for ignition existed.

8.12.7.2 An analysis must be made of the mechanism by which static electricity was generated. This analysis should include the identification of the materials or implements that caused the static accumulation, the extent of their electrical conductivity, and their relative motion, contact, and separation, or means by which electrons are exchanged.

8.12.7.3 The means of accumulating charge to sufficient levels where it can discharge in the form of an incendive arc should be identified. The states of bonding, grounding, and conductance of the material that accumulates the charge or to which the arc discharges should be identified.

8.12.7.4 Local records of meteorological conditions, including relative humidity, should be obtained and the possible influence on static accumulation or dissipation (relaxation) considered.

8.12.7.5 The location of the static electric arc should be determined as exactly as possible. In doing so, there is seldom any direct physical evidence of the actual discharge arc, if it occurred. Occasionally, there are witness accounts that describe the arc taking place at the time of the ignition. However, the investigator should endeavor to verify witness accounts through analysis of physical and circumstantial evidence.

8.12.7.6 The investigator should determine whether the arc discharge could have been of sufficient energy to be a competent ignition source for the initial fuel.

8.12.7.7 The potential voltage and energy of the arc in relation to the size of the arc gap should be calculated to determine whether the incendive arc is feasible.

8.12.7.8 The possibility for the incendive arc and the initial fuel (in the proper configuration and mixture) to exist in the same place at the same time should be established.

8.12.8* Lightning.

8.12.8.1 General. Lightning is another form of static electricity in which the charge builds up on and in clouds and on the earth below. Movement of water droplets, dust, and ice particles in the violent winds and updrafts of a thunderstorm build up a polarized electrostatic charge in the clouds. When sufficient charge builds up, a discharge occurs in the form of a lightning stroke between the charged cloud and objects of different potential. Lightning strokes may occur between clouds or between clouds and the earth. In the latter, charges of opposite polarity are generated in the cloud, while the charge in the ground below the cloud is induced by the cloud charge. In effect, the result is a giant capacitor, and when the charge builds up sufficiently, a discharge occurs.

8.12.8.2 Lightning Bolt Characteristics. Typically lightning bolts have a core of energy plasma 12.7 mm to 19 mm (½ in. to ¾ in.)in diameter, surrounded by a 102 mm (4 in.) thick channel of superheated ionized air. Lightning bolts average 24,000 A but can exceed 200,000 A, and potentials can range up to 15,000,000 V.

8.12.8.3 Lightning Strikes.

8.12.8.3.1 Lightning tends to strike the tallest object on the ground in the path of its discharge. Lightning enters structures in four ways:

(1) By striking a metallic object like a TV antenna, a cupola, or an air-conditioning unit extending up and out from the building roof
(2) By directly striking the structure
(3) By hitting a nearby tree or other tall structure and moving horizontally to the building
(4) By striking nearby overhead conductors and by being conducted into buildings along the normal power lines

8.12.8.3.2 The bolt generally follows a conductive path to ground. At points along its path, the main bolt may divert, for example, from wiring to plumbing, particularly if underground water piping is used as a grounding device for the structure's electrical system.

8.12.8.4 Lightning Damage. Damage by lightning is caused by two characteristic properties: first, the extremely high electrical potentials and energy in a lightning stroke; and second, the extremely high heat energy and temperatures generated by the electrical discharge. (A) through (D) are examples of these effects.

(A) A tree may be shattered by the explosive action of the lightning stroke striking the tree and the heat immediately vaporizing the moisture in the tree into steam, causing explosive effects.

(B) Copper conductors not designed to carry the thousands of amperes of a lightning stroke may be melted, severed, or completely vaporized by the overcurrent effect of a lightning discharge. It is also characteristic for electrical conductors that have experienced significant overcurrents to become severed and disjointed at numerous locations along their length, due to the extremely powerful magnetic fields generated by such overcurrents.

(C) When lightning strikes a steel-reinforced concrete building, the electricity may follow the steel reinforcing rods as the least resistive conductive path. The high energy and high temperature may destroy the surrounding concrete with explosive forces.

(D) Lightning can also cause fires by damaging fuel gas systems. Fuel gas appliance connectors have been known to have their flared ends damaged by electrical currents induced by lightning and other forms of electrical discharge. When gas lines are damaged, fuel gas can leak, and the same arcing that caused the gas line to fail may also cause ignition of the fuel gas.

8.12.8.5 Lightning Detection Networks. Lightning detection networks exist that may assist in establishing time and location (to within 500 meters) of a lightning strike. Historical data is also available, including report of any lightning strikes detected within a specified time prior to a fire.

Chapter 9 Building Fuel Gas Systems

9.1* Introduction. Fuel gas systems are found in or near most dwelling, storage, commercial, or industrial use structures. These systems commonly provide fuel for environmental comfort, water heating, cooking, and manufacturing processes. They can also be fuel sources for fires and explosions in these structures. The fire investigator or analyst should have a basic understanding of fuel gases and the appliances and equipment that utilize them. NFPA 54, *National Fuel Gas Code*, 49 CFR Part 192, "Transportation of Natural and Other Gases by Pipeline: Minimum Safety Standards," and NFPA 58, *Liquefied Petroleum Gas Code*, are generally considered to be the leading standards on this topic.

9.1.1 Impact of Fuel Gases on Fire and Explosion Investigations. Building fuel gas systems can influence the way a building burns in the following four ways: as an initial fuel source, as an initial ignition source, as both fuel and ignition sources, and as factors influencing fire spread. These influences can complicate the investigative process. The investigator should know at least the rudiments of fuel gas systems, how they work, and how they fail.

9.1.1.1 Fuel Sources. Fuel gases that escape from their piping, storage, or utilization systems can serve as easily ignited fuels for fires and explosions. These gases are commonly referred to as *fugitive* gases.

9.1.1.2 Ignition Sources. Ignition temperatures for most fuel gases range from approximately 384°C to 632°C (723°F to 1170°F). Minimum ignition energies are as low as 0.2 mJ. Thus, they are easily ignited from most commonly encountered ignition sources.

9.1.1.2.1 The open flames of fuel gas burners or pilot lights can serve as competent ignition sources for fuel gases and other fuels, particularly flammable gases or the vapors of ignitible liquids and dusts.

9.1.1.2.2 Overheated fuel gas utilization equipment or improperly installed appliances or flue vents can cause the ignition of solid fuels, such as where wooden structural building components or improperly stored combustibles are involved, or where proper clearances are not maintained.

9.1.1.3 Both Fuel Source and Ignition Source. On many occasions, the fuel gas piping and utilization systems, including burners and pilot lights, can serve as both the source of fuel and the ignition source.

9.1.2 Additional Fire Spread.

9.1.2.1 During fire or explosion events, disrupted fuel gas systems can provide additional fuel and can greatly change or increase fire spread rates, or can spread fire to areas of the structure that would not normally be burned. The flames issuing from broken fuel gas lines (often called *flares*) can spread fire and burn through structural components.

9.1.2.2 Pockets of fuel gas that are ignited during the fire event can create evidence of separate fire origins, flash fires, or explosions, causing increased fire spread.

9.2* Fuel Gases. Fuel gases by definition include natural gas, liquefied petroleum gas in the vapor phase only, liquefied petroleum gas–air mixtures, manufactured gases, and mixtures of these gases, plus gas–air mixtures within the flammable range, with the fuel gas or the flammable component of a mixture being a commercially distributed product. The fuel gases most commonly encountered by the fire and explosion investigator will be natural gas and commercial propane. *(See* NFPA 54, *National Fuel Gas Code, and* NFPA 58, *Liquefied Petroleum Gas Code.)*

9.2.1 Natural Gas.

9.2.1.1 Natural gas is a naturally occurring largely hydrocarbon gas product recovered by drilling wells into underground pockets, often in association with crude petroleum. Although exact percentages differ with geographic areas and there are no standards that specify its composition, natural gas is mostly methane, with lesser amounts of nitrogen, ethane, propane, and with traces of butane, pentane, hexane, carbon dioxide, and oxygen. The percentages may vary widely and have been reported in mixtures that range from 72 percent to 95 percent methane, 3 percent to 13 percent ethane, <1 percent to 4 percent propane, and <1 percent to 18 percent nitrogen.

9.2.1.2 Undiluted natural gas is lighter than air. Depending on the exact composition, it has a specific gravity (air) (vapor density) of 0.59 to 0.72, a lower explosive limit (LEL) of 3.9 percent to 4.5 percent, and an upper explosive limit (UEL) of 14.5 percent to 15 percent. A flammable mixture has a vapor density 0.96 to 0.98. Its ignition temperature is 482°C to 632°C (900°F to 1170°F).

9.2.2 Commercial Propane. Propane is derived from the refining of petroleum. Liquefied petroleum gases can be liquefied under moderate pressure at normal temperatures. This ability to condense the LP-Gases makes them more convenient to store and ship than natural gas and thus makes propane particularly suitable for rural and relatively inaccessible areas or for use with portable equipment and appliances. In populated areas where natural gas is unavailable, propane is sometimes premixed with air and piped to consumers at relatively low pressures through central underground distribution systems similar to that of natural gas.

9.2.2.1 Commercial propane is a minimum of 95 percent propane and propylene and a maximum of 5 percent other gases. The average content of propylene in commercial propane is 5 percent to 10 percent.

9.2.2.2 Undiluted propane gas is heavier than air. It has a specific gravity (air) (vapor density) of approximately 1.5 to 2.0, a lower explosive limit (LEL) of 2.15 percent, and an upper explosive limit (UEL) of 9.6 percent. A flammable propane–air mixture has a vapor density of 1.01 to 1.10. Its ignition temperature is 493°C to 604°C (920°F to 1120°F).

9.2.3 Other Fuel Gases. Other fuel gases that may be encountered by the investigator, particularly in commercial, industrial, or nondwelling settings, include commercial butane, propane HD5, and manufactured gases.

9.2.3.1 Commercial Butane. Commercial butane is a minimum of 95 percent butane and butylene and a maximum of 5 percent other gases, with the butylene component usually kept below 5 percent.

9.2.3.2 Propane HD5. Propane HD5 is a special grade of propane for motor fuel and other uses requiring more restrictive specifications than regular commercial propane. It is 95 percent propane and a maximum of 5 percent other gases.

9.2.3.3 Manufactured Gases. Manufactured gases are combustible gases produced from coal, coke, or oil; chemical processes; or by reforming of natural gas or liquefied petroleum gases or mixtures of such gases. They are most commonly used in industrial applications. The most common of the manufactured gases are acetylene, coke oven gas, and hydrogen.

9.2.4 Odorization. LP-Gas and commercial natural gas may not have a readily identifiable odor in their natural state. To increase the detectability of natural gas, an odorant blend containing *t*-butyl mercaptan, thiophane, or other mercaptans is usually added. To increase the detectability of LP-Gas, ethyl mercaptan is usually added. These odorants are required by law and fire code; 49 CFR 192.625 states, "A combustible gas in a distribution line must contain a natural odorant or be odorized so that at concentration in air of one-fifth of the lower explosive limit, the gas is readily detectable by a person with a normal sense of smell." Subsection 4.2.1 of NFPA 58, *Liquefied Petroleum Gas Code*, states, "All LP-Gases shall be odorized prior to delivery to a bulk plant by the addition of a warning agent of such character that the gases are detectable by a distinct odor, to a concentration in air of not over one-fifth the lower limit of flammability." The odorant for natural gas is added by the local distribution company prior to the introduction of the gas into the distribution system. Natural gas in long-distance transmission pipelines is usually not odorized. The odorant in LP-Gas is added by the gas supplier prior to delivery to an LP-Gas distributor's bulk plant.

9.2.4.1 Odorant verification should be a part of any explosion investigation involving or potentially involving fuel gas if it appears that there were no indications of a leaking gas being detected by people present. The odorant's presence in the proper amount should be verified. Specialized chemical detectors called *stain tubes* can be used in the field, and gas chromatography can be used, on other than natural gas samples, as a lab test for more accurate results.

9.2.4.2 The utilization of stain tubes requires that the identity of the odorant in the gas be known, as there is no universal stain tube for all odorants. ASTM D 5305, *Standard Test Method for Determination of Ethyl Mercaptan in LP-Gas Vapor*, is a standard for propane odorant determination by stain tubes. There is no similar standard for natural gas odorants. A sample of the gas (or liquid for LP-Gas) requires that the sample be properly taken. ASTM D 1265, *Standard Practice for Sampling Liquefied Petroleum (LP) Gases (Manual Method)*, covers the proper method for sampling LP-Gas. The utilization of Tedlar® bags is suggested for natural gas samples that are to be used for odorant verification. Gas samples taken from a propane tank give only a fraction of the information that can be obtained from a liquid sample. Not all laboratories can analyze gases or liquids for odorant content. The ability of the lab to analyze this should be verified prior to sending the sample, as these samples should be analyzed as rapidly as possible.

9.2.4.3 Laboratory testing of gas samples is not generally adequate to determine the effective level of odorant in a natural gas sample. Natural gas should be tested for a sufficient level of odorant in the field using sensory odorant detection equipment capable of determining the percentage of gas in air at which the odor becomes readily detectable. Some individuals cannot detect these odorants for various reasons, and under certain conditions the odorant's effectiveness can be reduced to a point that it cannot be detected. Therefore, test results should always be corroborated by a minimum of two people.

9.3 Natural Gas Systems. A difference between natural gas systems and propane systems is that natural gas is typically piped directly to the consumers' buildings from centralized production and storage facilities. The piping systems that deliver natural gas to the customer are quite complex, with many intervening procedures and pressure changes from collection to ultimate use.

9.3.1* Transmission Pipelines. Pipelines used to convey natural gas from storage or production facilities to local utilities are called *transmission pipelines*. In long-distance transmission pipelines, natural gas companies use pressures up to 8275 kPa (1200 psi).

9.3.2 Main Pipelines (Mains). Pipelines used to distribute natural gas in centralized grid systems for use by residential and business customers are called *main pipelines* or *mains*. Normal operating pressures in main pipelines vary widely among gas utility companies in different geographical areas. Pressures in main pipelines seldom exceed 1035 kPa (150 psi) in high-pressure systems and are typically 414 kPa or less (60 psi or less). Rural main systems, which must deliver gas to more distant customers, are necessarily at higher pressures than urban systems. A main is a type of distribution line.

9.3.3 Service Lines. Natural gas service lines, sometimes called "service laterals," are piping systems connecting the gas company's mains to the individual customer. They typically terminate at the regulator and utility meter. The minimum and maximum pressures delivered to customers' services after final pressure regulation are generally in the range of 1.0 kPa to 2.5 kPa (4 in. w.c. to 10 in. w.c.). A service line is a type of distribution line.

9.3.4 Metering.

9.3.4.1 A gas meter is an instrument installed on a gas system to measure the volume of gas delivered through it.

9.3.4.2 49 CFR 192.353 and NFPA 54, *National Fuel Gas Code*, require that gas meters be installed at least 0.9 m (3 ft) from sources of ignition and be protected from physical damage, extremes of temperature, overpressure, back pressure, or vacuum.

9.4 LP-Gas Systems. One difference between LP-Gas systems and natural gas systems is the storage and delivery of the fuel gases to the user's service piping. Typically, propane is delivered to the service customer's system in a compressed (liquid) state. It is delivered to the consumer by tank truck, with liquid transfer to the consumer's tank. In some isolated areas, where natural gas service is not available, underground propane or propane–air transmission and distribution piping systems similar to those discussed in Section 9.3 are used, though at generally lower pressures. Propane is the most commonly used LP-Gas, but butane and other LP-Gases or blends are used in some warm climates.

9.4.1 LP-Gas Storage Containers. LP-Gas storage containers may be cylinders, tanks, portable tanks, or cargo tanks. Specific definitions for these can be found in the various regulatory standards and guides. Generally, "cylinders" refers to containers of 454 kg (1000 lb) water capacity or less and are governed by Department of Transportation (DOT) regulations. "Tanks" are usually larger and governed by the American Society of Mechanical Engineers (ASME) *Boiler and Pressure Vessel Code*. In storage, LP-Gas is kept under pressure in both liquid and gaseous states. Generally, the maximum permissible amount of LP-Gas is 42 percent of the container's water capacity, by weight, or 80 percent of its volumetric capacity. Except for engine fuel or vaporizer applications, the LP-Gas is normally drawn from the vapor space of the storage container.

9.4.1.1* Tanks.

9.4.1.1.1 Residential and small commercial systems generally store LP-Gas in aboveground ASME stationary tanks of up to 3.786 m^3 (1000 gal) water capacity. These tanks are typically designed with a maximum working pressure of 1379 kPa to 1724 kPa (200 psi to 250 psi) and may not be transported with more than 5 percent liquid capacity. In some applications, tanks are located underground to minimize temperature changes.

9.4.1.1.2 ASME tanks for engine fuel applications are permanently mounted and designed to a maximum working pressure of 1724 kPa to 2150 kPa (250 psi to 312 psi). These tanks may be stored inside the vehicle, (e.g., trunk or other compartment) or outside, (e.g., saddle tanks). Forklifts with permanently mounted tanks typically have the tank exposed and mounted behind the operator.

9.4.1.1.3 Cargo tanks are those containers permanently mounted on a chassis, and are used for transporting LP-Gas. Cargo tanks are subject to DOT and ASME regulations and are typically designed with maximum working pressure of 1724 kPa (250 psi).

9.4.1.1.4 Portable tanks are used for transporting LP-Gas, are not mounted permanently on a chassis, and are in quantities over 454 kg (1000 lb) water capacity. These tanks are designed to ASME regulations and designed with a maximum working pressure of 1724 kPa (250 psi).

9.4.1.2* Cylinders. Cylinders must conform to DOT requirements 49 CFR 173 and 49 CFR 178. They are commonly used for rural homes and businesses, mobile homes, forklifts, recreational vehicles, and small appliances. Cylinders may be transported with their maximum LP-Gas capacity. They may be refillable, (e.g., DOT Specification 4BA or 4BW), or non-refillable, (e.g., DOT Specification 39, 2P, or 2Q). Nonrefillable cylinders are typically 1 lb LP-Gas capacity or less.

9.4.2 Container Appurtenances. Container appurtenances are items connected to container openings. These items include, but are not limited to, pressure relief devices, connections for flow control, liquid level gauges, pressure gauges, and plugs.

9.4.2.1 Pressure Relief Devices. Pressure relief devices are pressure or temperature activated to prevent the pressure from rising above a predetermined maximum and therefore prevent rupture of a normally charged container. LP-Gas containers are equipped with one or more relief devices that, except for certain DOT regulations, are designed to relieve vapor pressure. This venting of vapor results in the cooling of the remaining liquid and the associated reduction of pressure.

9.4.2.1.1 Fusible plug devices are thermally activated to open and vent the contents of the container. The activation temperature is 98°C (208°F) minimum to 104°C (220°F) maximum. They do not protect the container from overpressure due to improper filling. Once open, they do not reclose. Fusible plugs are not to be used on ASME containers of 544 kg (1200 lb) water capacity or greater.

9.4.2.1.2 Relief valves are activated by pressure. They maintain the pressure in the container as determined by the set pressure of the valve and thus do not protect against rupture of the container when the application of heat weakens the container to the point where its rupture pressure is less than the operating pressure of the valve. The set pressure for ASME tank applications is the design pressure for that tank, typically 1724 kPa or 2150 kPa (250 psi or 312 psi). The set pressure for refillable cylinders is 75 percent to 100 percent of the test pressure for that cylinder, typically 2481 kPa to 3308 kPa (360 psi to 480 psi).

9.4.2.2 Connections for Flow Control. Shutoff valves, excess-flow check valves, backflow valves, and quick-closing internal valves used individually or in combinations are utilized at container filling, withdrawal, and equalizing connections.

9.4.2.3 Liquid Level Gauging Devices. Gauges indicate the level of liquid propane within a container. Gauge types include fixed, such as fixed maximum level, and variable, such as float or magnetic, rotary, and slip tube.

9.4.2.3.1 Fixed level gauges (i.e., dip tubes) are primarily used to indicate when the filling of a tank or cylinder has reached its maximum allowable fill volume. They do not indicate liquid levels above or below their fixed lengths.

9.4.2.3.2 Variable gauges give readings of the liquid contents of containers, primarily tanks or large cylinders. They give readings at virtually any level of liquid volume.

9.4.2.4 Pressure Gauges. Pressure gauges, which are attached directly to a container opening or to a valve or fitting that is attached directly to a container opening, read the vapor space pressure of the container. Pressure gauges do not indicate the level of liquid within a container. Pressure gauges are also used in various areas of system piping, if needed.

9.4.3 Pressure Regulation.

9.4.3.1 Pressure in propane storage tanks and cylinders is the vapor pressure of the propane and is dependent on the temperature of the liquid propane. The vapor pressure gauge of propane ranges from 193 kPa (28 psi) at −18°C (0°F), to 876 kPa (127 psi) at 21°C (70°F), to 1972 kPa (286 psi) at 54°C (130°F).

9.4.3.2 For use with utilization equipment, the pressure is typically reduced in one or two stages by regulators to a working pressure of 2.74 kPa to 3.47 kPa (11 in. w.c. to 14 in. w.c.) before entering the service piping system.

9.4.4 Vaporizers. Where larger quantities of propane are required, such as for industrial applications, or where cold weather will hamper vaporization, specifically designed heaters called vaporizers are used to heat and vaporize the propane.

9.5 Common Fuel Gas System Components. Fuel gas delivery system components are common or similar for the various fuel gases. The following sections describe in general these commonly shared components.

9.5.1 Pressure Regulation (Reduction).

9.5.1.1 General. Pressure regulators are devices placed in a gas line system for reducing, controlling, and maintaining the pressure in that portion of the piping system downstream of the device. Regulators can be used singly or in combination to reduce the gas line pressures in stages.

9.5.1.1.1 The most common regulators in natural gas or propane consumer service are the diaphragm, or lever, type. In a diaphragm regulator, the flow of the inlet gas at high pressure is controlled by a shutoff disc, or seal, and gas at a specific lower pressure is discharged through the regulator outlet. The diaphragm is made of a rubberlike material, and its movement is controlled by adjustable spring pressure. Movement of the diaphragm controls the opening of the regulator inlet valve and its integral seal.

9.5.1.1.2 The proper operation of the regulator vent is important to the proper operation of diaphragm regulators. The vent equalizes the pressure above the diaphragm with atmospheric pressure and allows the diaphragm to move. If the vent becomes clogged or blocked by ice or debris, for example, the regulator may not operate properly or the rubber-like diaphragm material may be damaged, preventing proper operation.

9.5.1.1.3 In flood-prone areas, pressure regulators should be installed above the expected flood line, or the vent should be piped above the flood line. Flood waters filled with flood debris, such as mud, sticks, and trash, can easily clog or block the regulator vent. These conditions can result in an overpressure condition at downstream piping and at the gas appliance.

9.5.1.2 Normal Working Pressures. Normal working pressures in most structures and appliances are measured in inches of water column (w.c.), measured on a water-filled manometer. One pound per square inch gauge (psi) is equal to a 27.67 in. w.c. Normal inlet pressure for most nonindustrial natural gas appliances is 4 in. w.c. to 10 in. w.c. (1.0 kPa to 2.5 kPa). Normal inlet pressure for most nonindustrial propane appliances is 11 in. w.c. to 14 in. w.c. (2.74 kPa to 3.47 kPa).

9.5.1.3 Excess Pressures. Pressures significantly in excess of those for which appliances, equipment, devices, or piping systems are designed can cause gas leakage, damage to equipment, malfunction of equipment burners, or abnormally large flames.

9.5.2 Service Piping Systems.

9.5.2.1 Materials for Mains and Services. Fuel gas piping may properly be made of wrought iron, copper, brass, aluminum alloy, or plastic, as long as the material is used with gases that are not corrosive to them. Unapproved tubing or piping materials utilized in "not-to-code," homemade applications may lead to leaks and the release of fugitive gas.

9.5.2.2 Underground Piping. Improper installation of underground piping systems and use of unapproved materials may be a cause of gas leaks. Underground piping must be buried to a sufficient depth and in appropriate locations to be protected from physical damage. The pipe must be protected against corrosion. Underground piping under buildings may be acceptable, if unavoidable, but must be protected and encased in approved conduit designed to withstand the superimposed loads of the structure and contain any gas leakage.

9.5.3 Valves. Valves are devices used to control the gas flow to any section of a system or to an appliance. Examples of valve types include the following:

(1) *Automatic valves:* Devices consisting of a valve and operating mechanism that controls the gas supply to a burner during operation of an appliance. The operating mechanism may be activated by gas pressure, electrical means, or mechanical means.
(2) *Automatic gas shutoff valve:* A valve used in connection with an automatic gas shutoff device to shut off the gas supply to a fuel gas burning appliance.
(3) *Individual main burner valve:* A valve that controls the gas supply to an individual main burner.
(4) *Main burner control valve:* A valve that controls the gas supply to a main burner manifold.
(5) *Manual reset valve:* An automatic shutoff valve installed in the gas supply piping and set to shut off when unsafe conditions occur. The device remains closed until manually reset.
(6) *Relief valves:* A safety valve designed to prevent the rupture of a pressure vessel by relieving excess pressure.
(7) *Service shutoff valve:* A valve (usually installed by the utility gas supplier between the service meter or source of supply and the customer piping system) used to shut off gas to the entire piping system.
(8) *Shutoff valve:* A valve (located in the piping system and readily accessible and operable by the consumer) used to shut off individual appliances or equipment.

9.5.4 Gas Burners. Problems with fuel gas systems, including fires, are often caused by use of the inappropriate orifices or burners for natural gas or propane. Gas burners are devices for the final conveyance of the fuel gas, or a mixture of gas and air, to be burned. Although the several types of burners in common use are essentially the same in general design for both natural gas or propane, they may not be interchangeable from one gas usage to another. Physical differences between natural gas and propane require different-sized burner orifices.

9.5.4.1 Manual Ignition. Some gas appliances and equipment are designed to require the manual ignition of their burners when utilization of the appliances and equipment is desired. Some equipment furnished with a standing pilot light requires that the pilot light be ignited manually.

9.5.4.2 Pilot Lights. Automatic ignition of main burners on appliances is frequently accomplished by a pilot burner flame. For automatic operation by a thermostat, as on automatic water heaters and central heating appliances, the gas pilot flame must be burning and of sufficient size; otherwise, the valve controlling main burner gas flow will not open. In some designs, pilot lights themselves may be ignited automatically by electric arcs activated when gas is called for at the burner.

9.5.4.3 Pilotless Igniters. This type of ignition system consists of an electric ignition means, an electric arc, or a resistance heating element such as a glow plug or a glow bar that directly ignites the burner flame when gas flow begins. On failure to ignite, many, but not all, systems are designed to "lock out" the flow of gas to the burner.

9.6 Common Piping in Buildings. There are several considerations or requirements for the installation and use of fuel gas piping systems in buildings that are common no matter which fuel gases are used.

9.6.1 Size of Piping. The size of piping used is determined by the maximum flow requirements of the various appliances and equipment that it services.

9.6.2 Piping Materials. Piping may be made of wrought iron, copper, brass, aluminum alloy, or plastic, as long as the material is used with gases that are not corrosive to them. Flexible tubing may be of seamless copper, aluminum alloy, or steel. Aluminum alloy tubing is generally considered to be not suitable for underground or exterior use. Plastic pipe, tubing, and fittings are to be used for outside underground installations only.

9.6.3 Joints and Fittings. Piping joints may be screwed, flanged, or welded, and nonferrous pipes may be soldered or brazed. Tubing joints may be flared, soldered, or brazed. Special fittings such as compression fittings may be used under special circumstances. Outdoors, plastic piping joints and fittings may be made, with the appropriate adhesive method or by means of compression fittings, compatible with the piping materials. They are not to be threaded.

9.6.4 Piping Installation. When installed in buildings, gas piping should not weaken the building structure, should be supported with suitable devices (not other pipes), and should be protected from freezing. Drip pipes (drip legs) must be provided in any areas where condensation or debris may collect. Each gas outlet, including valve or cock outlets, should be securely capped whenever no appliance is connected.

9.6.5 Main Shutoff Valves. Accessible shutoff valves must be placed upstream of each service regulator in order to provide for a total shutdown of an entire piping system.

9.6.6 Prohibited Locations. NFPA 54, *National Fuel Gas Code*, prohibits the running of gas piping in or through circulating air ducts, clothes chutes, chimneys or gas vents, ventilating ducts, dumbwaiters, or elevator shafts. Fugitive gas in these prohibited areas is particularly dangerous because of the chances for widespread distribution of the leaking gas and the increased possibilities for accidental ignition.

9.6.7 Electrical Bonding and Grounding. Every aboveground portion of the piping system must be electrically bonded and the system grounded.

9.7 Common Appliance and Equipment Requirements. There are several considerations or requirements for the installation and use of fuel gas appliances that are common, no matter which fuel gases are used in the appliances.

9.7.1 Installation. The basic requirements for the installation of domestic, commercial, and industrial fuel gas appliances and equipment supplied at gas pressures of 3.47 kPa [0.5 psi (14 in. w.c.)] or less are similar.

9.7.1.1 Approved Appliances, Accessories, and Equipment. Subsection 5.1.1 of NFPA 54, *National Fuel Gas Code*, requires that "gas appliances, accessories, and gas utilization equipment shall be approved . . . acceptable to the authority having jurisdiction."

9.7.1.2 Type of Gas. Fuel gas utilization equipment must be used with the specific type of gas for which it was designed. A particular appliance cannot be used interchangeably with natural gas and propane without appropriate alteration.

9.7.1.3* Areas of Flammable Vapors. Gas appliances are not to be installed in residential garage locations where flammable vapors are likely to be present, unless the design, operation, and installation are such as to eliminate the probability of ignition of such vapors. For example, NFPA 54, *National Fuel Gas Code*, requires gas utilization equipment installed in garages to be installed with the burners and ignition devices not less than 0.5 m (18 in.) above the floor. Although not directly prohibited by the codes, installations below 0.5 m (18 in.) in other areas of buildings and dwellings have been found to be responsible for fires and injuries.

9.7.1.4 Gas Appliance Pressure Regulators. When the building gas supply pressure is higher than that at which the gas utilization equipment is designed to operate or varies beyond the design pressure limits of the equipment, a gas appliance pressure regulator is installed in the appliance.

9.7.1.5 Accessibility for Service. All gas utilization equipment should be located so as to be accessible for maintenance, service, and emergency shutoff.

9.7.1.6 Clearance to Combustible Materials. Gas utilization appliances and their vents should be installed with sufficient clearance from combustible materials so that their operation will not create a fire hazard.

9.7.1.7 Electrical Connections. All electrical components of gas utilization equipment should be electrically safe and should comply with *NFPA 70, National Electrical Code.*

9.7.2 Venting and Air Supply.

9.7.2.1 Venting is the removal of combustion products as well as process fumes (e.g., flue gases) to outer air. Although most fuel gases are clean-burning fuels, the products of combustion must not be allowed to accumulate in dangerous concentrations inside a building. Therefore, venting to the exterior is required for most appliances. Examples of appliances that require venting include furnaces and water heaters. Some appliances, such as ranges, ovens, and small space heaters, are allowed to be vented directly into interior spaces. A properly installed venting system should convey all the products of combustion gases to the outside; should prevent damage from condensation to the gas equipment, vent, building, and furnishings; and should prevent overheating of nearby walls and framing.

9.7.2.2 Gas utilization equipment requires an air supply for combustion and ventilation. Restriction of this air supply, particularly when equipment is installed in confined spaces, can result in overheating, fires, or asphyxiation.

9.7.3 Appliance Controls. General categories of appliance controls are the same for nearly all fuel gas appliances. Failure in these controls can lead to the overheating of appliances or the uncontrolled release of gas or flame. These common controls include the following:

(1) Temperature controls
(2) Ignition and shutoff devices
(3) Gas appliance pressure regulators
(4) Gas flow control accessories

9.8* Common Fuel Gas Utilization Equipment. All fuel gas systems ultimately employ the gases by burning them. The utilization of fuel gases in structures falls into seven main areas of domestic, commercial, or industrial use, each of which involves the combustion of the fuel gas. Common fuel gas utilization equipment is described in 9.8.1 through 9.8.7.

9.8.1 Air Heating. Forced air furnaces, space heaters, floor furnaces, wall heaters, radiant heaters, duct furnaces, or boilers are used for heating of environmental air. Direct or indirect burners are used for industrial ovens and dryers, and for heating of processes and materials, such as clothing and fabric dryers, both domestic and industrial.

9.8.2 Water Heating. Direct flame burners are used for the heating of potable or industrial process water.

9.8.3 Cooking. Ranges, stove-top burners, broilers, and cooking ovens, domestic and industrial, are used for cooking.

9.8.4 Refrigeration and Cooling. Fuel gases are often used as the energy sources for absorption system refrigeration and cooling systems.

9.8.5 Engines. Fuel gases are commonly used as fuel sources for stationary and motor vehicle engines and for auxiliary power units on service vehicles, such as to power pumps on tank trucks. Stationary engines fueled by fuel gases are commonly used as auxiliary or emergency motors for electrical power generation or fire pumps.

9.8.6 Illumination. Though commonly used for illumination in the early part of the 20th century, most fuel gas illumination systems have been replaced by electricity. The most common exception is gas lamps used for outdoor lighting. Principal residential lighting applications include yards, patios, driveways, porches, play areas, and swimming pools. Commercial applications include streets, shopping centers, airfields, hotels, and restaurants. These illumination uses of fuel gases include ornamental gas flames for memorials or decorative effects. These systems often involve underground fuel lines, which may not be buried deep enough or protected sufficiently from external damage.

9.8.7 Incinerators, Toilets, and Exhaust Afterburners. Gas-fired domestic, commercial, industrial, and flue-fed incinerators, toilets, and exhaust afterburners are used to burn rubbish, refuse, garbage, animal solids, and organic waste, as well as gaseous, liquid, or semisolid waste from industrial processes.

9.9 Investigating Fuel Gas Systems Incidents. Once it has been determined that a fuel gas system has influenced the way a building has burned, either as a fuel source, as an ignition source, as both a fuel and ignition source, or by providing additional fire spread, the system should be analyzed. This analysis should provide information as to the manner of and extent to which the fuel gas system may have been involved in the origin or cause of the fire or explosion.

9.9.1 The investigation of building fuel gas incidents can be an extremely complicated, technical, scientific, and potentially dangerous task requiring specialized knowledge, training and experience. Investigators faced with the requirement to investigate a fuel gas incident scene that exceeds the resources available or is beyond their knowledge or expertise should secure the scene to preserve evidence and endeavor to obtain technical expertise and adequate resources to accomplish the scene investigation in a safe and correct manner.

9.9.2 Fuel Gas System Analysis. Such an analysis should be a detailed examination of the fuel gas system. Each component of the system should be evaluated to determine whether, and to what extent, it operated or failed, and to what extent it contributed to the fire or explosion.

9.9.2.1 Necessary Measurements and Diagrams. Measurements and diagrams necessary to an adequate analysis of a fuel gas system include details of the structure involved; the fuel gas delivery piping and equipment, including piping materials, lengths, and sizes; as well as valves, connectors, and fittings. These measurements and diagrams should include all of the piping system and components from the utilization equipment and appliances back to and including the fuel gas source (i.e., tank, cylinder, or main). Notations should be made of the various pressures and flow rates, obvious breaks in piping, and the positions and settings (open, closed) of valves and controls.

9.9.2.1.1 Diagrams can be made to an appropriate approximate scale or may be schematic or isometric in nature. *(See Figure 9.9.2.1.1(a) and Figure 9.9.2.1.1(b).)*

9.9.3 Compliance with Codes and Standards.

9.9.3.1 The NFPA *National Fire Codes*, as well as many other fire codes and gas industry standards, specify a wide variety of safety rules for the installation, maintenance, servicing, and filling of fuel gas systems. Failure to comply with one or more of these standards can cause or contribute to a fuel gas fire or explosion incident.

9.9.3.2 The design, manufacture, construction, installation, and use of the various components of the fuel gas system should be evaluated for compliance with the appropriate codes and standards. Any relationship between a violation of the accepted codes and standards and the fire or explosion should be noted.

9.9.4 Leakage. Leakage from piping and equipment is the main cause of gas-fueled fires and explosions. Commonly, leaks occur at pipe junctions, at unlit pilot lights or burners, at uncapped pipes and outlets, at areas of corrosion in pipes, or from physical damage to the gas lines.

9.9.4.1 Pipe Junctions. Improper connections between piping elements, such as inadequate threading (not enough turns for gas tightness), improper threading (cross-threading or right-hand threads merged to left-hand threads), or improper use of pipe joint compound (too much or too little) can cause gas leaks. Pipe junctions are also the most common locations that leak as a result of physical damage to fuel gas piping systems. *(See 9.9.4.8.)*

9.9.4.2 Pilot Lights. Modern pilot light systems are designed to prevent gas flow to appliance burners if the pilot lights are not burning by the use of a thermocouple to sense the pilot flame. Such pilots could remain open if their automatic shutoff mechanisms failed to close off the flow of gas. The escape of gas from unlit pilots is not large enough to produce gas volumes sufficient to fuel significant fires or explosions, except in spaces that have little or no ventilation. Many modern appliances have pilot light systems that will not allow a flow of gas unless the pilots are lit, or else they have electronic ignition systems not requiring pilot lights at all.

9.9.4.3 Unlit Burners. In some gas appliance systems, a burner may emit gas even if the pilot light is not lit. This emission of gas generally will produce enough gas to fuel an explosion or fire even in well-ventilated rooms or structures.

9.9.4.4 Uncapped Pipes and Outlets. A common source of large quantities of fugitive gas is open, uncapped pipes and outlets. Such situations occur when gas appliances are removed and their attendant outlets or piping are not capped as is required by NFPA 54, *National Fuel Gas Code.* Unsuspecting persons then turn on the gas from remote valves, which causes high-volume leaks.

9.9.4.5 Malfunctioning Appliances and Controls. The malfunction and leaking of gas appliances or gas utilization controls, such as valves, regulators, and meters, can also produce fugitive gas. Often fittings and piping junctions within appliances can be sources of leakage. Shutoff valves may leak fugitive gas through the packing materials that are designed to seal the valve bodies from the activation levers. Valves may allow gas to pass through them when they should be closed, due to dirt or debris in their operating mechanisms or due to physical damage or binding of the mechanisms.

9.9.4.6 Regulators. Failures in gas regulators most often fall into one of three categories: faults with the internal diaphragm, faults with the rubberlike seal that controls the input of gas into the regulator, or faults with vents. Each of these fault categories can result in the regulator's failing to reduce the outlet pressure to acceptable levels or producing fugitive gases.

9.9.4.7 Corrosion. Metal pipes are subject to corrosion. Corrosion has been reported to be the cause of as many as 30 percent of all known gas leaks. Corrosion can be caused by oxidation of ferrous metal pipes (rust); electrolysis between dissimilar metals, metal and water, metal and soil, or stray currents; or even microbiological organisms. Corrosion can take place either above or below ground level and can subsequently release fugitive gas.

9.9.4.7.1 Because the size of corrosion leaks is cumulative as the corrosion continues, it may take long periods of time for the corrosion leaks to develop to sufficient size to produce enough fugitive gas to overcome the dissipating effects of ventilation or dispersion through the soil into the air.

9.9.4.7.2 Stress corrosion cracking of flexible brass appliance connectors has been shown to be a factor in many residential fires and explosions.

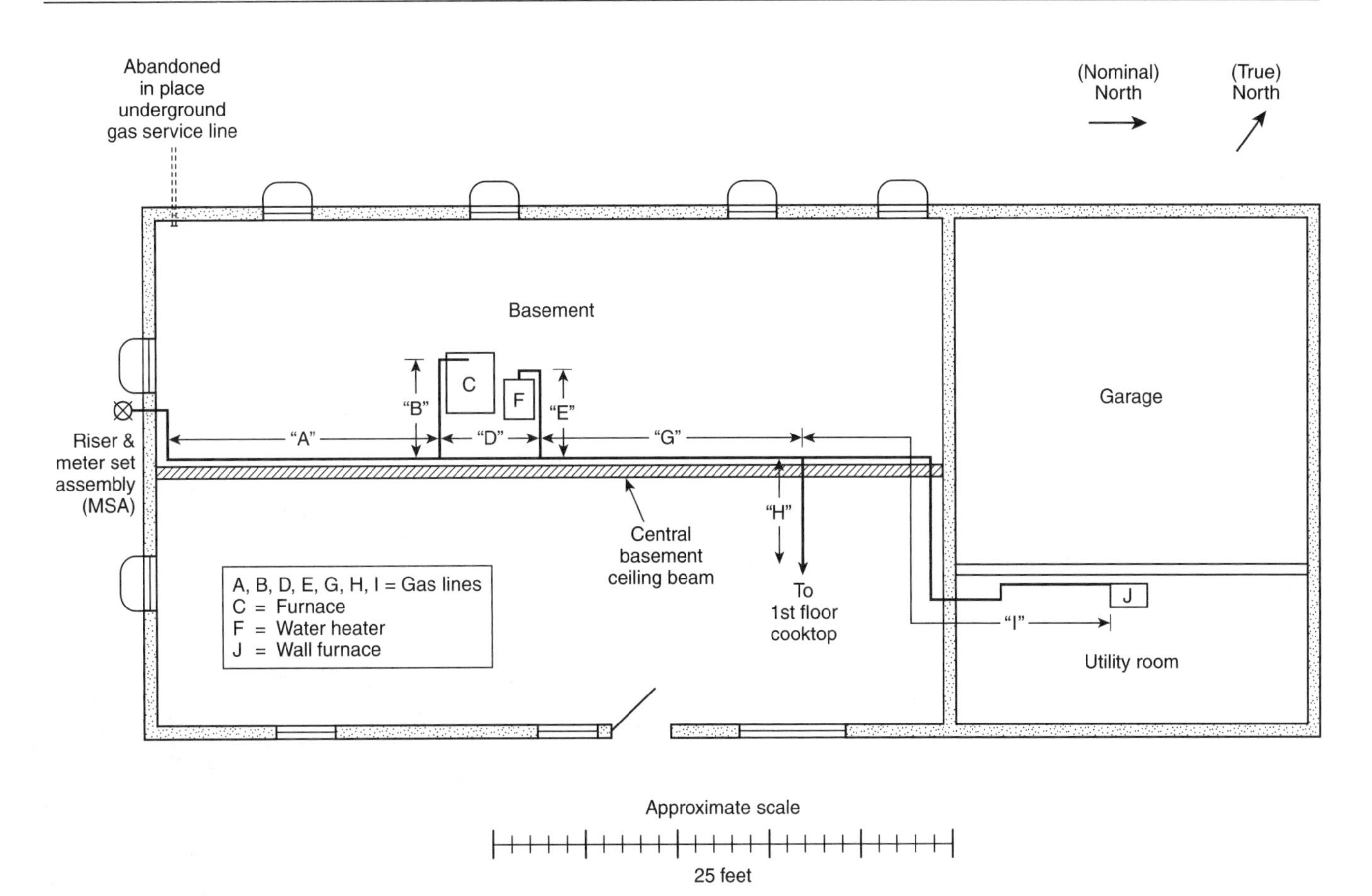

FIGURE 9.9.2.1.1(a) Example of a Scaled Fuel Gas Piping Diagram.

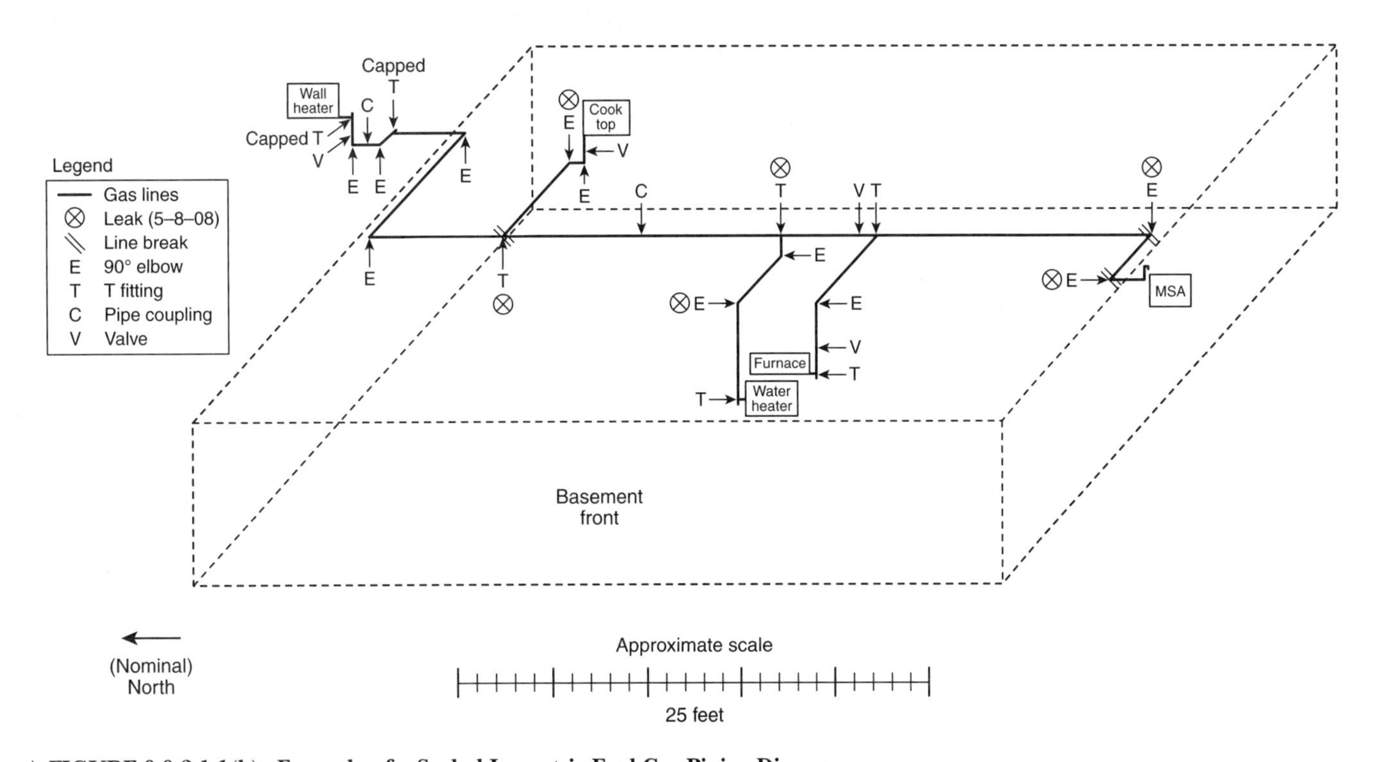

FIGURE 9.9.2.1.1(b) Example of a Scaled Isometric Fuel Gas Piping Diagram.

9.9.4.8* Physical Damage. Physical damage to fuel gas systems can cause leaks. Strain put on gas piping systems may manifest itself at the pipe junctions and unions. Because pipe elbows, T fittings, and couplings are more rigid and stronger than the pipes they connect, and because the threaded ends are weaker than the rest of the pipes, stress damage usually occurs in the threaded portions of the pipes immediately adjacent to the pipe fittings.

9.9.4.8.1 The leaks created by such strain may develop at junctions far removed from the actual point of physical contact. For example, if an automobile strikes a gas meter assembly, the strain on the underground piping of the system may cause a leak at a distant, underground pipe union many feet away. The movement of a gas range away from a wall may strain the gas piping system and cause a leak at the junction of the flexible tubing and the rigid main gas line or at a junction within the range itself.

9.9.4.8.2 Hidden pipes underground and in walls are often damaged by construction work. Pipes have been pierced by digging tools, nails, screws, drill bits, cutting tools, and other tools. When nails or screws pierce gas pipes, the resulting leak holes may be largely plugged and remain so until the nail or screw is removed or dislodged, such as by settling of the structure. Therefore, leaks from nails and screws may remain undetected until long after the original damage.

9.9.5 Pressure Testing.

9.9.5.1 General.

9.9.5.1.1 Fuel piping systems are designed to retain moderately pressurized gas. The presence of leaks in the system can be determined by detecting a drop in pressure within the closed system. Before the piping is used for such testing, any obviously damaged portions of the system should be isolated and capped. Sometimes it will be necessary to test the piping in two or more sections, one at a time.

9.9.5.1.2 When isolating portions of the piping that have been damaged by fire or explosion damage, it may be necessary to cut, rethread, and cap individual pipes. It may be possible to seal off damaged piping sections by the use of flexible tubing, hose clamps, and caps, without the necessity to rethread and cap lines. Screw junctions, unions, Ts, or elbows should not be unscrewed in order to isolate a section. Doing so may destroy evidence of previously existing poor connections.

9.9.5.2 Gas Meter Test.

9.9.5.2.1 If it is decided that it is safe to use the actual fuel gas of the system, the gas meter itself can be used to detect a flow of gas. After first checking that the meter is working properly and has not been damaged by the explosion or fire, or has not been bypassed, gas is reintroduced into the system through the meter, and the dial is observed to determine whether gas is escaping somewhere downstream. The meter should be observed with the needle on the upstroke, and observation should continue for up to 30 minutes.

9.9.5.2.2 NFPA 54, *National Fuel Gas Code*, recommends that if no leak is detected with the gas meter test, the test should be repeated with a small gas burner open and ignited, which will show whether the meter is working properly.

9.9.5.3* Pressure Drop Method. A gas piping system may also be tested by the pressure drop method. In this method, the system is pressurized with air or an inert gas such as nitrogen or carbon dioxide. For systems including appliances operating at gauge pressures of 3.4 kPa (0.5 psi) or less, the test can be conducted with the fuel gas itself at between 2.5 kPa and 3.5 kPa (10 in. w.c. and 14 in. w.c.) for 10 minutes. Test methods are listed in NFPA 54, *National Fuel Gas Code*, and NFPA 58, *Liquefied Petroleum Gas Code*.

9.9.6 Locating Leaks. Leaks in fuel gas piping may be located by one or more of the methods in 9.9.6.1 through 9.9.6.4.

9.9.6.1 Soap Bubble Test. Leaks at pipe junctions, fittings, and appliance connections can be detected by applying soap bubble solutions to the suspected leaking area. If the system is pressurized, the production of bubbles in the solution will disclose the leak. After testing, the area should be rinsed with water to prevent possible corrosion or stress cracking. Figure 9.9.6.1 illustrates a bubble test of leaking fuel gas from a piping T-fitting.

FIGURE 9.9.6.1 Illustrate a Bubble Test of Leaking Fuel Gas from a Piping T-Fitting.

9.9.6.2 Gas Detector Surveys. Gas detection instruments, known as flammable gas indicators, combustible gas indicators, explosion meters, or "sniffers," may be used as survey devices to detect the presence of fugitive fuel gases, hydrocarbon gases, or vapors in the atmosphere. Many instruments will also detect the presence of other combustible gases or vapors such as ammonia, carbon monoxide, and others, so the operator should be fully aware of the capabilities and limitations of the instrument being used.

9.9.6.2.1 On the outside of structures, the atmosphere is tested in every available opening in the pavement where gas that has escaped from mains and service lines may be present. Tests should be made along pavement cracks, curb lines, manholes and sewer openings, valve and curb boxes, catch basins, and in bar holes above and along gas piping runs.

9.9.6.2.2 The location of underground gas lines can be learned from utility company maps or by the use of an electronic locator device. These devices induce electromagnetic waves into the earth. Any metallic pipe in the wave field acts as a path for return current and is picked by the receiving unit of the device. An audio tone or analog meter needle then indicates the presence of an underground line. When plastic pipes are used, a metallic locating wire is usually buried along with the pipe to allow the locator devices to be utilized.

9.9.6.2.3 Within structures, the junctions and unions of gas piping can be tested. Spaces where fugitive gases may collect and pocket should also be tested. The relative vapor density of the fuel gases should be kept in mind. If lighter-than-air natural gas is suspected, upper areas of the rooms in structures should be tested. If heavier-than-air LP-Gas is suspected, lower levels should be checked.

9.9.6.3 Bar Holing. Bar holes are holes driven into the surface of the ground or pavement with either weighted metal bars or drills. Bar holing involves the systematic driving of holes at regular intervals along the path of and to either side of underground gas lines and the testing of the subsurface atmosphere with a gas detector. The results of these tests are recorded on a graph or chart known as a bar hole graph. A comparison of the readings of percentage of fugitive gas from each bar hole can indicate the location of an underground gas leak.

9.9.6.4 Vegetation Surveys. Over a period of time, some of the components of fugitive fuel gases from underground leaks can be harmful to grass, trees, shrubs, and other vegetation. When the root systems of plants are subjected to gas from underground leaks, the plants may turn brown, be stunted in their growth, or die. Long-existing underground leaks, which have been permeating the soil and dissipating into the air, may be located by the presence of dead grass or other vegetation over the area of the leak.

9.9.7 Testing Flow Rates and Pressures. If regulators or other gas appliance and service components have not been severely damaged by fire, they can be tested to see whether they are functioning correctly. These tests can be conducted with a variety of nonflammable and flammable gases, including air, nitrogen, helium, or the actual fuel gases for the system (i.e., natural gas, propane, or butane). When flammable gases are used, be sure to eliminate all ignition sources. With the use of proper laboratory or field equipment, both lockup and flow pressures, as well as normal or leak flow rates, can be determined. In such tests, the resultant data must be adjusted from the test medium gas to the gases for which the devices were designed. These adjustments are based on the relative vapor densities of the gases. *(See Figure 9.9.7.)*

9.9.8 Collection of Gas Piping. When collecting gas piping, steps should be taken to maintain evidentiary value. Screw junctions, unions, tees, or elbows should not be unscrewed or tightened in order to isolate a section. Doing so may destroy evidence of previously existing poor connections. Longitudinal witness marks should be placed on every joint and cut site to ensure accurate reconstruction of the spatial relationship of the piping. *(See 9.9.8.)*

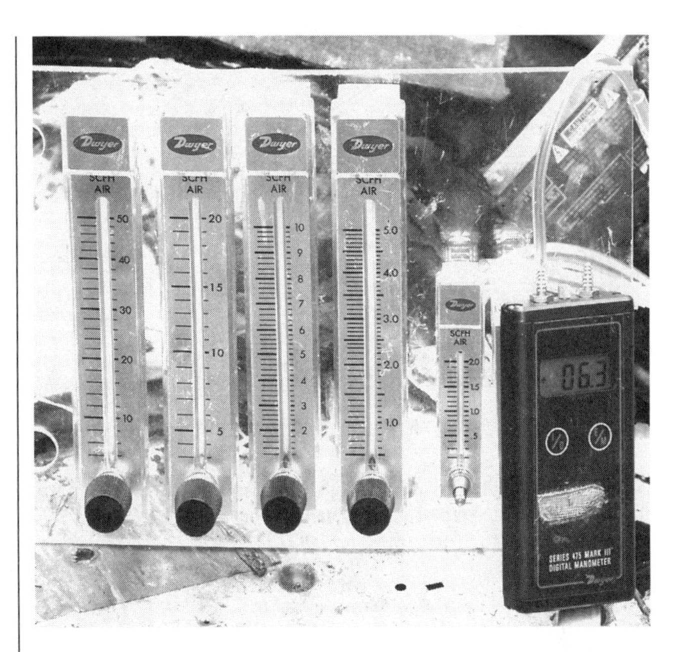

FIGURE 9.9.7 Example of Field Equipment for Testing Flow Rate (Five Meters on Left) and Line Pressure (Digital Manometer on Right).

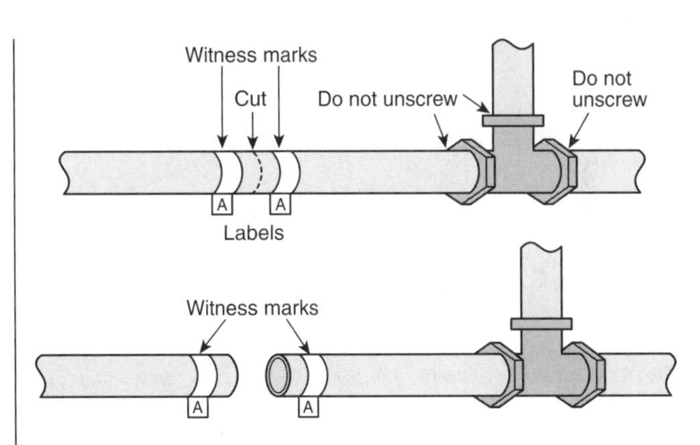

FIGURE 9.9.8 A Method of Intrusive, Nondestructive Marking and Cutting of Fuel Gas Pipes in a Manner That the Relationships of the Severed Cut Ends Will Be Recorded.

9.9.9* Underground Migration of Fuel Gases.

9.9.9.1 General.

9.9.9.1.1 It is common for fuel gases that have leaked from underground piping systems to migrate underground (sometimes for great distances), enter structures, and create flammable atmospheres. Both lighter-than-air and heavier-than-air fuel gases can migrate through soil; follow the exterior of underground lines; and seep into sewer lines, underground electrical or telephone conduits, drain tiles, or even directly through basement and foundation walls, none of which are as gastight as water or gas lines. *[See Figure 9.9.9.1.1(a) and Figure 9.9.9.1.1(b).]*

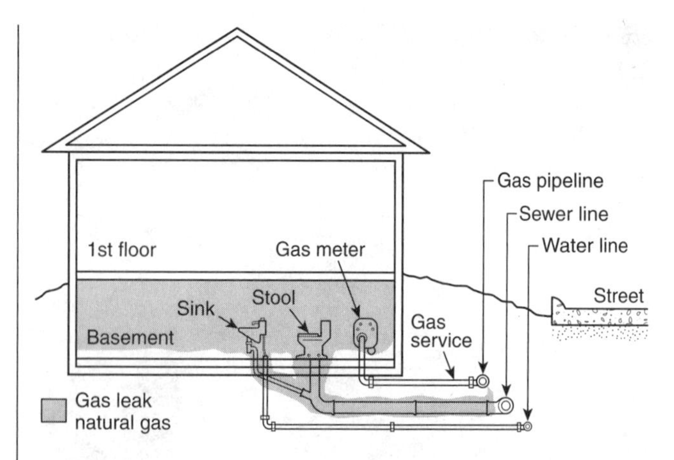

FIGURE 9.9.9.1.1(a) Gas Migrating Along the Sewer Line into the Home, After Leaking at the Service Tee. *[Source: The U.S. Department of Transportation (DOT) Pipeline Hazardous Material Safety Administration (PHMSA), Office of Pipeline Safety (OPS).]*

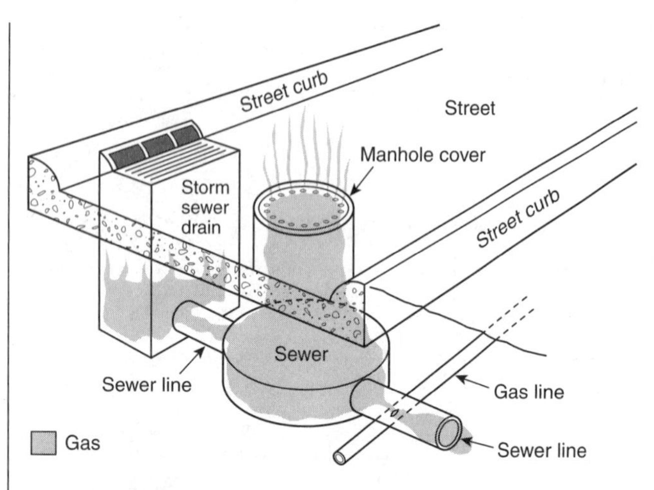

FIGURE 9.9.9.1.1(b) An Example of How a Gas Leak Can Get into a Sewer System. *[Source: The U.S. Department of Transportation (DOT) Pipeline Hazardous Material Safety Administration (PHMSA), Office of Pipeline Safety (OPS).]*

9.9.9.1.2 Such gases also tend to migrate upward, permeating the soil and dissipating harmlessly into the atmosphere. Whether the path of migration is lateral or upward is largely a matter of which path provides the least resistance to the travel of the fugitive gas, the depth at which the leak exists, the depth of any lateral buried lines that the gas might follow, and the nature of the surface of the ground. If the surface of the ground is obstructed by rain, snow, frozen earth, or paving, the gases may be forced to travel laterally. It is not uncommon for a long-existing leak to have been dissipating harmlessly into the air until the surface of the ground changes, such as by the installation of new paving or by heavy rains or freezing, and then be forced to migrate laterally and enter a structure, fueling a fire or explosion.

9.9.9.2* Odorant Removal from Gas. An odorized gas can lose odorant by a number of different mechanisms. This odorant loss has also been termed *odor fade.* This is a complex subject, and for a deeper understanding the reader is referred to the references cited in Annex B. Some of the important issues in odorant loss are summarized in 9.9.9.2(A) through 9.9.9.2(D).

(A) Loss of Odorant Due to Gas Migration in Soil. Gas odorants can be removed by dry, clay-type soils, and not by sand, loams, or heavily organic soils. Certain odorant components are better than others in terms of their ability to resist adsorption by clay-type soils. A large leak gives a lower contact time with the clay-type soil, and results in lower losses due to adsorption.

(B) Loss of Odorant Due to Adsorption of Odorant on Pipe and Container Walls. All odorant components are adsorbed by pipe or container walls to some extent. This is particularly true of new pipe (steel or plastic) and new propane containers.. Many natural gas companies treat the gas in new sections to a heavier dose of odorant after the section is placed in service. Propane industry practice, as found in National Propane Gas Association safety bulletin T133, calls for new propane containers to be purged of air and water vapor before being placed into service. Gas odorants can be adsorbed in gas pipe that has been in continuous service, if the flow rates of gas are lower than normal.

(C) Loss of Odorant Due to Oxidation of Odorant. Mercaptan odorants can be oxidized by ferric oxide (red rust), which can be found in new pipe and in new or out-of-service LP tanks.

(D) Loss of Odorant Due to Absorption. Absorption is a phenomenon that requires the dissolution of odorant in a liquid. It can occur in natural gas systems that have a problem with liquid condensates in their distribution lines. The most common liquid available in the environment is water. All odorants have a low solubility in water.

Chapter 10 Fire-Related Human Behavior

10.1* Introduction. The initiation, development, and consequences of many fires and explosions are either directly or indirectly related to the actions and omissions of people associated with the incident scene. As such, the analyses of fire-related human behavior will often be an integral part of the investigation.

10.1.1 This chapter discusses research findings associated with factors that contribute to fire-related human behavior: how people react to fire emergencies, both as individuals and in groups; factors related to fire initiation; factors related to fire spread and development; factors related to life safety; and factors related to fire safety.

10.1.2 The information discussed in this chapter is based on research conducted by specialists in the fire scene analysis and human behavior fields. The analysis of human behavior is not a substitute for a thorough and properly conducted investigation. While the analysis of human behavior will provide valuable investigative insights, such analysis must be integrated into the total investigation.

10.1.3 For more information on fire-related human behavior, see the *SFPE Engineering Guide to Human Behavior in Fire,* which provides a summary of research related to occupant characteristics, notification of occupants, decision making by occupants, and movement through egress paths.

10.2 History of Research. Fire-related human behavior began to emerge as a distinct field for study in the early 1970s. In 1972, English researcher Peter G. Wood, a pioneer in the field, completed a study of occupants in 952 fire incidents, published as

Fire Research Note #953. A few years later, John L. Bryan, a U.S. researcher and professor of fire protection engineering at the University of Maryland, published the results of his extensive studies on behavior in fires. Bryan has summarized both his work and much of the work of other researchers in this field. This summary is contained in *The SFPE Handbook of Fire Protection Engineering,* "Behavioral Response to Fire and Smoke."

10.3 General Considerations of Human Responses to Fires. Current accepted research indicates that there are myriad factors that affect an individual's or group's human behavior preceding, during, and following a fire or explosion incident. These factors can be broadly classified and evaluated as characteristics of the individual, characteristics of population groups, characteristics of the physical setting, and characteristics of the fire or explosion itself. A careful analysis and evaluation of these factors and their interaction with one another will provide valuable insight into the role of fire-related human behavior for any particular incident. These factors have been extensively examined in the U.S. Fire Administration publication, "Fire Related Human Behavior," (1994). This information is summarized in 10.3.1 through 10.3.2.4.

10.3.1 Individual. Fire-related human behavior is affected by characteristics of the individual in a variety of ways. These characteristics are comprised of physiological factors, including physical limitations, cognitive comprehension limitations, and knowledge of the physical setting. Each of these characteristics affect either an individual's ability to recognize and accurately assess the hazards presented by a fire or explosion incident or an individual's ability to respond appropriately to those hazards.

10.3.1.1 Physical Limitations. Physical limitations that may affect an individual's ability to recognize and react appropriately to the hazards presented by a fire or explosion incident include age (as it relates to mobility), physical disabilities, intoxication, incapacitating or limiting injuries or medical conditions, and other circumstances that limit an individual's mobility. Such limitations should be considered when evaluating an individual's fire-related human behavior, because they tend to restrict or limit one's ability to take appropriate action in response to a fire or explosion. The very old and very young are most affected by physical limitations.

10.3.1.2 Cognitive Comprehension Limitations. Cognitive comprehension limitations, which may affect an individual's ability to recognize and react appropriately to the hazards presented by a fire or explosion incident, include age (as it relates to mental comprehension), level of rest, alcohol use, drug use (legal or illegal), developmental disabilities, mental illness, and inhalation of smoke and toxic gases. These cognitive limitations are more likely to affect an individual's ability to accurately assess the hazards presented by a fire or explosion. Often such limitations account for delayed or inappropriate responses to such hazards. Children may fail to recognize the hazard and choose an inappropriate response, such as hiding or seeking a parent.

10.3.1.3 Familiarity with Physical Setting. An individual's familiarity with the physical setting in which a fire or explosion incident occurs may affect an individual's behavior. For example, a person would be more able to accurately judge a fire's development and progression in his or her own home than in a hotel. It is important to note, however, that physical and cognitive limitations may minimize the advantages of being familiar with the physical setting. Consequently, it may appear that a person has gotten "lost" in his or her home.

10.3.2 Groups. An individual's fire-related behavior is affected by more than his or her characteristics. When interacting with others, an individual's behavior will likely change and be further affected by his or her interaction with that population group and its characteristics. These group characteristics are related to the group size, structure, permanence, and its roles and norms.

10.3.2.1* Group Size. Research and experimental data indicate that when individuals are members of a group, they are less likely to acknowledge or react appropriately to the sensory cues that a fire or explosion incident presents. This tendency increases as the size of the group increases. Research suggests that this fire-related human behavior occurs because individuals in groups will delay their responses to such sensory cues until others in the group also acknowledge these cues and react. The same research suggests that this occurs because the responsibility for taking appropriate action is actually diffused among the group.

10.3.2.2* Group Structure. The structure of a group may also affect the fire-related behavior of both the group and its individual members. Generally, when the group has a formalized structure with defined and recognized leaders or authority figures, the group tends to react to fire and explosion incidents more quickly and in a more orderly manner. However, the reaction is not always appropriate. Examples of such groups include school populations, hospital populations, nursing home populations, and religious facility populations. The behaviors in 10.3.2.2.1 and 10.3.2.2.2 have been observed.

10.3.2.2.1 Research indicates that the interaction between individual members within such groups results in a sense of responsibility for the group as a whole. As such, an individual may be more likely to warn the others in his or her group of the threat than if he or she were interacting with a group of strangers.

10.3.2.2.2 When such a group does become aware of a fire or explosion, their requisite organization and cohesiveness will likely result in a more orderly response to any such threat.

10.3.2.3 Group Permanence.

10.3.2.3.1 Group permanence refers to how well established a group is or how long a particular group of individuals has interacted with one another. Closely related to the effects of group structure, the permanence of a group may also affect the behavior of both the group and its individual members.

10.3.2.3.2 Research indicates that more established groups (such as families, sports teams, or clubs) will be more formalized and structured and, therefore, will react differently during a fire or explosion incident than will a new or transient group (such as a shopping mall population). The latter is more likely to exhibit a multitude of conflicting individual behaviors as each group member responds and reacts on his or her own.

10.3.2.4 Roles and Norms.

10.3.2.4.1 The roles and norms of a group also affect its fire-related behavior. The norms of a group may be influenced by gender, social class, and occupational or educational makeup.

10.3.2.4.2 Gender roles are often a predominant factor during fire and explosion incidents. Research indicates, for example, that women are more likely to report a fire or explosion immediately, while their male counterparts may delay reporting the incident, opting rather to engage in suppression or other mitigation efforts.

10.3.3 Characteristics of the Physical Setting. The characteristics of the physical setting in which a fire occurs affect the development and the spread of a fire or explosion. The characteristics of the setting also affect fire-related behavior. Examples of these characteristics include location of exits, number of exits, the structure's height, fire-warning systems, and fire suppression systems.

10.3.3.1 Locations of Exits. The locations of available exits during a fire or explosion incident may affect the behavior of the occupants. If the locations of the available exits are not known to the occupants, or if they are not adequately identified, confusion and heightened levels of anxiety may be experienced.

10.3.3.2 Number of Exits. The number of available exits during a fire or explosion incident may affect the behavior of the occupants. An inadequate number of exits, blocked or restricted exits, and unprotected exits (i.e., nonpressurized or open interior stairwells) may result in the occupants being exposed to the fire and its by-products.

10.3.3.3 Height of Structure. The height of a structure may affect the behavior of its occupants during a fire or explosion incident. Some people believe that they are less safe in taller buildings during fires.

10.3.3.4* Fire Alarm Systems.

10.3.3.4.1 Fire alarm systems are among the variables of built-in fire safety that may be critical to an individual's awareness of a fire. Research has shown that verbal, directive messages may be most effective in creating response, compared to alarm bells and sounders alone.

10.3.3.4.2 Prior false alarms and alarm system malfunctions may reduce the positive effect of having an alarm system in the building, because the occupants may not respond appropriately to the alarm notification. Numerous false alarms reduce the occupants' appropriate responses to the alarm.

10.3.3.5 Fire Suppression Systems. The presence of automatic fire suppression systems, if known, may affect behavior. The effect may be positive or negative. A positive effect is that the increased margin of safety of such systems provides occupants of the involved structure more time to respond appropriately to the hazards presented by the incident. An example of a negative effect is possible decreased visibility caused by the discharge of the suppression agent, which may impede egress.

10.3.4 Characteristics of the Fire. Fire-related human behavior is directly related to an individual's or group's perception of the hazards or threats presented to them by a fire or explosion. Characteristics of the fire itself will tend to shape these perceptions and thereby affect fire-related human behavior. Examples of these characteristics include the presence of flames or smoke and the effects of toxic gases and oxygen depletion.

10.3.4.1 Presence of Flames. Most individuals have uneducated or uninformed perceptions of the hazards presented by fire and explosion incidents. This perception problem is especially true relative to an individual's observation of the presence of visible flames. The sight of flames makes the individual aware that it is not a false alarm and that some danger is present; however, because people do not understand fire dynamics and fire behavior, the presence of small flames may not be recognized as an immediate threat, and the resulting behavior is based on that belief. See Chapter 5 for further discussion on fire dynamics.

10.3.4.2 Presence of Smoke. Like visible flames, the presence of smoke may also affect fire-related behavior. A lack of knowledge regarding fire dynamics and fire behavior may result in erroneous perceptions relative to smoke. Individuals may perceive dense, black smoke as an immediate threat to their physical well being, while light, gray smoke may not be immediately perceived as a threat at all.

10.3.4.3 Effects of Toxic Gases and Oxygen Depletion. During fire and explosion incidents, individuals often inhale the by-products of combustion, including the toxic gases present in the smoke. Additionally, the development and progression of the fire, as well as the presence of these other gases, often results in a depletion of the oxygen that had originally been present in the ambient air. The inhalation of toxic gases, or low oxygen concentration levels below approximately 15 percent, may affect an individual's behavior and result in perceptual and behavioral changes. These changes may manifest themselves in delayed or inappropriate responses to the incident. Strength, stamina, mental acuity, and perceptual ability can all be severely decreased. See 23.5.5.

10.4 Factors Related to Fire Initiation. The initiation of many fire and explosion incidents is facilitated or fostered by the actions or omissions of people associated with the incident scene. Fire-related human behavior is often the reason that a competent source of ignition is present at the same time and place with a fuel in the presence of a sufficient amount of oxygen.

10.4.1 Factors Involved in Accidental Fires. The actions or inactions of people frequently result in accidental fires. Negligence, carelessness, lack of knowledge, disregard of fire safety principles, or an individual's failure to be cognizant of the ultimate results of such actions or inactions can be categorized into groups of similar behaviors. Examples of these behavior groupings are improper maintenance; poor housekeeping; issues involving product labels, instructions, and warnings; and violations of fire safety codes and standards.

10.4.1.1 Improper Maintenance and Operations.

10.4.1.1.1 Many types of equipment, systems, machinery, and appliances are potential ignition sources or fuel sources for fire and require some level of periodic maintenance or cleaning. Instructions pertaining to the type of maintenance or cleaning procedures, as well as a recommended schedule for maintenance or cleaning, are most often provided by the manufacturer or supplier. Failure to adhere to these recommendations may result in fire or explosion. It is often reported to the investigator that the accompanying maintenance or cleaning instructions to a specific piece of equipment are unavailable. In these instances, it is often possible to obtain this information directly from the manufacturer or supplier or from exemplar items. The investigator should, whenever possible, examine maintenance and cleaning instructions and records regarding equipment and appliances found in the area of origin. These records can prove helpful when a specific appliance or piece of equipment is being considered as an ignition source or fuel source.

10.4.1.1.2 Equipment and appliance operating procedures and cautions are also normally provided to the end user or consumer by the manufacturer or supplier. It may prove helpful to the investigator to obtain and review this type of information when accessing the condition of a specific piece of equipment or an appliance at the time of the fire.

10.4.1.2 Housekeeping. A lack of proper housekeeping measures can also contribute directly or indirectly to the occurrence of fire. Examples of lack of such measures include careless use or disposal of smoking materials, refuse and other combustibles allowed to accumulate too close to an ignition source, quantities of dust or other combustible particulate matter becoming suspended in air (due to dust collection equipment needing to be cleaned or emptied) in the same environment as open flame- or spark-producing equipment, lint in dryers, and grease buildup in cooking areas.

10.4.1.3 Product Labels, Instructions, and Warnings. A lack of awareness of, or a disregard for, warning labels and other safety instructions can also result in the accidental ignition of a fire. In many cases, the ignition factor of a fire is the result of the actions or omissions of the user of a product. The danger of improper actions or omissions may not always be obvious to a product user. Whenever a product has a hazard potential for supplying the ignition source, fuel, or oxygen portions of the ignition factor, it is incumbent upon the manufacturer and supplier to provide proper labeling, instructions, and warnings with the product. Likewise, it is incumbent upon the user to follow proper warnings and instructions.

10.4.1.4 Purpose of Labels. The purpose of labels is to provide the user with information about the product use at the closest possible point to its actual use. Labels can take several forms: printed labels attached to the product; labels printed on the packaging of the product; or molded, stamped, or engraved writing on the product or its container.

10.4.1.5 Purpose of Instructions. Instructions for a product are intended to inform the user on how the product is to be used safely, of the existence of any hazards, and how to minimize any risk to the user during the actual use of the product.

10.4.1.6 Purpose of Warnings. The overall purpose of a warning is to provide the user with information necessary to use a product safely or to make an informed decision not to use the product because of its hazard. Warnings on the labels or instructions of a product should serve two objectives: to inform an unknowing user of the dangers posed by the use or misuse of the product, or to remind the user of the hazards of the product.

10.4.1.7 Key Elements of a Proper Warning. According to federal regulations, in order for warnings to be appropriate and effective, there are certain key elements that must be present: an alert word, a statement of the danger, a statement of how to avoid the danger, and explanations of the consequences of the danger.

10.4.1.7.1 Alert Word. The alert word or signal word is the first sign to the user that there is a danger. "Caution," "Warning," or "Danger" are the most commonly used and approved alert words. Through its meaning, type style, type size, and contrast the alert word is designed to draw the user's attention to the warning that follows, and to give some concept of the degree of danger. Most standards hold that the alert words "Caution,""Warning," and "Danger" respectively signify increasing levels of hazard and risk. ANSI Z535.4, *Product Safety Signs and Labels*, provides the following definitions:

(1) CAUTION: Indicates a potentially hazardous situation that, if not avoided, may result in minor or moderate injury.
(2) WARNING: Indicates a potentially hazardous situation that, if not avoided, could result in death or serious injury.
(3) DANGER: Indicates an imminently hazardous situation that, if not avoided, will result in death or serious injury.

10.4.1.7.2 Statement of the Danger. The statement of the danger should identify the nature and extent of the danger and the gravity of the risk of injury, for example: "Combustible," "Flammable," or "Extremely Flammable." Risk is a function of the likelihood and severity of injury. Additional phrases such as "may explode," or "can cause serious burns," may also be necessary.

10.4.1.7.3 How to Avoid the Danger. Warnings should provide potential users with information on how the hazard can be avoided and should tell them what to do or refrain from doing to remain safe when using the product.

10.4.1.7.4 Consequences of the Danger. Warnings should also tell the user what would or could happen if the precautions listed are not followed.

10.4.1.8 Standards on Labels, Instructions, and Warnings. Over the years, government and industry have promulgated many standards, guidelines, and regulations dealing with safety warnings and safe product design. Among the standards that deal with labels, instructions, and warnings are the following:

(1) ANSI standards on labeling:
   (a) Z129.1, *Precautionary Labeling of Hazardous Industrial Chemicals*
   (b) Z400.1, *Material Safety Data Sheets — Preparation*
   (c) Z535.1, *Safety Color Code*
   (d) Z535.2, *Environmental and Facility Safety Signs*
   (e) Z535.3, *Criteria for Safety Symbols*
   (f) Z535.4, *Product Safety Signs and Labels*
   (g) Z535.5, *Accident Prevention Tags*
(2) UL standard on labeling:
   (a) ANSI/UL 969, *Standard for Marking and Labeling Systems*
(3) United States Federal Codes and Regulations:
   (a) "Consumer Safety Act" (15 USC Sections 2051–2084, and 16 CFR 1000)
   (b) "Hazardous Substances Act" (15 USC Sections 1261 et seq., and 16 CFR 1500)
   (c) "Federal Hazards Communication Standard" (29 CFR 1910)
   (d) "Flammable Fabrics Act" (15 USC Sections 1191–1204 and 16 CFR 1615, 1616, 1630–1632)
   (e) "Federal Food, Drug and Cosmetic Act" (15 USC Section 321 (m), and 21 CFR 600)
   (f) OSHA Regulations (29 CFR 1910)
(4) Industry standard:
   (a) *FMC Product Safety Sign and Label System Manual*

10.4.2 Recalls. Disregarding recall notices involving items that have the potential to become an ignition source can also result in a fire. Many times a recall notice is the result of reported fires, where a specific item has been identified as the ignition source.

10.4.3 Other Considerations. An individual's lack of knowledge, carelessness, willful disregard, or negligence are often unclear to the investigator who reviews the circumstances and events leading up to a fire. A review of training records and interviews conducted with persons occupying locations or spaces where a fire has occurred may provide the information needed to more clearly understand an individual's level of involvement regarding the initiation of a fire.

10.4.4 Violations of Fire Safety Codes and Standards. Failure to adhere to pertinent, established fire safety codes and standards, industry standards, or good practices may result in fires or explosions. Noncompliance with these various safety prescriptions may be deliberate or unintentional.

10.5 Children and Fire. Playing with fire is an activity that a large number of children participate in for many reasons. The most common reason for children playing with fire is curiosity. To a young child, fire is intriguing, very powerful, and too often accessible. Firesetting may be a child's means to express frustration or anger, to seek revenge, or to call attention to himself or herself or to difficult circumstances. Research has shown that the location where the fire is set and the motive for it often varies according to the age of the child. There are three recognized age groups, as follows.

10.5.1 Child Firesetters. Child firesetters (ages 2 to 6) are often responsible for fires in their homes or in the immediate area. Sometimes the fires are in areas that are hidden and out of sight of their guardian. These are usually curiosity firesetters.

10.5.2 Juvenile Firesetters. Juvenile firesetters (ages 7 to 13) are often responsible for fires that start in their homes or in the immediate environment. They may also start fires in their educational setting. These firesetting events are usually associated with some broken family environment or physical or emotional trauma.

10.5.3 Adolescent Firesetters. Adolescent firesetters (ages 14 to 16) are often responsible for fires that occur at places other than their homes. They target schools, churches, vacant buildings, fields, and vacant lots. These firesetters are often associated with a history of delinquency, disruptive rearing environment, poor social environment and emotional adjustments, peer pressure, and poor academic achievement. They sometimes work in pairs or small groups, with one dominant individual and others as followers. These fires are often set to express their stress, anxiety, and anger, or as a symptom of another problem.

10.6 Incendiary Fires. Human factors involved in the setting of incendiary fires are closely related to motives examined in Section 22.4. See Section 22.4 for additional information.

10.7 Human Factors Related to Fire Spread.

10.7.1 The spread of the fire can be affected significantly by the actions or omissions of the people present before or during the fire. These actions can either accelerate or retard the spread of the fire. The investigator may need to evaluate these actions to determine the effects these actions had on the fire. Some of these actions include opening and closing doors or windows, fire fighting, operation of fire protection systems, and rescue. Some of these actions are addressed in 17.6.1.

10.7.2 Pre-fire conditions, such as housekeeping, functioning alarms, and compartmentation, may be documented after the fire by inspecting unburned areas of the building, prior fire department inspection records for nonresidential buildings, as well as by post-fire interviews. The investigator should not presume the conditions in the building prior to the fire.

10.8 Recognition and Response to Fires. In a fire, survivability of an individual is based on the ability of him or her to recognize and safely respond to the hazard in several ways. The individual needs to perceive the danger, make a decision about some action to take, and carry out that action. These three basic concepts will be addressed in this section.

10.8.1 Perception of the Danger (Sensory Cues). People become aware of the fire by any one or combination of several sensory cues. The sensory perception can be affected by factors such as whether the person is awake, asleep, or impaired. Impairment may be physical, mental, or the result of effects of chemical agents (e.g., drugs, alcohol, or carbon monoxide). Sensory cues include the following:

(1) Sight: Direct view of flames, smoke, visual alarms, or flicker
(2) Sound: Crackling, failure of windows, audible alarms, dogs barking, children crying, voices, or shouts
(3) Feel: Temperature rise or structural failure
(4) Smell: Smoke odor

10.8.2 Decision to Act (Response). Once the danger has been perceived, a decision is made concerning how to respond. This decision is based on the severity of the danger perceived. The person's degree of impairment is a factor in the decision process.

10.8.3 Action Taken. The action taken by the individual can take any one or a combination of forms. These include the following:

(1) Ignore the problem
(2) Investigate
(3) Fight the fire
(4) Give alarm
(5) Rescue or aid others
(6) Re-enter after successful escape
(7) Flee (escape)
(8) Remain in place

10.8.4 Escape Factors. The success or failure of an attempt to escape a fire depends on a number of factors, including the following:

(1) Identifiability of escape routes
(2) Distance to a means of escape
(3) Fire conditions such as the presence of smoke, heat, or flames
(4) Presence of dead-end corridors
(5) Path blocked by obstacles or people
(6) Physical disabilities or impairments of occupants

10.8.5 Information Received from Survivors. Post-fire information obtained by interviewing witnesses (e.g., survivors, victims, occupants, passers-by, emergency responders) of a fire incident may be helpful in determining several factors. Such information may include the following:

(1) Pre-fire conditions
(2) Fire and smoke development
(3) Fuel packages and their location and orientation
(4) Victims' activities before, during, and after discovery of the fire
(5) Actions taken by survivors resulting in their survival; that is, escape or take refuge
(6) Decisions made by survivors and reasons for those decisions
(7) Critical fire events such as flashover, structural failure, window breakage, alarm sounding, first observation of smoke, first observation of flame, fire department arrival, and contact with others in the building

Chapter 11 Legal Considerations

11.1* Introduction. Legal considerations impact every phase of a fire investigation. Whatever the capacity in which a fire investigator functions (public or private), it is important that the investigator be informed regarding all relevant legal restrictions, requirements, obligations, standards, and duties. Failure to do so could jeopardize the reliability of any investigation and could subject the investigator to civil liability or criminal prosecution. It is the purpose of this chapter to alert the investigator to those areas that usually require legal advice, knowledge, or information. The legal considerations contained in this chapter and elsewhere in this guide pertain to the law in the United States. This chapter does not attempt to state the law as it is applied in each country or other jurisdiction. Such a task exceeds the scope of this guide. To the extent that statutes or case law are referred to, they are referred to by way of example only, and the user of this guide is reminded that "the law" is in a constant state of flux. As an analogy, both case law and statutory law can be compared to a living thing. They are constantly subject to creation (by new enactment or decision), change (by modification or amendment), and death (by being repealed, overruled, or vacated). It is recommended that the investigator seek legal counsel to assist in understanding and complying with the legal requirements of any particular jurisdiction. Recognition of applicable legal requirements and considerations will help to ensure the reliability and admissibility of the investigator's records, data, and opinions.

11.2 Constitutional Considerations. Within the United States and its territories, investigators should be aware of the constitutional safeguards that are generally applicable to criminal investigations and prosecutions as set forth in the Fourth, Fifth, and Sixth Amendments of the United States Constitution. The text of these three amendments is shown in Figure 11.2.

Amendment IV
The right of the people to be secure in their persons, houses, papers, and effects, against unreasonable searches and seizures, shall not be violated, and no Warrants shall issue, but upon probable cause, supported by Oath or affirmation, and particularly describing the place to be searched, and the persons or things to be seized.

Amendment V
No person shall be held to answer for a capital, or otherwise infamous crime, unless on a presentment or indictment of a Grand Jury, except in cases arising in the land or naval forces, or in the Militia, when in actual service in time of War or public danger; nor shall any person be subject for the same offence to be twice put in jeopardy of life or limb; nor shall be compelled in any criminal case to be a witness against himself, nor be deprived of life, liberty, or property, without due process of law; nor shall private property be taken for public use, without just compensation.

Amendment VI
In all criminal prosecutions, the accused shall enjoy the right to a speedy and public trial, by an impartial jury of the State and district wherein the crime shall have been committed, which district shall have been previously ascertained by law, and to be informed of the nature and cause of the accusation; to be confronted with the witnesses against him; to have compulsory process for obtaining witnesses in his favor, and to have the Assistance of Counsel for his defense.

FIGURE 11.2 Fourth, Fifth, and Sixth Amendments to the U.S. Constitution.

11.3 Legal Considerations During the Investigation.

11.3.1 Authority to Conduct the Investigation. The investigator should ascertain the basis and extent of his or her authority to conduct the investigation. The authority to investigate is given to police officers, fire fighters, and fire marshals according to the law of the jurisdiction. Private fire investigators receive their authority by contract or consent. Examples of contract consent are insurance contracts that obligate the insured to cooperate in the investigation. Also, a person having an interest in the property may retain (contract with) their own fire investigator. Consensual authority would include an investigator for another interested party being invited onto the premises to participate or observe the scene investigation. Proper identification of the basis of authority will assist the investigator in complying with applicable legal requirements and limitations. The scope of authority granted to investigators from the public or governmental sector is usually specified within the codified laws of each jurisdiction, as supplemented by applicable local, agency, and department rules and regulations. Many states and local jurisdictions (i.e., cities, towns, or counties) have licensing or certification requirements for investigators. If such requirements are not followed, the results of the investigation may not be admissible and the investigator may face sanctions.

11.3.2 Right of Entry. The fact that an investigator has authority to conduct an investigation does not necessarily mean that the investigator has the legal right to enter the property that was involved in the fire. Rights of entry are frequently enumerated by statutes, rules, and regulations. Illegal entry on the property could result in charges against the investigator (i.e., trespassing; breaking and entering; or obstructing, impeding, or hampering a criminal investigation). Once a legal right of entry onto the property has been established, the investigator should notify the officer or authority in control of the scene of the intent to enter. An otherwise legal right of entry does not authorize entry onto a crime scene investigation. Further authorization by the specific agency or officer in charge may be required. Care should be taken to avoid the spoliation of evidence.

11.3.2.1 Local code provisions designed to protect public safety may mandate that a building involved in a fire be demolished promptly to avoid danger to the public. This act can deny an investigator the only opportunity to examine the scene of a fire. When it is important to do so, court-ordered relief prohibiting the demolition until some later and specified date may be obtained, most typically by way of injunction, to allow for the investigator's presence at the scene. An injunction may prove costly, as the party seeking the delay may be required to post a bond, procure guards, and secure the property until the investigation is completed. Legal counsel should be able to anticipate needs in this regard and to respond to such needs promptly. The investigator may be required to produce evidence by order of court or pursuant to a subpoena. The investigator should exercise caution and not destroy, dispose of, or remove any evidence unless clearly and legally entitled to do so. Courts are becoming more willing to enter orders designed to preserve the fire scene, thereby preserving the rights of all interested parties to examine evidence in this post-fire location and condition.

11.3.2.2 In the event that destruction, disposal, or removal is authorized or necessary, the investigator should engage in such acts only after the scene has been properly recorded and the record has been verified as to accuracy and completeness. Care should be taken to avoid spoliation.

11.3.3 Method of Entry. Whereas "right of entry" refers to the legal authority to be on a given premise or fire scene, this section concerns itself with how that authority is obtained. There are four general methods by which entry may be obtained: consent, exigent circumstance, administrative search warrant, and criminal search warrant.

11.3.3.1 Consent. The person in lawful control of the property can grant the investigator permission or consent to enter and remain on the property. This is a voluntary act on the part of the responsible person and can be withdrawn at any time by that person. When consent is granted, the investigator should document it. One effective method is to have the person in lawful control sign a written consent form. The investigator may choose to make inquiries to ensure that the person giving consent has lawful control of the property. For example, if a tenant has rights to control leased property under a rental agreement, the property owner (landlord) may not have the immediate right to access that property, and may therefore lack the power to consent.

11.3.3.2 Exigent Circumstance.

11.3.3.2.1 It is generally recognized that the fire department has the legal authority to enter a property to control and extinguish a hostile fire. It also has been held that the fire department has an obligation to determine the origin and cause of the fire in the interest of the public good and general welfare.

11.3.3.2.2 The time period in which the investigation may continue or should conclude has been the subject of a Supreme Court decision (*Michigan v. Tyler*, 436 U.S. 499), when the Court held that the investigation may continue for a "reasonable period of time," which may depend on many variables. When the investigator is in doubt as to what is a "reasonable time," legal advice should be sought.

11.3.3.3 Administrative Search Warrant.

11.3.3.3.1 The purpose of an administrative search warrant is generally to allow those charged with the responsibility, by ordinance or statute, to investigate the origin and cause of a fire and to fulfill their obligation according to the law. An administrative search warrant may be obtained from a court of competent jurisdiction upon a showing that consent has not been granted or has been denied. It is not issued on the traditional showing of "probable cause," as is the criminal search warrant, although it is still necessary to demonstrate that the search is reasonable. The search should be justified by a showing of reasonable governmental interest, and supported by a statute, ordinance, or regulation. If a valid public interest justifies the intrusion, then valid and reasonable probable cause has been demonstrated.

11.3.3.3.2 The scope of an administrative search warrant is limited to the investigation of the origin and cause of the fire. If during the search permitted by an administrative search warrant, evidence of a crime is discovered, the search should be stopped and a criminal search warrant should be obtained (*Michigan v. Clifford*, 464 U.S. 287).

11.3.3.4 Criminal Search Warrant. The purpose of a criminal search warrant is to allow the entry of government agents to search for and collect evidence of a crime, as specified in the warrant. The warrant may authorize the search of the premises, a vehicle, or a person. Government agents with the authority to apply for a search warrant as well as the court to which the application must be made are specified by federal and state laws. A government agent authorized to apply for a warrant is not necessarily authorized by statute to execute the warrant. Seeming minor defects in the application or warrant can result in the suppression of evidence. The applicant should consider consulting with legal counsel when making an application.

11.3.3.4.1 The application for obtaining a criminal search warrant typically includes the following:

(1) The kind or character of the property sought
(2) The place or person to be searched
(3) Allegations of fact, based upon the personal knowledge of the applicant or upon information and belief of the applicant, with the grounds for such belief stated, that reasonable cause exists to support statements (1) and (2)

11.3.3.4.2 The application may also contain a request that it be executed at any time of the day or night and that entry may be made without giving notice, if supported by sufficient facts.

11.3.4 The Questioning of Suspects.

11.3.4.1 In light of the criminal charges that can be made as a result of a fire, the investigator should ascertain whether they are required to advise the person being questioned of his or her "Miranda rights" and if so, when and how to advise of those rights. The person being questioned should be advised of the following if the interrogation is conducted in a custodial setting by an investigator who represents a governmental agency or who is acting at the request of government investigators:

(1) They have a right to remain silent.
(2) Any statement they do make may be used as evidence against them.
(3) They have the right to the presence of an attorney.
(4) If they cannot afford an attorney, one will be appointed for them prior to any questioning if they so desire.
(5) Should they decide to waive the right to be silent, they can change their mind and stop the questioning or request an attorney at any time during the questioning.

11.3.4.2 Unless and until these warnings or a waiver of these rights are demonstrated at trial, no evidence obtained in the interrogation may be used against the accused (formerly the witness) (*Miranda v. Arizona*, 384 U.S. 436). Persons interviewed in "custodial settings" should be advised of their constitutional rights. Although interviews conducted on a fire scene are not generally considered to be custodial, they may be, depending on the circumstances. The custodial setting depends on many variables, including the location of the interview, the length of the interview, who is present and who participates, and the person's perception of whether they will be restrained if they attempt to leave. If there is any doubt in the mind of the investigator as to whether the person is being questioned in a custodial setting, the person should be advised of their constitutional rights. When a person is advised of their constitutional rights as required by the Miranda ruling, the rights may be listed on a written form that can be signed by the person.

11.3.5 Spoliation of Evidence. Spoliation of evidence refers to the loss, destruction, or material alteration of an object or document that is evidence or potential evidence in a legal proceeding by one who has the responsibility for its preservation. Spoliation of evidence may occur when the movement, change, or destruction of evidence, or the alteration of the scene significantly impairs the opportunity of other interested parties to obtain the same evidentiary value from the evidence, as did any prior investigator.

11.3.5.1 Responsibility. The responsibility of the investigator (or anyone who handles or examines evidence) is evidence preservation, and the scope of that responsibility varies according to such factors as the investigator's jurisdiction, whether he or she is a public official or private sector investigator, whether criminal conduct is indicated, and applicable laws and regulations. However, regardless of the scope and responsibility of the investigation, care should be taken to avoid destruction of evidence.

11.3.5.2 Documentation. Efforts to photograph, document, or preserve evidence should apply not only to evidence relevant to an investigator's opinions, but also to evidence of reasonable alternate hypotheses that were considered and ruled out.

11.3.5.3 Remedies for Spoliation. Criminal and civil courts have applied various remedies when there has been spoliation of evidence. Remedies employed by the courts may include discovery sanctions, monetary sanctions, application of evidentiary inferences, limitations under the rules of evidence, exclusion of expert testimony, dismissal of a claim or defense, independent tort actions for the intentional or negligent destruction of evidence, and even prosecution under criminal statutes relating to obstruction of justice. Investigators should conduct their investigations so as to minimize the loss or destruction of evidence and thereby to minimize allegations of spoliation.

11.3.5.4 Notification to Interested Parties. Claims of spoliation of evidence can be minimized when notice is given to all known interested parties that an investigation at the site of the incident is going to occur so as to allow all known interested parties the opportunity to retain experts and attend the investigation. Such notice may be made by telephone, letter, or e-mail. Oral notification should be confirmed in writing. Notification should include the date of the incident; the nature of the incident; the incident location; the nature and extent of loss; damage, death, or injury to the extent known; the interested party's potential connection to the incident; next action date; circumstances affecting the scene (such as pending demolition orders or environmental conditions); a request to reply by a certain date; contact information as to whom the notified person is to reply; and the identity of the individual or entity controlling the scene. The notification should also include a roster of all parties to whom notice has been provided. Public sector investigators may have different notification responsibilities than the private sector investigators. Responsibility for notification varies based on jurisdictions, scope, procedures, and the circumstances of the fire. Interested parties should make public officials aware of their interest. A private sector consent to search does not constitute notice unless it conforms with this section.

11.3.5.5 Documentation Prior to Alteration. Anytime the investigator determines that significant alteration of the fire scene will be necessary to complete the fire investigation, it should be done, only after notification to all known interested parties has been given, and the interested parties have been afforded the opportunity to be present. Special care should be taken to photograph and document the scene and preserve relevant evidence. The scene should be properly documented prior to any alteration, and relevant evidence should be preserved. Destructive disassembly of any suspected or potential ignition sources should be avoided whenever possible to permit later forensic examination after notice is given to all known interested parties.

11.3.5.6 Alteration and Movement of Evidence.

11.3.5.6.1 Fire investigation usually requires the movement of evidence or alteration of the scene. In and of itself, such movement of evidence or alteration of the scene should not be considered spoliation of evidence. Physical evidence may need to be moved prior to the discovery of the cause of the fire. Additionally, it is recognized that it is sometimes necessary to remove the potential causative agent from the scene and even to carry out some disassembly in order to determine whether the object did, in fact, cause the fire, and which parties may have contributed to that cause. For example, the manufacturer of an appliance may not be known until after the unit has been examined for identification. Such activities should not be considered spoliation.

11.3.5.6.2 Still another consideration is protection of the evidence. There may be cases where it is necessary to remove relevant evidence from a scene in order to ensure that it is protected from further damage or theft. Steps taken to protect evidence should also not be considered spoliation.

11.3.5.7 Notification Prior to Destructive Testing. Once evidence has been removed from the scene, it should be maintained and not be destroyed or altered until others who have a reasonable interest in the matter have been notified. Any destructive testing or destructive examination of the evidence that may be necessary should occur only after all reasonably known parties have been notified in advance and given the opportunity to participate in or observe the testing. This section is not intended to apply to evidence collected as part of a criminal investigation. Once the evidence is no longer required for a criminal investigation it should be appropriately released. Guidance regarding notification can be found in ASTM E 860, *Standard Practice for Examining and Testing Items That Are or May Become Involved in Litigation,* and ASTM E 1188, *Standard Practice for Collection and Preservation of Information and Physical Items by a Technical Investigator.* Guidance for disposal of evidence may be found in Section 16.11 of this guide. Guidance for labeling of evidence can be found in ASTM E 1459, *Standard Guide for Physical Evidence Labeling and Related Documentation.*

11.4 Pretrial Legal Considerations.

11.4.1 Introduction. Between the time an investigation is concluded and the time the matter comes to trial, there may be legal proceedings in which information and documents are exchanged between parties, testimony is taken, and admissions are requested. These proceedings can be categorized as "discovery." They serve to help the parties prepare their cases, understand the evidence and facts possessed by the other parties, and evaluate their cases for potential settlement when appropriate. These proceedings occur primarily in civil cases, but may be available in criminal cases in some jurisdictions. While discovery is governed by legal rules, there is usually not a judge involved in this part of the litigation, unless the parties are unable to resolve a particular issue. Other pre-trial issues may involve a judge or magistrate who may issue advance rulings on what objects, documents, facts, or opinions will be allowed as evidence at trial.

11.4.2 Forms of Discovery. Discovery is the process occurring during the pretrial phase of a legal proceeding where each party to the litigation obtains information, documents, and evidence from opposing parties or nonparties to be used in preparing for trial. Discovery, which is governed by court rules applicable to the jurisdiction in which the case is pending, can take several forms.

11.4.2.1 Request to Produce. A Request to Produce is a written demand on another party or witness requesting that certain documents be produced. Usually, when a request is made on a nonparty, it will be accompanied by a subpoena. The request or subpoena will identify specific documents or categories of documents to be produced and when and where the documents must be produced.

11.4.2.2 Interrogatories. Interrogatories are written questions one party serves on another party. Interrogatories must be answered in writing, under oath, and signed by a party or their representative.

11.4.2.3 Depositions. A deposition is a method of obtaining oral testimony under oath, whereby the witness (deponent) must answer questions of one or more of the attorneys representing the parties to a lawsuit. There are several purposes for taking a deposition. They include discovering what facts, opinions, or evidence a witness has and may offer at trial; obtaining testimony to be used in later court proceedings, such as motions; or to preserve the testimony of a witness who may be unavailable to testify at trial. A court stenographer (court reporter), who may later produce a transcript of the deposition proceedings, records depositions. It is common for depositions to also be videotaped.

11.4.2.3.1 Procedure. Regardless of the purpose of a deposition, the procedure for taking a deposition is almost always the same. In a deposition, the witness is obligated to swear or affirm under penalty of perjury that the testimony to be given will be the truth. The court reporter will administer the oath and record everything that is said by the witness and attorneys during the deposition. A deposition proceeds in a question-and-answer format. An attorney will ask questions of the witness, who is obligated to provide answers, unless otherwise instructed.

11.4.2.3.2 Discovery Depositions. A discovery deposition is one that is taken to learn or discover what facts, opinions, or information a witness has. The attorney who requested the deposition will begin the questioning. Often, but not always, after the first attorney is finished asking questions, the attorneys for the other parties may also ask questions. In general, the strategy in a discovery deposition is to learn all of the facts and opinions that a witness has, the contents of the witness's file, the bias, if any, of the witness, and what testimony the witness may offer at trial. If the witness later testifies at trial in a way different from or inconsistent with his deposition testimony, the deposition testimony may be used to impeach the witness. Discovery depositions may cover a wide range of topics, including the witness's background, training, experience, qualifications, and the methodologies used by the witness in formulating any expert opinions. In these situations, a fire investigator must communicate opinions clearly and understandably. The difficulty in communicating opinions in this setting is that the investigator must communicate facts and opinions in response to questions posed by an attorney who may represent an adverse party in a proceeding in which the investigator has little control. Therefore, the investigator should understand how deposition testimony may be used in the future, and the importance of creating a record that clearly establishes an opinion and a valid basis for that opinion.

11.4.2.3.3 Trial Depositions. A trial deposition is usually taken to preserve the testimony of a witness who may be unavailable to testify in person at the time of trial. Unlike a discovery deposition, a trial deposition is conducted by the attorney for the party wanting to offer the witness's testimony during trial, either in the case in chief, or rebuttal. Rather than taking the deposition to discover facts and opinions, the strategy in a trial deposition is to question the investigator to establish their credentials, to establish a foundation for what the investigator did, and to allow the investigator to render opinions.

11.4.2.4 Reports. The Federal Rules of Civil Procedure and some state courts may require that experts who will be called as trial witnesses prepare reports, which may form the basis for cross-examination during the witness's deposition. These reports contain the following information:

(1) A list of materials reviewed and investigative activities conducted
(2) A list of opinions the expert expects to express at trial
(3) The bases for those opinions
(4) A list of publications by the expert within the last ten years
(5) A list of testimony given either at trial or in deposition in the last four years
(6) The compensation the witness receives for his or her work

11.4.3 Motions. A motion is a request for the court to take action. Facts, documents, and evidence that come to light during the discovery phase often form the basis for motions to exclude certain items of evidence or witness statements, or to limit or exclude the testimony of certain individuals from the trial. Such motions may argue that constitutional rights were violated, or that evidence was obtained illegally, or that proposed expert witness testimony is not relevant or reliable. The investigator may be required to provide an affidavit or testimony concerning the motion.

11.5 Trials. If the parties are unable to resolve the matter in dispute through a plea in a criminal case or a settlement in a civil suit, the case proceeds to a trial. A trial is presided over by the judge, who acts as a finder of fact, but most trials involving fires employ juries as the finder of fact. The judge instructs the jury on the relevant law, and makes rulings on the admissibility of evidence based on the law.

11.5.1* Rules of Evidence.

11.5.1.1 Rules of evidence regulate the admissibility of proof at a trial. The purpose of rules of evidence is to ensure that the proof offered is reliable. A goal of every fire investigation is to produce reliable documents, samples, statements, information, data, and conclusions. It is not necessary that every fire investigator become an expert on rules of evidence. If the practices and procedures recommended within this guide are complied with, the results of the investigation should be admissible.

11.5.1.2 Evidentiary requirements, standards, and rules vary greatly from jurisdiction to jurisdiction. For this reason, those rules of evidence that are in effect in individual states, territories, provinces, and international jurisdictions should be consulted. The United States Federal Rules of Evidence have been relied on throughout this guide for guidance in promoting their general criteria of relevance and identification. The Federal Rules of Evidence became effective on January 2, 1975, and have been amended several times. The most recent amendment was January 28, 2002. The federal rules are applicable in all civil and criminal cases in all United States courts of appeal, district courts, courts of claims, and before United States magistrates. The federal rules are recognized as having essentially codified the well-established rules of evidence, and many states have adopted, in whole or in part, the federal rules.

11.5.2 Types of Evidence. There are basically three types of evidence, all of which in some manner relate to fire investigations. They are "demonstrative evidence," "documentary evidence," and "testimonial evidence." They are described in detail in 11.5.2.1, 11.5.2.2, and 11.5.2.3.

11.5.2.1 Demonstrative Evidence. Demonstrative evidence consists of tangible items as distinguished from testimony of witnesses about the items. It is evidence from which one can derive a relevant firsthand impression by seeing, touching, smelling, or hearing the evidence. Demonstrative evidence should be authenticated. Evidence is authenticated in one of two ways: through witness identification (i.e., recognition testimony), or by establishing a chain of custody (an unbroken chain of possession from the taking of the item from the fire scene to the exhibiting of the item).

11.5.2.1.1 Photographs/Illustrative Forms of Evidence. Among the most frequently utilized types of illustrative demonstrative evidence are maps, sketches, diagrams, and models. They are generally admissible on the basis of testimony that they are substantially accurate representations of what the witness is endeavoring to describe. Photographs and movies are viewed as graphic portrayals of oral testimony and become admissible when a witness has testified that they are correct and accurate representations of relevant facts observed by the witness. The witness often need not be the photographer, but he or she should know about the facts being represented or the scene or objects being photographed. Once this knowledge is demonstrated by the witness, he or she can state whether a photograph correctly and accurately portrays those facts.

11.5.2.1.2 Samples. Chain of custody is especially important regarding samples. To ensure admissibility of a sample, an unbroken chain of possession should be established.

11.5.2.2 Documentary Evidence. Documentary evidence is any evidence in written form. It may include business records such as sales receipts, inventory lists, invoices, and bank records, including checks and deposit slips; insurance policies; personal items such as diaries, calendars, and telephone records; fire department records such as the fire investigator's report, the investigator's notes, the fire incident report, and witness statement reduced to writing; or any law enforcement agency reports, including investigation reports, police officer operational reports, and fire or police department dispatcher logs; division of motor vehicle records; and written transcripts of audio- or videotape recordings. Any information in a written form related to the fire or explosion incident is considered documentary evidence. Documentary evidence is generally admissible if the documents are maintained in the normal course of business. All witness statements should be properly signed by the witness, dated, and witnessed by a third party when possible. It is important to obtain the full name, address, and telephone number of the witness. Any additional identifying information (e.g., date of birth, social security number, and automobile license number) may prove helpful in the event that difficulties are later encountered in locating the witness. Statements actually written by the witness may be required in certain jurisdictions.

11.5.2.3 Testimonial Evidence. Testimonial evidence is that given by a competent live witness speaking under oath or affirmation. Investigators are frequently called on to give testimonial evidence regarding the nature, scope, conduct, and results of their investigation. It is incumbent on all witnesses to respond completely and honestly to all questions. There are two types of witnesses that offer testimony in a legal proceeding: "fact witnesses" and "expert witnesses."

11.5.2.3.1 Fact Witnesses. A fact witness is one whose testimony is not based on scientific, technical, or other specialized knowledge. An example of a fact witness is a neighbor who discovered the fire and testifies about their observations. An investigator will often be called to give testimony before courts, administrative bodies, regulatory agencies, and related entities as a fact witness. In addition to giving factual testimony, an investigator can be called to give conclusions or opinions regarding a fire, as an expert witness. Opinion testimony by a fact witness is allowed in limited circumstances. The circumstances are governed by Federal Rules of Evidence, Rule 701, or state rules of evidence.

11.5.2.3.2 Expert Witnesses. An expert witness is generally defined as someone with sufficient skill, knowledge, or experience in a given field so as to be capable of drawing inferences or reaching conclusions or opinions that an average person would not be competent to reach. The expert's opinion testimony should aid the judge or jury in their understanding of the fact at issue and thereby aid in the search for truth.

(A) Prior to offering opinion testimony identifying the origin and cause, a fire investigator must be accepted by the court as an expert. The opinion or conclusion of the investigator testifying as an expert witness is of no greater value in ascertaining the truth of a matter than that warranted by the soundness of the investigator's underlying reasons and facts. The evidence that forms the basis of any opinion or conclusion should be relevant and reliable and, therefore, admissible. The proper conduct of an investigation will ensure that these indices of reliability and credibility are met. The rules governing the admissibility of expert witnesses are contained in Rules 702, 703, 704, and 705 of the Federal Rules of Evidence in Federal Court, or the rules of the particular jurisdiction in which the case is pending. Rule 702 is shown in 11.5.2.3.2(B).

(B) If scientific, technical, or other specialized knowledge will assist the trier of fact to understand the evidence or to determine a fact in issue, a witness qualified as an expert through knowledge, skill, experience, training, or education may testify thereto in the form of an opinion or otherwise, if (1) the testimony is based upon sufficient facts or data, (2) the testimony is the product of reliable principles and methods, and (3) the witness has applied the principles and methods reliably to the facts of the case.

11.5.2.3.3 Admissibility of Expert Testimony. In order for expert testimony to be admitted as evidence in a legal proceeding, the court must determine that the testimony is relevant, that a witness is qualified, and that the testimony is reliable.

11.5.2.3.4 Relevance. A court will find that expert testimony is relevant if scientific, technical, or other specialized knowledge will assist the court or jury in understanding the evidence or decide the facts in the case. For example, in a case where the origin and cause of a fire is at issue, the testimony of an expert in fire origin and cause issues will be relevant in assisting the court or jury in understanding the issues in the case.

11.5.2.3.5 Qualifications of Expert. The court determines if a witness that is going to give expert testimony possesses the necessary qualifications to give such opinions. Typically, the court will look at the education, training, experience, or skill of the expert.

11.5.2.3.6 Reliability of Opinions. If the court determines that the expert's testimony is relevant and that the expert has the necessary qualifications to give an opinion, there is yet a third requirement that must be met before the opinion can be admitted into evidence. The court must find that the expert's opinion is reliable.

(A) Oftentimes a challenge to the reliability of an expert's opinions is called a "Daubert" challenge, based upon the decision of the United States Supreme Court in *Daubert v. Merrell Dow* (509 U.S. 579, 113 S.Ct. 2786). The holding in *Daubert* applies in federal courts and in those state courts that have recognized the rules set forth in that case. Not all state courts follow the rules set forth in *Daubert*.

(B) The Supreme Court in *Daubert* set forth factors a court may use in evaluating whether or not an expert's opinion is sufficiently reliable to be admissible. Subsequent Supreme Court decisions make it clear that the test of reliability is flexible and that this list of specific factors neither necessarily nor exclusively applies to all experts or in every case. These factors are as follows:

(1) Whether a theory or technique can be (and has been) tested
(2) Whether a theory or technique has been subjected to peer review and publication (although publication, or the lack thereof, is not a dispositive consideration)
(3) The known or potential rate of error of a particular scientific technique and the existence and maintenance of standards controlling the technique's operation
(4) That a "reliability assessment" does not require, although it does permit, explicit identification of a relevant scientific community and an express determination of a particular degree of acceptance of a theory or technique within that community

(C) It is important to note that the United States Supreme Court has held that the *Daubert* factors used in determining reliability apply not only to scientific testimony, but to testimony based upon technical or other specialized knowledge. The court's inquiry into the reliability of proposed expert testimony may extend to an evaluation of the methodology on which the opinion is based. The methodology will be validated upon a showing that accepted investigative techniques were used and that the methodology and reasoning were correctly applied to the facts at issue. The potential witness can use this document, as well as others, to establish that the methodology used in reaching the opinion was reliable.

11.5.3 Forms of Examination. The examination of witnesses generally proceeds in one of two forms, direct examination and cross-examination.

11.5.3.1 Direct Examination. Direct examination is the first examination of a witness in a trial or legal proceeding conducted by the attorney for the party on whose behalf the witness is called.

11.5.3.2 Cross-Examination. Cross-examination is the examination of a witness in a trial or legal proceeding by the party opposed to the one who produced the witness. The purpose of cross-examination is to further develop testimony given on direct examination, or to test the truth or impeach the witness's testimony.

11.5.4 Forms of Testimony. Testimony may be given both orally and in writing. Whether given in written or oral form, testimony is always given under oath.

11.5.4.1 Affidavits. An affidavit is a written statement of fact or opinion made by the witness voluntarily and signed by the witness under oath.

11.5.4.2 Answers to Interrogatories. Interrogatories are written questions to be answered by a party witness or other person who may have information of interest to a party in a legal proceeding. The answers must be signed under oath.

11.5.4.3 Depositions and Trial Testimony. These are the oral statements of a witness given under oath at a deposition or trial.

11.5.5 Burden of Proof. The burdens of proof in civil cases differ from those in criminal cases. In a criminal case, because the civil liberties of the defendant are at stake, the prosecutor must prove the defendant's guilt beyond a reasonable doubt. Civil cases typically involve disputes over money. In most civil cases, the plaintiff must prove his or her claims by a preponderance of the evidence, which means, "more likely than not." In some jurisdictions, the burden of proof in certain kinds of civil trials (e.g., those involving a claim of fraud) the standard for proof is "clear and convincing." It means the trier of fact must be persuaded by the evidence that it is highly probable that the claim or affirmative defense is true. The clear and convincing evidence standard is a heavier burden than the preponderance of the evidence standard but less than beyond a reasonable doubt.

11.5.6 Criminal Prosecution. Although there are certain fire-related crimes that appear to exist in all jurisdictions (e.g., arson), the full scope of possible criminal charges is as varied as the jurisdictions themselves, their resources, histories, interests, and concerns.

11.5.6.1 Arson.

11.5.6.1.1 Arson is the most commonly recognized fire-related crime.

11.5.6.1.2 Arson. *Black's Law Dictionary* defines arson as follows: At common law, the malicious burning of the house of another. This definition, however, has been broadened by state statutes and criminal codes. For example, the *Model Penal Code*, Section 220.1(1), provides that a person is guilty of arson, a felony of the second degree, if he starts a fire or causes an explosion with the purpose of: (a) destroying a building or occupied structure of another; or (b) destroying or damaging any property, whether his own or another's, to collect insurance for such a loss. In several states, this crime is divided into arson in the first, second, and third degrees: the first degree including the burning of an inhabited dwelling-house in the nighttime; the second degree, the burning (at night) of a building other than a dwelling-house, but so situated with reference to a dwelling-house as to endanger it; the third degree, the burning of any building or structure not the subject of arson in the first or second degree, or the burning of property, his own or another's, with intent to defraud or prejudice the insurer thereof.

11.5.6.2 Arson Statutes. The laws of each jurisdiction should be carefully researched regarding the requirements, burden of proof, and penalties for the crime of arson. Arson generally, or in the first and second degrees (if so classified), is deemed a felony offense. Such felony offenses require proof that the person intentionally damaged property by starting or maintaining a fire or causing an explosion. Arson in the third degree (if so classified) generally requires only reckless conduct that results in the damage of property and is often a misdemeanor offense.

11.5.6.3 Factors to Be Considered. The following factors are of relevance to most investigations when there is a possibility that the criminal act of arson was committed:

(1) Was the building, starting, or maintaining of a fire or the causing of an explosion intentional?
(2) Was another person present in or on the property?
(3) Who owned the property?

(4) If the property involved was a building, what type of building and what type of occupancy was involved in the fire?
(5) Did the perpetrator act recklessly, though aware of the risk present?
(6) Was there actual presence of flame?
(7) Was actual damage to the property or bodily injury to a person caused by the fire or explosion?

11.5.6.4 Other Fire-Related Criminal Acts.

11.5.6.4.1 The bases of fire-related criminal prosecution vary greatly from jurisdiction to jurisdiction. It is impossible to list all possible offenses. The following nonexclusive list of sample acts that can result in criminal prosecution will alert the investigator to the possibilities in any given jurisdiction: insurance fraud; leaving fires unattended; allowing fires to burn uncontrolled; allowing fires to escape; burning without proper permits; reckless burning; negligent burning; reckless endangerment; criminal mischief; threatening a fire or bombing; failure to report a fire; failure to report smoldering conditions; tampering with machinery, equipment, or warning signs used for fire detection, prevention, or suppression; failure to assist in suppression or control of a fire; sale or installation of illegal or inoperative fire suppression or detection devices; and use of certain equipment or machinery without proper safety devices, without the presence of fire extinguishers, or without other precautions to prevent fires. Criminal sanctions are almost universally imposed for failures to obey orders of fire marshals, fire wardens, and other officials and agents of public sector entities created to promote, accomplish, or otherwise ensure fire prevention, protection, suppression, or safety.

11.5.6.4.2 Key industries or resources within a given jurisdiction often result in the enactment of special and detailed criminal provisions. By way of example, criminal statutes exist with specific reference to fires in coal mines, woods, prairie lands, forests, and parks, and during drought or emergency conditions. Special provisions also exist regarding the type of occupancy or use of a given structure (i.e., penal/correctional institutions, hospitals, nursing homes, day-care or child-care centers, and schools). The use or transportation of hazardous or explosive materials is regulated in nearly all jurisdictions.

11.5.6.5 Arson-Reporting Statutes. Many jurisdictions have enacted statutes requiring that information be released to public officials regarding fires that may have been the result of a criminal act. Commonly referred to as the "arson immunity acts," the arson-reporting statutes generally provide that an insurance company should, on written request from a designated public entity or official, release enumerated items of information and documentation regarding any loss or potential loss due to a fire of "suspicious" or incendiary origin. The information is held in confidence until its use is required in a civil or criminal proceeding. The insurance company is held immune from civil liability and criminal prosecution, premised upon its release of the information, pursuant to the statute. The number of jurisdictions with an arson-reporting act is growing, and it is anticipated that they will continue to grow. As enacted in each jurisdiction, the acts vary greatly as to both requirements and criminal sanctions. Each act does impose criminal sanctions for failure to comply. In order to avoid criminal prosecution, the insurance companies and investigators operating on its behalf should be aware of any applicable arson-reporting act. One should be alert to the variations in 11.5.6.5(A) through 11.5.6.5(E) that currently exist.

(A) In addition to the insurance company, some jurisdictions require compliance by its employees, agents, investigators, insureds, and attorneys.

(B) In addition to response to specific written requests for information or documentation, some jurisdictions state that an insurance company may inform the proper authorities whenever it suspects a fire was of "suspicious origin." Other jurisdictions state that an insurance company must inform the proper authorities whenever it suspects a fire was of "suspicious origin." Note that the term *suspicious origin*, as used within this section, refers to the actual language of some arson-reporting statutes. This guide does not recognize mere suspicion as an accurate or acceptable level of proof for making determinations of origin or cause, nor does it recognize "suspicious origin" as an accurate or acceptable description of cause or origin. This guide discourages the use of such terms.

(C) In addition to requiring production of specifically enumerated items of information and documentation, some jurisdictions require production of all information and documentation.

(D) Though most jurisdictions ensure absolute confidentiality of the information and documentation released, pending its use at a criminal or civil proceeding, other jurisdictions allow its release to other interested public entities and officials.

(E) In many jurisdictions, the immunity from civil liability and criminal prosecution is lost in the event that information was released maliciously or in bad faith.

11.5.7 Civil Litigation. Many fires result in civil litigation. These lawsuits typically involve claims of damages for death, injury, property damage, and financial loss caused by a fire or explosion. The majority of civil lawsuits are premised on allegations of negligence. A significant number of civil lawsuits are premised on the legal principle of product liability or alleged violations of applicable codes and standards.

11.5.7.1 Negligence.

11.5.7.1.1 Negligence generally applies to situations in which a person has not behaved in the manner of a reasonably prudent person in the same or similar circumstances. Liability for negligence requires more than conduct. The elements that traditionally should be established to impose legal liability for negligence may be stated briefly, as follows:

(1) Duty: A duty requiring a person to conform to a certain standard of conduct, for the protection of others against unreasonable risks
(2) Failure: A failure by the person to conform to the standard required
(3) Cause: A reasonably close causal connection between the conduct of the person and resulting injury to another (generally referred to as "legal cause" or "proximate cause")
(4) Loss: Actual loss or damage resulting to the interests of another

11.5.7.1.2 Hypothetical Example of the Elements of Negligence. The operator of a nursing home has a *duty* to install operable smoke detectors within the nursing home for the protection of the inhabitants of the nursing home. A reasonably prudent nursing home operator would have installed the smoke detectors. The operator of the nursing home *failed* to install operable smoke detectors. A fire began in a storage room. Because there were no smoke detectors, the staff and occupants of the nursing home were not alerted to the presence of the fire in time to allow the occupants to reach safety, and an occupant who could have otherwise been saved died as a result of the fire. The death of the occupant was proximately *caused* by the failure to install operable smoke detectors. The death constitutes actual *loss* or *damage* to the deceased occupant and his or her family. Once all four elements are established, liability for negligence may be imposed.

11.5.7.2 Codes, Regulations, and Standards. Various codes, regulations, and standards have evolved through the years to protect lives and property from fire. Violations of codes, regulations, rules, orders, or standards can establish a basis of civil liability in fire or explosion cases. Further, many jurisdictions have legislatively determined that such violations either establish negligence or raise a presumption of negligence. By statute, violation of criminal or penal code provisions may also entitle the injured party to double or triple damages.

11.5.7.3 Product Liability. Product liability refers to the legal liability of manufacturers and sellers to compensate buyers, users, and even bystanders for damages or injuries suffered because of defects in goods purchased. This tort makes manufacturers liable if their product has a defective condition that makes it unreasonably dangerous (unsafe) to the user or consumer. Although the ultimate responsibility for injury or damage most frequently rests with the manufacturer, liability may also be imposed upon a retailer, occasionally on a wholesaler or middleman, on a bailor or lessor, and infrequently on a party wholly outside the manufacturing and distributing process, such as a certifier. This ultimate responsibility may be imposed by an action by the plaintiff against the manufacturer directly, or by way of claims for indemnification or contribution against others who might be held liable for the injury caused by the defective product.

11.5.7.4 Strict Liability.

11.5.7.4.1 Courts apply the concept of strict liability in product liability cases in which a seller is liable for any and all defective or hazardous products that unduly threaten a consumer's personal safety. This concept applies to all members involved in the manufacturing and selling of any facet of the product. The concept of strict liability in tort is founded on the premise that when a manufacturer presents a product or good to the public for sale, the manufacturer represents that the product or good is suitable for its intended use. In order to recover in strict liability, it is essential to prove that the product was defective when placed in the stream of commerce and was, therefore, unreasonably dangerous.

11.5.7.4.2 The following types of defects have been recognized: design defects; manufacturing defects; failure to warn or inadequacy of warning; and failure to comply with applicable standards, codes, rules, or regulations. The three most commonly applied defects are described in 11.5.7.4.2(A) through 11.5.7.4.2(C).

(A) Design Defect. The basic design of the product contains a fault or flaw that has made the product unreasonably dangerous.

(B) Manufacturing Defect. The design of the product may have been adequate, but a fault or mistake in the manufacturing or assembly of the product has made it unsafe.

(C) Inadequate Warnings. The consumer was not properly instructed in the proper or safe use of the product; nor was the consumer warned of any inherent danger in the possession of, or any reasonably foreseeable use or misuse of, the product. Strict liability applies, although the seller has exercised all possible care in the preparation and sale of a product. It is not required that negligence be established.

Chapter 12 Safety

12.1* General. Fire scenes, by their nature, are dangerous places. Fire investigators have an obligation to themselves and perhaps to others (such as other investigators, equipment operators, laborers, property owners, attorneys) who may be endangered at fire scenes during the investigation process. This chapter will provide the investigator with some basic recommendations concerning a variety of safety issues, including personal protective equipment (PPE). It should be noted, however, that the investigator should be aware of and follow the applicable requirements of safety-related laws (OSHA, federal, or state) or those policies and procedures established by their agency, company, or organization.

12.1.1* General Injury/Health Statistics. The fireground atmosphere encountered by fire and explosion investigators as part of their normal work routine changes rapidly with time, may contain a combination of multiple respiratory hazards, and can be immediately dangerous to life and health (IDLH). The inhalation of harmful dusts, toxic gases, and vapors at fire and explosion scenes is a common hazard to investigators who typically arrive to initiate their investigation after fire suppression and overhaul operations are completed.

12.1.1.1* Many researchers have assessed the extent to which fire fighters are exposed to hazardous substances during extinguishment activities. These studies set the foundation upon which better protective standards for fire investigators can be developed and present issues related to short-term and long-term health effects.

12.1.1.2 Although limited, some research has attempted to quantify the hazards present during the investigation of fires. The National Institute for Occupational Safety and Health (NIOSH) in conjunction with the Bureau of Alcohol, Tobacco, Firearms and Explosives recognized the hazards of overhaul operations from fire investigators sifting through debris in a 1998 health hazard evaluation report, and in 2007, published the results of a study regarding contamination of clothing exposed at fire scenes.

12.1.1.2.1 The 1998 study quantified compounds present at fire scenes post-fire extinguishment. Although in low concentrations, compounds detected included dusts, aliphatic hydrocarbons, acetone, acetic acid, ethyl acetate, isopropanol, styrene, benzene, toluene, xylene, furfural, phenol, and naphthalene. Polycyclic aromatichydrocarbons (PAHs) with carcinogenic potential included benzo(a)anthracene, benzo(b)fluoranthene, and benzo(a)pyrene. While all of the previous compounds were found at levels below NIOSH recommended exposure limits, formaldehyde was found in concentrations nearly twice as high as its limit of 0.1 ppm.

12.1.1.2.2 The 2007 study quantified the hazards presented to investigators and their families due to contamination of their clothing during fire scene investigations. Researchers found a potential for contamination of other clothing washed with the soiled uniforms. Based on the report findings, the researchers recommended that protective clothing should be worn during fire scene investigations, and to reduce the potential for carrying contaminants home, investigators should use disposable coveralls, or use a professional laundry service for this purpose.

12.1.2 Health and Safety Programs All public and private sector employers have a responsibility to provide a "safe" workplace and to protect their employees from recognized hazards, as required under the General Duty Clause of the Occupational Safety and Health Administration (OSHA) Act of 1970. Investigators and their employers are expected to comply with all OSHA regulations, standards, and practices applicable to the tasks and activities conducted at their workplace, which most often will be at fire and explosion scenes. The key to compliance with occupational safety and health regulations and the foundation of an organization's standard operating procedures, policies, and employee training programs is a comprehensive written Occupational Safety and Health Program.

12.1.2.1 OSHA has identified five critical elements that have consistently proven successful in helping organizations reduce the incidence of occupational injuries, illnesses, and fatalities and that are necessary to develop and implement an effective fire investigator occupational safety and health program.

12.1.2.1.1 Management Commitment and Employee Participation. Organizations must have a clearly articulated written safety and health policy statement that is understood by all personnel. It is critical that everyone understand the priority of safety and health protection in relation to other organizational values.

12.1.2.1.2 Hazard and Risk Assessment. Identifying potential hazards at a fire or explosion scene requires an active, ongoing examination and analysis of work processes, practices, procedures, equipment, and working conditions. Identifying hazards not only helps to determine the appropriate level of personal protective clothing and equipment (PPE) needed to adequately protect investigators, but it also can be used to identify appropriate training and education needs.

12.1.2.1.3 Hazard Prevention and Control. Hazard prevention and control is based on the determination that a potential hazard always exists at every scene. Hazards are either eliminated or managed by the implementation of SOPs and work practices that outline effective engineering controls and PPE. This process provides for the systematic identification, evaluation, prevention, and control of general workplace hazards and less obvious hazards that may arise during on-site activities.

12.1.2.1.4 Safety and Health Training and Education. An effective training and education program addresses the safety and health responsibilities of all personnel throughout the organization, including supervisors. Agencies should consider integrating some aspect of safety and health training and education into all organizational training and education activities to reinforce the importance of safety.

12.1.2.1.5 Long-Term Commitment. Management and employees must make a serious commitment to sustain the organization's safety and health program and make it a key priority. Without this level of commitment, the safety and health program is doomed for failure. Organizations should reach out and continually look for new and improved practices, methods, programs, technology, and equipment specifically tailored to the duties and responsibilities of investigators.

12.1.2.2 An effective Fire Investigator Occupational Safety and Health Program includes provisions for the systematic identification, evaluation, and prevention or control of general workplace hazards and less obvious hazards that may arise during on-site activities. Investigators should refer to NFPA 1500, *Standard on Fire Department Occupational Safety and Health Program*, for specific guidance on developing an effective risk management/health and safety program for their organization.

12.1.3 Safety Clothing and Equipment. Proper personal protective equipment, including safety shoes or boots with a protective mid-sole, gloves, safety helmet, eye protection, and protective clothing, should be worn at all times while investigating the scene. The type of protective clothing will depend on the type and level of hazard present. When there is a potential for injuries from falling objects or potential cuts or scrapes from sharp objects, fire-fighting turnout gear or similar clothing that provides this type of protection may be the best choice. When an investigator is dealing with a potential exposure of toxic substances and debris, disposable coveralls as required by some safety-related regulations may be necessary. In high hazard atmospheres, *hazardous environmental suits* may be required. Whenever PPE is worn to provide protection from a hazardous environment, it should be properly decontaminated or disposed of in order to avoid subsequent exposure to residues. Even when choosing to wear standard cloth coveralls or fire-fighting turnout gear, consideration should be given to the safe handling of the clothing so as not to create additional exposure.

12.1.3.1 Appropriate respiratory protection is necessary at most fire scenes. Immediately following fire extinguishment there may be combustible gases and smoke, low oxygen concentrations, toxic or carcinogenic airborne particles, and high heat conditions present. In these atmospheres, the investigator should utilize SCBA and other PPE that are appropriate and should recognize that air-purifying respirators should not be utilized in atmospheres where the oxygen level is below 19.5 percent or Immediately Dangerous to Life and Health (IDLH) atmospheres are present. The act of disturbing the fire debris can create dust and release organic vapors, which should be considered hazardous, and the investigator should be wearing a filter mask and an air purifying respirator with appropriate cartridges. The decision to wear a full-face respirator versus a half-face respirator will be up to the employer and depends on the hazards present. In the respirator selection process, consideration should be given to eye protection, as many toxic substances can be absorbed through the sclera. If a half-face respirator is selected, then wearing a pair of vented goggles will provide protection from this type of hazard. If respiratory protection is worn, the investigator or other individual will need to be properly trained, medically and physically fit, and have been properly fit tested when required for the particular respiratory protection being worn. Additional guidance concerning respirators and the responsibilities of the employer and employee are contained in Occupational Safety and Health Administration (OSHA) Regulation 29 CFR, Section 1910.134 (Respiratory Protection).

12.1.3.2 The proper selection of gloves that provide puncture protection or protection from biological or chemical contamination should also be considered. When conducting scene excavation or debris removal, puncture-resistant fire-fighting gloves or lighter leather gloves should be selected. Additional protection from the leaching of toxic substances should be provided by wearing latex (or similar) gloves underneath the leather gloves, or the investigator may need to select gloves that would be more appropriate for the hazard present.

12.1.3.3 Certain other equipment might also be necessary to maintain safety. This equipment includes flashlights or portable lighting, fall protection equipment, environmental monitoring and sampling equipment, and other specialized tools and equipment. Some of this equipment requires special training in its use.

12.2 General Fire Scene Safety. The investigator should remain aware of the general and particular dangers of the scene under investigation. The investigator should keep in mind the potential for serious injury at any time and should not become complacent or take unnecessary risks.

12.2.1 Investigating the Scene Alone. Fire scene examinations should not be undertaken alone. A minimum of two individuals should be present to ensure that assistance is at hand if an investigator should become trapped or injured. If the fire scene is investigated by one investigator, a clear communications protocol needs to be established between the site investigator and an off-site contact person. An estimated completion time should be established, and periodic contacts between the scene investigator and off-site contact person should be made at regular intervals. If it is impossible for the investigator to be accompanied, he or she should at least notify a responsible person of where the investigator will be and of when he or she can reasonably be expected to return.

12.2.2 Investigator Fatigue.

12.2.2.1 It is common for investigators to put in long periods of strenuous personal labor during an incident scene investigation. This labor may result in fatigue, which can adversely influence an investigator's physical coordination, strength, or judgment to recognize or respond to hazardous conditions or situations. Keep in mind that the use of heavy safety clothing and respiratory protection will further increase fatigue.

12.2.2.2 Periodic rest, fluid replacement, and nourishment should be obtained in a safe atmosphere, remote from but convenient to the fire scene. Sanitation facilities that include a restroom and washing station are necessary on large or major incidents. The hazard to the fire investigator is not just through aspiration and absorption but also through ingestion, so it is essential that eating and drinking occur out of the scene after removal of contaminated gear and the washing of face and hands.

12.2.3 Working Above or Below Grade Level. Whenever the investigators have to work above or below grade level they should be aware of the special hazards that may be present.

12.2.3.1 Standing water can pose a variety of dangers to the investigator. Puddles of water in the presence of energized electrical systems can be lethal if the investigator should touch an energized wire, ungrounded appliance, or other piece of equipment while standing in water.

12.2.3.2 Pools of water that may appear to be only inches deep may in fact be well over the investigator's head. Pools of water may also conceal hidden danger such as holes or dangerous objects that may trip or otherwise injure the investigator.

12.2.3.2.1 Suppression foam is used by fire departments in both Class A and Class B fires. Foam can pose a hazard to any fire scene and the investigators. The foam can hide holes in the floor, tripping hazards, debris, sharp objects, tools, and various other items left at the fire scene. The foams can make walking surfaces slippery and can cause falls. If foam has been used, then it is recommended that the foam be allowed to dissipate, or the foam be carefully washed out of the scene prior to making entry so as to minimize the possibility of altering the scene or destroying evidence.

12.2.3.3 Air quality of basement or underground areas may require atmospheric testing. The testing should determine the oxygen concentration or evaluate other potential atmospheric conditions that are suspected.

12.2.3.4 When working above grade, the investigator should consider the need for appropriate fall protection equipment. Requirements of the OSHA (state or federal) regulations for fall protection and fall protection trigger heights should be consulted and followed.

12.2.3.5 When working from any aerial platform, the investigator should determine if that platform or piece of equipment has been designed (labeled) for use by people. Equipment not designated for use by people should not be used.

12.2.4 Working Around Mechanized Equipment. The utilization of heavy and mechanized equipment at the fire or explosion scene presents unique issues and concerns for both those investigators that are present and the overall scene safety.

12.2.4.1 When mechanized equipment is in use, the area should be isolated by barricades to prevent entry into that area or if investigators are required to be in that area, they should wear appropriate high-visibility vests or clothing, and a safety monitor that will communicate with the operator and warn investigators of changing hazards or conditions should be present.

12.2.4.2 The swing areas of cranes and the path that will be taken to remove debris should be identified and barricaded to prevent entry and potential injury. No one should work under a load that is being moved by a crane.

12.2.5 Safety of Bystanders.

12.2.5.1 Fire and explosion scenes often generate the interest of bystanders. Their safety, as well as the security of the scene and its evidence, should be addressed by the investigator.

12.2.5.2 The investigation scene should be secured from entry by curious bystanders. This security may be accomplished by merely roping off the area and posting "Keep Out" signs and barricade tape, or it may require the assistance of police officers, fire service personnel, or other persons serving as guards. Any unauthorized individuals found within the fire investigation scene area should be identified and their identity noted; then they should be escorted off the site to prevent potential injury.

12.2.6 Status of Suppression.

12.2.6.1 If the investigator is going to enter parts of the structure before the fire is completely extinguished, the investigator should receive permission from the fire ground commander. The investigator should coordinate his or her activities with the fire suppression personnel and keep the fire ground commander advised of the areas into which he or she will be entering and working. The investigator should not move into other areas of the structure without informing the fire ground commander. The investigator should not enter a burning structure unless accompanied by fire suppression personnel, and unless appropriately trained to do so.

12.2.6.2 When conducting an investigation in a structure soon after the fire is believed to be extinguished, the investigator should be mindful of the possibility of a rekindle. The investigator should be alert for continued burning or a rekindle and should remain aware at all times of the fastest or safest means of egress.

12.2.7 First Aid Kit and Emergency Notification Numbers. The controlling entity at a fire or explosion scene should at a minimum have a first aid kit and access to local emergency notification numbers and the location of emergency medical care in the event that an emergency arises during the investigative process.

12.2.8 Emergency Notification Signal. The controlling entity at a fire or explosion scene should have an established emergency evacuation signal and meeting place identified for other investigators that may be working on the scene. The type of signal and evacuation location should be discussed during the first safety meeting and at other times when new investigators arrive at the scene.

12.3 Fire Scene Hazards. The investigator should remain aware of the general and particular dangers of the scene under investigation. The investigator should keep in mind the potential for serious injury at any time and should not become complacent or take unnecessary risks. The need for this awareness is especially important when the structural stability of the scene is unknown or when the investigation requires that the investigator be working above or below ground level. Even in cases where the fire investigator believes the structure to be stable, caution should always be taken, as visual observations of the stability of the structure are not always consistent with the actual stability of the building. Heat and/or suppression activities can cause the structural components of the building to fail or weaken. It is recommended that investigators work in teams of two or more. By working in teams, the investigators can assist each other and help ensure each other's safety. Whereas working alone is not recommended, when instances arise that necessitate that an investigator work alone, information as to where, when, and for how long the investigator is working at the scene should be provided to someone in case of an accident or mishap.

12.3.1 Physical Hazards. Slip, trip, and fall hazards; holes in floors; sharp surfaces; broken glass; and other such hazards can cause injury to the investigator. Investigator fatigue increases the potential for physical injury while investigating the fire scene. When using hand tools, portable power tools, and ladders, care should be taken to observe all safety requirements and operational guidelines to lessen the potential for injury. The use of flashlights or portable lighting (intrinsically safe, if required) will reduce the potential of a slip or fall. Additionally, identification of, marking of, and covering holes and other items that can pose a physical hazard will reduce the potential for injury. Standing water and wet or slippery surfaces should be appropriately marked and barricaded to prevent investigators from entering the area.

12.3.2 Structural Stability Hazards. By their nature, most structures that have been involved in fires or explosions are structurally weakened. Roofs, ceilings, partitions, load-bearing walls, and floors may have been compromised by the fire or explosion.

12.3.2.1 Heat affects various building components in different ways, some of which may not be visible to the naked eye. Caution should always be taken to assess the stability of the structure prior to entering and throughout the processing of the scene. If the scene processing takes more than one day, stability assessments should be conducted numerous times, as the effects of the fire damage to the building may change throughout scene processing. Weather conditions can affect the stability of the building, necessitating constant reassessments throughout the entire scene processing.

12.3.2.2 The investigator's task requires that he or she enter these structures and often requires that he or she perform tasks of debris removal that may dislodge or further weaken these already unsound structures. Before entering such structures or beginning debris removal, the investigator should make a careful assessment of the stability and safety of the structure. If necessary, the investigator should seek the help of qualified structural experts to assess the need for the removal of dangerously weakened construction or should make provisions for shoring up load-bearing walls, floors, ceilings, or roofs.

12.3.3 Electrical Hazards. Electrical hazards at the investigation scene can come from the building's electrical utility service, emergency or standby power, or those tools and equipment the investigator brings on to the scene. The electrical service should be disconnected or the appropriate circuits isolated.

12.3.3.1 Although the fire investigators may arrive on the scene hours or even days later, they should recognize potential hazards in order to avoid injury or even death. Serious injury or death can result from electric shocks or burns. Investigators as well as fire officers should learn to protect themselves from the dangers of electricity while conducting fire scene examinations. The risk is particularly high during an examination of the scene immediately following the fire. When conditions warrant, the investigator should ensure that the power to the building or to the area affected has been disconnected prior to entering the hazardous area. The investigator should also recognize that buildings may have several utility feeds and should ensure that all feeds are disconnected prior to entering the hazardous area. The fire investigator should not disconnect the building's electric power but should ensure that the authorized utility does so.

12.3.3.2* Lockout/Tagout (LOTO). When electrical service has been interrupted and the power supply has been disconnected, a warning tag or lock should be attached to the appropriate disconnect, indicating that power has been shut off. If more than one person or group is investigating the scene, each person or group should attach their own warning tag or lock on the requisite appropriate disconnect meter. This may precludes the disconnect meter from being inadvertently switched to the "on" position by a person or group leaving the area while a second person or group is still processing the scene. In considering potential electrical hazards, always assume that danger is present. The investigator should personally verify that the power has been disconnected. This verification can be accomplished with the use of a voltmeter. Some meters allow the accurate measuring of volts, ohms, and resistance. Other devices are designed simply to indicate the presence of alternating current. These pencil-sized products give an audible or visual alarm when the device tip is placed on the wire (bare or jacketed). When utilizing voltage-testing equipment, it is imperative that the testing device be rated for the voltage supplied to the structure under investigation. Utilization of equipment that is not rated exposes the investigator to electrocutions and puts other investigators in the area of the testing at great risk. If any doubt exists as to whether the equipment is energized, the local electric utility should be called for verification.

12.3.3.3 The investigator may be working at a fire scenes that have has been equipped with temporary wiring. The investigator should be aware that temporary wiring for lighting or power arrangements is often not properly installed, grounded, or insulated and, therefore, may be unsafe.

12.3.3.4 The investigator should consider the electrical hazards shown in 12.3.3.4.1 through 12.3.3.4.12 when examining the fire scene.

12.3.3.4.1 All wires should be considered energized or "hot," even when the meter has been removed or disconnected.

12.3.3.4.2 When approaching a fire scene, the investigator should be alert to fallen electrical wires on the street; on the ground; or in contact with a metal fence, guard rail, or other conductive material, including water.

12.3.3.4.3 The investigator should look out for antennas that have fallen on existing power lines, for metal siding that has become energized, and for underground wiring.

12.3.3.4.4 The investigator should use caution when using or operating ladders or when elevating equipment in the vicinity of overhead electric lines.

12.3.3.4.5 It should be noted that building services are capable of delivering high amperage and that short circuiting can result in an intense electrical flash, with the possibility of serious physical injury and burns.

12.3.3.4.6 Rubber footwear should not be depended on as an insulator.

12.3.3.4.7 A flooded basement should not be entered if the electrical system is energized. Energized electrical equipment should not be turned off manually while standing in water.

12.3.3.4.8 Operation of any electrical switch or non-explosionproof equipment in the area that might cause an explosion if flammable gas or vapors are suspected of being present should be avoided. *(See 12.3.4.)* When electric power must be shut off, it should be done at a point remote from the explosive atmosphere.

12.3.3.4.9 Lines of communication and close cooperation with the utility company should be established. Power company personnel possess the expertise and equipment necessary to deal with electrical emergencies.

12.3.3.4.10 The investigator should locate and avoid underground electric supply cables before digging or excavating on the fire scene.

12.3.3.4.11 The investigator should be aware of multiple electrical services that may not be disconnected, extension cords from neighboring buildings, and similar installations.

12.3.3.4.12 A meter always should be used to determine whether the electricity is off.

12.3.4 Chemical Hazards. Fires and explosions often generate toxic gases. The presence of hazardous materials in the structure is certain. Homes contain chemicals in the kitchen, bath, and garage that can create great risk to the investigator if he or she is exposed to them. Commercial and business structures are generally more organized in the storage of hazardous materials, but the investigator cannot assume that the risk is less in such structures. Many buildings built prior to 1975 will contain asbestos. The investigator should be aware of the possibility that he or she could become exposed to dangerous atmospheres during the course of an investigation.

12.3.4.1 Fire scene atmospheres may contain ignitible gas, vapors, and liquids as well as low oxygen concentrations. The atmosphere should be tested using appropriate equipment to determine whether such hazards or conditions exist before working in or introducing ignition sources into the area. Such ignition sources may include electrical arcs from flashlights, radios, cameras and their flashes, and smoking materials.

12.3.4.2 The investigator should be aware that the atmosphere may change while processing a scene. As the investigator moves objects during the excavation of the scene, pockets of gases may escape or containers and pipes may be ruptured. Therefore, the atmosphere may need to be monitored.

12.3.4.3 Chemicals that are normally present at the scene or those that are a result of the incident should be considered. In commercial occupancies, the investigator may wish to obtain copies of material safety data sheets (MSDS) to determine the hazards of those products. The identification of chemical hazards that may be present as a result of the incident is more difficult. There are many reference documents the investigator may use to determine the hazards of suspected chemicals present at the investigation scene, including the National Institute of Safety and Health (NIOSH) *Pocket Guide to Chemical Hazards.*

12.3.4.4 Gas utilities that serve the structure or the industrial process should be identified and shut off at the meter and locked out or tagged out. If complete isolation of the building cannot be accomplished, the investigator should ensure that the area of the building or industrial process being excavated and examined be isolated from any connected gas utilities.

12.3.4.5 The presence of chemicals such as pesticides should also be considered in both residential and commercial occupancies. If they are properly contained, they generally will not pose a threat. However, if the container is broken prior to or while processing the scene, the investigator will need to take appropriate precautions such as avoiding the area or utilizing appropriate PPE.

12.3.5 Biological Hazards. Sources of biological hazards include bacteria, viruses, insects, plants, birds, animals, and humans. These sources can cause a variety of health effects ranging from skin irritation and allergies to infections (e.g., tuberculosis, communicable diseases), cancer, etc. Some of these hazards may not be recognized without specialized assistance.

12.3.5.1 There are common sources of biological hazards found in residential and commercial occupancies, including decomposing food, garbage, animals that did not survive the fire, and broken or damaged waste water pipes and systems. The investigator should not open refrigerators or freezers without considering the condition of the food, especially if the electrical service has been off for several days.

12.3.5.2 If the investigator is required to work around biological hazards, appropriate PPE should be worn and upon completion of the work, appropriate decontamination and disposal should occur. The use of disposable outer garments is helpful, as they can provide excellent protection and limit the need for decontamination of garments worn under them.

12.3.6 Mechanical Hazards. Machinery and equipment present on the scene may have stored energy. Prior to working around machinery and equipment, the investigator will need to determine if they are at zero mechanical state or if they are still operational or functional. For specialized machinery or equipment, the investigator may need to seek the assistance of the property owner or other technical resource to assist in controlling the stored energy.

12.3.7 Miscellaneous Hazards. In addition to the hazards previously listed, there are some hazards specific to the particular occupancy.

12.3.7.1 Radiological Hazards. Radiological hazards may be found in medical offices and some industrial occupancies. Medical offices may contain small amounts of radioactive materials.

12.3.7.2 Utilities. The investigator should determine the status of all utilities (i.e., electric, gas, and water) within the structure under investigation. Determine before entering if electric lines are energized (primary, secondary, or temporary electrical service), if fuel gas lines are charged, or if water mains and lines are operative. Determining the status of all utilities is necessary to prevent the possibility of electrical shock or inadvertent release of fuel gases or water during the course of the investigation.

12.3.7.3 Mechanized Equipment Hazards Using mechanized equipment during the fire scene processing brings additional dangers to the scene. Care should be taken while processing the scene during mechanized equipment usage. The investigator should be aware of the movement of equipment and materials and recognize that the operator of that equipment may not be aware that the investigator may be in danger.

12.4 Safety Plans. There could be a number of safety plans that the investigator may be required to develop as a part of the investigative process. The complexity of the plans and the topics included will vary depending on the hazards and risks identified on the scene. Other factors may need to be considered, including the number of investigators and support personnel, severity of the hazards and risks present, the use of specialized PPE, use of mechanized equipment, and government and organizational policies and procedures.

12.4.1* Hazard and Risk Assessment. One of the first tasks that should be completed before a fire or explosion scene investigation is begun is a Hazard and Risk Assessment. The investigator will be able to determine the hazards present and control those hazards by engineering or administrative control or through the selection and use of appropriate PPE.

12.4.1.1 Identify the Hazards. The hazard and risk assessment process begins with the identification of hazards. To simplify the hazard identification process and to allow for more systematic and complete identification, hazards can be grouped by type.

12.4.1.1.1 Physical Hazards. Physical hazards, such as slip, trip, and fall hazards, or sharp surfaces, broken glass, and other such hazards, can cause a physical hazard to the investigator.

12.4.1.1.2 Structural Hazards. Many structural hazards are easily identified without the need to have specialized technical assistance, but in complex scenes or heavily damaged scenes the investigator may want to consider the assistance of a structural engineer.

12.4.1.1.3 Electrical Hazards. Electrical hazards at the investigation scene can come from the building's electrical utility service, emergency or standby power, or those tools and equipment the investigator brings on to the scene. The electrical service should be disconnected or the appropriate circuits isolated.

12.4.1.1.4 Chemical Hazards. Chemicals that are normally present at the scene or those that are a result of the incident should be considered. In commercial occupancies, the investigator may wish to obtain copies of Material Safety Data Sheets (MSDS) to determine the hazards of those products. The identification of chemical hazards that may be present as a result of the incident is more difficult. There are many reference documents the investigator may use to determine the hazards of suspected chemicals present at the investigation scene, including the National Institute of Safety and Health (NIOSH) *Pocket Guide to Chemical Hazards*.

12.4.1.1.5 Biological Hazards. Sources of biological hazards include blood, body fluids, and human remains, bacteria, viruses, insects, plants, birds, animals, and humans. These sources can cause a variety of health effects ranging from skin irritation and allergies to infections (e.g., tuberculosis, communicable diseases), cancer, etc. Some of these hazards may not be recognized without specialized assistance.

12.4.1.1.6 Mechanical Hazards. Machinery and equipment present on the scene may have stored energy. Prior to working around machinery and equipment, the investigator will need to determine if they are at zero mechanical state or if they are still operational or functional. For specialized machinery or equipment, the investigator may need to seek the assistance of the property owner or other technical resource to assist in controlling the stored energy.

12.4.1.2 Determine the Risk of the Hazard. Depending on the specific hazard identified, the determination of the risks associated with the hazard could vary from simple qualitative assessments to complex quantitative assessments. Also, as a part of this analysis, the investigator will determine the likelihood that they will come in contact with that hazard. As an example, for a chemical (even if it is a chemical hazard) contained in a sealed drum, the risk is minimal. Given that example, a control mechanism may be to isolate the area where the container is in order to prevent damage and potential release.

12.4.1.3 Control the Hazard. Following the determination of the risk level, this level should be compared to a suitable benchmark or acceptance criteria. In some cases, the acceptance criteria has been established by regulators (OSHA). To control a hazard, the investigator can utilize several methodologies that include engineering controls, administrative controls, or the selection and use of appropriate PPE.

12.4.1.3.1 Engineering Controls. Engineering controls can be as simple as placing appropriate shoring to reinforce damaged structural elements or the demolition of those areas after they are properly documented. Or, they can be very complex solutions that will require a structural engineer to evaluate, design corrective measures, and manage the installation of the corrective measures.

12.4.1.3.2 Administrative Controls. Administrative controls can include the isolation of an area by the use of signs or barrier tape, by briefing of those that will be working in the area of the hazards and cautioning them that they are not to enter within the isolated area, by obtaining specialized resources that have expertise dealing with the hazard present, or by a combination of methodologies.

12.4.1.3.3 Proper Selection and Use of Personal Protective Equipment (PPE). The use of PPE is generally considered the least effective of the control measures. However, due to the conditions that the investigator may encounter at the scene and the duration of the work, PPE can be a suitable control mechanism. Care will need to be taken to determine the hazard present to ensure that the PPE selected is acceptable for the hazard present and that the user of the PPE is trained and capable to use it.

12.4.2 Site-Specific Safety Plans. After the Hazard and Risk Assessment process has been completed, specialized safety plans may need to be developed. If there are no hazards present that require a site-specific safety program, then one does not need to be developed. The controlling entity will need to determine the plans that are applicable, such as those listed below. Other investigators may also need to have a compatible program for their employees.

12.4.2.1* Hazard Communication Site Plan (HazCom Plan). The HazCom Plan includes the identification and location of hazardous materials, location of MSDSs, how exposure to the chemicals may occur, and labeling or identification of the materials. The HazCom plan also requires training and documentation of such training.

12.4.2.2 Confined Space Program. If the investigation requires entry into a confined space as defined by 29 CFR 1910.146, then a site program should be developed. Any persons not entering the confined space will not require training.

12.4.3 Management of Plans and Site Safety. Depending on the complexity of the scene, a formal organizational structure may need to be established for managing the safety component. For small scenes with very limited safety concerns, the safety component may be management in an informal manner only requiring the assessment of the scene and the development of a plan to control those hazards. For scenes of "complex investigations," safety will be a major function that will have to be assigned as a specific organizational function with direct input into the management of that investigation. Figure 12.4.3 is an example of how the safety function may be integrated into the management of a complex investigation.

12.4.4 Safety Meetings and Briefings. General safety meetings are conducted two or three times a workday. They can be conducted more frequently as need arises. Times for conducting such meetings that the investigator or person managing the safety task should consider are at the start of the day and after lunch. General safety meetings should be conducted as frequently as needed, often two or three times a day. A special safety meeting may be required prior to beginning a new phase or a new task. A safety de-briefing can also be used at the end of the special tasks or at the end of the investigation.

12.5 Chemical and Contaminant Exposure As part of the process of determining the type and level of PPE, the investigator should understand some basic terminology associated with the exposure to chemicals and substances they may encounter on the fire and explosion scene.

12.5.1 Types of Exposure Effects.

12.5.1.1 Local Effect. Local effects occur at the site of the contact, for example, an acid or caustic burn, or contamination by dusts or some liquids.

12.5.1.2 Systemic Effect. Systemic effects occur at a site that could be distant from the entry point of the substance, ultimately acting on a target organ or organ systems.

12.5.2 Routes of Exposure.

12.5.2.1 Inhalation. Inhalation is the most common route of entry for toxins in the workplace. Inhalation is also the most rapid and efficient route of exposure, immediately introducing toxic chemicals into the respiratory tissues and bloodstream. For this reason, it is considered the most important and potentially most serious method of exposure. This route can cause both local and systemic effects.

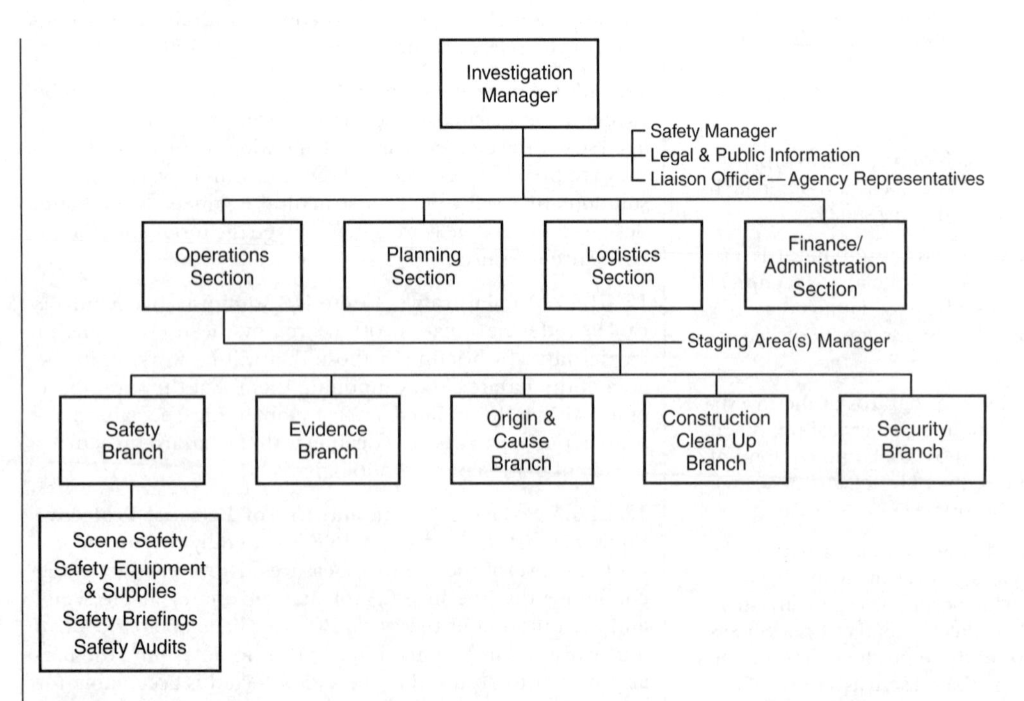

FIGURE 12.4.3 An Example of How the Safety Function May Be Completed at a Complex Investigation.

12.5.2.2 Cutaneous. Skin absorption occurs when the chemical passes directly through the skin. The primary function of the skin is to act as a barrier against the entry of foreign materials into the body. This barrier, which is effective against many chemicals, does allow some toxic materials to readily pass through. Usually the presence of cuts and scrapes will greatly increase the absorption rate. Absorption through the mucous membrane is more effective than through the skin. Absorption can lead to local and systemic effects.

12.5.2.3 Ingestion. The entry of toxic materials through ingestion usually occurs due to contamination of food and drinks, and to smoking. This type of exposure is most often associated with poor or no decontamination. Care taken to wash hands, arms, and face as well as the decontamination of clothing will limit the potential of ingesting a hazard. Also, moving to a place away from the potential contaminated area to eat or take rest breaks will assist in preventing cross-exposure.

12.5.2.4 Injection. Chemicals can enter the body when the skin is penetrated by a contaminated object. Minor lacerations such as a paper cut, if the paper is contaminated, can in fact cause problems. The contamination can be severe if the injury occurs around chemicals. Even small wounds should be treated to prevent contamination. Whenever possible, the proper protective clothing should be worn to lessen the potential of injury and contamination.

12.5.2.5 Ocular Exposure Route. Chemicals can be absorbed directly through the eyes. In some instances, the chemical cannot be detected, only the symptoms of the toxic effects can be noticed. In other situations the chemical causes an adverse effect on contact.

12.5.3 Toxicity Exposure Levels.

12.5.3.1 Acute Exposure. Acute exposures typically refer to a one-time high level of exposure of over a short period of time. This type of exposure is usually associated with inhalation of high concentrations or from direct skin contact by splash or immersion. The symptoms and effects are usually immediately apparent. However, in some situations the symptoms can be delayed until the chemical reaches a target organ. Effects can be reversible or irreversible.

12.5.3.2 Chronic Exposure. Chronic exposures typically refer to repetitive or continuous low-level exposures over a long period of time (weeks to years). In this type of exposure the inhalation concentrations are usually low, or direct skin contact involves substances that have a low potential for skin absorption. The symptoms and effects are usually delayed, in some instances 20 to 30 years. The effects can be reversible or irreversible.

12.5.3.3 Cumulative Exposure. Repeated exposure, either over a short period of time or longer periods, may allow the chemical exposure to add to the original dosage. Carbon monoxide is an example of a material for which the exposure dosage is cumulative. However, in this case, over time, the carbon monoxide is removed from the body. Lead, however, is cumulative and is not cleansed from the body through bodily functions.

12.5.3.4 Latency Period. Some chemical exposures will not cause symptoms until some time after exposure. This is called the latency period. Carcinogens are examples of products that have a latency period.

12.6 Personal Protective Equipment (PPE).

12.6.1 Proper Selection and Use of Personal Protective Equipment (PPE). The use of PPE is generally considered the least effective of the control measures. However, due to the conditions that the investigator may encounter at the scene and the duration of the work, PPE can be a suitable control mechanism. Care will need to be taken to determine the hazard present to ensure that the PPE selected is acceptable for the hazard present, and that the user of the PPE is trained and capable of using it, and understands the limitations of the equipment, the need for effective personal decontamination after using the equipment, and how to inspect and clean the equipment.

12.6.1.1 Safety Clothing and Equipment. Proper PPE, including safety shoes or boots with a puncture-resistant sole and steel toe, gloves, safety helmet, eye protection, and protective clothing, should be worn at all times while investigating the scene. The type of protective clothing will depend on the type and level of hazards present. When there is a potential for injuries from falling objects or potential cuts or scrapes from sharp objects, fire-fighting turnout gear or similar clothing that provides this type of protection may be the best choice. When an investigator is dealing with a potential exposure of toxic substances and debris, disposable coveralls may be necessary. In high hazard atmospheres, hazardous environmental suits may be required.

12.6.1.2 PPE Use. Whenever PPE is worn to provide protection from a hazardous environment, the wearer must be trained in the proper donning, doffing, limitations, use, and decontamination of such equipment to ensure that it is properly worn and functioning.

12.6.1.3* Decontamination. The investigator should be trained in the proper methodology to complete personal decontamination and in the proper method of decontamination or disposal of PPE worn in order to avoid subsequent exposure to residues still in the clothing and gear. The effort required to decontaminate clothing can be reduced through the use of outer disposable garments such as Tyvek® coveralls and latex booties over footware.

12.6.1.3.1 Standard Cloth Clothing. Even when choosing to wear standard cloth coveralls or fire-fighting turnout gear, consideration should be given to the safe handling of the clothing so as not to create additional exposure.

12.6.1.3.2 Investigators should decontaminate all potentially contaminated personal protective equipment (PPE) prior to leaving the scene to limit the potential for contaminating their vehicles, offices, and residences (or change their clothes to avoid spreading contamination to "clean" areas away from the scene).

12.6.1.3.3 If investigators opt to wash their clothing at home, contaminated clothing should not be washed with other "clean" clothing, to avoid the potential for cross-contamination. Investigators should also consider using a commercial specialty laundry service on a regular basis to ensure the greatest probability that their protective clothing does not contain potentially harmful contaminants that may lead to short-term and long-term health effects.

12.6.1.3.4 In those situations where these measures are not utilized or practical, investigators should employ a basic decontamination process that consists of scrubbing and rinsing contaminated gear and equipment with soap (detergent) and water. This process should be implemented in accordance with any specific manufacturer's recommendations for their respective equipment, such as respirators.

12.6.2* Examples of Personal Protective Equipment (PPE). Table 12.6.2 provides examples of PPE and the part of the body protected by that equipment.

Table 12.6.2 Protection of Body Part by Equipment Chart

Body Part	Example of PPE
Eye	Safety glasses, goggles, UV, welding and laser
Face	Face shield
Head	Hard hat, helmet
Feet	Safety shoes, boots
Hands and arms	Gloves
Body (torso)	Vests, aprons, chemical suits
Hearing	Earplugs, canal caps and earmuffs
Respiratory	APR, PAPR, SCBA, air supplied

12.7 Emergency Action Plans. A number of potential emergency situations could occur while processing a fire scene. Proper action by those on the scene will lessen the potential impact of those emergencies. Contained in the General Industry OSHA Standards 1910.38, there is a requirement to develop and implement an emergency action plan. While that standard does reference what is needed for a fire emergency, there is application for other emergencies to be included. An emergency action plan for two investigators processing the scene may be simple and communicated verbally to each other. On large or complex scenes where there will be a number of investigators present who may be working in different areas of the building or site, a more formalized set of emergency action plans may need to be developed. The plan examples in 12.7.1 through 12.7.5 are intended to provide the investigator with some basic information that should be included in the emergency action plans.

12.7.1 Emergency Evacuation Plans. In the event that the scene will need to be evacuated because of a change in structural conditions, accidental release of a hazardous material, severe weather, or some other unexpected condition, an emergency evacuation plan should be developed and implemented. The plan should include a method of notification, routes of escape, location for gathering, and method to account for all people who were working at the scene. This plan can be formal and written or discussed by all present during the investigation. This information should be communicated to additional investigators or groups as they arrive on scene.

12.7.2 Medical Emergency Plans. This plan can be written or simply communicated to participants during a safety meeting. The medical emergency plan should include locations of hospitals or other emergency facilities, emergency phone numbers for the local emergency medical system (EMS), the location of the first aid kit that is kept on scene, and notification of management controlling the scene. There may be additional items included as dictated by the condition or location of the scene.

12.7.3 Severe Weather Plans. As with the other plans previously discussed in 12.7.1 and 12.7.2, the severe weather plan will only need to be considered if there is a potential for severe weather. When conditions could change rapidly, it is advantageous to have organized and discussed the methodology of notification and where the meeting place is located.

12.7.4 Fire Emergency Plan. Assuming that the original fire has been extinguished, there are still situations that may occur on the fire or explosion scene during the investigative process that may cause a fire. The use of mechanized equipment, portable tools, and cutting and welding equipment can all be possible ignition sources. Additional sources of fuels other than ordinary combustibles would include hazardous materials and the building utilities. Included in the fire emergency plan would be the phone number and location of the nearest fire department, notification of others that are working on the scene, evacuation routes and meeting places, and a methodology for the accounting of personnel. This information should be communicated to all individuals who will be working on the scene, regardless of their responsibilities.

12.7.5 Additional Emergency Action Plans. There may be a need to develop additional emergency action plans based on issues identified that are specific to the scene. If additional plans are needed, then the controlling entity would be responsible for the development of the plans and the communication of the plan information to all others working at the scene.

12.8 Post-Scene Safety Activities. There are a number of safety-related items that may need to be completed after processing a fire or explosion scene. Two such activities are described in 12.8.1 and 12.8.2.

12.8.1 Decontamination. Decontamination of people, PPE, clothing, tools, and equipment used on scene should be completed in a manner that will not cause cross contamination or exposure to others. The amount and level of decontamination efforts should be commensurate with the hazards identified and the level of contamination exposed.

12.8.2 Medical Screening. Exposure to health hazards during the processing of the fire or explosion scene should be noted on the appropriate medical-related documents. If the investigator was exposed to health hazards during the investigation, additional medical screening may be required. Reporting of exposure and additional medical screening should be done in accordance with agency procedures and policies. The investigator should be aware that governmental reporting and documenting requirements may also apply. *(See OSHA 29 CFR 1910.120.)*

12.9 Safety in Off-Scene Investigation Activities.

12.9.1 Safety considerations also extend to ancillary fire investigation activities not directly related to the fire or explosion scene examination. Such ancillary investigation activities include physical evidence handling and storage, laboratory examinations and testing, and live fire or explosion recreations and demonstrations. The basic safety precautions dealing with use of safety clothing and equipment, and the proper storage and prominent labeling of hazardous materials evidence, thermal, inhalation, and electrical dangers of fire and explosion recreations or demonstrations should be included in evidence storage, examination, and testing protocols.

12.9.2 Off-Scene Interviews. The investigator may have to conduct witness interviews away from the fire scene in locations that are not totally controlled by the investigator. In that case, the investigator should be aware of surroundings and other actions that could cause harm to the investigator. Environmental hazards may include dogs or other dangerous animals, an armed witness, gang activities in the neighborhood, or any other situation that may put the investigator at risk.

12.9.3 Valuable safety information for those conducting ancillary fire investigation not directly related to the fire or explosion scene examination or witness interviews may be found in NFPA 30, *Flammable and Combustible Liquids Code*, NFPA 45, *Standard on Fire Protection for Laboratories Using Chemicals*, NFPA 1403, *Standard on Live Fire Training Evolutions*, and NFPA 1500, *Standard on Fire Department Occupational Safety and Health Program.* Additional information may also be obtained by the appropriate government agency regulations such as Occupational Safety and Health Administration (OSHA) documents, Environmental Protection Agency (EPA) documents, state and local regulations, and documents written by other standards-making organizations such as the Compressed Gas Association (CGA), American Petroleum Institute (API), American Society of Testing and Materials (ASTM), American National Standards Institute (ANSI), and others that may impact the investigation activities.

12.10 Special Hazards.

12.10.1 Criminal Acts or Acts of Terrorism. Fire is an event that can result from a criminal act. The initial incendiary device that created the fire or explosion may not be the only device left at the scene by the perpetrator. A secondary incendiary or explosive device may be left at the scene with the intent to harm fire, rescue, or investigative personnel. Of further concern are the chemicals used in the device that may leave a residue, creating an additional exposure.

12.10.1.1 Secondary Devices. The potential endangerment from a secondary incendiary or explosive device is remote compared to other hazards created at the scene from the initial device. However, the investigator should always be wary of any unusual packages or containers at the crime scene. If there is reason to believe that such a device may exist, it is necessary to contact the appropriate authorities to have specialists "sweep" the area. Close cooperation between investigative personnel and the explosive ordnance disposal (EOD) specialists can preclude the unnecessary destruction of the crime scene.

12.10.2 Residue Chemicals. If the incendiary or explosive device is rendered safe by the appropriate personnel, care should be taken when handling the rendered device or any residue from the device. Exposure to the chemical residue could endanger the investigator. Appropriate protective clothing and breathing apparatus should be worn while in the process of collecting such evidence.

12.10.3 Biological and Radiological Terrorism. There is a potential for a terrorist to release biological or radiological particulates as a part of his or her terrorist act. Usually the emergency response personnel will be aware of such an act while mitigating the emergency incident. If there is any suspicion that either type of hazardous substance has been released, the scene must be rendered safe prior to the entry of investigative personnel. If this rendering is not possible and the investigation is to go forward, only those investigative personnel trained to work in such atmospheres should be allowed to enter the scene.

12.10.4 Drug Labs. Completing an investigation at the scene of a drug lab can expose the investigator to hazardous chemicals. The investigator should take appropriate actions to prevent contamination, including the use of appropriate PPE and ensuring that appropriate decontamination is completed and that the scene is isolated to prevent exposure to others.

12.10.5 Exposure to Tools and Equipment. Many of the tools and equipment used in the process of conducting an investigation may be rendered unsafe after being used in hazardous atmospheres. The necessary procedures, equipment, tools, and supplies to render the equipment safe should be in place prior to the undertaking of the investigation. Precautions should also be in place to dispose of the tools safely should they be incapable of being rendered safe.

12.11 Factors Influencing Scene Safety. Many varying factors can influence the danger potential of a fire or explosion scene. The investigator should be constantly on the alert for these conditions and should ensure that appropriate safety precautions are taken by all persons working at the scene.

12.11.1 The investigator should determine the status of all utilities (i.e., electric, gas, and water) within the structure under investigation. Determine before entering if electric lines are energized (primary, secondary, or temporary electrical service), if fuel gas lines are charged, or if water mains and lines are operative. Determining the status of all utilities is necessary to prevent the possibility of electrical shock or inadvertent release of fuel gases or water during the course of the investigation.

12.11.2 Electrical Hazards. Although the fire investigators may arrive on the scene hours or even days later, they should recognize potential hazards in order to avoid injury or even death. Serious injury or death can result from electric shocks or burns. Investigators as well as fire officers should learn to protect themselves from the dangers of electricity while conducting fire scene examinations. The risk is particularly high during an examination of the scene immediately following the fire. When conditions warrant, the investigator should ensure that the power to the building or to the area affected has been disconnected prior to entering the hazardous area. The investigator should also recognize that buildings may have several utility feeds and should ensure that all feeds are disconnected prior to entering the hazardous area. The fire investigator should not disconnect the building's electric power but should ensure that the authorized utility does so.

12.11.2.1 When electrical service has been interrupted and the power supply has been disconnected, a tag or lock should be attached to the meter, indicating that power has been shut off. In considering potential electrical hazards, always assume that danger is present. The investigator should personally verify that the power has been disconnected. This verification can be accomplished with the use of a voltmeter. Some meters allow the accurate measuring of volts, ohms, and resistance. Other devices are designed simply to indicate the presence of alternating current. These pencil-sized products give an audible or visual alarm when the device tip is placed on the wire (bare or jacketed). When utilizing voltage-testing equipment, it is imperative that the testing device be rated for the voltage supplied to the structure under investigation. Utilization of equipment that is not rated properly exposes the investigator to electrocutions and puts other investigators in the area of the testing at great risk. If any doubt exists as to whether the equipment is energized, the local electric utility should be called for verification.

12.11.2.2 The investigator may be working at fire scenes that have been equipped with temporary wiring. The investigator should be aware that temporary wiring for lighting or power arrangements is often not properly installed, grounded, or insulated and, therefore, may be unsafe.

Chapter 13 Sources of Information

13.1 General.

13.1.1 Purpose of Obtaining Information. The thorough fire investigation always involves the examination of the fire scene, either by visiting the actual scene or by evaluating the prior documentation of that scene.

13.1.1.1 By necessity, the thorough fire investigation also encompasses interviewing and the research and analysis of other sources of information. These activities are not a substitute for the fire scene examination; they are a complement to it.

13.1.1.2 Examining the fire scene, interviewing, and conducting research and analysis of other sources of information all provide the fire investigator with an opportunity to establish the origin, cause, and responsibility for a particular fire.

13.1.2 Reliability of Information Obtained.

13.1.2.1 Generally, any information solicited or received by the fire investigator during a fire investigation is only as reliable as the source of that information. As such, it is essential that the fire investigator evaluate the accuracy of the information's source. Certainly, no information should be considered to be accurate or reliable without such an evaluation of its source.

13.1.2.2 This evaluation may be based on many varying factors, depending on the type and form of information. These factors may include the fire investigator's common sense, the fire investigator's personal knowledge and experience, the information source's reputation, or the source's particular interest in the results of the fire investigation.

13.2 Legal Considerations.

13.2.1 Freedom of Information Act.

13.2.1.1 The Freedom of Information Act provides for making information held by federal agencies available to the public unless it is specifically exempted from such disclosure by law. Most agencies of the federal government have implemented procedures designed to comply with the provisions of the act. These procedures inform the public where specific sources of information are available and what appeal rights are available to the public if requested information is not disclosed.

13.2.1.2 Like the federal government, most states have also enacted similar laws that provide the public with the opportunity to access sources of information concerning government operations and their work products. The fire investigator is cautioned, however, that the provisions of such state laws may vary greatly from state to state.

13.2.2 Privileged Communications.

13.2.2.1 Privileged communications are those statements made by certain persons within a protected relationship such as husband-wife, attorney-client, priest-penitent, and the like. Such communications are protected by law from forced disclosure on the witness stand at the option of the witness spouse, client, or penitent.

13.2.2.2 Privileged communications are generally defined by state law. As such, the fire investigator is cautioned that the provisions of such laws may vary greatly from state to state.

13.2.3 Confidential Communications. Closely related to privileged communications, confidential communications are those statements made under circumstances showing that the speaker intended the statements only for the ears of the person addressed.

13.3 Forms of Information. Sources of information will present themselves in differing forms. Generally, information is available to the fire investigator in four forms: verbal, written, visual, and electronic.

13.3.1 Verbal Information. Verbal sources of information, by definition, are limited to the spoken word. Such sources, which may be encountered by the fire investigator, may include, but are not limited to, verbal statements during interviews, telephone conversations, tape recordings, radio transmissions, commercial radio broadcasts, and the like.

13.3.2 Written Information. Written sources of information are likely to be encountered by the fire investigator during all stages of an investigation. Such sources may include, but are not limited to, written reports, written documents, reference materials, newspapers, and the like.

13.3.3 Visual Information. Visual sources of information, by definition, are limited to those that are gathered utilizing the sense of sight. Beginning first with the advent of still photography, such sources may include, but are not limited to, photographs, videotapes, motion pictures, and computer-generated animations.

13.3.4 Electronic Information. Computers have become an integral part of modern information and data systems. As such, the computer system maintained by any particular source of information may provide a wealth of information relevant to the fire investigation.

13.4 Interviews.

13.4.1 Purpose of Interviews. The purpose of any interview is to gather both useful and accurate information. Witnesses can provide such information about the fire and explosion incident even if they were not eyewitnesses to the incident.

13.4.1.1 The investigator should make every effort to identify the ignition sequence factors as soon as possible. These questions should address those issues covered in Section 18.3, Section 18.4, and Section 18.5.

13.4.1.2 It is the responsibility of the investigator to evaluate the quality of the data obtained from the witness at the time of the interview.

13.4.2 Preparation for the Interview. The fire investigator should be thoroughly prepared prior to conducting any type of interview, especially if the investigator intends to solicit relevant and useful information. The most important aspect of this preparation is a thorough understanding of all facets of the investigation.

13.4.2.1 The fire investigator should also carefully plan the setting of the interview, that is, when and where the interview will be held. Although the time that the interview is conducted may be determined by a variety of factors, the interview should generally be conducted as soon as possible after the fire or explosion incident.

13.4.2.2 It may be helpful to the investigator to conduct preliminary interviews before the fire scene examination commences, although there are many instances when this may be impractical.

13.4.2.3 The interviewer and the person being interviewed should be properly identified. The interview should, therefore, begin with the proper identification of the person conducting the interview. The date, time, and location of the interview, as well as any witnesses to it, should be documented.

13.4.2.4 The person being interviewed should also be completely and positively identified. Positive identification may include the person's full name, date of birth, social security number, driver's license number, physical description, home address, home telephone number, place of employment, business address, business telephone number, or other information that may be deemed pertinent to establish positive identification.

13.4.2.5 Lastly, the fire investigator should also establish a flexible plan or outline for the interview.

13.4.3 Documenting the Interview. All interviews, regardless of their type, should be documented. Tape recording the interview or taking written notes during the interview are two of the most common methods of documenting the interview. Both of these methods, however, often tend to distract or annoy the person being interviewed, resulting in some information not being solicited from them. An alternative method used to document interviews can be accomplished through the use of visual taping. All taping must be done in accordance with applicable laws and regulations. The investigator should obtain signed written statements from as many witnesses as possible to enhance their admissibility in court.

13.5 Governmental Sources of Information.

13.5.1 Municipal Government.

13.5.1.1 Municipal Clerk. The municipal clerk maintains public records regarding municipal licensing and general municipal business.

13.5.1.2 Municipal Assessor. The municipal assessor maintains public records regarding plats or maps of real property, including dimensions, addresses, owners, and taxable value of the real property and any improvements.

13.5.1.3 Municipal Treasurer. The municipal treasurer maintains public records regarding names and addresses of property owners, names and addresses of taxpayers, legal descriptions of property, amount of taxes paid or owed on real and personal property, and former owners of the property.

13.5.1.4 Municipal Street Department. The municipal street department maintains public records regarding maps of the streets; maps showing the locations of conduits, drains, sewers, and utility conduits; correct street numbers; old names of streets; abandoned streets and rights-of-way; and alleys, easements, and rights-of-way.

13.5.1.5 Municipal Building Department. The municipal building department maintains public records regarding building permits, electrical permits, plumbing permits, blueprints, and diagrams showing construction details and records of various municipal inspectors.

13.5.1.6 Municipal Health Department. The municipal health department maintains public records regarding birth certificates, death certificates, records of investigations related to pollution and other health hazards, and records of health inspectors.

13.5.1.7 Municipal Board of Education. The municipal board of education maintains public records regarding all aspects of the public school system.

13.5.1.8 Municipal Police Department. The municipal police department maintains public records regarding local criminal investigations and other aspects of the activities of that department.

13.5.1.9 Municipal Fire Department. The municipal fire department maintains public records regarding fire incident reports, emergency medical incident reports, records of fire inspections, and other aspects of the activities of that department.

13.5.1.10 Other Municipal Agencies. Many other offices, departments, and agencies typically exist at the municipal level of government. The fire investigator may encounter different governmental structuring in each municipality. As such, the fire investigator may need to solicit information from these additional sources.

13.5.2 County Government.

13.5.2.1 County Recorder. The county recorder's office maintains public records regarding documents relating to real estate transactions, mortgages, certificates of marriage and marriage contracts, divorces, wills admitted to probate, official bonds, notices of mechanics' liens, birth certificates, death certificates, papers in connection with bankruptcy, and other such writings as are required or permitted by law.

13.5.2.2 County Clerk. The county clerk maintains public records regarding naturalization records, civil litigation records, probate records, criminal litigation records, and records of general county business.

13.5.2.3 County Assessor. The county assessor maintains public records such as plats or maps of real property in the county, which include dimensions, addresses, owners, and taxable value.

13.5.2.4 County Treasurer. The county treasurer maintains public records regarding names and addresses of property owners, names and addresses of taxpayers, legal descriptions of property, amounts of taxes paid or owed on real and personal property, and all county fiscal transactions.

13.5.2.5 County Coroner/Medical Examiner. The county coroner/medical examiner maintains public records regarding the names or descriptions of the deceased, dates of inquests, property found on the deceased, causes and manners of death, and documents regarding the disposition of the deceased.

13.5.2.6 County Sheriff's Department. The county sheriff's department maintains public records regarding county criminal investigations and other aspects of the activities of that department.

13.5.2.7 Other County Agencies. Many other offices, departments, and agencies typically exist at the county level of government. The fire investigator may encounter different governmental structuring in each county. As such, the fire investigator may need to solicit information from these additional sources.

13.5.3 State Government.

13.5.3.1 Secretary of State. The secretary of state maintains public records regarding charters and annual reports of corporations, annexations, and charter ordinances of towns, villages, and cities; trade names and trademarks registration; notary public records; and Uniform Commercial Code (UCC) statements.

13.5.3.2 State Treasurer. The state treasurer maintains public records regarding all state fiscal transactions.

13.5.3.3 State Department of Vital Statistics. The state department of vital statistics maintains public records regarding births, deaths, and marriages.

13.5.3.4 State Department of Revenue. The state department of revenue maintains public records regarding individual state tax returns; corporate state tax returns; and past, present, and pending investigations.

13.5.3.5 State Department of Regulation. The state department of regulation maintains public records regarding names of professional occupation license holders and their backgrounds; results of licensing examinations; consumer complaints; past, present, or pending investigations; and the annual reports of charitable organizations.

13.5.3.6 State Department of Transportation. The state department of transportation maintains public records regarding highway construction and improvement projects, motor vehicle accident information, motor vehicle registrations, and driver's license testing and registration.

13.5.3.7 State Department of Natural Resources. The state department of natural resources maintains public records regarding fish and game regulations, fishing and hunting license data, recreational vehicles license data, waste disposal regulation, and environmental protection regulation.

13.5.3.8 State Insurance Commissioner's Office. The state insurance commissioner's office maintains public records regarding insurance companies licensed to transact business in the state; licensed insurance agents; consumer complaints; and records of past, present, or pending investigations.

13.5.3.9 State Police. The state police maintain public records regarding state criminal investigations and other aspects of the activities of that agency.

13.5.3.10 State Fire Marshal's Office. The state fire marshal's office maintains public records regarding fire inspection and prevention activities, fire incident databases, and fire investigation activities.

13.5.3.11 Other State Agencies. Many other offices, departments, and agencies typically exist at the state level of government. The fire investigator may encounter different government structuring in each state. As such, the fire investigator may need to solicit information from these additional sources.

13.5.4 Federal Government.

13.5.4.1 Department of Agriculture. Under this department, the Food Stamps and Nutrition Services Agency maintains public records regarding food stamps and their issuance.

13.5.4.1.1 The Consumer and Marketing Service maintains public records regarding meat inspection, meat packers and stockyards, poultry inspection, and dairy product inspection.

13.5.4.1.2 The U.S. Forest Service maintains public records regarding forestry and mining activities.

13.5.4.1.3 The investigative activities of the Department of Agriculture are contained in the Office of the Inspector General. The investigative area of the Secretary of Agriculture is the Office of Investigations.

13.5.4.2 Department of Commerce. Under this department, the Bureau of Public Roads maintains public records regarding all highway programs in which federal assistance was given.

13.5.4.2.1 The National Marine Fisheries Service maintains public records regarding the names, addresses, and registration of all ships fishing in local waters.

13.5.4.2.2 The Commercial Intelligence Division Office maintains public records regarding trade lists, trade contract surveys, and world trade directory reports.

13.5.4.2.3 The U.S. Patent Office maintains public records regarding all patents issued in the United States, as well as a roster of attorneys and agents registered to practice before that office.

13.5.4.2.4 The Trade Mission Division maintains public records regarding information on members of trade missions.

13.5.4.2.5 The investigative activities of the Department of Commerce are contained in the Office of Investigations and Security.

13.5.4.3 Department of Defense. The Department of Defense oversees all of the military branches of the armed services including the Army, the Navy, the Marine Corps, the Air Force, and the Coast Guard. Each of these branches of the military maintains public records regarding its activities and personnel. Each of these branches has offices that conduct criminal investigations within its specific branch of armed service.

13.5.4.4 Department of Health and Human Services. Under this department, the Food and Drug Administration maintains public records regarding its enforcement of federal laws under its jurisdiction.

13.5.4.4.1 The Social Security Administration maintains public records with regard to its activities.

13.5.4.4.2 The investigative activities of the Department of Health and Human Services are contained in the Office of Security and Investigations.

13.5.4.5 Department of Housing and Urban Development. The Department of Housing and Urban Development maintains public records regarding all public housing programs in which federal assistance has been given. The investigative activities of the Department of Housing and Urban Development are contained in the compliance division.

13.5.4.6 Department of the Interior. Under this department, the Fish and Wildlife Service maintains public records regarding violations of federal laws related to fish and game.

13.5.4.6.1 The Bureau of Indian Affairs maintains public records regarding censuses of Indian reservations, names, degree of Indian blood, tribe, family background, and current addresses of all Indians, especially those residing on federal Indian reservations.

13.5.4.6.2 The National Park Service maintains public records regarding all federally owned or federally maintained parks and lands.

13.5.4.6.3 Each division of the Department of the Interior has its own investigative office.

13.5.4.7 Department of Labor. Under this department, the Labor Management Services Administration maintains public records regarding information on labor and management organizations and their officials.

13.5.4.7.1 The Employment Standards Administration maintains public records regarding federal laws related to minimum wage, overtime standards, equal pay, and age discrimination in employment.

13.5.4.7.2 The investigative activities of the Department of Labor are contained in the Labor Pension Reports Office Division.

13.5.4.8 Department of State.

13.5.4.8.1 The Department of State maintains public records regarding passports, visas, and import/export licenses.

13.5.4.8.2 The investigative activities of the Department of State are contained in the Visa Office.

13.5.4.9 Department of Transportation. Under this department, the Environmental Safety and Consumer Affairs Office maintains public records regarding its programs to protect the environment, to enhance the safety and security of passengers and cargo in domestic and international transport, and to monitor the transportation of hazardous and dangerous materials.

13.5.4.10 Internal Revenue Service. The Internal Revenue Service maintains public records regarding compliance with all federal tax laws.

13.5.4.11 Department of Justice. Under this department, the Antitrust Division maintains public records regarding federal sources of information relating to antitrust matters.

13.5.4.11.1 The Bureau of Alcohol, Tobacco, Firearms and Explosives (ATF) maintains public records regarding distillers, brewers, and persons or firms that manufacture or handle alcohol; retail liquor dealers; manufacturers and distributors of tobacco products; firearms registration; federal firearms license holders, including manufacturers, importers, and dealers; federal explosive license holders, including manufacturers, importers, and dealers; and the origin of all firearms manufactured and imported after 1968.

13.5.4.11.2 The Civil Rights Division maintains public records regarding its enforcement of all federal civil rights laws that prohibit discrimination on the basis of race, color, religion, or national origin in the areas of education, employment, and housing, and the use of public facilities and public accommodations.

13.5.4.11.3 The Criminal Division maintains public records regarding its enforcement of all federal criminal laws except those specifically assigned to the Antitrust, Civil Rights, or Tax Divisions.

13.5.4.11.4 The Drug Enforcement Administration maintains public records regarding all licensed handlers of narcotics, the legal trade of narcotics and dangerous drugs, and its enforcement of federal laws relating to narcotics and other drugs.

13.5.4.11.5 The Federal Bureau of Investigation maintains public records regarding criminal records, fingerprints, and its enforcement of federal criminal laws.

13.5.4.11.6 The Immigration and Naturalization Service maintains public records regarding immigrants, aliens, passengers and crews on vessels from foreign ports, naturalization records, deportation proceedings, and the financial statements of aliens and persons sponsoring their entry into the United States.

13.5.4.12 U.S. Postal Service. The U.S. Postal Service maintains public records regarding all of its activities. The investigative activities of the U.S. Postal Service are contained in the Office of the Postal Inspector.

13.5.4.13 Department of Energy. The Department of Energy is an executive department of the U.S. government that works to meet the nation's energy needs. The department develops and coordinates national energy policies and programs. It promotes conservation of fuel and electricity. It also conducts research to develop new energy sources and more efficient ways to use present supplies. The secretary of energy, a member of the president's cabinet, heads the department.

13.5.4.14 United States Department of Homeland Security. The Department of Homeland Security, established after the terrorist attacks of 9/11/2001, is an executive department of the U.S. government that works to maintain the security of the nation's needs. The department develops and coordinates national security policies and programs through a variety of border, transportation, and infrastructure protection. The Secretary of Homeland Security, a member of the President's cabinet, heads the department.

13.5.4.14.1 U.S. Customs and Border Protection (CBP) maintains public records regarding importers; exporters; customhouse brokers; customhouse truckers; and the registry, enrollment, and licensing of vessels not licensed by the Coast Guard or the United States that transport goods to and from the United States.

13.5.4.14.2 The U.S. Secret Service maintains public records regarding counterfeiting and forgery of U.S. coins and currencies and records of all threats on the life of the president and his immediate family, the vice president, former presidents and their wives, wives of deceased presidents, children of deceased presidents until age sixteen, president- and vice president–elect, major candidates for the office of president and vice president, and heads of states representing foreign countries visiting in the United States.

13.5.4.14.3 The United States Coast Guard maintains public records regarding persons serving on U.S.–registered ships, vessels equipped with permanently installed motors, vessels over 4.9 m (16 ft) long equipped with detachable motors, information on where and when ships departed or returned from U.S. ports, and violations of environmental laws.

13.5.4.14.4 The Federal Emergency Management Agency, a component of the Department of Homeland Security, provides the planning, preparation, response to and recovery from all types of natural and man-made disasters; provides federal support of disaster relief response, assistance, support and recovery to state and local entities impacted by federally declared disasters.

13.5.4.14.5 The U.S. Fire Administration maintains a wide array of fire service-based programs, training, education, and technical and statistical information for the overall planning/prevention/control of fire issues within the United States, including alerts, advisories, arson, juvenile fire setting, communications, critical infrastructure protection (EMR-ISAC), emergency medical services, rescue, fire service administration, fire fighter health and safety, hazardous materials, incident management, professional development, terrorism, and wildland fire.

13.5.4.14.5.1 The United States Fire Administration maintains an extensive database of information related to fire incidents through its administration of the National Fire Incident Reporting System (NFIRS).

13.5.4.14.5.2 In addition, the administration maintains records of ongoing research in fire investigation, information regarding arson awareness programs, and technical and reference materials focusing on fire investigation, and it coordinates the distribution of the Arson Information Management System (AIMS) software.

13.5.4.15 National Oceanic and Atmospheric Administration (NOAA). Weather data, past or present, for all reporting stations in the United States are available from the National Climatic Data Center in Asheville, North Carolina. Local NOAA weather stations can provide data for their areas.

13.5.4.16 Other Federal Agencies. There is a variety of other federal agencies and commissions that are part of the federal level of government. These federal agencies and commissions all maintain a variety of public records. As such, the fire investigator may need to solicit information from these additional sources. The U.S. Senate Committee on Government Operations publishes a handy reference entitled *Chart of the Organization of Federal Executive Departments and Agencies.* This chart provides the exact name of an office, division, or bureau, and the place it occupies in the organizational structure in a department or agency. With this reference, it should not be difficult for the fire investigator to determine the jurisdiction of a federal government agency or commission.

13.6 Private Sources of Information.

13.6.1 National Fire Protection Association (NFPA).

13.6.1.1 NFPA was organized in 1896 to promote the science and improve the methods of fire protection and prevention, to obtain and circulate information on these subjects, and to secure the cooperation of its members in establishing proper safeguards against loss of life and property. NFPA is an international, charitable, technical, and educational organization.

13.6.1.2 NFPA is responsible for the development and distribution of the *National Fire Codes*®. In addition to these, NFPA has developed and distributed a wealth of technical information, much of which is of significant interest to the fire investigator.

13.6.1.3* Every year NFPA sponsors or co-sponsors fire investigation training programs in various cities and countries.

13.6.2 Society of Fire Protection Engineers (SFPE). Organized in 1950, the Society of Fire Protection Engineers is a professional organization for engineers involved in the multifaceted field of fire protection. The society works to advance fire protection engineering and its allied fields, to maintain a high ethical standard among its members, and to foster fire protection engineering education.

13.6.3 American Society for Testing and Materials (ASTM).

13.6.3.1 The American Society for Testing and Materials, founded in 1898, is a scientific and technical organization formed for "the development of standards on characteristics and performance of materials, products, systems, and services, and the promotion of related knowledge." It is the world's largest source of voluntary consensus standards.

13.6.3.2 Many of these standards focus on acceptable test methods for conducting a variety of fire-related tests often requested by fire investigators. Those standards that outline fire tests are discussed in Chapter 16 of this document.

13.6.4 Founded in 1918, the American National Standards Institute (ANSI) is a private, non-profit organization that facilitates the development of American National Standards (ANSs) by accrediting the procedures of standards developing organizations (SDOs). These groups work cooperatively to develop voluntary national consensus standards. Accreditation by ANSI signifies that the procedures used by the standards body in connection with the development of American National Standards meet the Institute's essential requirements for openness, balance, consensus, and due process.

13.6.5 National Association of Fire Investigators (NAFI).

13.6.5.1 The National Association of Fire Investigators was organized in 1961. Its primary purposes are to increase the knowledge of and improve the skills of persons engaged in the investigation and analysis of fires and explosions, or in the litigation that ensues from such investigations.

13.6.5.2 The Association also originated and implemented the National Certification Board. Each year, the board certifies fire and explosion investigators, fire investigation instructors, and vehicle fire investigators. Through this program, those certified are recognized for their knowledge, training, and experience and accepted for their expertise.

13.6.6 International Association of Arson Investigators (IAAI).

13.6.6.1 The International Association of Arson Investigators was founded in 1949 by a group of public and private officials to address fire and arson issues. The purpose of the association is to strive to control arson and other related crimes, through education and training, in addition to providing basic and advanced fire investigator training. The IAAI has chapters located throughout the world.

13.6.6.2 In addition to an annual seminar, there are also regional seminars focusing on fire investigator training and education. The Association publishes the *Fire and Arson Investigator,* a quarterly magazine. The IAAI offers a written examination for investigators meeting IAAI minimum qualifications to become an IAAI–certified fire investigator (CFI).

13.6.7 Regional Fire Investigations Organizations. In addition to the National Association of Fire Investigators, the International Association of Arson Investigators, and its state chapters, many regional fire investigation organizations exist. These organizations generally exist as state or local fire/arson task forces, professional societies or groups of fire investigators, or mutual aid fire investigation teams.

13.6.8 Real Estate Industry. The real estate industry maintains certain records that may prove beneficial to the fire investigator during the investigation. Besides records of persons and businesses that are selling or purchasing property, real estate offices often maintain extensive libraries of photographs of homes and businesses located in their sales territory. These photographs may be of interest to the fire investigator.

13.6.9 Abstract and Title Companies. Abstract and title companies are another valuable source of information. Records maintained by such companies include maps and tract books; escrow indexes of purchasers and sellers of real estate; escrow files containing escrow instructions, agreements, and settlements; and abstract and title policies.

13.6.10 Financial Institutions. Financial institutions, including banks, savings and loan associations, brokers, transfer agents, dividend disbursing agents, and commercial lending services, all maintain records that serve as sources of valuable information. Besides the financial information about a particular person or business, the records of financial institutions contain other information about all facets of a person's life or a business's history.

13.6.11 Insurance Industry.

13.6.11.1 The insurance industry certainly has an interest in the results of most fire and explosion incidents. The industry's primary interest in such investigations is the detection of the crime of arson and other fraud offenses. The insurance industry can, however, also provide the fire investigator with a diverse amount of information concerning the structure involved or vehicle and the person(s) who have insured it. *(See 11.5.6.5.)*

13.6.11.2 The insurance industry also funds the Property Insurance Loss Register (PILR), which receives reports of property losses through fire, burglaries, and thefts. It is a computerized index of the insurance companies that paid the claims, the person to whom the claim was paid, the type of claim, and the like. It can serve as a valuable source of information to the fire investigator.

13.6.12 Educational Institutions. Educational institutions are not often considered as a source of information by fire investigators. The records maintained by such institutions can, however, provide an insight into a person's background and interests.

13.6.13 Utility Companies. During the normal course of business, utility companies maintain extensive databases, particularly concerning their customers. The fire investigator should not overlook that these companies, whether publicly or privately owned, also maintain records concerning the quality of and problems associated with the distribution of their products or services.

13.6.14 Trade Organizations. Trade organizations are often one of the most valuable sources of information available to the fire investigator. These organizations promote the interest of many of the prominent trades. Their value to the fire investigator is that each organization focuses on a specific trade or discipline. As such, they often function as clearinghouses for knowledge in their area of expertise. Besides this expertise, most trade organizations develop and distribute publications that serve as important reference materials to the fire investigator.

13.6.15 Local Television Stations. Local television stations often send camera crews to newsworthy fires. Copies of their videotape coverage may be obtained, if still available. Television stations also have records of the weather in the area and often have limited data from local amateur weather watchers from areas away from the airports.

13.6.16 Lightning Detection Networks. Lightning detection networks exist that may assist in the establishing of the time and location [to within 500 m (1640 ft)] of a lightning strike. Historical data is also available, including reports of any lightning strikes detected within a specified time prior to a fire.

13.6.17 Other Private Sources. There are a variety of other private sources of information. These private sources all maintain a variety of records. As such, the fire investigator may need to solicit information from these additional sources.

13.7 Conclusion. The number and diversity of governmental and private sources of information for the fire investigator are unlimited. While not a comprehensive listing, by any means, those sources of information enumerated in this chapter should provide the fire investigator with a realization that his or her ability to solicit information pertinent to a particular fire investigator is also unlimited.

Chapter 14 Planning the Investigation

14.1* Introduction. The intent of this chapter is to identify basic considerations of concern to the investigator prior to beginning the incident scene investigation.

14.1.1 Regardless of the number of people involved, the need to preplan investigations remains constant. Considerations for determining the number of investigators assigned include budgetary constraints, available staffing, complexity, loss of life, and size of the scene to be investigated.

14.1.2 The person responsible for the investigation of the incident should identify the resources at his or her disposal and those available from outside sources before those resources are needed. It is his or her responsibility to acquire additional resources as needed. Assistance can be gained from local or state building officials, universities and state colleges, and numerous other public and private agencies.

14.1.3 The "team concept" of investigating an incident is recommended. It is understood that at many incident scenes, the investigator may have to photograph or sketch the scene, collect evidence, interview, and be responsible for the entire scene investigation without other assistance. These functions and others described in this document should be performed regardless of the number of people involved with the investigation.

14.2 Basic Incident Information. Prior to beginning the incident scene investigation, numerous events, facts, and circumstances should be identified. Accuracy is important, because a mistake at this point could jeopardize the subsequent investigation results.

14.2.1 Location. The investigator, once notified of an incident, should obtain as much background information as possible relative to the incident from the requester. If the travel distance is great, arrangements may be required to transport the investigation team to the incident scene. The location of the incident may also dictate the need for specialized equipment and facilities. *(See 14.4.1.)*

14.2.2 Date and Time of Incident. The investigator should accurately determine the day, date, and time of the incident. The age of the scene may have an effect on the planning of the investigation. The greater the delay between the incident and the investigation, the more important it becomes to review pre-existing documentation and information such as incident reports, photographs, building plans, and diagrams.

14.2.3 Weather Conditions. Weather at the time of the investigation may necessitate special clothing and equipment. Weather may also determine the amount of time the team members can work an incident scene. Extreme weather may require that greater safety precautions be taken on behalf of the team members, for example, when the weight of snow on a structure weakens it. Weather conditions such as wind direction and velocity, temperature, and rain during a fire should be noted because all can have an effect on the ignition and fire spread.

14.2.4 Size and Complexity of Incident.

14.2.4.1 The size and complexity of the incident scene may suggest the need for assistance for the investigator. A large incident scene area may create communication problems for investigators, and arrangements for efficient communications should be made.

14.2.4.2 The size and complexity of the scene will also affect the length of the investigation, and preparations may be needed for housing and feeding the team members. Generally, the larger the incident scene, the greater the length of time required to conduct the investigation.

14.2.5 Type and Use of Structure.

14.2.5.1 The investigator should identify the type and use of the incident structure. The use or occupancy of the structure (e.g., industrial plant, chemical processing plant, storage warehouse, nuclear facility, or radiological waste storage) may necessitate special containment of debris, contamination, or radiation, including water runoff at the scene. Additionally, appropriate hazardous materials or contamination clothing, breathing apparatus, and other protective devices and equipment may be necessary to ensure safety at the incident scene. Conditions at certain scenes may be so hazardous that the investigators should work within monitored stay times.

14.2.5.2 Knowledge of the type of construction and construction materials will provide the investigator with valuable background information and allow anticipation of circumstances and problems to be encountered by the investigation team.

14.2.6 Nature and Extent of Damage.

14.2.6.1 Information on the condition of the scene may alert the investigator to special requirements for the investigation, such as utility testing equipment, specialized expertise, additional staffing, and special safety equipment. The investigator may be operating under time constraints and should plan accordingly.

14.2.6.2 The investigator should ensure that initiation of the investigation will not be contrary to post-incident orders issued by local, state, or federal regulatory agencies. Issues that often lead to such orders may involve structural stability and the presence of hazardous materials.

14.2.7 Security of Scene. The investigator should promptly determine the identity of the individual, authority, or entity that has possession or control of the scene. Right of access and means of access should be established. Scene security is a consideration. If possible, arrangements should be made to preserve the scene until the arrival of the investigator(s). If this is not possible, arrangements should be made to photograph and document existing conditions prior to disturbance or demolition.

14.2.8 Purpose of Investigation.

14.2.8.1 While planning the investigation, the investigator should remain aware of his or her role, the scope of the investigation, and areas of responsibility. Numerous investigators may be involved, from both the private and public sectors. Mutual respect and cooperation in the investigation is required.

14.2.8.2 The investigator, particularly the private sector investigator, may need to make a reasonable effort to notify all parties, identifiable at that time, who may have a legal interest in the investigation. *(See Section 11.3.)*

14.3 Organizing the Investigation Functions.

14.3.1 There are basic functions that are commonly performed in each investigation. These are the leadership/coordinating function; photography, note taking, mapping, and diagramming *(see Chapter 15)*; interviewing witnesses *(see Chapter 13)*; searching the scene *(see Chapter 17)*; evidence collection and preservation *(see Chapter 16)*; and safety assessment *(see Chapter 12)*.

14.3.2 In addition, specialized expertise in such fields as electrical, heating and air conditioning, or other engineering fields is often needed. The investigator should, if possible, fulfill these functions with the personnel available. In assigning functions, those special talents or training that individual members possess should be utilized.

14.4 Pre-Investigation Team Meeting. If the investigator has established a team, a meeting should take place prior to the on-scene investigation. The team leader or investigator should address questions of jurisdictional boundaries and assign specific responsibilities to the team members. Personnel should be advised of the condition of the scene and the safety precautions required.

14.4.1 Equipment and Facilities. Each person on the fire scene should be equipped with appropriate safety equipment, as required. A complement of basic tools should also be available. The tools and equipment listed in 14.4.2 and 14.4.3 may not be needed on every scene, but in planning the investigation, the investigator should know where to obtain these tools and equipment if the investigator does not carry them.

14.4.2 Personal Safety Equipment. Recommended personal safety equipment includes the following:

(1) Eye protection
(2) Flashlight
(3) Gloves
(4) Helmet or hard hat
(5) Respiratory protection (type depending on exposure)
(6) Safety boots or shoes
(7) Turnout gear or coveralls

14.4.3 Tools and Equipment. Recommended tools and equipment include the following:

(1) Absorption material
(2) Axe
(3) Broom
(4) Camera and film *(See 15.2.3.2 and 15.2.3.3 for recommendations.)*
(5) Claw hammer
(6) Directional compass
(7) Evidence-collecting container *(See Section 16.5 for recommendations.)*
(8) Evidence labels (sticky)
(9) Hand towels
(10) Hatchet
(11) Hydrocarbon detector
(12) Ladder
(13) Lighting
(14) Magnet
(15) Marking pens
(16) Paint brushes
(17) Paper towels/wiping cloths
(18) Pen knife
(19) Pliers/wire cutters
(20) Pry bar
(21) Rake
(22) Rope
(23) Rulers
(24) Saw
(25) Screwdrivers (multiple types)
(26) Shovel
(27) Sieve
(28) Soap and hand cleaner
(29) Styrofoam cups
(30) Tape measure
(31) Tape recorder
(32) Tongs
(33) Tweezers
(34) Twine
(35) Voltmeter/ohmmeter
(36) Water
(37) Writing/drawing equipment

14.5 Specialized Personnel and Technical Consultants.

14.5.1 General. During the planning of a fire investigation, specialized personnel may be needed to provide technical assistance. There are many different facets to fire investigation. If unfamiliar with a particular aspect, the investigator should never hesitate to call in another fire investigative expert who has more knowledge or experience in a particular aspect of the investigation. For example, there are some experts who specialize in explosions.

14.5.1.1 Sources for these specialized personnel/experts include colleges or universities, government agencies (federal, state, and local), societies or trade groups, consulting firms, and others. When specialized personnel are brought in, it is important to remember that conflict of interest should be avoided. Identification of special personnel in advance is recommended. Subsections 14.5.2 through 14.5.10 list examples of professional or specific engineering and scientific disciplines, along with areas where these personnel may help the fire investigator. This section is not intended to list all sources for these specialized personnel and technical consultants.

14.5.1.2 It should be kept in mind that fire investigation is a specialized field. Those individuals not specifically trained and experienced in the discipline of fire investigation and analysis, even though they may be experts in related fields, may not be well qualified to render opinions regarding fire origin and cause. In order to offer origin and cause opinions, additional training or experience is generally necessary.

14.5.1.3 The descriptions in 14.5.2 through 14.5.10 are general and do not imply that the presence or absence of a referenced area of training affects the qualifications of a particular specialist.

14.5.2 Materials Engineer or Scientist. A person in this field can provide specialized knowledge about how materials react to different conditions, including heat and fire. In the case of metals, someone with a metallurgical background may be able to answer questions about corrosion, stress, failure or fatigue, heating, or melting. A polymer scientist or chemist may offer assistance regarding how plastics react to heat and other conditions present during a fire and regarding the combustion and flammability properties of plastics.

14.5.3 Mechanical Engineer. A mechanical engineer may be needed to analyze complex mechanical systems or equipment, including heating, ventilation, and air-conditioning (HVAC) systems, especially how these systems may have affected the movement of smoke within a building. The mechanical engineer may also be able to perform strength-of-material tests.

14.5.4 Electrical Engineer. An electrical engineer may provide information regarding building fire alarm systems, energy systems, power supplies, or other electrical systems or components. An electrical engineer may assist by quantifying the normal operating parameters of a particular system and determining failure modes.

14.5.5 Chemical Engineer/Chemist.

14.5.5.1 A chemical engineer has education in chemical processes, fluid dynamics, and heat transfer. When a fire involves chemicals, a chemical process, or a chemical plant, the chemical engineer may help the investigator identify and analyze possible failure modes.

14.5.5.2 A chemist has extensive education in the identification and analysis of chemicals and may be used by the investigator in identifying a particular substance found at a fire scene. The chemist may be able to test a substance to determine its chemical and physical reaction to heat. When there are concerns about toxicity or the human reaction to chemicals or chemical decomposition products, a chemist, biochemist, or microbiologist should be consulted by the investigator.

14.5.6 Fire Science and Engineering. Within the field of fire science and engineering, there are a number of areas of special expertise that can provide advice and assistance to the investigator.

14.5.6.1 Fire Protection Engineer. Fire protection engineering encompasses all the traditional engineering disciplines in the science and technology of fire and explosions. The fire protection engineer deals with the relationship of ignition sources to materials in determination of what may have started the fire. He or she is also concerned with the dynamics of fire, and how it affects various types of materials and structures. The fire protection engineer should also have knowledge of how fire detection and suppression systems (e.g., smoke detectors, automatic sprinklers, or Halon systems) function and should be able to assist in the analysis of how a system may have failed to detect or extinguish a fire. The complexity of fire often requires the fire protection engineer to use many of the other engineering and scientific disciplines to study how a fire starts, grows, and goes out. Additionally, a fire protection engineer should be able to provide knowledge of building and fire codes, fire test methods, fire performance of materials, computer modeling of fires, and failure analysis.

14.5.6.2 Fire Engineering Technologist. Individuals with bachelor of science degrees in fire engineering technology, fire and safety engineering technology, or a similar discipline, or recognized equivalent, typically have studied fire dynamics and fire science; fire and arson investigation, fire suppression technology, fire extinguishment tactics, and fire department management; fire protection; fire protection structures and systems design; fire prevention; hazardous materials; applied upper-level mathematics and computer science; fire-related human behavior; safety and loss management; fire and safety codes and standards; and fire science research.

14.5.6.3 Fire Engineering Technician. Individuals with associate of science–level degrees in fire and safety engineering technology or similar disciplines, or recognized equivalent, typically have studied fire dynamics and fire science; fire and arson investigation; fire suppression technology, tactics, and management; fire protection; fire protection structures and systems design; fire prevention; hazardous materials; mathematics and computer science topics; fire-related human behavior; safety and loss management; fire and safety codes and standards; or fire science research.

14.5.7 Industry Expert. When the investigation involves a specialized industry, piece of equipment, or processing system, an expert in that field may be needed to fully understand the processes involved. Experience with the specific fire hazards involved and the standards or regulations associated with the industry and its equipment and processes can provide valuable information to the investigator. Industry experts can be found within companies, trade groups, or associations.

14.5.8 Attorney. An attorney can provide needed legal assistance with regard to rules of evidence, search and seizure laws, gaining access to a fire scene, and obtaining court orders.

14.5.9 Insurance Agent/Adjuster. An insurance agent or adjuster may be able to provide the investigator with information concerning the building and its contents prior to the fire, fire protection systems in the building, and the condition of those systems. Additional information regarding insurance coverage and prior losses may be available.

14.5.10 Canine Teams. Trained canine/handler teams may assist investigators in locating areas for collection of samples for laboratory analysis to identify the presence of ignitible liquids.

14.6 Case Management. A method should be employed to organize the information generated throughout the investigation and to coordinate the efforts of the various people involved. The topic of case management is addressed in the context of major loss investigations in Chapter 27 of this guide. It is also the focus of some of the reference material listed at the back of this guide.

Chapter 15 Documentation of the Investigation

15.1* Introduction.

15.1.1 The goal in documenting any fire or explosion investigation is to accurately record the investigation through media that will allow investigators to recall and communicate their observations at a later date. Common methods of accomplishing this goal include the use of photographs, videotapes, diagrams, maps, overlays, tape recordings, notes, and reports.

15.1.2 Thorough and accurate documentation of the investigation is critical because it is from this compilation of factual data that investigative opinions and conclusions can be supported and verified. There are a number of resources to assist the investigator in documenting the investigation.

15.2 Photography.

15.2.1 General. A visual documentation of the fire scene can be made using either film or video photography. Images can portray the scene better than words. They are the most efficient reminders of what the investigator saw while at the scene. Patterns and items may become evident that were overlooked at the time the photographs or videos were made. They can also substantiate reports and statements of the investigator.

15.2.1.1 For fire scene and investigation-related photography, color images are recommended.

15.2.1.2 Taking a basic photography or video course through a vocational school, camera club, or camera store would be most helpful in getting the photographer familiar with the equipment.

15.2.1.3 As many photographs should be taken as are necessary to document and record the fire scene adequately. It is recognized that time and expense considerations may impact the number of photographs taken, and the photographer should exercise discretion. It is far preferable to err on the side of taking too many photographs rather than too few.

15.2.1.4 The exclusive use of videotapes, motion pictures, or slides is not recommended. They are more effective when used in conjunction with still photographs. Also, additional equipment is obviously required to review and utilize videos, films, and slides.

15.2.2 Timing. Taking photographs during or as soon as possible after a fire is important when recording the fire scene, as the scene may become altered, disturbed, or even destroyed. Some reasons why time is important include the following:

(1) The building is in danger of imminent collapse or the structure must be demolished for safety reasons.
(2) The condition of the building contents creates an environmental hazard that needs immediate attention.
(3) Evidence should be documented when discovered as layers of debris are removed, as is done at an archaeological dig. Documenting the layers can also assist in understanding the course of the fire.

15.2.3 Basics.

15.2.3.1 General. The most fundamental aspect of photography that an investigator should comprehend is how a camera works. The easiest way to learn how a camera works is to compare the camera to the human eye.

15.2.3.1.1 One of the most important aspects to remember about fire investigation photography is light. The average fire scene consists of blackened subjects and blackened background, creating much less than ideal conditions for taking a photograph. As one can imagine, walking into a dark room causes the human eye to expand its pupil in order to gather more light; likewise, the camera requires similar operation. The person in a dark room normally turns on the light to enhance vision, just as a photographer uses flash or floodlight to enhance the imitated vision of the camera.

15.2.3.1.2 Both the human eye and the camera project an inverted image on the light-sensitive surface: the film in the camera, and the retina in the eye. The amount of light admitted is regulated by the iris (eye) or diaphragm (camera). In both, the chamber through which the light passes is coated with a black lining to absorb the stray light and avoid reflection.

15.2.3.1.3 Regardless of camera type, film speed, or whether slides or prints are being taken, it is recommended that the investigator use color film. The advantage of color film is that the final product can more realistically depict the fire scene by showing color variations between objects and smoke stains.

15.2.3.2 Types of Cameras. There is a multitude of camera types available to the investigator, from small, inexpensive models to elaborate versions with a wide range of attachments.

15.2.3.2.1 Some cameras are fully automatic, giving some investigators a sense of comfort knowing that all they need to do is point and shoot. These cameras will set the film speed from a code on the film canister, adjust the lens opening (f-stop), and focus the lens by means of a beam of infrared light.

15.2.3.2.2 Manual operation is sometimes desired by the investigator so that specialty photographs can be obtained that the automatic camera, with its built-in options, cannot perform. For example, with a manual camera, bracketing (taking a series of photographs with sequentially adjusted exposures) can be performed to ensure at least one properly exposed photograph when the correct exposure is difficult to measure. There are some cameras that can be operated in a manual as well as an automatic mode, providing a choice from the same camera. Most investigators prefer an automatic camera.

15.2.3.2.3 A 35 mm single-lens reflex camera is preferred over other formats, but the investigator who has a non–35 mm camera should continue to take photographs as recommended. A backup camera that instantly develops prints can be advantageous, especially for an important photograph of a valuable piece of evidence.

15.2.3.3 Film. There are many types of film and film speeds available in both slide and print film. There are numerous speeds of film (ASA ratings), especially in the 35 mm range. Because 35 mm (which designates the size of the film) is most recognized and utilized by fire investigators, film speeds will be discussed using this size only. The common speeds range from 25 to 1600 in color and to 6400 in black and white. The numbers are merely a rating system. As the numbers get larger, the film requires less light. While the higher ASA-rated (faster) film is better in low light conditions with no flash, a drawback is that it will produce poorer-quality enlargements, which will have a grainy appearance. The film with the lowest rating that the investigator is comfortable with should be used because of the potential need for enlargements. Most investigators use a film with an ASA rating between 100 and 400. Fire investigators should practice and become familiar with the type and speed of film they intend to use on a regular basis.

15.2.3.4 Digital Photography. With the advancements of computer-based technology and the improvement of digital cameras and technology, there are a number of issues that have been raised regarding the acceptance of the photograph during testimony. At this time, there is no established case law that bars the use of digital images in the courtroom. With digital photographs, as with all photographs, the tests of "a true and accurate representation" and "relevance to the testimony" must still be met.

15.2.3.4.1 Digital images can be manipulated and altered using readily available computer technology. Very often, legitimate alterations of the image can be used to further interpret or understand the image. Examples of this type of alteration include digital enhancement, brightening, color adjustment, and contrast adjustment. Similar techniques have also been used when developing prints from negatives; however, the film negative remains as a permanent record. If an image has been enhanced, it is incumbent upon the investigator to preserve the original image and to clearly state the extent to which the image was enhanced so as to not mislead the trier of fact.

15.2.3.4.2 When the investigator chooses to utilize digital photography, steps should be taken to preserve the original image and establish a methodology to allow authentication. An agency procedure should be established for the storage of images, such as placement on a storage medium (CD ROM) that will not allow them to be altered, or the utilization of a computer software program that does not allow the original image to be altered and saved using the original file name, or other programs that may be developed in the future.

15.2.3.5 Lenses. The camera lens is used to gather light and to focus the image on the surface of the film. Most of today's lenses are compound, meaning that multiple lenses are located in the same housing. The fire investigator needs a basic understanding of the lens function to obtain quality photographs. The convex surface of the lens collects the light and sends it to the back of the camera, where the film lies. The aperture is an adjustable opening in the lens that controls the amount of light admitted. The adjustments of this opening are sectioned into measurements called f-stops. As the f-stop numbers get larger, the opening gets smaller, admitting less light. These f-stop numbers are listed on the movable ring of the adjustable lenses. Normally, the higher the f-stop that can be used, the better the quality of the photograph.

15.2.3.5.1 Focal lengths in lenses range from a normal lens (50 mm, which is most similar to the human eye) to the wide angle (28 mm or less) lenses, to telephoto and zoom lenses (typically 100 mm or greater). The investigator needs to determine what focal lengths will be used regularly and become familiar with the abilities of each.

15.2.3.5.2 The area of clear definition or depth of field is the distance between the farthest and nearest objects that will be in focus at any given time. The depth of field depends on the distance to the object being photographed, the lens opening, and the focal length of the lens being used. The depth of field will also determine the quality of detail in the investigator's photographs. For a given f-stop, the shorter the focal length of the lens, the greater the depth of field. For a given focal length lens, a larger f-stop (smaller opening) will provide a greater depth of field. The more the depth of field, the more minute are the details that will be seen. This is an important technique to master. These are the most common lens factors with which the fire investigator needs to be familiar. If a fixed-lens camera is used, the investigator need not be concerned with adjustments, because the manufacturer has preset the lens. A recommended lens is a medium range zoom, such as the 35 mm to 70 mm, providing a wide angle with a good depth of field and the ability to take high-magnification close-ups (macros).

15.2.3.6 Filters. The investigator should know that problems can occur with the use of colored filters. Unless the end results of colored filter use are known, it is recommended that they not be used. If colored filters are used, the investigator should take a photograph with a clear filter also. The clear filter can be used continually and is a good means of protecting the lens.

15.2.3.7 Lighting. The most usable light source known is the sun. No artificial light source can compare realistically in terms of color, definition, and clarity. At the beginning and end of the day, inside a structure or an enclosure, or on an overcast day, a substitute light source will most likely be needed. This light can be obtained from a floodlight or from a strobe or flash unit integrated with the camera.

15.2.3.7.1 Because a burned area has poor reflective properties, artificial lighting using floodlights is useful. Floodlights, however, will need a power source either from a portable generator or from a source within reach by extension cord.

15.2.3.7.2 Flash units are necessary for the fire investigator's work. The flash unit should be removable from the camera body so that it can be operated at an angle oblique to that of the lens view. This practice is valuable in reducing the amount of reflection, exposing more depth perception, and amplifying the texture of the heat- and flame-damaged surfaces. Another advantage to a detachable flash unit is that, if the desired composition is over a larger area, the angle and distance between the flash and the subject can be more balanced.

15.2.3.7.3 A technique that will cover a large scene is called photo painting. It can be accomplished by placing the camera in a fixed position with the shutter locked open. A flash unit can be fired from multiple angles, to illuminate multiple subjects or large areas from all angles. The same general effect can be obtained by the use of multiple flash units and remote operating devices called slaves.

15.2.3.7.4 For close-up work, a ring flash will reduce glare and give adequate lighting for the subject matter. Multiple flash units can also be used to give a similar effect to the ring flash by placing them to flash at oblique angles.

15.2.3.7.5 The investigator should be sure that glare from a flash or floodlight does not distort the actual appearance of an object. For example, smoke stains could appear lighter or nonexistent. In addition, shadows created could be interpreted as burn patterns. Movie lights used with videotapes can cause the same problems as still camera flash units. Using bounce flash, light diffusers, or other techniques could alleviate this problem.

15.2.3.7.6 The investigator concerned with the potential outcome of a photograph can bracket the exposure. Bracketing is the process of taking the same subject matter at slightly different exposure settings to ensure at least one correct exposure.

15.2.3.8 Special Types of Photography. Today's technology has produced some specialty types of photography. Infrared, laser, and microscopic photography can be used under controlled circumstances. An example is the ability of laser photography to document a latent fingerprint found on a body.

15.2.4 Composition and Techniques.

15.2.4.1 Photographs may be the most persuasive factor in the acceptance of the fire investigator's theory of the fire's evolution.

15.2.4.1.1 In fire investigation, a series of photographs should be taken to portray the structure and contents that remain at the fire scene. The investigator generally takes a series of photographs, working from the outside toward the inside of a structure, as well as from the unburned toward the most heavily burned areas. The concluding photographs are usually of the area and point of origin, as well as any elements of the cause of the fire. Deviations from the general photography sequence described in this section do not necessarily indicate faulty investigative methodology.

15.2.4.1.2 It can be useful for the photographer to record, and thereby document, the entire fire scene and not just the suspected point of origin, as it may be necessary to show the degree of smoke spread or evidence of undamaged areas.

15.2.4.2 Sequential Photos. Sequential photographs, shown in Figure 15.2.4.2, are helpful in understanding the relationship of a small subject to its relative position in a known area. The small subject is first photographed from a distant position, where it is shown in context with its surroundings. Additional photographs are then taken increasingly closer until the subject is the focus of the entire frame.

FIGURE 15.2.4.2 Sequential Photographs of a Chair.

15.2.4.3 Mosaics. A mosaic or collage of photographs can be useful at times when a sufficiently wide angle lens is not available and a panoramic view is desired. A mosaic is created by assembling a number of photographs in overlay form to give a more-than-peripheral view of an area, as shown in Figure 15.2.4.3. An investigator needs to identify items (e.g., benchmarks) in the edge of the view finder that will appear in the print and take the next photograph with that same reference point on the opposite side of the view finder. The two prints can then be combined to obtain a wider view than the camera is capable of taking in a single shot.

FIGURE 15.2.4.3 Mosaic of Warehouse Burn Scene from Aerial Truck.

15.2.4.4 Photo Diagram. A photo diagram can be useful to the investigator. When the finished product of a floor plan is complete, it can be copied, and directional arrows can be drawn to indicate the direction from which each of the photographs was taken. Numbers corresponding to the film frame and roll are then placed on the photographs. This diagram will assist in orienting a viewer who is unfamiliar with the fire scene. A diagram prepared to log a set of photographs might appear as shown in Figure 15.2.4.4.

15.2.4.4.1 Recommended documentation includes identification of the photographer, identification of the fire scene (i.e., address or incident number), and the date that the photographs were taken. A title form can be used for the first image to record this photo documentation.

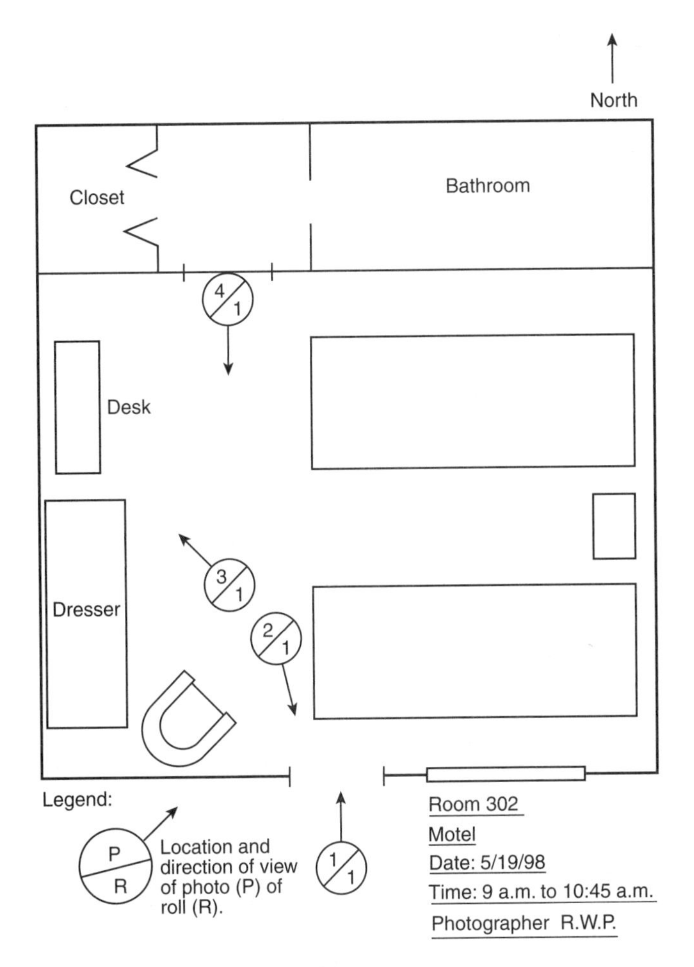

FIGURE 15.2.4.4 Diagram Showing Photo Locations.

15.2.4.4.2 The exact time a photograph is taken does not always need to be recorded. There are instances, however, when the time period during which a photograph was taken will be important to an understanding of what the photograph depicts. In photographs of identical subject, natural lighting conditions that exist at noon may result in a significantly different photographic image than natural lighting conditions that exist at dusk. When lighting is a factor, the approximate time or period of day should be noted. Also, the specific time should be noted for any photograph taken prior to extinguishment of the fire, as these often help establish time lines in the fire's progress.

15.2.4.5 Assisting Photographer. If a person other than the fire investigator is taking the photographs, the angles and composition should be supervised by the fire investigator to ensure that shots needed to document the fire are obtained. Investigators should communicate their needs to the photographer, as they may not have a chance to return to the fire scene. The investigators should not assume that the photographer understands what essential photographs are needed without discussing the content of each photo.

15.2.4.6 Photography and the Courts. For the fire investigator to weave photographs and testimony together in the courtroom, one requirement in all jurisdictions is that the photograph should be relevant to the testimony. There are other requirements that may exist in other jurisdictions, including noninflammatory content, clarity of the photograph, or lack of distortion. In most courts, if the relevancy exists, the photograph will usually withstand objections. Since the first color photographs were introduced into evidence in a fire trial, most jurisdictions have not distinguished between color or black and white photographs, if the photographs met all other jurisdictional criteria.

15.2.5 Video. In recent years, advancements have made motion pictures more available to the nonprofessional through the use of video cameras. There are different formats available for video cameras, including VHS, Beta, and 8 mm. Video is a very useful tool to the fire investigator. A great advantage to video is the ability to orient the fire scene by progressive movement of the viewing angle. In some ways, it combines the use of the photo diagram, photo indexing, floor plan diagram, and still photos into a single operation.

15.2.5.1 When taking videos or movies, "zooming-in" or otherwise exaggerating an object should be avoided, as it can be considered to present a dramatic effect rather than the objective effect that is sometimes required for evidence in litigation work.

15.2.5.2 Another use of video is for interviews of witnesses, owners, occupants, or suspects when the documentation of their testimony is of prime importance. If demeanor is important to an investigator or to a jury, the video can be helpful in revealing that.

15.2.5.3 The exclusive use of videotape or movies is not recommended, because such types of photography are often considered less objective and less reliable than still photographs. Video should be used in conjunction with still photographs.

15.2.5.4 Videotape recording of the fire scene can be a method of recording and documenting the fire scene. The investigator can narrate observations, similar to an audio (only) tape recorder, while videorecording the fire scene. The added benefit of video recording is that the investigator can better recall the fire scene, specifically fire patterns or artifact evidence, their location, and other important elements of the fire scene. Utilized in this method, the recording is not necessarily for the purpose of later presentation, but is simply another method by which the investigator can record and document the fire scene.

15.2.5.5 Video recording can also be effective to document the examination of evidence, especially destructive examination. By videotaping the examination, the condition and position of particular elements of evidence can be documented.

15.2.6 Suggested Activities to Be Documented. An investigation may be enhanced if as many aspects of the fire ground activities can be documented as possible or practical. Such documentation may include the suppression activities, overhaul, and the cause and origin investigation.

15.2.6.1 During the Fire. Photographs of the fire in progress should be taken if the opportunity exists. These help show the fire's progression as well as fire department operations. As the overhaul phase often involves moving the contents and sometimes structural elements, photographing the overhaul phase will assist in understanding the scene before the fire.

15.2.6.2 Crowd or People Photographs. Photographs of people in a crowd are often valuable for identifying individuals who may have additional knowledge that can be valuable to the overall investigation.

15.2.6.3 Fire Suppression Photographs. Fire suppression activities pertinent to the investigation include the operation of automatic systems as well as the activities of the responding fire services, whenever possible. All aspects pertinent to these, such as hydrant locations, engine company positions, hose lays, attack line locations, and so forth, play a role in the eventual outcome of the fire. Therefore, all components of those systems should be photographed.

15.2.6.4 Exterior Photographs. A series of exterior shots should be taken to establish the location of a fire scene. These shots could include street signs or access streets, numerical addresses, or landmarks that can be readily identified and are likely to remain for some time. Surrounding areas that would represent remote evidence, such as fire protection and exposure damage, should also be photographed. Exterior photographs should also be taken of all sides and corners of a structure to reveal all structural members and their relationships with each other. *(See Figure 15.2.6.4.)*

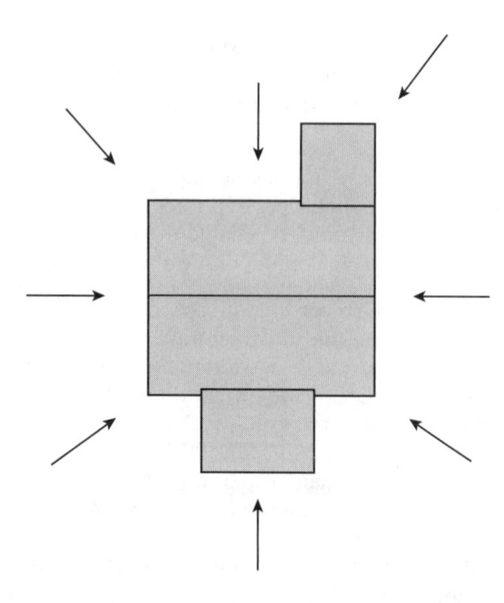

FIGURE 15.2.6.4 Photographing the Scene from All Angles and Corners.

15.2.6.5 Structural Photographs. Structural photographs document the damage to the structure after heat and flame exposure. Structural photos can expose burn patterns that can track the evolution of the fire and can assist in understanding the fire's origin.

15.2.6.5.1 A recommended procedure is to include as much as possible all exterior angles and views of the structure. Oblique corner shots can give reference points for orientation. Photographs should show all angles necessary for a full explanation of a condition.

15.2.6.5.2 Photographs should be taken of structural failures such as windows, roofs, or walls, because such failures can change the route of fire travel and can play a significant role in the eventual outcome of the fire. Code violations or structural deficiencies should also be photographed because fire travel patterns may have resulted from those deficiencies.

15.2.6.6 Interior Photographs. Interior photographs are equally important. Lighting conditions will likely change from the exterior, calling for the need to adjust technique, but the concerns (tracking and documenting fire travel backward toward the fire origin) are the same. All significant ventilation points accessed or created by the fire should be photographed, as well as all significant smoke, heat, and burn patterns. Figure 15.2.6.6 provides a diagram of basic shots.

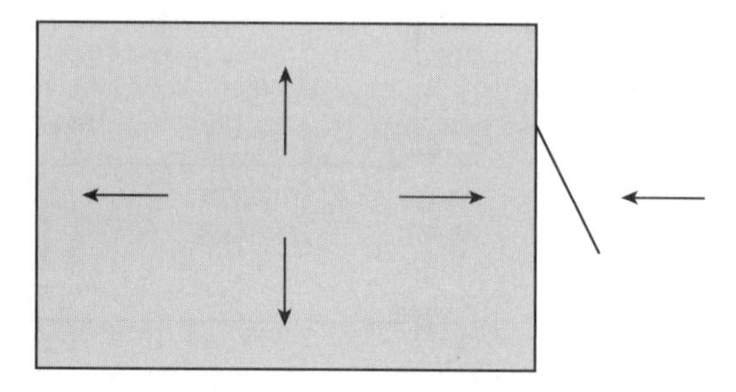

FIGURE 15.2.6.6 Photographing All Four Walls and Both Sides of Each Door.

15.2.6.6.1 Rooms within the immediate area of the fire origin should be photographed, even if there is no damage. If warranted, closets and cabinet interiors should also be documented. In small buildings, this documentation could involve all rooms; but in large buildings, it may not be necessary to photograph all rooms unless there is a need to document the presence, absence, or condition of contents.

15.2.6.6.2 All heat-producing appliances or equipment, such as furnaces, in the immediate area of the origin or connected to the area of origin should be photographed to document their role, if any, in the fire cause.

15.2.6.6.3 All furniture or other contents within the area of origin should be photographed as found and again after reconstruction. Protected areas left by any furnishings or other contents should also be photographed, as in the example shown in Figure 15.2.6.6.3.

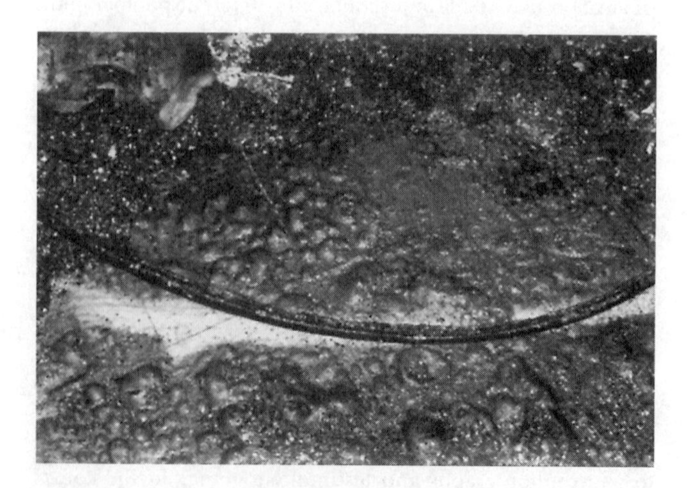

FIGURE 15.2.6.6.3 Floor Tile Protected from Radiant Heat by Wire.

15.2.6.6.4 The position of doors and windows during a fire is important, so photographs should be taken that would document those indications and resulting patterns.

15.2.6.6.5 Interior fire protection devices such as detectors, sprinklers, extinguishers used, door closers, or dampers should be photographed.

15.2.6.6.6 Clocks may indicate the time power was discontinued to them or the time in which fire or heat physically stopped their movement.

15.2.6.7 Utility and Appliance Photographs. The utility (gas, electric) entrances and controls both inside and outside a structure should be photographed. Photos should include gas and electric meters, gas regulators, and their location relative to the structure. The electric utility pole(s) near the structure that is equipped with the transformer serving the structure, and the electrical services coming into the structure, as well as the fuse or circuit breaker panels, should also be photographed. If there are gas appliances in the fire area of origin, the position of all controls on the gas appliances should be photographed. When photographing electrical circuit breaker panels, the position of all circuit breaker handles and the panel's schedule indicating what electrical equipment is supplied by each breaker, when available, should be photographed. Likewise, all electrical cords and convenience outlets pertinent to the fire's location should be photographed.

15.2.6.8 Evidence Photographs. Items of evidentiary value should be photographed at the scene and can be rephotographed at the investigator's office or laboratory if a more detailed view is needed. During the excavation of the debris strata, articles in the debris may or may not be recognized as evidence. If photographs are taken in an archaeological manner, the location and position of evidence that can be of vital importance will be documented permanently. Photographs orient the articles of evidence in their original location as well as show their condition when found. Evidence is essential in any court case, and the photographs of evidence stand strong with proper identification. In an evidentiary photograph, a ruler can be used to identify relative size of the evidence. Other items can also be used to identify the size of evidence as long as the item is readily identifiable and of constant size (e.g., a penny). A photograph should be taken of the evidence without the ruler or marker prior to taking a photograph with the marker *(see 16.5.2.1)*.

15.2.6.9 Victim Photographs. The locations of occupants should be documented, and any evidence of actions taken or performed by those occupants should be photographed. This documentation should include marks on walls, beds victims were occupying, or protected areas where a body was located. *(See Figure 15.2.6.9.)* If there is a death involved, the body should be photographed. Surviving victims' injuries and their clothing worn should also be photographed.

FIGURE 15.2.6.9 Protected Area Where Body Was Located.

15.2.6.10 Witness Viewpoint Photographs. During an investigation, if witnesses surface and give testimony as to what they observed from a certain vantage point, a photograph should be taken from the most identical view available. This photograph will orient all persons involved with the investigation, as well as a jury, to the direction of the witnesses' observations and could support or refute the possibility of their seeing what they said they saw.

15.2.6.11 Aerial Photographs. Views from a high vantage point, which can be an aerial fire apparatus, adjacent building, or hill, or from an airplane or helicopter can often reveal fire spread patterns. Aerial photography can be expensive, and a number of special problems exist that can affect the quality of the results. It is suggested that the investigator seek the advice or assistance of an experienced aerial photographer when such photographs are desired.

15.2.7 Photography Tips. Investigators may help themselves by applying some or all of the photography tips in 15.2.7.1 through 15.2.7.9.

15.2.7.1 Upon arrival at a fire scene and after shooting an 18 percent gray card, a written "title sheet" that shows identifying information (i.e., location, date, or situational information) should be photographed.

15.2.7.2 The film canister should be labeled after each use to prevent confusion or loss.

15.2.7.3 If the investigator's budget will allow, bulk film can be purchased and loaded into individual canisters that can allow for specific needs in multiple roll sizes and can be less expensive in certain situations.

15.2.7.4 A tripod that will allow for a more consistent mosaic pattern, alleviate movement and blurred photographs, and assist in keeping the camera free of fire debris should be available. A quick-release shoe on the tripod will save time.

15.2.7.5 Multiple fire incidents should not be combined on one roll of film. The last roll should be removed from the camera before leaving the scene. This will eliminate potential confusion and problems later on.

15.2.7.6 Extra batteries should be carried, especially in cold weather when they can be drained quickly. Larger and longer-life battery packs and battery styles are available.

15.2.7.7 Batteries should not be left in the photography equipment for an extended period of time. Leaking batteries can cause a multitude of problems to electrical and mechanical parts.

15.2.7.8 Obstruction of the flash or lens by hands, camera strap, or parts of the fire scene should be avoided. Additionally, when the camera is focused and ready to shoot, both eyes of the photographer should be opened to determine whether the flash went off.

15.2.7.9 In the event that prints from a single roll of film may have become out of sequence, examination of the numbering on the film negatives provides a permanent record of the sequence in which photographs were taken on that particular roll.

15.2.8 Presentation of Photograph. There is a variety of methodologies available to the investigator for the presentation of reports, diagrams, and photographs. A key to the decision-making process is, "What method of presentation shows or presents the item with the greatest clarity?" A secondary consideration to assist in the preparation of the presentation is to follow guidelines or practices that are used for instructional presentations, specifically in the area of instructional aids. The investigator should determine what methods of presentation and types of photographs currently are acceptable to the court. Additionally, the investigator should identify and obtain equipment that may be needed to support the presentation, oversee the setup, and test the equipment prior to use. Preparation is one of the most important aspects of presenting demonstrative evidence.

15.2.8.1 Prints versus Slides.

15.2.8.1.1 There are advantages and disadvantages to both prints and slides. A benefit of slides over prints is that large size images may be displayed at no additional cost. When showing slides in court, the investigator can keep every juror's attention on what the investigator is testifying about. If prints are utilized, the investigator's testimony may be recalled only vaguely, if the jury member is busy looking at photographs that are passed among the jurors as testimony continues. The use of poster-sized enlargements can help.

15.2.8.1.2 Conversely, during testimony of a long duration or during detailed explanations of the scene, slides are a burden to refer to without the use of a projector. In this case, photographs are easier to handle and analyze. When slides are used, problems can occur, such as the slides jamming or a lamp burning out in the projector; thus, there may be no alternate way to display the scene to the jurors without delay. Prints require no mechanical devices to display them, and notations for purposes of identification, documentation, or description are easily affixed on or adjacent to a still photograph.

15.2.8.2 Video Presentation.

15.2.8.2.1 The use of video to present important information in testimony is an excellent methodology. Key to proper use of video presentation is to ensure that the size of the screen is sufficient to allow all interested parties to see the material adequately. The use of additional monitors may assist in overcoming this problem.

15.2.8.2.2 The investigator should be aware of quality issues when preparing the video presentation, as those that will be viewing the presentation are accustomed to broadcast-quality video.

15.2.8.3 Computer-Based Presentations. The advancement and increased use of computer-based presentations provides the investigator with an excellent tool for presentation. As with other presentation formats, there are inherent advantages and disadvantages to those programs.

15.2.8.3.1 Computer-based presentations provide the user with the ability to put drawings and photographs on the same slide, as well as to provide other highlighting or information that may enhance the observer's ability to understand relationships or information being presented.

15.2.8.3.2 The investigator should have backup resources available, such as the original photographs and drawings, in the event that hardware incompatibility or software problems prevent the presentation from being viewed or reduce the effectiveness of the presentation.

15.3 Note Taking. Note taking is a method of documentation in addition to drawings and photographs. Items that may need to be documented in notes may include the following:

(1) Names and addresses
(2) Model/serial numbers
(3) Statements and interviews
(4) Photo log
(5) Identification of items
(6) Types of materials (e.g., wood paneling, foam plastic, carpet)
(7) Data that was needed to produce an accurate computer model (see Section 20.6)
(8) Investigator observations (e.g., burn patterns, building conditions, position of switches and controls)

15.3.1 Forms of Incident Field Notes. The collection of data concerning an investigation is important in the analysis of any incident. The use of forms is not required in data collection; however, some forms have been developed to assist the investigator in the collection of data. These example forms and the information documented are not designed to constitute the report but, rather, they provide a means to gather data that may be helpful in reaching conclusions so that a report can be prepared.

15.3.2* Forms for Collecting Data. Some forms have been developed to assist the investigator in the collection of data. These forms and the information documented on them are not designed to constitute the incident report. They provide a means to gather data that may be helpful in reaching conclusions so that the incident report or the investigation report can be prepared. See Table 15.3.2.

Table 15.3.2 Field Notes and Forms

Form	Purpose
Fire incident field notes	Any fire investigation to collect general incident data
Casualty field notes	Collection of general data on any victim killed or injured
Wildfire field notes	Data collection specifically for wildfire
Evidence form	Documentation of evidence collection and chain of custody
Vehicle inspection form	Data collection of incidents specifically involving motor vehicles
Photograph log	Documentation of photographs taken during the investigation
Electrical panel documentation	Collection of data specifically relating to electrical panels
Structure fire notes	Collection of data concerning structure fires
Compartment fire modeling	Collection of data necessary for compartment fire modeling

15.3.3 Dictation of Field Notes. Many investigators dictate their notes using portable tape recorders. Investigators should be careful not to rely solely on tape recorders or any single piece of equipment when documenting critical information or evidence.

15.3.4 The retention of original notes, diagrams, photographs, and measurements such as detailed in Section 15.3 is the best practice. Unless otherwise required by a written policy or regulation, such data should be retained. These data constitute a body of factual information that should be retained until all reasonably perceived litigation processes are resolved. Information collected during the investigation may become significant long after it is collected and after the initial report is written. For example, notes or a diagram of a circuit breaker panel showing the status of the breakers may not be pertinent for a fire where the origin is in upholstered furniture, but may be of value regarding the status of the circuit powering a smoke detector. The retention of notes is not necessarily intended to apply to the fire fighter completing the required data fields in a fire incident reporting system.

15.4 Diagrams and Drawings. Clear and concise sketches and diagrams can assist the investigator in documenting evidence of fire growth, scene conditions, and other details of the fire scene. Diagrams are also useful in providing support and understanding of the investigator's photographs. Diagrams may also be useful in conducting witness interviews. However, no matter how professional a diagram may appear, it is only as useful as the accuracy of the data used in its creation. Various types of drawings, including sketches, diagrams, and plans can be made or obtained to assist the investigator in documenting and analyzing the fire scene.

15.4.1 Types of Drawings. Fire investigations that can be reasonably expected to be involved in criminal or civil litigation should be sketched and diagrammed.

15.4.1.1 Sketches. Sketches are generally freehand diagrams or diagrams drawn with minimal tools that are completed at the scene and can be either three-dimensional or two-dimensional representations of features found at the fire scene.

15.4.1.2 Diagrams. Diagrams are generally more formal drawings that are completed after the scene investigation is completed. Diagrams are completed using the scene sketches and can be drawn using traditional methodologies or computer-based drawing programs. It should be noted that the completion of formal scene diagrams may not be required in some instances. The decision to complete a more formal scene diagram will be determined by the investigator, agency policies, and scope of the investigation.

15.4.2 Selection of Drawings. It is recommended that original sketches and finalized diagrams be retained throughout the life of the investigation and any resulting litigation. See examples of various types of drawings in Figure 15.4.2(a) through Figure 15.4.2(f).

15.4.3 Drawing Tools and Equipment. Depending on the size or complexity of the fire, various techniques can be used to prepare the drawings. As with photographs, drawings are used to support memory, as the investigator may only get one chance to inspect the fire scene. As with the other methods of documenting the scene, the investigator will need to determine the type and detail of the diagrams developed and the type of drawings that may be requested from the building or equipment designer or manufacturer. During the course of the investigation, the investigator may have available a variety of drawings. These drawings may have been prepared by a building or equipment designer or manufacturer, may have been drawn by the investigator, or may have been developed by other investigators documenting conditions found at the time of their investigation.

15.4.3.1 Computer software is available that allows the investigator to prepare high-quality diagrams from scene-generated sketches. The fire investigator should look to several features in deciding on the best computer drawing tool to meet the intended goals. The fire investigator must first decide whether or not three-dimensional (3D) capability is required. In making that decision, the fire investigator must also decide whether or not the time invested in learning the package and additional complexity warrant the investment in such a tool. However, 3D drawings can yield great benefits in the investigation in determining and demonstrating such issues as the physical interrelationships of building components or the available view to witness. Regardless of the package selected, the most important criterion is the fire investigator's ability to create, modify, produce output, and manipulate a drawing within the selected package. Another consideration would be the compatibility of the Computer Aided Drawing (CAD) output to provide computer fire models input.

15.4.3.2 A good drawing package should also allow for the drawing on separate "layers" that can be turned on and off for different display purposes, such as pre-fire layout and post-fire debris. The package should also provide for automatic dimensioning and various dimensioning styles (i.e., decimal 1.5 ft, architectural 1 ft 6 in., and 457 mm). The package should also come with a wide variety of dimensioned "parts libraries." This component of the package provides the pre-drawn details, such as kitchen and bathroom fixtures, for placement by the investigator in the drawing.

15.4.4 Diagram Elements. The investigator, depending on the scope and complexity of the investigation, and on agency procedures, will decide on what elements to include on sketches and diagrams; however, there are a number of key elements that should be on all sketches and diagrams, as outlined in 15.4.4(A) through 15.4.4(D).

(A) General Information. Identification of the individual who prepared the diagram, diagram title, date of preparation, and other pertinent information should be included.

(B) Identification of Compass Orientation. Identification of compass orientation should be included on sketches and diagrams of fire scenes.

(C) Scale. The drawing should be drawn approximately to scale. The scale should be identified or indicated "Approximate Scale" or "Not to Exact Scale," and a graphic scale or approximate scale may be provided on the drawing.

(D)* Symbols. Symbols are commonly utilized on sketches and drawings to denote certain features; for example, a door symbol is used to indicate that there is a door in the wall and is drawn in the direction of swing. To facilitate understanding, it is recommended that the investigator utilize standard drawing symbols commonly found in the architectural or engineering community. For fire protection symbols, the investigator may utilize the symbols contained in NFPA 170, *Standard for Fire Safety and Emergency Symbols.*

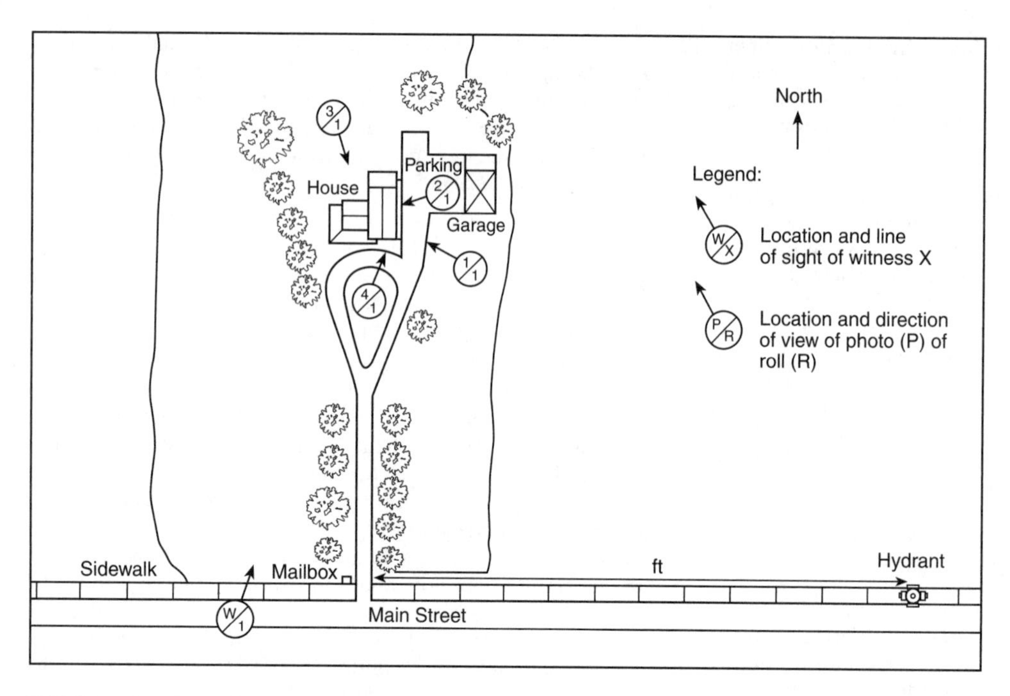

FIGURE 15.4.2(a) Site Plan Showing Photo and Witness Locations.

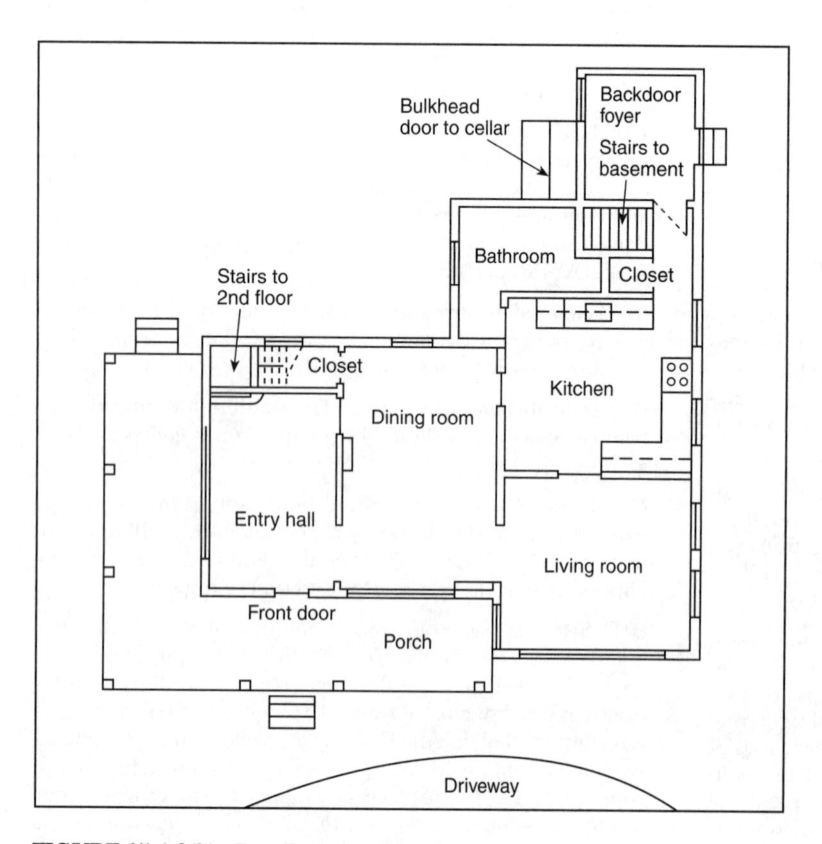

FIGURE 15.4.2(b) Detailed Floor Plan.

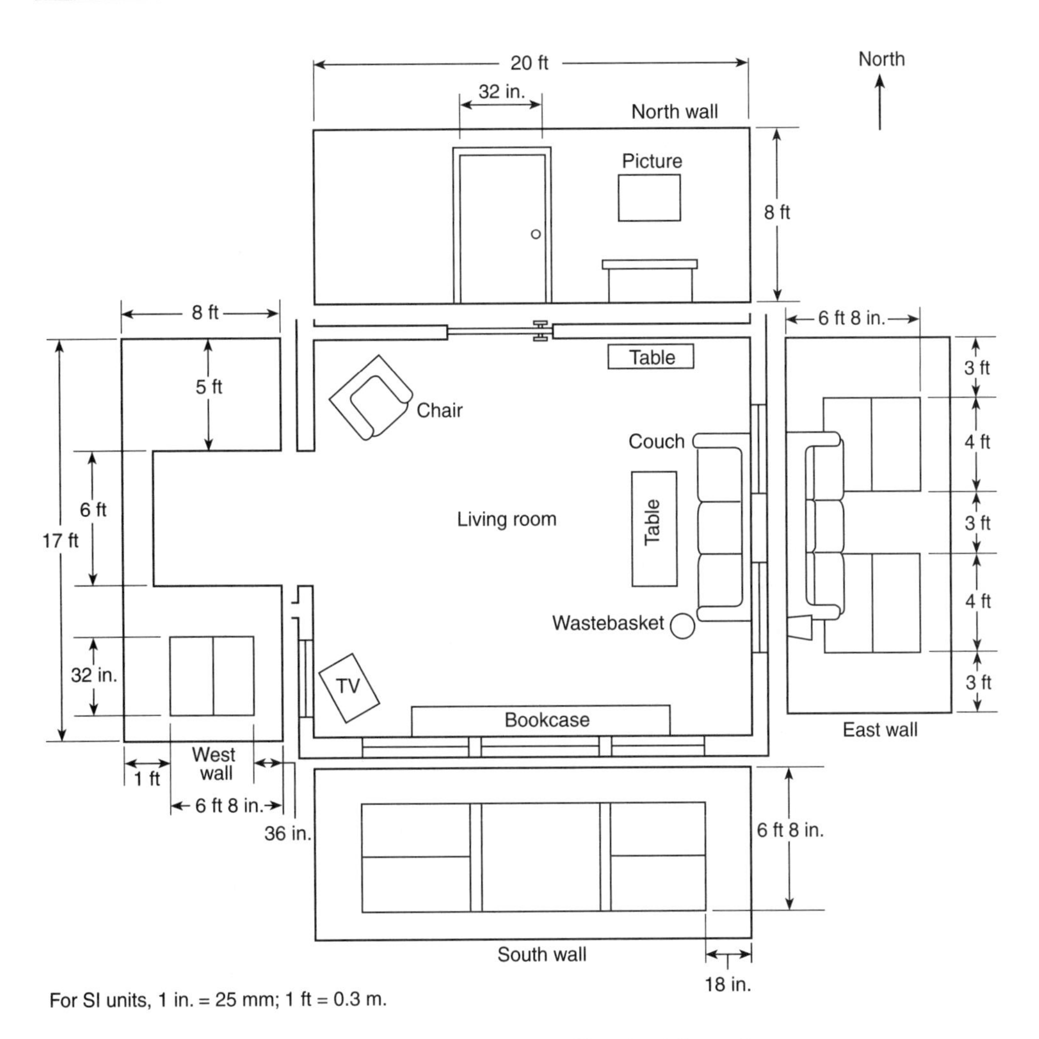

FIGURE 15.4.2(c) Diagram of Room and Contents Showing Dimensions.

(E) Legend. If symbols are utilized that are not readily identifiable, the investigator should use a legend on the drawing to eliminate the potential for confusion as to what the symbol represents.

15.4.5 Drawings. Generally, a simple sketch of the room of origin or immediate scene should be prepared and should include items such as furniture, windows and doors, and other useful data. A typical building sketch can show the relative locations of rooms, stairs, windows, doors, and associated fire damage. These drawings can be done freehand with approximate dimensions. This type of drawing should suffice on fire cases where the fire analysis and conclusions are simple. More complex scenes or litigation cases will often require developing or acquiring actual building plans and detailed documentation of construction, equipment, furnishings, witness location, and damage.

15.4.5.1 Site or Area Plans. Plot or area diagrams may be needed to show the placement of apparatus, the location of the fire scene to other buildings, water supplies, or similar information. Diagrams of this nature assist in documenting important factors outside of the structure. See Figure 15.4.2(a).

15.4.5.2 Floor Plans. Floor plans of a building identify the locations of rooms, stairs, windows, doors, and other features of the structure. See Figure 15.4.2(b).

15.4.5.3 Elevations. Elevation drawings are single plane diagrams that show a wall, either interior or exterior, and specific information about the wall. See elevation portion of Figure 15.4.5.3.

15.4.5.4 Details and Sections. Details and sections are drawn to show specific features of an item. There is a wide variety of information that can be represented in a detail or section diagram, such as the position of switches or controls, damage to an item, location of an item, construction features, and many more.

15.4.5.5 Exploded View Diagrams. Exploded view diagrams are often used to show assembly of components or parts lists. The investigator may also utilize this format to show all surfaces inside of a room or compartment on the same diagram. See Figure 15.4.2(d).

15.4.5.6 Three-Dimensional Representations. In many cases, it will be desirable, if not necessary, for the investigator to obtain sufficient dimensional data to develop a three-dimensional representation of the fire scene.

15.4.5.6.1 Structural Dimensions. The investigator should measure and document dimensions that would be required to develop an accurate three-dimensional representation of the structure, as illustrated in Figure 15.4.2(c). Consideration

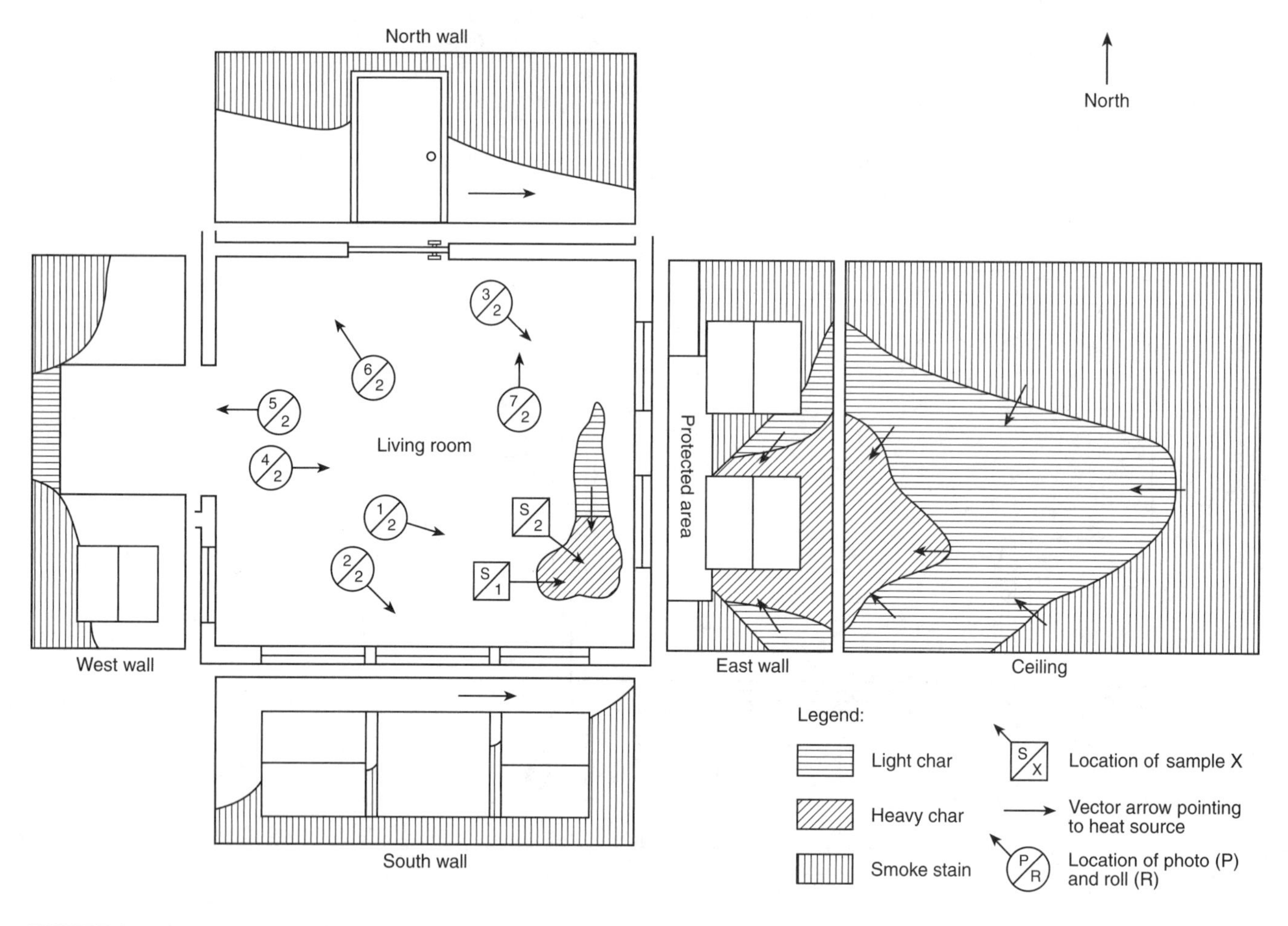

FIGURE 15.4.2(d) Exploded Room Diagram Showing Damage Patterns, Sample Locations, and Photo Locations.

should be given to the documentation of such often overlooked dimensions as the thickness of walls, air gaps in doors, and the slope of floors, walls, and ceilings. Such representative geometry may be required if subsequent fire modeling and/or experimental tests are to be conducted as part of the incident investigation.

15.4.5.6.2 Availability of Dimensional Data. While dimensional data may be found in building plans, layouts, or as-built drawings, it may not be known at the time of the scene investigation if such sources of information exist, especially in the case of older structures. Thus, it is prudent for the investigator to collect the physical dimensions independent of the existence of plans, layouts, or drawings.

15.4.5.7 Specialized Fire Investigation Diagrams.

15.4.5.7.1 In addition to a basic floor plan diagram, it is recommended that the fire investigator utilize specialized sketches and diagrams to assist in documenting specifics of the fire investigation. The decision to utilize these or other specialized sketches and diagrams is dependent on the decision of the investigator and the need to represent a complex fact or issue. These types of specialized investigation diagrams will include electrical, mechanical, process system, and fuel gas piping schematics, fire pattern, depth of char survey, depth or calcination survey, witness line of sight, heat and flame vector analysis, and others, as required.

15.4.5.7.2 The use of a computer-drawing program facilitates the development of many different specialized drawings from the initial floor plan. The use of layers or overlays can assist in the understanding of specific features and prevent the drawing from becoming overly complicated.

15.4.6 Prepared Design and Construction Drawings.

15.4.6.1 General. Prepared design and construction diagrams are those diagrams that were developed for the design and construction of buildings, equipment, appliances, and similar items by the design professional. These diagrams are often useful to the investigator to assist in determining components, design features, specifications, and other items.

15.4.6.1.1 The availability and complexity of prepared design and construction drawings will vary, depending on the type and size of the occupancy for structures or the ability to identify a specific manufactured item.

15.4.6.1.2 During or after building construction or as a result of occupancy changes, modifications may occur. These modifications may not be reflected on any existing drawings. When using prepared building diagrams, the investigator will need to compare the drawing to the actual building layout.

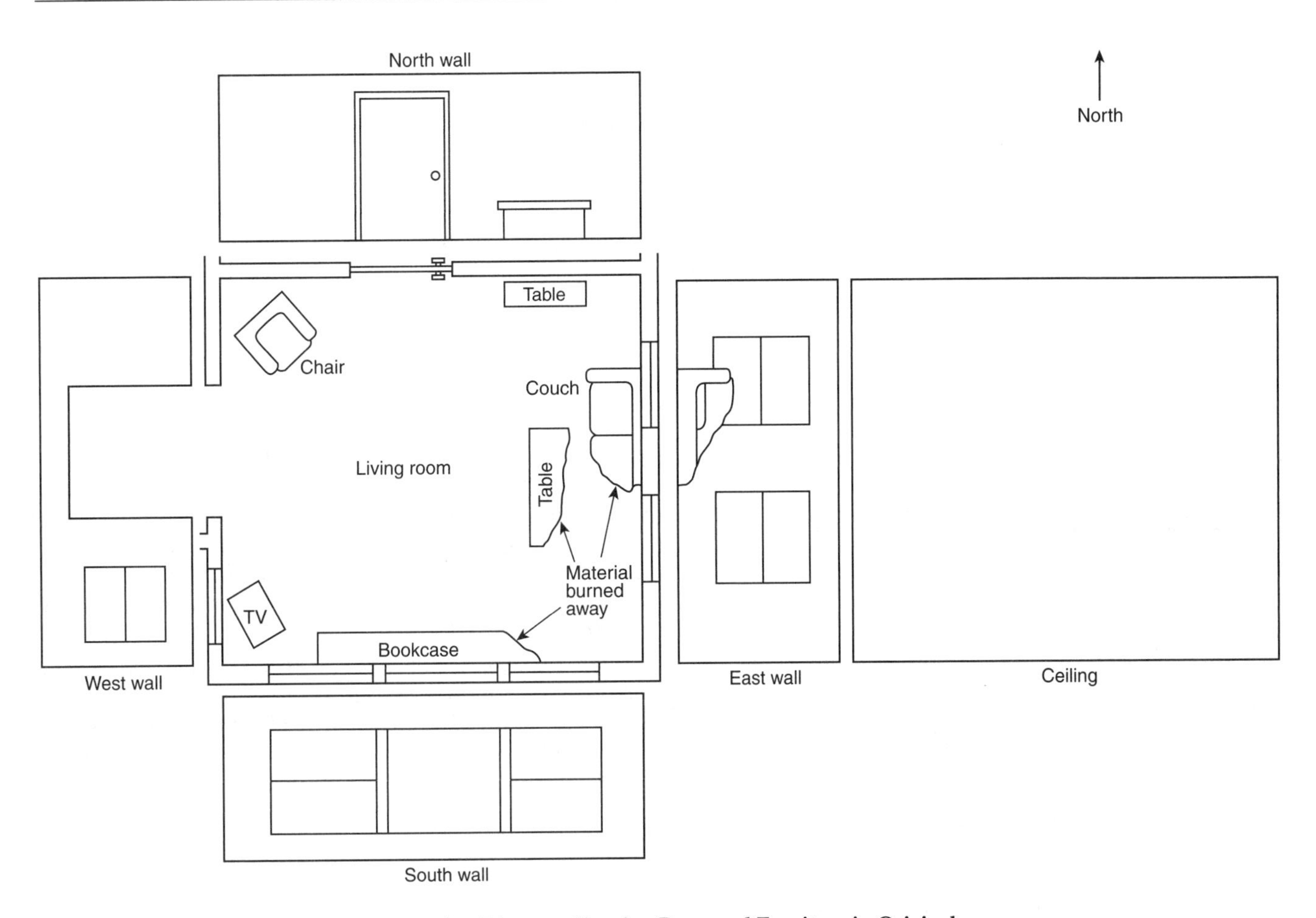

FIGURE 15.4.2(e) Contents Reconstruction Diagram Showing Damaged Furniture in Original Positions.

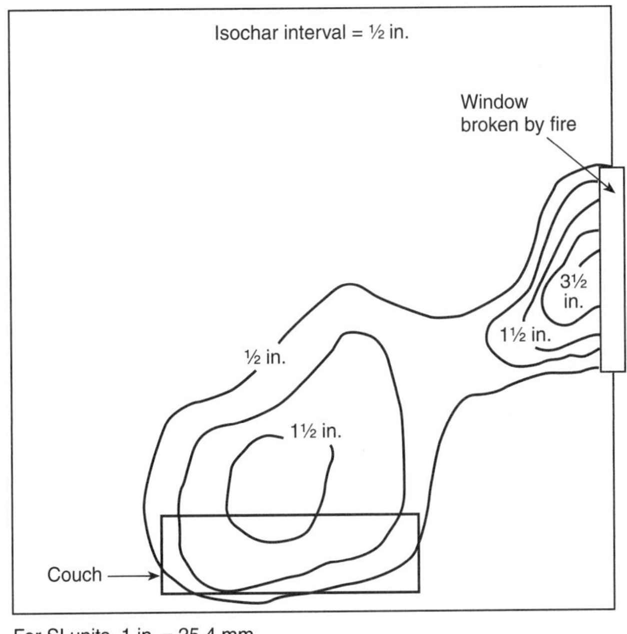

FIGURE 15.4.2(f) An Isochar Diagram Showing Lines of Equal Char Depth on Exposed Ceiling Joists.

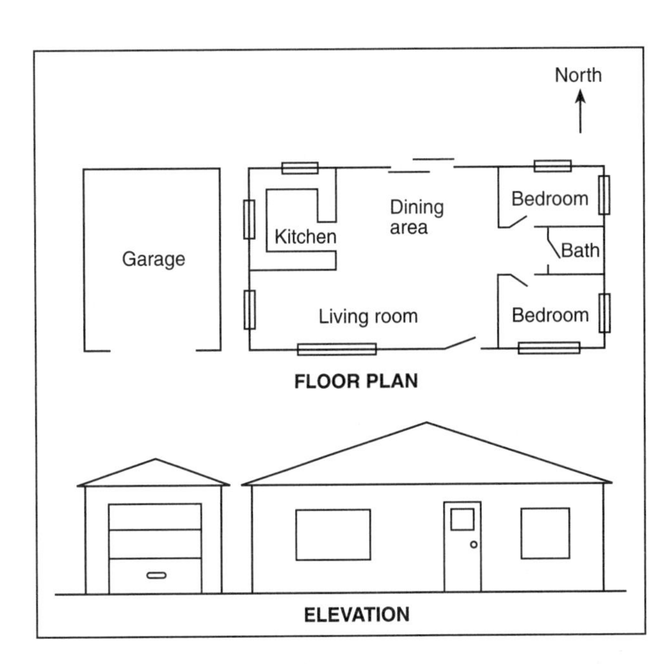

FIGURE 15.4.5.3 Minimum Drawing for Simple Fire Analysis.

15.4.6.2 Architectural and Engineering Drawings. Within the design and construction process, there are several types of drawings with which the investigator should be familiar. The most common drawings along with the discipline that generally prepares them are shown in Table 15.4.6.2.

15.4.6.3 Architectural and Engineering Schedules. On larger projects, it may be necessary to detail the types of equipment in lists that are called *schedules*. Where many components are specified in great detail, a schedule will usually exist. Typical schedules are as follows:

(1) Door and window schedule
(2) Interior finish schedule
(3) Electrical schedule
(4) HVAC schedule
(5) Plumbing schedule
(6) Lighting schedule

15.4.6.4* Specifications. Architects and engineers prepare specifications to accompany their drawings. While the drawings show the geometry of the project, the specifications detail the quality of the materials, responsibilities of various contractors, and the general administration of the project. Specifications are usually divided into sections for the various components of the building. For the fire investigator, the properties of materials can be identified through a specification review and may assist in the analysis of the fire scene.

15.4.6.5 Appliances and Building Equipment. Parts diagrams and shop drawings may be available for appliances and equipment that may have been involved in a fire scenario. These diagrams may assist the investigator in determining or obtaining specific information about components or other features.

15.5* Reports. The purpose of a report is to effectively communicate the observations, analyses, and conclusions made during an investigation. The specific format of a report is not prescribed. For guidance on court-mandated reports, see Chapter 11.

15.5.1 Descriptive Information. Generally, reports should contain the following information, preferably in the introduction:

(1) Date, time, and location of incident
(2) Date and location of examination
(3) Date the report was prepared
(4) Name of the person or entity requesting the report
(5) The scope of the investigation (tasks completed)
(6) Nature of the report (preliminary, interim, final, summary, supplementary)

15.5.2 Pertinent Facts. A description of the incident scene, items examined, and evidence collected should be provided. The report should contain observations and information relevant to the opinions. Photographs, diagrams, and laboratory reports may be referenced.

15.5.3 Opinions and Conclusions. The report should contain the opinions and conclusions rendered by the investigator. The report should also contain the foundation(s) on which the opinion and conclusions are based. The name, address, and affiliation of each person who has rendered an opinion contained in the report should be provided.

Table 15.4.6.2 Design and Construction Drawings That May Be Available

Type	Information	Discipline
Topographical	Varying grade of the land	Surveyor
Site plan	Structure on the property with sewer, water, electrical distributions to the structure	Civil engineer
Floor plan	Walls and rooms of structure as if you were looking down on it	Architect
Plumbing	Layout and size of piping for fresh water and wastewater	Mechanical engineer
Electrical	Size and arrangement of service entrance, switches and outlets, fixed electrical appliances	Electrical engineer
Mechanical	HVAC system	Mechanical engineer
Sprinkler/ fire alarm	Self-explanatory	Fire protection engineer
Structural	Frame of building	Structural engineer
Elevations	Shows interior/exterior walls	Architect
Cross-section	Shows what the inside of components look like if cut through	Architect
Details	Show close-ups of complex areas	All disciplines

Chapter 16 Physical Evidence

16.1* General. During the course of any fire investigation, the fire investigator is likely to be responsible for locating, collecting, identifying, storing, examining, and arranging for testing of physical evidence. The fire investigator should be thoroughly familiar with the recommended and accepted methods of processing such physical evidence.

16.2 Physical Evidence.

16.2.1 Physical evidence, defined generally, is any physical or tangible item that tends to prove or disprove a particular fact or issue. Physical evidence at the fire scene may be relevant to the issues of the origin, cause, spread, or the responsibility for the fire.

16.2.2* The decision on what physical evidence to collect at the incident scene for submission to a laboratory or other testing facility for examination and testing, or for support of a fact or opinion, rests with the fire investigator. This decision may be based on a variety of considerations, such as the scope of the investigation, legal requirements, or prohibition. *(See Section 13.2.)* Additional evidence may also be collected by others, including other investigators, insurance company representatives, manufacturer's representatives, owners, and occupants. The investigator should also be aware of standards and procedures relating to evidentiary issues and those issues related to spoliation of evidence.

16.3* Preservation of the Fire Scene and Physical Evidence.

16.3.1 General. Every attempt should be made to protect and preserve the fire scene as intact and undisturbed as possible, with the structure, contents, fixtures, and furnishings remaining in their pre-fire locations. Evidence such as the small paper match shown in Figure 16.3.1 could easily be destroyed or lost in an improperly preserved fire scene.

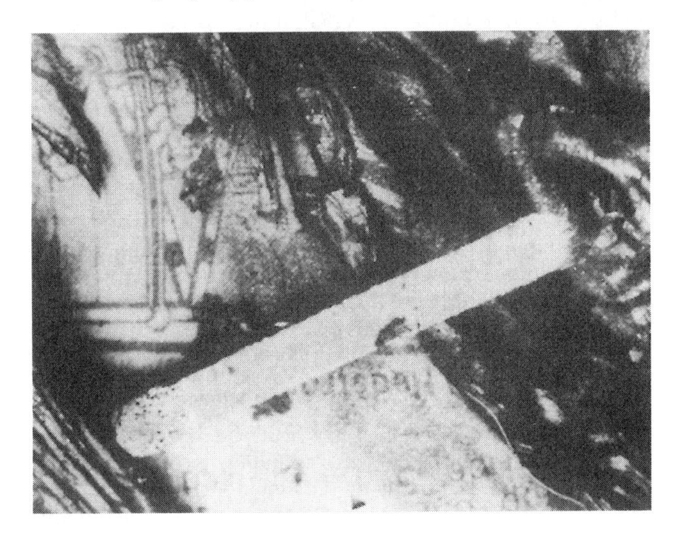

FIGURE 16.3.1 Physical Evidence at a Fire Scene.

16.3.1.1 Generally, the cause of a fire or explosion is not known until near the end of the investigation. Therefore, the evidentiary or interpretative value of various pieces of physical evidence observed at the scene may not be known until, at, or near the end of the fire scene examination, or until the end of the complete investigation. As a result, the entire fire scene should be considered physical evidence and should be protected and preserved. Consideration should be given to temporarily placing removed ash and debris into bags, tarps, or other suitable containers labeled as to the location from which it was removed. This way, if components from an appliance or an incendiary device are found to be missing they can be more easily found in a labeled container.

16.3.1.2 The responsibility for the preservation of the fire scene and physical evidence does not lie solely with the fire investigator, but should begin with arriving fire-fighting units or police authorities. Lack of preservation may result in the destruction, contamination, loss, or unnecessary movement of physical evidence. Initially, the incident commander and, later, the fire investigator should secure or ensure the security of the fire scene from unnecessary and unauthorized intrusions and should limit fire suppression activities to those that are necessary.

16.3.1.3 Evidence at the fire scene should be considered not only in a criminal context, such as in traditional forensic evidence (e.g., weapons, bodily fluids, footprints), nor should it be limited to arson-related evidence, items, or artifacts, such as incendiary devices or containers. Potential evidence at the fire scene and surrounding areas can include the physical structure, the contents, the artifacts, and any materials ignited or any material on which fire patterns appear.

16.3.2 Fire Patterns as Physical Evidence. The evidentiary and interpretative use of fire patterns may be valuable in the identification of a potential ignition source, such as an incendiary device in an arson fire or an appliance in an accidental fire. Fire patterns are the visible or measurable physical effects that remain after a fire. These include thermal effects on materials, such as charring, oxidation, consumption of combustibles, smoke and soot deposits, distortion, melting, color changes, changes in the character of materials, structural collapse, and other effects. *(See Section 6.3.)*

16.3.3 Artifact Evidence. Artifacts can be the remains of the material first ignited, the ignition source, or other items or components in some way related to the fire ignition, development, or spread. An artifact may also be an item on which fire patterns are present, in which case the preservation of the artifact is not for the item itself but for the fire pattern that is contained thereon.

16.3.4 Protecting Evidence.

16.3.4.1 There are a number of methods that can be utilized to protect evidence from destruction. Some methods include posting a fire fighter or police officer as a sentry to prevent or limit access to a building, a room, or an area; use of traffic cones or numerical markers to identify evidence or areas that warrant further examination; covering the area or evidence with tarpaulins prior to overhaul; or isolating the room or area with rope, caution tape, or police line tape. The investigator may benefit from supervising overhaul and salvage operations.

16.3.4.2 Items found at the fire scene, such as empty boxes or buckets, may be placed over an artifact. However, these items may not clearly identify the artifact as evidence that should be preserved by fire fighters or others at the fire scene. If evidence is not clearly identified, it may be susceptible to movement or destruction at the scene.

16.3.5 Role and Responsibilities of Fire Suppression Personnel in Preserving the Fire Scene.

16.3.5.1 Generally, fire officers and fire fighters have been instructed during basic fire training that they have a responsibility at the fire scene regarding fire investigation.

16.3.5.1.1 In most cases, this responsibility is identified as recognizing the indicators of incendiarism, such as multiple fires, the presence of incendiary devices or trailers, and the presence of ignitible liquids at the area of origin *(see Section 22.2)*. While this is an important aspect of their responsibilities in the investigation of the fire cause, it is only a small part.

16.3.5.1.2 Prompt control and extinguishment of the fire protects evidence. The ability to preserve the fire scene is often an important element in the investigation. Even when fire officers and fire fighters are not responsible for actually determining the origin or cause of the fire, they play an integral part in the investigation by preserving the fire scene and physical evidence.

16.3.5.2 Preservation. Once an artifact or other evidence has been discovered, preliminary steps should be taken to preserve and protect the item from loss, destruction, or movement. The person making the discovery should notify the incident commander as soon as practical. The incident commander should notify the fire investigator or other appropriate individual or agency with the authority and responsibility for the documentation and collection of the evidence.

16.3.5.3 Caution in Fire Suppression Operations. Fire crews should avoid causing unnecessary damage to evidence when using straight-stream hoselines, pulling ceilings, breaking windows, collapsing walls, and performing overhaul and salvage.

16.3.5.3.1 Use of Water Lines and Hose Streams. When possible, fire fighters should use caution with straight-stream applications, particularly at the base of the fire, because the base of the fire may be the area of origin. Evidence of the ignition source can sometimes be found at the area of origin. The use of hoselines, particularly straight-stream applications, can move, damage, or destroy physical evidence that may be present.

(A) The use of water hoselines for overhaul operations such as washing down, or for opening up walls or ceilings, should also be restricted to areas away from possible areas of origin.

(B) The use of water should be controlled in areas where the investigator may wish to look at the floor for possible fire patterns. When draining the floor of standing water, the drain hole should be located so as to have the least impact on the fire scene and fire patterns.

16.3.5.3.2 Overhaul.

(A) It is during overhaul that any remaining evidence not damaged by the fire is susceptible to being destroyed or displaced. Excessive overhaul of the fire scene prior to the documentation and analysis of fire patterns can affect the investigation, including failure to determine the area of origin.

(B) While the fire fighters have a responsibility to control and extinguish the fire and then check for fire extension, they are also responsible for the preservation of evidence. These two responsibilities may appear to be in conflict and, as a result, it is usually the evidence that is affected during the search for hidden fire. However, if overhaul operations are performed in a systematic manner, both responsibilities can be met successfully.

16.3.5.3.3 Salvage. The movement or removal of artifacts from a fire scene can make the reconstruction difficult for the investigator. If the investigator cannot determine the pre-fire location of the evidence, the analytical or interpretative value of the evidence may be lost. Moving, and particularly removing, contents and furnishings or other evidences at the fire scene should be avoided until the documentation, reconstruction, and analysis are completed.

16.3.5.3.4 Movement of Knobs and Switches. Fire fighters should refrain from turning knobs and operating switches on any equipment, appliances, or utility services at the fire scene. The position of components, such as the knobs and switches, may be a necessary element in the investigation, particularly in developing fire ignition scenarios or hypotheses. These components, which are often constructed of plastics, can become very brittle when subjected to heating. Their movement may alter the original post-fire state and may cause the switch to break or to become impossible to relocate in its original post-fire position. *(See 24.5.3.)*

16.3.5.3.5 Use of Power Tools. The use of gasoline- or diesel-powered tools and equipment should be controlled carefully in certain locations. The refueling of any fuel-powered equipment or tools should be done outside the perimeter of the fire scene. Whenever fuel-powered equipment is used on the fire scene, its use and location should be documented and the investigator advised.

16.3.5.3.6 Limiting Access of Fire Fighters and Other Emergency Personnel. Access to the fire scene should be limited to those persons who need to be there. This precaution includes limiting fire fighters and other emergency or rescue personnel to those necessary for the task at hand. When possible, the activity or operation should be postponed until the evidence has been documented, protected, evaluated, and collected.

16.3.6 Role and Responsibilities of the Fire Investigator. If the fire fighters have not taken the preliminary steps to preserve or protect the fire scene, then the fire investigator should assume the responsibility for doing so. Then, depending on the individual's authority and responsibility, the investigator should document, analyze, and collect the evidence.

16.3.7 Practical Considerations. The precautions in this section should not be interpreted as requiring the unsafe or infinite preservation of the fire scene. It may be necessary to repair or demolish the scene for safety or for other practical reasons. Once the scene has been documented by interested parties and the relevant evidence removed, there is no reason to continue to preserve the scene. The decision as to when sufficient steps have been taken to allow the resumption of normal activities should be made by all interested parties known at that time.

16.4 Contamination of Physical Evidence. Contamination of physical evidence can occur from improper methods of collection, storage, or shipment. Like improper preservation of the fire scene, any contamination of physical evidence may reduce the evidentiary value of the physical evidence.

16.4.1 Contamination of Evidence Containers.

16.4.1.1 Unless care is taken, physical evidence may become contaminated through the use of contaminated evidence containers. For this reason, the fire investigator should take every reasonable precaution to ensure that new and uncontaminated evidence containers are stored separately from used containers or contaminated areas.

16.4.1.2 One practice that may help to limit a possible source of cross-contamination of evidence collection containers, including steel paint cans or glass jars, is to seal them immediately after receipt from the supplier. The containers should remain sealed during storage and transportation to the evidence collection site. An evidence collection container should be opened only to receive evidence at the collection point, at which time it should be resealed pending laboratory examination.

16.4.2* Contamination During Collection. Most contamination of physical evidence occurs during its collection. This is especially true during the collection of liquid and solid accelerant evidence. The liquid and solid accelerant may be absorbed by the fire investigator's gloves or may be transferred onto the collection tools and instruments.

16.4.2.1 Avoiding cross-contamination of any subsequent physical evidence, therefore, becomes critical to the fire investigator. To prevent such cross-contamination, the fire investigator can wear disposable plastic gloves or place his or her hands into plastic bags during the collection of the liquid or solid accelerant evidence. New gloves or bags should always be used during the collection of each subsequent item of liquid or solid accelerant evidence.

16.4.2.2 An alternative method to limit contamination during collection is to utilize the evidence container itself as the collection tool. For example, the lid of a metal can may be used to scoop the physical evidence into the can, thereby eliminating any cross-contamination from the fire investigator's hands, gloves, or tools.

16.4.2.3 Similarly, any collection tools or overhaul equipment such as brooms, shovels, or squeegees utilized by the fire investigator need to be cleaned thoroughly between the collection of each item of liquid or solid accelerant evidence to prevent similar cross-contamination. The fire investigator should be careful, however, not to use waterless or other types of cleaners that may contain volatile solvents.

16.4.3 Contamination by Fire Fighters. Contamination is possible when fire fighters are using or refilling fuel-powered tools and equipment in an area where an investigator later tests for the presence or omission of an ignitible liquid. Fire fighters should take the necessary precautions to ensure that the possibility of contamination is kept to a minimum, and the investigator should be informed when the possibility of contamination exists.

16.5 Methods of Collection.

16.5.1 General. The collection of physical evidence is an integral part of a properly conducted fire investigation.

16.5.1.1 The method of collection of the physical evidence is determined by many factors, including the following:

(1) *Physical State.* Whether the physical evidence is a solid, liquid, or gas
(2) *Physical Characteristics.* The size, shape, and weight of the physical evidence
(3) *Fragility.* How easily the physical evidence may be broken, damaged, or altered
(4) *Volatility.* How easily the physical evidence may evaporate

16.5.1.2* Regardless of which method of collection is employed, the fire investigator should be guided by ASTM standards as well as by the policies and procedures of the laboratory that will examine or test the physical evidence.

16.5.2 Documenting the Collection of Physical Evidence.

16.5.2.1 Physical evidence should be thoroughly documented before it is moved. This documentation can be best accomplished through field notes, written reports, sketches, and diagrams, with accurate measurements and photography. The diagramming and photography should always be accomplished before the physical evidence is moved or disturbed. The investigator should strive to maintain a list of all evidence removed and of who removed it.

16.5.2.2 The purpose of such documentation is twofold. First, the documentation should assist the fire investigator in establishing the origin of the physical evidence, including not only its location at the time of discovery, but also its condition and relationship to the fire investigation. Second, the documentation should also assist the fire investigator in establishing that the physical evidence has not been contaminated or altered. *(See 15.2.6.8.)*

16.5.3 Collection of Traditional Forensic Physical Evidence. Traditional forensic physical evidence includes, but is not limited to, finger and palm prints, bodily fluids such as blood and saliva, hair and fibers, footwear impressions, tool marks, soils and sand, woods and sawdust, glass, paint, metals, handwriting, questioned documents, and general types of trace evidence. Although usually associated with other types of investigations, these types of physical evidence may also become part of a fire investigation. The recommended methods of collection of such traditional forensic physical evidence vary greatly. As such, the fire investigator should consult with the forensic laboratory that will examine or test the physical evidence.

16.5.4 Collection of Evidence for Accelerant Testing. An accelerant is any fuel or oxidizer, often an ignitible liquid, used to initiate a fire or increase the rate of growth or speed the spread of fire. Accelerant may be found in any state: gas, liquid, or solid. Evidence for accelerant testing should be collected and tested in accordance with ASTM E 1387, *Standard Test Method for Ignitible Liquid Residues in Extracts from Fire Debris Samples by Gas Chromatography*, or with ASTM E 1618, *Standard Test Method for Ignitible Liquid Residues in Extracts from Fire Debris by Gas Chromatography–Mass Spectrometry.*

16.5.4.1 Liquid Accelerant Characteristics. Liquid accelerants have unique characteristics that are directly related to their collection as physical evidence. These characteristics include the following:

(1) Liquid accelerants are readily absorbed by most structural components, interior furnishings, and other fire debris.
(2) Generally, liquid accelerants float when in contact with water (alcohol is a noted exception).
(3) Liquid accelerants have remarkable persistence (survivability) when trapped within porous material.

16.5.4.2 Canine/Handler Teams. When a canine/handler team is used to detect possible evidence of accelerant use, the handler should be allowed to decide what areas (if any) of a building or site to examine. Prior to any search, the handler should carefully evaluate the site for safety and health risks such as collapse, falling, toxic materials, residual heat, and vapors, and should be the final arbiter of whether the canine is allowed to search. It should also be the handler's decision whether to search all of a building or site, even areas not involved in the fire. The canine/handler team can assist with the examination of debris (loose or packaged) removed from the immediate scene as a screening step to confirm whether the appropriate debris has been recovered for laboratory analysis.

16.5.4.3 Collection of Liquid Samples for Ignitible Liquid Testing. When a possible ignitible liquid is found in a liquid state, it can be collected using any one of a variety of methods. Whichever method is employed, however, the fire investigator should be certain that the evidence does not become contaminated. If readily accessible, the liquid may be collected with a new syringe, eye dropper, pipette, siphoning device, or the evidence container itself. Sterile cotton balls or gauze pads may also be used to absorb the liquid. This method of collection results in the liquid becoming absorbed by the cotton balls or gauze pads. The cotton balls or gauze pads and their absorbed contents then become the physical evidence that should be sealed in an airtight container and submitted to the laboratory for examination and testing.

16.5.4.4 Collection of Liquid Evidence Absorbed by Solid Materials. Often, liquid accelerant evidence may be found only where the liquid accelerant has been absorbed by solid materials, including soils and sands. This method of collection merely involves the collection of these solid materials with their absorbed contents. The collection of these solid materials may be accomplished by scooping them with the evidence container itself or by cutting, sawing, or scraping. Raw, unsealed, or sawed edges, ends, nail holes, cracks, knot holes, and other similar areas of wood, plaster, sheet rock, mortar, or even concrete are particularly good areas to sample. If deep penetration is suspected, the entire cross-section of material should be removed and preserved for laboratory evaluation. In some solid material, such as soil or sand, the liquid accelerant may absorb deeply into the material. The investigator should therefore remove samples from a greater depth. In those situations where liquid accelerants are believed to have become trapped in porous material, such as a concrete floor, the fire investigator may use absorbent materials such as lime, diatomaceous earth, or non-self-rising flour. This method of collection involves spreading the absorbent onto the concrete surface, allowing it to stand for 20 to 30 minutes, and securing it in a clean, airtight container. The absorbent is then extracted in the laboratory. The investigator should be careful to use clean tools and containers for the recovery step, because the absorbent is easily contaminated. A sample of the unused absorbent should be preserved separately for analysis as a comparison sample.

16.5.4.5 Collection of Solid Samples for Accelerant Testing. Solid accelerant may be common household materials and compounds or dangerous chemicals. Because some incendiary materials remain corrosive or reactive, care should be taken in packaging to ensure that the corrosive residues do not attack the packaging container. In addition, such materials should be handled carefully by personnel for their own safety.

16.5.4.6* Comparison Samples. When physical evidence is collected for examination and testing, it is often necessary to also collect comparison samples.

16.5.4.6.1 The collection of comparison samples is especially important in the collection of materials that are believed to contain liquid or solid accelerant. For example, the comparison sample for physical evidence consisting of a piece of carpeting believed to contain a liquid accelerant would be a piece of the same carpeting that does not contain any of the liquid accelerant. Comparison samples allow the laboratory to evaluate the possible contributions of volatile pyrolysis products to the analysis and also to estimate the flammability properties of the normal fuel present.

16.5.4.6.2 When collected for the purpose of identifying the presence of accelerant residue, the comparison sample should be collected from an area that the investigator believes is free of such accelerants, such as under furniture or in areas that have not been involved in the fire. Assuming that the comparison sample tests negative for ignitible liquids, any ignitible liquids that are found in the suspect sample can be shown to be foreign to the area when the suspect sample was taken.

16.5.4.6.3 It is recognized that comparison samples may be unavailable due to the condition of the fire scene. It is also recognized that comparison samples are frequently unnecessary for the valid identification of ignitible liquid residue. The determination of whether comparison samples are necessary is made by the laboratory analyst, but because it is usually impossible for an investigator to return to a scene to collect comparison samples, they should be collected at the time of the initial investigation.

16.5.4.6.4 If mechanical or electrical equipment is suspected in the fire ignition, exemplar equipment may be identified and collected or purchased as a comparison sample.

16.5.4.7* Canine Teams. Properly trained and validated ignitible liquid detection canine/handler teams have proven their ability to improve fire investigations by assisting in the location and collection of samples for laboratory analysis for the presence of ignitible liquids. The proper use of detection canines is to assist with the location and selection of samples.

16.5.4.7.1 In order for the presence or absence of an ignitible liquid to be scientifically confirmed in a sample, that sample should be analyzed by a laboratory in accordance with 16.5.3. Any canine alert not confirmed by laboratory analysis should not be considered validated.

16.5.4.7.2* Research has shown that canines have responded or have been alerted to pyrolysis products that are not produced by an ignitible liquid and have not always responded when an ignitible liquid accelerant was known to be present. If an investigator feels that there are indicators of an accelerant, samples should be taken even in the absence of a canine alert.

16.5.4.7.3 The canine olfactory system is believed capable of detecting gasoline at concentrations below those normally cited for laboratory methods. The detection limit, however, is not the sole criterion or even the most important criterion for any forensic technique. Specificity, the ability to distinguish between ignitible liquids and background materials, is even more important than sensitivity for detection of any ignitible liquid residues. Unlike explosive- or drug-detecting dogs, these canines are trained to detect substances that are common to our everyday environment. The techniques exist today for forensic laboratories to detect submicroliter quantities of ignitible liquids, but because these substances are intrinsic to our mechanized world, merely detecting such quantities is of limited evidential value.

16.5.4.7.4* Current research does not indicate which individual chemical compounds or classes of chemical compounds are the key "triggers" for canine alerts. Research reveals that most classes of compounds contained in ignitible liquids may be produced from the burning of common synthetic materials. Laboratories that use ASTM standards (*see Section 16.10*) have minimum standards that define those chemical compounds that must be present in order to make a positive determination. The sheer variety of pyrolysis products present in fire scenes suggests possible reasons for some unconfirmed alerts by canines. The discriminatory ability of the canine to distinguish between pyrolysis products and ignitible liquids is remarkable but not infallible.

16.5.4.7.5 The proper objective of the use of canine/handler teams is to assist with the selection of samples that have a higher probability of laboratory confirmation than samples selected without the canine's assistance.

16.5.4.7.6 Canine ignitible liquid detection should be used in conjunction with, and not in place of, the other fire investigation and analysis methods described in this guide.

16.5.5 Collection of Gaseous Samples. During certain types of fire and explosion investigations, especially those involving fuel gases, it may become necessary for the fire investigator to collect a gaseous sample. The collection of gaseous samples may be accomplished by several methods.

16.5.5.1 The first method involves the use of commercially available mechanical sampling devices. These devices merely draw a sample of the gaseous atmosphere and contain it in a sample chamber or draw it through a trap of charcoal- or polymer-adsorbing material for later analysis.

16.5.5.2 Another method is the utilization of evacuated air-sampling cans. These cans are specifically designed for taking gaseous samples.

16.5.6 Collection of Electrical Equipment and System Components. Before attempting to collect electrical equipment or components from circuits of a power distribution system, the fire investigator should verify that all sources of electricity are off or disconnected. All safety procedures described in Chapter 12 should be followed. Electrical equipment and components may be collected as physical evidence to assist the fire investigator in determining whether the component was related to the cause of the fire.

16.5.6.1 Electrical components, after being involved in a fire, may become brittle and subject to damage if mishandled. Therefore, methods and procedures used in collection should preserve, as far as practical, the condition in which the physical evidence was found. Before any electrical component is collected as physical evidence, it should be thoroughly documented, including being photographed and diagrammed. Electrical wiring can usually be cut easily and removed. This type of evidence may consist of a short piece, a severed or melted end, or it might be a much longer piece, including an unburned section where the wiring's insulation is still intact. The fire investigator should collect the longest section of wiring practicable so that any remaining insulation can also be examined. Before wires are cut, a photograph should be taken of the wire(s), and then both ends of the wire should be tagged and cut so that they can be identified as one of the following:

(1) The device or appliance to which it was attached or from which it was severed
(2) The circuit breaker or fuse number or location to which the wire was attached or from which it was severed
(3) The wire's path or the route it took between the device and the circuit protector

16.5.6.2 Electrical switches, receptacles, thermostats, relays, junction boxes, electrical distribution panels, and similar equipment and components are often collected as physical evidence. It is recommended that these types of electrical evidence be removed intact, in the condition in which they were found.

16.5.6.3 When practical, it is recommended that any fixtures housing such equipment and components be removed without disturbing the components within them. Electrical distribution panels, for example, should be removed intact. An alternative method, however, would be the removal of individual fuse holders or circuit breakers from the panel. If the removal of individual components becomes necessary, the fire investigator should be careful not to operate or manipulate them while being careful to document their position and their function in the overall electrical distribution system.

16.5.6.4 If the investigator is unfamiliar with the equipment, he or she should obtain assistance from someone knowledgeable regarding the equipment, prior to disassembly or on-scene testing, to prevent damage to the equipment or components.

16.5.7 Collection of Appliances or Small Electrical Equipment. Whenever an appliance or other type of equipment is believed to be part of the ignition scenario, it is recommended that the fire investigator have it examined or tested. Appliances may be collected as physical evidence to support the fire investigator's determination that the appliance was or was not the cause of the fire. This type of physical evidence may include many diverse items, from the large (e.g., furnaces, water heaters, stoves, washers, dryers) to the small (e.g., toasters, coffee pots, radios, irons, lamps).

16.5.7.1 Where practical, the entire appliance or item of equipment should be collected intact as physical evidence. This includes any electrical power cords or fuel lines supplying or controlling it.

16.5.7.2 Where the size or damaged condition of an appliance or item of equipment makes it impractical to be removed in its entirety, it is recommended that it be secured in place for examination and testing. Often, however, only a single component or group of components in an appliance or item of equipment may be collected as physical evidence. The fire investigator should strive to ensure that the removal, transportation, and storage of such evidence maintains the physical evidence in its originally discovered condition. *(See 11.3.5 on spoliation.)*

16.6 Evidence Containers.

16.6.1 General. Once collected, physical evidence should be placed and stored in an appropriate evidence container.

16.6.1.1 Like the collection of the physical evidence itself, the selection of an appropriate evidence container also depends on the physical state, physical characteristics, fragility, and volatility of the physical evidence. The evidence container should preserve the integrity of the evidence and should prevent any change to or contamination of the evidence. Evidence should not be packed directly in loose packing materials, such as "peanuts" or shredded paper, but first placed in a proper bag or container to avoid the loss of small items. Alternatively, the packing material can be placed in bags and packed around the evidence.

16.6.1.2 Evidence containers may be common items, such as envelopes, paper bags, plastic bags, glass containers, or metal cans, or they may be containers specifically designed for certain types of physical evidence. The investigator's selection of an appropriate evidence container should be guided by the policies and procedures of the laboratory that will examine or test the physical evidence or the use to which the evidence will be subjected.

16.6.2 Liquid and Solid Accelerant Evidence Containers. It is recommended that containers used for the collection of liquid and solid accelerant evidence be limited to four types. These include metal cans, glass jars, special evidence bags, and common plastic evidence bags. The fire investigator should be concerned with preventing the evaporation of the accelerant and preventing its contamination. It is important, therefore, that the container used be completely sealed to prohibit such evaporation or contamination.

16.6.2.1 Metal Cans. The recommended container for the collection of liquid and solid accelerant evidence is an unused, clean metal can, as shown in Figure 16.6.2.1. In order to allow space for vapors to collect, the can should be not more than two-thirds full.

FIGURE 16.6.2.1 Various Types of Metal Cans.

16.6.2.1.1 The advantages of using metal cans include their availability, economic price, durability, and ability to prevent the evaporation of volatile liquids.

16.6.2.1.2 The disadvantages, however, include the inability to view the evidence without opening the container, the space requirements for storage, and the tendency of the container to rust when stored for long periods of time. If metal cans are used to store bulk quantities of volatile liquids, such as gasoline, high storage temperatures [above 38°C (100°F)] can produce sufficient vapor pressure to force the lid open and cause loss of sample. For such samples, glass jars may be more appropriate.

16.6.2.2 Glass Jars. Glass jars can also be used for the collection of liquid and solid accelerant evidence. It is important that the jars not have glued cap liners or rubber seals, especially when bulk liquids are collected. The glue often contains traces of solvent that can contaminate the sample, and rubber seals can soften or even dissolve in the presence of liquid accelerants or their vapors, allowing leakage or loss of the sample. In order to allow space for vapor samples to be taken during examination and testing, the glass jar should be not more than two-thirds full.

16.6.2.2.1 The advantages of using glass jars include their availability, their low price, the ability to view the evidence without opening the jar, the ability to prevent the evaporation of volatile liquids, and their lack of deterioration when stored for long periods of time.

16.6.2.2.2 The disadvantages, however, include their tendency to break easily and their physical size, which often prohibits the storage of large quantities of physical evidence.

16.6.2.3 Special Evidence Bags. Special bags, designed specifically for liquid and solid accelerant evidence, can also be used for collection. Unlike common plastic evidence bags, these special evidence bags do not have a chemical composition that can cause erroneous test results during laboratory examination and during testing of the physical evidence contained in such bags.

16.6.2.3.1 The advantages of using special evidence bags include their availability in a variety of shapes and sizes, their economic price, the ability to view the evidence without opening the bag, their ease of storage, and the ability to prevent the evaporation of volatile liquids.

16.6.2.3.2 The disadvantages, however, are that they are susceptible to being damaged easily, resulting in the contamination of the physical evidence contained in them, and they may be difficult to seal adequately.

16.6.2.4 Common Plastic Bags. While they are not generally usable for volatile evidence, common (polyethylene) plastic bags can be used for some evidence packaging. They can be used for packaging incendiary devices or solid accelerant residues, but they could be permeable, allowing for loss and contamination.

16.6.2.4.1 The advantages of using common plastic bags include their availability in a variety of shapes and sizes, their economic price, the ability to view the evidence without opening the bag, and their ease of storage.

16.6.2.4.2 The disadvantages, however, are their susceptibility to easy damage (tearing and penetration), resulting in the contamination of the physical evidence contained in them, and their marked inability to retain light hydrocarbons and alcohols, resulting in loss of the sample, misidentification, or cross-contamination between containers in the same box.

16.7 Identification of Physical Evidence. All evidence should be marked or labeled for identification at the time of collection, as required by ASTM E 1188, *Standard Practice for Collection and Preservation of Information and Physical Items by a Technical Investigator*, and ASTM E 1459, *Standard Guide for Physical Evidence Labeling and Related Documentation.*

16.7.1 Recommended identification includes the name of the fire investigator collecting the physical evidence, the date and time of collection, an identification name or number, the case number and item designation, a description of the physical evidence, and where the physical evidence was located. This can be accomplished directly on the container (*see Figure 16.7.1*) or on a preprinted tag or label that is then securely fastened to the container.

FIGURE 16.7.1 Marking of the Evidence Container.

16.7.2 The fire investigator should be careful that the identification of the physical evidence cannot be easily damaged, lost, removed, or altered. The fire investigator also should be careful that the placement of the identification, especially adhesive labels, does not interfere with subsequent examination or testing of the physical evidence at the laboratory.

16.8 Transportation and Storage of Physical Evidence. Transportation of physical evidence to the laboratory or testing facility can be done either by hand delivery or by shipment.

16.8.1 Hand Delivery. Whenever possible, it is recommended that physical evidence be hand delivered for examination and testing. Hand delivery minimizes the potential of the physical evidence becoming damaged, misplaced, or stolen.

16.8.1.1 During such hand delivery, the fire investigator should take every precaution to preserve the integrity of the physical evidence. It is recommended that the physical evidence remain in the immediate possession and control of the fire investigator until arrival and transfer of custody at the laboratory or testing facility.

16.8.1.2 The fire investigator should define in writing the scope of the examination or testing desired. This request should include the name, address, and telephone number of the fire investigator; a detailed listing of the physical evidence being submitted for examination and testing; and any other information required, depending on the nature and scope of the examination and testing requested. This request may also include the facts and circumstances of the incident yielding the physical evidence.

16.8.2 Shipment.

16.8.2.1 General. It may sometimes become necessary to ship physical evidence to a laboratory or testing facility for examination and testing. When shipping becomes necessary, the fire investigator should take every precaution to preserve the integrity of that physical evidence.

16.8.2.1.1 The fire investigator should choose a container of sufficient size to adequately hold all of the individual evidence containers from a single investigation. Physical evidence from more than one investigation should never be placed in the same shipment.

16.8.2.1.2 The individual evidence container should be packaged securely within the shipping container. A letter of transmittal should be included. The letter of transmittal is a written request for laboratory examination and testing. It should include the name, address, and telephone number of the fire investigator; a detailed listing of the physical evidence being submitted for examination and testing; the nature and scope of the examination and testing desired; and any other information required, depending on the nature and scope of the examination and testing requested. This letter of transmittal may also include the facts and circumstances of the incident yielding the physical evidence.

16.8.2.1.3 The sealed package should be shipped by registered United States mail or any commercial courier service. The fire investigator should, however, always request return receipts and signature surveillance.

16.8.2.2 Shipping Electrical Evidence. In addition to the procedures described in 16.8.2.1, the investigator should be aware that some electrical equipment with sensitive electromechanical components may not be suitable for shipment. Examples include certain circuit breakers, relays, or thermostats. The fire investigator should consult personnel at laboratory or testing facilities for advice on how to transport the evidence.

16.8.2.3 Shipping of Volatile or Hazardous Materials. The fire investigator is cautioned about shipping volatile or hazardous materials. The investigator should ensure that such shipments are made in accordance with applicable federal, state, and local laws. When dealing with volatile evidence, it is important that the evidence be protected from extremes of temperature. Freezing or heating of the volatile materials may affect lab test results. Generally, the lower the temperature at which the evidence is stored, the better the volatile sample will be preserved, but it should not be allowed to freeze.

16.8.3 Storage of Evidence. Physical evidence should be maintained in the best possible condition until it is no longer needed. It should always be protected from loss, contamination, and degradation. Heat, sunlight, and moisture are the chief sources of degradation of most kinds of evidence. Dry and dark conditions are preferred, and the cooler the better. Opening of sealed evidence bags containing evidence not intended for accelerant testing will allow moisture to evaporate, will better preserve metallic items, and can prevent molding of organic items such as wet clothing. Refrigeration of volatile evidence is strongly recommended. If a sample is being collected for fire-debris analysis, it may be frozen, since freezing will prevent microbial and other biological degradation. However, freezing may interfere with flash point or other physical tests and may burst water-filled containers.

16.9 Chain of Custody of Physical Evidence.

16.9.1 The value of physical evidence entirely depends on the fire investigator's efforts to maintain the security and integrity of that physical evidence from the time of its initial discovery and collection to its subsequent examination and testing. At all times after its discovery and collection, physical evidence should be stored in a secured location that is designed and designated for this purpose. Access to this storage location should be limited in order to limit the chain of custody to as few persons as possible. Wherever possible, the desired storage location is one that is under the sole control of the fire investigator.

16.9.2 When it is necessary to pass chain of custody from one person to another, it should be done using a form on which the receiving person signs for the physical evidence. Figure 16.9.2 shows an example of such a form.

16.10 Examination and Testing of Physical Evidence. Once collected, physical evidence is usually examined and tested in a laboratory or other testing facility. Physical evidence may be examined and tested to identify its chemical composition; to establish its physical properties; to determine its conformity or lack of conformity to certain legal standards; to establish its operation, inoperation, or malfunction; to determine its design sufficiency or deficiency, or other issues that will provide the fire investigator with an opportunity to understand and determine the origin of a fire, the specific cause of a fire, the contributing factors to a fire's spread, or the responsibility for a fire. The investigator should consult with the laboratory or other testing facility to determine what specific services are provided and what limitations are in effect.

16.10.1 Evidence Collection or Inspections Involving Alteration Without Changes to the Evidentiary Value of the Artifacts.

16.10.1.1 Frequently, the efficient collection and examination of items of physical evidence require that certain alterations be made to the evidence, without which the evidence could not be properly preserved, examined, or evaluated.

Crime Scene Search Evidence Report

Name of subject ____________________

Offense ____________________

Date of incident __________ Time __________ a.m. p.m.

Search officer ____________________

Evidence description ____________________

____

Location ____________________

____

Chain of Possession

Received from ____________________

By ____________________

Date __________ Time __________ a.m. p.m.

Received from ____________________

By ____________________

Date __________ Time __________ a.m. p.m.

Received from ____________________

By ____________________

Date __________ Time __________ a.m. p.m.

Received from ____________________

By ____________________

Date __________ Time __________ a.m. p.m.

FIGURE 16.9.2 Chain of Custody Form.

16.10.1.2 These inspections commonly include photography, x-ray, and diagramming; physical measurement of the size, weight, or density of the artifacts (i.e., depth of char and depth of calcination measurements); and testing of such properties as electrical resistance and continuity and ferromagnetism (nature of a metal).

16.10.1.3 On many occasions the effective collection, inspection, or examination of evidence first requires that unimportant portions of, or debris adhering to an evidence item must be removed or altered in order to gain access to the important evidentiary information, which, by definition, do not change the evidentiary value of the evidence.

16.10.1.4 Common procedures include the opening or removal of an artifact's cover, outer case, or door; the cutting away or cleaning off of charred or melted material or debris that is obstructing access to an area or information of interest; or the cutting of electrical wires or piping between connections for more efficient examination, collection, or testing *[see Figure 16.10.1.4(a) through Figure 16.10.1.4(c)]*. Such procedures, though they may in fact alter portions of the physical evidence, typically do not effectively alter the overall evidentiary value of the item.

16.10.1.5 For example, Figure 16.10.1.4(b) and Figure 16.10.1.4(c) display a portion of fire scene floor debris, which could not be x-rayed without removal from the much larger original evidence configuration. Such removal required that four electrical conductors be cut, and their respective end pieces marked.

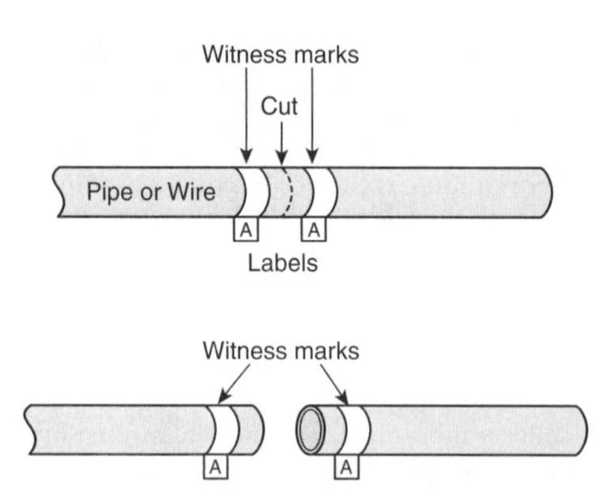

FIGURE 16.10.1.4(a) A Method of Intrusive, Nondestructive Marking and Cutting of Wires, Electrical Conductors, or Pipes in a Manner that the Relationships of the Severed Ends Will Be Recorded.

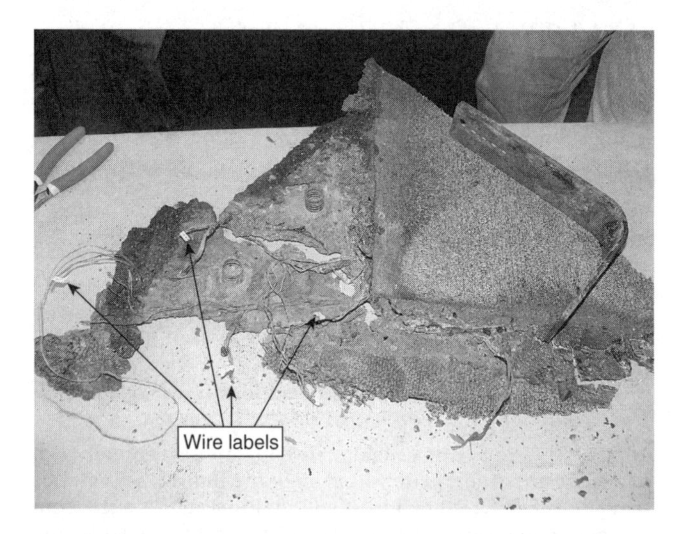

FIGURE 16.10.1.4(b) Partially Melted Carpet Sample with Cut and Marked Embedded Wires (Note Wire Labels), and Other Evidence Obscured by Melted and Resolidified Carpet Material.

16.10.1.6 Laboratory Examination and Testing. A wide variety of standardized tests are available, depending on the physical evidence and the issue or hypothesis being examined or tested. Whenever possible, such tests should be performed and carried out by procedures that have been standardized by some recognized group. Such conformance better ensures that the results are valid and that they will be comparable to results from other laboratories or testing facilities.

16.10.1.6.1 It should be noted that the results of many laboratory examinations and tests may be affected by a variety of factors. These factors include the abilities of the person conducting or interpreting the test, the capabilities of the particular test apparatus, the maintenance or condition of the particular test apparatus, sufficiency of the test protocol, and the quality of the sample or specimen being tested. Fire investigators should be aware of these factors when using the interpretations of test results.

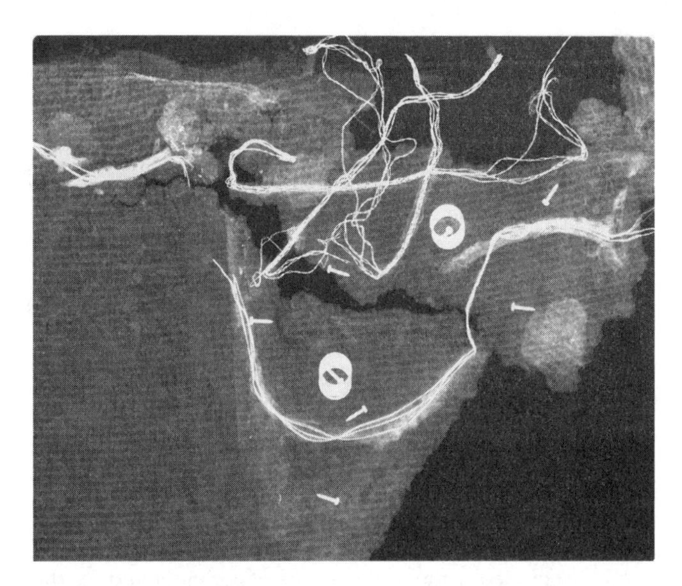

FIGURE 16.10.1.4(c) X-Ray Image of the Partially Melted Carpet Sample Shown in Figure 16.10.1.4(b), Displaying Hidden Screws and Electrical Conductors.

16.10.1.6.2 If it is determined that testing might alter the evidence, interested parties should be notified prior to testing to allow them an opportunity to object or be present at the testing. Guidance regarding notification can be found in ASTM E 860, *Standard Practice for Examining and Testing Items That Are or May Become Involved in Product Liability Litigation. (See also 16.5.4.6.)*

16.10.1.6.3 While standard tests are recommended and preferred, it is recognized that not every situation in need of testing falls precisely into a previously developed standard test category. In those instances, the fire investigator may need to develop "specialized tests." When specialized tests are needed, they should be performed within the scientific principles of experimental tests. Every step of the specialized test should be documented such that any other competent investigator could reproduce the experiment and obtain the same results within experimental error. The specialized test to evaluate some unknown should change only one variable at a time. If more than one variable is changed at a time, it may not be possible to determine the effect or cause. Control samples should be included to form a basis for comparison. Key steps in the experiment should be photographed and/or videotaped. The design and results of a specialized test should assist the investigator in reaching a scientific and valid conclusion.

16.10.2 Test Methods. The following is a listing of selected analytical methods and tests that are applicable to certain fire investigations. When utilizing laboratories to perform any of these tests, investigators should be aware of the quality of the laboratory results that can be expected.

16.10.2.1 Gas Chromatography (GC). This test method separates the mixtures into their individual components and then provides a graphical representation of each component and its relative amount. This method is useful for mixtures of gases or liquids that can be vaporized without decomposition. GC is sometimes a preliminary test that may indicate the need for additional testing to specifically identify the components. For most petroleum distillate accelerants, GC provides adequate characterization if conducted according to accepted methods. These methods are described in ASTM E 1387, *Standard Test Method for Ignitible Liquid Residues in Extracts from Fire Debris Samples by Gas Chromatography.*

16.10.2.2 Mass Spectrometry (MS). This test method is usually employed in conjunction with gas chromatography. This method further analyzes the individual components that have been separated during gas chromatography. Methods of GC/MS analysis are described in ASTM E 1618, *Standard Test Method for Ignitible Liquid Residues in Extracts from Fire Debris by Gas Chromatography–Mass Spectrometry.*

16.10.2.3 Infrared Spectrophotometer (IR). This test method can identify some chemical species by their ability to absorb infrared light in specific wavelength regions.

16.10.2.4 Atomic Absorption (AA). This test method identifies the individual elements in nonvolatile substances such as metals, ceramics, or soils.

16.10.2.5 X-Ray Fluorescence. This test analyzes for metallic elements by evaluating an element's response to X-ray photons.

16.10.2.6 Flash Point by Tag Closed Tester (ASTM D 56). This test method, from ASTM D 56, *Standard Test Method for Flash Point by Tag Closed Tester,* covers the determination of the flash point, by tag closed tester, of liquids having low viscosity and a flash point below 93°C (200°F). Asphalt and those liquids that tend to form a surface film under test conditions and materials that contain suspended solids are tested using the Pensky-Martens (*see 16.10.2.8*) closed tester.

16.10.2.7 Flash and Fire Points by Cleveland Open Cup (ASTM D 92). This test method, from ASTM D 92, *Standard Test Method for Flash and Fire Points by Cleveland Open Cup,* covers determination of the flash and fire points of all petroleum products (except oils) and those products having an open-cup flash point below 79°C (175°F).

16.10.2.8 Flash Point by Pensky-Martens Closed Tester (ASTM D 93). This test method, from ASTM D 93, *Standard Test Method for Flash Point by Pensky-Martens Closed Cup Tester,* covers the determination of the flash point by Pensky-Martens closed-cup tester of fuel oils, lubricating oils, suspensions of solids, liquids that tend to form a surface film under test conditions, and other liquids.

16.10.2.9 Flash Point and Fire Point of Liquids by Tag Open-Cup Apparatus (ASTM D 1310). This test method, from ASTM D 1310, *Standard Test Method for Flash Point and Fire Point of Liquids by Tag Open-Cup Apparatus,* covers the determination by tag open-cup apparatus of the flash point and fire point of liquids having flash points between −18°C and 163°C (0°F and 325°F) and fire points up to 163°C (325°F).

16.10.2.10 Flash Point by Setaflash Closed Tester (ASTM D 3828). This test method, from ASTM D 3828, *Standard Test Methods for Flash Point by Small Scale Closed Tester,* covers procedures for the determination of flash point by a Setaflash closed tester. Setaflash methods require smaller specimens than the other flash point tests.

16.10.2.11 Autoignition Temperature of Liquid Chemicals (ASTM E 659). This test method, from ASTM E 659, *Standard Test Method for Autoignition Temperature of Liquid Chemicals,* covers the determination of hot- and cool-flame autoignition temperatures of a liquid chemical in air at atmospheric pressure in a uniformly heated vessel.

16.10.2.12 Heat of Combustion of Hydrocarbon Fuels by Bomb Calorimeter (Precision Method) (ASTM D 4809). This test method, from ASTM D 4809, *Standard Test Method for Heat of Combustion of Liquid Hydrocarbon Fuels by Bomb Calorimeter (Precision Method)*, covers the determination of the heat of combustion of hydrocarbon fuels. It is designed specifically for use with aviation fuels when the permissible difference between duplicate determinations is of the order of 0.1 percent. It can be used for a wide range of volatile and nonvolatile materials where slightly greater differences in precision can be tolerated.

16.10.2.13 Flammability of Apparel Textiles (ASTM D 1230). This test method, from ASTM D 1230, *Standard Test Method for Flammability of Apparel Textiles*, covers the evaluation of the flammability of textile fabrics as they reach the consumer for or from apparel other than children's sleepwear or protective clothing.

16.10.2.14 Cigarette Ignition Resistance of Mock-up Upholstered Furniture Assemblies (ASTM E 1352). This test method, from ASTM E 1352, *Standard Test Method for Cigarette Ignition Resistance of Mock-up Upholstered Furniture Assemblies*, is intended to cover the assessment of the resistance of upholstered furniture mock-up assemblies to combustion after exposure to smoldering cigarettes under specified conditions.

16.10.2.15 Cigarette Ignition Resistance of Components of Upholstered Furniture (ASTM E 1353). This test method, from ASTM E 1353, *Standard Test Methods for Cigarette Ignition Resistance of Components of Upholstered Furniture*, is intended to evaluate the ignition resistance of upholstered furniture component assemblies when exposed to smoldering cigarettes under specified conditions.

16.10.2.16 Flammability of Finished Textile Floor-Covering Materials (ASTM D 2859). This test method, from ASTM D 2859, *Standard Test Method for Flammability of Finished Textile Floor Covering Materials*, covers the determination of the flammability of finished textile floor covering materials when exposed to an ignition source under controlled laboratory conditions. It is applicable to all types of textile floor coverings regardless of the method of fabrication or whether they are made from natural or manmade fibers. Although this test method may be applied to unfinished material, such a test is not considered satisfactory for the evaluation of a textile floor-covering material for ultimate consumer use.

16.10.2.17 Flammability of Aerosol Products (ASTM D 3065). This test method, from ASTM D 3065, *Standard Test Methods for Flammability of Aerosol Products*, covers the determination of flammability hazards for aerosol products.

16.10.2.18 Surface Burning Characteristics of Building Materials (ASTM E 84). This test method, from ASTM E 84, *Standard Test Method for Surface Burning Characteristics of Building Materials*, for the comparative surface burning behavior of building materials, is applicable to exposed surfaces, such as ceilings or walls, provided that the material or assembly of materials, by its own structural quality or the manner in which it is tested and intended for use, is capable of supporting itself in position or being supported during the test period. This test is conducted with the material in the ceiling position. This test is not recommended for use with cellular plastic.

16.10.2.19 Fire Tests of Roof Coverings (ASTM E 108). This test method, from ASTM E 108, *Standard Test Method for Fire Tests of Roof Coverings*, covers the measurement of relative fire characteristics of roof coverings under simulated fire originating outside the building. It is applicable to roof coverings intended for installation on either combustible or noncombustible decks, when applied as intended for use.

16.10.2.20 Critical Radiant Flux of Floor-Covering Systems Using a Radiant Heat Energy Source (ASTM E 648). This test method, from ASTM E 648, *Standard Test Method for Critical Radiant Flux of Floor-Covering Systems Using a Radiant Heat Energy Source*, describes a procedure for measuring the critical radiant flux of horizontally mounted floor covering systems exposed to a flaming ignition source in graded radiant heat energy environment in a test chamber. The specimen can be mounted over underlayment or to a simulated concrete structural floor, bonded to a simulated structural floor, or otherwise mounted in a typical and representative way.

16.10.2.21 Room Fire Experiments (ASTM E 603). This guide, ASTM E 603, *Standard Guide for Room Fire Experiments*, covers full-scale compartment fire experiments that are designed to evaluate the fire characteristics of materials, products, or systems under actual fire conditions. It is intended to serve as a guide for the design of the experiment and for the interpretation of its results. ASTM E 603 may be used as a guide for establishing laboratory conditions that simulate a given set of fire conditions to the greatest extent possible.

16.10.2.22 Concentration Limits of Flammability of Chemicals (ASTM E 681). This test method, from ASTM E 681, *Standard Test Method for Concentration Limits of Flammability of Chemicals*, covers the determination of the lower and upper concentration limits of flammability of chemicals having sufficient vapor pressure to form flammable mixtures in air at 1 atmosphere pressure at the test temperature. This method may be used to determine these limits in the presence of inert dilution gases. No oxidant stronger than air should be used.

16.10.2.23 Measurement of Gases Present or Generated During Fires (ASTM E 800). Analytical methods for the measurement of carbon monoxide, carbon dioxide, oxygen, nitrogen oxides, sulfur oxides, carbonyl sulfide, hydrogen halide, hydrogen cyanide, aldehydes, and hydrocarbons are described in ASTM E 800, *Standard Guide for Measurement of Gases Present or Generated During Fires*, along with sampling considerations. Many of these gases may be present in any fire environment. Several analytical techniques are described for each gaseous species, together with advantages and disadvantages of each. The test environment, sampling constraints, analytical range, and accuracy often dictate use of one analytical method over another.

16.10.2.24 Heat and Visible Smoke Release Rates for Materials and Products (ASTM E 906). This test method, from ASTM E 906, *Standard Test Method for Heat and Visible Smoke Release Rates for Materials and Products*, can be used to determine the release rates of heat and visible smoke from materials and products when exposed to different levels of radiant heat using the test apparatus, specimen configurations, and procedures described in this test method.

16.10.2.25 Pressure and Rate of Pressure Rise for Combustible Dusts (ASTM E 1226). This test method, from ASTM E 1226, *Test Method for Pressure and Rate of Pressure Rise for Combustible Dusts*, can be used to measure composition limits of explosibility, ease of ignition, and explosion pressures of dusts and gases.

16.10.2.26 Heat and Visible Smoke Release Rates for Materials and Products Using an Oxygen Consumption Calorimeter (ASTM E 1354). This test method, from ASTM E 1354, *Standard Test Method for Heat and Visible Smoke Release Rates for Materials and*

Products Using an Oxygen Consumption Calorimeter, is a bench-scale laboratory instrument for measuring heat release rate, radiant ignitibility, smoke production, mass loss rate, and certain toxic gases of materials.

16.10.2.27 Ignition Properties of Plastics (ASTM D 1929). This test method, from ASTM D 1929, *Standard Test Method for Determining Ignition Temperature of Plastics*, covers a laboratory determination of the self-ignition and flash-ignition temperatures of plastics using a hot-air ignition furnace.

16.10.2.28 Dielectric Withstand Voltage (Mil-Std–202F Method 301). This test method, from Mil-Std–202F, *Test Method for Electronic and Electrical Components*, also called high-potential, overpotential, voltage-breakdown, or dielectric-strength test, consists of the application of a voltage higher than rated voltage for a specific time between mutually insulated portions of a component part or between insulated portions and ground.

16.10.2.29 Insulation Resistance (Mil-Std–202F Method 302). This test, from Mil-Std–202F, *Test Method for Electronic and Electrical Components*, measures the resistance offered by the insulating members of a component part to an impressed direct voltage tending to produce a leakage current through or on the surface of these members.

16.10.3 Sufficiency of Samples. Fire investigators often misunderstand the abilities of laboratory personnel and the capabilities of their scientific laboratory equipment. These misconceptions usually result in the fire investigator's collecting a quantity of physical evidence that is too small to examine or test.

16.10.3.1 Certainly, the fire investigator will not always have the opportunity to determine the quantity of physical evidence he or she can collect. Often, the fire investigator can collect only that quantity that is discovered during his or her investigation.

16.10.3.2 Each laboratory examination or test requires a certain minimum quantity of physical evidence to facilitate proper and accurate results. The fire investigator should be familiar with these minimum requirements. The laboratory that examines or tests the physical evidence should be consulted concerning these minimum quantities.

16.10.4 Comparative Examination and Testing.

16.10.4.1 During the course of certain fire investigations, the fire investigator may wish to have appliances, electrical equipment, or other products examined to determine their compliance with recognized standards. Such standards are published by the American Society for Testing and Materials, Underwriters Laboratories Inc., and other agencies.

16.10.4.2 Another method of comparative examination and testing involves the use of an exemplar appliance or product. Utilizing an exemplar allows the testing of an undamaged example of a particular appliance or product to determine whether it was capable of causing the fire. The sample should be the same make and model as the product involved in the fire.

16.11 Evidence Disposition.

16.11.1 The fire investigator is often faced with disposing of evidence after an investigation has been completed. The investigator should not destroy or discard evidence unless proper authorization is received. Circumstances may require that evidence be retained for many years and ultimately may be returned to the owner.

16.11.2 Criminal cases such as arson require that the evidence be kept until the case is adjudicated. During the trial, evidence submitted — such as reports, photographs, diagrams, and items of physical evidence — will become part of the court record and will be kept by the courts. Volatile or large physical items may be returned to the investigator by the court. There may be other evidence still in the investigator's possession that was not used in the trial. Once all appeals have been exhausted, the investigator may petition the court to either destroy or distribute all of the evidence accordingly. A written record of authorization to dispose of the evidence should be kept. The criminal investigator should be mindful of potential civil cases resulting from this incident, which may require retention of the evidence beyond the criminal proceedings.

Chapter 17 Origin Determination

17.1 Introduction. This chapter recommends a methodology to follow in determining the origin of a fire. The *area of origin* is defined as a structure, part of a structure, or general geographic location within a fire scene, in which the "point of origin" of a fire or explosion is reasonably believed to be located. The *point of origin* is defined as the smallest location a fire investigator can define, within an "area of origin," in which a heat source, source of oxygen, and a fuel interacted with each other and a fire or explosion began. *(See 3.3.9, Area of Origin.)* For example, in practical use in the field a point of origin may be as large as a single chair or electrical appliance or as small as a specific location on the chair or within the appliance, depending upon the smallest location the investigator can define. The origin of a fire is one of the most important hypotheses that an investigator develops and tests during the investigation. Generally, if the origin cannot be determined, the cause cannot be determined, and generally, if the correct origin is not identified, the subsequent cause determination will also be incorrect. The purpose of determining the origin of the fire is to identify in three dimensions the location at which the fire began.

17.1.1 This chapter deals primarily with the determination of origin involving structures; however, the methodology generally applies to all origin determinations. Separate chapters address the particular requirements for determining origin in nonstructure fire incidents (motor vehicles, boats, wildfire, etc.).

17.1.2 Determination of the origin of the fire involves the coordination of information derived from one or more of the following:

(1) *Witness Information.* The analysis of observations reported by persons who witnessed the fire or were aware of conditions present at the time of the fire
(2) *Fire Patterns.* The analysis of effects and patterns left by the fire *(See Chapter 6.)*
(3) *Arc Mapping.* The analysis of the locations where electrical arcing has caused damage and the documentation of the involved electrical circuits *(See Section 8.10.)*
(4) *Fire Dynamics.* The analysis of the fire dynamics, that is, the physics and chemistry of fire initiation and growth *(see Chapter 5)*, and the interaction between the fire and the building's systems *(See Chapter 7.)*

17.2 Overall Methodology. The overall methodology for determining the origin of the fire is the scientific method as described in Chapter 4. This methodology includes recognizing and defining the problem to be solved, collecting data, analyzing the data, developing a hypothesis or hypotheses, and most importantly, testing the hypothesis or hypotheses. In order to use the scientific method, the investigator must develop at least one hypothesis based on the data available at the time. These hypotheses should be considered "working hypotheses," which upon testing may be discarded, revised, or expanded in detail as new data is collected during the investigation and new analyses are applied. This process is repeated as new information becomes available. *(See Figure 17.2.)*

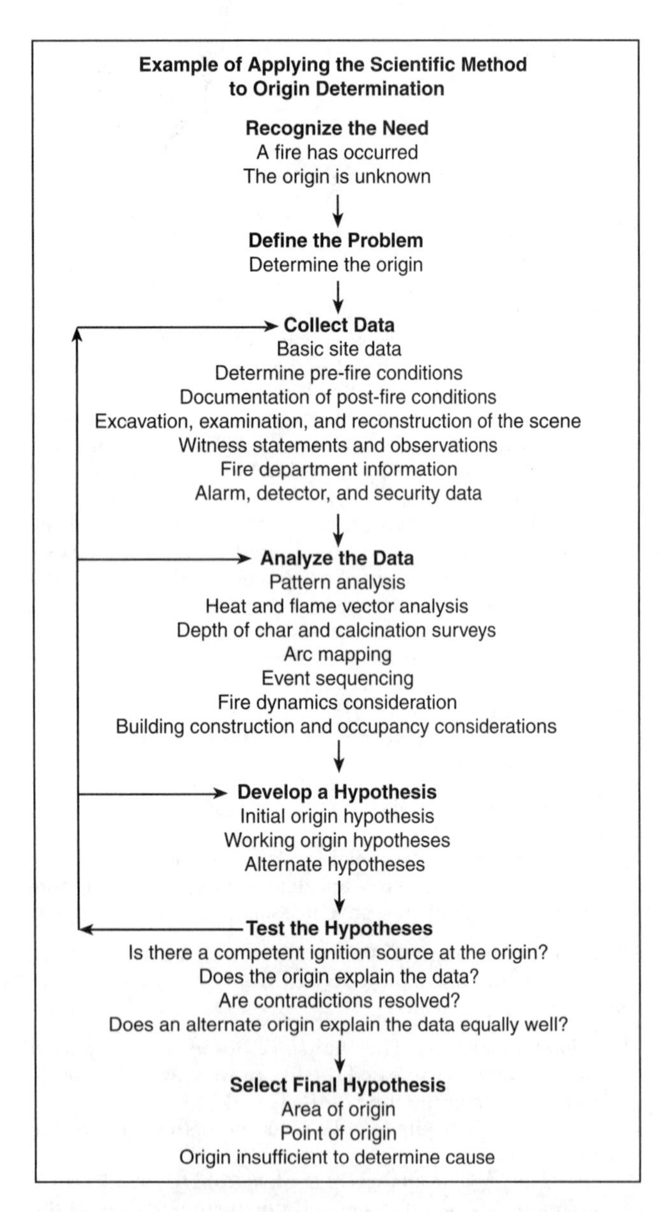

FIGURE 17.2 An Example of Applying the Scientific Method to Origin Determination.

17.2.1 Testing any origin hypothesis requires an understanding of the associated fire events as well as the growth of the fire and how the fire spread through the structure. A narrow focus on only identifying the first item ignited and a competent ignition source fails to take into account important data that can be used to test any origin hypothesis. In such a narrow focus, the growth and spread of the fire and the resulting fire damage are not well considered.

17.2.1.1 The purpose of the fire spread analysis is to determine whether the resulting physical damage and available data are consistent with the area of origin hypothesis. For example, a fire starting in a wastebasket is a plausible working hypothesis, but the resulting fire damage would be highly dependent on the position of the initial fuel and any subsequently ignited fuels. If the wastebasket had been located in an area with no adjacent fuel, then the results may be significantly different than if the wastebasket had been located next to a polyurethane sofa. Both hypotheses posit the same first item ignited, but the outcome is very different. Thus, if the origin hypothesis is not consistent with the resulting growth and spread of the fire, it is not a valid hypothesis. Fire spread scenarios within a compartment or building should be analyzed using the principles of fire dynamics presented in Chapter 5 and fire pattern development in Chapter 6.

17.2.1.2 In some instances, a single item, such as an irrefutable article of physical evidence or a credible eyewitness to the ignition, or a video recording, may be the basis for a determination of origin. In most cases, however, no single item is sufficient in itself. The investigator should use all available resources to develop origin and spread hypotheses and to determine which hypotheses fit all of the evidence available. When an apparently plausible hypothesis fails to fit some item of evidence, the investigator should try to reconcile the two and determine whether the hypothesis or the evidence is erroneous.

17.2.1.3 In some cases, it will be impossible to fix the point of origin of a fire. Where a single point cannot be identified, it can still be valuable for many purposes to identify the area(s) of origin. In such instances, the investigator should be able to provide plausible explanations for the area of origin with the supporting evidence for each option. Not identifying a point of origin will not necessarily preclude determining an origin and cause. In some situations, the extent of the damage may reduce the ability to specifically identify the point of origin, without removing the ability to put forward credible origin and cause hypotheses.

17.2.2 Sequence of Activities. The various activities required to determine the origin using the scientific method (data collection, analysis, hypothesis development, and hypothesis testing) occur continuously. Likewise, recording the scene, note taking, photography, evidence identification, witness interviews, cause investigation, failure analysis, and other data collection activities may be performed simultaneously with these efforts. Generally, the various activities of origin determination will follow a routine sequence, while the specific actions within each activity may be taking place at the same time.

17.2.3 Sequential Pattern Analysis. The area of origin may be determined by examining the fire effects and fire patterns. The surfaces of the fire scene record all of the fire patterns generated during the lifetime of the event, from ignition through suppression, although these patterns may be altered, overwritten, or obliterated after they are produced. The key to determining the origin of a fire is to determine the sequence in which these patterns were produced. Investigators should strive to identify and collect sequential data and, once collected, organize the information into a sequential format. Sequential data not only indicate what happened, but the order in which it happened. Identifiable fire spread patterns should be traced back to an area or point of origin. Once the area of origin has been established, the investigator should be able to understand and explain the fire spread.

17.2.4 Systematic Procedure. Investigators should establish a systematic procedure to follow for each type of incident. By following a familiar procedure, the investigator can concentrate on the incident at hand and need not dwell on the details of what the next step in the procedure will be. By doing so, the investigator may avoid overlooking significant evidence and will avoid forming premature conclusions about the origin.

17.2.5 Recommended Methodology. This chapter discusses a recommended methodology for the examination of the fire scene. This methodology consists of an initial scene assessment, development of a preliminary fire spread hypothesis, an in-depth examination of the fire scene and reconstruction of the fire scene, development of a final fire spread hypothesis, and identification of the fire's origin. Origin identification may occur earlier in the process, depending on the types and time of arrival of various data. This recommended methodology serves to inform the investigator but is not meant to limit the origin determination to only this procedure. Investigators, within the scope of their investigations, should consider all aspects of the fire event during the investigation. Witness statements, the investigator's expertise, and fire-fighting procedures play important roles in the determination of the fire origin.

17.3 Data Collection for Origin Determination. This section describes the data collection process for origin determination, including initial scene assessment, excavation and reconstruction, and collection of additional data from witnesses and other sources.

17.3.1 Initial Scene Assessment. An initial assessment should be made of the fire scene. As the investigator starts the initial scene assessment, data collection for the determination of the origin begins. Care should be taken during each of the steps within the initial scene assessment to protect the investigator from scene hazards and preserve the scene. The purpose of this initial examination is to determine the scope of the investigation, such as equipment and manpower needed, to determine the safety of the fire scene, and to determine the areas that warrant further study.

17.3.1.1 Safety Assessment. The investigator should first make an initial safety assessment. The investigator should determine if it is safe to enter the scene. If it is not safe to enter, the investigator must determine what steps are required to provide for personal safety or to render the scene safe to enter. Each of the hazards described in Chapter 12 should be assessed. There is no reason the investigator should compromise safety.

17.3.1.2 Scope of the Examination. After safety issues have been addressed, the investigator may begin the initial scene assessment. The purpose of this initial assessment is to determine the complexity and extent of the investigation, identify required equipment and staffing, and determine the areas that warrant further analysis.

17.3.1.3 Order of the Examination. This assessment may take place concurrently with the initial documentation of the scene. The assessment should include an overall look at the entire scene or structure, both exterior and interior, and all pertinent areas. The order in which the assessment takes place may vary, depending on scene conditions. Some investigators prefer to start with the least damaged area and move toward the most damaged area. Some investigators prefer to start at the highest point in a scene and work downward. Whatever order is selected for a particular scene, the point is to assess all areas that are pertinent to the origin and spread of the fire.

17.3.1.4 Surrounding Areas. Investigators should include in their examination the site or areas around the scene. These areas may exhibit significant evidence or fire patterns, away from the main body of the scene, that may enable the investigator to better define the site and the investigation. Anything of interest should be documented as to its location in reference to the scene. This phase of the examination can be used to canvass the neighborhood for witnesses to the fire and for persons who could provide information about the incident.

17.3.1.5 Structure Exterior. An inspection of the entire perimeter of the structure may reveal the extent and location of damage and may help determine the size and complexity of the scene. The general construction method and occupancy classification should be noted. The construction refers to how the building was built, types of materials used, exterior surfaces, previous remodeling, and any unusual features that may have affected how the fire began and spread. A significant consideration is the degree of destruction that can occur in a structure consisting of mixed types and methods of construction.

17.3.1.5.1 The occupancy classification refers to the current use of the building. Use is defined as the activities conducted and the manner in which such activities are undertaken. The number and circumstances of those individuals occupying the space may also be relevant. If the occupancy classification or use of the structure has changed, this should be considered and noted. Changes in use and occupancy classification sometimes require modifications to the structural, architectural, or fire protection features in accordance with applicable codes. Such modifications may or may not have been undertaken.

17.3.1.5.2 The fire damage and evidence of significant smoke, heat, and flame venting on the exterior should be documented and considered to assist in determining those areas that warrant further study. An in-depth examination of fire effects and patterns is not necessary at this point in the investigation.

17.3.1.6 Structure Interior. On the initial assessment, investigators should examine all rooms and other areas that may be relevant to the investigation, including those areas that are fire damaged or adjacent to the fire and smoke damaged areas. The primary purpose of this assessment is to identify the areas that require more detailed examination. The investigator should be observant of conditions of occupancy, including methods of storage, nature of contents, housekeeping, and maintenance. The type of construction, interior finish(es), and furnishings should be noted. Areas of damage, and extent of damage in each area (severe, minor, or none) should be noted. This damage should be compared with the damage seen on the exterior. During this examination, the investigator should reassess the soundness of the structure.

17.3.1.7 Post-Fire Alterations. During this assessment, the investigator should document any indication of post-fire alterations. Such alterations may affect the investigator's interpretation of the physical evidence. Alterations may include debris removal or movement, contents removal or movement, electrical service panel alterations, changes in valve positions on automatic sprinkler systems, and changes to fuel gas systems. If alterations are indicated, attempts should be made to contact the person(s) who altered the scene. They should be interviewed as to the extent of the alterations and the documentation they may have of the unaltered site.

17.3.1.8 At the conclusion of the preliminary scene assessment, the investigator should have determined the safety of the fire scene, the probable staffing and equipment requirements, and the areas around and in the structure that will require a detailed examination. The preliminary scene assessment is an important aspect of the investigation. The investigator should take as much time in this assessment as is needed to make these determinations. Time spent in this endeavor may save significant time and effort in later stages of the investigation.

17.3.2 Excavation and Reconstruction. Fire scene excavation and reconstruction allows the investigator to observe patterns on the exposed surfaces and to locate other evidence that can assist the investigator to make an accurate origin analysis. The purpose of fire scene reconstruction is to recreate as nearly as practicable the pre-fire positions of contents and structural components. Interviews, diagrams, photographs, and other means can be helpful in establishing pre-fire conditions.

17.3.2.1 Scope of Excavation and Reconstruction. Because the preliminary scene assessment has identified the areas warranting further examination, the task of fire scene reconstruction may not require the removal of debris and the replacement of the contents throughout the entire structure. As mentioned previously, the preliminary scene assessment should not be done hastily. Careful analysis of the fire scene may help to reduce to a practical level the strenuous task of debris removal. If the area to be reconstructed cannot be reduced, then the investigator should accept the necessity of removing the debris from the entire area of interest.

17.3.2.2 Safety. Safe work practices throughout this effort are required. Debris excavation and removal can weaken a structure and cause it to collapse. Debris removal can also expose hazardous substances, uncover holes in the floor, and can expose energized electrical wiring. All significant risks that may be encountered during an investigation should be minimized before the investigation continues. See Chapter 12 for a detailed discussion on safety.

17.3.2.3 Excavation. Adequate debris removal is essential for a thorough fire investigation. Inadequate removal of debris and the resultant exposure of limited portions of the fire patterns and other evidence may lead to an incorrect analysis. A fire scene investigation normally involves dirty and strenuous work. Acceptance of this fact is essential in conducting a proper fire investigation.

17.3.2.3.1 The removal of debris during the overhaul stage of fire suppression operations is an area of concern for the fire investigator. Firefighters may disturb the scene, thus making origin determination more difficult. Only those overhaul and suppression operations necessary to ensure complete extinguishment should be conducted. When these operations call for substantial scene alterations, an attempt should be made to document the fire scene with photography and notes prior to the alterations, if practical.

17.3.2.3.2 Investigators should consider where debris will be placed during removal. In some instances, it may be desirable to move the debris to a secure location. Debris should only be moved to an area that has already been examined or has no future need for examination or documentation. Moving debris twice is counterproductive. Debris removal should be performed in a planned and systematic fashion. This means that debris should be removed in layers, with adequate documentation as the process continues. If more than one investigator is doing the removal, they should discuss the purpose for the debris removal prior to starting. A discussion may prevent one investigator from discarding something the other investigator considers important. Each layer should be examined for significant artifacts as the debris is being removed.

17.3.2.3.3 During the excavation of a scene there exists a danger of evidence destruction. Although it is desirable to work efficiently, the use of heavy equipment to remove debris should only be undertaken if it is not practical to use hand tools to accomplish the task. Inappropriate use of mechanized excavating equipment can potentially destroy more evidence than it reveals.

17.3.2.4 Heavy Equipment. An investigation may require the use of heavy equipment such as cranes, backhoes, or front-end loaders. The condition of the fire scene may necessitate removal of building components or contents, because they constitute a safety hazard, are blocking access, or need to be removed during the systematic examination of the scene. Before using heavy equipment, the scene should be documented in the same manner as in all investigations. Documentation should also be conducted at frequent intervals when heavy equipment is being used.

17.3.2.4.1 Working with heavy equipment can be dangerous and noisy. One investigator should be appointed to communicate with the heavy equipment operator. A briefing session including the goals for this section should be held with the crew/operator as the investigation advances into new physical areas. The speed at which fire investigations occur is substantially slower than the ordinary speed at which heavy equipment operates. A commonly understood set of hand signals should be confirmed before beginning to use heavy equipment so that the crew/operator is clear from where direction comes and about the meaning of hand signals indicating specific actions. To reduce the risk of personal injury, the operation of heavy equipment should cease whenever a person enters the area of hazard where the machine is operating.

17.3.2.4.2 Prior to the use of heavy equipment, areas where the presence of ignitible liquid residue is suspected should be identified, if practicable. When possible, the samples should be collected prior to the use of equipment in those areas. Prior to use, heavy equipment should be inspected and any leaks of petroleum products should be noted. Fueling of the equipment should occur in a designated area, removed from the areas of interest, that will not contaminate the scene or items entering the scene. If contamination is of concern, samples of the equipment fluids should be collected for comparison purposes.

17.3.2.4.3 It is preferable to utilize the heavy equipment in a manner that has the least impact on the scene. The least destruction is usually accomplished by positioning the equipment outside the building and using the equipment to lift or move items from the inside to the outside of the building. When lifting a potential piece of evidence, rigging can be used to minimize damage to the removed item. If large components such as walls need to be demolished, it is preferable to do so in a manner that does not change or add debris to the underlying area to be examined. Safety concerns and site constraints can preclude this practice. Shoring should be considered as an alternative to demolition when possible and appropriate.

17.3.2.4.4 Items that are removed from the building can be examined and documented, and can remain at the fire scene if needed for further investigation. The removal of debris offsite can limit further investigation, but is sometimes necessary. Removal off site should be well documented.

17.3.2.4.5 Some investigations will require the use of heavy equipment inside the scene. In such cases, equipment should be used in a manner that is the least disruptive of the scene. One method involves using the equipment to systematically progress into a building. The equipment is initially positioned outside the building and used to aid the examination of a portion of the interior of the building. Once that area is examined, documented, and the debris removed, the heavy equipment is brought onto that cleared location. From that new position, the equipment is used to aid in the examination of adjacent areas. This progression is repeated as many times as needed. At all times, the equipment is only located in areas that have been previously examined and documented. At all times, the equipment operator should be under the direction of a fire investigator. The collection and unloading of each load should be observed for relevant evidence. Preplanning the progression of the examination can help to limit unnecessary alteration of the scene.

17.3.2.5 Avoiding Spoliation. During the excavation, care should be taken to avoid damaging ignition sources, fuels, or other potentially important evidence within the scene. If the investigator's area of interest contains important evidence, consideration should be given to suspending the investigation and putting potentially interested parties on notice, to give them an opportunity to see the evidence in place. For more guidance on the subject of avoiding spoliation, refer to Chapter 11.

17.3.2.6 Avoiding Contamination. To avoid scene contamination, extreme care should be taken with respect to the use of portable liquid-fueled equipment, such as gasoline-powered saws. Re-fueling of such equipment should be done away from the structure.

17.3.2.7 Washing Floors. After adequate debris removal has occurred, necessary samples have been taken for examination or testing, and proper scene documentation is completed, it may be useful to flush the floor or surface with water. This flushing may help to better reveal fire patterns. The use of high pressure and straight streams should be used with caution because such activities may harm significant evidence.

17.3.2.8 Contents. Any contents, or their remains, uncovered during debris removal should be documented as to their location, condition, and orientation. Once the debris has been removed, the contents can be placed in their pre-fire positions for analysis.

17.3.2.8.1 When the contents have been displaced during fire suppression or overhaul, post-fire reconstruction becomes much more difficult. The position where the item was located may exhibit a protected area or other indicator from the item, such as table legs leaving small clear spots on the floor. The problem is knowing which leg goes to which spot. If a definite determination is not possible by scene analysis or witness identification, then all potential orientations should be considered. Otherwise, the orientation should not be included in the fire scene reconstruction. A guess as to how contents were oriented may be wrong, thereby contributing false data to the analysis process. An alternative is to document the contents in all probable positions in the hope that later information will pinpoint the accurate location.

17.3.2.8.2 In addition to the replacement of contents, reconstruction may also include the replacement of structural elements (e.g., doors, joists, studs, sections of walls and floors) that may have recorded fire patterns.

17.3.3 Additional Data Collection Activities for Origin Determination. The following paragraphs describe activities in addition to the scene examination and reconstruction, which will lead to the development of data useful in the determination of the origin.

17.3.3.1 Pre-Fire Conditions. The pre-fire conditions of the structure should be determined to the extent practicable. Details such as the state of repair, condition of foundations and chimneys, insect damage, the presence and condition of fire protection systems, and so forth may prove to be significant data. Obtaining pre-fire photographs or video may be beneficial, but be aware that changes may have occurred between the time the photograph was taken and the time of the fire. Owners, employees, or occupants may be able to provide information and diagrams of pre-fire conditions. Checking with neighbors may also provide photographs showing the structure. Some jurisdictions now offer pre-fire photographs of the structure on tax record websites. There are companies specializing in aerial photography, primarily of commercial structures, which may offer pre-fire views. Satellite images are also available for many areas and may offer pre-fire documentation of both the site and the surrounding area. Additional information may be available from the fire department, which could include photographs, structure diagrams, special hazards, and fire protection systems. Other government agencies may also have pre-fire information and records.

17.3.3.2 Description of Fuels. The investigator should identify the fuels present in the building or area of interest and the characteristics of those fuels. When considering the area of origin, the type, quantity, and specific location of structural and content fuels should be identified to assist in the analysis of the fire patterns, fire growth, and spread characteristics. In this process, it is not only important to identify the potential first fuel ignited, but also to identify subsequent fuels involved.

17.3.3.3 Structure Dimensions. The physical dimensions of a structure are important data. In many instances, the post-fire structure provides the only means of obtaining dimensions. Dimensions should be recorded for all areas of the structure that may be used to understand fire growth, and smoke and fire spread. Dimensions should include the width, length, and height of a room or structure. The location, size, and condition (opened/closed) of all openings should be recorded, as well as any structures or obstructions that would affect the flow of fire gases. The effort associated with obtaining the dimensions can be time consuming and the amount of information collected may be limited depending on the extent of destruction. Specific dimensional information is necessary to reconstruct the fire event via a fire model or hand calculations *(see Chapter 20)*. When the scene is no longer available, information may be obtained from the investigative photographs, notes and diagrams of previous investigators, or from architects, engineers, contractors, insurance companies, or government offices such as building departments. The investigator should assess the accuracy of the plans and whether the plans actually represent the "as-built" structure. *(See the discussion of building design in 7.2.2.6.)*

17.3.3.4 Building Systems and Ventilation. Building systems may cause fires or influence the fire spread. The investigator should consider collecting information regarding the pre-fire conditions of the electrical, HVAC, fuel gas, and fire protection systems in a structure. This collection should be performed in all cases when a system is believed to be involved in the cause, detection, spread, or extinguishment of a fire.

17.3.3.5 Weather Conditions. The investigator should document weather factors that may have influenced the fire. The surrounding area may provide evidence of the weather conditions. Wind direction may be indicated by smoke movement or by fire damage sustained by structures or vegetation. Additionally, post-fire weather may cause changes to the physical condition of the scene.

17.3.3.6 Electrical Systems. The electrical system should be documented. The means used to distribute electricity should be determined, and damage to the systems should be documented. The documentation process should begin with the incoming electrical service. The main panel amperage and voltage input should be noted. The type, rating, position (on/tripped/off), and condition of the circuit protection devices may be relevant to the investigation and should be documented.

17.3.3.7 Electrical Loads. Note the location of electrical receptacles and switches within the room or area of origin. Electrical items plugged into the receptacles should be identified and documented. The investigative process may involve the tracing of circuits throughout a structure. The purpose for tracing these circuits is to identify the switches, receptacles, and fixtures on a particular circuit, as well as which overcurrent device protects that circuit, and its position and condition. Electrical appliances and loads should be noted. A more detailed documentation of electrical systems and devices may be necessary where they are believed to be the fire cause or a contributing factor, or when arc mapping is used. Use caution when interpreting damage to electrical wiring and equipment because it may be difficult to distinguish cause from effect. For a more detailed explanation of electrical systems, see Chapter 8.

17.3.3.8 HVAC Systems. The air movement through HVAC systems can affect the growth and spread of a fire and can transport combustion products throughout a structure. The investigator should record the location, size, and function (supply/return/exhaust) of vents in the area of interest, and whether the vent was open, closed, or covered at the time of the fire. Checking filters may provide evidence of heat or smoke damage and soot deposition to determine whether the HVAC system was operating at the time of the fire. Some HVAC systems are equipped with detectors designed to change the operation of the system in case of fire. Some systems are equipped with manual or automatic dampers designed to control fire spread, smoke movement, or airflow. Where these devices are present, their specific location and condition should be noted and any activation records should be obtained. The location and setting of any thermostats, switches, or controls for the HVAC system should be identified and documented.

17.3.3.9 Fuel Gas Systems. The fuel gas supply should be identified and documented. The purpose of this examination is to assist in determining whether the fuel gas contributed to the fire. If the examination reveals that fuel gases may have been a contributing factor, then the system should be examined and documented in detail. This examination should include testing for leaks, if possible, and determining the supply pressure, if possible. As with electrical systems, it may be difficult to distinguish between evidence of cause and effect. Fires can, and frequently do, compromise the integrity of gas distribution systems. The investigator should document the condition and position (open/closed) of system valves. Valves are frequently turned off during fires, so an attempt should be made to ascertain if anyone operated any valves during the event.

17.3.3.10 Liquid Fuel Systems. A variety of liquid fuel systems and appliances exists. These may be permanent systems, such as oil-fired space heaters and water heaters, or portable systems, such as kerosene or white gas heaters. In either case, the location and quantity of fuel present should be documented. Supply lines and valves to connected fuel supplies in remote tanks should also be documented. If the device contains an attached or integral tank, the amount of fuel remaining in the device should be estimated or measured. If the heating device is a suspected cause, or its fuel is believed to have contributed to the spread of the fire, a sample of the fuel should be preserved.

17.3.3.11 Fire Protection Systems. The examination of all involved fire protection systems (fire detection, fire alarm, and fire suppression systems) is important in determining if each system functioned properly, and can assist in tracking the growth and spread of a fire. If the system was monitored, records should be obtained from the monitoring service. In some instances, information can be downloaded from the central panel to indicate alarm and trouble signal locations and times. This is volatile data and care must be taken in extracting it from the alarm panel. Extracting this data generally requires specific knowledge and equipment. A qualified technician should be employed for downloading the data as substantial permanent loss of data can occur if this is done incorrectly. In many cases, building electrical power may be discontinued after a fire so that a limited amount of time is available for recovering the data while the system is operating on its backup battery. This limited time window should be taken into account when ordering scene activities.

17.3.3.12 Fire Protection System Data. Device locations and conditions should be documented, including the height of wall-mounted devices or the distance of ceiling-mounted devices from walls. Which sprinklers activated should be considered when examining fire spread patterns. Both detector activation and sprinkler activation may provide sequential data. In some cases, the specific location or zone of the first activating detector or sprinkler can be used to narrow down an area of origin, allowing an investigator to assess specific ignition sources in that area. Some systems provide only alarm or water flow data, and do not specify a particular zone. This information can be helpful in comparing the time of system activation to the time and observations of first arriving fire fighters or other witness, in assessment of the growth and spread of the fire.

17.3.3.13 Security Cameras. Security cameras that monitor buildings or ATMs may be very useful, particularly for providing "hard" times *(see the discussion of timelines in Chapter 20)*. Events before or during the fire including, in some cases, the actual ignition and development of the fire may have been recorded. The video recorder may be found in a secure area or a remote location. It should be recovered and reviewed even if damaged.

17.3.3.14 Intrusion Alarm Systems. An intrusion system may activate during a fire due to heat, smoke movement, the destruction of wiring, or loss of power. A monitored intrusion system may send a trouble signal to the monitoring station if a transmission line is compromised or power is lost. As with fire alarm systems, attempts should be made to recover the alarm panel history before the alarm system is reset. This frequently requires special expertise. Some alarm systems may record the identity of persons entering and leaving the building.

17.3.3.15 Witness Observations. Observations by witnesses are data that can be used in the context of determining the origin. Such witnesses can provide knowledge of conditions prior to, during, and after the fire event. Witnesses may be able to provide photographs or videotapes of the scene before or during the fire. Observations are not necessarily limited to visual observations. Sounds, smells, and perceptions of heat may shed light on the origin. Witness statements regarding the location of the origin create a need for the fire investigator to conduct as thorough an investigation as possible to collect data that can support or refute the witness statements. When witness statements are not supported by the investigator's interpretation of the physical evidence, the investigator should evaluate each separately.

17.4 Analyze the Data. The scientific method requires that all data collected that bears upon the origin be analyzed. This is an essential step that must take place before the formation of any hypotheses. The identification, gathering, and cataloging of data does not equate to data analysis. Analysis of the data is based on the knowledge, training, experience, and expertise of the individual doing the analysis. If the investigator lacks the knowledge to properly attribute meaning to a piece of data, then assistance should be sought from someone with the necessary knowledge. Understanding the meaning of the data will enable the investigator to form hypotheses based on the evidence, rather than on speculation or subjective belief.

17.4.1 Fire Pattern Analysis. An investigator should read and understand the concepts of fire effects, fire dynamics, and fire pattern development described in Chapters 5 and 6. This knowledge is essential in the analysis of a scene to determine the origin of the fire.

17.4.1.1 Consideration of All Patterns. All observed patterns should be considered in the analysis. Accurate determination of the origin of a fire by a single dominant fire pattern is rare, as in the case of very limited fire damage where there may be only one fire pattern.

17.4.1.2 Sequence of Patterns. While fire patterns may be the most readily available data for origin determination, the investigator should keep in mind that the damage and burn patterns observed after a fire represent the total history of the fire. A major challenge in the analysis of fire pattern data is to determine the sequence of pattern formation. Patterns observed in fires that are extinguished early in their development can present different data than those remaining after full room involvement or significant building destruction. Patterns generated as a result of a rekindle may impact the perception of the fire's history or sequence of pattern production.

17.4.1.3 Pattern Generation. The investigator should not assume that the fire at the origin burned the longest and therefore fire patterns showing the greatest damage must be at the area of origin. Greater damage in one place than in another may be the result of differences in thermal exposure due to differences in fuel loading, the location of the fuel package in the compartment, increased ventilation, or fire-fighting tactics. For similar reasons, a fire investigator should consider these factors when there is a possibility of multiple origins.

17.4.1.3.1 The size, location, and heat release rate of a fuel package may have as much effect on the extent of damage as the length of time the fuel package was burning. An area of extensive damage may simply mean that there was a significant fuel package at that location. The investigator should consider whether the fire at such a location might have spread there from another location where the fuel load was smaller.

17.4.1.3.2 Fuel packages of identical composition and equal size may burn very differently, depending on their location in a compartment. The possible effect of the location of walls relative to the fire should be considered in interpreting the extent of damage as it relates to fire origin. In making the determination, the possibility that the fuel in the suspected area of origin was not the first material ignited and that the great degree of damage was the result of wall or corner effects should be considered.

17.4.1.4 Ventilation. Ventilation, or lack thereof, during a fire has a significant impact on the heat release rate and consequently on the extent of observable burn damage. The analysis of fire pattern data should, therefore, include consideration that ventilation influenced the production of the pattern. Ventilation-controlled fires tend to burn more intensely near open windows or other vents, thereby producing greater damage. Knowledge of the location and type of fuel is important in fire pattern analysis. During full room involvement conditions, the development of fire patterns is significantly influenced by ventilation. Full room involvement conditions can cause fire patterns that developed during the earlier fuel-controlled phase of the fire to evolve and change. In addition, fires can produce unburned hydrocarbons that can be driven outside the compartment through ventilation openings. This unburned fuel can mix with air and burn on the exterior of the compartment, producing additional fire patterns that indicate the fire spread out of the original compartment. Thus, knowledge of changes in ventilation (e.g., forced ventilation from building systems, window breakage, opening or closing of doors, burn-through of compartment boundaries) is important to understand in the context of fire pattern analysis. Determination of what patterns were produced at the point of origin by the first item ignited usually becomes more difficult as the size and duration of the fire increases. This is especially true if the compartment has achieved full room involvement.

17.4.1.5 Movement and Intensity Patterns. As discussed in Chapter 6, fire patterns are generated by one of two mechanisms: the spread of the fire or the intensity of burning. As discussed above, fuel composition, rate of heat release, location, and ventilation differences may lead to differences in the intensity patterns that do not necessarily point to the area where the first fuel was ignited. Patterns that arise from the growth and movement (spread) of the fire are invariably better indicators of the area of origin. It may be difficult, however, to distinguish movement patterns from intensity patterns. Further, some patterns display a combination of intensity and movement (spread) indicators.

17.4.2* Heat and Flame Vector Analysis. Heat and flame vector analysis, along with accompanying diagram(s), is a tool for fire pattern analysis. Heat and flame vectoring is applied by constructing a diagram of the scene. The diagram should include walls, doorways and doors, windows, and any pertinent furnishings or contents. Then, through the use of arrows, the investigator notes the interpretations of the direction of heat or flame spread based upon the identifiable fire patterns present. The size of the arrows should reflect the scaled magnitude (actual size) of the individual patterns depicted. The arrows can point in the direction of fire travel from the heat source, or point back toward the heat source, as long as the direction of the vectors is consistent throughout the diagram. The investigator should identify each vector as to the respective fire pattern it represents. In a legend accompanying the diagram the investigator may give details of the corresponding fire pattern, such as height above the floor, height of the vertex of the pattern, the nature of the surface upon which the pattern appears, the pattern geometry, the particular fire effect which constitutes the pattern, and the direction(s) of fire spread which the pattern(s) represent. For example, as shown in Figure 17.4.2, Vector #7 represents burn damage on the carpet with decreasing fire damage as one moves northeast, Vector #8 represents comparison of the burn damage differences between the two sides of the chair with the south side displaying more severe damage, and Vector #10 represents a truncated cone pattern with decreasing fire damage and increasing height to the line of demarcation as one moves north.

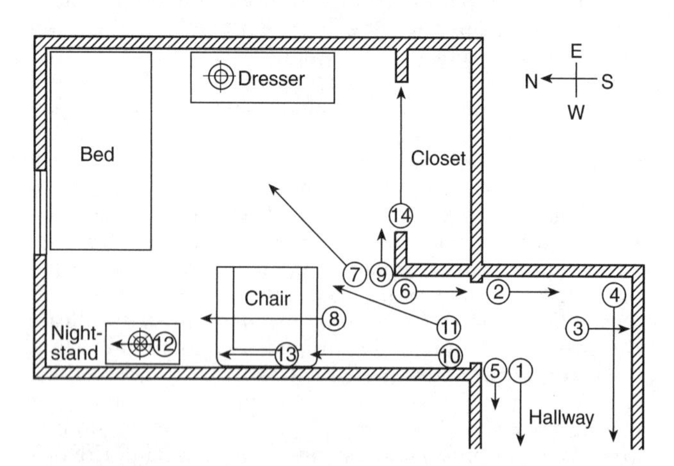

FIGURE 17.4.2 Heat and Vector Analysis Diagram Showing Vectors of the Physical Size and Direction of Heat Travel of the Fire Patterns and Demonstrating a Fire Origin in the Area of Vectors 6, 10, and 11. (Source: Kennedy and Shanley, "USFA Fire Burn Pattern Tests — Program for the Study of Fire Patterns.")

17.4.2.1 Complementary Vectors. Complementary vectors can be considered together to show actual heat and flame spread directions. In that case, the investigator should clearly identify which vectors represent actual fire patterns and which vectors represent heat flow derived from the investigator's analysis of these patterns. An important point to be made regarding this discussion is the terminology heat source and source of heat. These terms are not synonymous with the origin of the fire. Instead, these terms relate to any heat source that creates an identifiable fire pattern. The heat source may or may not be generated by the initial fuel. It is imperative that the use of heat and flame vector analysis be tempered by an accurate understanding of the progress of the fire and basic fire dynamics. A vector diagram can give the investigator an overall viewpoint to analyze. The diagram can also be used to identify any conflicting patterns that need to be explained. The ultimate purpose of the vector analysis is to discuss and graphically document the investigator's interpretation of the fire patterns.

17.4.2.2 Heat Source. A heat source can be any fuel package that creates an identifiable fire pattern. The pattern may or may not be produced by the initial fuel. Consider a fire that spreads into a garage and ignites flammable liquids stored there. The burning liquid represents a new heat source that leaves fire patterns on the garage's surfaces. Therefore, it is imperative that fire pattern analysis be tempered by an accurate understanding of the progress of the fire and basic fire dynamics.

17.4.2.3 Additional Tools for Pattern Visualization. When fire patterns are not visually obvious, a depth of char or depth of calcination survey may help the investigator to locate areas of greater or lesser heat damage and recognized lines of demarcation defining patterns. Survey results should be plotted on a diagram. On such diagrams, the depth of char or calcination measurements are recorded to a convenient scale. Once the depth of char or calcination measurements have been recorded on the diagram, lines can be drawn connecting points of equal, or nearly equal, char or calcination depths. The resulting lines may reveal identifiable patterns. *[See Figure 15.4.2(f).]*

17.4.3 Depth of Char Analysis. Analysis of the depth of charring is most reliable for evaluating fire spread, rather than for the establishment of specific burn times or intensity of heat from adjacent burning materials. By measuring the relative depth and extent of charring, the investigator may be able to determine what portions of a material or construction were exposed the longest to a heat source. The relative depth of char from point to point is the key to appropriate use of charring — locating the places where the damage was most severe due to exposure, ventilation, or fuel placement. The investigator may then deduce the direction of fire spread, with decreasing char depths being farther away from a heat source. Certain key variables affect the validity of depth of char pattern analysis. These factors include the following:

(1) Single versus multiple heat or fuel sources creating the char patterns being measured. Depth of char measurements may be useful in determining more than one fire or heat source.
(2) Comparison of char measurements, which should be done only for identical materials. It would not be valid to compare the depth of char from a wall stud to the depth of char of an adjacent wooden wall panel.
(3) Ventilation factors influencing the rate of burning. Wood can exhibit deeper charring when adjacent to a ventilation source or an opening where hot fire gases can escape.
(4) Consistency of measuring technique and method. Each comparable depth of char measurement should be made with the same tool and same technique. *[See Figure 15.4.2(f).]*

17.4.3.1 Depth of Char Diagram. Lines of demarcation that may not be visually obvious can often be identified for analysis by a process of measuring and charting depths of char on a grid diagram. By drawing lines connecting points of equal char depth (isochars) on the grid diagram, lines of demarcation may be identified.

17.4.3.2 Measuring Depth of Char. Consistency in the method of measuring the depth of char is the key to generating reliable data. Sharp pointed instruments, such as pocket knives, are not suitable for accurate measurements because the sharp end of the knife will have a tendency to cut into the noncharred wood beneath. Thin, blunt-ended probes, such as certain types of calipers, tire tread depth gauges, or specifically modified metal rulers are best. Dial calipers with depth probes of round cross-section, shown in Figure 17.4.3.2(a), are excellent depth of char measurement tools. Figure 17.4.3.2(b) illustrates their use. The same measuring tool should be used for any set of comparable measurements. Consistent pressure for each measurement while inserting the measuring device is also necessary for accurate results.

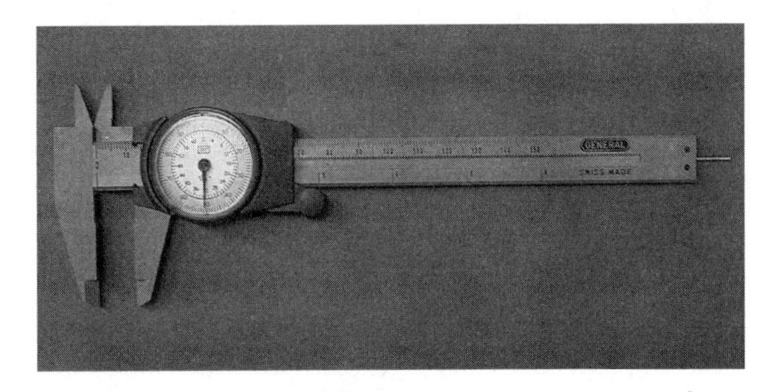

FIGURE 17.4.3.2(a) Dial Calipers with Depth Probes.

FIGURE 17.4.3.2(b) Using Dial Calipers to Measure Depth of Char.

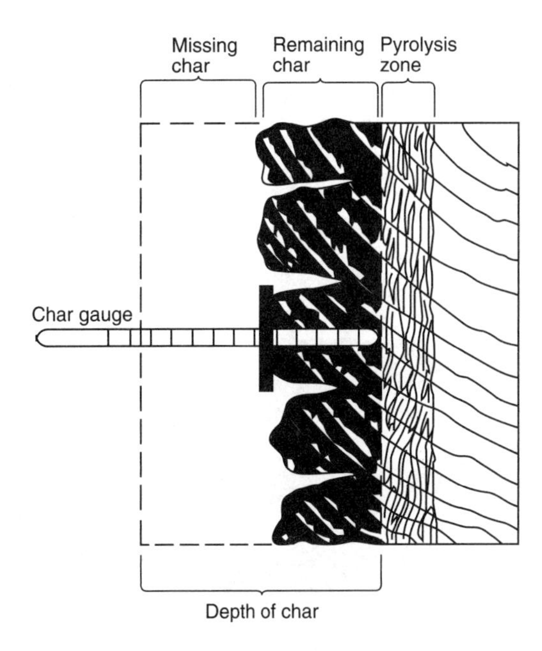

FIGURE 17.4.3.3 Measuring Depth of Char.

17.4.3.3 Char depth measurements, illustrated in Figure 17.4.3.3, should be made at the center of char blisters, rather than in or near the crevasses between blisters.

17.4.3.4 Missing Wood. When determining the depth of charring, the investigator should take into consideration any burned wood that may have been completely destroyed by the fire and add that missing depth of wood to the overall depth measurement.

17.4.3.5 Depth of Char Surveys with Fuel Gases. When fugitive fuel gases are the initial fuel sources for fires, they produce relatively even depths of char over the often wide areas that they cover. Progressive changes in depth of char that are used by investigators to trace fire spread might exist only in those areas to which the fire spreads from the initial locations of the pocketed fuel gases. Deeper charring might exist in close proximity to the point of gas leakage, as burning might continue there after the original quantity of gas is consumed. This charring may be highly localized because of the pressurized gas jets that can exist at the immediate point of leakage and may assist the investigator in locating the leak.

17.4.4 Depth of Calcination Survey. Relative depth of calcination can indicate differences in total heating of the fire-exposed gypsum wallboard. Deeper calcination readings indicate longer or more intense heating (heat flux), and the higher temperatures that those areas of the wallboard achieved during the fire.

Certain key variables affect the validity of depth of calcination analysis. These factors include the following:

(1) Single versus multiple heat or fuel sources, creating the calcination patterns being measured, should be considered. Depth of calcination patterns may be useful in determining multiple heat or fire sources.
(2) Comparisons of depth of calcination measurements should be made only from the same material. It should be recognized that gypsum wallboard comes in different thicknesses, is made of different materials of construction, and changes with time. The investigator should carefully consider sections of walls or ceilings that may have had new sections inserted as a repair.
(3) The finish of the gypsum wallboard (e.g., paint, wallpaper, stucco) should be considered when evaluating depth of calcination. The investigator should recognize that some of these finishes are combustible and may affect the patterns if they are ignited.
(4) Measurements should be made in a consistent fashion to reduce errors in this data collection.
(5) Gypsum wallboard can be damaged during suppression, during overhaul, and post-fire by hose streams and standing water. Wetting of the calcined wallboard can soften the gypsum to the point where no reliable measurements can be made.

17.4.4.1 Depth of Calcination Diagram. A depth of calcination diagram can be produced in the same manner as that for depth of char.

17.4.4.2 Measuring Depth of Calcination. The technique for measuring and analyzing depth of calcination can use a visual observation of cross-sections or a probe survey. The visual method requires careful removal of small, full-thickness sections [minimum approximately 50 mm (2 in.) diameter] of walls or ceilings to observe and measure the thickness of the calcined layer. The probe method requires that a survey of the depth of calcination be undertaken by inserting a small cross-section probe device, such as illustrated in Figure 17.4.4.2(a) and Figure 17.4.4.2(b), and recording the depth at which a relative difference in resistance of the calcined gypsum is felt. When using the probe method the investigator should conduct the survey at regular lateral and vertical grid intervals along the surface of the involved wallboard, usually in increments of 0.3 m (1 ft) or less. Care should be taken to use approximately the same insertion pressure for each measurement. Such surveys can be made on either wall or ceiling installations of wallboard.

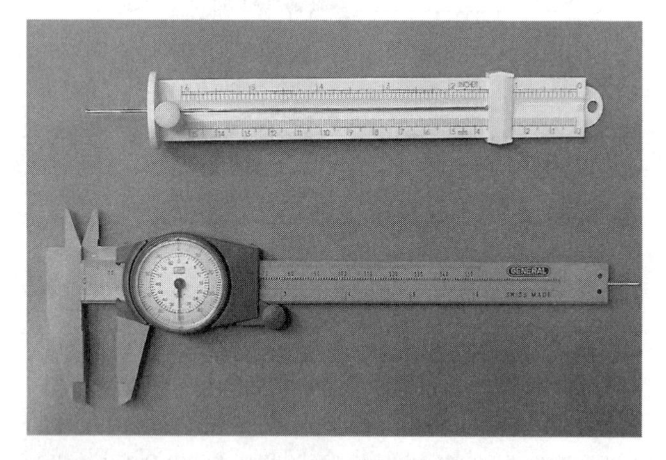

FIGURE 17.4.4.2(a) Two Instruments That Can be Used to Measure the Depth of Calcination.

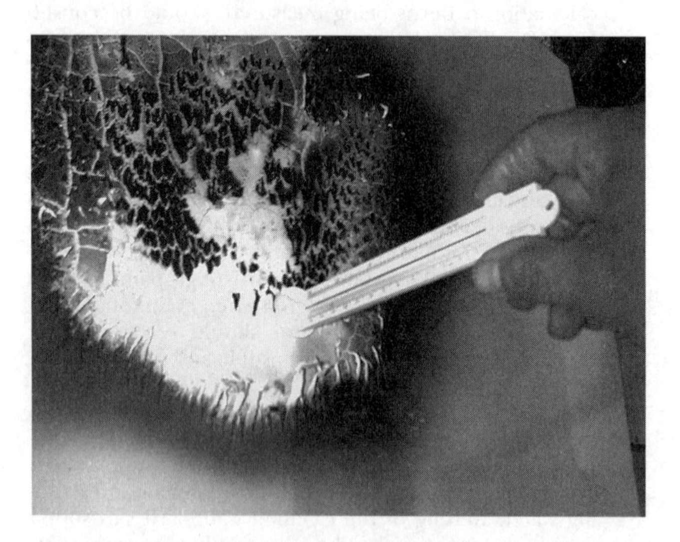

FIGURE 17.4.4.2(b) Measuring Depth of Calcination on a Piece of Gypsum Wallboard.

17.4.5 Arc Surveys or Arc Mapping. Arc surveys (also known as arc mapping) is a technique in which the investigator uses the identification of arc locations or "sites" to aid in determining the area of fire origin. This technique is based on the predictable behavior of energized electrical circuits exposed to a spreading fire. The spatial relationship of the arc sites to the structure and to each other can be a pattern, which can be used in an analysis of the sequence in which the affected parts of the electrical system were compromised. This sequential data can be used in combination with other data to more clearly define the area of origin. There are circumstances, such as complete destruction of the branch circuits, melting of conductors from fire exposure, post-fire re-energizing of the electrical system, or the inability to recognize arc damage on conductors, that make it more difficult or impossible to use this technique. The identification of arc sites in the area of fire origin may help to identify a potential ignition source(s) for consideration. The investigator is cautioned to consider that conductors only pass through certain areas, and therefore, the amount of information available will be limited by the spatial distribution of the conductors available for arcing. Mapping of arcs on conductors that are exposed to the developing fire, such as appliance cords or branch circuit conductors that are in the compartment of the fire, may provide the most useful information. Branch circuits that are located behind some type of thermal barrier, such as gypsum board or plywood, may not provide useful arc mapping information, as the circuit protection may open prior to the fire damaging the conductors located behind ceilings and walls.

17.4.5.1* Suggested Procedure. One procedure to perform an arc survey is as follows:

(1) Identify the area that will be surveyed.
(2) Sketch and diagram the area as completely and accurately as possible.
(3) Identify zones within the survey area, such as ceiling, floor, north wall, south wall, etc.
(4) Identify all conductors of the electrical circuits passing through the zone, noting, when possible, loads on each circuit, direction of power flow (upstream versus downstream), locations of junction boxes, outlets, switches (or any such control), size of each conductor, and the over-current protection size, type, and status.
(5) Select a zone for examination and begin the process of systematically examining each of the conductors in that zone.
(6) Examine and feel each conductor, for the purpose of identifying surface anomalies or damage, such as beads and notches. When it is necessary to remove conductors from conduits, take care to prevent damage to the conductors.
(7) Determine if the surface anomaly occurred from arcing, environmental heat, or eutectic melting (alloying of metals).
(8) Locate the arc site on the sketch and document its physical characteristics (faulted to another conductor in same cable, faulted to conductor from another cable, completely severed conductor, partially severed conductor, faulted to grounded metallic conduit, or a conductive building element).
(9) Flag the location of the arc site(s) with a suitable marking and document such location(s).
(10) Preserve the items as evidence, when warranted.

17.4.5.2 Arc Survey Diagram. The drawing used to plot the arc sites should be as detailed as possible. Precision in sketching the drawing will aid in reducing errors in the subsequent analysis. When setting the boundaries of each zone, keep in mind that some or all of the conductors may be routed through other zones as well. Having a compass or reference direction on the drawing is particularly useful. When analyzing the status of the overcurrent protection device for each circuit, note the type of device such as fuse, circuit breaker, ground fault circuit interrupter (GFCI), or arc fault circuit interrupter (AFCI).

17.4.5.3 Locating Arc Sites. Typically, examining each conductor can be accomplished by passing the conductor through one's fingertips, feeling the conductor's surface for imperfections or anomalies, either convex or concave. Dragging a cotton ball along the conductor is another way to detect surface imperfections. Once the imperfection is detected, the location is examined visually to assess if the imperfection is the result of conductor metal loss or deposit (indications of possible arc sites), or from some other reason, such as dirt accumulation, oxidation, or charred insulation scale. Search the entire length of the damaged conductor in case there are multiple sites where arcing occurred on the same conductor. Locate each arc site on the plan and elevation drawings as precisely as possible.

17.4.5.3.1 The identification of arc damage is not a simple task. Fire melting of conductors, also called environmental melting, can be difficult to distinguish from melting caused by arcing. The primary distinction between the two is the relative area where the melting is found. Arcing can create highly localized damage where the temperatures exceed the melting point of the conductor metal. Arcing can create complementary damage to adjacent conductors or grounded surfaces. The edges of the melted areas are generally quite distinct. Melting from environmental heating can be relatively widespread and may involve numerous conductors in an area. Chapter 8 provides several photographic examples of the two types of damage. Some locations of melting on the conductors may not indicate what caused the melting. Such locations can be noted on the documentation as possibilities or as unknowns.

17.4.5.3.2 The visibility of arc damage is related to the duration of the arc, as well as the timing of the arc. In alternating current circuits, the potential difference between two conductors depends on the point in the AC cycle when the arc occurs. Circuit breakers or fuses may operate to cut off power in as little as one-half cycle (0.8 ms in 60 cycle circuits, 1.0 ms in 50 cycle circuits). During each typical 120 VAC cycle, the potential between the "hot" conductor and the neutral conductor or ground ranges between +120V and −120V. As the potential approaches 0V, arc damage will tend to become less severe, and therefore less visible.

17.4.5.4 Documenting Arc Sites. To document arc sites, attach visible markers such as colored ribbon, colored cable ties, or tape to the conductors and document using photographs or videotape. Should the need exist to take as evidence the proof of the arc survey, then one can collect the electrical circuits in the structure. Collecting each circuit, both those that arced and those that did not, will be of value, only if the spatial relationship of the circuits is maintained. The relationship in space of the arc sites to those conductors that did not arc, and the individual arc sites, not the fire-damaged conductors taken out of context, are the significant evidence.

17.4.5.5 Arc Survey Evidence Collection. Take care to properly identify, tag, and collect electrical conductors. The conductors are sometimes brittle and can be quite fragile. Handling may result in fractures, which make labeling and tagging much more tedious. Attachments to the conductors, such as junction box remains, may become loose and fall away, which can potentially prevent any future circuit tracing.

17.4.5.6 Arc Survey Utilization. The utility of arc mapping is primarily the analysis of the data to determine the sequence of events, but it should be noted that arc mapping can be useful in both formulating and testing hypotheses. Clearly, if a conductor is arc-severed, it can be correctly concluded that any arcing events electrically downstream of the arc-severed point happened before the severance. Exceptions to this rule exist when the conductors are back-fed through a pre-fire wiring error, uninterruptible power supply (UPS) systems, or generators. Further, if an area is hypothesized as the origin of a fire, then, in the absence of contrary evidence, one would expect that the origin area would be one of the first areas where the electrical system was compromised.

17.4.5.7 Arc Survey Limitations. Arc surveys can identify areas where the fire had damaged energized electrical conductors early in the fire's development. Likewise, the spatial relationship of arc sites can identify a specific space where the fire occurred before electrical energy to that space was cut off. Both of these investigative tools can be helpful in the origin determination. The accuracy of the effort, however, is directly dependent upon the investigator correctly identifying arc damage on the wires. Fire damage to copper conductors can mimic arc damage, and visual inspection at the fire scene site may not be sufficient to correctly identify validate arc sites. If the analysis of the circuits incorrectly identifies damage on the conductors as arcing, hypotheses formed from the analyses will be based on flawed data and will be incorrect. The investigator may want to collect each perceived arc site for more detailed evaluation and verification.

17.4.6 Analysis of Sequential Events. The analysis of the timing or sequence of events during a fire can be useful in determining the origin. Much of the data for this analysis will come from witnesses. In some instances, a witness may be found who saw the fire in its incipient stage and can provide the investigator with an area of fire origin. Such circumstances create a burden on the fire investigator to conduct as thorough an investigation as possible to find facts that can support or refute the witness's statements. Means to verify such statements could include patterns analysis, arc mapping, or matching smoke detector, heat detector, and security detector activation times with the witness's observations. This analysis can identify gaps or inconsistencies in information, assist in developing questions for additional witness interviews, and provide support in the analysis and reconstruction of the progression of the fire. A more detailed discussion of time lines is included in Section 20.2.

17.4.7 Fire Dynamics. Fundamentals of fire dynamics can be used to analyze the data to aid in the development of origin hypotheses and to complement other origin determination techniques. Such analyses can help in the identification of potential fuels that may have been the first item to ignite, the sequence of subsequent fuel involvement, the recognition of other data that may need to be collected, the analysis of fire patterns, and the identification of potential competent ignition sources.

17.5 Developing an Origin Hypothesis. Based on the data analysis, the investigator should now produce a hypothesis or group of hypotheses to explain the origin and development of the fire. This hypothesis should be based solely on the empirical data that the investigator has collected. It is understood that when using the scientific method, an investigator may continuously be engaged in data collection, data analysis, hypothesis development, and hypothesis testing. An investigator may develop an origin hypothesis early in the investigative process, but when the process is completed, regardless of the order of steps followed, the investigator should be able to describe how those steps conform to the scientific method. Figure 17.2 shows how the procedures set forth in this chapter follow the scientific method.

17.5.1 Initial Hypothesis. The initial origin hypothesis is developed by considering witness observations, by conducting an initial scene assessment, and by attempting to explain the fire's movement through the structure. This process is accomplished using the methods described in earlier sections of this chapter. The initial hypothesis allows the investigator to organize and plan the remainder of the origin investigation. The development of the initial hypothesis is a critical point in the investigation. It is important at this stage that the investigator attempt to identify other feasible origins, and to keep all reasonable origin hypotheses under consideration until sufficient evidence is developed to justify discarding them.

17.5.2 Modifying the Initial Hypothesis. The investigation should not be planned solely to prove the initial hypothesis. It is important to maintain an open mind. The investigative effort may cause the initial hypothesis to change many times before the investigation is complete. The investigator should continue to reevaluate potential areas of origin by considering the additional data accumulated as the investigation progresses.

17.6 Testing of Origin Hypotheses. In order to conform to the scientific method, once a hypothesis is developed, the investigator must test it using deductive reasoning. A test using deductive reasoning is based on the premise that *if* the hypothesis is true, *then* the fire scene should exhibit certain characteristics, assuming that the fire did not subsequently obliterate those characteristics. For example, if a witness stated that a specific door was closed during the fire, then there should be a protected area on the door jamb, which would tend to prove the hypothesis that the door was closed. *(See Chapter 4 and A.4.3.6.)*

17.6.1 Means of Hypothesis Testing. During the investigation, the investigator may develop and test many hypotheses about the progress of the fire. For example, the investigator often has to determine whether a door or window was open or closed. Ultimately, the origin determination is arrived at through the testing of origin hypotheses. A technically valid origin determination is one that is consistent with the available data. In testing the hypothesis, the questions addressed in 17.6.1.1 through 17.6.1.3 should be answered.

17.6.1.1 Is there a competent ignition source at the hypothetical origin? The lack of a competent ignition source at the hypothesized origin should make the hypothesis subject to increased scrutiny. Investigators should be wary of the trap of circular logic. While the cause of the fire was at one time necessarily located at the point of origin, the investigator who eliminates a potential ignition source because it is "not in the area of the hypothesized origin," needs to be especially diligent in testing the origin hypothesis and in considering alternate hypotheses. *(See Section 18.2.)* This is particularly true in cases of full room involvement. Unless there is reliable evidence to narrow the origin to a particular portion of the room, every potential ignition source in the compartment of origin should be given consideration as a possible cause.

17.6.1.2 Can a fire starting at the hypothetical origin result in the observed damage? The investigator should be cautious about deciding on an origin just because a readily ignitible fuel and potential ignition source are present. The sequence of events that bring the ignition source and the fuel together and cause the observed damage indicates the origin, and ultimately the cause. The hypothetical origin should not only account for physical damage to the structure and contents, but also for the exposure of occupants to the fire environment.

17.6.1.3 Is the growth and development of a fire starting at the hypothetical origin consistent with available data at a specific point(s) in time? Few data are more damaging to an origin hypothesis than a contradictory observation by a credible eyewitness. Any data can be contradictory to the ultimate hypothesis. The data must be taken as a whole in considering the hypothesis, with each piece of data being analyzed for its reliability and value. Ultimately, the investigator should be able to explain how the growth and development of a fire, starting at the hypothesized origin, is consistent with the data.

17.6.2 Analytical Techniques and Tools. Analysis techniques and tools are available to test origin hypotheses. Using such tools and techniques to analyze the dynamics of the fire can provide an understanding of the fire that can enhance the technical basis for origin determinations. Such analyses can also identify gaps or inconsistencies in the data. The utility of fire dynamics tools is not limited to hypothesis testing. They may also be used for data analysis and hypothesis development. Techniques and tools include time line analysis, fire dynamics analysis, and experimentation.

17.6.2.1 Time Line Analysis. Time lines are an investigative tool that can show relationships between events and conditions associated with the fire. These events and conditions are generally time-dependent, and thus, the sequence of events can be used for testing origin hypotheses. Relevant events and conditions include ignition of additional fuel packages, changes in ventilation, activation of heat and smoke detectors, flashover, window breakage, and fire spread to adjacent compartments. Much of this information will come from witnesses. Fire dynamics analytical tools *(see 20.4.8)* can be used to estimate time-dependent events and fire conditions. A more detailed discussion of time lines is included in Section 20.2.

17.6.2.2 Fire Modeling. Fundamentals of fire dynamics can be used to test hypotheses regarding fire origin. Such fundamentals are described in the available scientific literature and are incorporated into fire models ranging from simple algebraic equations to more complex computer fire models *(see 20.4.8)*. The models use incident-specific data to predict the fire environment given a proposed hypothesis. The results can be compared to physical and eyewitness evidence to test the origin hypothesis. Models can address issues related to fire development, spread, and occupant exposure.

17.6.2.3 Experimental Testing. Experiments can be conducted to test the hypothesized origin. If the experimental results match the damage at the scene, the experiment can be said to support the hypothesis. If the experiment produces different results, a new origin hypothesis or additional data may need to be considered, taking into account potential differences between testing and actual fire conditions. The following is an example of such an experiment. The hypothesized origin is a wicker basket located in the corner of a wood-paneled room. The data from the actual fire shows the partial remains of the basket, undamaged carpet in the corner, and wood paneling still intact in the corner. A fire test replicating the hypothesized origin totally consumes the carpet, the wicker basket, and the wood paneling. Thus (assuming the test replicated the pre-fire conditions), testing revealed that this hypothesized origin is inconsistent with the damage that would be expected from such a fire.

17.7 Selecting the Final Hypothesis. Once the hypotheses regarding the origin of the fire have been tested, the investigator should review the entire process, to ensure that all credible data are accounted for and all credible alternate origin hypotheses have been considered and eliminated. When using the scientific method, the failure to consider alternate hypotheses is a serious error. A critical question to be answered by fire investigators is, "Are there any other origin hypotheses that are consistent with the data?" The investigator should document the facts that support the origin determination to the exclusion of all other potential origins.

17.7.1 Defining the Area of Origin. Although *area of origin* is common terminology used to describe the origin, the investigator should describe it in terms of the three-dimensional space where the fire began, including the boundaries of that space.

17.7.2 Inconsistent Data. It is unusual for a hypothesis to be totally consistent with all of the data. Each piece of data should be analyzed for its reliability and value — not all data in an alysis has the same value. Frequently, some fire pattern or witness statement will provide data that appears to be inconsistent. Contradictory data should be recognized and resolved. Incomplete data may make this difficult or impossible. If resolution is not possible, then the origin hypothesis should be re-evaluated.

17.7.3 Case File Review. Other investigators can assist in the evaluation of the origin hypothesis. An investigator should be able to provide the data and analyses to another investigator, who should be able to reach the same conclusion as to the origin. Review by other investigators is almost certain to happen in any significant fire case. Differences in opinions may arise from the weight given to certain data by different investigators or the application of differing theoretical explanations (fire dynamics) to the underlying facts in a particular case.

17.8 Origin Insufficiently Defined. There are occasions when it is not possible to form a testable hypothesis defining an area that is useful for identifying potential causes. The goal of origin investigation is to identify the precise location where the fire began. In practice, the investigator has an origin hypothesis when first arriving at a fire scene. The origin is the scene. Sometimes, it is not possible to find an area or volume that is any smaller than the entire scene. Thus, a conclusion of the origin investigation can be the identification of a volume of space too large to identify causal factors, or where no practical boundaries can be established around the volume of the origin. An example of such an origin can be a building that has been totally burned, with no eyewitnesses. Such fires are sometimes called total burns. The area of origin is the building, but in reality no further testable origin hypothesis can be developed because there is insufficient reliable data.

17.8.1 Large Area Adequate for Determination. There are cases in which a lack of an origin determination does not necessarily hinder the investigation. An example is a case in which a fire resulted from the ignition of fuel gas vapors inside a structure. The resulting damage may preclude the defining of the location where the fuel combined with the ignition source. However, probable ignition sources may still be hypothesized.

17.8.2 Justification of a Large Area of Origin. The origin analysis should identify the data that justify the conclusion that the area of fire origin cannot be reduced to a practical size. Examples of such data could include establishing the fact that there were no significant patterns to trace, that most or all combustible materials were consumed, or that other methods of origin determination were attempted but no reasonable conclusion could be established.

17.8.3 Eyewitness Evidence of Origin Area. If the origin is too large to be useful, then the determination of the fire's cause may become very difficult, or impossible. In some instances, where no further testable origin hypothesis can be developed by examination of the scene alone, a witness may be found who saw the fire in its incipient stage and can provide the investigator with an area of fire origin.

Chapter 18 Fire Cause Determination

18.1 Introduction. This chapter recommends a methodology to follow in determining the cause of a fire. Fire cause determination is the process of identifying the first fuel ignited, the ignition source, the oxidizing agent, and the circumstances that resulted in the fire. Fire cause determination generally follows origin determination (see Origin Determination chapter). Generally, a fire cause determination can be considered reliable only if the origin has been correctly determined.

18.1.1 Fire Cause Factors. The determination of the cause of a fire requires the identification of those factors that were necessary for the fire to have occurred. Those factors include the presence of a competent ignition source, the type and form of the first fuel ignited, and the circumstances, such as failures or human actions, that allowed the factors to come together and start the fire. Device or appliance failures can involve, for example, a high-temperature thermostat that fails to operate. The device may have failed due to a design defect. Human contributions to a fire can include a failure to monitor a cooking pot on the stove, failure to connect electrical wiring tightly resulting in a high-resistance connection, or intentional acts. For example, consider a fire that starts when a blanket is ignited by an incandescent lamp in a closet. The various factors include having a lamp hanging down too close to the shelf, putting combustibles too close to the lamp, and leaving the lamp on while not using the closet. The absence of any one of those factors would have prevented the fire. The function of the investigator is to identify those factors that contribute to the fire.

18.1.2 First Fuel Ignited. The first fuel ignited is that which first sustains combustion beyond the ignition source. For example, the wood of the match would not be the first fuel ignited, but paper, ignitable liquid, or draperies would be, if the match were used to ignite them.

18.1.3 Ignition Source. The ignition source will be at or near the point of origin at the time of ignition, although in some circumstances, such as the ignition of flammable vapors, the two may not appear to coincide. Sometimes the source of ignition will remain at the point of origin in recognizable form, whereas other times the source may be altered, destroyed, consumed, moved, or removed. Nevertheless, the source should be identified in order for the cause to be proven. There are, however, occasions when there is no physical evidence of the ignition source, but an ignition sequence can be hypothesized based on other data.

18.1.4 Oxidant. Generally, the oxidant is the oxygen in the earth's atmosphere. Medical oxygen, such as that stored in cylinders or produced by oxygen concentrators, and certain chemical compounds may support or enhance combustion reactions (see Oxidizing Agents in the Basic Fire Science chapter).

18.1.5 Ignition Sequence. A fuel by itself or an ignition source by itself does not create a fire. Fire results from the combination of fuel, an oxidant, and an ignition source. The investigator's description of events, including the ignition sequence, (the factors that allowed the ignition source, fuel, and oxidant to react) can help establish the fire cause.

18.2 Overall Methodology. The overall methodology for determining the cause of the fire is the scientific method as described in the Basic Methodology chapter. This methodology includes recognizing and defining the problem to be solved, collecting data, analyzing the data, developing a hypothesis or hypotheses, and, most importantly, testing the hypothesis or hypotheses. In order to use the scientific method, the investigator must develop at least one hypothesis based on the data available at the time. These initial hypotheses should be considered "working hypotheses," which upon testing may be discarded, revised, or expanded in detail as new data is collected during the investigation and new analyses are applied. This process is repeated as new information becomes available. *(See Figure 18.2).*

18.2.1 Consideration of Data. In some instances, a single item, such as an irrefutable article of physical evidence or a credible eyewitness to the ignition, or a video recording, may be the basis for a determination of cause. In most cases, however, no single item is sufficient in itself to allow determination of the fire cause. The investigator should use all available resources to develop fire cause hypotheses and to determine which hypotheses fit all of the credible data available. When an apparently plausible hypothesis fails to fit some item of data, the investigator should try to reconcile the two and determine whether the hypothesis or the data is erroneous.

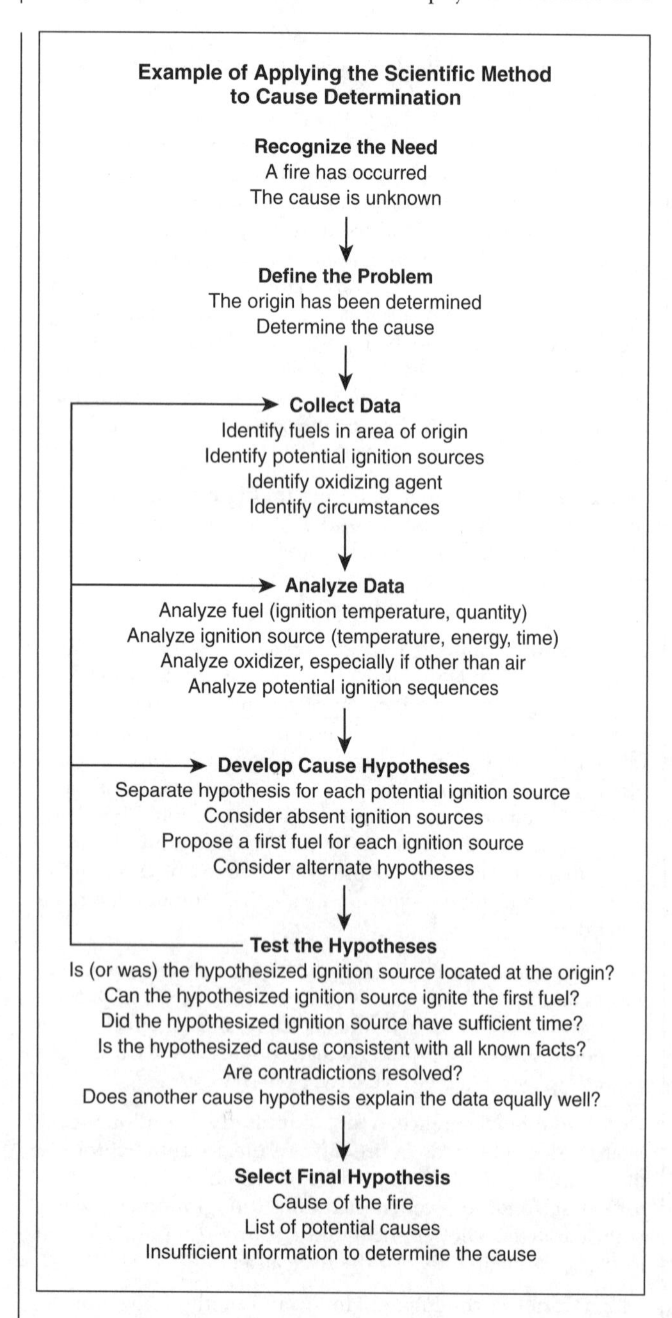

FIGURE 18.2 An Example of Applying the Scientific Method to Cause Determination.

18.2.2 Sequence of Activities. The various activities required to determine the cause using the scientific method (data collection, analysis, hypothesis development, and hypothesis testing) occur continuously. Likewise, recording the scene, note taking, photography, evidence identification, witness interviews, origin investigation, failure analysis, and other data collection activities may be performed simultaneously with these efforts. Investigators should refer to the other sections of this guide that deal with these specific activities. Similarly, investigators need to remain aware of potential spoliation and scene contamination issues and should refer to the chapters on Legal Considerations and Physical Evidence.

18.2.3 Point and Area of Origin. In some cases, it will be impossible to determine the point of origin of a fire within the area of origin. Where a single point cannot be identified, it can still be valuable for many purposes to identify the area(s) of origin. In such instances, the investigator should be able to provide reliable explanations for the area of origin with the supporting evidence for each option. In some situations, the extent of the damage may reduce the ability to specifically identify the point of origin, without removing the ability to put forward credible origin and cause hypotheses.

18.3 Data Collection for Fire Cause Determination. Data collection processes for cause determination includes identification of fuel packages, ignition sources, oxidizers, and circumstances. Data should be collected to identify all potential fuels, ignition sources, and oxidants within the area or areas of origin. Data may also need to be collected from outside the area of origin. Examples of this would be unburned fuel samples or exemplar ignition sources located in other areas. Data on the circumstances bringing the fuel, ignition sources, and oxidizer together may come from many different sources. If available, a review of pre-fire documentation of possible areas of origin can be of value.

18.3.1 Identify Fuels in the Area of Origin. The investigator should identify the fuels present in the area of origin at the time of ignition. One of these fuels will be the first fuel ignited. The type, quantity, and specific location of structural and content fuels should be identified.

18.3.1.1 Identifying the initial fuel is necessary for evaluating the competency of potential ignition sources and understanding the events that caused the fire. Sometimes a portion of the first ignited fuel will survive the fire, but often it does not. The initial fuel must be capable of being ignited within the limitations of the ignition source. The components in most buildings are not susceptible to ignition by heat sources having low energy, low temperature, or short duration. For example, flooring, structural lumber, wood cabinets, and carpeting do not ignite unless they are exposed to a substantial heat source. The investigator should identify easily ignited items that, once ignited, could provide the heat source to damage or involve these harder-to-ignite items. *(See Basic Fire Science chapter, Fuel Load).*

18.3.1.2 The initial fuel could be part of a device that malfunctions or fails. Examples include insulation on a wire that is heated to its ignition temperature by excessive current, or the plastic housing on an overheating coffee maker.

18.3.1.3 The initial fuel might be something too close to a heat-producing device. Examples are clothing against an incandescent lamp or a radiant heater, wood framing too close to a wood stove or fireplace, or combustibles too close to an engine exhaust manifold or catalytic converter.

18.3.1.4 Certain fuels produce residues not typically found after a fire. These residues differ from construction and contents materials that are normally present in the area of origin. Examples include residues of ignitable liquids or pyrotechnic materials, such as flares.

18.3.1.5 Gases, vapors, and combustible dusts can be the initial fuel and can cause confusion about the location of the point of origin, because the point of ignition can be some distance away from where sustained fire starts in the structure or furnishings. Also, flash fire may occur with sustained burning of light density materials, such as curtains, that are located away from the initial vapor-fuel source.

18.3.1.6 Information should be sought from persons having knowledge (such as occupants) about recent activities in the area of origin and what fuel items should or should not have been present. Information should also be obtained about the construction of the structure in the origin area. Construction details could include information about the floor, ceiling and wall coverings, type of doors, type of windows, or other information necessary for the analysis. The age of construction materials and attachment methodologies may be relevant. This information could reveal the initial fuel for the fire. This information would also be helpful to an investigator to prevent overlooking secondary and subsequent fuels that were present in the origin area that would contribute to fire growth. The investigator should refer to the chapters on Basic Fire Science, Fire Patterns, Building Systems, and Sources of Information when analyzing an origin area for the initial fuel.

18.3.2 Identify Source and Form of the Heat of Ignition. The investigator should identify and document all heat-producing items in the area of origin. Heat-producing items include devices, appliances, equipment, and self-heating and reactive materials. The investigator should also identify devices or equipment that are not normally heat producing, but may produce enough heat for ignition through misuse or malfunction.

18.3.2.1 Potential sources of ignition for gases, vapors, or dusts include open flames, arcs from motors and switches, electric igniters, standing pilots or flames in gas appliances, hot surfaces, and static electricity.

18.3.3 Identify Items and Activities in Area of Origin. Information should be obtained from owners and occupants about recent activities in the area of origin and what appliances, equipment, or heat-producing devices were present. This information is especially important when potential ignition sources are not identifiable post-fire. The information would also be helpful in alerting an investigator to small or easily overlooked items when examining the area of origin. When electrical energy sources are considered as potential ignition sources, the investigator should refer to the chapters on Electricity and Appliances. Information on purchase, such as new or used, how and when they were used, repair history, and problems should also be gathered.

18.3.4* Identify the Oxidant. The most common oxidant (oxidizer or oxidizing agent) within a fire is the oxygen in earth's atmosphere and no special documentation is required. However, other oxidants, as described in 18.3.4.1 through 18.3.4.3, should be identified and documented when they are in or near the area of origin.

18.3.4.1 Sometimes oxygen exists at greater than the normal atmospheric concentration, such as in hyperbaric chambers, oxygen tents, or around oxygen generation and storage equipment.

18.3.4.2 Some chemicals other than molecular oxygen are classified as oxidants. Certain common chemicals, such as pool sanitizers, may also act as oxidants.

18.3.4.3 Some chemical mixtures, such as solid rocket fuel, contain an oxidizer as well as a fuel and require no external oxidizing source.

18.3.5 Identify Ignition Sequence Data. The investigator should develop data that can be used to analyze the events that brought the fuel and ignition source together (ignition sequence). This information on the conditions surrounding the coincidence of fuel, ignition source, and oxidizer may be available through observations, witness accounts, or weather data. Time lines can be useful in organizing and analyzing this data. *(See chapter on Failure Analysis and Analytical Tools.)* Additional data collection may be necessary in order to determine the circumstances that brought the fuel, ignition source, and oxidizer together. Data collection may continue even after the fire scene has been processed and could require specialized laboratory equipment. Such additional data may result in modification or rejection of previously developed hypotheses or reconsideration of previous rejected hypotheses.

18.4 Analyze the Data. The scientific method requires that all data collected that bears upon the fire cause be analyzed. Analyzing the data requires the examination and interpretation of each component of data collected that bears upon the fire cause. This is an essential step that must take place before the formation of any hypotheses. The purpose of the analysis is to attribute specific meaning to the results of the examination and interpretation process, which will ultimately play a role in hypothesis development and testing. The identification, gathering, and cataloging of data does not equate to data analysis. Analysis of the data is based on the knowledge, training, experience, and expertise of the individual doing the analysis. If the investigator lacks the knowledge to properly attribute meaning to a piece of data, then assistance should be sought from someone with the necessary knowledge. Understanding the meaning of the data will enable the investigator to form hypotheses based on the evidence, rather than on speculation or subjective belief.

18.4.1 Fuel Analysis. Fuel analysis is the process of identifying the first (initial) fuel item or package that sustains combustion beyond the ignition source and identifying subsequent target fuels beyond the initial fuel.

18.4.1.1 Geometry and Orientation. An understanding of the geometry and orientation of the fuel is important in determining if the fuel was the first material ignited. The physical configuration of the fuel plays a significant role in its ability to be ignited. A nongaseous fuel with a high surface-to-mass ratio is much more readily ignitable than a fuel with a low surface-to-mass ratio. Examples of high surface-to-mass fuels include dusts, fibers, and paper. As the surface-to-mass ratio increases, the heat energy or time required to ignite the fuel decreases. Gases and vapors are fully dispersed (in effect, an extremely high surface-to-mass ratio) and can be ignited by a low heat energy source instantly.

18.4.1.2 Ignition Temperature. The fuel must be capable of being ignited by the hypothesized ignition source. The ignition temperature of the fuel should be understood. It is important to understand the difference between piloted ignition and autoignition temperatures. The components in most buildings are not susceptible to ignition by heat sources of low energy, low temperature, or short duration. For example, flooring, structural lumber, wood cabinets, and carpeting do not ignite unless they are exposed to a substantial heat source.

18.4.1.3 Quantity of Fuel. The first material ignited may not result in fire growth and spread if a sufficient quantity of the fuel does not exist. For example, if the lighter fluid used to start a charcoal fire is consumed before enough heat is transferred to the briquettes, the fire goes out. The investigator should conduct an analysis of the quantity of fuels (primary, secondary, tertiary, etc.) to determine that it is sufficient to explain the resulting fire.

18.4.2 Ignition Source Analysis. The investigator should evaluate the potential ignition sources in the area of origin to determine if they are competent. A competent ignition source will have sufficient energy and be capable of transferring that energy to the fuel long enough to raise the fuel to its ignition temperature.

18.4.2.1 Heating of the potential fuel will occur by the energy that reaches it. Each fuel reacts differently to the energy that impacts on it based upon its thermal and physical properties. Energy can be reflected, transmitted, or dispersed through the material, with only the absorbed energy causing the fuel temperature to rise.

18.4.2.2 Flammable gases or liquid vapors, such as those from gasoline, may travel a considerable distance from their original point of release before reaching a competent ignition source. Only under specific conditions will ignition take place, the most important condition being concentration within the flammable limits and an ignition source of sufficient energy located in the flammable mixture.

18.4.3 Oxidant. The oxidant is usually the oxygen in the atmosphere. In some cases alternate or additional oxidants may have been present and the investigator should consider this and the role of such conditions in ignition and spread.

18.4.3.1 If the existence of an oxidant other than atmospheric oxygen is suspected based upon the presence of residue, that residue should be collected and analyzed in a laboratory. Typically the oxidant does not survive in its original form, but may leave characteristic residues.

18.4.4 Ignition Sequence.

18.4.4.1 The ignition sequence of a fire event is defined as the succession of events and conditions that allow the source of ignition, the fuel, and the oxidant to interact in the appropriate quantities and circumstance for combustion to begin. Simply identifying a fuel or an ignition source by itself does not and cannot describe how a fire came to be. Fire results from the interaction of fuel, an oxidant, and an ignition source. Therefore, the investigator should be cautious about deciding on a cause of a fire just because a readily ignitable fuel, potential ignition source, or any other of an ignition sequence's elements is identified. The sequence of events that allow the source of ignition, the fuel, and the oxidant to interact in the appropriate quantities and circumstances for combustion to begin, is essential in establishing the cause.

18.4.4.2 Analyzing the ignition sequence requires determining events and conditions that occurred or were logically necessary to have occurred, in order for the fire to have begun. Additionally, in describing an ignition sequence, the order in which those events occurred should be determined.

18.4.4.2.1 In each fire investigation, the various contributing factors to ignition should be investigated and included in the ultimate explanation of the ignition sequence. These factors should include:

(1) How and sequentially when the first fuel ignited came to be present in the appropriate shape, phase, configuration, and condition to be capable of being ignited (a competent fuel);
(2) How and sequentially when the oxidant came to be present in the right form and quantity to interact with the first fuel ignited and ignition source and allow the combustion reaction;
(3) How and sequentially when the competent ignition source came to be present and interact with the fuel;
(4) How and sequentially when the competent ignition source transferred its heat energy to the fuel, causing ignition;
(5) How safety devices and features designed to prevent fire from occurring or becoming a hostile fire operated or failed to operate. *(See Appliance chapter for additional discussion)*;
(6) How and sequentially when any acts, omissions, outside agencies, or conditions brought the fuel, oxidant, and competent ignition source together at the time and place for ignition to occur;
(7) How the first fuel subsequently ignited any secondary, tertiary, and successive fuels which resulted in any fire spread.

18.4.4.3 There are times when there is no physical evidence of the ignition source found at the origin, but where an ignition sequence can logically be inferred using other data. Any determination of fire cause should be based on evidence rather than on the absence of evidence; however, there are limited circumstances when the ignition source cannot be identified, but the ignition sequence can logically be inferred. This inference may be arrived at through the testing of alternate hypotheses involving potential ignition sequences, provided that the conclusion regarding the remaining ignition sequence is consistent with all known facts *(see Basic Methodology chapter)*. The following are examples of situations that lend themselves to formulating an ignition scenario when the ignition source is not found during the examination of the fire scene. The list is not exclusive and the fire investigator is cautioned not to hypothesize an ignition sequence without data that logically supports the hypothesis.

(A) Diffuse fuel explosions and flash fires.

(B) When an ignitable liquid residue (confirmed by laboratory analysis) is found at one or more locations within the fire scene and its presence at that location(s) does not have an innocent explanation. *(See Incendiary Fires chapter)*.

(C) When there are multiple fires *(See Incendiary Fires chapter)*.

(D) When trailers are observed. *(See Incendiary Fires chapter)*.

(E) The fire was observed or recorded at or near the time of inception or before it spread to a secondary fuel.

18.5 Developing a Cause Hypothesis. The investigator should use the scientific method *(see the Basic Methodology chapter)* as the method for data gathering, hypothesis development, and hypothesis testing regarding the consideration of potential ignition sequences. This process of consideration actually involves the development and testing of alternate hypotheses. In this case, a separate hypothesis is developed considering each individual competent ignition source at the origin as a potential ignition source. Systematic evaluation (hypothesis testing) is then conducted with the elimination of those hypotheses that are not supportable (or refuted) by the facts discovered through further examination. The investigator is cautioned not to eliminate a potential ignition source merely because there is no obvious evidence for it. For example, the investigator should not eliminate the electric heater because there is no arcing in the wires or because the contacts are not stuck. There may be other methods by which the heater could have been the ignition source other than a system failure, such as combustible materials being stored too close to it. Potential ignition sources should be eliminated from consideration only if there is reliable evidence that they could not be the ignition source for the fire. For example, an electric heater can easily be eliminated from consideration if it was not energized.

18.5.1 Devices present at the point/area of origin which are either heat-producing, or are capable of heat production when they sustain a fault or failure (e.g., electrical devices of various kinds) should always be placed on the list of hypotheses, even if, for some reason, they are easy to eliminate.

18.5.2 The investigator should carefully consider potential ignition sources which do not correspond to a physical device that can be recovered. Such potential ignition sources include open flames where the device does not remain (e.g., a cigarette lighter was used, but not left at the scene) and static electricity discharges (including lightning). Given the lack of a physical device, other evidence is needed to establish the presence or absence of an ignition source.

18.5.3 For each potential ignition source in the area of origin, it must be established that there existed a fuel or fuels, in an appropriate form and configuration, for which the potential ignition source could be considered a competent ignition source. A cause hypothesis can be developed even in the absence of being able to state specifically which of these fuels was the first ignited.

18.5.4 There may be multiple competent ignition sources in the area of origin with a known first fuel. A cause hypothesis can be developed in the absence of being able to state specifically which of these competent ignition sources ignited the known first fuel. Where propane leaks into a cellar, the standing pilot on either the water heater or the furnace may have been the ignition source, however post-fire it may not be possible to definitively determine which of the two ignited the gas.

18.6 Testing the Cause Hypothesis. Each of the alternate hypotheses that were developed must then be tested using the Scientific Method. If one remaining hypothesis is tested using the "scientific method" and is determined to be probable, then the cause of the fire is identified.

18.6.1 Scientific Method. Use of the Scientific Method dictates that any hypothesis formed from analysis of the data collected in an investigation must stand the test of careful and serious challenge, by the investigator testing the hypothesis or by examination by others. *[See the Basic Methodology chapter and Daubert v. Merrell Dow Pharmaceuticals, Inc. 509 U.S. 579, 113 S. Ct. 2786 (1993).]*

18.6.2 Deductive Reasoning. Testing of the hypothesis is done by the principle of deductive reasoning, in which the investigator compares the hypothesis to all the known facts as well as the body of scientific knowledge associated with the phenomena relevant to the specific incident. Ultimately, the cause determination is arrived at through the testing of cause hypotheses.

18.6.3 Hypotheses Testing Questions. In testing a cause hypothesis, the following questions should be answered:

(1) Is the hypothesized ignition source a competent ignition source for the first fuel ignited?
(2) Is the required time for ignition consistent with the time line associated with the cause hypothesis and facts of the incident?
(3) What were the circumstances that brought the ignition source in contact with the first fuel ignited?
(4) What, if any, were the failure modes required for ignition to occur?

18.6.4 Means of Hypothesis Testing. When testing a hypothesis, the investigator should attempt to disprove, rather than to confirm, the hypothesis. If the hypothesis cannot be disproved, then it may be accepted as either possible or probable. Hypothesis testing may include: any application of fundamental principles of science, physical experiments or testing, cognitive experiments, analytical techniques and tools, and systems analysis.

18.6.4.1* Scientific Literature. The use of the scientific literature is an important means to develop information that can be used in hypothesis testing. A review of the literature may include descriptions of experiments and testing that can also be applied to the investigator's specific case. "Gateways" to the scientific literature can include Internet databases, technical libraries, textbooks, and handbooks. The validity of the information in the literature should be considered by the investigator.

18.6.4.2 Fundamental Principles of Science. A cause hypothesis is disproved if it violates the fundamental laws of physics or thermodynamics. Water does not burn — a hypothesis positing the ignition of water would be wrong.

18.6.4.3 Physical Experiments or Testing. Experiments can be conducted to test the hypothesized cause. Care must be exercised in developing an experimental protocol that will produce reliable and applicable results for the specific fire or explosion incident. For more information, see the section on Fire Testing in the chapter on Failure Analysis and Analytical Tools.

18.6.4.4 Cognitive Experiments. In a cognitive experiment, one sets up a premise and tests it against the data. An example of a cognitive experiment is, "If it were posited that the door was open during the fire and the hinges were found with mirror image patterns, then the hypothesis would be disproved." For more information see the chapter on Basic Methodology.

18.6.4.5 Time Lines. In the context of testing a cause hypothesis, the time frame may be a discriminator for determining if an ignition scenario is consistent with the available data as it related to time frames.

18.6.4.6 Fault Trees. Fault trees can be used to test the possibility of a hypothesized fire cause. Fault trees are developed by breaking down an event into causal component parts. These components are then placed in a logical sequence of events or conditions necessary to produce the event. If the conditions or sequence are not present then the hypothesis is disproved.

18.6.4.7 Additional Techniques. Additional analytical techniques and tools in the chapter on Failure Analysis and Analytical Tools can be helpful in hypothesis testing.

18.6.5* Inappropriate Use of the Process of Elimination. The process of determining the ignition source for a fire, by eliminating all ignition sources found, known, or believed to have been present in the area of origin, and then claiming such methodology is proof of an ignition source for which there is no evidence of its existence, is referred to by some investigators as "negative corpus." Negative corpus has typically been used in classifying fires as incendiary, although the process has also been used to characterize fires classified as accidental. This process is not consistent with the Scientific Method, is inappropriate, and should not be used because it generates un-testable hypotheses, and may result in incorrect determinations of the ignition source and first fuel ignited. Any hypothesis formulated for the causal factors (e.g., first fuel, ignition source, and ignition sequence), must be based on facts. Those facts are derived from evidence, observations, calculations, experiments, and the laws of science. Speculative information cannot be included in the analysis.

18.6.5.1 Cause Undetermined. In the circumstance where all hypothesized fire causes have been eliminated and the investigator is left with no hypothesis that is evidenced by the facts of the investigation, the only choice for the investigator is to opine that the fire cause, or specific causal factors, remains undetermined. It is improper to base hypotheses on the absence of any supportive evidence *(see 11.5.2, Types of Evidence)*. That is, it is improper to opine a specific ignition source that has no evidence to support it even though all other hypothesized sources were eliminated.

18.6.5.2* Ignition Source vs Fire Cause. The investigator should remember that the cause of a fire is defined as "the circumstances, conditions, or agencies that bring together a fuel, ignition source, and oxidizer (such as air or oxygen) resulting in a fire or a combustion explosion" *(see the Definitions Chapter, Fire Cause)*. The identification of an ignition source and a first fuel is not sufficient to determine a cause. Determining a fire cause and ignition sequence requires that any proposed hypothesis include consideration of the relationship between the competency of the ignition source and the first fuel ignited. The investigator should determine if the proposed ignition source is a competent ignition source for the proposed first fuel ignited *(see 18.4.2, Ignition Source Analysis)*.

18.7 Selecting the Final Hypothesis. Once the hypotheses regarding the "cause" of the fire have been tested, the investigator should review the entire process, to ensure that all credible data are accounted for and all credible alternate cause hypotheses have been considered and eliminated. When using the Scientific Method, the failure to consider alternate hypotheses is a serious error. A critical question to be answered by fire investigators is, "Are there any other cause hypotheses that are consistent with the data?" The investigator should document the facts that support the cause determination to the exclusion of all other reasonable causes.

18.7.1 Establishing the Cause. Although cause is common terminology, the investigator should describe it in terms of the competent ignition source providing enough heat to ignite the first fuel, and the circumstances of how they came together. The fuels involved after the first fuel should be noted, this may be especially true when the first fuel is part of the source, such as an appliance. In such a case the subsequent fuels may be the combustibles that are located near the appliance where the fire originated.

18.7.2 Inconsistent Data. It is unusual for all data items to be totally consistent with the selected hypothesis. Each piece of data should be analyzed for its reliability and value. Not all data in an analysis has the same value. Frequently, some analysis or witness statement will provide data that appears to be inconsistent. Contradictory data should be recognized and resolved. Incomplete data may make this difficult or impossible. If resolution is not possible, then the cause hypothesis should be re-evaluated.

18.7.3 Safety Devices and Features. Safety devices and features are often engineered and built to prevent fires from occurring or becoming a hostile fire. The cause determination will need to account for the actions of safety devices.

18.7.4 Undetermined Fire Cause. The final opinion is only as good as the quality of the data used in reaching that opinion. If the level of certainty of the opinion is only "possible" or "suspected," the fire cause is unresolved and should be classified as "undetermined." This decision as to the level of certainty in data collected in the investigation or of any hypothesis drawn from an analysis of the data rests with the investigator.

Chapter 19 Analyzing the Incident for Cause and Responsibility

19.1* General.

19.1.1 The purpose of fire and explosion investigations is often much broader than just determining the cause of a fire or explosion incident. The goal of any particular fire investigation is to come to a correct conclusion about the features of a particular fire or explosion incident that resulted in death, injury, damage, or other unwanted outcome. The features can be grouped under the following four headings:

(1) *The cause of the fire or explosion.* This feature involves a consideration of the circumstances, conditions, or agencies that bring together a fuel, ignition source, and oxidizer (such as air or oxygen), resulting in a fire or a combustion explosion.
(2) *The cause of damage to property resulting from the incident.* This feature involves a consideration of those factors that were responsible for the spread of the fire and for the extent of the loss, including the adequacy of fire protection, the sufficiency of building construction, and the contribution of any products to flame spread and to smoke propagation.
(3) *The cause of bodily injury or loss of life.* This feature addresses life safety components such as the adequacy of alarm systems, sufficiency of means of egress or in-place protective confinement, the role of materials that emit toxic by-products that endanger human life, and the reason for fire fighter injuries or fatalities.
(4) *The degree to which human fault contributed to any one or more of the causal issues described in (1), (2), and (3).* This feature deals with the human factor in the cause or spread of fire or in bodily injury and loss of life. It encompasses acts and omissions that contribute to a loss (responsibility), such as incendiarism and negligence.

19.1.2 The cause of a fire or the causes of damage or casualties may be grouped in broad categories for general discussion, for assignment of legal responsibility or culpability, or for reporting purposes. Local, state, or federal reporting systems or legal systems may have alternative definitions that should be applied as required.

19.2 The Cause of the Fire or Explosion. The determination of the cause of a fire requires the identification of those circumstances and factors that were necessary for the fire to have occurred. Those circumstances and factors include, but are not limited to, the device or equipment involved in the ignition, the presence of a competent ignition source, the type and form of the material first ignited, and the circumstances or human actions that allowed the factors to come together to allow the fire to occur. An individual investigator may not have responsibility for, or be required to address, all of these issues.

19.2.1 Classification of the Cause. Classification of a fire cause may be used for assignment of responsibility *(see Section 18.6)*, reporting purposes, or compilation of statistics. Different jurisdictions may have alternative definitions that should be applied as required. The cause of a fire may be classified as accidental, natural, incendiary, or undetermined. Use of the term *suspicious* is not an accurate description of a fire cause. Suspicion refers to a level of proof, or level of certainty, and is not a classification for a fire cause. Suspicion is not an acceptable level of proof for making a determination of cause within the scope of this guide and should be avoided. Fires in which the level of certainty is possible or suspected, or in which there is only suspicion of that cause, should be classified as undetermined. Determining the cause of a fire and classifying the cause of the fire are two separate processes that should not be confused with each other. *(See Section 18.6.)*

19.2.1.1 Accidental Fire Cause. Accidental fires involve all those for which the proven cause does not involve an intentional human act to ignite or spread fire into an area where the fire should not be. When the intent of the person's action cannot be determined or proven to an acceptable level of certainty, the correct classification is undetermined. *(See Section 18.6.)* In most cases, this classification will be clear, but some deliberately ignited fires can still be accidental. For example, in a legal setting, a trash fire might be spread by a sudden gust of wind. The spread of fire was accidental even though the initial fire was deliberate.

19.2.1.2 Natural Fire Cause. Natural fire causes involve fires caused without direct human intervention or action, such as fires resulting from lightning, earthquake, and wind.

19.2.1.3 Incendiary Fire Cause. The incendiary fire is one intentionally ignited under circumstances in which the person igniting the fire knows the fire should not be ignited. When the intent of the person's action cannot be determined or proven to an acceptable level of certainty, the correct classification is undetermined. *(See Section 18.6.)*

19.2.1.4 Undetermined Fire Cause. Whenever the cause cannot be proven to an acceptable level of certainty, the proper classification is undetermined. *(See Section 18.6.)*

(A) Undetermined fire causes include those fires that have not yet been investigated or those that have been investigated, or are under investigation, and have insufficient information to classify further. However, the fire might still be under investigation and the cause may be determined later with the introduction or discovery of new information.

(B) In the instance in which the investigator fails to identify the ignition source, the fire need not always be classified as undetermined *(see 18.6.5.1)*. If the physical evidence established one factor, such as the use of an accelerant, that evidence may be sufficient to establish an incendiary fire cause classification even where other factors such as ignition source cannot be identified. Determinations in the absence of physical evidence of an ignition source may be more difficult to substantiate. Therefore, investigators should strive to remain objective throughout the investigation.

19.3 The Cause of Damage to Property Resulting from the Incident.

19.3.1 The following are considerations that may be utilized in establishing cause of property damage. These factors are divided into two major categories: fire and smoke spread damage and other types of damage.

19.3.2 Fire/Smoke Spread. Elements of damage caused by fire or its products of combustion can further be affected by the following conditions:

(1) *Compartmentation.* Effectiveness or failure of confining the fire and smoke by methods of construction or specific passive fire protection assemblies.
(2) *Change of occupancy/hazard.* Change from the original design and use from a lower hazard to a greater hazard without appropriate changes to fire protection, structural, or means of egress features.
(3) *Detection/alarm systems.* Failure to provide timely and effective notice of a fire, resulting from a delay in detection or a delay in notification.
(4) *Human behavior.* This includes intentional or unintentional acts or omissions by people.
(5) *Fire suppression.* Failure of the building fire suppression systems to properly control or mitigate the fire and fire suppression activities may also be factors in fire development or spread.
(6) *Fuel loads.* The type, amount, and configuration of fuels affect fire development and spread.
(7) *Housekeeping.* Poor housekeeping can contribute to fire damage by providing a more easily ignited fuel configuration and may allow more rapid fire/smoke spread. Poor housekeeping can also obstruct access of fire fighters during suppression activities.
(8) *Ventilation.* Fire department ventilation operations, HVAC, and open windows and doors may affect fire growth. Ventilation can also cause smoke and hot gases to move within or from compartments.
(9) *Code violations.* Violations of fire safety codes and standards can increase or cause damage (e.g., leaving fire doors open or penetrations of fire walls).
(10) *Structural failure.* Failure of building components or systems (utility, fire protection, compartmentation, etc.) can increase damage by allowing or causing fire and products of combustion to spread, providing additional fuels, or impeding fire suppression.

19.3.3 Other Consequential Damage. Other consequential damage can include loss of utilities, subsequent weather damage, corrosion, contamination, water damage, mold damage, or theft.

19.4 The Cause of Bodily Injury or Loss of Life. *(See Chapter 10 and Chapter 23.)*

19.4.1 Fire/Smoke Spread. The following are considerations that may be utilized in establishing the fire related cause of deaths or injuries:

(1) *Toxicity.* Deaths or injuries resulting from exposure to products of combustion.
(2) *Hazardous materials.* Deaths or injuries resulting from exposure to hazardous materials released as a result of the fire or explosion incident not directly related to products of combustion.
(3) *Compartmentation.* Casualties resulting from improper design or poor performance of compartmentalization features.
(4) *Change of occupancy/hazard.* Change from the original design and use from a lower hazard to a greater hazard without appropriate changes to fire protection, structural, or means of egress features.
(5) *Detection/alarm systems.* Failure to provide timely and effective notice of a fire, resulting from a delay in detection or a delay in notification, or ineffective notification.

(6) *Human behavior.* Human behavior resulting in casualties may include failure to react, inappropriate reaction to notification, or delay in evacuation *(see Chapter 10).*
(7) *Fire suppression.* Failure of the building fire suppression systems to properly control or mitigate the fire and fire suppression activities may also be factors in fire injuries or deaths.
(8) *Housekeeping.* Poor housekeeping can contribute to fire injuries by providing a more easily ignited fuel configuration and may allow for more rapid fire/smoke spread. Poor housekeeping can also obstruct access of fire fighters during suppression activities or egress of occupants.
(9) *Fuel loads.* The type, amount and configuration of fuels affect fire development and spread.
(10) *Ventilation.* Fire department ventilation operations, HVAC, and open windows and doors may affect fire growth. Ventilation can also cause smoke and hazardous gases to move within structures, endangering occupants.
(11) *Code violations.* Violations of fire safety codes and standards can increase or cause hazards (e.g., exceeding occupancy limits).
(12) *Means of egress/refuge.* The inability of occupants to escape or find refuge as a result of improper design, installation, maintenance, operation, or non–code compliance may increase the potential for injuries or death.
(13) *Structural failure.* Failure of building components or systems (utility, fire protection, compartmentation, etc.) can increase danger by allowing or causing fire and products of combustion to spread, providing additional fuels, or impeding fire suppression. Structural failure or collapse itself can cause injuries or death.
(14) *Intentional acts.* Many of the above conditions can be the result of intentional acts, with or without the intent to cause injury or death.

19.4.2 Emergency Preparedness. Failure to properly prepare, test, and implement an appropriate emergency plan can result in casualties.

19.5 Determining Responsibility. After determining the origin, cause, and development of a fire or explosion incident, the fire investigator may be required to do a failure analysis and to determine responsibility. It is only through the determination of such responsibility for the fire that remedial codes and standards, fire safety, or civil or criminal litigation actions can be undertaken.

19.5.1 Nature of Responsibility. The nature of responsibility in a fire or explosion incident may be in the form of an act or omission. It may be something that was done, accidentally or intentionally, that ultimately brought about the fire or explosion, or it may be some failure to act to correct or prevent a condition that caused the incident, fire/smoke spread, injuries, or damage. Responsibility may be attributed to a fire or explosion event notwithstanding the classification of the fire cause: natural, accidental, incendiary, or undetermined. Responsibility may be attributed to the accountable person or other entity because of negligence, reckless conduct, product liability, arson, violations of codes or standards, or other means.

19.5.2 Definition of Responsibility. Responsibility for a fire or explosion incident is the accountability of a person or other entity for the event or sequence of events that caused the fire or explosion, spread of the fire, bodily injuries, loss of life, or property damage.

19.5.3 Assessing Responsibility. While it is frequently a court's role to affix a final finding of responsibility and to assign liability, remedial measures, compensation, or punishment, it is the role of the person who performs the analysis to identify responsibility so that fire safety, code enforcement, or litigation processes can be undertaken.

19.5.4 Degrees of Responsibility. A series or sequence of events or conditions often causes a fire or explosion and the resulting spread, injuries, and damage. A failure analysis often shows that a change to any one or more of these conditions, acts, or omissions could have prevented or mitigated the incident. In this way, responsibility may fall on more than one person or entity. In such a case, multiple or various degrees of responsibility may be assessed.

Chapter 20 Failure Analysis and Analytical Tools

20.1* Introduction. This chapter identifies methods available to assist the investigator in the analysis of a fire/explosion incident. Additional tools requiring special expertise are also discussed. These methods can be used to analyze fires of any size or complexity. In many cases, the methods are used to organize information collected during the documentation of the incident into a rational and logical format. They can also be used to identify aspects of the investigation needing additional information and where future efforts should be directed.

20.2 Time Lines.

20.2.1 General. A time line is a graphic or narrative representation of events related to the fire incident, arranged in chronological order.

20.2.1.1 The events included in the time line may occur before, during, or after the fire incident. This investigative tool can show relationships between events, identify gaps or inconsistencies in information and sources, assist in witness interviews, and otherwise assist in the analysis and investigation of the incident. A graphic time line is useful as a demonstrative document. The value of a time line is dependent upon the accuracy of the information used to develop the time line.

20.2.1.2 Estimates of fire size or fire conditions are frequently valuable in developing time lines. Using the tools of fire dynamics analysis *(see 20.4.8)*, fire conditions can be related to specific events. If there are sufficient events it may be possible to develop an estimate of the heat release history for at least the early stages of a fire.

20.2.1.2.1 For example, the observed height of flames, relative to the height of known objects, can be used to estimate the rate of heat release of the fire. Given the response characteristics of detectors and sprinklers, the size of a fire at the time of operation of such devices can be estimated. If the time of operation is recorded at an alarm panel or remote location such as a central alarm service, the estimated size of the fire in the area of the building where the alarm occurred can become part of the time line. Where the detection or suppression systems have multiple zones, the time of operation in each zone can be used to track the spread of the fire through a building and the events can be added to the time line. If the heat release can be estimated for several points in time, a possible heat release history may be postulated and used as one means to assist in testing various hypotheses for the cause and growth scenarios in a given fire.

20.2.1.2.2 Fire dynamics analysis can also be used to provide estimated times for relevant events to occur where limited eyewitness observations or hard times are available. Such events include ignition of additional fuels, detector activation, flashover, window breakage, fire spread to adjacent compartments, and occupant incapacitation and death. These analysis tools have acknowledged limitations, and the associated input data are subject to uncertainties. Therefore, estimates of fire conditions and related events in the time line may require to be described as a time interval (e.g., flashover between 10:46 and 10:48) as opposed to a single and specific time (e.g., 10:46).

20.2.1.3 In order to construct a time line, it is necessary to relate events or activities to the time of their occurrence. In assigning time to events or activities, it is important to identify the confidence the investigator has in the assigned time. One means of doing this is to identify the quality of the data as hard time (actual) or soft time (estimated or relative).

20.2.2 Hard Time (Actual).

20.2.2.1 Hard time identifies a specific point in time that is directly or indirectly linked to a reliable clock or timing device of known accuracy. It is possible to have a time line with no hard times. Hard times can be obtained from sources such as the following:

(1) Fire department dispatch telephone or radio logs
(2) Police department dispatch and radio logs
(3) Emergency Medical Service reports
(4) Alarm system records (on-site, central station, fire dispatch, etc.)
(5) Building inspection report(s)
(6) Health inspection report(s)
(7) Fire inspection report(s)
(8) Utility company records (maintenance/emergency/repair records)
(9) Private videos/photos (check with local film developers)
(10) Media coverage (newspaper photographer, radio, television, magazines)
(11) Timers (clocks, time clocks, security timers, water softeners, lawn sprinkler systems)
(12) Weather reports (NOAA, airports, lightning tracking services)
(13) Current and/or prior owner/tenant records (re: maintenance)
(14) Interviews
(15) Computer-based fire department alarms, communications audio tapes, and transcripts
(16) Building or systems installation permits

20.2.2.2 All clocks and timing devices are usually not synchronized. Discrepancies between different clocks should be recorded and adjustments made where necessary.

20.2.3 Soft Time (Estimated).

20.2.3.1 Soft time can be either estimated or relative time. *Relative time* is the chronological order of events or activities that can be identified in relation to other events or activities. *Estimated time* is an approximation based on information or calculations that may or may not be relative to other events or activities. Often, relative or estimated times can be determined within a known degree of accuracy. For example, they may be bound by two known events or within a time range. It may be desirable to report them as a time range rather than as discrete time.

20.2.3.2 Relative time can be very subjective in nature. The concept of elapsed time varies with the individual and the stress caused by the incident. It is important for witnesses to be as specific as possible by having them refer to their actions and observations in relation to each other and to other events. All relative time is based on an estimate. It is also possible to have events for which the estimated time cannot be related to a hard time but that are valuable to the analysis. These are referred to as *estimated times*. Relative or estimated times are generally provided by witnesses.

20.2.3.3 Potential sources of soft times include those sources for hard times listed in 20.2.2.1, along with estimation of times for an activity to be performed or an event to occur.

20.2.4 Benchmark Events. Some events are particularly valuable as a foundation for the time line or may have significant relation to the cause, spread, detection, or extinguishment of a fire. These are referred to as *benchmark events*. An example of a benchmark event could be the dispatch and arrival times of the fire fighters as recorded on the fire department incident report. Other examples may include events such as a roof collapsing, a window breaking out, or an explosion.

20.2.5 Multiple Time Lines. It is quite possible that two or more time lines will be required to effectively evaluate and document the sequence of events precipitating the fire, the actual fire incident, and post-fire activity. These time lines can be called *macro* and *micro*.

20.2.5.1 A macro evaluation of events may incorporate activity that occurred months before the fire and that terminated on the demolition of the building. As an example, this activity might include renovations that altered the building's electrical system and that may be attributable as the ignition source.

20.2.5.2 A micro evaluation of events focuses on some discrete segment of the total time line for which the investigator has a particular interest. For example, it may consist of an evaluation of events during the time period immediately prior to ignition, during initial fire fighting, during fire growth, or from ignition to extinguishment.

20.2.5.3 Parallel time lines can be presented to demonstrate two or more series of events. The purpose of such a presentation may be to show whether or not they are related in some manner.

20.2.5.4 Various tools are available to assist in the development of time lines. Although a simple time line can be constructed with pencil and paper, there are software packages available as well, from simple word processing or database, to sophisticated scheduling software. See Figure 20.2.5.4 for an example of a simple chronology time line that does not identify hard, soft, or benchmark times and does not graphically display the temporal relationships between events.

20.2.5.5 Scaled Time Line. A scaled time line displays a list of events that appear on the time line in both chronological and elapsed time relationship to each other. Individual events that are closer together to each other in time are drawn physically closer together on the time line, while events that are further apart in time are further apart on the

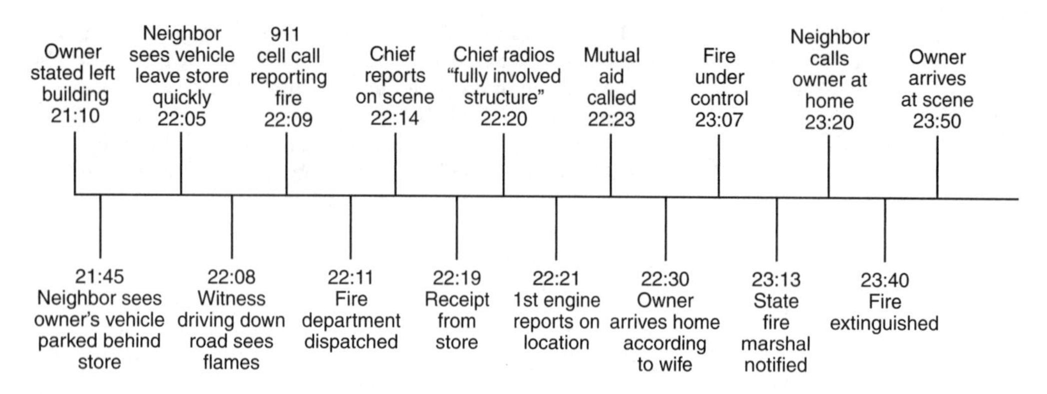

FIGURE 20.2.5.4 Illustration of a Simple Chronology Time Line.

time line and there is a definite scale or ratio of time to distance along the length of the line. See Figure 20.2.5.5. In addition, Figure 20.2.5.5 identifies hard and soft times by listing hard times above the time line and soft or estimated times below. Hard times in Figure 20.2.5.5 include when John Doe punched the time clock at work and the reported alarm and arrival times of the fire department, all of which can be correlated to known clocks. Soft times listed in Figure 20.2.5.5 include estimated times for ignition, smoke alarm activation, and flashover derived from computer modeling, and the estimated time that the eyewitness first became aware of the fire.

20.3 Systems Analysis. Systems analysis techniques are important tools in identifying when and how engineering analysis and modeling may be useful. These techniques, developed for use in system safety analyses, include failure modes and effects analysis, fault tree analysis, HAZOP analysis, and what-if analysis. These tools provide a systematic method for analyzing systems to determine hazards or faults. The tools can utilize either qualitative or quantitative formats. Hazard probabilities or failure rates can be factored in when using quantitative formats. Some of the more common techniques — fault tree analysis and failure mode and effects analysis — are described in 20.3.1 through 20.3.2. Several other systems analyses are available, each with its inherent advantages and limitations.

20.3.1* Fault Trees. A fault tree is a logic diagram that can be used to analyze a fire or explosion. A fault tree is developed using deductive reasoning. The diagram places, in logical sequence and position, the conditions and chains of events that are necessary for a given fire or explosion to occur.

20.3.1.1 Fault trees can be used to test the possibility of a proposed fire cause or spread scenario and to identify or evaluate possible alternative scenarios. Fault trees are developed by breaking down an undesired event into its causal elements or component parts. The components are then placed in logical sequences of events or conditions necessary to produce the fire or explosion, or into categories of specific aspect of associated damage, death, or injury. If the conditions are not present or if the events did not occur in the necessary sequence, then the proposed scenario is not possible. For example, if the proposed scenario required a live electrical circuit and there was no electrical service, the scenario would be incorrect unless an alternative source for the electricity could

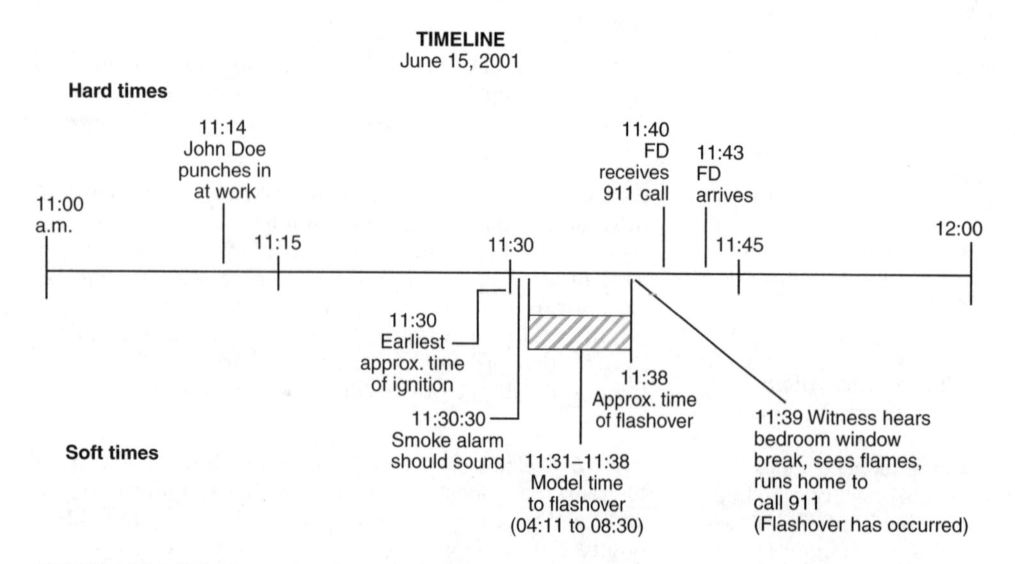

FIGURE 20.2.5.5 Example of a Scaled Time Line (displaying hard times above the line and soft, or estimated, times below).

be shown. The logic for evaluating the events and conditions that control undesired events is represented by "and" decisions and "or" decisions. In a graphic representation of a fault tree, these decision points are called *gates*. In most cases, fault trees involve combinations of "and" gates and "or" gates, as shown in Figure 20.3.1.1.

20.3.1.2 For an "and" controlled event to occur, all the elements and conditions must be present. An example using an "and" gate is the set of conditions that must be present for a flashlight to work and produce light. There must be good batteries, the bulb must be good, and the switch must work to produce light. A fault tree for this process is shown in Figure 20.3.1.2.

20.3.1.3 For an "or" controlled event, any one of several series of elements and conditions may result in the subject event. An example using an "or" gate would be a flashlight that does not work when the switch is operated. The failure might be due to a switch failure, a blown bulb, or battery problems. Figure 20.3.1.3 shows the fault tree for this example.

20.3.1.4 Fault tree analysis may be used to estimate the probability of an undesired event by assigning probabilities to the conditions and events. Assigning reliable probabilities to events or conditions is often difficult and may not be possible.

20.3.1.5 All system components, their relationships, and the validity of data used need to be identified. In order to construct a fault tree properly, it may be necessary to consult people with special expertise regarding the equipment, materials, or processes involved.

20.3.1.6 The fault tree method of analysis may produce multiple feasible scenarios for a given undesirable event. As a result of insufficient data, it may not be possible to establish which scenario is most likely.

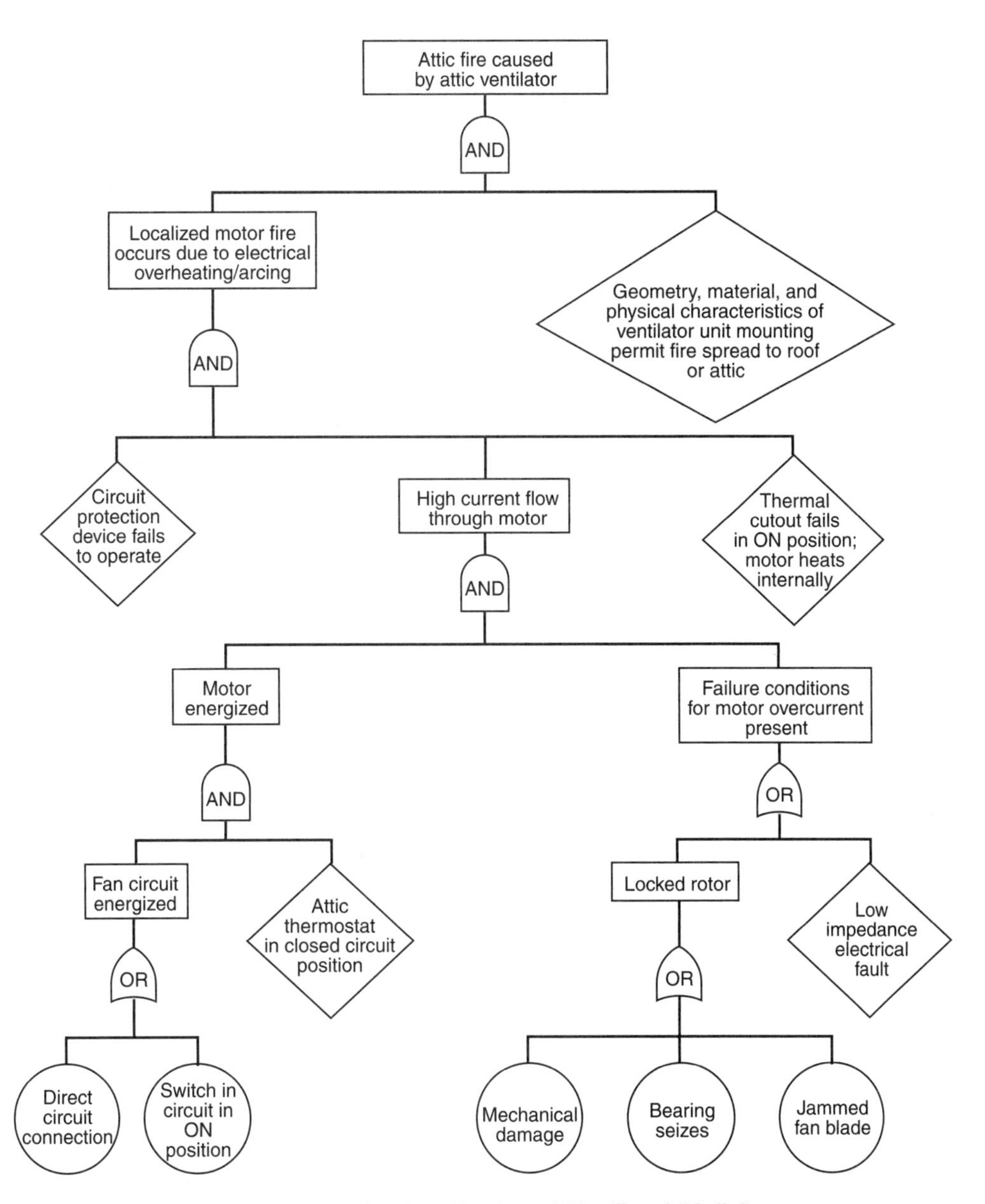

FIGURE 20.3.1.1 Fault Tree Showing Combination of "And" and "Or" Gates.

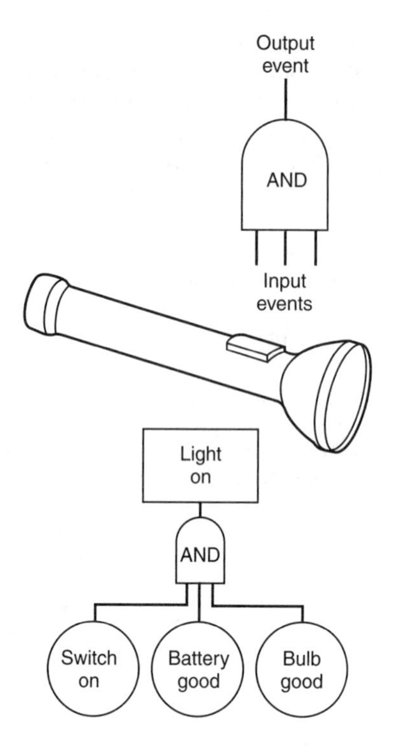

FIGURE 20.3.1.2 Example of Fault Tree Showing "And" Gate.

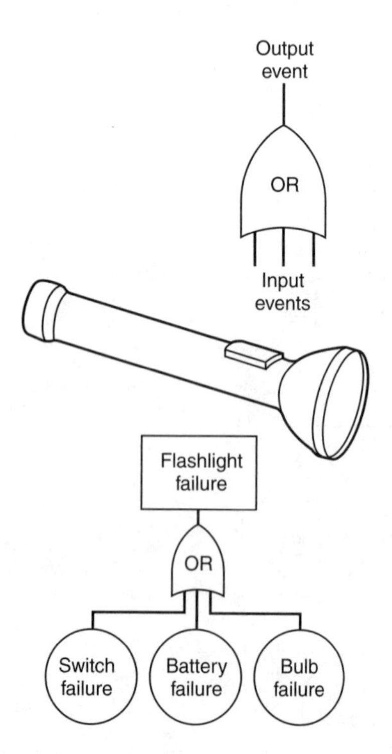

FIGURE 20.3.1.3 Example of Fault Tree Showing "Or" Gate.

20.3.1.7 Suggested sources for data to be used in fault tree analysis include the following:

(1) Operations and maintenance manuals
(2) Maintenance records
(3) Parts replacement and repair records
(4) Design documents
(5) Services of expert with knowledge of system
(6) Examination and testing of exemplar equipment or materials
(7) Component reliability databases
(8) Building plans and specifications
(9) Fire department reports
(10) Incident scene documentation
(11) Witness statements
(12) Medical records of victims
(13) Human behavior information

20.3.1.8 Fault trees are constructed using a standard format familiar to the technical community. Software is available for assisting the user in developing and analyzing fault trees.

20.3.2 Failure Mode and Effects Analysis (FMEA). FMEA is a technique used to identify basic sources of failure within a system, and to follow the consequences of these failures in a systematic fashion. In fire/explosion investigations, FMEA is a systematic evaluation of all equipment and/or actions that could have contributed to the cause of an incident. FMEA is prepared by filling in a table with column headings such as are shown in Figure 20.3.2. The column headings and format of the table are flexible, but at least the following three items are common:

(1) Item (or action) being analyzed
(2) Basic fault (failure) or error that created the hazard
(3) Consequence of the failure

20.3.2.1 FMEA can help identify potential causes of a fire or explosion and can indicate where further analysis could be beneficial. FMEA is particularly useful in a large or complex incident. It can be effective in identifying factors, both physical and human, that could have contributed to the cause of the fire/explosion. Similarly, it can be helpful in eliminating potential causes of a fire/explosion.

20.3.2.2 Additional columns are added by the investigator as appropriate, to address the needs of the particular investigation. An assessment of the likelihood of each individual failure mode is frequently included. It is helpful to assess the consequence of a given failure relative to the fire/explosion. FMEA tables can be cataloged by item and can serve as reference material for further investigations. FMEA tables can be developed using computer spreadsheets or specialized software.

20.3.2.3 When filling out the table, the investigator should consider the range of environmental conditions and the process status (i.e., normal operation, shutdown, and startup) for each item or action. Probabilities or degrees of likelihood can be assigned to each occurrence. When a sequence of failures is required for the incident to occur, the probabilities or degrees of likelihood can be combined to assess the likelihood that any given sequence of events led to the incident.

20.3.2.4 All known system components and human actions that may have contributed to the incident need to be identified. The accuracy of the determination of the sequence of the events is dependent on the accuracy assigned to each of the individual failure modes.

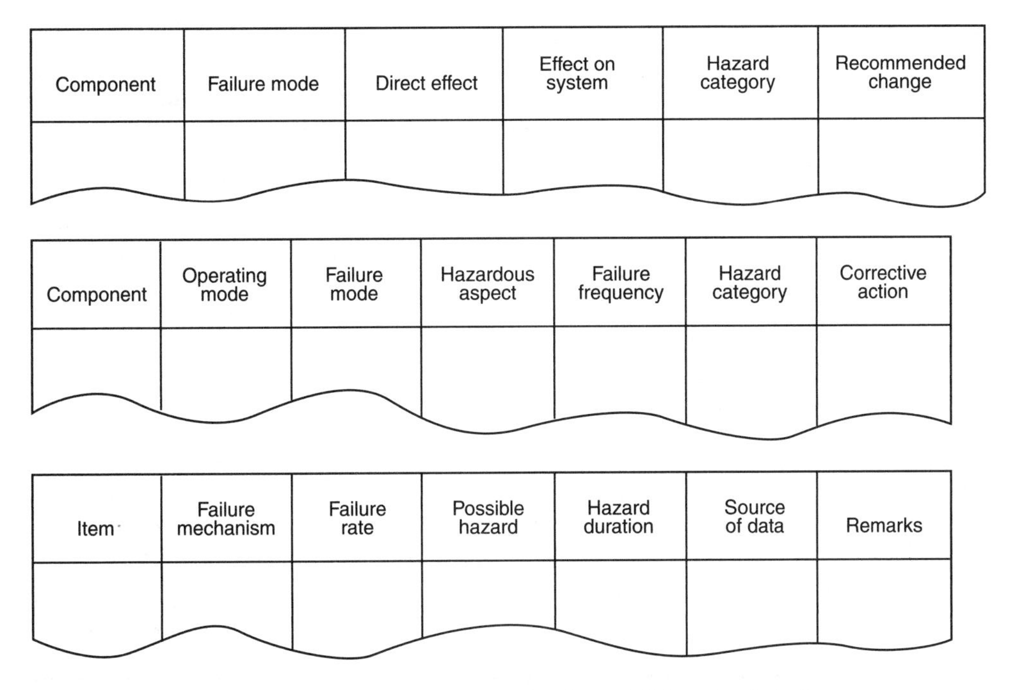

FIGURE 20.3.2 Simplified Examples of Failure Mode and Effects Analysis Forms.

20.3.2.5 The data required for an FMEA depend on the extent of the analysis desired. Minimum information typically includes a list of all system components and human actions that may have led to the incident, possible failure modes for each component and action, and the immediate consequences of each failure. It is important to recognize that many system components will have more than one failure mode, so each possible failure mode and its particular consequences should be listed for each component or action.

20.3.2.6 Data for systems and components can be obtained from many sources, including the following:

(1) Operations and maintenance manuals
(2) Maintenance records
(3) Parts replacement and repair records
(4) Design documents
(5) Services of expert with knowledge of system
(6) Examination and testing of exemplar equipment or materials
(7) Component reliability databases
(8) Building plans and specifications
(9) Fire department reports
(10) Incident scene documentation
(11) Witness statements
(12) Medical records of victims
(13) Human behavior information

20.3.2.7 Table 20.3.2.7 shows a hypothetical example of an FMEA applied to a particular fire scenario, in the determination of a cause of that fire.

20.4 Mathematical Modeling.

20.4.1 General. Mathematical modeling techniques provide the investigator with tools for testing hypotheses regarding the origin and cause of the fire/explosion and the cause of the resulting damage to property or injury to people. Even when the origin and cause are not issues, it is often possible and important to establish the cause of the resulting damage to property or injury to people.

20.4.1.1 The scope of this discussion emphasizes models and analyses that can be exercised using hand or computer-aided calculations. Usage of these analytical tools depends on the scope of the investigator's assignment, the particular incident, and the practical purpose of the investigation. A special expert may be needed to complete the analysis.

20.4.1.2 Mathematical models are intended to simulate or predict real-world phenomena using scientific principles and empirical data. There are numerous fields and specialty disciplines that use models. Some that have proven useful in fire and explosion investigations are discussed in 20.4.2 through 20.4.8.

20.4.1.3 Limitations of Mathematical Modeling. Mathematical modeling, whether simplified hand calculations or computer fire models, has inherent limitations and assumptions that should be considered. Models generally rely upon empirical data and are validated via comparison with other empirical data. Care must be taken to assure that the model is being used with due regard for limitations, assumptions, and validation. While computational models can be used to test hypotheses, models should not be utilized as the sole basis of a fire origin and cause determination.

20.4.1.4 In the selection of a mathematical model for use in hypothesis testing, the scope, applicability, basis, and validation and verification of the model should be considered. Models selected should be known to be capable of addressing the technical issues posed in the hypothesis testing. Use of proprietary software may create issues that will need to be addressed, with respect to the ability of other parties to examine the results and use the software.

20.4.1.5 Inputs to mathematical models are subject to uncertainties that should be considered in the evaluation of model results. The effect of input uncertainties on model results should be assessed through the use of sensitivity analysis. Uncertainties in model inputs may be significantly increased if standardized

Table 20.3.2.7 Sample Failure Mode and Effects Analysis for Lunchroom Fire

Component Item	Failure Mode	Cause of Failure	Effects of Failure	Hazard Created	Necessary Conditions	Indication of Failure
Coffee maker	Heater current flows without shutoff	Switch left on and controls fail	"Boils" out any water in reservoir; thermal runaway of heating element; local temperature increases above 600°C	Ignition of plastic housing	Power on; switch on or fails closed; thermostat fails in ON position; both thermal fuses fail to open	Melting of aluminum housing around heating element; condensed aluminum at base of maker; thermostat closed circuit; both fuses closed circuit
Range (electric)	Autoignition of cooking oil	Unattended cooking Control failure	Oil temperature raised above autoignition temperature	Burning oil fire and large amount of smoke	Unit on; switch on or fails in closed position; no temperature regulation	Burner control in ON position; melted aluminum pan; oil consumed or spilled on unit; contacts fused or welded

Note: The data and conclusions presented in this table are hypothetical and used for example purposes only.

methods for input determination have not been developed in conjunction with the selected model. Other sources of uncertainty in inputs can result from the use of generic data from the fire science literature or use of exemplar materials for experimental determination of model inputs.

20.4.1.6* Results of mathematical models are subject to uncertainties resulting from approximations made within the model. The uncertainties introduced by modeling approximations should be considered in hypothesis testing. Information on modeling uncertainties is typically included in validation and verification (V & V) documents. Additional comparisons of model results with relevant existing experimental results may be useful to further establish the V & V basis of the model for use in the investigation.

20.4.1.7 Input files and output files from computer models should be retained in their original form as part of the investigative record. When input and output files are provided to other parties, the files should be provided electronically in their original form so that they can be used and examined via software intended for that purpose. Files should not be provided as scans or facsimiles, where the ability to use or examine the results via software is limited. As with all discovery issues, parties are free to litigate the reasonableness of the scope of discovery, cost sharing, the burden of production, and the protection of proprietary or trade secrets. Courts can fashion remedies that accommodate and protect the interests of all parties.

20.4.2 Heat Transfer Analysis.

20.4.2.1 Heat transfer models allow quantitative analysis of conduction, convection, and radiation in fire scenarios. These models are then used to test hypotheses regarding fire causation, fire spread, and resultant damage to property and injury to people. Heat transfer models are often incorporated into other models, including structural and fire dynamics analysis. Various general texts on heat transfer analysis are available.

20.4.2.2 Heat transfer models and analyses can be used to evaluate various hypotheses, including those relating to the following:

(1) Competency of ignition source *(See Section 18.3.)*
(2) Damage or ignition to adjacent building(s)
(3) Ignition of secondary fuel items
(4) Thermal transmission through building elements

20.4.3 Flammable Gas Concentrations. Models can be used to calculate gas concentrations as a function of time and elevation in the space and can assist in identifying ignition sources. Flammable gas concentration modeling, combined with an evaluation of explosion or fire damage and the location of possible ignition sources, can be used (a) to establish whether or not a suspected or alleged leak could have been the cause of an explosion or fire, and (b) to determine what source(s) of gas or fuel vapor were consistent with the explosion or fire scenario, damage, and possible ignition sources.

20.4.4 Hydraulic Analysis.

20.4.4.1 Analysis of automatic sprinkler and water supply systems is often required in the evaluation of the cause of loss. The same mathematical models and computer codes used to design these systems can be used in loss analysis. However, the methods of application are different for design than they are for forensic analysis.

20.4.4.2 A common application of hydraulic analysis is to determine why a sprinkler system did not control a fire. Modeling can also be used to investigate the loss associated with a single sprinkler head opening, to investigate the effect of fouling in the piping, and to determine the effect of valve position on system performance at the time of loss. There are also models and methods available to analyze flow through systems other than water-based systems, such as carbon dioxide, gaseous suppression agents, dry chemicals, and fuels.

20.4.5 Thermodynamic Chemical Equilibrium Analysis. Fires and explosions believed to be caused by reactions of known or suspected chemical mixtures can be investigated by a thermodynamics analysis of the probable chemical mixtures and potential contaminants.

20.4.5.1 Thermodynamic chemical equilibrium analysis can be used to evaluate various hypotheses, including those relating to the following:

(1) Reaction(s) that could have caused the fire/explosion
(2) Improper mixture of chemicals
(3) Role of contamination
(4) Role of ambient conditions
(5) Potential of a chemical or chemical mixture to overheat
(6) Potential for a chemical or chemical mixture to produce flammable vapors or gases
(7) Role of human action on process failures

20.4.5.2 Thermodynamic reaction equilibrium analysis traditionally required tedious hand calculations. Currently available computer programs make this analysis much easier to perform. The computer programs typically require several material properties as inputs, including chemical formula, mass, density, entropy, and heat of formation.

20.4.5.3 Chemical reactions that are shown not to be favored by thermodynamics can be eliminated from consideration as the cause of a fire. Thermodynamically favored reactions must be further analyzed to determine whether the kinetic rate of the considered reactions is fast enough to have caused ignition, given the particular circumstances of the fire.

20.4.6 Structural Analysis. Structural analysis techniques can be utilized to determine reasons for structural failure or change during a fire or explosion. Numerous references can be found in engineering libraries, addressing matters such as strength of materials, formulas for simple structural elements, and structural analysis of assemblies.

20.4.7* Egress Analysis. The failure of occupants to escape may be one of the critical issues that an investigator needs to address. Egress models can be utilized to analyze movement of occupants under fire conditions. Integrating egress models with a fire dynamics model is often necessary to evaluate the effect of the fire environment on the occupants. See Section 10.3 on human factors.

20.4.8* Fire Dynamics Analysis. Fire dynamics analyses consist of mathematical equations derived from fundamental scientific principles or from empirical data. They range from simple algebraic equations to computer models incorporating many individual fire dynamics equations. Fire dynamics analysis can be used to predict fire phenomena and characteristics of the environment such as the following:

(1) Time to flashover
(2) Gas temperatures
(3) Gas concentrations (oxygen, carbon monoxide, carbon dioxide, and others)
(4) Smoke concentrations
(5) Flow rates of smoke, gases, and unburned fuel
(6) Temperatures of the walls, ceiling, and floor
(7) Time of activation of smoke detectors, heat detectors, and sprinklers
(8) Effects of opening or closing doors, breakage of windows, or other physical events

20.4.8.1 Fire dynamics analyses can be used to evaluate hypotheses regarding fire origin and fire development. The analyses use building data and fire dynamics principles and data to predict the environment created by the fire under a proposed hypothesis. The results can be compared to physical and eyewitness evidence to support or refute the hypothesis.

20.4.8.2 Building, contents, and fire dynamics data are subject to uncertainties. The effects of these uncertainties should be assessed through a sensitivity analysis and should be incorporated in hypothesis testing. Uncertainties may include the condition of openings (open or closed), the fire load characteristics, HVAC flow rates, and the heat release rate of the fuel packages. See Section 20.6 for recommended data-collection procedures.

20.4.8.3 Fire dynamics analyses can generally be classified into three categories: specialized fire dynamic analyses, zone models, and field models. They are listed in order of increasing complexity and required computational power.

20.4.8.3.1 Specialized Fire Dynamics Routines. Specialized fire dynamics routines are simplified procedures designed to solve a single, narrowly focused question. In many cases, these routines can answer questions related to a fire reconstruction without the use of a fire model. Much less data is typically required for these routines than is required to run a fire model. Examples of available routines can be found in the FIREFORM section of the FPETOOL program.

20.4.8.3.2 Zone Models. Most of the fire growth models that can be run on personal computers are zone models. Zone models usually divide each room into two spaces or zones, an upper zone that contains the hot gases produced by the fire, and a lower zone that is the source of the air for combustion. Zone sizes change during the course of the fire. The upper zone can expand to occupy virtually all the space in the room.

20.4.8.3.3 Field, Computational Fluid Dynamics (CFD) Models. CFD models usually require large-capacity computer work stations or mainframe computers. By dividing the space into many small cells (frequently tens of thousands), CFD models can examine gas flows in much greater detail than zone models. Where such detail is needed, it is often necessary to use the sophistication of a field model. In general, however, field models are much more expensive to use, require more time to set up and run, and often require a high level of expertise to make the decisions required in setting up the problem and interpreting the output produced by the model. The use of CFD models in fire investigation and related litigation, however, is increasing. CFD models are particularly well suited to situations where the space or fuel configuration is irregular, where turbulence is a critical element, or where very fine detail is sought.

20.5 Fire Testing.

20.5.1 Role of Fire Testing. Fire testing is a tool that can provide data that complement data collected at the fire scene *(see 4.3.3)*, or can be used to test hypotheses *(see 4.3.6)*. Such fire testing can range in scope from bench scale testing to full-scale recreations of the entire event. These tests may relate to the origin and cause of the fire, or to fire spread and development. The components and subsystems to be tested may include building contents, building systems, and architectural and structural elements of the building itself.

20.5.1.1 Used as a part of data collection, fire testing can provide insights into the characteristics of fuels or items consumed in the fire, into the characteristics of materials or assemblies affected by the fire, or into fire processes that may have played a role in the fire. This information is valuable in the analysis of data and the formation of hypotheses. *(See also Section 16.10.)*

20.5.1.2 Used as a part of hypothesis testing, fire testing can assist in evaluating whether a hypothesis is consistent with the case facts and the laws of fire science. In this manner, fire testing is used in much the same way as fire modeling. In addition, fire testing may support modeling by providing input data for models or by providing benchmark data that can be used to assess the accuracy and applicability of a model.

20.5.2* Fire Test Methods. To the extent possible, fire test methods, procedures, and instrumentation should follow or be modeled after standard tests or test methods that have been reported in the fire science literature. Tests consistent with standard test methods or the fire science literature will contribute to the scientific credibility of the results. Testing not performed to a recognized standard should be consistent with the relevant facts of the case. Credible testing includes the use of materials and assemblies that are suitable exemplars of actual materials and assemblies, as well as conducting experiments that reflect the relevant conditions of the scene at the time of the fire. Valuable data may be obtained from testing that addresses limited aspects of a fire incident.

20.5.3 Limitations of Fire Testing. While fire testing can provide useful information, it is not possible to perfectly recreate all of the conditions of a specific fire that may affect the results of a full-scale test fire. Weather conditions are an example of a parameter that may not be reproduced readily and that may affect the results of a test fire. These conditions should be considered in reaching conclusions that are based on the test results.

20.6 Data Required for Modeling and Testing. Scene data required for modeling and testing, typically obtained by the fire investigator, are used to quantify or characterize the physical scene. Relevant scene data include structural dimensions; type of building materials; size, location, and type of contents; and size, location, and type of sources of ventilation.

20.6.1 Materials and Contents.

20.6.1.1 A meaningful analysis of a fire requires understanding of the heat release rate, fire growth rate, and total heat released. The determination of these parameters requires identification of the types, quantities, location, and configuration of fuel actually involved in the fire. For example, a vertical configuration will burn faster than a horizontal configuration of the same fuel.

20.6.1.2 The composition, thickness, condition, and layers of the materials comprising the walls, floors, windows, doors, and ceiling should be documented. The ceiling, wall, and decorative finishes, as well as the type, configuration, and condition of contents, should be documented.

20.6.2 Ventilation. Understanding ventilation conditions is important to the validity of a fire test or model. The position and condition of doors, windows, skylights, and other sources of ventilation, such as thermostatically controlled exhaust fans, should be determined. Determining when ventilation sources were opened or closed is important. Ventilation effects may include wind, fire department ventilation, and HVAC operation and should be considered.

Chapter 21 Explosions

21.1* General.

21.1.1 Historically, the term explosion has been difficult to define precisely. The evidence that indicates an explosion occurred includes damage or change brought about by blast overpressure as an integral element, producing physical effects on structures, equipment, and other objects.

21.1.2 This effect can result from the confinement of the blast overpressure or the impact of an unconfined pressure or shock wave on an object, such as a person or structure. Overpressure is the pressure generated or released in excess of the surrounding ambient pressure. (*See Section 21.11*)

21.1.3 Explosion Definition. For fire and explosion investigations, an explosion is the sudden conversion of potential energy (chemical or mechanical) into kinetic energy with the production and release of gas(es) under pressure. These gases then do mechanical work, such as defeating their confining vessel or moving, changing, or shattering nearby materials.

21.1.3.1 Hydrostatic Vessel Failure. The failure and bursting of a tank or vessel from hydrostatic pressure of a non-compressible fluid such as water is not an explosion, because the pressure is not created by gas. Explosions are gas dynamic.

21.1.3.2 Flash Fires. A flash fire is a fire that spreads rapidly through a diffuse fuel, such as dust, gas, or the vapors of an ignitible liquid, without the production of damaging pressure. The ignition of diffuse fuels does not necessarily always cause explosions. Whether an explosion occurs depends on the location and concentration of diffuse fuels and on the geometry, venting, and strength of the confining structure or vessel, if present, and the presence of obstacles.

21.1.4 Although an explosion is almost always accompanied by the production of a loud noise, the noise itself is not an essential element in the definition of an explosion. The generation and violent escape of gases are the primary criteria of an explosion.

21.1.5 The ignition of a flammable vapor–air mixture within a can, which bursts the can or even only pops off the lid, is considered an explosion. The ignition of the same mixture in an open field, while it is a deflagration, may not be an explosion as defined in this document, even though there may be the release of gas under pressure.

21.1.6 In applying this chapter, the investigator should keep in mind that there are numerous factors that control the effects of explosions and the nature of the damage produced. These factors include the type, quantity, and configuration of the fuel, the size and shape of the containment vessel or structure, obstacles located within the structure, the type and strength of the materials of construction of the containment vessel or structure, and the type and amount of venting present. *(See Section 21.5.)*

21.1.7 Sections of this chapter present explosion analysis techniques and terms that have been developed primarily from the analysis of explosions involving diffuse fuel sources, such as ignitable gases, dusts, and the vapors from ignitable liquids in buildings as described in the Building Systems chapter section on Types of Construction. The scope of this chapter covers all structures in general. However, many of the aids developed covering structural damage are most appropriate for residential and commercial frame structures, e.g., high/low order damage. The reader is cautioned that application of these principles to structures of other construction types may require additional research and/or guidance from other references on explosion effects. The analysis of explosions involving condensed-phase (solid or liquid) explosives, particularly detonating (high) explosives, will also require specialized knowledge that goes beyond the scope of this text.

21.1.8 The investigation of explosion events can be an extremely complicated, technical, scientific, and potentially dangerous task. Investigators faced with the requirement to investigate an explosion scene that exceeds the resources available or is beyond their knowledge or expertise, the investigator should secure the scene to preserve evidence and endeavor to obtain technical expertise and adequate resources to accomplish the scene investigation in a safe and correct manner.

21.2* Types of Explosions. There are two major types of explosions with which investigators are routinely involved: mechanical and chemical, with several subtypes within these types. These types are differentiated by the source or mechanism by which the blast overpressure is produced.

21.2.1 Mechanical Explosions. A mechanical explosion is the rupture of a closed container, cylinder, tank, boiler, or similar storage vessel resulting in the release of pressurized gas or vapor. The pressure within the confining container, structure, or vessel is not due to a chemical reaction or change in chemical composition of the substances in the container.

21.2.2* BLEVEs. The boiling liquid expanding vapor explosion (BLEVE) is the type of mechanical explosion that will be encountered most frequently by the fire investigator. These are explosions involving vessels that contain liquids under pressure at temperatures above their atmospheric boiling points. The liquid need not be flammable. BLEVEs are a subtype of mechanical explosions but are so common that they are treated here as a separate explosion type. A BLEVE can occur in vessels as small as disposable lighters or aerosol cans and as large as tank cars or industrial storage tanks. While the initiating event can be caused by a vessel failure, the explosion and overpressure associated with a BLEVE is due to expansion of pressurized gas or vapor in the ullage (vapor space) combined with the rapidly boiling liquid liberating vapor.

21.2.2.1 A BLEVE frequently occurs when the temperature of the liquid and vapor within a confining tank or vessel is raised by an exposure fire to the point where the increasing internal pressure can no longer be contained and the vessel explodes. *(See Figure 21.2.2.1).* This rupture of the confining vessel subjects the pressurized liquid to a sudden drop in pressure and allows it to vaporize almost instantaneously contributing to the overpressure and explosion. If the contents are ignitible, there is almost always a fire. If the contents are noncombustible, there can still be a BLEVE, but no ignition of the vapors. Ignition usually occurs either from the original external heat that caused the BLEVE or from some electrical or friction source created by the blast or fragments of the vessel.

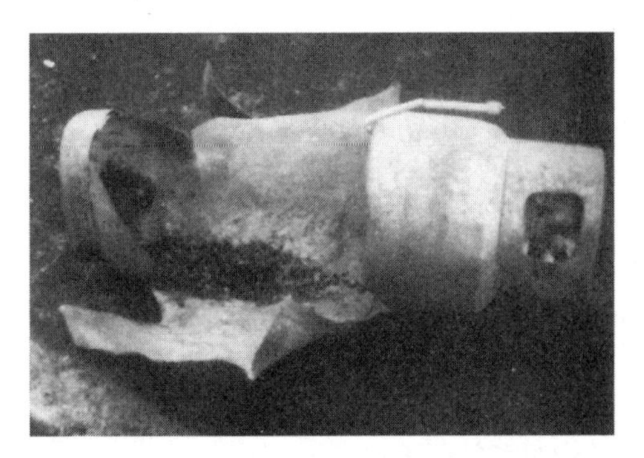

FIGURE 21.2.2.1 An LP-Gas Cylinder That Suffered a BLEVE as a Result of Exposure to an External Fire.

21.2.2.2 A BLEVE may also result from a reduction in the strength of a container as a result of mechanical damage or localized heating above the liquid level. This rupture of the confining vessel subjects the pressurized liquid to a sudden drop in pressure and allows it to vaporize almost instantaneously, contributing to the overpressure and explosion. A common example of a BLEVE not involving an ignitable liquid is the bursting of a steam boiler. The source of overpressure is the steam created by the sudden vaporization of the water. This overpressure and explosion can catastrophically fail the boiler. No chemical, combustion, or nuclear reaction is necessary. This contributes to the energy source, overpressure, and explosion. The chemical nature of the steam (H_2O) is not changed.

21.2.2.3 BLEVEs may also result from mechanical damage, overfilling, runaway reaction, overheating vapor-space explosion, and mechanical failure. See Figure 21.2.2.3, which shows the extent of possible damage from a BLEVE.

FIGURE 21.2.2.3 A Railroad Tank Car of Butadiene That Suffered a BLEVE as a Result of Heating Created by an Internal Chemical Reaction.

21.2.3* Chemical Explosions.

21.2.3.1 In chemical explosions, the generation of the overpressure is the result of exothermic reactions wherein the fundamental chemical nature of the fuel is changed. Chemical reactions of the type involved in an explosion usually propagate in a reaction front away from the point of initiation.

21.2.3.1.1 Combustion Explosions. The most common of the chemical explosions are those caused by the burning of combustible hydrocarbon fuels. These are combustion explosions and are frequently characterized by the presence of a fuel with air as an oxidizer. A combustion explosion may also involve dusts. In combustion explosions, overpressures are caused by the rapid volume production of heated combustion products as the fuel burns.

21.2.3.1.2 Chemical explosions can involve solid combustibles or explosive mixtures of fuel and oxidizer, but more common to the fire investigator will be reactions involving gases, vapors, or dusts mixed with air. Such combustion reactions are called propagation reactions because they occur progressively through the reactant (fuel), with a definable flame front or reaction zone separating the reacted and unreacted fuel.

21.2.3.1.3 Combustion explosions are classified as either deflagrations or detonations, depending on the velocity of the flame front propagation through the fuel air mixture. This should not be confused with the flame speed, which is the speed of the flame propagation relative to a fixed point. The regimes of propagating flame fronts are more accurately described as a deflagration or a detonation. *(See Figure 21.2.3.1.3.)*

21.2.3.1.3.1 Detonations are combustion reactions in which the velocity of the reaction zone relative to the unreacted flammables is faster than the speed of sound. Detonations are reaction zones that are autoignited by the shock waves ahead of the flame. Unlike fast deflagrations, detonations are self-sustaining processes and do not require other driving mechanisms. To initiate a detonation a very strong release of energy is required (i.e., deflagration to detonation transition (DDT) or the detonation of explosives).

21.2.3.1.4 Several subtypes of combustion explosions can be classified according to the types of fuels involved. The most common of these fuels are as follows:

(1) Flammable gases
(2) Vapors of ignitable (flammable and combustible) liquids
(3) Combustible dusts
(4) Smoke and flammable products of incomplete combustion (backdraft explosions)
(5) Aerosols

21.2.4 Electrical Explosions. High-energy electrical arcs may generate sufficient heat to cause an explosion. The rapid heating of the surrounding gases results in a mechanical explosion that may or may not cause a fire. The clap of thunder accompanying a lightning bolt is an example of an electrical explosion effect. High-energy electrical arcs require high voltage and are not covered in this chapter.

21.2.5 Nuclear Explosions. In nuclear explosions, the high pressure is created by the enormous quantities of heat produced by the fusion or fission of the nuclei of atoms. The investigation of nuclear explosions is not covered by this document.

21.3 Characterization of Explosion Damage. or descriptive and investigative purposes, it can be helpful to characterize incidents, particularly in structures, on the basis of the type of damage noted. The terms high-order damage and low-order damage have been used by the fire investigation community to characterize explosion damage. Use of the terms high-order damage and low-order damage is recommended to reduce confusion with similar terms used to describe the energy release from explosives. *(See Section 21.12.)* The differences in damage are a function of the blast load applied to surfaces (rate of pressure rise, peak pressure, and impulse, achieved in the incident) and the strength of the confining or restricting structure, or vessel, rather than the maximum pressures being reached. It should be recognized that the use of the terms low-order damage and high-order damage may not always be appropriate, and a site may contain evidence spanning both categories.

21.3.1 Low-Order Damage. Low-order damage is characterized by walls bulged out or laid down, virtually intact, next to

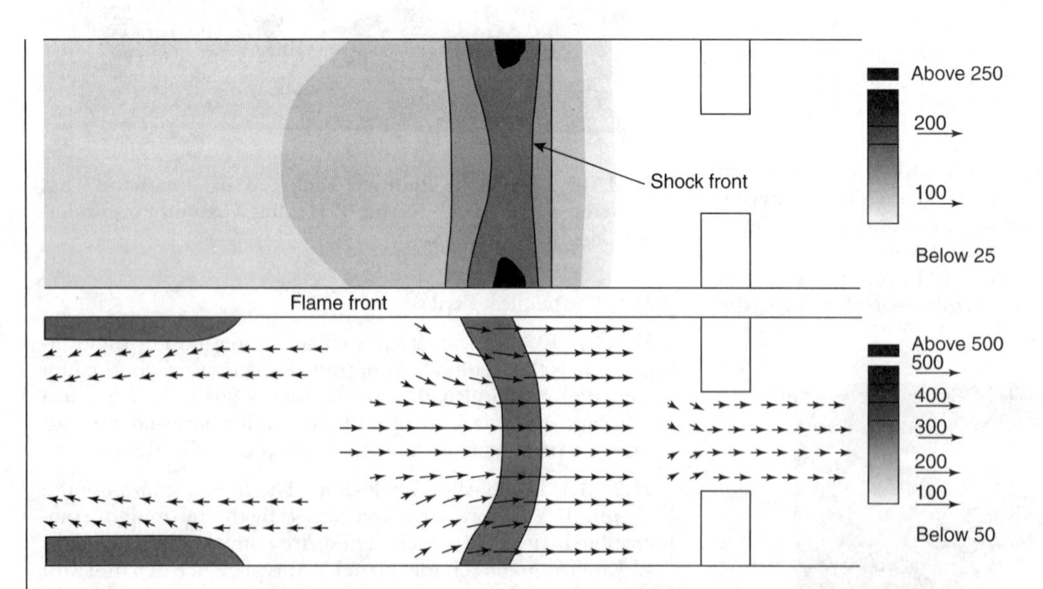

FIGURE 21.2.3.1.3 CFD-simulations of a typical pressure in psi (upper) and flame/velocity in m/s (lower) for a detonation flame. Overpressures greater than 16 times the ambient pressure are typically observed the flame front, which can propagate at typical speeds of 1500-2000 m/s. There is no region of overpressure ahead of the flame. [Simulations result in smoothing of the shock front and small velocities ahead of the flame due to the run-up period. Actual shock fronts will be even sharper and the velocities ahead of the front will no longer be present when it is fully developed.]

the structure. Roofs may be lifted slightly and returned to their approximate original position. Windows may be dislodged, sometimes without glass being broken. Debris produced is generally large and is moved short distances. Low-order damage is produced when the blast load is sufficient to fail structural connections of large surfaces, such as walls or roof, but insufficient to break up larger surfaces and accelerate debris to significant velocities. *(See Figure 21.3.1.)*

FIGURE 21.3.1 Low-Order Damage in a Dwelling.

21.3.2* High-Order Damage. High-order damage is characterized by shattering of the structure, producing small debris pieces. Walls, roofs, and structural members are broken apart with some members splintered or shattered, and with the building completely demolished. Debris is thrown considerable distances, possibly hundreds of feet. High-order damage is the result of relatively high blast loads. *[See Figure 21.3.2(a), Figure 21.3.2(b), and Figure 21.3.2(c).]*

FIGURE 21.3.2(a) High-Order Damage of a Four-Bedroom Single Story House without Ensuing Fire.

21.4 Effects of Explosions. An explosion is a gas dynamic phenomenon that, under ideal theoretical circumstances, will manifest itself as an expanding spherical heat and pressure wave front. The heat, overpressure, and pressure waves produce the damage characteristic of explosions. The effects of explosions can be observed in five major groups: blast overpressure and wave effect, dynamic drag loads, projected fragment effect, thermal effect, and seismic effect (ground shock).

FIGURE 21.3.2(b) High-Order Damage of a Three-Bedroom House with Ensuing Fire.

FIGURE 21.3.2(c) High-Order Damage Remains of a Commercial Structure.

21.4.1 Blast Overpressure and Wave Effect.

21.4.1.1 General. Certain explosions produce significant volumes of gases. As these gases are generated, the pressure in the confining vessel increases and can significantly damage the containing vessel. In addition, the expanding gases and the displaced air moved by the gases produce a pressure front that is primarily responsible for the damage and injuries associated with explosions. *(See Figure 21.4.1.1.)*

21.4.1.1.1 For smaller cylindrical objects, such as pipes, supports or lamp posts, damage is largely due to the dynamic drag loads induced by the displaced gases (explosion wind) flowing past the object. The effects of the explosion wind can even be greater after the blast wave and peak overpressures have passed the object. *(See Figure 21.4.1.1.1.)*

21.4.1.1.2 The blast pressure front occurs in two distinct phases, based on the direction of the forces in relation to the point of origin of the explosion. These are the positive pressure phase and the negative pressure phase.

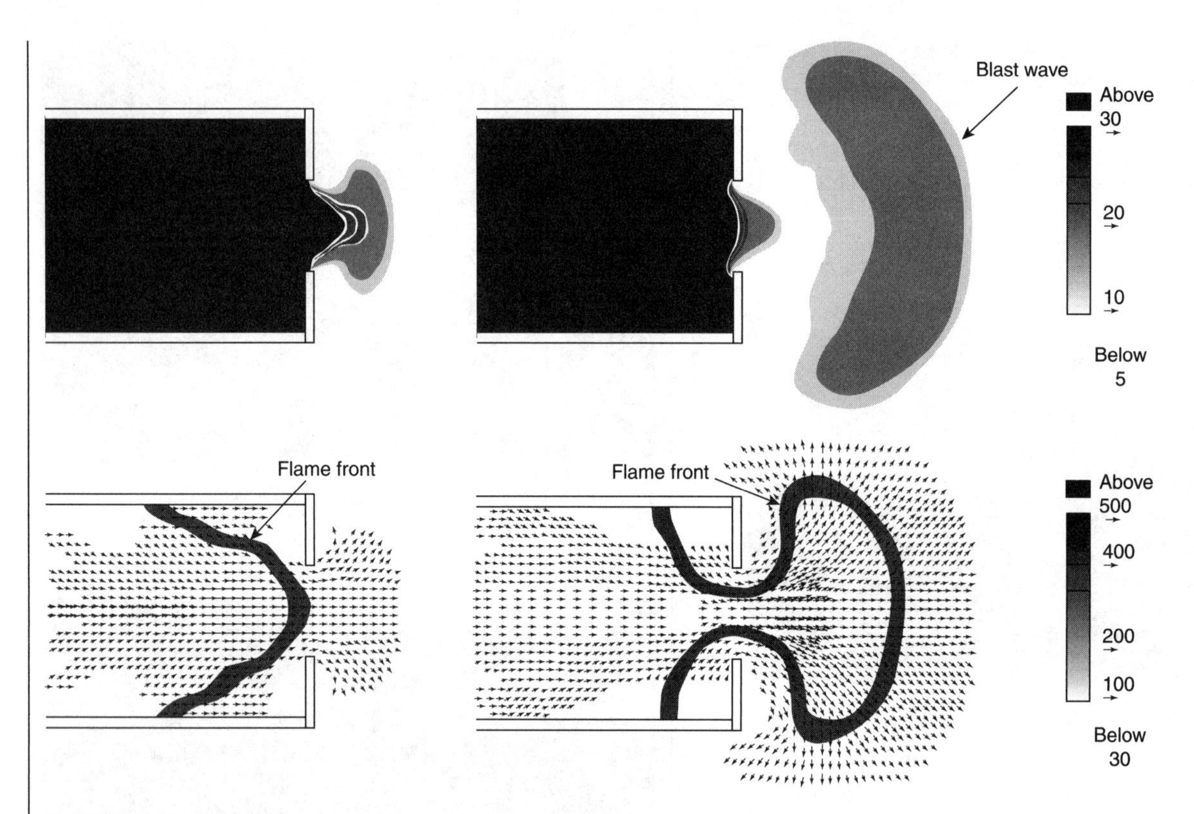

FIGURE 21.4.1.1 Illustration of pressure fronts in psi (upper) and flame front/velocities in m/s (lower) from a vented ethylene/air explosion just after initial venting 132 msec after ignition (left) and 142 msec after ignition (right).

FIGURE 21.4.1.1.1 Lamp Post Damage due to Dynamic Drag Loads Imparted by the Displaced Gases (Explosion Wind) at Flixborough, England.

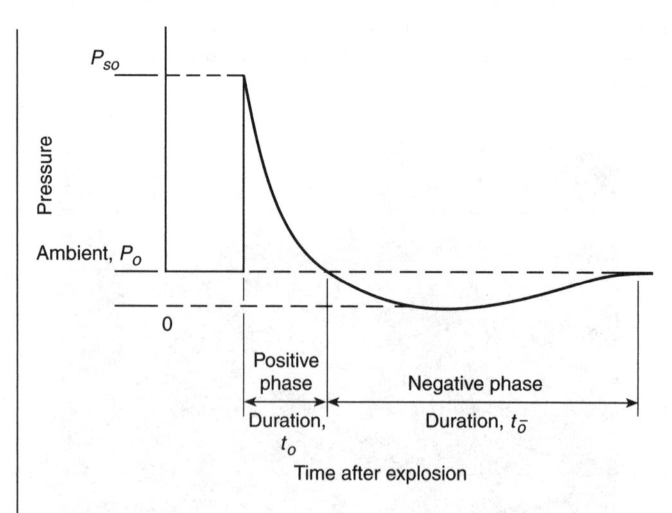

FIGURE 21.4.1.1.3 Typical Pressure History from an Idealized Detonation, Measured at a Point Away from the Point of Detonation.

21.4.1.1.3 A typical pressure history from an idealized detonation, measured at a point away from the point of detonation, is shown in Figure 21.4.1.1.3 and consists of positive and negative phases. The area under the pressure–time curve is called the impulse of the explosion.

21.4.1.2 Positive Pressure Phase. The positive pressure phase is that portion of the blast pressure front in which the expanding gases are moving away from the point of origin. The positive pressure phase is more powerful than the negative pressure phase and is responsible for the majority of pressure damage. This damage can include weakening of the structure such that the structure can be further damaged by the negative pressure phase.

21.4.1.3 Negative Pressure Phase.

21.4.1.3.1 As the extremely rapid expansion of the positive pressure phase of the explosion moves outward from the origin of the explosion, it displaces, compresses, and heats the ambient surrounding air. A low air pressure condition (relative to ambient) is created at the epicenter or origin. Due to the negative pressure condition (relative to ambient), air rushes back to the area of the origin to equilibrate this low-pressure condition.

21.4.1.3.2 The negative pressure phase can cause secondary damage and move items of physical evidence toward the point of origin. Movement of debris during the negative pressure phase may conceal the point of origin. The negative pressure phase is usually of considerably less power than the positive pressure phase but may be of sufficient strength to cause collapse of structural features already weakened by the positive pressure phase. The negative pressure phase may be difficult to detect by witnesses or by post-blast examination in diffuse-phase (gas/vapor) explosions.

21.4.1.4 Shape of Blast Wave (Front). The shape of the blast front from an idealized explosion would be spherical. It would expand evenly in all directions from the epicenter. In the real world, confinement, obstruction, ignition position, cloud shape, or concentration distribution at the source of the blast pressure wave changes and modifies the direction, shape, and force of the front. *[See Figure 21.4.1.4(a) and Figure 21.4.1.4(b).]*

21.4.1.4.1 Venting of the confining vessel or structure may cause damage outside of the vessel or structure. The most damage can be expected to be in the path of the venting as a result of the blast wave and expelled hot products. For example, the blast pressure front in a room may travel through a doorway and damage items or materials directly in line with the doorway in the adjacent room. The same relative effect may be seen directly in line with the structural seam of a tank or drum that fails before the sidewalls. As the blast wave travels radially from the venting source, damage can also be observed at locations not in line with the vent. *(See Figure 21.4.1.1.)*

21.4.1.4.2 The blast pressure front may also be reflected off solid obstacles and redirected, resulting in a substantial increase or possible decrease in pressure, depending on the characteristics of the obstacle struck.

21.4.1.4.3 After propagating reactions have consumed their available fuel, the force of the expanding blast pressure front decreases with the increase in distance from the epicenter of the explosion. *(See Figure 21.4.1.4.3.)*

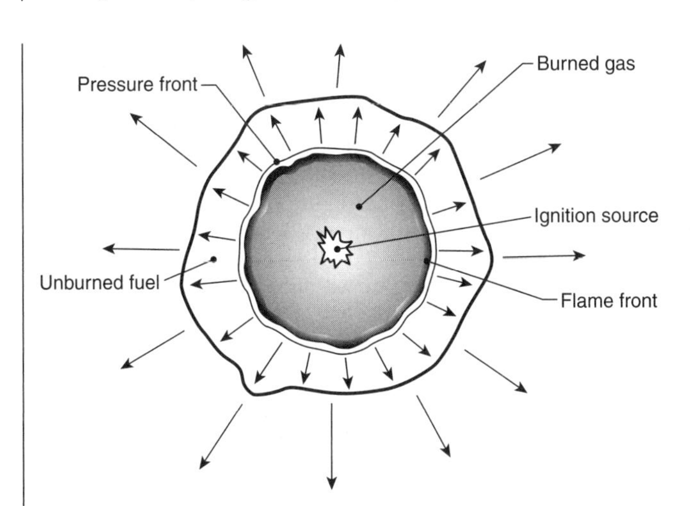

FIGURE 21.4.1.4(a) Idealized Propagating Flame and Pressure Fronts [After Harris (1983) p.3]

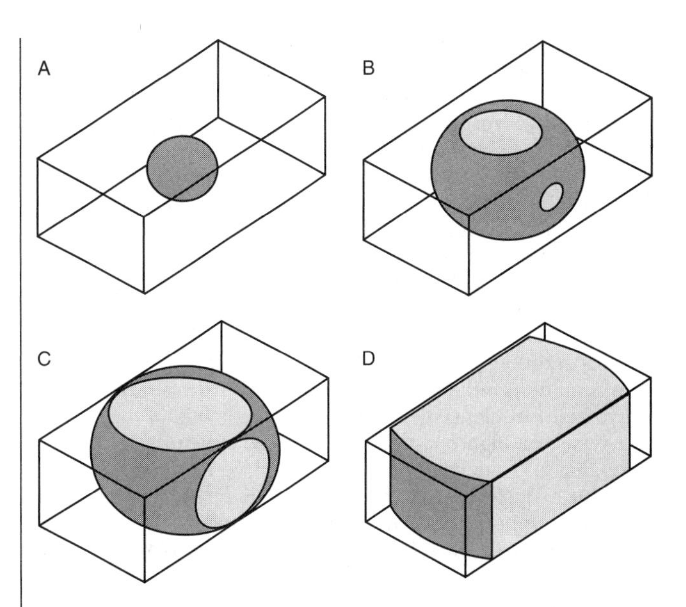

FIGURE 21.4.1.4(b) Figure 21.4.1.4 (b) Idealized Representation of a Flame Front in a Cuboid Vessel (From Harris, 1983)

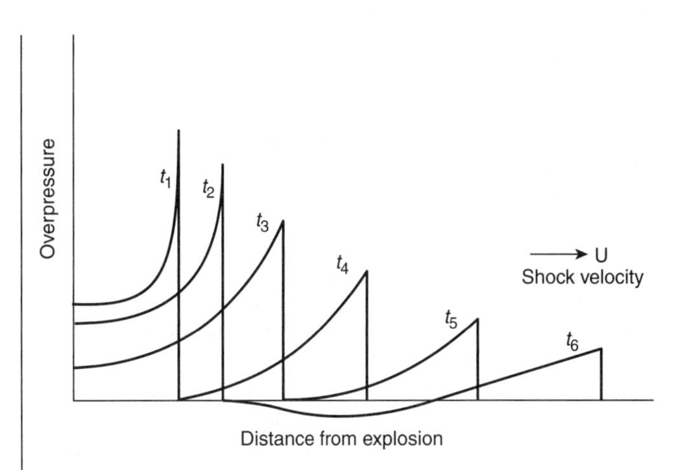

FIGURE 21.4.1.4.3 Typical Overpressure History at Locations Distant from Center of Explosion.

21.4.1.5 Rate of Pressure Rise versus Maximum Pressure. The type of damage caused by the blast pressure front of an explosion is dependent not only on the total amount of energy generated but also, and often to a larger degree, on the rate of energy release and the resulting rate of pressure rise.

21.4.1.5.1 Relatively slow rates of pressure rise will produce the pushing or bulging type of damage effects seen in low-order damage. The weaker parts of the confining structure or vessel, such as windows or structural seams, will rupture first; thereby venting the blast pressure wave and reducing the total damage effects of the explosion.

21.4.1.5.2 In explosions of structures where the rate of pressure rise is very rapid, faster than the structure can respond to it, there will be more shattering of the confining vessel or container, and debris will be thrown great distances, as the venting effects are not allowed sufficient time to develop. This is characteristic of high-order damage.

21.4.1.5.3 Where the rate of pressure rise is less rapid, but faster than the structure can respond to it, the venting effect will have an important impact on the maximum pressure developed. See NFPA 68, *Standard on Explosion Protection by Deflagration Venting,* for equations, data, and guidance on calculating the theoretical effect of venting on pressure during a deflagration. Such calculations assume a structure or vessel that can sustain such a high pressure. The maximum theoretical pressure that can be developed by a slow deflagration can, under some circumstances, be as high as 7 to 9 times higher than the initial pressure. These pressures however are achieved in testing vessels which do not allow the pressure wave to vent, and so are not readily transferred to actual conditions encountered at an explosion scene. Under certain conditions with fuels of high reactivity (hydrogen, acetylene, ethylene), turbulence generated by the geometry of the confining vessel, or high congestion within the structure, local pressures due to fast deflagrations can exceed 7 times the initial pressure due to dynamic effects such as pressure focusing, reflections, and pre-compression. Specialized experiments or Computational Fluid Dynamics (CFD) tools can be used to analyze such effects.

21.4.1.5.4 In commonly encountered situations, such as fugitive gas explosions in residential or commercial buildings, the maximum pressure will be limited to a level slightly higher than the pressure that major elements of the building enclosure (e.g., walls, roof, and large windows) can sustain without rupture (minimum failure pressure). In a well-built residence, this pressure will seldom exceed 21 kPa (3 psi).

21.4.2 Shrapnel Effect (Projectiles).

21.4.2.1 When the containers, structures, or vessels that contain or restrict the blast overpressures are ruptured, they are often broken into pieces or fragments that may be thrown over great distances with great force. These fragments are also called missiles, shrapnel, debris, or projectiles. They can cause great damage and personal injury, often far from the source of the explosion. In addition, fragments can sever electric utility lines, fuel gas or other flammable fuel lines, or storage containers, thereby adding to the size and intensity of post-explosion fires or causing additional explosions.

21.4.2.2 The distance to which missiles can be propelled outward from an explosion depends greatly on their initial direction, velocity, mass, and aerodynamic characteristics. An idealized diagram for missile trajectories is shown in Figure 21.4.2.2 for several different initial directions. The actual distances that missiles can travel depend greatly on aerodynamic conditions, obstructions, and occurrences of ricochet impacts. As illustrated in Figure 21.4.2.2, the investigator should be mindful that the total distance of travel through the air for a missile, may not be represented by the actual linear distance from the missile's original location.

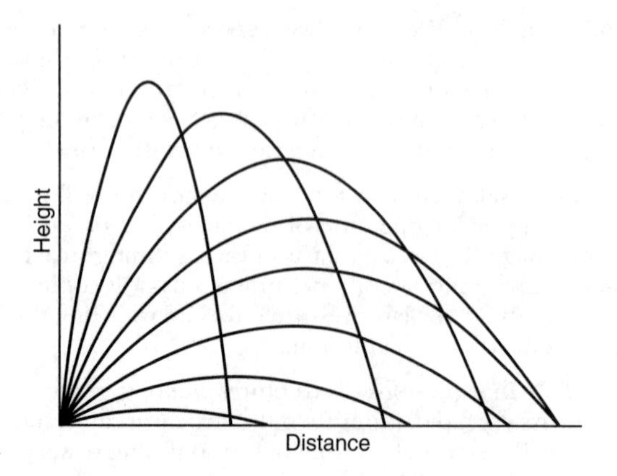

FIGURE 21.4.2.2 Idealized Missile Trajectories for Several Initial Flight Directions.

21.4.3 Thermal Effect. Combustion explosions release quantities of energy that heat combustion gases and ambient air to high temperatures. This heat can ignite nearby thermally thin, and low thermal inertia combustibles or can cause burn injuries to anyone nearby. *(See Basic Fire Science chapter, section on Heat Transfer.)* These secondary fires increase the damage and injury from the explosion and complicate the investigation process. In some cases the fire may actually occur as the primary event; it may be difficult to determine which occurred first, the fire or the explosion.

21.4.3.1 All chemical explosions liberate heat because of the chemical changes in the fuel that occur. The thermal damage *(See section on Effective Temperatures in Fire Patterns chapter)* depends on the nature of the fuel as well as the duration of the high temperatures.

21.4.3.2* Fireballs and firebrands are possible thermal effects of explosions, particularly BLEVEs involving liquefied gas. Fireballs are the momentary ball of flame present during or after the explosive event. As the outer envelope of the gas cloud burns, it lifts and forms the fireball. As fireballs rise they produce mushroom clouds, in which violent convection currents can form. A fireball may produce high-intensity, short-duration thermal radiation. Fireballs can be the result of momentum-driven forces such as a BLEVE or burst vessel, or from buoyancy driven forces resulting from a combusting vapor cloud. Firebrands are hot or burning fragments propelled from the explosion. All these effects may serve to initiate fires away from the center of the explosion.

21.4.4 Seismic Effect (Ground Shock). For ground shock to occur, the explosion must transmit significant energy into the ground, causing soil motion. Or as damaged portions of large structures are knocked to the ground, localized kinetic energy can be transmitted into the ground. These ground motion effects, usually negligible for small explosions and diffuse fuel explosions, can sometimes produce additional damage to structures, underground utility services, pipelines, tanks, and cables.

21.5 Factors Controlling Explosion Effects. Factors that can control the effects of explosions include the type and configuration of the fuel; nature, size, volume, and shape of any containment vessel or object affected; level of congestion and obstacles within the vessel; location and magnitude of ignition source; venting of the containment vessel; relative maximum pressure; and rate of pressure rise. The nature of these factors and their various combinations in any one explosion incident can produce a wide variety of physical effects with which the investigator will be confronted. Various phenomena affect the characteristics of a blast pressure front as it travels away from the source.

21.5.1 Fuel. The nature of the fuel, whether it is a dust, gas, vapor or aerosol of a flammable or combustible liquid, or explosive, will have a profound impact upon the effects that the explosion will produce. Airborne fuels produce remarkably different effects that those from explosives or pyrotechnics. (See the Fuels section of the Basic Fire Science chapter.)

21.5.2 Turbulence. Turbulence within a fuel-air mixture increases the flame speed and, therefore, greatly increases the rate of combustion and the rate of pressure rise. Turbulence can produce rates of pressure rise with relatively small amounts of fuel that can result in high-order damage even though the mixture was above or below stoichiometric conditions. The shape, size, and location of obstacles within the confining vessel can have a profound effect on the severity of the explosion by affecting the nature of turbulence. Congestion from an abundance of obstacles in the path of the combustion wave has been shown to increase turbulence and greatly increase the severity of the explosion, mainly due to increasing the flame speed of the mixture involved. Other mixing and turbulence sources, such as fans and forced-air ventilation, may increase the explosion effects. *(See Figure 21.8.2.1.7.)*

21.5.3* Nature of Confining Space. The nature of containment—its size, shape, construction, volume, materials, design, and internal obstacles—will also greatly change the effects of the explosion. For example, a specific percentage by volume of natural gas mixed with air will produce a different rate-of-pressure rise if it is contained in a 28.3 m^3 (1000 ft^3) room than if it is contained in a 283.2 m^3 (10,000 ft^3) room at the time of ignition. (*See* NFPA 68, *Standard on Explosion Protection by Deflagration Venting.*)This variation in effects is true even though the maximum overpressure achieved will be essentially the same.

21.5.3.1 A long, narrow corridor filled with a combustible vapor–air mixture, when ignited at one end, will be very different in its pressure distribution, rate of pressure rise, and its effects on the structure than if the same volume of fuel–air were ignited in a cubical compartment.

21.5.3.2 During the explosion, turbulence caused by obstructions within the containment vessel can increase the damage effects. This turbulence can be caused by solid obstructions, such as columns or posts, machinery, pipes, racks, etc., which increase flame speed, and thus increase the rate of pressure rise.

21.5.4* Location and Magnitude of Ignition Source. The highest rate of pressure rise will occur if the ignition source is in the center of an uncongested, cubic confining structure. The closer the ignition source is to the walls of such a confining vessel or structure, the sooner the flame front will reach the wall and extinguish causing a reduction in the flame surface and reaction zone. This results in the loss of energy and a corresponding lower rate of pressure rise and a less violent explosion. *(See Figure 21.5.4 for experimental test results.)*

21.5.4.1 In commonly encountered structures, which include areas of high aspect ratio, semi-confinement, venting, or partial congestion, the location of the ignition source is very important to the development of the overpressure (maximum pressure and rate of pressure rise) and the corresponding damage level and distribution. Specialized experiments or Computational Fluid Dynamics (CFD) tools can be used to analyze such incidents.

21.5.4.2 The energy of the ignition source generally has a minimal effect on the course of an explosion, but unusually large ignition sources (e.g., blasting caps or explosive devices) can significantly increase the speed of pressure development and, in some instances, can cause a deflagration to transition into a detonation.

Spark Position	Maximum Pressure (psia)	Maximum Rate of Pressure Rise (psi/sec)	Time to Maximum Pressure (sec)
Center	117	4960	0.095
Bottom	113	1870	0.133
Side	109	1750	0.147
Top	108	1620	0.148

* Reported by Harris (1983).

FIGURE 21.5.4 Laboratory Test Results for Varying Ignition Source Positions* (10.15 Percent Methane/Air Mixture, 0.28 m^3 Spherical Vessel).

21.5.5 Venting. With diffuse fuel (i.e., gas, vapor, or dust) explosions, the venting of the containment vessel will also have a profound effect on the nature of explosion damage. For example, it may be possible to cause a length of steel pipe to burst in the center if it is sufficiently long, in spite of the fact that it may be open at both ends. The number, size, and location of doors and windows in a room may determine whether the room experiences complete destruction, merely a slight movement of the walls and ceiling, or no damage to walls and ceiling.

21.5.5.1 Venting of a confining vessel or structure may also cause damage outside of the vessel or structure. The most damage can be expected in the path of venting as a result of the blast wave and expelled hot products. For example, the blast pressure front in a room may travel through a doorway and may damage items or materials directly in line with the doorway in the adjacent room. The same relative effect may be seen directly in line with the structural seam of a tank or drum that fails before the sidewalls. As the blast wave travels radially from the venting source, damage can also be observed at locations not in line with the vent.

21.5.5.2 With detonations, venting effects are considerably less than deflagrations, as the high speeds of the blast pressure fronts are too fast for any venting to effectively relieve the peak pressures.

21.5.6 Blast Pressure Wave (Blast Pressure Front) Modification by Reflection. As a blast pressure front encounters objects in its path, the blast pressure front may amplify due to its reflection. This reflection will cause the overpressure to increase with the amplification, depending on the angle of incidence and the incident overpressure. This is further exacerbated in corners where pressure may be locally focused due to the reflections.

21.5.7 Blast Pressure Front Modification by Refraction and Blast Focusing. At times a lack of homogeneity in the affected atmosphere can cause anomalies in the behavior of the expected blast pressure front. When a blast pressure front encounters a layer of air at a significantly different temperature or density, it may cause the blast pressure front to bend, or refract. This occurs because the speed of sound is proportional to the square root of temperature, and thereby the density, of the air. A low-level temperature inversion can cause an initially hemispherical blast front to refract and to focus on the ground around the center of the explosion. Severe weather-related wind shear can cause focusing in the downwind direction.

21.6 Seated Explosions.

21.6.1 General. The seat of an explosion is defined as the crater or concentrated area of great damage, frequently roughly circular or spheroid in shape. The seat of an explosion may not always be located at the point of initiation (epicenter). Material may be thrown out of the crater. This material is called ejecta and may range from larger pieces of shattered debris to fine dust. The presence of a seat indicates the explosion of a concentrated fuel source in contact with or in close proximity to the seat. *(See Figure 21.6.1.)*

FIGURE 21.6.1 A 3 ft diameter explosion seat from an explosive detonated on the ground. Soil ejecta can be seen in the lower right quadrant of the photo.

21.6.1.1 These seats can be of any size, depending on the size and strength of the explosive material involved. They typically range in size from a few centimeters (inches) to 7.6 m (25 ft) in diameter. They display an easily recognizable crater of pulverized soil, floors, or walls located at the center of otherwise less damaged areas. Seated explosions are generally characterized by high pressure and rapid rates of pressure rise.

21.6.1.2 Only specific types or configurations of explosive fuels can produce seated explosions. These include explosives, steam boilers, tightly confined gaseous fuels or liquid fuel vapors, and BLEVEs occurring in relatively small containers, such as cans or barrels.

21.6.1.3 In general, it is accepted that reaction velocities should exceed the speed of sound (detonations) to produce seated explosions, unless the damage is produced by shrapnel (projectiles) from a failing vessel. Each of these explosions involves a rapid release of energy from a containment vessel, resulting in a pressure wave that decays with distance.

21.6.2 Explosives. Explosions fueled by many explosives are often most easily identified by their highly centralized epicenters, or seats. High explosives especially produce such high-velocity, positive pressure phases at detonation that they often shatter their immediate surroundings and produce craters or highly localized areas of great damage. However, if an explosion occurs where the explosive material is not in contact with a surface (suspended or at some distance above ground), a crater may not be present and the investigator would have to consider just the area of greatest damage and the types of fuels that may be present to make a determination.

21.6.3 Boiler and Pressure Vessels. Vessels. A boiler explosion often creates a seated explosion because of its high energy, rapid rate of pressure release, and confined area of origin. Boiler and pressure vessel explosions will exhibit effects similar to explosives, though with lesser localized overpressure near the source.

21.6.4 Confined Fuel Gas and Liquid Vapor. Fuel gases or ignitible liquid vapors when confined to such small vessels as cylinders, small tanks, barrels, or other containers can also produce seated explosions.

21.6.5 BLEVE. A boiling liquid expanding vapor explosion will produce a seated explosion if the confining vessel (e.g., a cylinder, drum, or tank) is of a small size and if the rate of pressure release when the vessel fails is rapid enough (mechanical explosion).

21.7 Nonseated Explosions. Nonseated explosions occur most often when the fuels are dispersed or diffused at the time of the explosion because the rates of pressure rise are moderate and because the explosive velocities are subsonic. It should be kept in mind that even detonations of diffuse fuels and condensed phase explosives may produce nonseated explosion damage under certain conditions, such as an elevated explosion.

21.7.1 Fuel Gases. Fuel gases, such as commercial natural gas and liquefied petroleum (LP) gases, most often produce nonseated explosions. This is because these gases often are confined in large vessels, such as individual rooms or structures, and their deflagration speeds are subsonic.

21.7.2 Pooled Flammable/Combustible Liquids. Explosions from the ignition of vapors of pooled flammable or combustible liquids are typically nonseated explosions. The large areas that they cover and from which they evolve their ignitable vapors, and their subsonic explosive speeds preclude the production of small, concentrated, high-damage seats.

21.7.3* Dusts. Dust explosions most often occur in confined areas of relatively wide dispersal, such as grain elevators, materials-processing plants, and coal mines, where combustible dust can accumulate in sufficient quantity to support a propagating reaction. These large areas of origin preclude the production of pronounced seats.

21.7.4 Backdraft (Smoke Explosion). Backdrafts involve a widely diffused volume of particulate matter and combustible gases. Their explosive velocities are subsonic, thereby precluding the production of pronounced seats.

21.8 Gas/Vapor Combustion Explosions. The most commonly encountered explosions are those involving gases or vapors, especially fuel gases or the vapors of ignitible liquids. Table 21.8 provides some useful properties of common flammable gases. NFPA 68, *Standard on Explosion Protection by Deflagration Venting*, provides a more complete introduction to the fundamentals of these explosions. Table 21.8 lists only single values of lower and upper limits, those limits do in fact change with temperature. Figure 21.8 shows how the flammable limits might vary with temperature.

21.8.1* Ignition of Gases and Vapors. Gaseous fuel–air mixtures are the most easily ignitible fuels capable of causing an explosion. Minimum ignition temperatures in the 370°C to 590°C (700°F to 1100°F) range are common, and they can be even lower for heavier hydrocarbons, such as n-heptane at 215°C (419°F). Minimum ignition energies of some selected fuels are shown in Table 21.8.1. While Table 21.8.1 shows single values, minimum ignition energy varies with fuel/air ratio as shown in Figure 21.8.1 for methane.

Table 21.8 Combustion Properties of Common Flammable Gases

Gas	Comparative Units MJ/m^3 (gross)	Comparative Units Btu/ft^3 (gross)	Limits of Flammability Percent by Volume in Air Lower	Limits of Flammability Percent by Volume in Air Upper	Specific Gravity (air = 1.0)	Air Needed to Burn 1 m^3 of Gas (m^3)	Air Needed to Burn 1 ft^3 of Gas (ft^3)	Ignition Temperature °C	Ignition Temperature °F
Natural gas									
High inert type[a]	35.7–39.2	958–1051	4.5	14.0	0.660–0.708	9.2	9.2	—	—
High methane type[b]	37.6–39.9	1008–1071	4.7	15.0	0.590–0.614	10.2	10.2	482–632	900–1170
High Btu type[c]	39.9–41.9	1071–1124	4.7	14.5	0.620–0.719	9.4	9.4	—	—
Blast furnace gas	3.0–4.1	81–111	33.2	71.3	1.04–1.00	0.8	0.8	—	—
Coke oven gas	21.4	575	4.4	34.0	0.38	4.7	4.7	—	—
Propane (commercial)	93.7	2516	2.15	9.6	1.52	24.0	24.0	493–604	920–1120
Butane (commercial)	122.9	3300	1.9	8.5	2.0	31.0	31.0	482–538	900–1000
Sewage gas	24.9	670	6.0	17.0	0.79	6.5	6.5	—	—
Acetylene	208.1	1499	2.5	81.0	0.91	11.9	11.9	305	581
Hydrogen	12.1	325	4.0	75.0	0.07	2.4	2.4	500	932
Anhydrous ammonia	14.4	386	16.0	25.0	0.60	8.3	8.3	651	1204
Carbon monoxide	11.7	314	12.5	74.0	0.97	2.4	2.4	609	1128
Ethylene	59.6	1600	2.7	36.0	0.98	14.3	14.3	490	914
Methyl acetylene, propadiene, stabilized[d]	91.3	2450	3.4	10.8	1.48	—	—	454	850

[a]Typical composition CH_4 71.9–83.2%; N_2 6.3–16.20%.
[b]Typical composition CH_4 87.6–95.7%; N_2 0.1–2.39%.
[c]Typical composition CH_4 85.0–90.1%; N_2 1.2–7.5%.
[d]MAPP® Gas from the NFPA *Fire Protection Handbook*, 19th ed., Table 8.7.3.

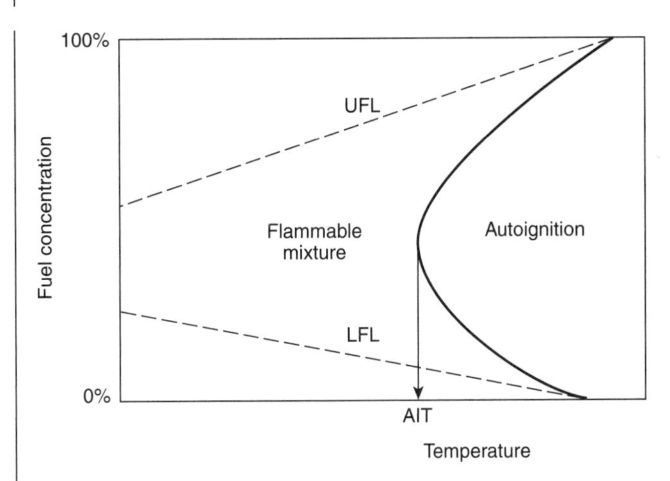

FIGURE 21.8 Chart Demonstrating the Effect of Temperature on the Flammable/Explosive Range of Gases. *(Source: GexCon—Gas Explosion Handbook, Figure 4.5)*

Table 21.8.1 Minimum Ignition Energies of Selected Fuels*

Gas/Vapor	Minimum Ignition Energy (mJ)
Acetylene	0.02
Benzene	0.225
Butane	0.26
Ethane	0.24
Heptane	0.24
Hexane	0.248
Hydrogen	0.018
Methane	0.28
Methanol	0.14
Methyl ethyl ketone	0.28
Pentane	0.22
Propane	0.25

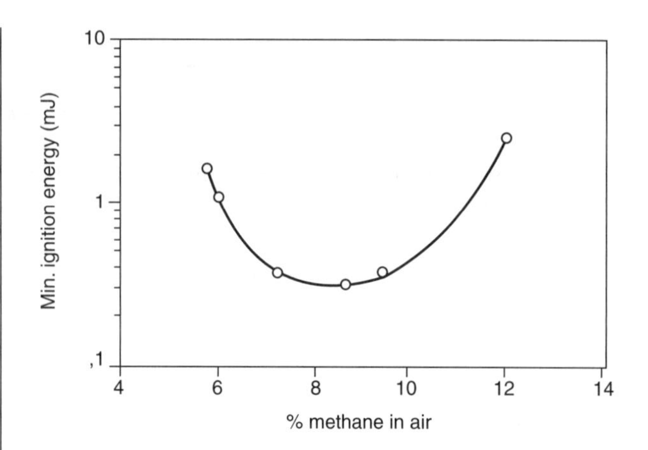

FIGURE 21.8.1 An Experimental Comparison of the Minimum Ignition Energy of Methane as a Function of Percentage of Methane in the Air *(Source: GexCon—Gas Explosion Handbook, Figure 4.5)*.

21.8.2 Interpretation of Explosion Damage. The explosion damage to structures (low-order and high-order) is related to a number of factors. These include the fuel-to-air ratio, vapor density of the fuel, turbulence effects, volume of the confining space, location and magnitude of the ignition source, venting, and the characteristic strength of the structure.

21.8.2.1* Fuel-to-Air Ratio. The nature of damage to the confining structure can be an indicator of the fuel–air mixture at the time of ignition.

21.8.2.1.1 Adiabatic flame temperatures are related to the concentration of fuel as illustrated in Figure 21.8.2.1.1. The theoretical peak pressure of a fuel is largely related to its adiabatic flame temperatures, and, to a lesser extent, the net mole production/consumption in the reaction.

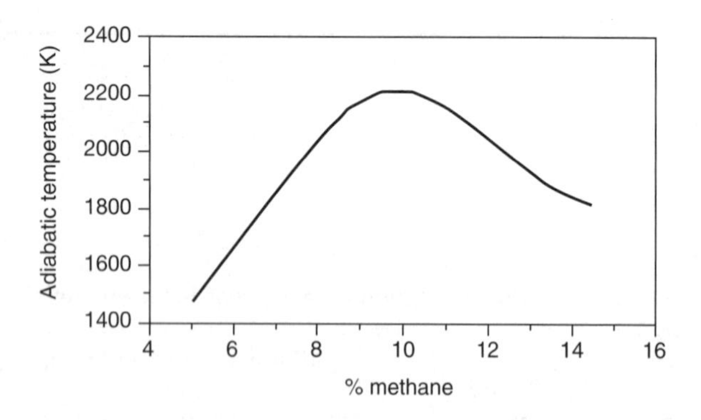

FIGURE 21.8.2.1.1 Adiabatic Flame Temperature for Initial Conditions 1 atm. and 25°C *(Source: GexCon—Gas Explosion Handbook, Figure 4.7.)*

21.8.2.1.2 It is not necessary for an entire volume to be occupied by a ignitable mixture of fuel and air for there to be an explosion. Relatively small volumes of ignitable mixtures capable of causing damage may result from gases or vapors collecting in a given area and being ignited before having migrated to all areas of a room or confining vessel. Depending upon the fuel's properties and the geometry of the confining structure, an explosion or flash fire may result. The absence of explosion damage does not preclude the presence of an ignitable fuel/air mixture.

21.8.2.1.3 Explosions that occur in mixtures at or near the lower explosive limit (LEL) or upper explosive limit (UEL) of a gas or vapor produce less violent explosions than those near the optimum concentration (i.e., usually just slightly rich of stoichiometric). This is because the less-than-optimum ratio of fuel and air results in lower flame speeds, lower rates of pressure rise, and lower maximum pressure. In general, these explosions tend to push and heave at the confining structure, producing low-order damage. However, there are cases when stratified levels of fuel-rich mixtures are pushed out a vent opening and mix with the ambient air, creating flammable mixtures and potentially strong explosions.

21.8.2.1.4* Laminar Burning Velocity. In laminar flame propagation, the flame propagation rate relative to the unburned gas is termed the laminar burning velocity, SL. The burning velocity is the rate of flame propagation relative to the velocity of the unburned gas ahead of it. The fundamental burning velocity is the burning velocity for laminar flame under stated conditions of composition, temperature, and pressure of the unburned gas. Fundamental burning velocity is an inherent characteristic of a combustible and is a fixed value, whereas actual flame speed can vary widely, depending on the existing parameters of temperature, pressure, confining volume and configuration, combustible concentration, and turbulence. The laminar burning velocity is the velocity component that is normal to the flame surface and is determined experimentally by a wide variety of techniques including: (1) Bunsen flames: ratio of the volume flow rate, divided by the actual flame front area, (2) flat flame burners, (3) spherically propagating flames, and (4) counterflow flame configurations. This is often not the velocity which is of actual interest to the investigator. *(See Table 21.8.2.1.4.)*

21.8.2.1.4.1 The laminar burning velocity is the velocity at which a flame reaction front moves into the unburned mixture as it chemically transforms the fuel and oxidant into combustion products. It is only a fraction of the flame speed. The flame speed is the product of the velocity of the flame front caused by the volume expansion of the combustion products due to the increase in temperature and any increase in the number of moles and any flow velocity due to motion of the gas mixture prior to ignition. The burning velocity of the flame front can be calculated from the fundamental burning velocity, which is reported in NFPA 68, *Standard on Explosion Protection by Deflagration Venting*, at standardized conditions of temperature, pressure, and composition of unburned gas. As pressure and turbulence increase substantially during an explosion, the fundamental burning velocity will increase, further accelerating the rate of pressure increase. NFPA 68 lists data on the various materials.

21.8.2.1.5 Expansion Ratio. The expansion ratio is the post-ignition rate of expansion of a fuel/air mixture's products of combustion behind the expanding flame front. The expanding products of combustion propel the unreacted fuel/air mixture ahead of the flame front. Expansion ratio is generally a specific value for each fuel at a defined temperature, pressure and fuel/air concentration. The density ratio, is the expansion ratio for the mixture, where ρu is the density of unburned gas, and ρb is the density of burned gas. If the chemical composition of the fuel and the oxidizer are known, the expansion ratio can be computed for stoichiometric combustion under ideal conditions. The expansion ratio is highest

Table 21.8.2.1.4 Typical Combustion Properties of Common Gaseous Fuels

Gas	% Gas at Stochiometric Ratio	% Gas at Maximum Burning Velocity (Optimum Mixture)	Maximum Laminar Burning Velocity m/sec. (ft/sec.)	Adiabatic flame temperature ° kelvin (°F)	Expansion Ratio** (T_f/T_i)	Max. Laminar Flame Speed m/sec. (ft/sec.)
Acetylene	7.7	9.3	1.58(5.18)	2598(4217)	9.0	14.2(46.6)
Benzene	2.7	3.3	0.62(2.03)	2287(3657)	7.9	4.9(16.08)
Butane	3.1	3.5	0.50(1.64)	2168(3443)	7.5	3.7(12.13)
Ethane	5.6	6.3	0.53(1.74)	2168(3443)	7.5	4.0(13.12)
Heptane	1.9	2.3	0.53(1.71)	2196(3493)	7.6	4.0(13.12)
Hexane	2.2	2.5	0.53(1.71)	2221(3538)	7.7	4.0(13.12)
Hydrogen	30.0	54	3.5 (11.48)	2318(3713)	8.0	28.0(91.86)
Methane	9.5	10.0	0.45(1.48)	2148(3407)	7.4	3.5(11.48)
Pentane	2.6	2.9	0.52(1.71)	2232(3558)	7.7	4.0(13.12)
Propane	4.0	4.5	0.52(1.71)	2198(3497)	7.6	4.0(13.12)

*after Harris (1983)

** The actual Expansion Ratio is given by ρ_u/ρ_b the density of the reactants (ρ_u) divided by the density of the products (ρ_b), or equivalently in near constant pressure conditions (T_f/T_i) (N_b/N_u), the product of the ratio of absolute flame temperature (T_f) to absolute initial gas temperature (T_i) and the ratio of the number of moles of the products (N_b) to the number of moles of the reactants (N_u). Harris assumes the number of moles of the products (N_b) is equal to the number of moles of the reactants (N_u) and computes the expansion ratio based on the absolute temperature ratio alone. As such, the values in the table will be slightly different than actual values.

in fuel/air mixtures slightly above the stochiometric concentration. This expansion ratio is directly proportional to the volume and temperature of the products of combustion. For most commonly encountered gaseous fuels the expansion factors are between 7 and 8. *(See Table 21.8.2.1.4.)*

21.8.2.1.6 Laminar Flame Speed. The laminar flame speed is the local speed of a freely propagating flame relative to a fixed point without the effect of turbulence in the fuel/air mixture. *(See 21.8.2.1.7.)* It is the product of the burning velocity and the expansion ration of the flame front and is given by: $S_b = S_L \frac{\rho_u}{\rho_b}$. For hydrocarbon-air mixtures one may say that the higher the laminar flame speed, the more reactive is the mixture. This means that the flame can propagate fast through a cloud and thereby cause flame acceleration and pressure build-up. The maximum laminar flame speeds for methane and propane are 3.5 m/sec (11.5 ft/sec) and 4 m/sec (13.1 ft/sec), respectively. For certain mixture concentrations instabilities can occur causing the flame to wrinkle and increase the flame surface. This can further increase the laminar flame speed higher than reported above. *(See Figure 21.8.2.1.6.)*

21.8.2.1.7 Turbulent Flame Speed. Explosions will rarely involve laminar (non-turbulent) combustion. Turbulence greatly increases flame speed and the turbulent flame speed will generally be the flame speed which is relevant to a real explosion. It is difficult to compute the turbulent flame speed for an accidental explosion, since local flow conditions have to be known, including factors such as obstructions and their congestion, which increase the turbulent flame speed. However, it is worthy to note that turbulent flame speeds can be higher than the speed of sound in the unreacted mixture (i.e., fast deflagrations). *(See Figure 21.8.2.1.7.)*

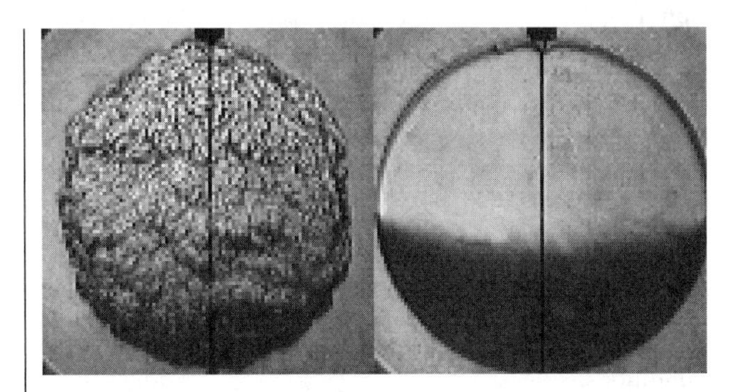

FIGURE 21.8.2.1.6 Expanding Spherical $H_2/O_2/N_2$ Flame Front at 3 atm. Left-Wrinkled Equivalence Ratio 0.7, Right-Non-wrinkled Equivalence Ratio 2.25. Burning Velocity, (S_L) × Expansion Ratio, (E) = Laminar Flame Speed, (S_b For Methane: 1.47 ft./sec. × 7.4 = 10.87 ft./sec.

21.8.2.1.8 Explosions of mixtures near the LEL do not tend to produce large quantities of post-explosion fire, as nearly all of the available fuel is consumed during the explosive propagation.

21.8.2.1.9 Explosions of mixtures near the UEL tend to produce post-explosion fires because of the fuel-rich mixtures. The delayed combustion of the remaining fuel produces the post-explosion fire. Often, a portion of the mixture over the UEL has fuel that does not burn until it is mixed with air during the explosion's venting phase or negative-pressure phase, thereby producing the characteristic post-explosion fire or secondary explosion.

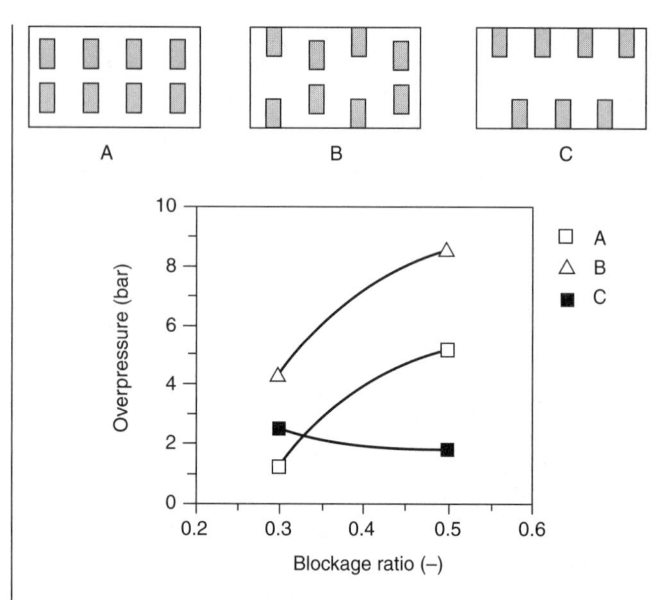

FIGURE 21.8.2.1.7 Influence of Obstacle Arrangement on Flame Propagation in Ethylene-Air Mixtures in a Vessel (van Wingerden et al., 1991).

21.8.2.1.10 When optimum (i.e., most violent) explosions occur; it is almost always at mixtures near or just above the stoichiometric mixture (i.e., slightly fuel rich). This is the optimum mixture. These mixtures produce the most efficient combustion and, therefore, the highest flame speeds, rates of pressure rise, maximum pressures, and consequently the most damage. Post-explosion fires and secondary explosions can occur if there are pockets of overly rich mixtures that further mix with air (oxygen) resulting in mixtures within the flammable/explosive range and subsequent burning. Hydrogen and certain other fuels, however, can produce explosions for a much larger range of concentrations away from stoichiometry. Furthermore, explosions can also result from structures with a high level of congestion, even when mixtures are well away from stoichiometric conditions.

21.8.2.2* Specific Gravity (air) (Vapor Density). The specific gravity (air) (vapor density) of the gas or vapor fuel can have a marked effect on the nature of the explosion damage to the confining structure, such as dwellings and other buildings. While air movement from both natural and forced convection is the dominant mechanism for moving gases in a structure, the specific gravity (air) (vapor density) can affect the movement of a gas or vapor as it escapes from its container or delivery system.

21.8.2.2.1 Fugitive gas from a natural gas leak in the first story of a multistory structure, if it is not ignited promptly, may be manifested in an explosion with an epicenter in an upper story. The natural gas, being lighter than air, will initially rise through natural openings and may even migrate inside walls. However, as natural gas mixes with air to a flammable mixture, the mixture is essentially the same density as air, and continued migration is governed largely by air movement and diffusion. The gas will continue to disperse in the structure until an ignition source is encountered. However, eventually in the absence of a continuing leak, any gas will be diluted and will dissipate.

21.8.2.2.2 Fugitive gas from an LP-Gas leak, if it is not ignited promptly, can travel away from the source and, due to its initial density, will tend to migrate downward over time. However, as LP-gas mixes with air to a flammable mixture, the mixture is essentially the same density as air, and continued migration is governed by air movement and diffusion. During the leak and for some time thereafter, the gas may be at a higher concentration in low area. However, in the absence of a continuing leak, any gas will be diluted and eventually will dissipate.

21.8.2.2.3 Ignition of the gas will occur only if the concentration is within the flammable/explosive limits and in contact with a competent ignition source.

21.8.2.2.4 Whether lighter- or heavier-than-air gases are involved, there may be evidence of the passage of flame where the fuel–air layer was. Scorching and blistering of paintwork and thermally thin materials (e.g., lampshades) are indicators of this phenomenon. The operation of heating and air-conditioning systems, temperature gradients, and the effects of wind on a building can cause mixing and movement that can reduce the effects of vapor density. Vapor density effects are greatest in still-air conditions.

21.8.2.2.5* Full-scale testing of the distribution of flammable gas concentrations from turbulent jets into rooms has shown that near stoichiometric concentrations of gas will develop between the location of the leak and either the ceiling for lighter-than-air gases or the floor for heavier-than-air gases. It was also reported that a heavier-than-air gas that leaked at floor level will create a greater concentration at floor level and that the gas will slowly diffuse upward. A similar but inverse relationship is true for a lighter-than-air gas leaked at ceiling height. In both cases this testing indicated the gas/air mixtures would become fairly homogeneous and that for turbulent jets the previous general belief was incorrect that lighter-than-air gas leaks would fill a compartment from the top down. However, for low-momentum leaks in still air, lighter-than-air gas leaks will fill a compartment from the top down *(see Figure 21.8.2.2.5).* Ventilation, both natural and mechanical, can change the movement and mixing of the gas and can result in gas spreading to adjacent rooms.

21.8.2.2.6* The specific gravity (air) (vapor density) of the fuel is not necessarily indicated by the relative elevation of the structural explosion damage above floor level. It was once widely thought that if the walls of a particular structure were blown out at floor level, the gaseous fuel would be heavier than air, and, conversely, if the walls were blown out at ceiling level, the fuel would be lighter than air. Since explosive pressure within a room equilibrates at the speed of sound, a wall will experience a similar pressure-time history across its entire height, unless significant flame acceleration occurs due to very reactive fuels (i.e., hydrogen, acetylene, ethylene) or significant turbulence generation from the structure geometry and congestion. The level of the explosion damage within a conventional room is a function of the construction strength of the wall headers and bottom plates, the least resistive giving way first.

21.8.2.2.7 Specific Gravity (air) (Vapor Density) of Ignitable Gaseous Mixtures. The specific gravity (air) (vapor density) of gases vapors in their 100 percent undiluted states can be considerably different than when they are mixed with air to a mixture within their flammable/explosive ranges. Ignitable mixtures of gases and vapors are a combination of the fuel components and air. In general, these ignitable mixtures are

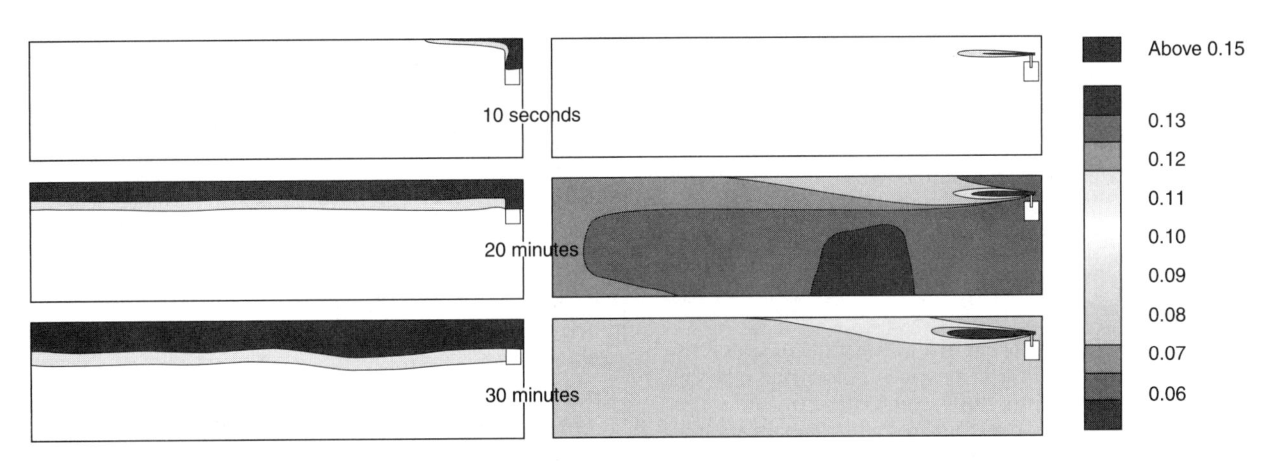

FIGURE 21.8.2.2.5 The plot is showing percent methane in air for a natural gas release involving a high momentum turbulent jet (right) and low-momentum jet (left) due to flow impingement or partial confinement. Each leak is approximately 100 ft^3/hr and the room is approximately 5300 ft^3.

predominantly air, being 85 percent to 99 percent air (notable exceptions include hydrogen, acetylene, ethylene). The specific gravity (air) (vapor density) of most ignitable mixtures are much closer to the actual specific gravity (air) (vapor density) of air (1.0) than to the specific gravity (air) (vapor density) of the 100 percent undiluted fuel components which are commonly referenced in fire and explosion investigation texts. The calculations for determining the actual specific gravity (air) (vapor density) of a given fuel/air mixture at the minimum ignitable mixture are accomplished by a mixing equation as follows:

$$\frac{VD_m = VD_f\left(C_f\right)+\left(100-C_f\right)}{100}$$

where:

VD_m = is the specific gravity (air) (vapor density) of the fuel/air mixture

VD_f = is the specific gravity (air) (vapor density) of the gaseous fuel component

C_f = is the concentration by volume percentage of the fuel within the fuel/air mixture

21.8.3 Underground Migration of Fuel Gases. Underground fuel gas leaks have migrated underground, entered structures, and fueled fires or explosions. Because the soil surrounding underground pipes and utility lines may have been more disturbed than adjacent soil or may contain special backfill material, the soil may be less dense and more porous. Both lighter-than-air and heavier-than-air fugitive fuel gases may follow these soil pathways, or the annular spaces around the exterior of such underground constructions and can enter structures through openings such as those found in foundations, floors, or walls. Often, these fugitive gases will permeate the soil, migrate upward, and dissipate harmlessly into the air. This dissipation may take place unnoticed for months or years. However, if the surface of the ground is then obstructed by rain, frozen soil, snow, ice, new paving, or other impervious cover, the gases may then begin to migrate laterally and enter structures.

21.8.3.1 Fuel gases migrating underground have been known to enter buildings by seeping into sewer lines, underground electrical or telephone conduits, drain tiles, or even directly through basement and foundation walls, none of which are as gastight as water or gas lines.

21.8.3.2 In addition, gases can move through underground conduits for hundreds of feet and then fuel explosions or fires in distant structures.

21.8.3.3 Natural gas and propane have little or no natural odors of their own. In order for them to be readily detected when leaking, foul-smelling malodorant compounds are added to the gases. Odorant verification should be a part of any explosion investigation involving or potentially involving fuel gas, especially if it appears that there were no indications of a leaking gas being detected by witnesses. The odorant's presence, in the proper amount, should be verified.

21.8.4* Multiple Explosions. A migration and pocketing effect can be manifested by the production of multiple explosions, generally referred to as secondary explosions (and sometimes cascade explosions). Gas and vapors that have migrated to adjacent stories or rooms can collect or pocket on each level. When an ignition and explosion takes place in one story or room, subsequent explosions can occur in adjoining areas or stories.

21.8.4.1 The migration and accumulation of gases often produces areas or pockets with different fuel–air mixtures. One pocket could be within the explosive range of the fuel, while a pocket in an adjoining room, story, or interstitial space could be over the upper explosive limit (UEL). When the first mixture is ignited and explodes, damaging the structure, the dynamic forces of the explosion, including the positive and negative pressure phases, tend to mix air into the fuel-rich mixture and bring it into the explosive range. This mixture in turn will explode if an ignition source of sufficient energy is present. In this way, a series of vapor/gas explosions is possible.

21.8.4.2 Multiple explosions are a very common occurrence. However, often the explosions occur so rapidly that witnesses report hearing only one, but the physical evidence, including multiple epicenters, indicates more than one explosion.

21.8.4.3 A secondary or cascade explosion in an adjacent compartment can be more violent than the primary explosion in certain situations. This violence is generally due to the first explosion acting as a very strong ignition source, creating additional turbulence and possible pre-compression in the compartment.

21.9 Dust Explosions.

21.9.1 General. Finely divided solid materials (e.g., dusts and fines), when dispersed in the air, can fuel particularly violent and destructive explosions. Even materials that are not normally considered to be combustible, such as aspirin, aluminum, sugar, or milk powders, can produce explosions when burned as dispersed dusts.

21.9.1.1* Dust explosions occur in a wide variety of materials, such as grain dusts; sawdust; carbonaceous materials, such as coal and charcoal; chemicals; drugs, such as aspirin and ascorbic acid (i.e., vitamin C); dyes and pigments; metals, such as aluminum, magnesium, and titanium; plastics; and resins, such as synthetic rubber.

21.9.2* Particle Size. Since the combustion reaction takes place at the surface of the dust particle, the rates of pressure rise generated by combustion are largely dependent on the surface area of the dispersed dust particles. For a given mass of dust material, the total surface area, and consequently the violence of the explosion, increases as the particle size decreases. The finer the dust, the more violent is the explosion. In general, an explosion hazard concentration of combustible dusts can exist when the particles are 500 microns or less in diameter, and when fibers are 500 microns or less in their smallest dimension. *(See NFPA 654.)*

21.9.2.1 There are some cases where certain particle sizes may exceed 500 microns and still present an explosion hazard. This is the case for some fibrous materials such as flock, which have very high length-to-diameter ratios.

21.9.3* Concentration. The concentration of the dust in air has a profound effect on its ignitibility and violence of the blast pressure wave. As with ignitible vapors and gases, there are minimum explosive concentrations of specific dusts required for a propagating combustion reaction to occur. Minimum explosive concentrations (MEC) can vary with the specific dust from as low as 20 g/m^3 to 2000 g/m^3 (0.015 oz/ft^3 to 2.0 oz/ft^3) with the most common concentrations being less than 1000 g/m^3 (1.0 oz/ft^3). Particle size greatly influences minimum explosive concentration. Minimum concentration values vary widely depending upon the type of dust. Published values have to be used with caution, since the literature contains values which are unreliability low due to measurement error. The Minimum Explosive Concentration (MEC) values measured in the commonly used Hartman Apparatus are roughly too low by a factor of two, while the degree of error from other apparatus has not yet been quantified. *(See Babrauskas' Ignition Handbook.)* The chemical reactivity of dust clouds is lower than that of gases, so MEC values necessarily have to be higher. Apart from exceptionally reactive gases MEC values are all greater that 30 gms/m^3, thus reported dust cloud MEC values that are lower are unreliable. Lacking more specific information a rough estimate can be made by multiplying reported Hartman Apparatus values by 2.

21.9.3.1 Unlike most gases and vapors, however, there is generally no reliable maximum limit of concentration. The reaction rate is controlled more by the surface-area-to-mass ratio than by a maximum concentration.

21.9.3.2 Similar to gases and vapors, the rate of pressure rise and the maximum pressure that occur in the dust explosion are higher if the pre-explosion dust concentration is at or close to the optimum mixture. The combustion rate and maximum pressure decrease if the mixture is fuel rich or fuel lean. The rate of pressure rise and total explosion pressure are very low at the lower explosive limit and at very high fuel-rich concentrations.

21.9.4 Turbulence in Dust Explosions. Turbulence within the suspended dust–air mixture greatly increases the rate of combustion and thereby, the rate of pressure rise. The shape and size of the confining vessel can have a profound effect on the severity of the dust explosion by affecting the degree of turbulence, such as the pouring of grain from a great height into a largely empty storage bin.

21.9.5* Moisture. Generally, increasing the moisture content of the dust particles increases the minimum energy required for ignition and the ignition temperature of the dust suspension. The initial increase in ignition energy and temperature is generally low, but, as the limiting value of moisture concentration is approached, the rate of increase in ignition energy and temperature becomes high. Above the limiting values of moisture, suspensions of the dust will not ignite. The moisture content of the surrounding air, however, has little effect on the propagation reaction once ignition has occurred.

21.9.6 Minimum Ignition Energy for Dust.

21.9.6.1 Dust explosions have been ignited by open flames, smoking materials, light bulb filaments, welding and cutting, electric arcs, static electric discharges, friction sparks, heated surfaces, and spontaneous heating.

21.9.6.2 Ignition temperatures for most material dusts range from 320°C to 590°C (600°F to 1100°F). Layered dusts generally have lower ignition temperatures than the same dusts suspended in air. Minimum ignition energies are higher for dusts than for gas or vapor fuels and generally fall within the range of 10 mJ to 40 mJ, considerably higher than most flammable gases or vapors (~.02-.29 mJ).

21.9.7 Multiple Explosions. Dust explosions in industrial scenarios usually occur in a series. The initial ignition and explosion are most often less severe than subsequent secondary explosions. However, the first explosion puts additional dust into suspension, which results in additional explosions. The mechanism for this is that acoustical and structural vibrations and the blast front from one explosion will propagate faster than the flame front, lofting dust ahead of it and entraining it in the air. In facilities such as grain elevators, these secondary explosions often progress from one area to another or from building to building.

21.10* Backdraft (Smoke Explosions).

21.10.1 When fires occur within rooms or structures that are relatively airtight, it is common for fires to become oxygen depleted. In these cases, high concentrations of heated airborne particulates and aerosols, and other flammable gases can be generated due to incomplete combustion. These heated fuels will collect in a structure where there is insufficient oxygen to allow combustion to occur and insufficient ventilation to allow them to escape.

21.10.2 When this accumulation of fuels mixes with air, such as by the opening of a window or door, they can ignite and burn sufficiently fast to produce low-order damage, though usually with less than 13.8 kPa (2 psi) overpressure in conventional structures. These are called backdrafts (smoke explosions).

21.11 Outdoor Vapor Cloud Explosions. An outdoor vapor cloud explosion is the result of the release of gas, vapor, or mist into the atmosphere, forming a cloud within the fuel's flammable limits and subsequently becoming ignited. An essential element of a vapor cloud explosion is flame acceleration to velocities that produce a blast wave. The principal characteristic of the event is potential airblast damage within and beyond the boundary of the cloud due to deflagration or detonation phenomena.

21.11.1 A vapor cloud deflagration can only occur when obstacles and confining surfaces are in the area of the flammable cloud. Congestion from an abundance of obstacles creates turbulence in the combustion zone, which increases the combustion rate. Confining surfaces in the congested region limit flame expansion in certain directions, causing an increase in pressure and flame speed, which may also enhance the combustion rate. A sufficient number of obstacles and a large enough congested volume are needed to generate a blast wave. The explosion energy in a deflagration is limited by the size of the congested zone, outside of which the flame speed decelerates rapidly and is no longer a contributor to blast pressure generation. In certain cases where the flame speed is higher than the speed of sound in the unburned mixture (fast deflagrations), unburned gases immediately outside the congestion zone may need to be considered as they may contribute to the energy in the blast wave.

21.11.2 Fuel properties affect the flame speed and overpressure achieved in a vapor cloud explosion. More reactive fuels such as hydrogen, ethylene, and acetylene achieve much higher flame speeds than lower reactivity fuels. Ammonia and methane are among the low-reactivity fuels. Propane, butane and other longer chain alkanes are among the medium-reactivity fuels.

21.11.3 Blast waves produced by a slow deflagration do not move at supersonic speeds and therefore are not shock waves. Rather, slow deflagrations usually produce blast waves with a relatively slow rate of pressure rise, which is less damaging than a shock wave of comparable pressure and impulse. A shock wave with an instantaneous rise in pressure is produced from a fast deflagrations or detonation.

21.11.4 This phenomenon also has historically been referred to as an unconfined vapor air explosion or unconfined vapor cloud explosion, but it is now recognized that congestion and confinement are needed to cause a vapor cloud explosion. Absent congestion and confinement, ignition of a flammable cloud results in a flash fire.

21.11.5 Outdoor vapor cloud explosions have generally occurred at chemical processing plants, wherein large amounts of fuel (thousands of pounds or more) are generally involved. However, congestion from natural sources (i.e., trees, vegetation) has been known to accelerate flames and cause significant overpressures.

21.12* Explosives. Explosives are any chemical compound, mixture, or device, the primary purpose of which is to function by explosion. Explosives are categorized into two main types: low explosives and high explosives (not to be confused with low-order damage or high-order damage and low-order detonation or high-order detonation).

21.12.1 Low Explosives.

21.12.1.1 Low explosives are characterized by deflagration (subsonic blast pressure wave) or a relatively slow rate of reaction and the development of low pressure when initiated. Common low explosives are smokeless gunpowder, flash powders, solid rocket fuels, and black powder. Low explosives are designed to work by the pushing or heaving effects of the rapidly produced hot reaction gases.

21.12.1.2 It should be noted that some low explosives (i.e., double-base smokeless powder) can achieve detonation under circumstances where confinement is adequate to produce sufficient reaction speed, when the initiation source is very strong, or where instabilities in combustion occur.

21.12.2 High Explosives.

21.12.2.1 High explosives are characterized by a detonation propagation mechanism. Common high explosives are dynamites, water gel, TNT, ANFO, RDX, and PETN. High explosives are designed to produce shattering effects by virtue of their high rate-of-pressure rise and extremely high detonation pressure [on the order of 6,900,000 kPa (1,000,000 psi)]. These high, localized pressures are responsible for cratering and localized damage (seats) near the center of the explosion.

21.12.2.2 The effects produced by diffuse phase (i.e., fuel–air) explosions and solid explosives are very different. In a diffuse phase explosion (usually slow deflagration), structural damage to the original confining structure will tend to be uniform and omnidirectional (unless significant flame acceleration occurs due to very reactive fuels such as hydrogen, acetylene, and ethylene, turbulence generated by the geometry of the confining vessel, or high levels of congestion such as fast deflagration), and there may be evidence of burning, scorching, or blistering. In contrast, the rate of combustion of a condensed phase explosive is extremely fast in comparison to the speed of sound. Therefore, pressure does not equalize through the explosion volume and extremely high pressures are generated near the explosive. The pressure and the resultant level of damage rapidly decay with distance away from the center of the explosion. At the location of the condensed phase fuel explosion, there should be evidence of crushing, splintering, and shattering effects produced by the higher pressures. This frequently is manifested by a pronounced "seat." Away from the source of the explosion, there is usually very little evidence of intense burning or scorching, except where hot fragments or firebrands have landed on combustible materials.

21.13 Investigation of Explosive Incidents. The investigation of incidents involving explosives requires specialized training. Explosives are strictly regulated by local and federal laws, so most explosives incidents will be investigated by law enforcement or regulatory agencies. It is suggested that only investigators with the appropriate training endeavor to conduct such investigations. Those without this training should contact law enforcement or other agencies for assistance.

21.14 Investigating the Explosion Scene.

21.14.1 General. The general objectives of the explosion scene investigation are no different from those for a regular fire investigation: determine the origin, identify the fuel and ignition source, describe the ignition sequence, determine the cause, and establish the responsibility for the incident. A systematic approach to the scene examination is at least equally and possibly more important in an explosion investigation than in a fire investigation. Explosion scenes are often larger and more disturbed than fire scenes. Without a preplanned, systematic approach, explosion investigations become even more difficult or impossible to conduct effectively.

21.14.1.1 Typical explosion incidents can range from a small pipe bomb in a dwelling to a large process explosion encompassing an entire facility. While the investigative procedures described in 21.14.1.2 through 21.14.4 are more comprehensive for the large incidents, the same principles should be applied to small incidents, with appropriate simplification.

21.14.1.2 When damage is very extensive and includes much structural damage, an explosion dynamics expert and a structural expert should be consulted early in the investigation to aid in the complex issues involved.

21.14.2 Securing the Scene. The first duty of the investigator is to secure the scene of the explosion. First responders to the explosion should establish and maintain physical control of the structure and surrounding areas. Unauthorized persons should be prevented from entering the scene or touching blast debris remote from the scene itself because the critical evidence from an explosion (whether accidental or criminal) may be very small and may be easily disturbed or moved by people passing through. Evidence is also easily picked up on shoes and tracked out. Properly securing the scene also tends to prevent additional injuries to unauthorized persons or to the curious who may attempt to enter an unsafe area.

21.14.2.1 Establishing the Scene. As a general rule, it is desirable for the outer perimeter of the incident scene to be established at 1½ times the distance of the farthest piece of debris evidence found. Significant pieces of blast debris can be propelled great distances or into nearby buildings or vehicles, and these areas should be included in the scene perimeter. If additional pieces of debris are found, the scene perimeter should be widened. It may be more practical for spot debris documentation and collection to be implemented in conditions of extreme distances from the explosion.

21.14.2.2 Obtain Background Information. Before beginning any search, all relevant information should be obtained pertaining to the incident. This information should include a description of the incident site and systems or operations involved and of conditions and events that led to the incident. The locations of any combustibles and oxidants that were present and what abnormal or hazardous conditions existed that might account for the incident need to be determined. Any pertinent information regarding suspected explosive materials and causes will be of interest and will aid in the search as well.

21.14.2.2.1 In developing the evidence, the investigator should examine witness accounts, maintenance records, operational logs, manuals, weather reports, previous incident reports, and other relevant records. Recent changes in equipment, procedures, and operating conditions can be especially significant.

21.14.2.2.2 Obtaining pre-explosion drawings or photographs of the building or process will greatly improve documentation of the scene.

21.14.2.3 Establish a Scene Search Pattern. The investigator should establish a scene search pattern. With the assistance of investigation team members, the scene should be searched from the outer perimeter inward toward the area of greatest damage. The final determination of the location of the explosion's origin (epicenter) should be made only after the entire scene has been examined.

21.14.2.3.1 The search pattern itself may be spiral, circular, or grid shaped. Often, the particular circumstances of the scene and number of available searchers will dictate the nature of the pattern. In any case, the assigned areas of the search pattern should overlap so that no evidence will be lost at the edge of any search area. It is often useful to search areas more than once. When this search pattern is done, a different searcher should be used to repeat the search to help ensure that evidence is not overlooked.

21.14.2.3.2 The number of actual searchers will depend on the physical size and complexity of the scene. The investigator in charge should keep in mind, however, that too many searchers can often be as counterproductive as too few. Searchers should be briefed as to the proper procedures for identifying, logging, photographing, and marking and mapping the location of evidence. Consistent procedures are imperative whenever there are several searchers involved.

21.14.2.3.3 The location of evidence may be marked with chalk marks, spray paint, flags, stakes, or other marking means. After photographing, the evidence may be tagged, moved, and secured.

21.14.2.3.4 Evidence at or near the center of an explosion is frequently among the most helpful to an investigation. However, damage near or inside the explosion area may be significant, and evidence may degrade with time and exposure to weather. Searching the explosion area as early as can safely be accomplished can lead to collection of evidence that may degrade or be lost with time. Such evidence includes chemical and metallurgical samples, paper records, soot patterns, electronic data, and positions of light weight materials.

21.14.2.4 Safety at the Explosion Scene. All of the fire investigation safety recommendations listed in the Safety chapter also apply to the investigation of explosions. In addition, there are some special safety considerations at an explosion scene.

21.14.2.4.1 Structures that have suffered explosions are often more structurally damaged than merely burned buildings. The possibility of floor, wall, ceiling, roof, or entire building collapse is great and should always be considered.

21.14.2.4.2 In the case of fuel gas or dust explosions, secondary explosions are the rule rather than the exception. Early responders need to remain alert to that possibility. Leaking gas or pools of flammable liquids need to be made safe before the investigation is begun. Toxic materials in the air or on material surfaces need to be neutralized. The use of appropriate personal safety equipment is recommended.

21.14.2.4.3 Explosion scenes that involve bombings or explosives have added dangers. Investigators should be on the lookout for additional devices and undetonated explosives. The modus operandi of some bombers or arsonists includes using secondary explosive devices specifically targeted for the law enforcement or fire service personnel who will be responding to the bombing incident.

21.14.2.4.4 A thorough search of the scene should be conducted for any secondary devices prior to the initiation of the post-blast investigation. If undetonated explosive devices or explosives are found, it is imperative that they not be moved or touched. The area should be evacuated and isolated, and explosives disposal authorities summoned.

21.14.3 Initial Scene Assessment.

21.14.3.1 General. Once the explosion scene has been established, the investigator should make an initial assessment of the type of incident with which he or she is dealing. The initial assessment also includes scene safety. Scene safety often precludes investigation activities until hazards are rendered safe and the level of personal protective equipment has been established. (*See Chapter 12.*) If at any time during the investigation the investigator determines that the explosion was fueled by explosives or involved an explosive device, the investigator should discontinue the scene investigation, secure the scene, and contact the appropriate law enforcement agency.

21.14.3.2 Identify Explosion or Fire. An early task in the initial assessment is to determine whether the incident was a fire, explosion, or both. It may be a lengthy process to determine what type of event occurred and which came first. Often the evidence of an explosion is not obvious, for example, where a weak explosion of a gaseous fuel is involved.

21.14.3.2.1 The investigator should look for signs of an overpressure condition existing within the structure, including displacement or bulging of walls, floors, ceilings, doors and windows, roofs, other structural members, nails, screws, utility service lines, panels, and boxes. Localized fragmentation and pressure damage should be noted as attributable to condensed phase explosive fuel reaction.

21.14.3.2.2 The investigator should look for and assess the nature and extent of heat damage to the structure and its components and decide whether it can be attributed to fire alone or a propagating flame front.

21.14.3.3 Document Damage. Explosion damage may be used to help classify the type, quantity, and mixture of the fuel involved. Damage should be documented before the scene is disturbed and as the scene is altered to uncover or gain access to areas. In ordinarily dwellings or commercial structures the investigator may find it helpful to determine whether the nature of damage indicates high-order or low-order damage. *(See Section 21.3.)*

21.14.3.4 Seated or Nonseated Explosion. The investigator should determine whether the explosion was seated or nonseated. This will help classify the type of explosion (e.g., condensed phase detonation, mechanical explosion, diffuse fuel, etc.). *(See Section 21.6.)*

21.14.3.5 Identify Type of Explosion. The investigator should identify the type of explosion involved (e.g., mechanical, combustion, other chemical reaction, or BLEVE).

21.14.3.6 Identify Potential General Fuel Type.

21.14.3.6.1 The investigator should identify which types of fuel were potentially available at the explosion scene by identifying the condition and location of utility services including fuel gases, and sources of other fuels such as ignitible dusts or liquids.

21.14.3.6.2 The investigator should analyze the nature of damage in comparison to the typical damage patterns available from the following:

(1) Gases
(2) Liquid vapors
(3) Dusts
(4) Explosives
(5) Backdrafts
(6) BLEVEs

21.14.3.7 Establish the Origin. The investigator should attempt early on to establish the origin of the explosion, the location where the ignition occurred. The origin may not be the area of most damage. The origin may include a crater or other localized area of severe damage in the case of a seated explosion. In the case of a diffuse fuel explosion, the origin will be the location where the flammable mixture came in contact with the competent ignition source at the time of ignition. In case of a diffuse fuel explosion inside a structure, the smallest identifiable area of origin may be the confining volume or room. For many explosions, it is also necessary to understand how the gas cloud or flammable mixture accumulates within the structure from a given leak source. The size, shape and concentration and distribution within the structure may be very important in determining the origin.

21.14.3.8 Establish Ignition Source.

21.14.3.8.1 The investigator should attempt to identify the ignition source involved. At times, this can be very difficult. Examination should be made for potential sources, such as hot surfaces, electrical arcing, static electricity, open flames, sparks, chemicals, etc., where fuel–air mixtures are involved. Specialized experiments or Computational Fluid Dynamics (CFD) tools can be used to evaluate the location of ignition sources based on the observed damage and the initiating dynamics of the explosion.

21.14.3.8.2 Where explosives are involved, the initiation source may be a blasting cap or other pyrotechnic device. Wires and device components will often survive, but finding and identifying such components requires detailed examination of the scene and debris.

21.14.4 Detailed Scene Assessment. Armed with general information from the initial scene assessment, the investigator may now begin a more detailed study of the blast damage and debris. As in any fire incident investigation, the investigator should record his or her investigation and findings by accurate note taking, photography, and diagramming. It is important to use proper collection and preservation techniques.

21.14.4.1 Identify Damage Effects of Explosion. The investigator should make a detailed examination and analysis of the specific explosion or overpressure damage. Damaged articles should be identified as having been affected by one or more of the following typical explosion forces:

(1) Blast overpressure and wave (Blast Pressure Front) — positive phase
(2) Blast pressure and wave (Blast Pressure Front) — negative phase
(3) Fragment impact
(4) Thermal energy
(5) Ground shock
(6) Dynamic drag loads (explosion wind)

21.14.4.1.1 The investigator should examine and classify the type of damage present, whether it was shattered, bent, broken, or flattened, and also look for changes in the nature of damage. For example, as the distance increases from a fast deflagration or detonation, the pressure will decrease and the effects may resemble those of a slow deflagration explosion, while materials in the immediate vicinity of the shock front will exhibit splintering and shattering. Significant overpressures and shock waves can be generated by fast deflagrations or detonations.
Upper H2+NO2+1.33N2, Lower C3H8+5O2+9N2 (Shepherd, Proc. Combust. Inst.32, pp. 83–98, 2009)

21.14.4.1.2 The investigator should make a detailed examination and analysis of the specific explosion or overpressure damage. Damaged articles should be identified as having been affected by one or more of the damaging effects of explosions: blast overpressure and wave, dynamic drag loads, fragment impact, thermal effects, and ground shock effects.

21.14.4.1.3 The scene should be examined carefully and fragments and foreign material should be recovered, as well as debris from the seat itself. Fragments may require forensic laboratory analysis for their identification to determine if they are fragments of the original vessel or container or portions of an explosive device.

21.14.4.1.4 Debris Field Diagram. A detailed diagram of the debris field, as shown in Figure 21.14.4.1.4, denoting the material, size, mass, direction, distance, physical characteristic, and identity can be helpful in determining the origin of the explosion.

21.14.4.1.5 Table 21.14.4.1.5(a) and Table 21.14.4.1.5(b) provide examples of injury and building damage, respectively for various reported overpressure levels. These data are from peak overpressure applied to the structure's exterior. The effects of explosion overpressure on the inside of the structure need not be as great to produce similar damage. Actual injury and building damage level may vary from Table 21.14.4.1.5(a) and Table 21.14.4.1.5(b) due to differences in the size, health, and age of people, structural design and condition of buildings, and age of the data reported in the tables.

21.14.4.1.6 It should be noted that the estimation of structural damage from an explosion is a very complex topic. A thorough treatment involves maximum pressure and impulse of the explosion, as well as the natural period and strength characteristics of the confining structure. Generally, one can expect a peak overpressure of 6.9 kPa to 13.8 kPa (1 psi to 2 psi) to cause the failure of most light structural assemblies, such as doors, non-reinforced wood siding, corrugated steel panels, or masonry block walls, and conventional windows can break at less than 3.4 kPa (0.5 psi). In comparison, much higher overpressures can be tolerated when the structural design is reinforced, particularly with materials of good ductility (e.g., steel).

21.14.4.2 Identify Pre-Blast and Post-Blast Fire Damage. Fire or heat damage should be identified as having been caused by a pre-existing fire or by the thermal effect of the explosion. Debris that has been propelled away from the point of origin should be examined to determine whether it has been burned. Debris of this nature that is burned may be an indicator that a fire preceded the explosion.

21.14.4.2.1 Probably the most common sign of an overpressure condition is window glass thrown some distance from the windows of the structure. The residue of smoke or soot on fragments of window glass or other structural debris reveals that the explosion followed a fire by some time, whereas perfectly clean pieces of glass or debris thrown large distances from the structure can indicate an explosion preceding the fire or that the fire did not affect the glass prior to the explosion.

21.14.4.2.2 The direction of flow of melted and resolidified debris may tell the investigator the position or orientation of the debris at the time of heat exposure.

21.14.4.3 Locate and Identify Articles of Evidence. Investigators should locate, identify, note, log, photograph, and diagram any of the articles of physical evidence. Because of the propelling nature of explosions, the investigator should keep in mind that significant pieces of evidence may be found in a wide variety of locations, such as outside the exploded structure, embedded in the walls or other structural members of the exploded structure, on or in nearby vegetation, inside adjacent structures or vehicles, or embedded in these adjacent structures. In the case of bombing incidents or incidents involving the explosion of tanks, appliances, or equipment, significant pieces of evidence debris may have pierced the bodies of victims or be contained in their clothing.

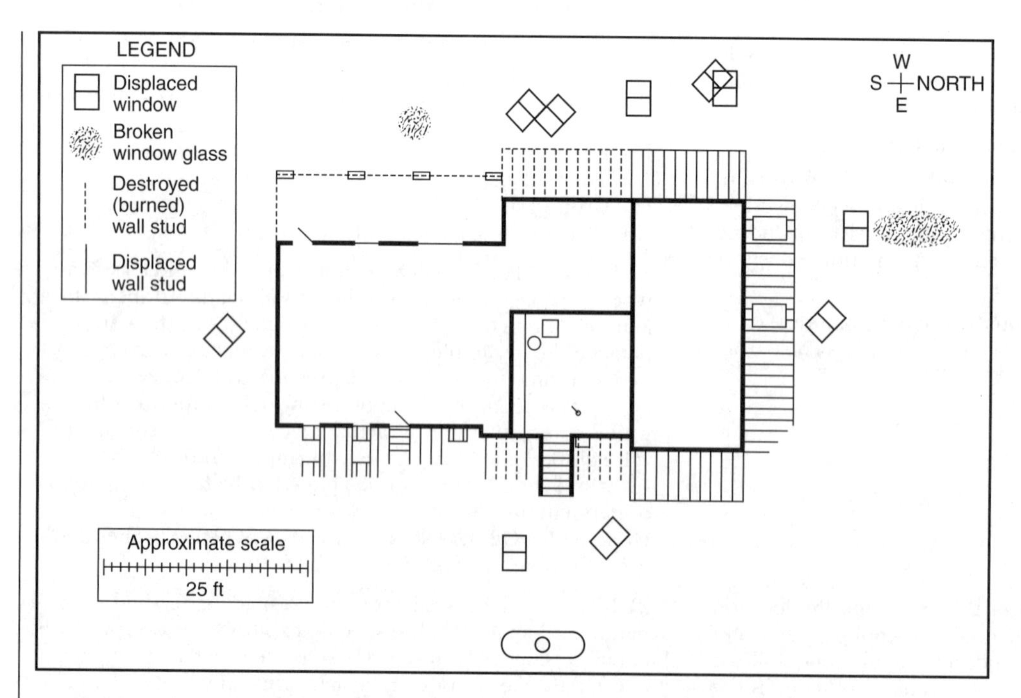

FIGURE 21.14.4.1.4 A Simple Debris Field Diagram from a Dwelling Gas Explosion.

Table 21.14.4.1.5(a) Human Injury Criteria (Includes Injury from Flying Glass and Direct Overpressure Effects)

Overpressure (psi)	Injury	Comments	Source
0.6	Threshold for injury from flying glass*	Based on studies using sheep and dogs	a
1.0–2.0	Threshold for skin laceration from flying glass	Based on U.S. Army data	b
1.5	Threshold for multiple skin penetrations from flying glass (bare skin)*	Based on studies using sheep and dogs	a
2.0–3.0	Threshold for serious wounds from flying glass	Based on U.S. Army data	b
2.4	Threshold for eardrum rupture	Conflicting data on eardrum rupture	b
2.8	10% probability of eardrum rupture	Conflicting data on eardrum rupture	b
3.0	Overpressure will hurl a person to the ground	One source suggested an overpressure of 1.0 psi for this effect	c
3.4	1% eardrum rupture	Not a serious lesion	d
4.0–5.0	Serious wounds from flying glass near 50% probability	Based on U.S. Army data	b
5.8	Threshold for body-wall penetration from flying glass (bare skin)*	Based on studies using sheep and dogs	a
6.3	50% probability of eardrum rupture	Conflicting data on eardrum rupture	b
7.0–8.0	Serious wounds from flying glass near 100% probability	Based on U.S. Army data	b
10.0	Threshold lung hemorrhage	Not a serious lesion [applies to a blast of long duration [over 50 m/sec (164 ft/sec)]; 20–30 psi required for 3 m/sec (9 ft/sec) duration waves	d
14.5	Fatality threshold for direct blast effects	Fatality primarily from lung hemorrhage	b
16	50% eardrum rupture	Some of the ear injuries would be severe	d
17.5	10% probability of fatality from direct blast effects	Conflicting data on mortality	b
20.5	50% probability of fatality from direct blast effects	Conflicting data on mortality	b
25.5	90% probability of fatality from direct blast effects	Conflicting data on mortality	b
27	1% mortality	A high incidence of severe lung injuries [applies to a blast of long duration (over 50 m/sec)]; 60–70 psi required for 3 m/sec duration waves	d
29	99% probability of fatality from direct blast effects	Conflicting data on mortality	b

For SI units, 6.9 kPa = 1 psi.
*Interpretation of tables of data presented in reference.
Source a = Fletcher, Richmond, and Yelverton, 1980.
Source b = F. Lees, Loss Prevention in the Process Industries, 1996.
Source c = Brasie and Simpson, 1968.
Source d= U.S. Department of Transportation, 1988.

21.14.4.3.1 The clothing of anyone injured in an explosion should be obtained and preserved for examination and possible analysis. The investigator should ensure that photographs are taken of the injuries and that any material removed from the victims during medical treatment or surgery is preserved. This is true whether the person survives or not.

21.14.4.3.2 Investigators should note the condition and position of any damaged and displaced structural components, such as walls, ceilings, floors, roofs, foundations, support columns, doors, windows, sidewalks, driveways, and patios.

21.14.4.3.3 Investigators should note the condition and position of any damaged and displaced building contents, such as furnishings, appliances, heating or cooking equipment, manufacturing equipment, victims' clothing, and personal effects.

21.14.4.3.4 Investigators should note the condition and position of any damaged and displaced utility equipment, such as fuel gas meters and regulators, fuel gas piping and tanks, electrical boxes and meters, electrical conduits and conductors, heating oil tanks, parts of explosive devices, or fuel vessels.

21.14.4.3.5 If it is believed that fugitive gas from fuel gas systems is involved in an explosion incident, the relevant portions of the piping system should be examined and documented, and if necessary, tested. For residential and light commercial structures, this examination and documentation should include the customer service lateral in the case of commercial natural gas and the storage tank or cylinder in the case of LP Gas, The examination and documentation should also include any underground piping, interior house piping, equipment, and appliances, including valves and

Table 21.14.4.1.5(b) Property Damage Criteria

Overpressure (psi)	Damage	Source
0.03	Occasional breaking of large glass windows already under strain	a
0.04	Loud noise (143 dB). Sonic boom glass failure	a
0.1	Breakage of small windows, under strain	a
0.15	Typical pressure for glass failure	a
0.30	"Safe distance" (probability 0.95 no serious damage beyond this value)	a
	Missile limit	
	Some damage to house ceilings	
	10% window glass broken	
0.4	Minor structural damage	a, c
0.5–1.0	Shattering of glass windows, occasional damage to window frames. One source reported glass failure at 1 kPa (0.147 psi)	a, c, d, e
0.7	Minor damage to house structures	a
1.0	Partial demolition of houses, made uninhabitable	a
1.0–2.0	Shattering of corrugated asbestos siding	a, b, d, e
	Failure of corrugated aluminum–steel paneling	
	Failure of wood siding panels (standard housing construction)	
1.3	Steel frame of clad building slightly distorted	a
2.0	Partial collapse of walls and roofs of houses	a
2.0–3.0	Shattering of nonreinforced concrete or cinder block wall panels [10.3 kPa (1.5 psi) according to another source]	a, b, c, d
2.3	Lower limit of serious structural damage	a
2.5	50% destruction of brickwork of house	a
3	Steel frame building distorted and pulled away from foundations	a
3.0–4.10	Collapse of self-framing steel panel buildings	a, b, c
	Rupture of oil storage tanks	
	Snapping failure — wooden utility tanks	
4.0	Cladding of light industrial buildings ruptured	a
4.8	Failure of reinforced concrete structures	e
5.0	Snapping failure — wooden utility poles	a, b
5.0–7.0	Nearly complete destruction of houses	a
7.0	Loaded train wagons overturned	a
7.0–8.0	Shearing/flexure failure of brick wall panels [20.3 cm to 30.5 cm (8 in. to 12 in.) thick, not reinforced]	a, b, c, d
	Sides of steel frame buildings blown in	d
	Overturning of loaded rail cars	b, c
9.0	Loaded train boxcars completely demolished	a
10.0	Probable total destruction of buildings	a
30.0	Steel towers blown down	b, c
88.0	Crater damage	e

For SI units, 6.9 kPa = 1 psi.
Source a = F. Lees, Loss Prevention in the Process Industries, 1996.
Source b = Brasie and Simpson, 1968.
Source c = U.S. Department of Transportation, 1988.
Source d = U.S. Air Force, 1983.
Source e = McRae, 1984.

fittings. This documentation should include photography and notes, as well as diagrams.

21.14.4.4 Identify Force Vectors. Investigators should identify, diagram, photograph, and note those pieces of debris that indicate the direction and relative force of the explosion. Keep in mind that the force necessary to shatter a wall is more than that necessary to merely dislodge or displace it. The force necessary to shatter a window is more than that necessary to blow out a window intact. The greater the force, the farther is the distance that similar pieces of debris will be thrown from the epicenter.

21.14.4.4.1 The investigator should log, diagram, and photograph varying missile distances and directions of travel for similar debris, such as window glass. Larger, more massive missiles should be measured and weighed for comparison of the forces necessary to propel them.

21.14.4.4.2 The distance as well as the direction of significant pieces of evidence from the apparent epicenter of the explosion may be critical. The location of all significant pieces should be completely documented on the explosion scene diagram, along with notes as to both distance and direction. This procedure allows the investigator to reconstruct the displacement of various components.

21.14.4.4.3 Other directional indicators may be present in some cases. Fuel gas explosions may exhibit heat damage impingement effects on articles in the path of the flame front. Clothing and skin of personnel may show directional patterns as well. Dust explosions may exhibit similar phenomena, and may also include imbedded burned or unburned dust particles on the articles of impingement.

21.15 Analyze Origin (Epicenter). After identifying the force vectors, the investigator should trace backward from the least to the most damaged areas, following the general path of the explosion force vectors. This process is known as an explosion dynamics analysis. It can be accomplished most efficiently by plotting on a diagram of the exploded structure the various directions of debris movement and, if possible, an estimate of the relative force necessary for the damage or movement of each significant piece of debris, as indicated in Figure 21.15. A dimensional diagram is desirable in cases where an engineering analysis of the damage effects is anticipated.

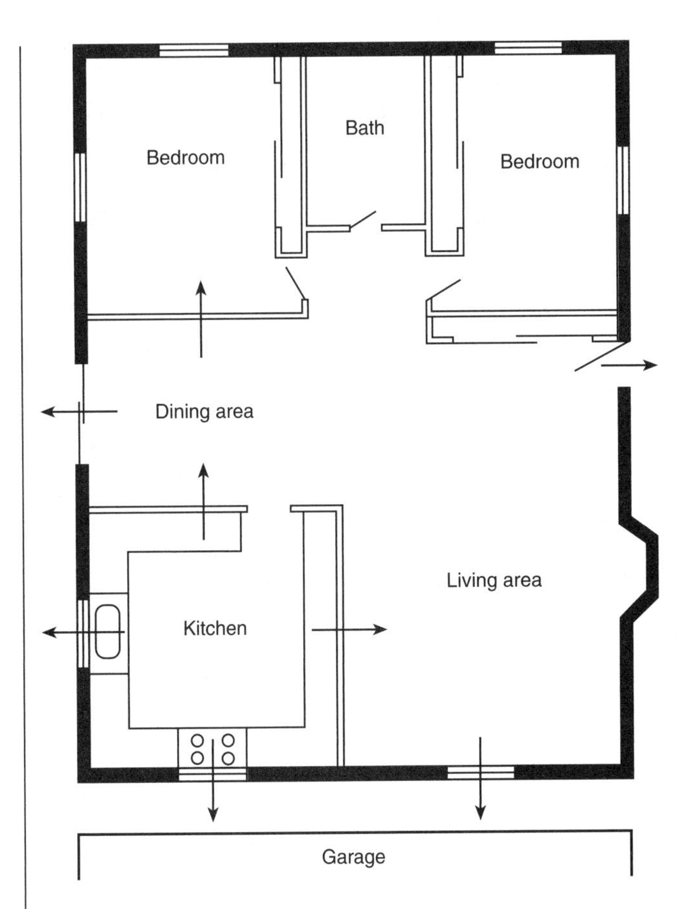

FIGURE 21.15 Diagram Showing Displacement of Walls, Doors, and Windows Due to Explosion.

21.15.1 The analysis of the explosion dynamics is based on the debris movement away from the epicenter of the explosion in a roughly spherical pattern and on the decreasing force of the explosion as the distance from the epicenter increases. Figure 21.15 illustrates a general origin for the explosion in the kitchen.

21.15.2 Often, more than one explosion dynamics diagram is necessary. The first diagram might show a relatively large area that may indicate a specific area or room for further study as the origin (as in Figure 21.15). A second, smaller-scale diagram might then be constructed to analyze the explosion dynamics of the room or compartment of origin itself. This smaller-scale diagram is especially useful when dealing with a seated explosion.

21.15.3 Often, especially when dealing with non-seated explosions, such as fugitive fuel gas explosions, the investigator may be unable to pinpoint the origin (epicenter) any more precisely than to a specific room or area.

21.15.4 The explosion dynamics analysis is often complicated by evidence of a series of explosions, each with its own epicenter. This situation calls for a detailed comparison of the force vectors. Movement of more solid debris, such as walls, floors, and roofs, is generally less in subsequent explosions than in the first. The first forceful explosion tends to vent the structure, allowing more of the positive pressure phase of subsequent explosions to be released.

21.15.5 This finding is true, however, only when the secondary explosions are of the same or lesser force than the first. Dust explosions are a notable exception to this phenomenon, with subsequent explosions frequently being more powerful than the first.

21.16 Analyze Fuel Source. Once the origin or epicenter of the explosion has been identified, the investigator should determine the fuel. This determination is made by a comparison of the nature and type of damage to the known available fuels at the scene.

21.16.1 All available fuel sources should be considered and eliminated until one fuel can be identified as meeting all of the physical damage criteria as well as any other significant data. For example, if the epicenter of the explosion is identified as a 1.8 m (6 ft) crater of pulverized concrete in the center of the floor, fugitive natural gas can be eliminated as the fuel, and only fuels that can create seated explosions should be considered.

21.16.2 Chemical analysis of debris, soot, soil, or air samples can be helpful in identifying the fuel. With explosives or liquid fuels, gas chromatography, mass spectroscopy, or other chemical tests of properly collected samples may be able to identify their presence.

21.16.3 Air samples taken in the vicinity of the area of origin can be used in identifying gases or the vapors of liquid fuels. For example, commercial "natural gas" is a mixture of methane, ethane, propane, nitrogen, and butane. The presence of ethane in an air sample may show that commercial "natural gas" was there rather than naturally occurring "swamp, marsh, or sewer" gas, of which methane is the only ignitable component.

21.16.4 Once a fuel is identified, the investigator should determine its source. For example, if the fuel is identified as a lighter-than-air gas and the structure is serviced by natural gas, the investigator should locate the source of gas that will most likely be at or below the epicenter, possibly from a leaking service line or malfunctioning gas appliance.

21.16.5 All gas piping, including from the street mains or LP-Gas storage tanks, up to and through the service regulator and meter, up to and including all appliances, should be examined and leak tested if possible. (*See* NFPA 54, *National Fuel Gas Code, and the National Fuel Gas Code Handbook.*) Unless it has been determined that the use of Fuel Gases for testing would be safe, leak testing inside a building that has had a fire or gas explosion, and which remains essentially intact, should be performed using air or an inert gas (e.g. carbon dioxide, nitrogen or helium).

21.16.5.1 In no case should oxygen been used as the testing medium.

21.16.5.2 The leakage shall be located by means of an approved gas detector, a noncorrosive leak detection fluid, or other approved leak detection methods. Matches, candles, open flames, or other methods that provide a source of ignition shall not be used. (*See* NFPA 54, *National Fuel Gas Code.*)

21.16.6 Odorant verification should be part of any explosion investigation involving, or potentially involving, fuel gas(es), especially if there are indications that leaking gas was not detected by people present. Its presence should be verified. Stain tubes can be used in the field, or gas samples can be collected for laboratory analysis and gas chromatography can be used as a lab test for more accurate results.

21.17 Analyze Ignition Source. When the area of origin and fuel are identified, the means of ignition should be analyzed. This analysis is often the most difficult part of the overall explosion investigation, especially with fugitive fuel gases, because multiple competent ignition sources may be present. In the event of multiple possible ignition sources, the investigator should take into consideration all the available information, including witness statements. A careful evaluation of every possible ignition source should be made. Factors to consider include the following:

(1) Minimum ignition energy of the fuel
(2) Ignition energy of the potential ignition source
(3) Ignition temperature of the fuel
(4) Temperature of the ignition source
(5) Location of the ignition source in relation to the fuel
(6) Simultaneous presence of the fuel and ignition source at the time of ignition
(7) Witness accounts of conditions and actions immediately prior to and at the time of the explosion

21.18 Analyze to Establish Cause.

21.18.1 General. Having identified the origin, fuel, and ignition source, the investigator should analyze and determine what brought together the fuel and ignition at the origin. The circumstances that brought these elements together at that time and place are the cause. This involves a careful analysis of the probable ignition sequence.

21.18.1.1 Part of this analysis may include considerations of how the explosion could have been prevented, such as failure to conform to existing codes or standards. It should be noted that, due to the destructive effects of fire and explosions, the cause cannot always be determined.

21.18.1.2 Many techniques are suggested in 21.18.2 through 21.18.6 to aid in establishing causation. The choice of the technique(s) used will depend on the unique circumstances of the incident.

21.18.2 Time Line Analysis. Based on the background information gathered (e.g., statements and logs), a sequence of events should be tabulated for the time both prior to the explosion and during the explosion. Consistencies and inconsistencies with causation theories can then be examined and a "best fit" hypothesis established.

21.18.3 Damage Pattern Analysis. Various types of damage patterns, principally debris and structural damage, should be documented for further analysis.

21.18.3.1 Debris Analysis.

21.18.3.1.1 Investigators should identify, diagram, photograph, and note those pieces of debris that indicate the direction and relative force of the explosion. In general, the greater the explosive energy, the farther similar pieces of debris will be thrown from the center of the explosion. However, different drag and lift (i.e., aerodynamic) characteristics of various fragment shapes will tend to favor some pieces going farther. Also, construction details affect debris throw, including mass of various surfaces, strength of connections, window and door area, below grade rooms, and other details.

21.18.3.1.2 The distance as well as the direction of significant pieces of evidence from the apparent center of the explosion may be critical. The location of all significant pieces should be completely documented on the explosion scene diagram, along with notes as to both distance and direction. This procedure allows the investigator to reconstruct the trajectories of various components. In some cases, it is desirable to weigh and make geometric measurements of significant missiles, especially large ones. These measurements can then be used in a more complete engineering analysis of trajectories.

21.18.3.2 Relative Structural Damage Analysis. Investigators should diagram the damage to the areas surrounding the explosion site. Analysis of damage to surrounding structures may be used to estimate blast loads and develop a pressure contour map. Such an analysis will give additional clues to explosion propagation and can be used for further input to a more complete engineering analysis.

21.18.4* Correlation of Explosion Type and Energy with Damage Incurred. There are several methods that analysts use to correlate the degree of damage and projectile distance with the type and amount of fuel involved. Due to the great differences in chemical dynamics between solid explosives and gas/vapor deflagrations, it is not possible to directly correlate the amount of fuel involved in one to the weight of explosive used in the other. Condensed phase materials that detonate may be compared to high explosive blast data, but differences in performance may occur. Weight equivalencies for common condensed-phase explosives can be found in the literature.

21.18.5 Analysis of Damaged Items and Structures. Frequently, the determination of the cause in explosion incidents requires a multidisciplinary approach to relate damage to the fuels involved. The use of special experts may be necessary.

21.18.6 Correlation of Thermal Effects. A collection of artifacts exhibiting heat damage from an explosive event may provide evidence of a fireball or fire during the sequence of events. These artifacts may be further proof that the explosion event may have involved a BLEVE, a fuel jet fire (flare), or other phenomenon, depending on the character of those articles. Specialized analysis of thermal damage effects can be conducted by a person trained in this area. From this material, an isothermal diagram (i.e., heat damage map) can be developed.

Chapter 22 Incendiary Fires

22.1* Introduction. An incendiary fire is a fire that has been deliberately ignited under circumstances in which the person knows the fire should not be ignited. This chapter provides guidance to assist the investigator in identifying incendiary fires and documenting evidence regarding their origin and cause. In the event the investigator concludes that a fire was incendiary, other evidentiary factors are addressed regarding suspect development and identification. The existence of a single indicator or a combination of indicators is not necessarily conclusive proof that a fire is of incendiary cause. However, the presence of indicators may suggest that the fire deserves further investigation.

22.2 Incendiary Fire Indicators. There are a number of conditions related to fire origin and spread that may provide physical evidence of an incendiary fire cause.

22.2.1 Multiple Fires. Multiple fires are two or more separate, nonrelated, simultaneously burning fires. The investigator should search to uncover any additional fire sets or points of origin that may exist. In order to conclude that there are multiple fires, the investigator should determine that any "separate" fire was not the natural outgrowth of the initial fire.

22.2.1.1 Fires in different rooms, fires on different stories with no connecting fire, or separate fires inside and outside a building are examples of multiple fires. A search of the fire building and its surrounding areas should be conducted to determine whether there are multiple fires.

22.2.1.2 Separate fires that are not caused by multiple deliberate ignitions can result from the following:

(1) Fire spread by conduction, convection, or radiation
(2) Fire spread by flying brands
(3) Fire spread by direct flame impingement
(4) Fire spread by falling flaming materials (i.e., drop down) such as curtains
(5) Fire spread through shafts, such as pipe chases or air-conditioning ducts
(6) Fire spread within wall or floor cavities within "balloon construction"
(7) Overloaded electrical wiring
(8) Utility system failures
(9) Lightning

22.2.1.3 Apparent multiple points of origin can also result from continued burning at remote parts of a building during fire suppression and overhaul, particularly when building collapse or partial building collapse is involved.

22.2.1.4 The earlier a fire is extinguished, the easier it is to identify multiple points of origin. Once full-room involvement or room-to-room extension has occurred, identifying multiple fires becomes more difficult and a complete burnout or "black hole" may make identification impossible.

22.2.1.5 If there has been a previous fire in the building, care should be taken not to confuse earlier damage with a multiple fire situation.

22.2.1.6 Fire scene reconstruction *(see Section 17.7)*, an important aspect of the fire scene examination, is especially important when multiple fires are suspected.

22.2.1.7 A careful examination of the fire scene may reveal additional fire sets that are intended to ignite additional fires, particularly in the same type of area. For example, if the investigator observes or discovers an area of origin in a closet, an examination of other closets for additional fires or fire sets is prudent. The investigator may be required to obtain legal authority to conduct a search in areas not affected or involved in the discovered fire. *(See 13.3.2 and 13.3.3.)*

22.2.2 Trailers. A trailer is a deliberately introduced fuel or manipulation of existing fuel(s) used to aid the spread of a fire from one area to another. Trailers can be used along floors to connect separate fire sets, or on stairways to move fires from one story or level to another. Fuels used for trailers may be ignitible liquids, solids, or combinations of these. Trailers are frequently indicated by elongated fire patterns. *(See Figure 6.3.7.10.1.)*

22.2.2.1 Confirmation of the presence of a trailer is a compelling indication that the fire was incendiary.

22.2.2.2 Materials such as clothing, paper, straw, and ignitible liquids are often used. Remnants of solid materials frequently are left behind and should be collected and documented.

22.2.2.3 Ignitible liquids may leave linear patterns, particularly when the fires are extinguished early. Radiant energy from the extension of flame or hot gases through corridors or up stairways can also produce linear patterns. As with suspected solid accelerants, samples of possible liquid accelerants should be collected and analyzed. *(See Section 16.5.)*

22.2.2.4 Often, when the floor area is cleared of debris to examine damage, long, wide, straight patterns will be found showing areas of extensive heat damage, bound on each side by undamaged or less damaged areas. These patterns have often been interpreted to be trailers. While this conclusion is possible, the presence of furniture, stock, counters, or storage may result in these linear patterns. These patterns may also result from fire impact on worn areas of floors and the floor coverings. Irregularly shaped objects on the floor, such as clothing or bedding, may provide protection to the floor, resulting in patterns that may be inaccurately interpreted.

22.2.2.5 For example, gasoline itself poured out to assist the fire is an accelerant. It is the deliberate use of the gasoline to spread the fire from one location to another that causes the stream of gasoline to be a trailer. Trailing gasoline from one room to another and up the staircase constitutes laying a trailer. Dousing a building with gasoline from cellar to rooftop or over a widespread area does not constitute laying a trailer; instead, it is considered using an accelerant. So it can be seen that the fuel does not constitute a trailer, but rather the manner in which the fuel or accelerant is used. This distinction is similar to the "use" requirement in the definition of an accelerant. The burning action has no effect on whether there is a trailer. Gasoline, rags, or newspapers can all be used as trailers, but they burn differently. The pattern that is left by a trailer is evidence of the trailer; the pattern is not the trailer. If an arsonist lays a trailer but is arrested prior to ignition, there is still a trailer.

22.2.3 Lack of Expected Fuel Load or Ignition Sources. When the fire damage at the origin is inconsistent with the known or reported fuel load, limited rates of heat release, or limited potentially accidental ignition sources, the fire may be incendiary. An example of all three is an isolated burn at floor level in a large, empty room. Examples of limited fire load areas include corridors and stairways. Stairways, while usually having limited fire loads, may promote rapid fire spread by allowing flames or hot gases to travel vertically to other areas. This action may cause severe damage on exposed stairway surfaces. Additional examples of areas with limited potentially accidental ignition sources include closets, crawl spaces, and attics.

22.2.4* **Exotic Accelerants.** Mixtures of fuels and Class 3 or Class 4 oxidizers *(see NFPA 430, Code for the Storage of Liquid and Solid Oxidizers)* may produce an exceedingly hot fire and may be used to start or accelerate a fire. Some of these oxidizers, depending on various conditions, can self ignite and will cause the same type of fire growth. Thermite mixtures also produce exceedingly hot fires. Such accelerants generally leave residues that may be visually or chemically identifiable. Presence of remains from the oxidizers does not in itself constitute an intentionally set fire. *(See 5.7.4.1.5.)*

22.2.4.1 Exotic accelerants have been hypothesized as having been used to start or accelerate some rapidly growing fires and were referred to in these particular instances as *high temperature accelerants (HTA)*. Indicators of exotic accelerants include an exceedingly rapid rate of fire growth, brilliant flares (particularly at the start of the fire), and melted steel or concrete. A study of 25 fires suspected of being associated with HTAs during the 1981–1991 period revealed that there was no conclusive scientific proof of the use of such HTA.

22.2.4.2 In any fire where the rate of fire growth is considered exceedingly rapid, other reasons for this should be considered in addition to the use of an accelerant, exotic or otherwise. These reasons include ventilation, fire suppression tactics, and the type and configuration of the fuels.

22.2.5 Unusual Fuel Load or Configuration. If the investigation reveals the presence of an unusually large fuel load in the area of origin, or a fuel load in the area of origin that either would normally not be expected in that area or would not be expected to be in the configuration in which it was found, the fire may be incendiary. An example of an unusual configuration is where furniture, stock, or contents are deliberately stacked or piled in a configuration to encourage rapid or complete fire development. An example of an unusually large fuel load is where accumulations of trash, debris, or cardboard cartons are deliberately introduced into a room or space in order to encourage greater fire involvement.

22.2.6 Burn Injuries. The manner and extent of burn injuries may provide clues to the origin, cause, or spread of the fire. Burn injuries may be sustained while setting an incendiary fire. The investigator should ascertain whether the fire victim's burns and the nature and extent of the injuries are consistent with the investigative hypothesis regarding fire cause and spread. The investigator should check the local hospitals for the identification of any persons admitted or treated for burn injuries.

22.2.7 Incendiary Devices. *Incendiary device* is a term used to describe a wide range of mechanisms used to initiate an incendiary fire. In some cases, the firesetter may have used more than one incendiary device. Frequently, remains of the fuel used will be found with the ignition device. If an incendiary fire is suspected, the investigator should search for other fire sets that may have burned out or failed to operate.

> **WARNING:** When an incendiary device is discovered that has not activated, *do not move it!* Such devices must be handled by specially trained explosive ordnance disposal personnel. Touching or moving such devices is extremely dangerous and can result in an ignition or explosion.

22.2.7.1 Examples of Incendiary Devices. Examples of some incendiary devices, and the evidence that may establish their presence or use, are as follows:

(1) Books of paper matches and cigarettes from which the striker from the matchbook, cigarette filters, remaining cigarette ash, and the combustible materials ignited by the matches or cigarettes may be found in the area of origin
(2) Candles from which their wax and the remains of any combustible material ignited by the candles may be found in the area of origin
(3) Wiring systems or electric heating appliances to initiate a fire (which may be evidenced by indications of tampering or modification of the wiring system, by the movement or arrangement of heat-producing appliances to locations near combustible materials, or by evidence of combustible materials being placed on or near heat-producing appliances)
(4) Fire bombs, commonly called Molotov cocktails (which leave evidence in the form of the ignitible liquid, chemicals, or compounds used within them, the broken or intact containers, and wicks)
(5) Paraffin wax–sawdust incendiary device [which can be evidenced by remains of wax impregnated with sawdust (e.g., artificial fire logs)]

22.2.7.2 Delay Devices. Timers or delay devices can be employed to allow the firesetter an opportunity to leave the scene and to possibly establish an alibi prior to the ignition. Common delay devices include candles, cigarettes, and mechanical or electrical timers.

22.2.7.3 Presence of Ignitible Liquids in Area of Origin. The use or presence of ignitible liquids is generally referred to as a *liquid accelerant* when used in conjunction with an incendiary fire.

22.2.7.3.1 The presence of ignitible liquids may indicate that a fire was incendiary, especially when the ignitible liquids are found in areas in which they are not normally expected. Containers of ignitible liquids in an automobile garage may not be unusual, but a container of ignitible liquids found in a bedroom may be unusual. In either case, the presence of ignitible liquids near the area of origin should be fully investigated.

22.2.7.3.2 "Irregular patterns" *(see 6.3.7.8)* may indicate the presence of an ignitible liquid. If the investigator observes patterns associated with a liquid accelerant, he or she may also observe the remains of a container used to hold the liquid. The investigator should ensure that samples are taken from any area where ignitible liquids are suspected to be present.

22.2.8 Assessment of Fire Growth and Fire Damage. Investigators may form an opinion that the speed of fire growth or the extent of damage was greater than would be expected for the "normal" fuels believed to be present and for the building configuration. However, these opinions are subjective. Fire growth and damage are related to a large number of variables, and the assumptions made by the investigator are based on that investigator's individual training and experience. If subjective language is used, the investigator should be able to explain specifically why the fire was "excessive," "unnatural," or "abnormal."

22.2.8.1 What an investigator may consider as "excessive," "unnatural," or "abnormal" can actually occur in an accidental fire, depending on the geometry of the space, the fuel characteristics, and the ventilation of the compartment *(see 5.5.4)*. Some plastic fuels that are difficult to burn in the open may burn vigorously when subjected to thermal radiation from other burning materials in the area. This type of burning might occur in the conditions during or after flashover.

22.2.8.2 The investigator is strongly cautioned against using subjective opinions to support an incendiary cause determination in the absence of physical evidence.

22.2.8.3 Mathematical models of fire growth exist that can provide assistance, if used properly, in assessing the potential accuracy of these subjective observations.

22.3 Potential Indicators Not Directly Related to Combustion. These indicators are generally conditions or circumstances that, in and of themselves, are not directly related to the fire or explosion cause, but that may be used by the investigator to develop ignition hypotheses, to select witnesses for interviewing, to develop suspects, and to develop avenues for further investigation. The indicators in this section are those that tend to show that somebody had prior knowledge of the fire.

22.3.1 Remote Locations with View Blocked or Obscured. A fire in a secluded location or where the view is hidden from observation may indicate a firesetter who did not want to be seen or caught. Fires at such locations would also allow the fire to develop before it was discovered. Examples include situations where windows are painted over or paper covers the windows for no apparent reason other than to conceal the fire. The fact that a fire origin is in a remote location, or in a place that is difficult to observe, should not, in and of itself, be construed as indicating prior knowledge.

22.3.2 Fires Near Service Equipment and Appliances. A fire near gas or electrical equipment, appliances, or fireplaces may be intended to make the fire appear to be from an accidental cause. The investigator should examine the fuel supply or service connections to determine whether they were loose or disconnected and then should determine whether tampering or sabotage of the equipment or appliances has occurred. If the investigator does not have sufficient knowledge regarding the equipment or appliance, it should be examined by qualified personnel.

22.3.3 Removal or Replacement of Contents Prior to the Fire. Through the course of the investigation, the investigator may believe that prior to the fire, the contents of a building have been removed or replaced with less-valued or more-valued items.

22.3.3.1 Replacement. When the investigator believes the contents have been replaced, as complete an inventory as possible of the contents should be made prior to release of the building. The inventory of the pre-fire structure should be obtained and corroborated through witness statements, invoice and inventory receipt, and so forth. The insurance proof of loss and underwriting file will provide a list of what was claimed to have been present. If contents that are abnormal to the occupancy are found, this can be another indicator that someone had prior knowledge of the fire. The items and contents that may be replaced depend on the occupancy of the building or space. Consider the following examples:

(1) Residential occupancy — furniture, clothing
(2) Industrial/commercial occupancy — machinery, equipment, stock, merchandise
(3) Vehicles — tires, batteries

22.3.3.2 Removal. Fire scenes or fire buildings that are devoid of the "normal" contents reasonably expected (or identified through witness statements, etc.) to be in the structure prior to the fire should be investigated and explained. The items removed are generally valuable items, such as television sets, VCRs, stereo systems, computers, camera equipment, stock, or equipment, or items that are difficult to replace, including files, business records, and so forth. Other items that may be removed prior to a fire may be those incriminating to a firesetter.

22.3.3.3 Absence of Personal Items Prior to the Fire. The absence of items that are personal, irreplaceable, or difficult items to replace should be investigated. Examples of those items include jewelry, photographs, awards, certificates, trophies, art, pets, sports and hobby equipment, and so forth. Also, the removal of the important documents (e.g., fire insurance policy, business records, tax records) prior to the fire should be investigated and explained.

22.3.4 Entry Blocked or Obstructed.

22.3.4.1 The entrance to a structure or to the property may be blocked or obstructed to hamper fire fighters from extinguishing the fire. Obstructions to the property may include fallen trees, street barricades, or construction features that deny fire vehicle access, such as masonry columns, fences, and gates.

22.3.4.2 Obstructions to the structure may include what appear to be "security" measures, such as boarded up windows and doors, "security grilles," and chains and locks.

22.3.5 Sabotage to the Structure or Fire Protection Systems.

22.3.5.1 Introduction. Sabotage refers to intentional damage or destruction to the physical structure of the building, or intentional damage to a fire protection system or system components.

22.3.5.1.1 A firesetter is often intent on developing conditions that will lead to the rapid and complete destruction of the building or its contents. In order to fulfill this goal, the firesetter may sabotage the structure (fire-resistive assembly) or the fire protection systems.

22.3.5.1.2 Investigators should determine whether the failure of structural components or fire protection systems was the result of deliberate sabotage or other factors, such as improper construction, lack of maintenance, systems shutdown for maintenance, improper design, or equipment or structural assembly failure.

22.3.5.2 Damage to Fire-Resistive Assemblies. Fire-resistive design, accomplished through the construction of various fire resistance–rated assemblies (e.g., walls, ceilings, and floors) and the proper protection of openings (e.g., fire doors, windows and shutters, and fire dampers), is intended to separate portions of a structure into "compartments" or "fire areas" that confine a fire within the "compartment" in which the fire originated, preventing smoke and fire movement to other portions of the building.

22.3.5.2.1 Penetrations in fire-resistant assemblies may be an indication that the firesetter attempted to spread the fire from one area to another. The investigator should try to determine whether the penetrations occurred with the intent of spreading the fire. Penetrations of fire resistance–rated construction may be the result of poor initial construction, renovations, service wiring, or cables, or may be the result of fire-fighting activities, such as ventilation or overhaul.

22.3.5.2.2 Open doors are the most common method of fire travel through a structure. Sabotage to fire or smoke doors (e.g., wedging doors open) or fire shutters can increase fire and smoke spread throughout the structure. Sabotage to stairway doors can further increase rapid smoke and fire spread. However, frequently these doors are held or propped open by building occupants to improve ventilation or access during regular building operations. The investigator should determine whether the doors and other opening protection were intentionally opened by the firesetter or were open as a normal operational use of the building.

22.3.5.3 Damage to Fire Protection Systems. Fire protection systems include heat, smoke, or flame detection; alarm and signaling systems; sprinkler and standpipe systems; special extinguishing systems, such as those using carbon dioxide, foam, or halon; and private water mains and fire hydrants.

22.3.5.3.1 Sabotage to fire protection systems or components can delay notification to occupants and the fire department and can prevent the control or extinguishment of the fire. Such sabotage is intended to allow the fire to develop fully and to create greater destruction.

22.3.5.3.2 Sabotage can include removing or covering smoke detectors; obstructing sprinklers; shutting off control valves; damaging threads on standpipes, hose connections, or fire hydrants; and obstructing or placing debris in fire department Siamese connections or fire hydrants.

22.3.5.3.3 Another type of sabotage, although more subtle, is igniting multiple fires (*see 22.2.1*). In addition to increased destruction from additional fires, multiple fires can have the effect of overtaxing the fire suppression system beyond its design capabilities. Assistance may be needed to determine the design limitations on the fire protection system. *(See Section 14.5.)*

22.3.6 Open Windows and Exterior Doors. Open windows and exterior doors can speed the growth and spread of a fire. When these conditions exist during cold weather or in violation of normal building security, it may be an indicator that someone attempted to provide extra ventilation for the fire. Windows may have been broken out for the same purpose.

22.4 Other Evidentiary Factors.

22.4.1 Once the investigator has completed the fire scene examination and has concluded that the fire was incendiary, there are other evidentiary factors that should be recorded and examined, which may be critical regarding future suspect development and identification.

22.4.1.1 These evidentiary factors regarding the identification of a suspected firesetter, or the "motive" or opportunity for the fire, cannot be substituted for a properly conducted investigation and determination of the fire's origin and cause.

22.4.1.2 In the absence of physical evidence of an incendiary fire, the investigator is strongly cautioned against using the discovery or presence of these other evidentiary factors in developing a hypothesis, forming opinions, or drawing conclusions concerning the cause of the fire.

22.4.2 Analysis of Confirmed Incendiary Fires. It is through the analysis of confirmed incendiary fires that trends or patterns in repetitive firesetting behaviors may be detected. The key to this analysis is whether the firesetting is repetitive or not. This analysis may assist the investigator in the development and identification of possible suspects. Repetitive firesetting, sometimes called *serial fire setting* or *serial arson*, refers to a series of three or more (incendiary) fires, where the ignition is attributed to the individual or a group acting together. There are three principal trends that may be identified through analysis: geographic areas, or "clusters"; temporal frequency; and materials and methods.

22.4.2.1* Geographic Areas, or Clusters. Repetitive firesetting activities tend to group within the same geographic location (i.e., same neighborhood), or cluster. Locating incendiary fires by utilizing computer-assisted pattern recognition systems, such as the Arson Information Management System (AIMS) or by looking at a map of the local area, can assist the investigator in identifying clusters.

22.4.2.2 Temporal Frequency. Incendiary fires set by the same individual often occur during the same time period of the day or on the same day of the week. This occurrence may have several reasons, including the level of activity in the area, the firesetter's assessment of his or her chances of success, or the firesetter's routine. For example, the firesetter may pass the location (to or from work or to or from a bar) during a certain period of the day or on a certain day of the week.

22.4.2.3 Materials and Method. The method and material used in the ignition of incendiary fires vary according to the firesetter. Generally, however, once a firesetter begins repetitive firesetting behavior, the materials and method tend to remain similar, as do the locations of the incendiary fires.

22.4.3 Evidence of Other Crimes, Crime Concealment.

22.4.3.1 An incendiary fire may be an attempt to conceal other crimes, such as homicides and burglaries. In other cases, a staged burglary may occur to disguise an incendiary fire. The issue of which occurred first, the other crime or the fire, is more related to the motive for the fire and has little to do with the cause of the fire.

22.4.3.2 Although possible motives do not determine a fire's cause (i.e., if the motive was to burglarize the structure and to conceal the burglary with a fire, or to set a fire but make it appear as a burglary), motives may lead the investigator to approach the investigation (i.e., the search for evidence) and possible suspects differently.

22.4.4 Indications of Financial Stress. The investigation may reveal indicators of financial stress. These indicators may include the following: liens, attachments, unpaid taxes, mortgage payments in arrears, real estate for sale (inability to sell, property is nonmarketable, etc.), poor business location, or new competition.

22.4.4.1 Financial stress may also be indicated by factors associated with the use or type of occupancy of the building. For business occupancies, indicators may include periods of economic decline, particularly within that industry; changes within an industry, in either product or equipment; obsolescence of equipment; and new competition within the industry. Other indicators can include factors such as the need to relocate or new competition in the same geographic region or area.

22.4.4.2 Examples of financial stress for residential properties can include landlords who cannot collect rent or who cannot rent out vacant units, rent control, the owner's need to relocate, and mass loss of jobs within the region resulting from industrial cutbacks or closings.

22.4.5 Existing or History of Code Violations.

22.4.5.1 Closely related to, and possibly another indication of financial stress, is the existence of or a history of building, fire safety, housing, or maintenance code violations. This may indicate either the financial inability to maintain the building or the intentional choice to let the building deteriorate (refusal to reinvest in the structure).

22.4.5.2 Where the deterioration of a building is intentional, other indicators related to financial stress, such as overinsurance or the inability to sell the property, may be discovered during the investigation.

22.4.6 Owner with Fires at Other Properties. If a structure is owned by persons who have had incendiary fires at other properties, especially if they have collected insurance as a result of those fires, there is a possibility they will experience another incendiary fire.

22.4.7 Overinsurance. Another indicator closely related to financial stress is overinsurance. Overinsurance is a condition whereby the insurance coverage is greater than the value of the property in a valued policy state or whereby there are multiple insurance policies on the property.

22.4.8 Timed Opportunity. Timed opportunity refers to the indicators that a firesetter has timed the fire to coincide with conditions or circumstances that assist the chances of successful destruction of the target (property) or to utilize those conditions or circumstances to increase the chances of not being apprehended.

22.4.8.1 Fires During Severe Natural Conditions. Fires during periods of extreme natural conditions such as floods, snowstorms, hurricanes, or earthquakes may delay fire department response or hinder fire-fighting capabilities. Other natural conditions to note are electrical storms, periods of high winds, low humidity, and freezing or extremely high temperatures.

22.4.8.2 Fires During Civil Unrest. This is a type of opportunistic fire. Other indicators, such as financial stress, often accompany this indicator.

22.4.8.2.1 Also, incendiary fires during civil unrest usually do not involve elaborate ignition devices or materials, although "fire bombs" or liquid accelerants are sometimes used. More often, available materials are utilized as an initial fuel.

22.4.8.2.2 A similar pattern may develop when *repetitive fires* or a series of incendiary fires occur in the same geographic area (*see 22.4.2.1*). The owners or occupants may attempt to set a fire and have the cause attributed to another firesetter. In these instances, the investigator may discover a difference in the method (such as time of day, days of the week, location of the fire), the materials (such as different fuels) used, or ignition source that does not fit the established firesetting pattern. (*See 22.4.2.*)

22.4.8.3 Fire Department Unavailable. Fires may be set at times when the fire department is unavailable. Examples include deliberately calling in a false alarm to get the fire department away from the area or starting the fire while there is a working fire in progress or when the fire department is involved in a parade or other community function.

22.4.9* Motives for Firesetting Behavior.

22.4.9.1 General. Motive indicators should not be included or substituted as analytical elements of the fire scene for the purpose of determining or classifying the fire cause. The proper use of motive indicators in the fire investigation process is in identifying potential suspects only after the fire origin and cause have been determined and the fire has been classified as incendiary.

22.4.9.1.1 *Motive* is defined as an inner drive or impulse that is the cause, reason, or incentive that induces or prompts a specific behavior. The identification of an offender's motive is a key element in crime analysis. Crime analysis is a method of identifying personality traits and characteristics exhibited by an unknown offender. It is the identification and analysis of the personality traits that eventually lead to the classification of a motive. Once a possible motive is identified, the investigator can begin to evaluate potential suspects for the incendiary fire.

22.4.9.1.2 Behaviors related to the classifications of motive may not be exclusive to one motive classification but may appear to overlap categories and to be similar for different motives. In these instances, it is important to obtain additional information that may clarify the behaviors.

22.4.9.1.3 In addition to the identification of a motive, other analyses should be considered that might assist in determining if a serial firesetter exists. Through the analysis of confirmed incendiary fires, trends or patterns in repetitive firesetting behaviors may be detected. The three principal trends that may be identified are geographic clustering, temporal frequency, and methods and materials. (*See 22.4.1.*)

22.4.9.2 Motive Versus Intent. There is an important distinction to be made between motive and intent. *Intent* refers to the purposefulness or deliberateness of the person's actions or, in some instances, omissions. It also refers to the state of mind that exists at the time the person acts or fails to act. Intent is generally necessary to show proof of crime. The showing of intent generally means that some substantive steps have been taken in perpetuating the act. *Motive* is the reason that an individual or group may do something. It refers to what causes or moves a person to act or not to act and the stimulus that causes action or inaction. Motive is generally not a required element of a crime. For example, a person with indications of "financial difficulty" could experience a fire to his insured property that is ignited by his falling asleep with a lit cigarette. While this person may have motive to cause a fire, that person did not intend to have a fire. Thus, no element of intent existed.

22.4.9.3* Classifications of Motive.

22.4.9.3.1 Introduction. The classifications discussed in this chapter are those identified in Douglas et al., *Crime Classification Manual* (CCM). The CCM uses a diagnostic system intended to standardize terminology and formally classify the critical characteristics of the perpetrators and the victims of the three major violent crimes: murder, arson, and sexual assault.

22.4.9.3.1.1 The CCM identifies analytical factors that have been identified as essential elements in order to classify the motive of an offense. These factors include information about the victim, the crime scene, and the nature of the victim–offender exchange. Not all the information will be, or should be expected to be, present in every case. The intent is to provide the fire investigator with as much information as possible.

22.4.9.3.1.2 The behaviors that may identify a possible motive, and thus a possible suspect, apply whether the fire is the result of a one-time occurrence or multiple occurrences, such as with a repetitive or serial firesetter. There are three classifications of repetitive firesetting behavior. These are identified as serial arson, spree arson, and mass arson. The terminology used in classifying a repetitive firesetter is similar to the terminology used in murderers. *Serial arson* involves an offender who sets three or more fires, with a cooling-off period between the fires. *Spree arson* involves an arsonist who sets three or more fires at separate locations with no emotional cooling-off period between fires. *Mass arson* involves an offender who sets three or more fires at the same site or location during a limited period of time.

22.4.9.3.1.3 The National Center for the Analysis of Violent Crime (NCAVC) has identified the six motive classifications as the most effective in identifying offender characteristics for firesetting behavior, as follows:

(1) Vandalism
(2) Excitement
(3) Revenge
(4) Crime concealment
(5) Profit
(6) Extremism

22.4.9.3.2 Vandalism. Vandalism-motivated firesetting is defined as mischievous or malicious firesetting that results in damage to property. Common targets include educational facilities and abandoned structures, but also include trash fires and grass fires. Vandalism firesetting categories include willful and malicious mischief and peer or group pressure.

22.4.9.3.2.1 Willful and Malicious Mischief. These are incendiary fires that have no apparent motive or those that seemingly are set at random and have no identifiable purpose. These are fires that are often attributed to juveniles or adolescents.

22.4.9.3.2.2 Peer or Group Pressure. Recognition or pressure from peers is sometimes regarded as a reason for firesetting, particularly among juveniles.

22.4.9.3.3* Excitement. The excitement-motivated firesetter may enjoy the excitement that is provided by actual firesetting or the activities surrounding the fire suppression efforts, or may have a psychological need for attention. The excitement-motivated offender is often a serial firesetter. This firesetter will generally remain at the scene during the fire and will often get in position to respond to, or view the fire and the surrounding activities. The excitement-motivated firesetter's targets range from small trash and grass fires to occupied buildings.

22.4.9.3.3.1 The excitement-motivated firesetter includes the following subcategories.

(A) Thrill Seeking. Setting a fire provides this offender with feelings of excitement and power. The thrill-seeking firesetter is often a repetitive firesetter, who compulsively sets fires to satisfy some psychological desire or need.

(B) Attention Seeking. These firesetters have a need to feel important, and they set fires in order to satisfy a psychological need.

(C) Recognition. These firesetters are sometimes described as the hero or vanity firesetter. These firesetters often remain at the fire scene to warn others, report the fire, or assist in firefighting efforts. They enjoy or may seek the recognition and praise they receive for their efforts. Typical among these firesetters are security guards and fire fighters. Occasionally these firesetters may even take responsibility for setting the fire.

(D) Sexual Gratification or Perversion. These are firesetters who set fires as a means of sexual release. The firesetter in this category is considered rare.

22.4.9.3.3.2 Attention-seeking, recognition, and sexual-gratification firesetters rarely attempt or intend to harm people, but these firesetters may disregard the safety of innocent bystanders or occupants. However, the thrill-seeking offender, whose compulsion requires the inherent sense of satisfaction, will often set a big fire or a series of fires. Fires will typically involve structures, but when vegetation is involved, these fires are also large.

22.4.9.3.4* Revenge.

22.4.9.3.4.1 The revenge-motivated firesetter retaliates for some real or perceived injustice. An important aspect is that a sense of injustice is perceived by the offender. The event or circumstance that is perceived may have occurred months or years before the firesetting activity. A fire by the revenge-motivated offender may be a well-planned, one-time event or may represent serial firesetting, with little or no pre-planning. Serial offenders may direct their retaliation at individuals, institutions, or society in general.

22.4.9.3.4.2 Subcategories of revenge firesetting include personal, societal, institutional, and group retaliation.

(A) Personal Retaliation. The triggering event for this motive may be an argument, a fight, a personal affront, or any event perceived by the offender to warrant retaliation. Favorite targets include the victim's vehicle, home, or personal possessions. The specific location and the materials involved in the fire may be a significant factor in identifying the offender. Igniting clothing or other personal possessions is seen as a more personal affront to the victim than simply setting a fire in a common area. The fire scene may also be vandalized. These fires may be a one-time event or an act of serial arson.

(B) Societal Retaliation. This offender usually suffers from a feeling of inadequacy, loneliness, persecution, or abuse. The societal retaliation offender generally is not satisfied with a single fire or even a series of fires. Therefore, this serial offender is likely to set many more fires than other revenge-motivated firesetters.

(C) Institutional Retaliation. This classification of offender targets institutions such as religious, medical, governmental, and educational institutions, or corporations. The firesetter may be a disgruntled employee, a former employee, a customer, or a patient.

(D) Group Retaliation. Targets for group retaliation may be religious, racial, fraternal, or other groups, including gangs. Graffiti, symbols or markings, and other vandalism may accompany the fire.

22.4.9.3.5* Crime Concealment. This category involves firesetting that is a secondary or a collateral criminal activity, perpetrated for the purpose of concealing the primary criminal activity. In some cases, however, the fire may actually be part of the intended crime, such as revenge. Many people erroneously believe that a fire will destroy all physical evidence at the crime scene. Categories for crime concealment firesetting include murder or burglary concealment and destruction of records or documents.

(A) Murder Concealment. This scenario is where a fire is set in an attempt to conceal the fact that a homicide has been committed, to destroy forensic evidence that may identify the offender, or to conceal the identity of the victim.

(B) Burglary Concealment. This is a fire that is set in an attempt to conceal the fact that a burglary has occurred or to destroy forensic evidence that may identify the offender.

(C) Destruction of Records or Documents. This is a fire that targets records or documents. These fires may involve files ignited still in their folders or an origin inside a file cabinet. It may involve ordinary combustibles located in an exposure position to the files, such as a trash can moved adjacent to the files. Potential suspects in these incidents involve those who have some interest in the documents or records that were targeted.

22.4.9.3.6* Profit. Fires set for profit involve those set for material or monetary gain, either directly or indirectly. The direct gain may come from insurance fraud, eliminating or intimidating business competition, extortion, removing unwanted structures to increase property values, or from escaping financial obligations.

22.4.9.3.6.1 The broad category of fraud is frequently identified as an arson motive. However, fraud is classified as a subcategory in the profit motive category. Fraud-motivated fires may include commercial or residential properties. Commercial fraud fires may be set or arranged by an owner to destroy old or antiquated equipment, to destroy records to avoid taxes or audits, or for the purpose of obtaining insurance money. Fires may be set by a competitor to gain market advantage, or by agents of organized crime for purposes of extortion, protection rackets, or intimidation. Residential fraud may include an owner intending to defraud an insurance carrier, or a tenant defrauding an owner or a welfare agency. Increasing taxes, physical deterioration (and legally mandated repairs such as by code enforcement agencies), vacancy or inability to rent, or statutory rent-control measures may be reasons for a landlord to consider burning the structure.

22.4.9.3.6.2 There are several subcategories that further identify fraud as a motive. These include fraud to collect insurance, fraud to liquidate property, fraud to dissolve a business, and fraud to conceal a loss or liquidate inventory. The other categories include employment, parcel/property clearance, and competition.

22.4.9.3.7* Extremism. Extremist-motivated firesetting is committed to further a social, political, or religious cause. Fires have been used as a weapon of social protest since revolutions first began. Extremist firesetters may work in groups or as individuals. Also, due to planning aspects and the selection of their targets, extremist firesetters generally have a great degree of organization, as reflected in their use of more elaborate ignition or incendiary devices. Subcategories of extremist firesetting are terrorism and riot/civil disturbance.

(A) Terrorism. The targets set by terrorists may appear to be at random; however, target locations are generally selected with some degree of political or economic significance. Political targets generally include government offices, newspapers, universities, political party headquarters, and military or law enforcement installations. Political terrorists may also target diverse properties such as animal research facilities or abortion clinics. Economic targets may include business offices, distribution facilities of utility providers (e.g., atomic generation plants), banks, or companies thought to have an adverse impact on the environment. Fires or explosions become a means of creating confusion, fear, or anarchy. The terrorist may include fire as but one of a variety of weapons, along with explosives, used in furthering his or her goal.

(B) Riot/Civil Disturbance. Intentionally set fires during riots or civil disturbances may be accompanied by vandalism and looting. It is worth noting that all fires ignited during periods of civil unrest may not be the result of the extremist firesetter but may be set by others, such as owners, hoping that the fire is attributed to the extremist firesetter and the circumstances surrounding the civil disturbance.

Chapter 23 Fire and Explosion Deaths and Injuries

23.1 General. Each year, thousands of people are injured or die in fire- and explosion-related incidents. The investigation of fire- and explosion-related deaths and injuries require the utilization of specialized skills that are not typically used during routine fire investigations. These skills may include subsets of toxicology, pathology, and human behavior, among others. The data from the deaths and injuries of fire victims may provide information related to the nature and development of a fire.

23.2* Mechanisms of Death and Injury. The combustion products arising from a fire are many, and their effects on healthy individuals vary; however, none are without toxicological effects. The inhalation of these products or contact with skin or eyes can result in deleterious biological effects, such as immediate irritation of the eyes and respiratory tract or systemic effects that influence other functions of the body. These combustion products include carbon monoxide, hydrogen cyanide, carbon dioxide, nitrogen oxides, halogen acids (hydrochloric, hydrofluoric, and hydrobromic acids), acrolein, benzene, particulates (ash, soot), and aerosols (complex organic molecules resulting from pyrolysis products).

23.2.1* Carbon Monoxide. Carbon monoxide (CO) is produced at some level in virtually every fire. All carbon-based fuels (e.g., wood, paper products, plastics) produce carbon monoxide as a result of incomplete combustion. During burning of organic fuels, CO is initially formed and then subsequently oxidized to carbon dioxide (CO_2). In underventilated fires or in fires where the initial products of combustion mix with colder gases (such as in smoldering fires), conversion of CO to CO_2 can be halted, and CO can become a major product of combustion. In well-ventilated fires, the level of CO produced may be as little as a few hundred parts per million (i.e., 0.02 percent). However, in under ventilated, smoldering, or post flashover fires, CO concentrations of 1 percent to 10 percent (10,000 ppm to 100,000 ppm) can be produced. Elevated CO concentrations can also develop during fire suppression.

23.2.1.1* Carbon monoxide acts as a central nervous system depressant. When inhaled, CO binds with hemoglobin in the blood, creating carboxyhemoglobin (COH_b). The affinity of carbon monoxide for hemoglobin is approximately 240–250 times more than the affinity of oxygen for hemoglobin. Therefore, the blood can accumulate dangerous levels of COH_b from even low CO concentrations in the air.

23.2.1.2 Although carbon monoxide has asphyxiating effects, it is more powerful than an asphyxiant. Carbon monoxide not only reduces the oxygen-carrying capacity of the blood, as it binds with hemoglobin, it causes a shift in the dissociation curve which affects the ability of the blood to release bound oxygen to the tissues. CO delivered to the cells interferes with cellular respiration, causing incapacitation and death. The binding of carbon monoxide to hemoglobin is reversible. Carbon monoxide can be eliminated from the body by breathing fresh air or oxygen. In cases of high carboxyhemoglobin levels, however, hyperbaric chambers are used to expedite the elimination of the toxicant from the body. The rate of elimination of carbon monoxide from the body, therefore, is dependent on the oxygen concentration in the air and pressure. For examples, individuals treated with 100 percent oxygen will eliminate the toxicant more quickly than those breathing room air (21 percent oxygen).

23.2.1.3* Because carboxyhemoglobin is so stable, it can be readily measured in the blood of fire victims, even long after death. The average fatal level of blood CO is widely accepted as 50 percent COH_b. However, fire victims have died from CO exposure with a blood COH_b level as low as 20 percent and as high as 90 percent. Victims with less that 20 percent COH_b most likely died from other causes, such as a lack of oxygen, cardiac arrest, or thermal burns. In contrast, victims with COH_b concentrations of 40 percent or higher are likely to have died from carbon monoxide alone or in combination with other factors (such as age, alcohol, or a heart condition) or may simply have been incapacitated sufficiently by carbon monoxide poisoning to be unable to flee the fire.

23.2.1.4* In assessing the significance of a victim's COH_b level, it should be noted that carbon monoxide is produced endogenously in the body due to the break down of hemoglobin to bile pigments in the liver. CO can also enter the body from environmental and habitual sources. A test of 3022 transfusion blood samples for COH_b levels found that 65 percent were below 1.5 percent, 26.5 percent were between 1.5 percent and 5 percent, 6.7 percent were between 5 percent and 10 percent, and 0.3 percent were in excess of 10 percent. A smoker may be exposed to CO levels in the range of 400–500 ppm during the average 6 minute duration that it takes to smoke a cigarette. As such, smokers typically have a COH_b range of 3 percent–8 percent with an average baseline of 4 percent. Heavy smokers (greater than 2 packs per day) can reach levels as high as 15 percent. Nonsmokers average approximately 1 percent COH_b in their blood.

23.2.1.5* The COH_b level of a fire survivor begins to decrease as soon as the person is removed from the fire environment. The rate at which CO is eliminated from the body is dependent on the oxygen concentration of the air being breathed. The concentration of CO in the blood (COH_b saturation) will be decreased by one-half (COH_b half-life), for example, reducing COH_b from 45 percent to 22 percent in 250 minutes to 320 minutes at ambient O_2 levels in air (21 percent). COH_b half-life is approximately 45 minutes to 90 minutes when a near 100 percent oxygen concentration is administered during emergency medical treatment. Hyperbaric oxygen treatment can reduce COH_b half-life to approximately 20–30 minutes.

23.2.1.6* In determining the significance of the victim's COH_b level, the investigator should consider the effects of any medical treatment that the victim received prior to succumbing to their injuries. Individuals that received pulmonary resuscitation or were breathing and administered oxygen therapy prior to death will have a lower COH_b level than was present when they were removed from the fire environment. Therefore, knowledge of the time elapsed between removal from the fire and death and the duration and type of oxygen therapy used, e.g., non-rebreather, nasal cannula, hyperbaric chamber, prior to blood sampling are important data to gather.

23.2.1.7* Studies have shown that 60 percent or more fire victims die from carbon monoxide poisoning. Because of the transport of CO, most of these people are found outside the room of origin. Most fires do not produce lethal levels of CO until the fire becomes ventilation controlled (the exception is smoldering fires). Thus, victims of carbon monoxide inhalation are typically found outside the room of origin unless the fire resulted from a locally under-ventilated or smoldering ignition. However, during flashover, thermal injury and lack of oxygen can cause death before substantial concentrations of COH_b are developed. The same can occur if the victim is involved in a flash fire involving fuel gases or vapors.

23.2.2* Cyanide. Hydrogen cyanide (HCN), in addition to carbon monoxide, is another key toxicant that is produced in fires. Hydrogen cyanide gas is produced during the combustion of wool, nylon, and plastics, such as polyurethane and ABS, which are commonly found in household furniture and interior finishes. HCN can act alone or in combination with CO to cause incapacitation and death. HCN can contribute to smoke inhalation deaths by quickly incapacitating fire victims, thereby preventing them from escaping.

23.2.2.1* The role of cyanide in fire deaths has been highlighted in several landmark fires, such as the 1990 Happy Land Social Club Fire in New York City, the 1986 DuPont Plaza Hotel Fire in Puerto Rico, the 1977 Maury County Jail Fire in Tennessee, and the 1983 Ramada Inn Central Fire in Texas. When HCN is inhaled during respiration, it is rapidly absorbed into the blood. The diffusion capabilities of cyanide are more significant than those of CO, which explains the rapid lethal effects of the toxicant.

23.2.2.2 While carbon monoxide affects the binding of oxygen to hemoglobin, cyanide effects the utilization of oxygen. Once it diffuses into the body from inhalation, cyanide affects cellular respiration by interfering with the last step in the electron transport chain, cytochrome c oxidase. This cellular interference hampers the transport of electrons necessary to form water and the resulting ATP molecules needed for energy production. As a result, the Kreb's cycle ceases, causing metabolic acidosis, histotoxic anoxia, paralysis of the respiratory center in the brain, and ultimately death.

23.2.2.3 Although cyanide is most commonly measured in the blood, techniques are available for quantification of cyanide in various other biological samples, such as the liver, brain, lung, kidney, and stomach contents. Investigators should take caution, however, that the distribution of cyanide to these various organs and the stability of cyanide in these organs are variable. Cyanide does not show the same stability as carbon monoxide in post-mortem samples. The dominant and most consistently reported changes in cyanide concentration consist of decreases with time in whole cadavers and in postmortem specimens of blood and other tissues. In some cases, however, cyanide concentrations have been shown to increase over time due to bacterial production in the sample.

23.2.2.4* The rate at which cyanide changes in biological samples is dependent on four main criteria. These criteria include the sample concentration at the time of death, the length of time between death and removal of the sample from the cadaver, the length of time after removal prior to toxicological testing, and the storage (e.g., refrigerated, frozen, ambient) and conditioning (addition of preservatives) of the sample. The investigator should attempt to gather this information, as it will be important in the overall interpretation of the victim's cyanide concentration. The addition of 1 percent–2 percent sodium fluoride to the sample has been shown to aid in the reduction of cyanide concentration changes over time.

23.2.2.5* Incapacitating levels of cyanide are between 2.0 to 2.5 μg/mL, and lethal levels are reported to be 3.0 μg/mL or greater in whole blood. In terms of parts per million, approximately 280 ppm exposure will cause immediate death, and 100 ppm will result in death after one hour of exposure. Endogenous cyanide levels of less than 0.26 mg/L are found in normal healthy adults due to the metabolism of Vitamin B12 (cyanocobalamin). Therefore, the levels of cyanide found in fire victims are generally at least one order of magnitude higher (i.e., 2–3 mg/L) than normal level.

23.2.2.6* Research indicates that the effects of CO and cyanide are at least additive and may be synergistic. The additive effect is reflected in the use of fractional effective dose models (i.e., a fractional effective dose of 100 percent) which assume, for example, if a victim has 30 percent COH_b (60 percent of a typical lethal dose) and 1.2 mg/l of HCN (40 percent of a typical lethal dose), that the combination is generally considered lethal.

23.2.3 Other Toxic Gases. There are many toxic gases found in fire environments that can cause irritation and swelling (edema) sufficient to interfere with breathing. Hydrogen chloride and acrolein are two commonly found irritant gases in fires. Hydrogen chloride (HCl) can be produced during the combustion of polyvinyl chloride (PVC) plastics. Acrolein is produced during the combustion of wood and other cellulosic products. The investigator should consider the effect of irritant gases on the ability of the victim to move throughout the fire compartment or exit the structure.

23.2.4 Hyperthermia. Hyperthermia is the condition of overheating of the body. Victims exposed to the hot environment of a fire, including high moisture content, are subject to incapacitation or death due to hyperthermia, especially if the person is active. The time duration and type of exposure can lead to either simple hyperthermia or acute hyperthermia.

23.2.4.1* Simple hyperthermia results from prolonged exposures (typically more than 15 minutes) to hot environments where the ambient temperature is too low to cause burns. Such conditions range from 80°C to 120°C (176°F to 248°F) depending on the relative humidity, and usually result in a gradual increase in the body core temperature. High humidity makes it harder for the body to dispel excess heat by evaporation and thereby accelerates the heating process. Core body temperatures above approximately 43°C (109°F) are generally fatal within minutes unless treated.

23.2.4.2 Acute hyperthermia involves exposure to high temperatures for short periods of time (less than 15 minutes). This type of hyperthermia is accompanied by burns. However, when death occurs shortly after exposure to severe heat, the cause of death is generally considered to be from a rise in blood temperatures rather than from burns.

23.2.5* Skin Burns. When the temperature of the skin reaches approximately 45°C (113°F), pain will result and an additional increase in temperature will cause thermal burns. Thermal burns can result from conductive, convective, or radiative heat exposure. Researchers found that a 60°C (140°F) brass block applied directly to the skin caused pain within 1 second of contact, partial thickness burns within 10 seconds, and full thickness burns within 100 seconds. Clothing, especially heavier cellulosic fabrics like denim or canvas, can transmit enough heat by conduction to cause skin burns even though the fabric does not exhibit any burning or charring. When skin was exposed to convective heat, pain and the onset of burns occurred at air temperatures above 120°C (248°F).

23.2.5.1* When radiant heating raises the temperature of the skin, the higher the radiant flux, the faster damage will occur. For instance, a heat flux of 2 kW/m_2 will cause pain after a 30-second exposure, while a heat flux of 10 kW/m_2 will cause pain after just 5 seconds. A flux of 2 kW/m_2 will not cause blisters, while 10 kW/m_2 will blister in 12 seconds. *(See Figure 23.2.5.1.)* A radiant heat flux of 20 kW/m_2, typically associated with flashover, is sufficient to ignite clothing or cause severe burns or death by brief thermal exposure. Radiant heat, sufficient to cause burns, can be reflected from some surfaces. Heat can be transferred through clothing, causing burns to the underlying skin, without any readily identifiable damage to the clothing.

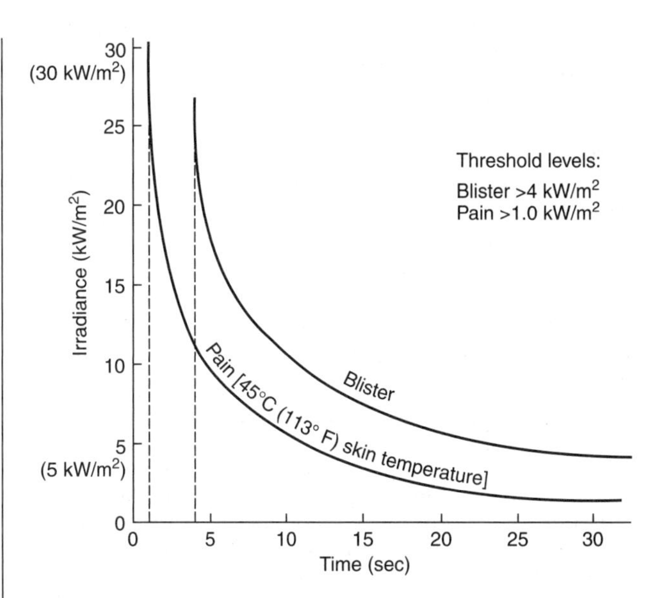

FIGURE 23.2.5.1 Diagram Showing Incident Radiant Heat Flux Effect on Bard Skin (Based on Stoll and Greene, 1959).

23.2.5.2 Burns induced by chemicals or contact with hot liquids (scalds) may not be distinguishable from those induced by hot gases or flames.

23.2.6* Inhalation of Hot Gases. Inhalation of hot fire gases can result in death or injury. However, it is difficult to distinguish the effects of thermal inhalation burns from edema and inflammation caused by chemical irritants in smoke. A distinguishing characteristic of thermal inhalation burns is that they are accompanied by external facial burns, as the temperatures are sufficient to burn skin and facial hair. Research on animals determined that dry air at 350°C (662°F)and flame gases at 500°C (932°F)resulted in larynx and trachea damage, while exposure to steam at 100°C (212°F) resulted in burns down to the deep lung.

23.2.7 Soot and Smoke. Soot and smoke can contribute to fire deaths and injuries through several mechanisms. Hot soot particles can be inhaled and can cause thermal injuries leading to edema in the respiratory system. Soot particulate can also contain toxic chemicals and can provide inhalation and ingestion pathways for these toxins. Excessive soot can also physically block the airways, causing asphyxiation. Liquid aerosols (mists) of pyrolysis products are often acidic, causing chemical edema, and are often very toxic, causing systemic failures upon inhalation.

23.2.8 Hypoxia. Hypoxia is a condition caused by breathing a reduced oxygen atmosphere. A reduced oxygen environment occurs in an enclosure fire as a natural consequence of the combustion process. There is little effect of reduced environmental oxygen down to 15 percent oxygen in air. However, as the oxygen concentration in inhaled air decreases from 15 percent to 10 percent a gradual increase in respiration occurs, followed by disorientation and loss of judgment. As the oxygen concentration in the ambient environment decreases below 10 percent, unconsciousness occurs, followed rapidly by cessation of breathing and death. This situation is aggravated by a high level of carbon dioxide in the air, which causes a substantial increase in the rate and depth of respiration. The increase in respiration rate due to the lower oxygen and higher carbon dioxide concentrations can result in an increased rate of CO uptake.

23.2.9 Sublethal Inhalation Exposure Effects on the Individual. The discussion in this section is limited to the specific effects of exposure to sublethal concentrations of narcotic gases [carbon monoxide, hydrogen cyanide, oxygen-depleted air (hypoxia)], irritant gases (hydrogen chloride, acrolein, etc.), and smoke.

23.2.9.1 Narcotic Gases. Carbon monoxide, hydrogen cyanide, and oxygen-depleted air (causing hypoxia) are all narcotic gases. Narcotic gases cause loss of alertness (intoxication), mental function, and psychomotor ability (the ability to carry out simple coordinated movements as are required in exiting a building). Carbon monoxide acts without the subject being aware of the extent of exposure and impairment. Hypoxia as a result of reduced oxygen concentration has a similar effect. Conversely, while hydrogen cyanide will ultimately result in mental depression and unconsciousness just as other narcotic gases, the effects of HCN exposure are more rapid and dramatic. At sublethal conditions, all these gases will reduce the ability of an individual to make decisions and carry out intended actions.

23.2.9.2 Irritant Gases. Irritant gases can alert people to the presence of a fire, even at low concentrations. Because of the unpleasant aspects of irritation of the eyes and respiratory tract, individuals may become aware of a fire earlier than would otherwise be the case, and may be motivated to escape. As the irritant effects become more pronounced, irritancy can have a direct impact on the ability of individuals to see and in this way may interfere with exiting behaviors. Post-fire effects of these irritants can be lung edema and inflammation.

23.2.9.3 Smoke. Visible products of combustion will impair the ability of individuals to see, and this in turn will reduce the speed of movement of escaping individuals. Sufficiently reduced visibility can cause individuals to not use an exit path. The reduced visibility that will cause an individual to abandon an exit path is dependent upon many factors, including the individual's familiarity with the building.

23.2.10 Explosion-Related Injuries. The location and distribution of explosion injuries to a victim can be useful in the reconstruction of the incident. These findings may indicate the location and activity of the victim at the time of the explosion, and they may help establish the location, orientation, energy, and function of the exploding mechanism or device. Explosion injuries can be divided into four categories based largely upon the explosion effect that caused them: blast pressure, shrapnel, thermal, and seismic.

23.2.10.1 Blast Pressure Injuries. The concussive effect upon a victim can cause internal injuries to various organs and body systems such as the gastrointestinal tract, lungs, eardrums, and blood vessels.

23.2.10.1.1 Frequently, the blast pressure front is strong enough to violently move or even propel the victim into solid objects, or conversely can violently move or propel large solid objects (walls, doors, etc.) into victims. These actions can cause blunt trauma injuries, fractures, lacerations, amputations, contusions, and abrasions.

23.2.10.1.2 Dirt, sand, explosive powders and other fine particles can be blasted into unprotected skin, causing a type of injury commonly called tattooing.

23.2.10.1.3 With detonations, there may be violent amputations or dismemberment of the body caused by the blast pressure wave. Parts of the body or its clothing may be propelled great distances and should be searched for and documented.

23.2.10.2 Shrapnel Injuries. Shrapnel (solid fragments) traveling at high speeds from the epicenter of an explosion can cause amputations, dismemberment, lacerations or perforations resembling stab wounds, localized blunt trauma such as broken and crushed bones, and soft tissue damage. Bruising, abrasions, and lacerations caused by the impact of shrapnel is referred to as body stippling.

23.2.10.3 Thermal Injuries. Thermal injuries associated with explosion flame fronts (and not the following fires, which often accompany low-order explosions) are usually of the first- and second-degree types because of their very short duration. Third-degree burns can also be encountered in these situations, but with much less frequency. These burns can be fatal. Brief exposure to the high-temperature expanding flame front causes burn damage to the exposed skin surfaces. Often even a thin layer of clothing can protect the underlying skin from injury. Synthetic-fabric clothing may be melted by exposure to flash flames from deflagrations, where cotton fabrics may only be scorched. Frequently, the burn injuries can be localized to the side of the body that is facing the expanding flame front. This finding can be used by the investigator as a heat and flame or explosion dynamics vector.

23.2.10.4 Building Collapse Injuries. The seismic effects of explosions are most dangerously manifested in the collapse of buildings and their structural elements. Injuries and deaths resulting from such occurrences are similar to what might be encountered by building damage from blast pressure waves. Collapse of buildings can cause blunt trauma injuries, lacerations, fractures, amputations, contusions, and abrasions.

23.3* Consumption of the Body by Fire. The investigator should be aware that, in some cases, the body is part of the fuel load of a burning room. That is why the burn patterns on the body and any consumption of it should be considered within the context of the entire scene and not in isolation. Exposure to fire results in a predictable progression of effects to the body and its components. The body reacts to heat and parts of the body can move in a predictable response due to contraction or destruction of countering forces of the muscles and ligaments. Aside from the clothing, there are four major combustible components of the body; skin, fat, muscle, and bone. The significant difference between these components is moisture content.

23.3.1 Skin. Skin will change color, blister, dehydrate, and split as it responds to heat. The splitting of the skin does not extend into the muscle but does expose the underlying fat layer. Skin is not a good fuel, but it will burn when dehydrated and exposed to sufficient heat.

23.3.2 Muscle. During exposure to fire, the muscle tissue shrinks due to dehydration. This shrinkage causes flexion. The flexion occurs to the fingers, hands, wrists, elbows, shoulders, toes, ankles, knees, and hips. This flexion can produce the so-called pugilistic attitude or pugilistic posture. The crouching stance with flexed arms, legs, and fingers is not the result of any pre-fire physical activity (such as self-defense or escape), but is a direct result of the fire. Bone fractures can result from such muscle contraction or as the result of extensive direct exposure to heat and flames. Muscle tissue is not a good fuel, but it will burn when dehydrated and exposed to sufficient heat.

23.3.3 Bone. The living bone will shrink, fracture and change color when heated. While its surface undergoes degradation to a flaky or powdery form, it does not readily oxidize to calcium oxide. The damaged bones can be fragile and may fracture during recovery and transport of the body. While not readily combustible, bone adds to the fuel load by supplying marrow and tissue. The skull can fracture (typically along the suture lines) or disintegrate when heated. In significantly burned bodies, it is common to see consumption of extremities and partial or full consumption of the skull cap.

23.3.4 Fat. Animal fat has a heat of combustion (ΔHc) of over 30 MJ/kg. It can be dehydrated by a modest flame, and then melted or rendered to sustain combustion. Human bodies do not combust spontaneously. Under certain conditions, the fat from a body can sustain a small but persistent flaming fire. If the body fat can be absorbed onto the rigid char of upholstery, clothing, bedding, or carpet, the flames can be sustained by the wicking action of the fat on the material in the manner of an oil lamp. The flames then promote dehydration and combustion of muscle tissues and internal organs and reduce bones to a flaky mass over a period of many hours. The fire thus sustained is small enough that other combustible fuels in the vicinity may not ignite by radiant or convective heating. The end result is a body most heavily burned away in the area where the most body fat is located (the torso and thighs), leaving the lower legs, arms, and often the head relatively unburned.

23.4 Postmortem Changes. The deceased body will undergo physical changes as a reaction to exposure to heat and to death. The investigator may encounter lividity and rigor mortis as postmortem physical changes of the body.

23.4.1 Lividity. Upon death, the circulation of blood through the vessels and capillaries ceases and the blood begins to settle, due to gravity, into the lowest available portions of the body. This process occurs over a period of hours. The settling of blood produces a purple or red coloration in the tissues, called lividity or livor mortis. In the first few hours after death, if the body is moved and its position altered, lividity disappears from one area and will develop in the new lowest area. After 6 to 9 hours, lividity becomes fixed and no longer shifts if the body is moved. The areas of lividity can appear red if the victim died with a significant COH_b level, because of the bright red color of blood with a high COH_b saturation. The presence, absence, and pattern of areas of lividity can help establish the position of the body after death and can reveal whether it has been moved or repositioned after death.

23.4.2 Rigor Mortis. Over a period of hours after death, chemical changes in the muscle tissue cause it and the joints to stiffen in place. This is called rigor mortis. It develops first in the hands and feet, progressively involving the limbs, torso, and head. Its onset depends on the temperature of the body (and its environment) and the physical activity of the victim just before death. After 12 to 24 hours, the rigor passes, leaving the joints and muscles limber. Loss of the rigor proceeds from extremities to torso and head over a several-hour period. Extreme muscular activity just prior to death and high environmental temperatures may hasten the onset (and often the loss) of rigor. Experienced forensic pathologists may use the progressive onset and loss of rigor to help establish an approximate time of death. Rigidity (and contraction) of muscles caused by exposure to fire is not the same as rigor mortis and does not leave the body over time.

23.5 Investigating Fire Scenes with Fatalities. There are a number of considerations to be made before the investigation of a fatal fire. Preplanning to ensure that appropriate procedures are followed during the investigation can have a significant impact on the length and success of the investigation. Authority notification, scene documentation, and body recovery are important aspects of fire death investigation. Collaboration with the fire department and forensic medical community is essential to ensure that the investigation is conducted properly and all factual data and evidence is protected and secured.

23.5.1 Notification. In death investigations, there are legal and procedural requirements for notifying the authorities, including police, coroner, medical examiner, and forensic lab. These procedures may vary from jurisdiction to jurisdiction, and may involve both civil and criminal agencies. It is the responsibility of the investigator to understand these requirements prior to beginning their investigation.

23.5.2 The Fire Department. In the process of extinguishing a fire, valuable evidence can be disturbed. Fire suppression personnel should be made aware that the use of hoses with straight-stream nozzle or nozzle patterns can unsettle fragile evidence, such as a badly charred body. As soon as a body is discovered, and it is determined that the victim is beyond medical aid, the fire department should not move the body. It might be thought advantageous to remove a body so that operations are not impeded, but it is beneficial to the entire fire-death investigation if the body is left in place until it can be properly documented and examined. Every effort should be made to minimize fire-fighting operations in close proximity to the victim, including foot traffic, hoselines, and equipment. If there is any chance of resuscitation, however, the survival of the victim must take priority. Only severe emergency conditions, such as imminent collapse of the building or uncontrollable fire in the vicinity, should force premature removal of the body.

23.5.3 Team Investigation. A proper death investigation is a team effort, and may involve the investigator, homicide detective, and coroner or medical examiner. All parties should be prepared to work side by side at the scene to ensure that all critical evidence is recovered, whether the death is determined to be accidental or otherwise. If there are indications of foul play or if the body is very badly burned, the investigator may need the special assistance of a criminalist (forensic scientist) with crime scene experience, a forensic odontologist, or a forensic anthropologist.

23.5.4 Safety. The investigation of a fatality involves work around and with a body, in a scene possibly contaminated by that body. Burned bodies can be in a condition that exposes the scene and investigators to body parts and body fluids. Investigators should take all appropriate safety precautions for dealing with a potential biohazard. The investigator should be aware of the unique requirements for a biohazard scene to include appropriate personal protective equipment, decontamination procedures, disposal procedures, and marking of all evidence that might be a biohazard.

23.5.5 Scene Documentation. As soon as conditions permit, photographic documentation of the body or body parts and their surroundings should be carried out. Video recordings or instant-photo films may not provide adequate detail but may be used as a supplement. If the body has to be moved due to emergency considerations, a few photographs may make the difference between a successful investigation and failure. Fire patterns or blast effects on clothing and on the body may be important evidence. Photographs should be taken of all exposed surfaces of the body before debris is disturbed, and again during examination and layering operations. The body should be photographed while it is being moved, to record any changes incurred during the removal process. In this situation, supplemental videotaping may be beneficial. The body should also be photographed once inside the body bag. After the body is removed, the location where it was found should be photographed. Clothing on the body is best collected at the post scene examination of the body, preferably at autopsy. If clothing on the body needs to be collected at the scene to preserve evidence, such as possible presence of ignitable liquid, photographs should be taken of the body before it is unclothed, as well as, after the removal of clothing.

23.5.5.1 Diagrams and sketches should supplement photos. Measurements and sketches should be performed prior to the movement of the body. Diagrams can show hidden details. They can record the dimensions of features of the scene and can document distances between the body (and its extremities) and furnishings, walls, doors, windows, and other features *(see Figure 23.5.5.1)*. The outline of the body should be recorded on a diagram and may be traced on the floor in chalk, tape, or string so that it can be referred to in later stages of the scene examination.

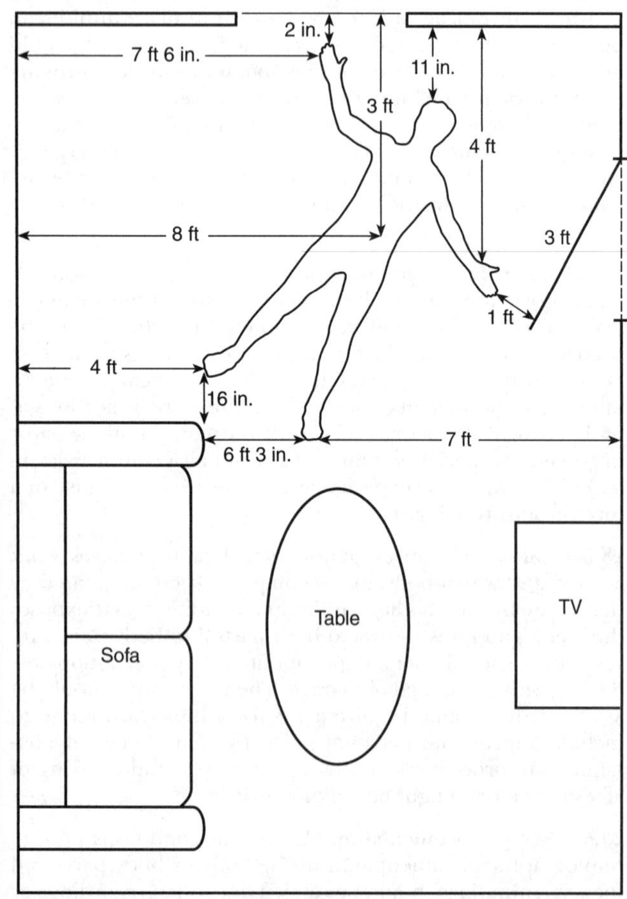

FIGURE 23.5.5.1 Diagram Showing Location of Body in Relationship to Room and Furnishings.

23.5.5.2 The body is unique in the fire scene as its final position can be radically different from its position at the start of the incident. While many items in a fire can be relocated during the event due to building collapse or suppression activity only the body has the capacity to actively move during the event. The victim may have engaged in starting the fire, fleeing the scene, fire suppression activity, rescue activity, or refuge from the fire. There may be evidence related to the victim's activity and movement that should be recorded on the diagram. Often, soot deposition in the compartment may provide an indication of the victim's path of travel or actions prior to being overcome. Doors, windows, and walls should be assessed for handprints, fingernail markings, fragments of clothing, blood, sloughed skin, or indications of the victim's initial location and subsequent movement. The investigator should take caution, however, that wall markings are not the result of fire department search operations.

23.5.6 Victim Documentation. The location, distribution, and degree of burns or other injuries should be shown on a body diagram such as shown in Figure A.15.3.2(b) Casualty Field. Such documentation of the burn patterns may assist the investigator in determining the victims' activities and location during the fire. Major physical trauma and wounds to the body, such as gunshot, fractures, blunt trauma, and knife wounds, should be viewed and documented. A detailed examination of the trauma and wounds should occur at the autopsy.

23.5.7 Recovery of Bodies and Evidence. At a fatal fire the search for evidence tends to focus on the body. Critical evidence is often recovered within arm's reach of the body. The body is a convenient reference point, but it should be remembered that evidence may be elsewhere in the vicinity, so a careful search must be made of the entire room or area. To aid in the search, this area can be marked off into sectors. A grid system may be developed to conduct the investigation by dividing the scene into specific areas as shown in Figure 23.5.7. The search in each grid needs to be documented and the evidence from each grid identified. The geometry of the scene may determine the grid system. Other search methods include strip or area searches. Regardless of the method used, the assigned search areas should overlap to ensure complete coverage.

23.5.7.1 Layering of Debris. The sequence of events of the death and fire or explosion may be revealed by the sequence of layers in the debris (ceiling, furniture, body, floor covering, etc.) and by noting where the fire damage has occurred. An unburned body found on an unburned sofa beneath a collapsed ceiling is a lot different than a burned body found on a burned sofa with a ceiling collapsed on top.

23.5.7.2 Sifting of Debris. The search proceeds through the layers of debris as they are found. The debris from each sector can be removed to a location where a more detailed search can be carried out by other searchers sieving through a series of sifting screens. Such screens are typically made of 1 in., 0.50 in., 0.25 in. in wire mesh fitted to wooden or metal frames. The scene should be photographically documented during each phase of layer removal.

23.5.7.3 Body Removal. Examining the debris in a path to the body and around the body prior to removing the body can prevent destroying potential evidence during body removal. When the body is removed the body should be placed in a new, unused, sealed body bag. All debris associated with or adhering to the body should be transported in the body bag and preserved for trace evidence, volatiles, weapons, projectiles, and the like. The

FIGURE 23.5.7 An Example of a Room That Has Been Marked Off into Sectors.

area under the body should then be carefully searched for evidence that has fallen loose while the body was being moved. Any evidence that has fallen loose from the body, with the exception of body parts which should be placed in the body bag, should be retained in appropriate evidence containers.

23.5.7.4 Victim Clothing. In cases where self-immolation is suspected or the victim was exposed to an ignitable liquid, the investigator should preserve the victim's clothing. In cases where the victim was transported to the hospital, the investigator should notify the hospital staff that all clothing must be retained as evidence. The collection of the clothing should be performed as soon as possible after removal. In cases where flammable or ignitable liquid may be present on the clothing, appropriate evidence retention procedures should be followed.

23.5.8 Collection of Other Physical Evidence. There are other kinds of physical evidence that need to be considered.

23.5.8.1 When possible, the investigator should be present when the postmortem examination (autopsy) is conducted. This will help to ensure complete documentation, which may include injuries, smoke inhalation, and toxicology. Additionally, the investigator may be able to answer questions that arise during the examination. Pathologists and medical examiners may have limited knowledge of fire chemistry, fire dynamics, or blast effects. In such instances, the investigator can advise as to fire conditions in the vicinity of the body. The investigator should ensure that physical evidence such as any foreign objects including bullets, casings, explosive residue, knives, and other weapons found with the body, as well as body fluids, recorded fingerprints, and dental records, are appropriately collected, preserved, and analyzed.

23.5.8.2* Animals that die in a fire may provide valuable information. Similar postmortem examinations that can be performed on humans can also be performed on animals.

23.6 Investigating Fire Scenes with Injuries. The injuries sustained by an individual exposed to a fire event may constitute important physical evidence. Fire injuries can lead to death hours, days, or even weeks after the event. Many of the investigative techniques discussed in Section 23.3 are also applicable at fire scenes with injuries; however there are additional considerations that should be made when investigating fire scenes where occupants were injured.

23.6.1* Notification Laws. Many jurisdictions have reporting laws that require emergency or medical personnel to notify police or fire authorities when a person suffering from significant burns is treated. These laws are patterned after gunshot wound notification laws and have been found to be successful in identifying both victims of assault and abuse, as well as perpetrators of arson who are burned in the execution of their crime.

23.6.2 Scene Documentation. Seriously injured occupants may not be able to provide statements regarding their observations until weeks to months after the fire incident. Therefore, it is important to gather information regarding the occupant's location in the structure during the fire from indicators in the structure.

23.6.2.1 Even though the injured person was removed from the fire, soot patterns, heat demarcation, and protected areas may still exist in the area where they were discovered. Diagrams, sketches, and photographs should be taken of any area where soot outlines are found. This area should be diagramed as if the body was still in its original location.

23.6.2.2 The location of the injured victim when discovered may not be the location of the victim throughout the incident. The scene should be examined for evidence related to the location of the victim prior to the fire and for movement during the fire. The evidence could include items such as burned fragments of clothing, blood, or even sloughed burned skin. These items should be fully documented with photographs and measurements for diagrams.

23.6.3 Victim Documentation. Documentation should include diagrams of distribution of burn injuries and photographs. Photos should be taken as soon as possible after the injury preferably, before significant treatment is under way. Medical treatment and healing will affect appearance; therefore, photos taken later in the healing process may be difficult to interpret. Removal of scar tissue formed over healing burn wounds, skin grafts, and incisions made to relieve pressure and allow flexibility can make burn areas look different (better or worse) than the original burns.

23.6.4 Victim Timeline. In addition to the development of a timeline of fire events, the investigator should develop an occupant timeline. The occupant timeline should include the time the injured person was discovered, the time the injured person was removed from the scene, the time the injured person began transport, and the time the injured person arrived at the medical facility. The investigator should make special efforts, when possible, to gather information related to the type of treatment that the patient received during transport to the hospital and while at the hospital. This treatment information should include the time and type of drugs administered, the time and type of oxygen therapy, and the time at which toxicological sampling and measurements were conducted. *See time lines in the chapter on Failure Analysis and Analytical Tools.*

23.6.5 Physical Evidence. Clothing of people injured in fires or explosions is likely to be removed by emergency personnel or by emergency room staff and may be discarded. The clothing, including outer clothing, undergarments, shoes, and socks, should be collected and preserved. If there is a suspicion that ignitible liquids or explosives were involved, there may be residues present on the garments. The clothing should be collected and preserved in accordance with information provided in the chapters on Physical Evidence and Fire & Explosion Investigation for later analysis.

23.7 Explosion Deaths and Injuries. The documentation of an explosion incident where injuries and/or deaths have occurred requires many of the same techniques discussed in Sections 23.4 and 23.5 with some additional considerations. When examining victims of explosions, the investigator should take extreme care to scrutinize the body parts, clothing, and associated debris to find, document, and preserve items of evidence, such as clothing and any foreign objects found.

23.7.1 Collecting Physical Evidence from Explosions. Physical evidence from an explosion may extend beyond the body itself. Such evidence may be visible to the naked eye (e.g., blood stains) or may be microscopic (e.g., hairs or fibers). This evidence may be found on such things as clothing or furnishings. All shrapnel evidence should be collected regardless of minute size or meaningless appearance as the significance of the item might not be determined until it is examined by a laboratory.

23.7.1.1 As in the case of fire injuries, clothing of people injured in an explosion is likely to be removed by emergency personnel or by emergency room staff and may be discarded. The clothing, including outer clothing, undergarments, shoes, and socks, should be collected and preserved. These garments may contain explosive residues, shrapnel, or shrapnel patterns. The clothing items may indicate the activity of the wearer at the time of the explosion. The clothing should be collected and preserved for later analysis in accordance with the Chapters on Physical Evidence and Explosions.

23.8 Post Scene Investigation of Injuries.

23.8.1 Burns. Evidence of burn injuries is often recorded in medical reports using terms with which the investigator should be familiar.

23.8.1.1* Degree of Burns. Degree of burn describes the depth and seriousness of injury as follows: Alternate descriptions of degrees of burn to skin are superficial, partial, and full-thickness burns.

(1) First degree: reddened skin only (like simple sunburn)
(2) Second degree: blistering
(3) Third degree: full-thickness damage to skin
(4) Fourth degree: damage to underlying tissue, charring

23.8.1.2 Body Area (Distribution). Burn damage to the body is often estimated by the medical community by the "rule of nines," where the major areas are represented by increments of 9 percent (%) as follows:

(1) Front of torso, 18 percent
(2) Right arm, 9 percent
(3) Front of right leg, 9 percent
(4) Rear of right leg, 9 percent
(5) Head, 9 percent
(6) Rear of torso, 18 percent
(7) Left arm, 9 percent
(8) Front of left leg, 9 percent
(9) Rear of left leg, 9 percent
(10) Genitals, 1 percent

23.8.1.3 A more precise distribution of skin surface to body area, which reflects the true proportions of the body, and which is sometimes used, is provided in Table 23.8.1.3.

Table 23.8.1.3 Percentage of Body Surface Area

Body Part	Infant	Child	Adult
Front of head	9.5	8.5	3.5
Rear of head	9.5	8.5	3.5
Front of neck	1	1	1
Rear of neck	1	1	1
Chest and abdomen	13	13	13
Genitalia	1	1	1
Back and buttocks	17	17	17
Front of arm and hand	4.25	4.25	4.75
Rear of arm and hand	4.25	4.25	4.75
Front of leg and foot	6.25	6.75	10
Rear of leg and foot	6.25	6.75	10

Infant: Up to age 4. Child: Age 5 to 10. Adult: Age 11 and above.

23.8.1.4 The total burned area of the body is sometimes used as a predictor of survivability, as indicated in Figure 23.8.1.4. Whether the victim survives or not may dictate further investigation. This figure can be used for assessing the likelihood of survivability.

23.8.2 Inhalation Medical Evidence. Like medical evidence concerning skin burns, medical evidence concerning inhalation exposure to toxic gases and heat can provide important information to the investigator to understand both the actions of the injured individual and the fire environment to which the individual was exposed.

23.8.3 Hospital Tests and Documentation. Normally, upon hospital entry of a patient with fire-related injuries, a blood sample should be taken and analyzed for percent saturation of carboxyhemoglobin (percent of COH_b), HCN concentration, blood alcohol, drugs, and blood pH to aid in the diagnosis and treatment of the individual. These measurements may be valuable in assessing the conditions of the individual at the fire scene and the fire environment to which the individual was exposed. In particular, the percent of COH_b is a valuable indicator. However, since the percent of COH_b begins to be reduced as soon as the individual is removed from the fire environment, it is important that the blood sample be taken as soon as possible.

23.8.3.1 The rate at which CO is eliminated from the body is dependent on the oxygen concentration of the inhaled air. The half-life of CO in surviving victims is described as the time period over which the quantity of CO in the blood decreases to half its original value. Typical literature values for the half-life of CO

Body Area Burned (%)	Age (year)																
	0–4	5–9	10–14	15–19	20–24	25–29	30–34	35–39	40–44	45–49	50–54	55–59	60–64	65–69	70–74	75–79	80 +
93+	1	1	1	1	1	1	1	1	1	1	1	1	1	1	1	1	1
88–92	0.9	0.9	0.9	0.9	1	1	1	1	1	1	1	1	1	1	1	1	1
83–87	0.9	0.9	0.9	0.9	0.9	0.9	1	1	1	1	1	1	1	1	1	1	1
78–82	0.8	0.8	0.8	0.8	0.9	0.9	0.9	0.9	1	1	1	1	1	1	1	1	1
73–77	0.7	0.7	0.8	0.8	0.8	0.8	0.9	0.9	0.9	1	1	1	1	1	1	1	1
68–72	0.6	0.6	0.7	0.7	0.7	0.8	0.8	0.8	0.9	0.9	0.9	1	1	1	1	1	1
63–67	0.5	0.5	0.6	0.6	0.6	0.7	0.7	0.8	0.8	0.9	0.9	1	1	1	1	1	1
58–62	0.4	0.4	0.4	0.5	0.5	0.6	0.6	0.7	0.7	0.8	0.9	0.9	1	1	1	1	1
53–57	0.3	0.3	0.3	0.4	0.4	0.5	0.5	0.6	0.7	0.7	0.8	0.9	1	1	1	1	1
48–52	0.2	0.2	0.2	0.3	0.3	0.3	0.3	0.5	0.6	0.6	0.7	0.8	0.9	1	1	1	1
43–47	0.2	0.2	0.2	0.2	0.2	0.3	0.3	0.4	0.4	0.5	0.6	0.7	0.8	1	1	1	1
38–42	0.1	0.1	0.1	0.1	0.2	0.2	0.2	0.3	0.3	0.4	0.5	0.6	0.8	0.9	1	1	1
33–37	0.1	0.1	0.1	0.1	0.1	0.1	0.2	0.2	0.3	0.3	0.4	0.5	0.7	0.8	0.9	1	1
28–32	0	0	0	0	0.1	0.1	0.1	0.1	0.2	0.2	0.3	0.4	0.6	0.7	0.9	1	1
23–27	0	0	0	0	0	0	0.1	0.1	0.1	0.2	0.2	0.3	0.4	0.6	0.7	0.9	1
18–22	0	0	0	0	0	0	0	0.1	0.1	0.1	0.1	0.2	0.3	0.4	0.6	0.8	0.9
13–17	0	0	0	0	0	0	0	0	0	0.1	0.1	0.1	0.2	0.3	0.5	0.6	0.7
8–12	0	0	0	0	0	0	0	0	0	0	0.1	0.1	0.1	0.2	0.3	0.5	0.5
3–7	0	0	0	0	0	0	0	0	0	0	0	0	0.1	0.1	0.2	0.3	0.4
0–2	0	0	0	0	0	0	0	0	0	0	0	0	0	0.1	0.1	0.2	0.2

1 = 100% mortality; 0.1 = 10% mortality (from Bull, 1979).

FIGURE 23.8.1.4 Mortality by Percentage of Body Burned and Age

when air (approximately 21 percent O_2) is inhaled after exposure are 3–5 hours. Half-life values with 100 percent O_2 therapy can range from 45 to 90 minutes. As with uptake, the rate of elimination is dependent on the activity level of the individual. Because treatment and time can significantly reduce the measured percentage of COH_b, the time from fire exposure to sampling of the blood for analysis and the treatment of the individual with oxygen by ambulance and hospital caregivers prior to sampling are important information, which the investigator should determine. Hospital records may provide information regarding multiple measurements of COH_b levels taken from the patient over time. In such cases, the actual patient-specific COH_b half-life can be calculated when accurate documentation of the sample extraction times is available.

23.8.3.2 Other information of importance to the fire investigator is the condition of the airways. The presence of soot or thermal damage in the upper airways provides information about the fire environment to which the individual was exposed. Lung edema and inflammation can be indications of exposure to irritant gases.

23.8.4 Access to Medical Evidence. The fire investigator should be aware of the applicable legal protections regarding the confidentiality of medical records, and the appropriate methods for obtaining and safeguarding this confidential information.

23.9* Fire Death Pathological and Toxicological Examination. Any time a fire death occurs, whether the death is immediate or sometime after the fire event, a pathological and toxicological examination of the victim should be performed.

23.9.1 The Coroner or Medical Examiner. The coroner's or medical examiner's analysis can provide valuable factual data which can aid in the investigation of a fire or explosion. Unfortunately, there is no standardized system for conducting autopsies; therefore, the completeness of the pathological and toxicological examination and the extent to which factual data is collected that will benefit the investigator may vary from jurisdiction to jurisdiction. The coroner or medical examiner may not be knowledgeable regarding the appropriate information to collect for the fire investigator. Therefore, the investigator should request that all fire-specific data and evidence be collected during the autopsy and toxicological analysis.

23.9.1.1 The coroner or medical examiner is tasked with determining the cause and manner of death of the victim. The cause of death may be defined as the event, injury, or illness that caused the sequence of changes that ultimately brought about death. Examples of causes of death include smoke inhalation, burn (incineration), gunshot, trauma (explosion, structural collapse), but may be heart attack or illness (chronic or acute). The manner of death describes the general course of events or circumstances that brought about the cause of death (accidental, homicidal, suicidal, natural, or undetermined).

23.9.1.2 The investigator should be familiar with the forensic medical system in their community, e.g., coroner or medical examiner. The educational background and level of training of forensic death investigators varies widely from jurisdiction to jurisdiction. In a coroner system, the death investigator is not required to be a medical doctor and is elected or appointed to the position by a governmental entity. In most medical examiner systems, the death investigator must be a medical doctor, typically a pathologist, and is appointed to the position by the health department. Additionally, the toxicological laboratory may be limited in their analytical equipment and analysis techniques; it may be necessary to send samples out for testing when certain toxicants cannot be measured by the in-house laboratory.

23.9.1.3 The investigator should stress the importance of performing a full autopsy to the coroner or medical examiner. For the purposes of fire victims, a full autopsy should include a detailed inspection of the respiratory system and collection of biological samples from various organs in the event that site-specific or duplicative toxicological measurements are necessary. In the event that a full autopsy is not possible or feasible, at minimum, toxicological analysis of the victim's blood and/or tissue should be performed. Additionally, at minimum, the victim's burn locations and severity (e.g., 2nd degree burns to 10 percent of abdomen) should be documented. The toxicological analysis of the victim's specimen should be performed as soon as possible after death and collection and should include quantification of carbon monoxide, cyanide levels, alcohol, and drugs.

23.9.1.4 Collection of Physical Evidence. When possible, the investigator should be present when the autopsy is conducted, not only to ensure that appropriate observations are made, but also to be on hand to answer any questions that arise during the examination. Pathologists and medical examiners may have limited knowledge of fire chemistry, fire dynamics, or blast effects. In such instances, the investigator can advise as to fire conditions in the vicinity of the body. The investigator should ensure that physical evidence such as bullets, casings, explosive residue, knives, and other weapons found with the body, as well as body fluids, recorded fingerprints, and dental records, are appropriately collected, preserved, and analyzed. Notification devices, such as smoke detection systems, should also be collected and preserved as evidence.

23.9.2 Identifying the Remains. There are a number of fundamental issues that may confront the investigator involved in a death related to fire or explosion. These may include remains identification, victim identification, cause and manner of death, victim activity during the fire, and postmortem changes. Although the cause and manner of death will ultimately be determined by the coroner or medical examiner, it may be the responsibility of the investigator, in some jurisdiction, to identify the victim.

23.9.2.1 Human vs. Animal Remains. In a very badly damaged body, it can be difficult to determine if the remains are human or animal. Some animals, such as pigs, deer, and even large dog breeds, have the same mass as an adult human and can be mistaken for human remains (or vice versa). Badly charred remains of children or infants are even harder to identify, because their smaller mass and reduced calcification allow for more thermal destruction. This critical identification may require the services of a physical or forensic anthropologist who is familiar with the anatomical characteristics of all species.

23.9.2.2* Visual Identification. The identification of victims can be carried out by a variety of means, depending on the extent of fire damage to the body. Identification by visual observation is most unreliable, because exposure to even a moderate fire induces tissue swelling and tightening of skin by shrinkage. Color changes to the face and hair can make identification of a person and sometimes even estimation of age and race difficult. Visual observation should be used only as a starting point.

23.9.2.3 Identification by Clothing and Personal Effects. Clothing and personal effects should be used, like visual identification, only as a starting point. It is far too easy for clothes, wallets, rings, watches, and other personal effects (even dental plates) to be substituted onto another person prior to a fire.

23.9.2.4 Fingerprint Identification. Fingerprints can be used with almost complete certainty if record prints are available for the person thought to be the victim. If even a small portion of unburned friction ridge skin remains on a fingertip, that may bear enough individual characteristics to permit comparison.

23.9.2.5 X-ray Identification. X-rays provide one of the surest means of identifying even badly burned bodies. The mass of the head tends to protect the teeth from most fire damage, and dental x-rays may be secured if even a tentative identification is made. The jaws must be resected and x-rays made by a qualified odontologist to replicate the positions and angles of whatever clinical antemortem x-rays are available. The shape and locations of fillings, bridges, and implants are then used to make the identification. In some cases, unusual root shapes or other irregularities have been used in the absence of dental work. X-rays of other parts of the body may yield previous fractures or other injuries or surgical procedures that can verify an identification. There are also custom-made joint implants, prostheses, and even pacemakers that can be used to identify a victim.

23.9.2.6 DNA Identification. Serological or DNA typing can be conducted if there are known DNA comparison samples, a known DNA profile or family members are available to provide reference samples. These techniques can be used on even fragmentary remains if they have not been completely charred and are a reliable form of identification.

23.9.3 X-ray Examination. Current pathological analysis techniques, such as x-rays, can reveal information about the victim that is not obvious during a simple visual examination. X-rays made of the entire body and all associated debris can be extremely beneficial. The x-ray examination may detect the presence of foreign objects in or on the body such as bullets, knife tips, and explosive device components. These can be supplemented with dental x-rays and detailed x-rays of anatomic features (broken bones, wounds, etc.). Depending on the jurisdiction, x-rays may not be routinely conducted on victims; therefore, it may be necessary to request this procedure.

23.9.4 Carbon Monoxide Levels. Fire victim blood and tissue specimens are most commonly assessed for carbon monoxide (CO) because it is simply measured and can provide an immediate indication of the cause of death. The percentage of hemoglobin that is saturation with the carbon monoxide as opposed to oxygen is referred to as carboxyhemoglobin or COH_b. Toxicological laboratories typically report carbon monoxide levels in fire victims as percent (%) COH_b. Carboxyhemoglobin levels should be measured for every fire victim.

23.9.4.1 Carbon monoxide causes a cherry-pink coloration to the skin that may not be visible in dark-skinned individuals or individuals that are heavily covered with soot. The cherry-pink coloration may be visible in the skin, lips, and nipples, as well as in the liquid blood, in areas of postmortem lividity, and in the internal organs. The coloration in internal organs will remain when the organ is preserved in formalin, when normal tissue turns a muddy gray-brown color.

23.9.5 Cyanide Levels. Another important toxicant in fire-related deaths is hydrogen cyanide (HCN). HCN can act alone or in combination with CO to cause incapacitation and death. Unlike CO, HCN is inherently unstable in biological samples. HCN concentrations can increase or decrease over time in samples, therefore, fire victim samples require special preservation techniques to ensure that the cyanide values measured at the time of autopsy are compatible to the values present at the time of death. Some toxicological laboratories do not quantify cyanide concentrations in fire victims because they believe it is not of value to them in their analysis or they do not have the appropriate analytical equipment. Cyanide concentrations, however, can be valuable to the fire investigator. When possible, blood cyanide levels should be measured for every fire victim.

23.9.6 Presence of Other Toxicants. The identification of other inhaled combustion products such as hydrogen bromide and hydrogen chloride or other organic or inorganic toxicant may be valuable to the investigator. Additionally, the victim's levels of alcohol, therapeutic drugs, or drugs of abuse should be quantified. Additionally, the victim's smoking, drinking, and drug habits should be documented as they may play a role in the toxicological findings of COH_b and HCN or may prove to be contributory to death.

23.9.7 Smoke and Soot Exposure. Evidence of smoke or soot in the lungs, bronchi, and trachea (even esophagus) is one of the most significant factors in confirming that the victim was alive and breathing smoke during the fire. This finding requires that the trachea be transected over its entire length. Soot in mouth or nasal openings alone may be the result of soot settling in openings and not from breathing. Additionally, knowing the position of the body when found may be critical to a correct interpretation of soot in the airways. Soot may also be swallowed and found in the esophagus and stomach. The investigator should request that a full autopsy be performed on all fire victims to establish these fact patterns. Depending on the rapidity of the exposure to the toxicants in the smoke, victims may have vomited or may have foam emanating from their mouth due to edema in the lungs.

23.9.8* Burns. Burns may be induced by antemortem or postmortem from exposure to radiant, convective, or conductive heat in the fire environment. Antemortem burns trigger a vital response, including color changes and blistering, which involves cellular and chemical changes that may be detected after death. The investigator should note, however, that blisters can naturally occur under certain conditions in decomposing bodies. Burns that occur immediately prior to death may not have time to exhibit a vital response and may not be distinguishable from postmortem burns. In some cases, the extent of burn injuries can be correlated with the level of heat exposure. A variety of research has been published in the fire science literature which correlates the duration and amount of conductive, convective, and radiative heat exposure to the onset of pain and the extent of thermal skin injuries. The Society of Fire Protection Engineers, Engineering Guide for Predicting 1st and 2nd Degree Skin Burns from Thermal Radiation, is a useful resource when assessing victim burn injuries.

23.9.9* Physical Trauma and Wounds. Major physical trauma and wounds to the body, such as gunshots, fractures, blunt trauma, and knife wounds, should be examined and thoroughly documented. Evidence of such pre-fire trauma will typically survive in a burned body. Prefire fractures are distinct from thermally induced fracturing of the bone. Prefire wounds will have a distinct appearance and will differ from thermally induced fissures in the skin. A detailed examination of the trauma and wounds should occur and should be photographed, including close-ups with a suitable scale in the field of view. Blood can seep from ears, nose, and mouth as a result of heating. Blood found external to the body can indicate antemortem physical trauma.

23.9.10 Stomach Contents. Activities prior to death, and possible time of death, may be established through assaying of stomach contents, which should be examined when indicated. Presence or absence of soot in the esophagus and stomach contents should be noted.

23.9.11 Internal Body Temperature. In some cases, the coroner or medical examiner may use the internal body temperature of the corpse to establish the time and mechanism of death. The internal temperature may be elevated due to hyperthermia, antemortem condition, or postmortem exposure to radiant and convective heat so the internal body temperature should be analyzed within that context.

23.9.12 Pre-Existing Medical Conditions. Pre-existing medical conditions and their associated medications can contribute to the behavior of the individual or the physiological response to the fire incident. Limited mobility, impaired vision, impaired hearing, decreased sensitivity to odors, decreased respiratory capacity, and cognitive impairment are just a few of the pre-existing conditions that can alter a behavior or physiological response.

23.9.13 Death Pre-Fire. The medical examination may determine that the individual died prior to the fire. The autopsy would show a general lack of soot within the airway and low COH_b consistent with ambient conditions. The investigator should determine the relationship between the death and the fire.

23.9.14 Death from a Medical Condition. There are instances where an individual will die in the course of a fire but the cause of death will be from a medical condition such as a heart attack or stroke instead of an effect of the fire. This condition may have been exacerbated by the fire and smoke conditions but not have been the actual cause of death.

23.10 Analysis of Data. The investigator should analyze the data developed from the death or injury investigation to correlate it with the other data from the investigation. Refer to the chapter on Failure Analysis and Analytical Tools for additional information.

23.10.1 Timeline Development. In addition to the development of a timeline of fire events, the investigator should develop a victim timeline. The victim timeline should include the time of victim discovery, the time of victim scene removal and transport, and the time of victim arrival at the examining facility. The investigator should make special efforts, when possible, to gather information related to the time of autopsy and toxicological analysis, as well as, the storage conditions and preservation of biological samples removed from the victim during autopsy. Due to post-mortem putrefactive effects and endogenous production, this information can become important in the interpretation of pathological and toxicological findings.

23.10.2 Victim Activity. An attempt should be made to determine the victim's activity before, during, and after the onset of the fire or explosion and at the time of death, including whether the person was alive and conscious. Factors that can assist the investigator in making these determinations include the following:

(1) Location of the body (in bed, at exit)
(2) Position of the body (in chair, hiding)
(3) Clothing on the body (pajamas, work clothes)
(4) Burn patterns on the clothing
(5) Burn patterns on the body
(6) Burn patterns including protected areas under the body
(7) Items found with the body (e.g., keys, telephone, flashlight, fire extinguisher, personal property)
(8) Blast damage to the body (e.g., pressure, impact, and shrapnel)

23.10.3 Pre-Fire Victim Impairment. Both alcohol and drugs (prescription and illegal) can lead to impairment of a victim. The impairment can decrease the response to fire indicators such as smoke, noise, flames, or alarm activation resulting in delayed or no notification of the adverse conditions. Refer to the toxicology report for pertinent information regarding alcohol and drugs. The toxicological report may report the blood alcohol content (BAC) in % ethanol or g/dL ethanol. As a reference point, 0.08 g/dL is considered by some states to indicate that a driver is impaired, although studies have reported impairment to alcohol at even lower blood concentrations.

23.10.4 Medical History. Pre-existing conditions, especially physical impairments related to vision, hearing, and mobility could have had a substantial impact on the victim's ability to detect and escape from a fire. Typically a medical examiner will gather information from existing medical records to match with the findings of the autopsy. The investigator may conduct interviews with the victim's family and associates to determine if there were medical conditions that were unreported or not discovered in the autopsy.

23.10.5 Fire Pattern. The patterns of damage on the clothing and the body should be considered in context with the total fire or explosion patterns in the room or area. Apparent inconsistencies should be examined. Burn patterns to the clothing (e.g., cigarette burns) may reveal a history of involvement with previous fires. Burn patterns to the clothing or the body may indicate that an attempt had been made to fight the fire or may be evidence of fire setting. The relationship between the death and the fire should be investigated, because not all fire-related deaths are directly caused by heat, flame, or smoke. Examples include a person smoking a cigarette on a sofa who dies of a heart attack, a person jumping from a window to escape a fire, fatal trauma from building collapse, suicide, and homicide prior to the fire.

23.10.6* Burns. The generally accepted criteria for the onset of radiant thermal injuries in a fire compartment is a thermal layer approximate temperature of 200°C (400°F) or a heat flux of 2.5 kW/m_2 at the floor. This layer temperature and flux is typically achieved when the layer is approximately 0.9 meters (3 feet) from the ceiling. This information in combination with fire modeling can provide the investigator with valuable information regarding the activities or movement of the victims prior to incapacitation or death.

23.10.7 Clothing. The clothing items may indicate the activity of the wearer at the time of the fire or explosion. What the clothing is made of and how it is made may play a role in its ignitibility by flaming or smoldering sources (loose long sleeves, fine fabrics, etc.). It may be important for the investigator to determine the ignitibility, burning properties (char, melt, or both), or heat release rate of the clothing involved.

23.10.8* Applications of Toxicology in Fire Investigation. A relationship exists between the nature of a fire, i.e., smoldering, flaming, post-flashover, and the production of toxic gases such as carbon monoxide (CO) and hydrogen cyanide (HCN). Because of this relationship, reliable toxicological data from a fire victim, in combination with fire models and other available tools, can provide the Fire Investigator with useful evidence.

23.10.8.1 Toxicological Analysis Techniques. There are numerous ways in which biological specimens can be analyzed for toxicants. Depending on the specific toxicant to be measured, laboratories may utilize any combination of spectrophotometry, chromatography, immunoassay, and mass spectrometry techniques for initial screening and confirmation. Depending on the preference of the laboratory and the validation of the method, these assays may be carried out on blood, liver, urine, spleen, or various other organs and tissues. In severely burned bodies, there may be limited blood, so other body fluids or tissue samples may be used. Some toxicological laboratories document the specimen and method that was used for toxicant quantification in their final reports; however, when this is not done, the investigator should consult with the toxicologist for this information and ensure that it has been documented.

23.10.8.2* The Colburn Forster Kane (CFK) Equation. The Colburn-Forster-Kane (CFK) equation describes the uptake of CO in humans and can be used as an investigative tool to estimate the quantity of carbon monoxide that a victim inhaled during a fire. The CFK model requires an input of the concentration of CO produced by the fire over time, in addition to the victim's weight and respiratory minute volume (RMV). The RMV is defined as the volume of air exchanged by breathing per minute (typically ranges from 8.5 l/min to 50 l/min depending on level of activity). The CFK model outputs a COH_b value; therefore, through iterations one can determine the average CO exposure needed to achieve the COH_b level found at autopsy.

23.10.8.2.1 The CFK equation is represented as

$$\frac{A[HbC0]_t - BV_{C0} - PI_{C0}}{A[HbC0]_0 - BV_{C0} - PI_{C0}} = e - tAV_bB$$

where:

$$A = \overline{P}_{C,02} / M[Hb0_2]$$
$$\overline{P}_{C,02} = PI_{02} - 49$$
$$PI_{02} = PI_{02} - 49$$
$$PI_{02} = 148.304 - 0.0208 \times PI_{C0}$$
$$M = 218$$
$$[HB0_2] = 0.22 - [HbC0]_t$$
$$[HbC0]t = [C0Hb\%t] \times 0.0022$$
$$B = 1/DL_{C0} + PL/VA$$
$$DL_{C0} = 35VO_2 \times e^{0.33}$$
$$VO_2 = RMV/22.274 - 0.0309$$
$$PL = 713$$
$$V_A = 0.933V_E - 132f$$
$$f = \exp[0.0165 \times RMV + 2.3293]$$

23.10.8.2.2 A fire modeling program, such as Fire Dynamic Simulator (FDS) or CFAST, can be used to correlate the amount of CO needed to achieve the victim's known COH_b level with the amount of CO estimated to be produced in the various, hypothesized fire scenarios. Therefore, the CO exposure quantity derived from the CFK equation can be related to a particular type of fire when compared with all the known facts of the case. Additionally, the victim's COH_b level in combination with burn injuries can provide details as to the location of the victim during the fire and/or proximity of the victim to the area of origin.

Chapter 24 Appliances

24.1* Scope.

24.1.1 This chapter covers the analysis of appliances as it relates to the investigation of the cause of fires. The chapter concentrates on appliances as ignition sources for fires but, where applicable, also discusses appliances as ignition sources for explosions. This chapter assumes that the origin of the fire has been determined and that an appliance at the origin is suspected of being an ignition source. Until an adequate origin determination has been done, it is not recommended that any appliances be explored as a possible ignition source.

24.1.2 Addressed in this chapter are appliance components, which are common to many appliances found in the home and business. Sections of this chapter also deal with specific but common residential-type appliances and with how they function.

24.2 Appliance Scene Recording. The material presented in Chapter 15 should be used where appropriate to record the scene involving an appliance. Material presented in this section is supplemental and has specific application to appliances.

24.2.1 Recording Specific Appliances. Once a specific appliance(s) has been identified in the area of origin, it should be carefully examined before it is disturbed in any way. The appliance should be photographed in place from as many angles as possible. Photographs should be close-ups of the appliance as well as more distant photographs that will show the appliance relative to the area of origin, the nearest combustible material(s), and a readily identified reference point (e.g., window, doorway, piece of furniture). This reference point will greatly aid later reconstruction efforts in placing the exact location of the appliance at the time of the fire. If an appliance has been moved since the start of the fire, then the same photographs should be taken where it was found. If it can be established where the appliance was located at the time of the fire, such as by observing a protected area that matches the appliance base, or by talking to someone familiar with the fire scene prior to the fire, the appliance should be moved to its pre-fire location and the same photographs taken. This movement by the investigator may not be done until all other necessary documentation is completed.

24.2.2 Measurements of the Location of the Appliance. The scene should be photographed and diagrammed as described in Section 15.4. The location of the appliance within the area of origin is particularly important. The investigator should take measurements that will establish the location of the appliance.

24.2.3 Positions of Appliance Controls. Special attention in the photography and diagramming should be paid to the position of all controls (e.g., dials, switches, power settings, thermostat setting, valve position), position of movable parts (e.g., doors, vents), analog clock hand position, power supply (e.g., battery and ac house current), fuel supply, and any other item that would affect the operation of the appliance or indicate its condition at the time of the fire.

24.2.4 Document Appliance Information. The manufacturer, model number, serial number, date of manufacture, warnings, recommendations, and any other data or labels located on the appliance should be documented. This information should be photographed, and notes should be taken, as these items may be difficult to photograph. Having notes will ensure that this valuable information is preserved. It is frequently necessary to move the appliance to obtain these data, and this should be done with minimal disturbance to the appliance and to the remainder of the fire scene. In no case should the appliance be moved prior to completion of the actions in 24.2.3.

24.2.5 Gathering All of the Parts from the Appliance. Where the appliance has been damaged by the fire or suppression activities, every effort should be made to gather all of the parts from the appliance and keep them together. After exposure to fire, many of the components may be brittle and may disintegrate with handling, which is why it is important to document their conditions at this point. Where it is considered helpful and will not result in significant damage to the remains of the appliance, some reconstruction of the parts may be done for documentation and analysis purposes. This reconstruction could include replacing detached parts and moving the appliance to its original location and position. Attempting to operate or test an appliance should not be done during the fire scene examination, as this may further damage the appliance, possibly destroying the critical clues within the appliance and its components. All testing at this point should be strictly nondestructive and only for the purpose of gathering data on the condition of the appliance after the fire. Examples of nondestructive testing include using a volt/ohmmeter to check resistance or continuity of appliance circuits.

24.3 Origin Analysis Involving Appliances. Chapter 6 and Chapter 17 deal with determining the origin of a fire in greater detail. The additional techniques and methodology presented here should be utilized when a fire involves an appliance. This is the case when the fire is confined to the appliance or when it is thought that a fire started by the appliance spread to involve other contents of the room.

24.3.1 Relationship of the Appliance to the Origin. It should be established that the appliance in question was in the area of origin. Those appliances that were clearly located outside the area of origin generally can be excluded as fire causes. In some cases, an appliance(s) remote from the area of origin may have something to do with the cause of the fire and should be included in the investigation. Examples of these are the use of an extension cord or the presence of a standing pilot on a gas appliance. Where doubt exists as to the area of origin, it should be classified as undetermined. When the origin is undetermined, the investigator should examine and document the appliances in any suspected areas of origin.

24.3.2 Fire Patterns. Fire patterns should be used carefully in establishing an appliance at the point of origin. Definite and unambiguous fire patterns help to show that the appliance was at the point of origin. Other causes of these patterns should be eliminated. The degree of damage to the appliance may or may not be an adequate indication of origin. Where the overall relative damage to the scene is light to moderate and the damage to the appliance is severe, then this may be an indicator of the origin. However, if there is widespread severe damage, other causes such as drop down, fuel load (i.e., fuel gas leak), ventilation, and other effects should be considered and eliminated. If the degree of damage to the appliance is not appreciably greater than the rest of the fire origin, then the appliance should not be chosen solely by virtue of its presence.

24.3.3 Plastic Appliance Components. Appliances that are constructed of plastic materials may be found at the fire scene with severe damage. The appliance may be severely distorted or deformed, or the combustible material may be burned away, leaving only wire and other metallic components. This condition of an appliance in and of itself is not an adequate indicator of the point of origin. This is especially true where there was sufficient energy from the fire in the room to cause this damage by radiant heating and ignition. Conditions approaching, or following, flashover can have sufficient energy to produce these effects some distance from the point of origin.

24.3.4 Reconstruction of the Area of Origin. Reconstruction of the area of origin may be necessary to locate and document those patterns and indicators that the investigator will be using to establish the area of origin. As much of the material from the appliance as possible should be returned to its original location and then recorded with photographs and a diagram. The help of a person familiar with the scene prior to the fire may be necessary.

24.4 Cause Analysis Involving Appliances. The material presented in Chapter 18 should be used where appropriate to analyze an appliance that may have caused a fire. Material presented in this section is supplemental or has specific application to appliances.

24.4.1 How the Appliance Generated Heat.

24.4.1.1 Before it can be concluded that a particular appliance has caused the fire, it should first be established how the appliance generated sufficient heat energy to cause ignition. The type of appliance will dictate whether this heat is possible under normal operating conditions or as a result of abnormal conditions. The next step is to determine the first material ignited and how ignition took place. The most likely ignition scenario(s) will remain after less likely or impossible ignition scenarios have been eliminated. If no likely ignition scenario exists, either accidental or intentional, then the cause should be classified as undetermined.

24.4.1.2 Patterns on the appliance may indicate the source of the ignition energy. However, hot spots or other burn patterns may be the result of other factors not related to the cause and need to be carefully considered. Patterns on nearby surfaces may provide information on the ignition source.

24.4.2 The Use and Design of the Appliance. The use and operation of an appliance should be well understood before it is identified as the fire cause. Some appliances are simple or very familiar to fire investigators and may not require in-depth study. However, appliance design can be changed by the manufacturer, or an appliance can be damaged or altered by the user, and, therefore, each appliance warrants investigation. More complicated appliances may require the help of specialized personnel to gain a full understanding of how they work and how they could generate sufficient energy for ignition.

24.4.3 Electrical Appliances as Ignition Sources. Many appliances use electricity as the power source, and electricity should be considered as a possible source for ignition. The material presented in Chapter 18 should be carefully considered and applied in this situation. Only under a specific set of conditions can sufficient heat be generated by electricity as a result of an overload or fault within or by an appliance and subsequently cause ignition.

24.4.4 Photographing Appliance Disassembly. When it is necessary to disassemble an appliance (or its remains) recovered from a fire scene, each step should be documented by photography. This is done to establish that the investigator did not haphazardly pull the artifact apart, causing pieces to be further damaged or lost. The documentation should show the artifact at the start and at each stage of disassembly, from multiple angles if possible, keeping careful track of loose pieces. Some investigators find it helpful to videotape this process. The investigator should have at least one specific reason for disassembling an artifact, and once an answer has been found, the disassembly process should stop. When an artifact cannot be easily disassembled or if the disassembly would be too destructive, the use of X-rays should be considered.

24.4.5 Obtaining Exemplar Appliances. To understand an appliance more fully, to test its operation, or to explore failure mechanisms, the investigator may need to obtain an exact duplicate (i.e., an exemplar). For this, the model and serial numbers may be required, and the manufacturer may need to be contacted to determine the history of this appliance. It may be that the manufacturer does not make the particular appliance any more or has changed it in some way. The investigator will need to determine whether the exemplar located is similar enough to the artifact to be useful.

24.4.6 Testing Exemplar Appliances. Exemplar appliances can be operated and tested to establish the validity of the proposed ignition scenario. If the ignition scenario requires the failure or malfunction of one or more appliance components, this can also be tested for validity on the exemplar. Where extensive or repeated testing is foreseen, the investigator will probably need more than one exemplar. The testing should show not just that the appliance is capable of generating heat, but that such heat is of sufficient magnitude and duration to ignite combustible material.

24.5 Appliance Components. Appliances are diverse in what they do and how they are constructed. Therefore, this section will provide a description of each of the common parts or components that might be found in various appliances. Where information is given in later sections about particular appliances, there will be references to the components that are used in those appliances.

24.5.1 Appliance Housings.

24.5.1.1 Introduction. Housings of appliances can be made of various materials. The nature of these materials can affect what happens to the appliances during fires and what the remains will look like after a fire. Most housings are made of metal or plastic, but other materials such as wood, glass, or ceramics might be found also.

24.5.1.1.1 Many appliances utilize painted steel finishes. This typically includes refrigerators, dryers, fluorescent fixtures, baseboard heaters, and the like.

24.5.1.1.2 Care should be exercised when evaluating heat damage patterns on painted steel surfaces. Many paints darken with heat exposure. Additional or greater heat exposure can cause some heat-darkened painted surfaces to lighten in color. Further heat exposure may cause the paint to decompose to a gray or white powder. This gray or white powder can be disturbed or removed by fire fighting, handling, or by the formation of rust. An apparently lighter surface color may reflect more thermal damage than a darker area.

24.5.1.2* Steel. Steel is used for the housings of many appliances because of its strength, durability, and ease of forming. Stainless steel is used where high luster and resistance to rusting is needed, such as in kitchen appliances or wherever appearance and sanitation are important. Other types of steel may be used and coated with plastic or enamel to achieve the desired appearance. Galvanized steel may be used where resistance to rusting is needed but appearance is not important, such as inside a washing machine.

24.5.1.2.1 Steel will not melt in fires except under very unusual circumstances of extremely high temperatures for extended times or by electrical arcs. Ordinarily, steel will be oxidized by fires, and the surface will be the dull blue-gray color resulting from ferrous oxide (FeO). The brown rust color does not appear until the steel item has been wet long enough to rust to the reddish color of ferric oxide (Fe_2O_3). When steel is deeply oxidized by long exposure in a fire, the oxide layer often will be thick enough to flake off. In severe cases, the flaking off may go through the steel and create a hole. In fires of short duration, the surfaces of polished or plated steel can show various color fringes, depending on the degree of heating. After a fire, bare galvanized steel will have a whitish coating from oxidation of the zinc. Often, the surfaces of steel housings will have a mottled appearance ranging from blue-gray to rust to white to black to reddish. The odd colors are usually from residues of decorative or protective coatings on the steel in addition to the oxides. The particular colors and the patterns depend on many factors, and not much importance should be put on the color and patterns without substantiating evidence.

24.5.1.2.2 On rare occasions, a steel housing may be found with a hole made by alloying with zinc or aluminum. Most of the time, when one of these metals drips onto steel during a fire, the surface oxides keep the metals separated. During a long fire, the molten metal might penetrate the oxide layers and alloy with the steel. If there is need to know the cause of the hole, analysis of the steel at the edge of the hole would show alloying elements or absence of them.

24.5.1.2.3 A steel housing does not necessarily keep internal components from reaching very high temperatures. If a closed steel box is exposed to a vigorous fire for a long enough time, the inside of the box can become hot enough to cook materials, to gray ashes, or to melt copper.

24.5.1.3 Aluminum. Aluminum housings are commonly made from formed sheets or castings. Extruded pieces might be found on or in the appliance as trim or supports for other components. Aluminum has a fairly low melting temperature of 660°C (1220°F) if pure; alloys melt at slightly lower temperatures. The extent of damage to the aluminum housing can indicate the severity of the fire or heat source at that point.

24.5.1.4 Other Metals. Other metals, such as zinc or brass, might be used in housings. They would be likely to be just decorative pieces or to be supports for other components. Zinc melts at the relatively low temperature of 419°C (786°F) and so is almost always found as a lump of gray metal. Brass is used in many electrical terminals. Brasses have ranges of melting temperatures in the neighborhood of 950°C (1740°F). Brass items are often found to be partly melted or just distorted after a fire. Because it is an alloy, brass softens over a range of temperatures rather than melting at a specific temperature.

24.5.1.5 Plastic. Plastic housings are used increasingly for a wide range of appliances that do not operate at high temperatures. Most plastics are made of carbon plus some other elements. Some plastics melt at low temperatures and then char and decompose at higher temperatures. Others do not melt but do char and decompose at higher temperatures. Nearly all plastics can form char when heated and will burn in existing fires. Many kinds of plastics will continue to burn by themselves if ignited. Other plastics will not continue to burn from a small ignition source at room temperature because of their chemical compositions or because of added fire retardants. Many plastic housings of recent manufacture have considerable fire retardant added, and they usually will not continue to burn from a small source of ignition. Each appliance in question would need to be checked for ease of burning of the plastic. In some cases, that check can be done by qualified personnel if enough of the material remains or if the identical appliance can be obtained.

24.5.1.5.1 After a brief fire, the plastic housing of an appliance may be melted and partially charred. If the pattern of damage shows that the heat source was inside, further examination of the remains is warranted. The plastics might show instead that the heat was from the outside and that the inside is less heated than the exterior. If the plastic housing has melted down to a partly charred mass, x-ray pictures can reveal encapsulated metal parts and wires. When a plastic housing has been mostly melted and burned by exterior fire, the underside of the appliance might still be intact or a metal base plate unheated.

24.5.1.5.2 When a fire is severe, all plastics might be consumed. Total consumption of the plastic does not by itself indicate that the fire started in the appliance.

24.5.1.5.3 Phenolic plastics are used for certain parts that must have resistance to heat, such as coffee pot handles and circuit breaker cases. Phenolics do not melt and will not burn by themselves. They can be consumed to a gray ash in a sustained fire. When a device that has been made with a molded phenolic body is moderately heated, the gray ash might be just a thin layer on the outside. Gray ash on the inside surfaces with little or no gray ash on the exterior may indicate internal heating.

24.5.1.5.4 When portions of an appliance melt and resolidify as a result of a fire, the direction of flow of the material can indicate the orientation of the appliance at the time that the melted component material cooled.

24.5.1.6 Wood. Wood still has occasional use in appliance housings. Wood can be fully consumed in a fire or can show a pattern of burning when only partly consumed. The pattern can help to show whether the fire came from inside or outside of the appliance.

24.5.1.7 Glass. Glass is used for transparent covers and doors on appliances. Glass might also be used in some decorations. Glass readily cracks when heated nonuniformly and can soften and sag or drip. Flame temperatures are higher than the softening temperatures of glass, so the degree of softening of glass is more a function of duration and continuity of exposure than of fire temperature.

24.5.1.8 Ceramics. Ceramics may be used for some novelty housings and are used as supports for some electrical components. Ceramics do not melt in fires, but a decorative glaze on them could melt.

24.5.2 Power Sources. Power sources for common appliances are usually the alternating current that is supplied by the power companies. There are a few other sources that will be considered. This section will not include voltages higher than 240 or three-phase power. For more detailed information on electrical power and devices, see Section 8.1 through 8.11.10.

24.5.2.1 Power Cords. Power companies in the United States supply electrical power at 60 Hz and 120/240 V ac (often called 110/220 V). Most appliances are designed to operate by plugging them into a 120 V outlet. Appliances that require more power, such as ranges and water heaters, operate at 240 V from the same electrical system in the structure.

24.5.2.1.1 Electrical cords that carry power to the appliance may be made of two or three conductors. The conductors are stranded to provide good flexibility. Some double-insulated appliances and most appliances made before 1962 had only two-conductor cords. Newer large appliances usually have three-conductor cords with the third conductor for grounding as a safety feature. The stranded conductors of cords usually survive fires, but the remains will usually be embrittled if the insulation was burned away during the fire. Careless handling of brittle stranded conductors can cause them to break apart. Cords should be checked for arcing damage. See Chapter 8 for information on electrical conductors and damage to them.

24.5.2.1.2 Plugs for connecting the power cord to the outlet have somewhat different designs, depending on the amperage of the appliance. Plugs made prior to 1987 for 20 A or less were two straight prongs of the same width. Newer plugs have the neutral prong wider than the "hot" prong. The plug may have a third prong for grounding. Factory-made plugs have the conductors attached to the prongs inside a molded plastic body. That body may melt or be entirely burned away in a fire. The conductors and brass prongs will usually survive a fire, but sometimes the brass parts may be melted. After a fire with only minor burning near the plug, the face of the plug will be nearly unheated because of being protected against the receptacle. That finding can show that the appliance was plugged in. Also, even after a more severe fire, the prongs may be less oxidized where they were protected in the receptacle during the fire.

24.5.2.1.3 Plugs for higher voltages or amperages will have larger prongs and different positioning.

24.5.2.2 Voltages Less than 120. Many appliances that plug into a wall receptacle actually operate at 6, 12, or other voltages less than 120 V. Normally, a step-down transformer is used to produce the lower voltage. The transformer will usually be part of the appliance, but sometimes it is a separate unit that plugs directly into the receptacle and feeds the appliance with a thin two-wire cord. Shorting of wiring at 6 V is not likely to cause a fire, but it can do so under circumstances where the energy (i.e., heat) can be concentrated in a small area close to a combustible material.

24.5.2.3 Batteries. Batteries are used for portable appliances and some security devices. Batteries can range from car batteries to common dry cells to small button batteries for cameras and watches. Batteries provide about 1.5 V of direct current. Batteries of 6 V or 9 V are actually made of four or six dry cells, respectively, in one package. Remains of batteries that were present in an appliance can usually be found after a fire. They usually will be damaged too much to indicate whether they provided power for ignition. However, what they were connected to could be important. One battery can provide enough power to ignite some materials under certain conditions. In most battery-powered devices, though, the normal circuitry will prevent the energy of the battery from being sufficiently concentrated at one spot at one time to achieve ignition.

24.5.2.4 Overcurrent Protection. Protection against excessive, damaging current is provided by fuses or circuit breakers in many appliances. After a minor fire, the remains of the protective device might show whether it operated. After a severe fire, the metal parts of the protective device might be found to show at least that it was present.

24.5.2.4.1 The fusing element in a fuse can be one of several metals. In all fuses, the element has the proper cross-section and electrical resistance for the temperature to rise to the melting point if current exceeds a specific level for a specified duration. If the excess current is moderate (e.g., less than twice the rating), the fuse element will melt without vaporizing. If the current is very high, as with a dead short, the element will usually partly vaporize to give an opaque deposit on a window or glass tube of the fuse.

24.5.2.4.2 Most circuit breakers operate thermally or magnetically, depending on the level of overcurrent. Above a specific current level, a bimetal strip deflects enough to let a spring pull the contacts apart. With an instantaneous high current, such as with a dead short, the magnetic field pulls the mechanism so that the contacts open. A circuit breaker that is in a fire environment can trip as the internal mechanism comes up to the activating temperature. Circuit breakers in appliances have a reset button.

24.5.3 Switches. Switches are used to turn appliances on or off and to change the operating conditions. Switches are found in a wide range of sizes, types, and modes of operation. Examination of switches after a fire can determine whether the appliance was on or off or other aspects of its operation. The remains of switches might be very delicate. Other than noting and documenting the positions of knobs, levers, or shafts or checking electrical continuity in place, it is recommended that the investigator not open, operate, or disassemble any switches. That job should be left to someone with technical expertise.

24.5.3.1 Manual Switches.

24.5.3.1.1 Many switches are intended for the user to operate. These include on–off switches and those to change functions, wattage, or other features of the appliance. The design of the switches can include moving lever (e.g., toggle), push button, turning knob, or sliding knob. They have metal parts that can be examined after a fire. Where lightly damaged, the switch might still electrically test on or off or show which position it was in. Where severely damaged, the remains might show only whether the contacts were welded together. Switches will create a parting arc when they open. Therefore, apparent damage to the switch surface may be normal.

24.5.3.1.2 Electronic switches in many appliances may be too damaged by even minor fires to determine their pre-fire position or whether they malfunctioned. Examples of those switches include touch pads on microwave ovens and remote-controlled TVs.

24.5.3.2 Automatic Switches. Many switches in appliances are automatic and are not intended for the user to operate. Those switches generally keep the appliance operating within its design parameters and prevent unsafe operation. Those kinds of switches may be operated by electrical current, temperature, or motion.

24.5.3.2.1 Fuses and Circuit Breakers. Fuses and circuit breakers are automatic switches that operate by overcurrent. Circuit breakers can be reset, but fuses and fusible links need to be replaced.

24.5.3.2.2 Temperature Switches. Automatic switches that operate by temperature and are intended to keep the appliance operating within certain temperature limits are called *thermostats.* Automatic switches that are intended to prevent the appliance from exceeding certain parameters are called *cutoffs, limit switches,* or *safeties.*

24.5.3.2.2.1 Switches that operate by temperature can be based on expanding metal, bimetal bending, fluid pressure, or melting. These switches are usually used to prevent an appliance from operating outside a fixed range of temperatures or to prevent it from exceeding a set temperature (cutoff switches). They ordinarily have enough metal parts to be recognizable after a severe fire, although it may not be possible to determine whether the switch was functional at the time of the fire.

24.5.3.2.2.2 A few switches use expanding metal, where a long rod is positioned in the warm area. If that area becomes too hot, the rod expands and pushes contacts open. More common is the bimetal type, where two dissimilar metals are bonded together in a flat piece. One metal expands more than the other with increasing temperature, so while the temperature rises, the piece bends. That motion can open contacts to turn off the appliance. These switches are slow make-and-break, which is more likely to cause either erosion or welding of the contacts. After a severe fire, the bimetal may be bent far out of position, which is a result of heating from the fire and does not indicate a defective thermostat.

24.5.3.2.2.3 A bimetal disc operates on differential expansion, but the disc snaps from a dish shape in one direction to a dish in the other direction. The edge of the circular disc is fixed, and so the center snaps back or forth at particular temperatures to open or close the contacts.

24.5.3.2.2.4 Some switches operate by expansion of a fluid in a bulb that is located in the hot area. The pressure of that fluid is passed to bellows, often back at a control panel, through a metal tube, commonly copper. The bellows push open the contacts.

24.5.3.2.2.5 These various mechanical switches can be arranged either to open contacts so as to shut the appliance off, or to close contacts so as to turn something else on, such as a cooling fan, that will counter the high temperature. High-temperature cutoff switches may be present in an appliance, but they should not open the circuit except when the temperature becomes too high in the appliance. The contacts of switches should be examined by competent persons. If contacts in cutoffs are eroded by arcing from repeated opening, that can indicate that the appliance was operating in an overheated condition for an extended time, which may indicate a defect in the appliance.

24.5.3.2.2.6 Mechanical switches can fail by overloads, which overheat certain internal parts, or by welding of the contacts. The latter can happen at normal currents as slow make-and-break contacts pass current without being firmly in contact. Poor connections internally, such as where wiring is attached or where brass parts are riveted, can cause destructive heating and failure of the switch. The faces of contacts of thermostats will normally be somewhat pitted because they open and close frequently. Faces of contacts in devices used as safety cutoffs should not be significantly pitted, because the devices should not operate except when there is overheating.

24.5.3.2.2.7 The contacts of a switch are more subject to surface pitting, erosion, and possibly welding when they slowly open and close. For that reason, most switches, especially for carrying substantial currents, are made to snap open or closed, which can be accomplished with a bimetal disc, a flat spring, or a magnet. When welded contacts are found after a fire, that fact does not by itself prove that failure of the switch caused the fire. Heat damage in the appliance could have caused a current surge if power were still available. Electrically welded contacts will have normal shapes, but the faces will be stuck together. If the contacts are found melted together into one lump, the cause is more likely to be severe fire exposure. The contacts are made of metals that have melting temperatures lower than that of copper, and they may melt together from fire exposure.

24.5.3.2.2.8 There are some cutoff devices that operate by internal melting of a material, which lets a spring push the contacts open. These are single-use devices that should be replaced if they operate, although they are sometimes deliberately bypassed, allowing the appliance to operate without protection. The appliance should be checked to verify the presence of such a device, and the device should be checked for signs of tampering or a history of previous replacement.

24.5.3.2.2.9 Many appliances have switches that operate from motion of some part of the appliance. Limit switches on appliances that have moving parts are intended to keep the part from moving too far. Forced-air furnaces may have a switch that operates by airflow pushing a vane up to allow the furnace to continue. Major appliances often have door switches, either to turn the appliance off as the door is opened, or to turn a light on. Motors in major appliances can usually have a centrifugal switch to disengage the starting winding as the motor comes up to speed. Those switches also may control a heating circuit so as not to allow heating unless the motor is running. As with all switches that operate from some mechanical action, these switches can fail to operate if the components that they depend on become misaligned or if the switch comes loose in its holder.

24.5.3.2.2.10 Many portable electric heaters have a tip-over switch that often is built into the thermostat. The switch has a weighted arm that hangs down and opens the contacts if the appliance is tipped so that the arm is not in its normal position.

24.5.4 Solenoids and Relays. Solenoids and relays are used in appliances to control a high-power circuit with one of lower power and often of low voltage. Activation of the low-power circuit energizes a coil or an electromagnet that causes an iron shaft or lever to move. That motion opens or closes the high-power circuit. Remains of solenoids or relays normally remain after a fire. Severe damage might make it impossible to determine whether they were operational or which position they were in at the time of the fire. The contacts should be examined to find whether they were stuck together during the fire.

24.5.5 Transformers. Transformers are used to reduce voltages from the normal 120 V and to isolate the rest of the appliance from the supply circuit. Some transformers are energized whenever the appliance is plugged in, so that the primary windings are always being heated by some amount of current. In other appliances, the transformer is not energized until the switch is turned on. The appliance is designed to keep heating of the transformer at a minimum under normal electrical loads. However, with long-term use, and if ventilation of the appliance is restricted, the temperature may increase and deteriorate the windings. As windings begin to short to each other, the impedance drops and more current flows, causing greater heating. More current flows can also lead to severe heating before the windings either fail by melting the wire or create a ground fault that could open the circuit protection. In some cases, the heated insulation or other combustibles in or on the transformer might be ignited before the electrical heating stops.

24.5.5.1 Appliance transformers are usually made of steel cores and copper windings, both of which will survive fires even when severely heated. Examination of a transformer from a burned appliance might show that the interior windings are less heated and might even be of bright copper color. That finding shows that the heating was external and not from the transformer itself. A transformer from a severely burned appliance might have the windings baked to where they have the appearance of oxidized copper, with no surviving insulation down to the core. The remains of the windings would be somewhat loose on the core. That can happen from long exposure in any fire and does not prove that the windings overheated and caused the fire. Overheating of the windings can be determined when there is a clear pattern of internal heating, arcing turn to turn, and a pattern of fire travel out from that source. It is possible for a transformer to overheat even when protected by a fuse, because the fuse should have a sufficient electrical rating to carry the operating currents plus a safety factor.

24.5.5.2 Some transformers may be totally enclosed in steel and would not be likely to be able to ignite adjacent combustibles before being turned off by protection or internal failure. Other transformers are open and often have paper and plastics, which can be ignited, in their construction.

24.5.6 Motors. Motors are common in appliances to provide mechanical action. They generally range from ⅓ hp to ¼ hp motors in washing machines or other large appliances to tiny motors in small devices. Most common motors are designed to operate at certain speeds. If the rotor is stopped while the motor is still energized, the impedance falls, and current flow increases. That can cause the motor windings to get hot enough to ignite the insulation and any plastics that are part of the construction.

24.5.6.1 Motors often have protection built into them that is intended to stop the current if the temperature gets too high for safe operation. That protection can be in the form of a fuse link, a single-acting thermal cutoff (TCO), a self-resetting thermal protector, or a manual resettable thermal protector. Some motors have both a resettable thermal protector and a single-acting TCO in them. A suspected motor and any protective device should be examined by competent personnel before deciding whether the motor caused the fire.

24.5.6.2 Windings of motors can be examined to find whether they are relatively unheated inside, which would indicate that heating came from the outside. If the windings are thoroughly baked, with oxidized strands all through, but materials around the motor are not so thoroughly heated, that indicates that the windings overheated. If there is much fire around the motor, the windings are likely to be thoroughly baked, whether the fire started in the motor or not.

24.5.6.3 Small motors that drive cooling fans or other devices are usually not sources of ignition. They do not have enough torque to generate much heat by friction. Some small motors are enclosed in metal cases, making ignition by internal heating unlikely. Shaded pole motors are often of open construction, and could ignite combustibles that are in contact with them if the windings get hot enough.

24.5.7 Heating Elements. Heating elements can be expected to get hot enough to ignite combustibles if the combustibles are in contact with the element. The design and construction of the appliance will usually keep combustibles away from the element. An exception is in cooking appliances, where the hot element is exposed for use. Elements can be sheathed, as is found in ovens and ranges, or they can be open wires that can get orange-hot during use. Open heating elements are usually wires or ribbons made from a nickel–chromium–iron alloy. Those that are designed to operate at glowing temperatures will get a dull gray surface oxide layer. In some appliances, a fan removes heat from the element fast enough to keep it from glowing. Those heating elements might retain their bright shiny surfaces after much use.

24.5.7.1 When a wire element burns out, the ends at the break might be left dangling. An end could contact the grounded metal of the appliance and form a new circuit. Depending on how much resistance was left in the segment of the element, the contact ground fault could allow the appliance to continue to function, to overheat, or to open the protection.

24.5.7.2 Sheathed elements consist of a resistance wire surrounded by an insulator (e.g., magnesium oxide) and encased in a metal sheath. The sheath is usually made from steel, but many baseboard or other space heaters have sheaths made from aluminum. Melting of an aluminum sheath is more likely to be a result of external fire than of internal heating; however, if melting and heating of the sheath, cooling fans, or adjacent materials show a clear pattern of coming from the element, that is good evidence that the element overheated. The element can be tested for electrical continuity and resistance. A burned-out element might indicate overheating or it might be simply old age. X-rays can assist in diagnosing the internal condition of the element.

24.5.7.3 A few electrical heaters have failed by ground faulting between the element and the sheath through the insulation, leaving characteristic eruptions of melted metal at various points along the sheath. Although heaters are normally designed so that no combustible materials are easily ignited by the element, the spatter from such arcing might ignite close combustibles if the spatters get through the protective grille.

24.5.8 Lighting. Lighting is used in many appliances to illuminate dials, work areas, or internal cavities. Lights are normally of low wattage and are not likely to be able to ignite anything ordinarily in or on the appliance. Most lighting will be incandescent, but fluorescent lights may be used to illuminate work spaces on the appliance. Fluorescent lights have ballasts (essentially a transformer) that can overheat. However, except for old ones, they have thermal protection and are usually enclosed in the appliance, where they are not likely to ignite anything. Fluorescent lamp tubes do not normally become hot enough to ignite adjacent combustibles, but some incandescent lamps may get hot enough to ignite combustibles that they touch.

24.5.8.1 Fluorescent Lighting Systems. Fluorescent lighting systems use one or more glass tubes filled with an auxiliary or starting gas and mercury at low pressure. An electrical discharge is created down the length of the glass tube. The mercury gas is excited, liberating UV light, which is converted to visible light by a coating on the inside of the lamp known as the phosphor or fluorescent powder.

24.5.8.1.1 Fluorescent lighting systems can be divided into two groups. The first group contains systems where the lamps have heated filaments at their ends. This group is composed of the preheat and rapid start systems. The second group is composed of the instant start system where the discharge is created by applying higher voltages across the lamp without separately heated filaments.

24.5.8.1.2 Fluorescent lamps require a ballast for operation. The ballast performs at least two functions. The first function is to generate sufficient voltage to cause the electrical discharge to progress down the lamp through the auxiliary gas and the mercury vapor. The second function is to limit the current flowing through the lamp in order to keep the lamp from burning out immediately.

24.5.8.1.3 There are two main types of fluorescent ballasts, the magnetic ballast and the electronic ballast. Magnetic ballasts typically incorporate either a reactor or a transformer, often with one or more capacitors. Since 1968, almost all fluorescent ballasts installed in new fixtures were required to incorporate protection against overheating. These ballasts were identified as being Class P ballasts. Replacement ballasts for non–Class P ballasts were not required to be Class P until 1984. Most Class P magnetic ballasts incorporate a self-resetting thermal protector next to the transformer. Some ballasts manufactured prior to 1984 incorporated a single-shot one-time operating thermal protector in them. Most magnetic fluorescent ballasts have a date code unique to each manufacturer stamped into the metal of their case.

24.5.8.1.4 Electronic ballasts typically use a printed circuit board with electronic components and smaller magnetic components to operate the fluorescent lamps at higher frequencies in order to provide higher efficiency. Some electronic ballasts incorporate resetting thermal protectors, while others incorporate fuses for protection.

24.5.8.1.5 Both magnetic and electronic fluorescent ballasts are commonly filled with an asphalt-based potting compound to provide better heat transfer, to reduce noise, and to hold the internal parts in place. The potting compound can soften and flow out of the ballast as the result of either internal heating or from external heat from a fire. The finding after a fire of evidence that the potting compound flowed out of the ballast enclosure is not proof that a ballast had a pre-fire failure or overheating. Ballasts are usually enclosed in a steel body of the light fixture. Any potting compound exiting the ballast due to internal heating is likely to be caught in the fixture enclosure. Potting compound that does drip out of the fixture will not ignite other materials unless the potting compound is already burning.

24.5.8.1.6 Some fire-producing failures of fluorescent ballasts include arc penetrations into combustible ceiling materials and extreme coil overheating conducting heat into adjacent combustibles. Fluorescent ballasts should not be disassembled at the fire scene. Suspect ballasts should be preserved along with their fixtures and wiring for laboratory examination by qualified experts. Commonly, ballasts are x-rayed prior to their disassembly. The fluorescent fixtures should also be examined for evidence of lampholder failures and arcing inside the fixture enclosure.

24.5.8.2* High Intensity Discharge Lighting Systems. High intensity discharge (H.I.D.) lighting systems are lighting sources that utilize a lamp that has a short tube filled with various metal vapors. An electric discharge is created along the length of the tube, exciting the atoms of the metal vapor, thereby creating light. High intensity discharge lamps operate at higher pressures than fluorescent lamps. High intensity discharge lighting systems include mercury vapor, metal halide, and high pressure sodium systems. High intensity discharge lighting systems are typically utilized in commercial structures such as warehouses, manufacturing areas, and retail occupancies.

24.5.8.2.1 High intensity discharge lighting systems typically utilize a ballast and a power capacitor to provide the starting voltage to the lamp, and to limit the current flowing through the lamp. These ballasts may be of the magnetic type or of the electronic type. The magnetic type of ballast is typically not surrounded by potting compound. Some high intensity discharge lighting ballasts are protected by fuses or thermal protectors. Most high intensity discharge ballasts are mounted inside a fixture enclosure above the lampholder and reflector assembly.

24.5.8.2.2 High intensity discharge lighting system ballasts can fail, resulting in arcing faults involving the ballast windings. Certain fixtures have sufficient voids and openings to allow metal droplets or sparks from the arcing faults to escape the fixture housing.

24.5.8.2.3 Most mercury and metal halide lamps use a cylinder of fused silica/quartz with electrodes at either end. This cylinder is called an arc tube. When at room temperature, this tube is typically below atmospheric pressure. This arc tube is supported in a frame inside a glass outer enclosure or jacket. The lamps are marked with a date code reflecting their date of manufacture. The date codes are unique to each manufacturer. During normal operation, the arc tube in a metal halide lamp can reach temperatures in the range of 900°C to 1100°C (1652°F to 2012°F) and internal pressures of 5 to 30 atmospheres. The mercury vapor arc tubes can operate in the range of 600°C to 800°C (1112°F to 1472°F) and pressures of about 3 to 5 atmospheres. High pressure sodium lamps operate at lower pressures typically close to 1 atmosphere.

24.5.8.2.4 Under certain conditions, including operating the lamp beyond rated life or during certain ballast failures, the arc tube of a metal halide lamp may fail during operation while at operating temperature and pressure. Hot pieces or particles from the arc tube may breach the outer glass jacket and escape the lamp. Some metal halide fixtures have lenses or shields designed to contain any escaping pieces or particles. Metal halide lamps are available with internal shields designed to stop the fractured pieces of the arc tube from breaching the outer glass jacket. Arc tube ruptures of mercury vapor lamps have been known to occur.

24.5.8.2.5 The investigation of a suspected high intensity discharge lighting fixture requires the evaluation of the fixture's electrical supply, wiring, ballast, capacitors, lamp, and any lens present. The fixtures should not be disassembled at the fire scene, but carefully preserved for laboratory examination. The evaluation of a metal halide or mercury vapor lamp requires as much of the lamp remains be found as possible. All pieces of the arc tube should be preserved and shielded from any additional damage or wear. The edges of the glass jacket and of the arc tube remains should not be handled, disturbed, or cleaned prior to laboratory examination. The lamp's frame pieces and lamp base should also be collected and carefully preserved. Since the quartz arc tube can be fractured mechanically by building collapse, firefighting, and overhaul, the presence of a fractured arc tube by itself is not proof that an arc tube rupture occurred prior to the ignition of the fire. It is often helpful to obtain additional exemplar lamps and fixtures from the scene for comparison purposes and to obtain the operating history of the lamps.

24.5.9 Miscellaneous Components. There are miscellaneous devices, such as dimmers and speed controllers, that might be found as components of appliances. Generally, many of these devices are now solid state, fully electronic devices. Older appliances may contain nonelectronic devices, such as rheostats or wire resistors. Electronic components are usually destroyed by fire unless the fire was brief. In most cases, the remains of dimmers or other electronic devices that use printed circuit boards will not be helpful in finding the cause because of their susceptibility to fire damage.

24.5.9.1 Timers can be built in or can be used as separate devices. They are driven by small clock motors with mechanical actuation of switches. Remains of any timers that were present can usually be found after a fire, but they may be badly damaged. Small timer motors last a long time and will not overheat to cause a fire. Failure of the timers is usually caused by the gears wearing out or loosing teeth. Electronic timers may not leave recognizable remains after being heated by fire.

24.5.9.2 Thermocouples are used to measure temperature differences. They function by creating a voltage at a junction of dissimilar metals, which is compared to the rest of the circuit or to a reference junction. The temperatures can be read on meters or on digital devices.

24.5.9.3 A thermopile is a series of thermocouples arranged so that the voltages at the series of junctions add to a large enough voltage to operate an electromagnet. Thermopiles have been used in gas appliances to keep a valve open when the pilot flame is burning but to let the valve close if the pilot flame is out. Newer gas appliances use electric igniters instead of standing pilots.

24.6 Common Residential Appliances. A brief description of the operation and components of common residential appliances is provided to assist the investigator in understanding how these appliances work.

24.6.1 Range or Oven. The heat is provided either by electricity passing through resistance heating coils or by burning natural gas or propane. In the oven, the interior temperature is controlled by a thermostat and a valve or switch on the fuel or power supply. On a gas range, the fuel flow rate and heat intensity is usually controlled with the burner fuel supply valve. An electric range typically utilizes a timing device that controls the cycling time of the heating element. This device is manually adjusted so that a high setting results in the longest (possibly continuous) on cycle. Ignition of the fuel gas in a gas range or oven may be by a standing pilot flame or by an electrical device that produces an arc for ignition.

24.6.2 Coffee Makers. The coffee maker design popular for home use consists of a water reservoir, heating tube, carafe, and housing. When started, the heating tube boils the water flowing through it from the reservoir. This boiling forces hot water to the area where the ground coffee is kept in a filter, and the coffee then drips into the carafe. The carafe in many designs sits on a warming plate. The warming plate is usually heated by the same resistance heater that heats the heating tube. The resistance heater is controlled by a thermostat that cycles it off when it reaches the upper limit of the thermostat. The heater will cycle on once it has cooled to a point determined by the thermostat. To prevent overheating by the heater, a TCO may be employed. If the maximum temperature of the TCO is reached, it is designed to open the heater circuit and prevent further heating. Some coffee maker designs may include multiple TCOs, automatic timing circuits that turn the coffee maker off after a fixed period, or a clock or automatic brew mode that turns the coffee maker on at a preset time. The TCO(s) should be checked to determine whether it has been bypassed.

24.6.3 Toaster. The toaster uses electrical resistance heaters to warm or toast food. It is a relatively simple appliance that utilizes an adjustable sensor to control the on time. By pushing down a lever, the food is lowered on a tray into the toaster, and the heaters are turned on. The sensor is usually a bimetal strip that might sense the temperature in the toaster, but more commonly the bimetal has its own heater and is nearly independent of the temperature in the toaster. At the conclusion of the heating cycle, a mechanical latch partly releases the tray and turns off the bimetal heater. As the bimetal cools, a second latch fully releases the tray, which then lifts the food. Some newer designs use an electronic timer that controls an electromagnetic latch.

24.6.4 Electric Can Opener. The electric can opener uses an electric motor to turn a can under a cutting wheel to open the can. Generally they will run only when a lever is manually held down. This seats and holds the cutting wheel in place and closes the power switch to the motor. The electric motor may or may not be protected against overheating by a thermal cutoff switch.

24.6.5 Refrigerator. The common refrigerator and freezer utilize a refrigeration cycle and ventilation system to keep the inside compartments at suitable temperatures. The refrigeration system consists of an evaporator (i.e., heat exchanger in the compartment), a condenser (i.e., heat exchanger outside the compartment), a compressor, a heat exchange medium (typically a fluorocarbon or Freon®), and tubing to connect these components. Warm air from inside the enclosure is used to evaporate the heat exchange medium; the coolant vapor moves to the compressor where it is compressed and condensed back to a liquid in the condenser. When the coolant condenses, it gives off the heat it picked up in the enclosure. As a result, the air around the evaporator is cooled and the air around the condenser is heated. The cool air is circulated within the refrigerator, and the hot air dissipates into the room in which the appliance is located. This cooling cycle is controlled by a timing device that regulates the length of the cycles, or it may have a thermostat device that controls the cycle.

24.6.5.1 The compressor is typically powered by an electric motor that is usually protected with a thermal cutoff. The compressor is usually located in a sealed container, which can prevent an overheated compressor from igniting nearby combustibles because it acts as a heat sink. Additional systems in a refrigerator include lighting, ice maker, ice and water dispenser, and a fan for the condenser and possibly one for the evaporator.

24.6.5.2 The refrigerator may also have heating coils in various areas for automatic defrosting and to prevent water condensation on outside surfaces. Automatic defrosters are designed to operate at regular intervals to prevent the accumulation of frost on inside surfaces, especially the inside of the freezer.

24.6.5.3 The antisweat (external condensation) heaters are located under exterior faces, and they operate at regular intervals to prevent condensation. Some models allow this feature to be disabled to conserve power. In both cases, these heaters are typically low-wattage electrical resistance heaters.

24.6.6 Dishwasher. A dishwasher uses a pump to spray and distribute hot water and soap onto the dishes. An electric resistance heater is typically located in the bottom of the unit, where it further heats the water being used. Once the washing and rinsing is complete, the water is drained, and the dishes are dried by the resistance heater, which is exposed to air after the water has drained. Some models allow the electric heater to be disabled during the drying cycle in order to save power. Other devices in the appliance include electrically operated valves and a timer control to regulate the various cycles. The electric pump motor may or may not be thermally protected. Some dishwashers have caused fires by electrical faulting in the push-button controls that then ignited the plastic housing.

24.6.7 Microwave Oven. A microwave oven utilizes a device known as a magnetron to generate and direct the radio waves (i.e., microwaves) into the enclosure. The frequency of these radio waves causes items placed in the oven to heat. To provide for even distribution of these waves, a device is used to scatter them inside the enclosure, and a food tray on the bottom may be rotated. The microwave oven will also have timing and control circuits, a transformer, and internal lighting. The transformer is used to produce the high voltage required by the magnetron. The magnetron will usually be provided with a TCO switch. There may be TCOs above the oven compartment to remove power in case of a fire in the oven.

24.6.8 Portable Space Heater. Portable space heaters for residential use have many designs but are generally divided into two groups: convective and radiant. Some convective heaters use a fan to force room air past a hot surface or element while others use natural convection. A natural convection or radiant heater does not have a fan. A complete discussion of the many heater designs is not appropriate here, but the investigator should become familiar with the particular design in question. Familiarization can be achieved by reviewing operating manuals and design drawings and by examining an exemplar heater. These heaters employ a variety of control methods and devices. Generally, these devices are present to control the heater, prevent overheating, or shut the heater off if it is upset from its normal position.

24.6.9 Electric Blanket. An electric blanket consists of an electric heating element within a blanket. The controls are typically located separate from the blanket, on the power cord. The control is typically manually adjusted to control the on–off cycle time. In one or more places near the heating elements within the blanket are located TCOs to prevent overheating of the appliance, and there may be as many as 12 or 15 of these, depending on the appliance. An electric blanket is designed not to overheat when spread out flat. If it is wadded or folded up, heat may accumulate in the blanket and get it hot enough to char and ignite. Normally, the cutoffs prevent overheating.

24.6.10 Window Air Conditioner Unit. A window air conditioner unit is designed to be placed in the window of a residence to cool the room. The unit does this by means of a refrigeration cycle very similar to that used by refrigerators. *(See 24.6.5.)* Air from the room is circulated through the unit, past the evaporator, which cools the air, and is then discharged to the room. A fan powered by an electric motor does the work of circulating the air. The fan motor is usually protected from overheating by a TCO. These units have controls for selecting fan speed, cooling capacity, and temperature. These units are powered by a nominal 120 V circuit, or larger units may require nominal 240 V service.

24.6.11 Hair Dryer and Hair Curler.

24.6.11.1 Typical residential hair dryers use a high-speed fan to direct air past an electric resistance heating coil. Controls are typically limited to on or off. Some units may have more than one heater power (wattage) and fan speed settings. One or more resettable TCO switches are typically provided near the heaters to prevent them from overheating.

24.6.11.2 Hair curlers or hair curling wands use an electric resistance heater within the wand, around which hair is wrapped to curl it. Some models allow the addition of water to a compartment that can be used to generate steam. Typical controls include an on–off switch and a power setting. Most models include a light to indicate that the unit is operating. Typically these units have one or more TCOs near the heating element, which may or may not be resettable.

24.6.12 Clothes Iron. A modern clothes iron uses an electrical resistance heater, located near the ironing surface, to heat that surface. Many models require the addition of water, which is used to distribute the heat and to produce steam. The controls on typical irons range from a simple temperature selector and an on–off switch to electronically controlled units that turn themselves off. Irons are designed to heat in both the vertical and horizontal position. Most irons are provided with one or more TCO to prevent overheating.

24.6.13 Clothes Dryer.

24.6.13.1 All clothes dryers use electricity to rotate the clothing drum and to circulate air with a blower. Energy for the heat source may be by the combustion of a fuel gas or by electricity. All electric dryers are powered by either a nominal 120 V or a 240 V source. The clothing is dried by spinning it in a drum, through which heated air is circulated. Air is discharged from the dryer via a duct that is typically directed to the exterior of the house. Most dryers have filters to trap lint, which can build up in the dryer. However, if the trap is clogged or not working or if the material being dried gives off a large quantity of lint, this material can accumulate in other areas of the dryer and its vent, which can be a fire hazard. Frictional heating sufficient to cause ignition can result if a piece of clothing or other material becomes trapped between the rotating drum and a stationary part. Fires have been reported in dryers when vegetable oil-soaked rags or plastic materials such as lightweight dry cleaner bags have been placed in the dryer.

24.6.13.2 Typical dryers have timing controls, humidity sensors, heat source selectors, and intensity selectors to control the operation of the dryer. Thermal cutoffs are provided to prevent overheating of the dryer and components such as the blower motor and heating elements.

24.6.14 Consumer Electronics. Consumer electronics include appliances such as televisions, VCRs, radios, CD players, video cameras, personal computers, and so forth. These devices are similar in their components in that they typically include a power supply, circuit boards with many electronic components attached, and a housing. Some of these appliances, such as televisions and CD players, have components that require high voltage. Additionally, many of these appliances can be operated via remote control. A complete discussion of the many designs of these appliances is not appropriate here, but the investigator should become familiar with the particular design in question. Familiarization can be achieved by reviewing operating manuals and design drawings and by examining an exemplar appliance.

24.6.15 Lighting. Typical residential lighting is either the incandescent or fluorescent type. Incandescent lighting uses a fine metal filament within the bulb, which has been filled with an inert gas such as argon, or the bulb is evacuated and sealed. When an incandescent bulb is working, a major by-product is the generation of heat. Fluorescent lightbulbs use high voltage from a transformer (the ballast) to initiate and maintain an electric discharge through the light tube. The interior of the tube is coated with a material that fluoresces or gives off light when exposed to the electrical discharge energy. The light-generating process in this case generates little heat as a by-product, but the ballast typically will give off heat. Thermally protected fluorescent light ballasts have a resettable thermal switch to prevent the ballast from overheating. *(See 24.5.5.)*

Chapter 25 Motor Vehicle Fires

25.1* Introduction. This chapter deals with factors related to the investigation of fires involving motor vehicles. Included in the discussion are automobiles, trucks, heavy equipment, farm implements, and recreational vehicles (motor homes). While vehicles that travel by air or on rails are not covered, there are many factors relating to incident scene documentation, fuels, ignition sources, and ignition scenarios that may apply. Marine fire investigations are described in Chapter 28.

25.1.1 The fire or damage patterns remaining on the body panels and vehicle frames, and in the interior of the vehicle are often used to locate the areas or point(s) of origin and for cause determination.

25.1.2 It was once felt that rapid-fire growth and extensive damage were indicative of an incendiary fire. However, the type and quantity of combustible materials found in motor vehicles today, when burned, can produce this degree of damage without the intentional addition of another fuel, such as gasoline. In the case of a total burnout, one cannot normally conclude whether or not the fire was incendiary on the basis of observations of the vehicle alone. The use of fire patterns to determine origin or cause should be used with caution. The interpretations drawn from these patterns may be verified by witness evidence, laboratory analysis, service records indicating mechanical or electrical faults, factory recall notices, or complaints and service bulletins that can be obtained from the National Highway Traffic Safety Administration (NHTSA), the Center for Auto Safety, and the Insurance Institute for Highway Safety (IIHS). The investigator should also be familiar with the composition of the vehicle and its normal operation.

25.1.3 The relatively small compartment sizes of vehicles may result in more rapid fire growth, given the same fuel and ignition source scenario, when compared to the larger compartments normally found in a structure fire. However, the principles of fire dynamics are the same in a vehicle as in a structure and, therefore, the investigative methodology should be the same. *(See Chapters 4 and 5.)*

25.2 Vehicle Investigation Safety. The completion of a thorough investigation of a burned vehicle may pose a variety of safety-related concerns that are different from those that may normally be found in a structure fire. See Chapter 12 for additional safety information

25.2.1 Before conducting an inspection of the vehicle undercarriage, the investigator should take care to prevent the vehicle from moving and causing investigator injury. The use of hydraulic lifts, jacks, or other lifting devices designed to hold the vehicle's weight should be used in conjunction with blocking or stands to prevent the vehicle's sudden movement or falling onto the investigator. Forklifts, tow trucks, or wreckers alone should not be used to support the vehicle for such inspections.

25.2.2 Undeployed airbags (supplemental restraint systems) may pose a serious potential safety concern for fire investigators. Sodium azide, the expelling agent for the airbag in older-model vehicles, is a hazardous material, and contact or inhalation can constitute a potential health hazard for the investigator. Due to the increased installation of airbags, some vehicles come equipped with multiple airbags and reactive occupant-restraint systems. The investigator will need to identify the systems that are present, the operational condition of those systems, and, if necessary, render those systems safe prior to disturbing the vehicle, in order to prevent accidental operation.

25.2.3 Vehicle fire inspections may present many other situations that pose safety hazards to the investigator. These can include fuel leaks or remaining fuel in fuel tanks posing a fire hazard; expelled lubricants, which may pose slip and fall hazards; electrical energy stored in the battery; or broken glass, which may pose puncture or cut hazards.

25.3 Fuels in Vehicle Fires. A wide variety of materials and substances may serve as the first materials ignited in motor vehicle fires. These include engine fuels; transmission, power steering, and brake fluids; coolants; lubricants; windshield wiper fluid; battery vapors; and the vehicle interior component materials, contents, or cargo. Once a fire is started, any of these materials may contribute as a secondary fuel, affecting the fire growth rate and the ultimate damage sustained.

25.3.1 Ignitible Liquids. Ignitible liquids are used in motor vehicles and can be associated with vehicle fires. These liquids may come in contact with an ignition source as a result of a malfunction of one of the vehicle systems, crash damage to one or more of the vehicle systems containing these liquids, or an incendiary act. Table 25.3.1 shows selected physical and fire properties of ignitible liquids used in motor vehicles. Whether a given liquid can ignite depends on the properties of the liquid, its physical state, the nature of the ignition source, and other variables related to the vehicle. The values of flash point and autoignition temperature in Table 25.3.1 were obtained in controlled laboratory tests, and are generally not applicable directly to ignition of these liquids in motor vehicles. The vehicle and the environment affect the surface temperature required to reach the autoignition temperature in a motor vehicle. These variables include airflow, the time of contact between the liquid and the heated surface, and the material composition, mass, shape, and surface texture of the heated surface among others. Autoignition of a liquid in contact with a heated surface generally requires a temperature substantially greater than published laboratory autoignition temperatures of that liquid.

25.3.1.1* Hot Surface Ignition. The hot surfaces which exist in a motor vehicle may be of sufficient temperature to autoignite ignitable liquids commonly found in these vehicles. This form of autoignition may be referred to as hot surface ignition. There is a difference between autoignition temperature, which is a property of a liquid determined through standard test methods, and hot surface ignition temperature, which is not a property. Experimental testing has shown that hot surface ignition temperatures for common automotive liquids may be substantially higher than reported autoignition temperatures. For example one study shows that the hot surface ignition temperature of gasoline was 354°C (670°F) whereas the reported autoignition temperature range for gasoline as shown in Table 25.3.1 is 257–280°C (495–536°F).

25.3.2 Gaseous Fuels. Alternative motor fuels, notably propane, hydrogen, and compressed natural gas, are finding increasing use in fleets of automobiles and in trucks as well as in some privately owned vehicles. Propane is also found aboard the majority of recreational vehicles as a cooking, heating, and refrigeration fuels. Wet-cell lead acid batteries can produce hydrogen which may be released during charging or as a consequence of a collision. Larger quantities of these gases may be found in larger vehicles, vehicles with batteries that support electric motor drives, or as cargo. Gases may be stored as a liquid under pressure, and become gaseous when released. Some properties of ignitible gaseous fuels are given in Table 25.3.2. The following publications provide more information on this topic:

(1) Crowl, D. and J. Louvar, *Chemical Process Safety Fundamentals with Applications* 2nd ed.
(2) NFPA 49, *Hazardous Chemical Data.*
(3) Glassman, I., *Combustion*, 3rd ed.
(4) Sax, N. I., and R. J. Lewis, *Dangerous Properties of Industrial Materials*, 7th ed.

Table 25.3.1 Properties of Ignitible Liquids

	Flash Point [a]		Autoignition Temperature [b]		Flammability Limits [c]		Boiling Point [d]				Density[e] Vapor
					LFL	UFL	IBP		FPB		
Liquid	°C	°F	°C	°F	%	%	°C	°F	°C	°F	(Air = 1)
Gasoline	−45 to −40	−49 to −40	257–280	495–536	1.4	7.6	26–49	78–120	171–233	339–452	3–4
Diesel fuel (fuel oil #2)	38–62	100–145	254–260	489–500	0.4	7	127–232	260–450	357–404	675–760	5–6
Brake fluid	110–171	230–340	300–319	572–606	1.2	8.5	232–288	111–142	460–550	238–288	5–6
Power steering fluid	175–180	347–356	360–>382	680–>720	1	7	309–348	588–658	507–523	945–973	>1
Motor oil	200–280	392–536	340–360	644–680	1	7	299–333	570–631	472–513	882–955	>1
Gear oil	150–270	302–510	>382	>716	1	7	316–371	601–700	>525	>977	>1
Automatic transmission fluid	150–280	302–536	330–>382	626–>716	1	7	239–242	462–468	507–523	945–973	>1
Ethylene glycol (antifreeze)	110–127	230–261	398–410	748–770	3.2	15.3	196–198	385–388			2.1
Propylene glycol (antifreeze)	93–107	199–225	371–421	700–790	2.6	12.5	187–188	369–370			2.6
Methanol (washer fluid)	11–15	52–55	464–484	867–903	6	36	65	149			1.1

[a]Flash point data was obtained from Technical Data Sheets and Material Safety Data Sheets from manufacturers and suppliers of the major brands of each type of fluid available in the United States. The flash points of gasolines reported in these sources were determined by ASTM D 56. The flash points for diesel fuels, brake fluids, power steering fluids, motor oils, transmission fluids, gear oils, ethylene glycol (antifreeze), propylene glycol (antifreeze), and methanol were determined by ASTM D 56, ASTM D 92, or ASTM D 93.

[b]Autoignition temperature data for gasoline, diesel fuel, brake fluid, ethylene glycol, propylene glycol, and methanol was obtained from Technical Data Sheets and Material Safety Data Sheets from manufacturers and suppliers of the major brands of each type of fluid available in the United States. These sources generally did not report the test method used to determine autoignition temperature; however, ASTM E 659 is the laboratory test method typically used to determine autoignition temperature. Autoignition temperature data for power steering fluid, motor oil, gear oil, and automatic transmission fluid were obtained using ASTM E 659.

[c]Flammability limit data was obtained from Technical Data Sheets and Material Safety Data Sheets from manufacturers and suppliers of the major brands of each in the United States. These sources generally did not specify the laboratory test method used to determine the reported flammability limits; however, ASTM E 681 is a laboratory test method typically used to determine the Lower Flammability Limit (LFL) and the Upper Flammability Limit (UFL).

[d]Boiling range data for gasolines was obtained from the Alliance of Automobile Manufacturers annual North American survey of gasoline properties for 2003. The boiling ranges reported in this survey were determined by ASTM D 86. Boiling range data for diesel fuel was obtained from Technical Data Sheets and Material Safety Data Sheets from manufacturers and suppliers of the major brands of diesel fuel in the Unites States. These sources generally did not report the laboratory test method used to determine the boiling range of diesel fuel. Boiling range data for brake fluid, power steering fluid, motor oil, gear oil, and automatic transmission fluid were determined by ASTM D 2887. Boiling point data for ethylene glycol, propylene glycol, and methanol were obtained from Material Safety Data Sheets from manufacturers and suppliers of these chemicals. These sources did not report the laboratory test method used to determine boiling point. In the table IBP and FBP are Initial Boiling Point and Final Boiling Point respectively.

[e]Vapor density data was obtained from Material Safety Data Sheets from manufacturers and suppliers of these materials.

*Studies include the following:

1. Arndt, S.M., Stevens, D.C., and Arndt, M.W., "The Motor Vehicle in the Post-Crash Environment, An Understanding of Ignition Properties of Spilled Fuels," SAE 1999-01-0086, International Congress and Exposition, Detroit, MI, March 1–4, 1999.

2. LaPointe, N.R., Adams, C.T., and Washington, J, "Autoignition of Gasoline on Hot Surfaces" *Fire and Arson Investigator,* Oct. 2005: pp. 18–21.

3. Colwell, J.D. and Reze, A. "Hot Surface Ignition of Automotive and Aviation Fluids," *Fire Technology,* Second Quarter 2005, pp. 105–123.

4. API PUBL 2216, *Ignition Risk of Hydrocarbon Vapors by Hot Surfaces in the Open Air.*

Table 25.3.2 Properties of Gaseous Fuels in Motor Vehicles

	Autoignition Temperature		Flammability Limits (Vol. % fuel in air)		Boiling Point		Specific Gravity (Air)	
Gas	°C	°F	LFL	UFL	°C	°F	Vapor Density (air =1)	Min. Ignition Energy (mJ)
Hydrogen	40–572	752–1061	4.0	75.0	−253	−422	0.07	0.018
Natural gas (methane)	632–650	1169–1202	5.3	15.0	−162	−259	0.60	0.280
Propane	450–493	842–919	2.2	9.5	−42	−44	1.56	0.250

Note: The data provided in this table are for generic or typical products and may not represent the values for a specific product. When possible, values specific to the product involved should be obtained from a material safety sheet, product specifications or by standard test methods.

25.3.3 Solid Fuels.

25.3.3.1 Many combustible plastics are easily ignited when exposed to an ignition source. These plastics can burn with heat release rates similar to those of ignitible liquids. Thermoplastic materials often sag or drop flaming pieces. Depending on the composition and orientation of the plastic, the flaming drips can contribute to the spread of fire. Table 25.3.3.1 lists and summarizes properties of typical plastics used in motor vehicles. Heat from friction may be sufficient to ignite drive belts, bearing, lubricant, or tires. Most combustible metals and their alloys need to be powdered or melted to burn. Solid magnesium, which is present in many motor vehicles, can ignite and burn vigorously. Given even a small initial fire, solid fuels may contribute significantly to the rate of fire growth and the extent of fire damage.

25.3.3.2 Investigators should not interpret the presence of melted metals to be an indicator of the use of an ignitible liquid as an accelerant, in the belief that only an ignitible liquid can produce sufficiently high temperatures. Common combustibles and ignitible liquids produce essentially the same flame temperature. In many cases, alloys are used rather than the pure metal. The melting temperature of an alloy is generally lower than that of its constituents. The actual composition of a metal part and its melting temperature should be determined before any conclusions are drawn from the fact that it has melted. Alloying may occur during a fire. For instance, low melting temperature zinc may drip onto a copper wire or tube and form a brass alloy, which melts at a lower temperature than copper. Likewise, molten aluminum can drip onto steel sheet metal, which can cause the appearance of melting of the sheet steel. Energy Dispersive Spectroscopy (EDS) can be used to nondestructively determine the composition of metal components from vehicle fires. Table 25.3.3.2 lists typical metals and alloys in motor vehicles and their melting temperature.

25.4 Ignition Sources. In most cases, the sources of ignition energy in motor vehicle fires are similar to those associated with structural fires such as arcs, mechanical sparks, overloaded wiring, open flames, and smoking materials. There are, however, some unique sources that should be considered, such as the hot surfaces of the engine exhaust system. This system may consist of the exhaust manifold, exhaust pipe, one or more catalytic converters, mufflers and tailpipes. Other hot surface ignition sources may include brakes, bearings, and turbochargers. Because some of these ignition sources may be difficult to identify following a fire, the description in 25.4.1 through 25.4.5 are provided to assist in their recognition

25.4.1 Open Flames. The most common source of an open flame in a vehicle is an exhaust system backfire out of the carburetor. Propagation will rarely occur if the air cleaner is properly in place. However, modern vehicles use fuel injection systems that eliminate the carburetor. Lit matches and other smoking materials may ignite debris in the ashtray, resulting in a fire. In recreational vehicles, appliance pilots or operating burners of ranges, ovens, water heaters furnaces, etc. are open-flame ignition sources.

25.4.2 Electrical Sources. When the engine is not running, the only available source of electrical power is the battery(s). A limited number of components remain electrically connected to the battery even though the ignition switch is off and the engine is not running. These components may include, the generator, (alternator) starter motor, under hood distribution panel (fuse box), remote car starter, some aftermarket accessories, cigarette lighter, power seats, power windows, power locks, lights, power side mirrors or ignition switch, can fail when the vehicle is off. A vehicle that is running has many more potential sources of electrical energy. Fuses, circuit breakers, and fusible links provide protection of electrical circuits in motor vehicles. In dc systems, the frame, metal body panels, and the engine are connected to the battery negative to create the ground system. The positive side of the battery supplies current to the fuse panel and to all electrical equipment. This means that an electrical device may have only one wire physically connected but it is attached to metal portions of the vehicle that complete the circuit back to the battery negative terminal (ground). It also means that the ground path may not be obvious. When an energized positive wire, terminal end, or conductive component touches a grounded surface a completed circuit can result.

25.4.2.1 Recreational vehicles (RVs) have both onboard auxiliary batteries and wiring like other motor vehicles. They may also have alternating current (ac) wiring and fixtures like a structure. Recreational vehicles can be equipped with a power converter, which changes 120-volt ac to 12-volt dc, as well as a power inverter, which inverts 12-volt dc to 120-volt ac. In some cases, converters/inverters are combined into a single unit. Recreational vehicles can have external 120-volt ac supplied by a power cord or generator. Additionally, RVs may have the ability to use 120-volt ac while moving.

Table 25.3.3.1 Common Plastics Found in Motor Vehicles

	Ignition Temperature*		Melting Temperature		ΔH_c	Peak Heat Release Rate†	
Material	°C	°F	°C	°F	kJ/g	kW/m²	Location in Vehicle
Acrylic fibers	560	1040	90–105	194–221	28	300	Floor covering
ABS	410	770	88–125	190–257	29	614–683	Body panels
Fiberglass (polyester resin)	560	1040	428–500	802–932			Resin burns but not glass body panels
Nylons	413–500	775–932	220–265	428–509	28	517–593	Trim, window gears, timing gears, HVAC unit, structure
Polycarbonate	440–522	824–972	265	510	22	16	Instrument panel, structure, headlights
Polyethylene	270–443	518–830	115–137	240–280	40	453–913	Wiring insulation, fuel tank, battery cover
Polypropylene	250–443	482–829	160–176	320–350	43	377–1170	Resonator, structure, air ducts, HVAC unit, battery cover
Polystyrene	346–365	655–689	120–240	248–465	36	723	Insulation, padding, trim
Polyurethanes	271–378	520–712	120–160	248–320	18	290	Seats, arm rests, padding, trim
Vinyl (PVC)	250–430	482–806	75–105	167–221	11	40–102	Wire insulation, upholstery, HVAC unit, bulkhead insulation

*Piloted ignition temperature. Typically, the autoignition temperature (AIT) is higher than the piloted ignition temperature.
† At an incident radiant heat flux of 20 kW/m² in the cone calorimeter by ASTM E 1354.
Lide (ed.), *Handbook of Chemistry and Physics.*
Hilado, *Flammability Handbook for Plastics.*
ASM Engineered Materials Reference Book, 2nd Ed., M. Bauccio ed. ASM Int'l, Materials Park, OH, 1994.
NFPA Fire Protection Handbook, Tables 3.13A and A.6 (17th edition).
Babrauskas, V. *Ignition Handbook*, Table 15, Fire Science Publishers, Issaquah, WA, 2003.
Cole, L., *Investigation of Motor Vehicle Fires.* Lee Books. Novato, CA: 1992.
Harper, Charles, *Handbook of Building Materials for Fire Protection*, McGraw-Hill Handbooks, 2004.
Note: The data provided in this table are generic and may not represent the values for a specific product. Thermal and combustion properties depend on composition, thickness and/or orientation. When possible, values specific to the product involved should be obtained from the manufacturer or by laboratory testing.

Table 25.3.3.2 Melting Temperatures of Metals Commonly Found in Motor Vehicles

	Melting Temperature	
	°C	°F
Zinc Die Cast (Pot Metal)	300–400	562–752
Magnesium	650	1202
Aluminum	660	1220
Aluminum alloys	566–650	1050–1200
Copper	1082	1981
Cast Iron		
Gray	1350–1400	2460–2550
White	1050–1100	1920–2010

See also Table 6.8.1.1, and Table 3.13A of the *NFPA Fire Protection Handbook*, (17th edition).

25.4.2.2 Overloaded Wiring. Overloaded wiring can present a problem when the current flowing through the wiring is greater than the conductor can safely carry. This can result in the conductor temperature rising to the ignition point of the insulation, particularly in bundled cables such as a multiple conductor wiring harnesses where the heat generated is not readily dissipated. This can sometimes occur without activating the circuit protection devices if these devices are not properly sized. Faults and mechanical failures of high current devices such as power seat motors, power window motors, and heaters used in seats and windows can also result in ignition of insulation, carpet materials, or combustible materials that may accumulate under seats. Some vehicles are equipped with seats that are heated by high resistance wires. In some seats a fault can cause a shorter-than-intended electrical path, resulting in an increase in surface temperature that may result in a fire. The addition of aftermarket accessories may also result in an overload of the original equipment wiring. Equipment of higher electrical demand may be installed in a vehicle, and may also have a higher capacity fuse or may have omitted the fuse.

25.4.2.3 Electrical High Resistance Connections. High resistance connections can occur when there is a poor or ineffective electrical connection between wiring conductors and terminations or other junctions between wiring. A loose connection can result in intermittent arcing when a load is placed on the circuit. High resistance results in a reduced or current limited circuit. It produces high temperatures but may not produce a condition that activates circuit protection devices such as fuses or circuit breakers. If a high resistance connection has occurred, the evidence of this type of condition will often exist after the fire. The temperatures at a high resistance location are often sufficiently high enough to eventually ignite the wiring insulating materials.

25.4.2.4 Electrical Short Circuits and Arcs (Electric Discharge). Short circuits can result when the wiring conductor insulation becomes abraded, cut, brittle, fractured, or otherwise damaged, allowing it to contact a grounded surface. A shorted circuit, if not properly fused, can result in excess heating and arcing or electrical discharging. During an impact shorts and arcs (electrical discharge) can be created as a result of the crushing, stretching, and cutting of wiring conductors. Some wiring conductors such as battery output and starter circuits are not overcurrent protected and carry very high amperage. An examination of the battery for indications of impact damage should also be performed.

25.4.2.5* Arc (Carbon) Tracking. An electrical potential applied across an insulating material can result in a short circuit by a phenomenon called arc tracking, see 8.9.4.5. The process occurs more quickly if the surface is contaminated with road salt or other conductive materials. The process can occur with 12-volt electrical systems.

25.4.2.6* Lamp Bulbs and Filaments. Bulb surfaces can produce sufficient heat to ignite some combustible materials that may be in contact with them. Lamp filaments of broken bulbs can also be a source of ignition for some vapors especially gasoline. Normally operating headlamp filaments have temperatures of approximately 1400°C (2550°F). However, most filaments operate in a vacuum or inert atmosphere. When the filament is exposed to ambient air, it will typically operate for only a few seconds, then burn open. Once the filament opens, the source of ignition is gone.

25.4.2.7 External Electrical Sources Used in Vehicles. While most electrical sources in vehicles are contained within the vehicle, there are situations where external electrical power is supplied to the vehicle. Examples of these sources are electrical hook-ups used in recreational vehicles and trailers, electric block heaters for engines, vehicle interior heaters, and battery chargers. Many vehicles used in colder climates have an electric block heater to warm the engine oil or coolant to ease starting. This type of heater is typically a permanent installation on the vehicle and will be equipped with a power cord. Inspection of electrical power cords should be made when applicable, because an overload of, or damage to, the cord or a failure of the appliance could be the cause of the fire. Improper application of external electricity can damage vehicle components, resulting in failure, and possibly in fire. Where recreational vehicles are connected to external power, the circuit wiring can be inspected for indications of ignition involvement.

25.4.3 Hot Surfaces.

25.4.3.1* Exhaust systems can generate sufficiently high temperatures to ignite combustible material, including ignitable liquids in the engine compartment. Automatic transmission fluid, particularly if heated due to an overloaded transmission, can ignite on a hot manifold. Engine oil and certain brake fluids (DOT 3 and 4) dropping on a hot manifold can also ignite. When a vehicle is suddenly brought to rest and shut off, the time for the exhaust manifold temperature to span 80 percent of the temperature difference between the initial temperatures and the ambient temperature is typically 20 to 30 minutes. While exhaust manifolds cool immediately when the vehicle is suddenly brought to rest and shut off, underbody catalytic converters generally experience a temperature increase lasting several minutes and then begin to cool. The time for underbody catalytic converters to span 80 percent of the temperature range from the steady-state temperature to ambient temperature typically range from 45 minutes to more than 90 minutes. If unburned fuel flows through the exhaust system, the catalytic converter temperatures can increase as this fuel is oxidized within the catalytic converters. Those fluids may ignite only shortly after the vehicle is shut off. This ignition is due to the loss of airflow through the engine compartment, which disperses these vapors and cools hot engine surfaces. In most vehicles, the pipe surface just upstream of the catalytic converter will operate hotter than the converter itself.

25.4.3.2 Typically, gasoline will not be ignited by a hot surface, but requires an arc, spark, or open flame for ignition. While ignition of gasoline vapor by a hot surface is difficult to reproduce, such ignitions should not be dismissed out of hand. As reported in LaPointe, et al., ignition of liquids by hot surface in the open air was not observed until the surface temperature was several hundred degrees above the published ignition temperature. The ignition of liquids by hot surfaces is influenced and determined by many factors, not just ignition temperature. These factors include ventilation; environmental conditions, such as humidity, air temperature and airflow; and fluids' physical properties, such as autoignition point, liquid flash point, liquid boiling point, liquid vapor pressure, liquid vaporization rate, and misting of liquid. Other factors include hot surface roughness, material type, and residence time of the liquid on the hot surface.

25.4.4 Mechanical Sparks. Metal-to-metal contact (steel, iron, or magnesium) or metal-to-road surface contact can create frictional contact sparks, with enough energy to ignite gases, vapors, and/or liquids that are in an atomized state. Metal-to-metal contact may occur at drive pulleys, drive shafts, or bearings, for example. Metal-to-road surface contact typically involves a broken component, such as a drive shaft, exhaust system, or wheel rim after the loss of a tire or in a crash. All metal-to-metal or metal-to-road surface sparking requires that the vehicle be running and/or in motion. Sparks generated at speeds as low as 8 kmh (5 mph) have been determined to reach temperatures of 800°C (1470°F) (orange sparks). Higher speeds have produced white sparks in the 1200°C (2190°F) range. Aluminum-to-road surface sparks are not a competent ignition source for most materials because of the relatively low melting temperature of aluminum. The small particle size (mass) of sparks limits the quantity of energy available from them to ignite materials they contact. Also, sparks cool rapidly, especially when moving through air, which further limits the rate of heat transfer to materials they contact. For these reasons, it is difficult for sparks to ignite solid materials.

25.4.5 Smoking Materials. Modern vehicle upholstery fabrics and materials are generally difficult to ignite with a lighted cigarette. Ignition may occur if a lighted cigarette is buried in paper, tissue, and/or other combustible debris, or if the seat material comes into contact with an open flame. Urethane foam, typical seat cushion material, burns readily once ignited with a flame, and adds substantially to the intensity of a vehicle fire.

25.5 System Identification and Function. Each system of a motor vehicle has a specific function. Not all motor vehicles have the same systems; however, many systems operate in a similar manner. System familiarity is essential to the proper investigation of a motor vehicle fire. If the investigator does not know how a system operates, he or she will not be able to determine if a system malfunctioned or has been altered, or whether such malfunction or alteration could be responsible for the fire. Almost every motor vehicle has some form of repair publication. Most of these repair manuals explain the systems unique to that particular vehicle. Many publications also have troubleshooting guides to diagnose potential problems. These publications can be found on the internet, bookstores, public libraries, auto parts stores, and the manufacturer. The manufacturer's service publications should be used whenever possible as this provides the most comprehensive information available.

25.5.1 Fuel Systems. There are two basic fuel systems used in gasoline powered motor vehicles: the vacuum/low-pressure carbureted system, and the high-pressure, fuel-injected system. Both systems use a fuel storage tank, tubing, and fuel pump to deliver fuel to the engine. Once the fuel is delivered to the engine, it is atomized, compressed, and burned.

25.5.1.1 Vacuum/Low-Pressure Carbureted Systems. In the vacuum/low-pressure carbureted system, the fuel is typically drawn from the fuel storage tank by a mechanical pump attached to the engine. This pump operates only when the engine is rotating. The fuel is then pumped under approximately 20 kPa to 35 kPa (3 psi to 5 psi) to the carburetor fuel bowl. The fuel is then drawn from the bowl into the venturi due to the vacuum created by the velocity of the air accelerated as it passes through the throat of the venturi. In the venturi, the fuel is mixed with air in a ratio of approximately 15 to 1. The fuel–air mix is then drawn into the combustion chamber though the intake valve. The piston compresses the mixture and a spark ignites it. Potential problems with this system typically occur on the pressure side. If a leak develops on the vacuum side, air is drawn into the system, and the vehicle generally will not run. Leaks on the pressure side, however, may range from a fine mist to a heavy stream. Leaks can occur in a fuel line or at the carburetor and generally occur within the confines of the engine compartment. If an ignition source is available, a fire could result. If a fuel line is compromised by a fire from another source, the fuel may contribute significantly to fire growth, and the result may be a fuel system that appears to be the cause of the fire. Note that some vehicles, originally sold with a mechanical pump, may have had the mechanical pump replaced by an electric pump. If the mechanical pump does not appear to be connected to a fuel line, an electric pump may be in use.

25.5.1.2* High-Pressure Fuel-Injected Systems. In the high-pressure fuel-injected system, the fuel is typically pumped from the fuel storage tank under pressures of 240 kPa to 480 kPa (35 psi to 70 psi). This pumping is accomplished with an electric pump in or near the fuel storage tank. This pump is typically energized any time the ignition key is in the run position. Most late model vehicles have electric fuel pumps that are integrated with the engine's computer. This computer is designed to shut the pump off if the engine is not running. Some vehicles are also equipped with an inertial switch. This switch will shut the fuel pump down in case of an impact of sufficient severity. The fuel is pumped to either a single venturi-mounted fuel injector or a fuel rail assembly on the engine. The single venturi-mounted fuel injector is typically located inside a device that resembles a carburetor. Fuel is injected (atomized) in a much m
in a carburetor, increasing fuel eco
ally, by providing a stable fuel–air m
duces pollution. The fuel rail assembly is
of the engine to supply constant fuel p
cylinder-mounted fuel injectors. Both of th
pressure systems may also include a fuel ret
fuel return system carries the unused fuel bac
storage tank at generally much lower pressure than
line, typically 20 kPa to 35 kPa (3 psi to 5 psi). Some
are equipped with *components* in the supply and retu
lines that *are* designed to prevent fuel from the fuel tank e
ing the lines with the fuel pump off. Potential problems w
this system include leaks at fittings. When a leak develops, the system pressure can propel a stream of fuel several feet. If the leak develops on the supply side of the system, operational problems may be noticed by the operator. These leaks can result in poor performance of the vehicle, including starting difficulties, erratic operation, and stalling. A fuel leak involving the return side of the system, downstream from the fuel pressure regulator, may not have a noticeable effect on the engine operation; therefore, a leak here may go undetected. Even vehicles that are not running typically have residual pressure in the fuel system. Many vehicles are being equipped with flow detection equipment that monitors fuel usage and fuel return. If this system detects an imbalance, the system will shut down. A fire originating in a system other than the fuel system, if allowed to progress unchecked, can compromise the fuel lines of a vehicle. Residual pressure within the delivery fuel line can result in fuel spillage once the line is compromised, however this spillage is generally minimal. Fuel lines running to and from the fuel storage tank can create a siphoning effect if compromised; however, check valves placed within the fuel delivery system can act to minimize or eliminate this conditioning from contributing to further fuel leakage. Some vehicles may incorporate a two-pump system consisting of a low-pressure lift pump (in the tank) and a high-pressure delivery pump (located outside the tank). During an impact, if a fuel line is compromised, fuel siphoning may occur. Siphoning is dependent upon several factors to permit a continued flow of fuel, including: fuel line opening, fuel level in the tank, and vehicle orientation. Factors that can reduce or prevent siphoning include: fuel tank pressure or vacuum, and flow restrictions such as fuel pumps, check valves, and fuel line kinking. The investigator should be careful to consider all of these elements before rendering the opinion that fuel siphoned from a damaged fuel line. The investigator should also recognize that internal fuel tank pressure created when the tank is exposed to heat from a fire can force fuel from lines that have been previously damaged by either the impact or the fire.

25.5.1.3 Diesel Fuel Systems. Diesel-powered vehicles typically use a combination of pumps to deliver the fuel from the fuel storage tank to the engine. Sometimes, electric lift pumps are used in or near the fuel storage tank, or engine-mounted mechanical pumps are used. These lift pumps are typically high-volume, low-pressure devices. They supply fuel to the engine-mounted fuel injector pump. The fuel injector pump meters and delivers the fuel into each cylinder at the appropriate time for combustion. Combustion air is brought in through natural aspiration or under pressure from a turbocharger. Diesel fuel delivery systems are typically very robust. Leaks may develop at fittings loosened by the vibration of the engine. While difficult, under the right conditions, diesel fuel may ignite on contact with a hot surface. "Runaway" diesel n occasionally if This may occur the intake sys- rspeed or me- osphere. The nstances is to

natural gas, t 20.7 MPa safety vent usible link t is usually r-than-air reduced d shutoff valves are ge cylinders and one before the low gulator. Both valves are closed when the engine is not running. More information about natural gas systems can be found in NFPA 54, *National Fuel Gas Code.*

25.5.1.5 Propane Fuel. Propane is stored as a liquid under pressure. Fuel for an engine is drawn off the tank as a liquid whereas fuel for appliances is drawn off as a vapor. As an engine fuel, the propane flows through the fuel lines as a liquid. Any leak or break in an engine supply line will result in the rapid evaporation of the released liquid, usually accompanied by a mist of condensed water vapor. As an appliance fuel, propane is typically found in recreational vehicles for cooking, heating, and refrigeration. From the tank, propane vapor flows immediately into a pressure regulator which significantly reduces the pressure in the distribution piping to the appliances. The investigation of appliance fuel fires requires many of the same techniques as that of a structure investigation. More information about propane gas systems can be found in NFPA 58, *Liquefied Petroleum Gas Code.*

25.5.1.6 Turbochargers. Turbocharging is the utilization of a turbine to add to the power output of an engine by increasing the amount of air being forced into the cylinders. The turbine used to drive the compressor can turn up to 100,000 rpm under full load. The turbocharger uses exhaust gases for propulsion, and in many cases, the turbocharger and exhaust manifold are the hottest external points on an engine. The heat created can ignite fuels and other combustible materials. Both gasoline- and diesel-fueled engines can be turbocharged, with almost all new diesel fueled engines being turbocharged. Some engines use two turbochargers mounted in series or parallel. The center bearings for the main shaft are lubricated by engine oil from a separate tube from the pressurized side of the engine lubricating system. A leak from this tube or fitting can spray oil onto the hot turbocharger housing and exhaust manifold, resulting in ignition of the oil. A broken shaft may allow a large quantity of oil to leak into the exhaust pipe causing a severe overheating of the catalytic converter if so equipped. Oil may also leak into the intake and be burned by the engine.

25.5.2 Emission Control System. Most motor vehicles are equipped with some form of emission control. This system controls exhaust tailpipe emissions and evaporative emissions of fuels. To control tailpipe emissions, the system typically regulates fuel input, spark timing, exhaust gas recirculation (EGR), idle speed, and transmission shift schedule (if automatic transmission). The major components of a gasoline engine system are the sensor to measure airflow into the engine, the heated exhaust gas oxygen sensors, the EGR valve, and the idle speed control.

25.5.2.1 A sensor measures airflow, into the engine. Based on that airflow, the vehicle's computer will calculate the correct of amount of fuel to add to meet driver demand and to minimize tailpipe emissions. The oxygen sensors are located in the exhaust system and measure the amount of oxygen in the exhaust gas. This provides feedback to the vehicle's computer to tell it if the correct amount of fuel is being added. The vehicle's computer can then make adjustments to either increase or decrease flow rate of fuel. Some exhaust gases may be routed back through the engine through the EGR valve. The purpose of the EGR valve is to reduce the combustion chamber temperature. Lower combustion temperatures result in lower production of oxides of Nitrogen (NOx), an unwanted tailpipe emission. The idle speed control valve is located near the throttle body and allows air to go around the throttle plates when the throttle plates are closed. By varying this flow, the engine's computer can control engine idle speed.

25.5.2.2 The air/fuel ratio and spark timing are also adjusted to control exhaust temperatures. The gasoline stoichiometric air/fuel ratio is approximately 14.6 pounds of air per pound of fuel. Operating at richer air fuel ratios will allow increased spark retard and result in lower exhaust gas temperatures. This is typically done during extreme driving conditions such as extended wide-open throttle or trailer towing.

25.5.2.3 A vehicle's evaporative emissions are also controlled. Evaporative emissions are gasoline vapors that escape from the vehicle's fuel system. The main component of the system is the carbon canister, which may be located in the engine compartment or near the fuel tank. The carbon canister contains activated charcoal that traps excess fuel tank vapors. Then, when the vehicle is running, those vapors are sent to the engine to be burned. An in-line valve allows fuel vapor to flow when the vehicle's computer commands it. A potential problem is vapor leakage from the charcoal canister or vapor hoses. If the vapors exist in a concentration above the lower flammability range and if an ignition source is present, a fire could result.

25.5.2.4 Exhaust System. The exhaust system does more than carry the exhaust gases from the engine. It also serves as a part of the emission control system. The exhaust manifold is bolted directly to the engine. It is the collector for the exhaust gases coming from each cylinder. This manifold is generally located below the valve cover. A leak in the valve cover or gasket may allow engine oil to contact the manifold, which may result in ignition. *(See 25.4.3.)* In the inlet header pipe (the first section of pipe from the engine), there is usually an oxygen sensor. This sensor detects the oxygen content in the exhaust stream and sends a signal to the onboard computers to make fuel delivery and timing adjustments. The next device downstream, in the exhaust system is the catalytic converter. Surface temperatures range between 316°C to 538°C (600°F to 1000°F) under normal operations, depending on the engine design. However, temperatures may exceed 538°C (1000°F) due to high engine loads or during abnormal engine conditions (i.e., misfires), or rich fuel concentrations due to other sensor malfunctions.

25.5.2.4.1 Diesel Particulate Filters (DPF). Diesel particulate filters are unique to diesel exhaust systems. The DPF is a device designed to remove diesel particulate matter or soot from the exhaust gas of a diesel engine. In addition to collecting the particulate, a method must exist to clean the filter. Some filters are single use disposable filters, while others are designed to burn off the accumulated particulate, referred to as regeneration. Regeneration can be passive, (through the use of a catalyst), or active, which may use an injected fuel to heat the filter to a temperature sufficient to burn carbon soot. In either case, high surface temperatures exist, creating an ignition hazard. Failure to maintain manufacturer-required shielding and clearance between these hot external surfaces and combustible materials may result in ignition of combustible materials.

25.5.2.5 A potential danger in a normally operating motor vehicle is contact with combustible items, such as tall grass or debris, with exhaust and catalytic converter surfaces. A malfunction in the engine can cause the catalytic converter to run hot enough to ignite undercoating and interior carpeting.

25.5.3 Motor Vehicle Electrical Systems.

25.5.3.1 Motor vehicles typically use a 12-volt dc or 24-volt dc electric power system with a negative ground. These systems typically include a lead-acid or gel-type batteries for energy storage and a generator (alternator) for production of electric power when the vehicle's engine is running. Some modern vehicles are designed using electric motor drives to supply the motive power to the vehicle (these are referred to as EVs or Electric Vehicles) and some are designed using a combination of electric motor and internal combustion engine systems to provide that power (these are referred to as HEVs or Hybrid Electric Vehicles). EVs and HEVs use higher voltages and in some cases ac power for the drive motors. These types of vehicle systems are discussed in 25.5.3.3 and 25.5.3.4. In a conventional system, if either battery cable is disconnected, no electrical activity can occur in the vehicle when it is not running. If a vehicle is running, it can continue to run using the generator power. The generator outputs electric power at a rate proportional to the speed of the engine, so there are times at engine idle speeds where the generator output is supplemented by power from the battery to operate all the systems on the vehicle. The battery power is used to maintain computer memories in the vehicle when the engine is turned off. The battery must also be large enough to operate the starter motor. The large amount of power required to start the engine usually means the starter motor cable is the largest conductor on the vehicle. As in structures, the more electric current a wire must carry, the larger the conductor must be. The starter cable is also usually one of the few wires in a vehicle that does not have an overcurrent protection device in it; it is sized to withstand the large current draw. The electric power system of a conventional vehicle is completed with connections from the battery and generator to a system of fuses or other overcurrent protection devices in various power distribution boxes in the vehicle, via other large conductors, much like the electrical service drop and panel in a structure.

25.5.3.2 In addition to providing electrical energy, the lead–acid battery can also be a source of fuel in the form of hydrogen gas. Hydrogen can be released during charging operations. Small amounts of hydrogen and oxygen are also present inside sealed (no maintenance) batteries and can be released due to mechanical damage during a collision. Hydrogen gas has a very wide explosive range and is easily ignited by low energy sources. However, it is unusual for a hot surface to cause hydrogen ignition.

25.5.3.3 Twelve-Volt Electrical Systems. A major difference between vehicles and structures is that the entire frame, body, and engine may be used as the ground path. This means that during an inspection, only one conductor may be found connected to a device.

25.5.3.3.1 If an energized conductor should chafe or rub on a frame, body, or engine surface, it could complete a short circuit fault, much the same way as a structure wire might chafe and ground out against a properly grounded device such as a fuse panel or metal switch box. In most cases, this will cause the overcurrent protection device to operate and de-energize the circuit. Unlike structure electric power systems that use standard sized circuits (i.e., 15A or 20A circuit breakers in each branch circuit) that are open to the user to plug into, a vehicle electric power system is mainly a closed system with most branch circuits designed for a specific use and without any user access. This means that if someone services an open overcurrent protection device by replacing it with a larger-value device, the larger fuse can allow overloading of the conductors in the circuit, which may cause overheating and lead to a fire.

25.5.3.3.2 There are circuits in vehicles that are connected to battery power that are not shut off, such as the starter solenoid cable at the starter motor, the generator output post, the ignition switch to use in starting the vehicle, the emergency flasher circuit, the brake lamps, and devices with memories in them such as radios, clocks, and powertrain control computers, to name a few. Circuits supplying aftermarket accessories connected directly to the battery or generator (with or without overcurrent protection) may also be energized when the engine is not running. Wiring diagrams will typically show conductor insulation color and device locations, and describe functionality.

25.5.3.3.3 As with any electrical appliance, electric devices in vehicles could fail or malfunction in ways that could generate heat. Motor vehicle conductors utilize stranded copper wiring that will become annealed or tempered if overheating occurs. This overheating can also be caused by flame impingement during a fire or by electrical activity before or during a fire. Damaged wires should be carefully inspected and documented before any substantial movement of the vehicle wiring is done, whenever possible.

25.5.3.4 Other Electrical Systems. Vehicles built before approximately 1956 may have a 6-volt dc system rather than a 12-volt dc system, and some vehicles of that era also utilized a positive ground system. Such vehicles are rare. Larger vehicles such as medium-duty and heavy trucks, buses, and heavy equipment may utilize a 24-volt dc system. Recreational vehicles, discussed in detail in Section 25.12 also may contain 120-volt ac systems. Also refer to the section on electric vehicles and hybrid electric vehicles in this chapter for further details of these different electrical systems.

25.5.3.5 Event Data Recorders (EDRs). Many modern vehicles are equipped with EDRs. Multiple vehicle parameters are recorded before and just after a crash, including vehicle speed, engine RPM, braking, airbag deployment, crash pulse, etc. The investigator should determine whether an EDR exists. Downloading data from EDRs requires special skills and equipment to avoid loss of the data. Postfire data may be corrupted by the action of the fire. There may also be legal issues associated with the ownership of the data.

25.5.4 Mechanical Power Systems. Motor vehicles are usually powered by an internal combustion engine that may run on a variety of fuels. An internal combustion engine is a complex machine with many components working in synchronization. To maintain proper function, all of these parts must remain securely attached, and many require lubrication or cooling. Mechanical failure of the engine can result in components, parts, or pieces being propelled at high speeds away from the engine. Pistons and connecting rods have been known to break apart within the engine block, break through the side or top of the engine block, and damage other components outside the engine. Liquid oil, crankcase vapors, and coolant can escape through a hole created by such a failure and ignite on a hot surface. Mechanical failure can also result in the failure of fuel lines. A piece of piston cutting through the fuel line may lead to a fire. Most engines will not operate after a catastrophic mechanical failure or will operate only with seriously diminished performance.

25.5.4.1 Lubrication Systems. Most engines use hydrocarbon oil for lubrication. Oil collects in a pan at the bottom of the engine and is distributed by a mechanical pump. The oil returns to the pan by gravity flow. Most oil pumps operate at 240 kPa to 415 kPa (35 psi to 60 psi). Oil can leak at the many joints or mating surfaces in the engine. Leaks around the pan gasket, for instance, can allow oil to be blown onto the hot exhaust components, or oil leaking from a valve cover gasket can flow down onto a hot exhaust manifold. Oil leaks may result in fires after the vehicle is turned off (see 25.4.3). Lack of oil in the engine can cause the failure of internal components. These failures can lead to catastrophic mechanical failure, which may, in turn, cause a fire.

25.5.4.2 Liquid Cooling Systems. Engine cooling is accomplished with liquid or air. Liquid-cooled engines use a closed system consisting of passageways through the entire block, a water pump, hoses, thermostat, and a radiator. As the engine reaches its normal operating temperature, the coolant in the engine also warms. This coolant is circulated through the system by the water pump. When the coolant reaches a pre-set temperature [typically in the range of 80°C to 90°C (180°F to 195°F)], the thermostat opens, allowing the coolant to flow through the radiator. The radiator is a heat exchanger, composed of a series of small tubes, connected between two tanks. A fan is used to provide airflow at low speeds or when the vehicle is stopped. Airflow around the tubes reduces coolant temperature before it is sent back to the engine. A normal cooling system will operate at approximately 110 kPa (16 psi). Most cooling systems are designed to use a mixture of 50 percent glycol antifreeze and 50 percent water. In colder climates, higher percentages of antifreeze are sometimes used.

25.5.4.3 Air-Cooled Systems. Air-cooled engines use a fan and ductwork to move air around the engine. These fans are typically belt driven. Engine overheating can occur if the belt breaks. Catastrophic engine failure may result.

25.5.4.4 Electric Motors. Vehicles such as pure electrics and hybrids (internal combustion-electric) will have one or more high-power electric motors to provide or augment the torque to drive the vehicle. These motors will be connected to power electronics, which in turn will be powered by a battery or ultra capacitors. The power supply providing energy to the motor(s) may range up to 600 volts. Initially, electric motors were dc, but the trend is toward variable frequency ac motors. However, small electric vehicles, such as golf carts, still operate on dc.

25.5.5 Mechanical Power Distribution. Mechanical power produced by the engine must be delivered to the drive wheels. This delivery is accomplished through a transmission. The transmission can consist of mechanical gearing or hydraulically driven gearing. The moving parts within the transmissions require lubrication.

25.5.5.1 Mechanically Geared Transmissions. Mechanically geared transmissions, commonly referred to as manual transmissions, receive engine power through the clutch assembly. The clutch assembly is located between the engine and transmission and consists of a pressure plate bolted to the engine flywheel and a clutch disc on the transmission front shaft. The clutch may be hydraulically actuated and the clutch fluid is also ignitible. The selection of drive gears is mechanically performed by rods and levers. The gears in the transmission are lubricated in an oil bath. The oil can be as heavy as 120- or 130-weight gear oil, or as light as automatic transmission fluid. Mechanical failures can be as catastrophic as those in the engine. Leaks of transmission lubricant may occur in the area of the exhaust system. Transmission fluid can typically be added only from under the vehicle, directly through openings in the side of the transmission. Overfilling of a manually geared transmission is rare.

25.5.5.2 Hydraulically Geared Transmission. Commonly referred to as automatic transmissions, these devices receive engine power through a torque converter. A torque converter allows the engine to run at idle without the vehicle moving. This is accomplished by vanes in the torque converter allowing a small quantity of transmission fluid to flow around the vanes. As engine speed increases, the fluid can no longer flow around the vanes and power is transferred to the transmission. Gear selection is performed by the transmission, depending on factors such as engine speed and load demands.

25.5.5.2.1 During operation of the transmission, heat is generated in the fluid as it flows throughout the transmission. This fluid requires cooling. Cooling is typically accomplished by routing the fluid through a heat exchanger located in the radiator end tank. If additional cooling is required, auxiliary heat exchangers may be necessary. They can be located in front of the radiator or HVAC condenser and will require additional fluid plumbing.

25.5.5.2.2 Fluid can be added by opening the engine compartment, removing the transmission fluid level indicator (dipstick), and pouring fluid down the tube. Most transmission level indicators specify that the transmission must be idling at normal operating temperatures before the fluid level is checked. Overfilling can cause the transmission to overheat the fluid and may allow the fluid to be expelled from the dipstick tube. During severe operation of the vehicle (high ambient temperatures, grade towing or high load towing) a transmission can expel the fluid out of the dipstick tube or the transmission vent tube. The dipstick tube and the vent tube are often located near or over the exhaust manifold or components of the exhaust system. Fluid expulsion can also result if the wrong type of fluid is added to the transmission. Some manufacturers have designed a locking dipstick that will prevent fluid expulsion from the dipstick tube after re-installing the dipstick properly. Leaks from the pan gasket may occur in the area of the exhaust system.

25.5.6 Accessories to the Mechanical Power System. Commonly installed accessories are air-conditioning compressors, power steering pumps, water pumps, generators (alternators), air pumps, and vacuum pumps. These mechanical devices may be attached to the engine and have their own potential to malfunction. These accessories are typically run by one or more drive belts. If a malfunction is suspected in one of these devices, an appropriate reference manual should be consulted.

25.5.7 Hydraulic Braking System. The hydraulic braking system is activated by a foot pedal connected, through the power brake booster, to a master cylinder. This master cylinder is connected with steel tubing to the wheel cylinders on drum brakes and calipers on disc brakes. Pressing the brake pedal produces higher pressure in the braking system, which is transmitted to the wheel cylinders and calipers. This pressure brings the brake linings (shoes and pads) in contact with the brake drums or rotors, creating friction, which stops the vehicle. Brake systems work under high pressures, and even a small leak can produce a spray that can be ignited if it contacts an ignition source. Brake systems also contain spare fluid that is held in a reservoir that is never pressurized. The purpose of the spare fluid is to replenish the master cylinder as the system takes in fluid during normal wear of the brake linings. The reservoirs on most vehicles are made of a plastic material that can melt during an engine compartment fire. If that occurs, the brake fluid in the reservoir will serve as a secondary fuel and accelerate the fire in the area of the master cylinder. Brake fluid can be expelled due to damage to the reservoir or a missing or dislodged cap and may provide the first fuel ignited.

25.5.8 Windshield Washer Systems. Windshield washer fluid or solvent is a mixture of water and methyl alcohol or methanol. Methyl alcohol is added to prevent the fluid or solvent from freezing. The content of methyl alcohol ranges from 20 percent to 60 percent. Windshield washer fluid or solvent, if sprayed on a hot surface, can form a vapor that may become ignited. The windshield washer solvent reservoir is usually a plastic material and may be consumed in a fire. If the reservoir is consumed, note whether body parts of the vehicle have penetrated the space that would have been occupied by the reservoir. Some systems use heated nozzles to keep the washer fluid from freezing. A malfunction in the heater may cause a fire.

25.6 Body Systems. Many body panels of modern motor vehicles are made of plastic, polymers, or fiberglass materials and will burn during a fire. These materials by themselves do not pose a significant ignition hazard; however, after a fire begins, they may significantly add to the fuel load. Often the entire cab of a tractor will be consumed, or door and hood panels will be gone. The inside panels of the front fenders in many cars are plastic; when they burn through, additional ventilation will be available for engine compartment fires. It should be noted that aluminum, magnesium, and their alloys are being used in panels in some vehicles. These panels will burn, often with great intensity. The partition between the engine compartment and the passenger compartment is commonly referred to as either the *fire wall*, *cowling*, or *bulkhead*. In modern motor vehicles, this partition may have numerous penetrations, some associated with the heating and air-conditioning system ducts. The ducts are usually made of reinforced plastic and can burn, resulting in a path for fire spread into the passenger compartment. Fire can also spread by conduction through the metal partition to combustibles under the dash.

25.6.1 Interior Finishes and Accessories. The interior finishes and furnishings (seats and padding) of most motor vehicles represent a significant fuel load. If the vehicle is burned out, try to determine what the original interior fuel load was and how it was arranged. Document the presence (or absence) of seats and accessory equipment such as radios, CD players, or telephones.

25.6.2 Cargo Areas. A motor vehicle fire may involve the trunk of an automobile, the storage areas of a motor home, or the cargo compartment of a truck. It is important to determine whether the fire originated in or spread to these areas. The investigator should make an inventory of the materials that were present in these areas. Inspection of the debris may be sufficient, or it may be necessary to interview owners or occupants to obtain the information needed.

25.7 Recording Motor Vehicle Fire Scenes. The same general techniques are employed for vehicles as are used for structure fires. Whenever possible, the vehicle should be examined in place at the scene. In many cases, however, the investigator may not have the opportunity to view the vehicle in place. For many reasons, the vehicle may have to be moved before the investigator reaches the scene. Frequently, part of the documentation takes place at a salvage yard, repair facility, or warehouse. A vehicle fire inspection should include the procedures in 25.7.1 through 25.7.5. Post crash vehicle fires may require the assistance of an accident reconstructionist.

25.7.1 Vehicle Identification. Identify the vehicle to be inspected and record the information. This will entail describing it by make, model, model year, and including any other identifying features. The vehicle should be accurately identified by means of the vehicle identification number (VIN). The composition of the VIN provides information on such things as the manufacturer, country of origin, body style, engine type, model year, assembly plant, and production number. The VIN plate is most commonly placed on the dash panel in front of the driver's position. It may be affixed with rivets. If this plate survives the fire, the number should be recorded accurately. Manufacturing information is also frequently found on a combustible label on the driver's door pillar. Many manufacturers also stamp at least a partial VIN on the engine side of the bulkhead. The bulkhead may be referred to as the dash and toe panel or the fire wall. It is the partition separating the machinery compartment from the passenger compartment. A brass bristle brush will clean this number so that it is legible. If the number is rendered unreadable or appears to have been tampered with, then the assistance of one of the following should be requested:

(1) A police auto theft unit
(2) A member of the National Insurance Crime Bureau in the United States
(3) A member of the Canadian Automobile Theft Bureau in Canada

25.7.1.1 These persons have the necessary expertise to identify the vehicle by means of confidential numbers located elsewhere on the vehicle. The VIN should be checked on either the National Crime Information Center (NCIC) or the Canadian Police Information Centre (CPIC) to ensure that there is no record outstanding on it.

25.7.2 Vehicle Fire Scene History. One of the initial steps to be undertaken in a vehicle fire investigation is the collection of information relative to the condition and use of the vehicle at the time of and prior to the fire. Pertinent information may be obtained through interviews with various people such as the vehicle owner, service/repair persons, the most recent operator, the person or persons who discovered the fire, bystanders, fire-fighting personnel, and police officers. It is important to record the information that is obtained so that it can be taken into consideration later during the forming of a fire cause hypothesis. For more information on recording information obtained during interviews, see 13.4.3. Examples of information that should be obtained include the following:

(1) Last use
(2) Mileage
(3) Operation
(4) Service
(5) Fuel
(6) Equipment
(7) Personal effects
(8) Photographs prior to the fire
(9) Photographs during the fire
(10) Photographs after the fire

25.7.2.1 The investigator should keep in mind that the above list represents only a sampling of the questions that may be relevant in any given investigation and is not meant to be all-inclusive.

25.7.3 Vehicle Particulars.

25.7.3.1 Once the vehicle has been positively identified as being the subject of the investigation, the mechanical functions of that particular vehicle, its composition, and its fire susceptibility should be reviewed. To ensure that no details are overlooked, the investigator may examine a vehicle of similar year, make, model, and equipment, or the appropriate service manuals. The use of a checklist may assist the investigator in the complete inspection of the vehicle. *[See Figure A.15.3.2(e) for an example of the checklist.]* It is not intended to be all-inclusive, and it may not apply to a given investigation and additionally may not cover all items that an investigator may identify at the scene.

25.7.3.2 Information regarding fires and fire causes in vehicles of the same make, model, and year can be obtained from The National Highway Traffic Safety Administration, *www.nhtsa.gov* (1-800-424-9393 or 1-202-366-0123), from the Insurance Institute for Highway Safety, *www.hwysafety.org* (202-328-7700), or from the Center for Auto Safety, *www.autosafety.org* (1-202-328-7700). In Canada, contact Transport Canada, *www.tc.gc.ca/* (1-613-993-9851).

25.7.4 Documenting the Scene. The investigator should make a diagram of the fire scene, showing points of reference and distances relative to the vehicle. The diagram should be of sufficient detail to pinpoint the location of the vehicle before its removal. The overall scene should be photographed, showing surrounding buildings, highway structures, vegetation, other vehicles, and impressions left by tires or footprints. All fire damage to any of the above items or signs of fuel discharge should be photographed and documented to help in the analysis of the fire spread. The location and condition of any parts or debris that are detached should be documented. Post-crash fire scenes may require additional documentation such as digital surveys and aerial surveys.

25.7.4.1 The vehicle should be photographed in an orderly and consistent manner. The photographs should include all surfaces, including the top and underside. Both the damaged and undamaged areas, including the interior and exterior damage, should be photographed. Documentation of the vehicle should illustrate the condition, position, and contents of the vehicle at the time of the inspection. Any changes to the vehicle during the inspection should be documented with a series of photographs that reasonably show those changes.

25.7.4.2 Any evidence showing the path of fire spread either into or out of any compartment (engine, passenger, trunk, cargo, etc.) or within any compartment should be photographed. As with structure fires, the path of fire travel may be difficult to determine in a totally burned-out vehicle.

25.7.4.3 The cargo spaces should be photographed. The type and quantity of cargo and any involvement in the fire should be noted. If possible, the removal of the vehicle(s) and any damage that results from the removal process should be documented. Also, after removal of the vehicle(s), the scene should be photographed while noting burns on the earth or roadway, and the location of glass and other debris. Drawings and notes should be prepared to augment the photographs.

25.7.5 Documenting the Vehicle Away from the Scene.

25.7.5.1 If the vehicle has been removed from the scene, a visit to the scene may be helpful. Photographs that were taken at the scene should be reviewed. Prior to the vehicle inspection, as much background information should be gathered as possible. This information should include date and time of loss; location of loss; operator, passenger, or witness statements; police and fire department reports; the vehicle's current location; and method of transportation (roll-back, tow truck, driven, etc.). The basic process of documenting the condition of the vehicle is the same regardless of where it is. When the inspection is delayed and the vehicle is located at a remote location, parts may be missing or damaged. Additionally, the vehicle(s) may have been damaged by the elements, and fire patterns, most notably those on metal surfaces, may be obscured. If outdoor storage is likely, arrangements should be made for the vehicle to be covered with a tarp or other suitable material. Often, rust developing on body panels within a few days of the fire can make burn patterns more visible, but subsequent rusting can obscure burn patterns.

25.7.5.2 Even if the vehicle was examined at the scene, there are advantages to inspecting a vehicle away from the scene. For example, it is easier to move or remove body panels that may be blocking a view of critical parts. Power is often available, as are tools for disassembly, if needed. Frequently, arrangements can be made to have equipment such as a forklift available to raise the vehicle for a more detailed inspection. The vehicle should be thoroughly photographed as it is examined at locations away from the scene.

25.8 Motor Vehicle Examinations.

25.8.1 General. The examination of a motor vehicle after it has burned is a complex and varied task. As with structure fires, the first step is to determine an area of origin. Most motor vehicles can be divided into three major compartments: the engine compartment, the passenger compartment or interior, and the cargo compartment. The size, construction, and fuel load of these compartments can vary considerably. The use of vehicle inspection field notes *[see Figure A.15.3.2(e)]* may assist the investigator in recording information.

25.8.1.1 Examination of the exterior may reveal significant fire patterns. The location of the fire, and the way that the windshield reacts to it, may allow a determination of the compartment of the origin. Diagrams illustrating potential fire pattern development as a function of compartment of origin are shown in Figure 25.8.1.1(a) and Figure 25.8.1.1(b). A passenger compartment fire will frequently cause failure at the top of the windshield and will leave radial fire patterns (fire patterns that appear to radiate from an area) on the hood, as shown in Figure 25.8.1.1(c) and Figure 25.8.1.1(d).

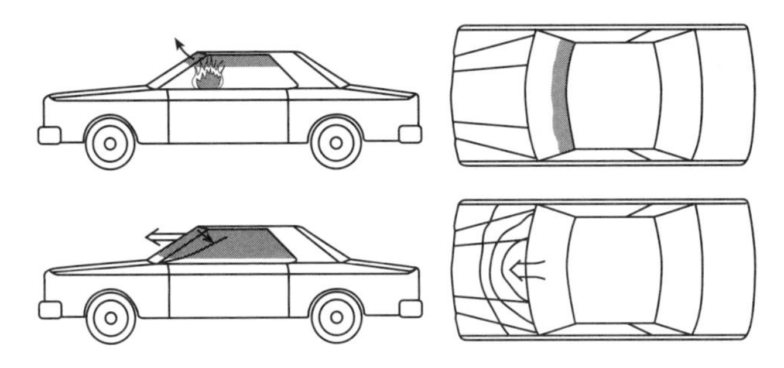

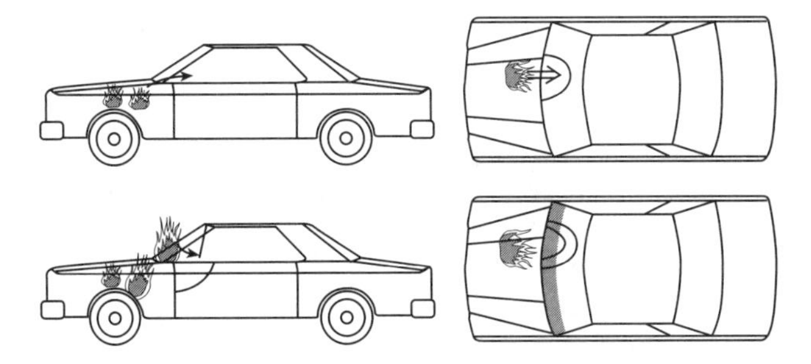

FIGURE 25.8.1.1(a) Fire Pattern Development from an Interior Origin.

FIGURE 25.8.1.1(b) Fire Pattern Development from an Engine Compartment Origin.

FIGURE 25.8.1.1(c) Radial Fire Pattern Produced by a Passenger Compartment Fire.

25.8.1.2 Engine compartment fires may spread into the passenger compartment through preexisting penetrations in the bulkhead, as described in paragraph 25.6, and typically cause failure at the bottom of the windshield often on the passenger side. Radial patterns from an engine fire may be observed on the doors. Examples of typical patterns produced by engine fires are shown in Figure 25.8.1.2(a) and Figure 25.8.1.2(b).

25.8.2 Examination of Vehicle Systems. After establishing a compartment of origin, the individual systems in that area can be examined. Using the system identifications and functions described in Section 25.5 through 25.5.8, an investigator should inspect each system and determine its condition and the possible involvement of that system in the fire.

FIGURE 25.8.1.1(d) Another Radial Fire Pattern Produced by a Passenger Compartment Fire.

FIGURE 25.8.1.2(a) Incipient Windshield Failure Caused by an Engine Compartment Fire.

FIGURE 25.8.1.2(b) Radial Pattern on the Driver's Side Door Produced by an Engine Compartment Fire.

25.8.2.1 Inspect the fuel tank for crushing or penetrations. Note the condition of the fuel filler system. Fuel fill and vent tubes and hoses are often multi-piece systems with elastomeric connection(s). Some filler systems are inserted into the tank through an elastomeric bushing or gasket. These connections may release fuel by mechanically separating during a crash or may burn through from an exposure fire.

25.8.2.2 The presence or absence of the fuel tank fill cap and any fire or mechanical damage at the end of the filler system should be noted and recorded. Many fuel tank caps have plastic or low-melting-temperature metal components that may be destroyed during the fire, with the result that the metal parts may become dislodged, missing, or found in the fuel tank. Fuel tanks exposed to heat or flame generally exhibit a "liquid" line of demarcation that represents the fuel level at the time the fire was extinguished.

25.8.2.3 Fuel supply, return, and vapor tubes and hoses should be inspected for leakage or indications of fire damage. These tubes and hoses usually have elastomeric connections at one or more points along their length that may release fuel. The condition of these tubes and hoses and any evidence of fuel release in relation to areas of hot surface or mechanical damage should be examined and recorded.

25.8.3 Switches, Handles, and Levers. During inspection of the vehicle interior, the position of switches should be noted to determine their respective positions (on/off, up/down, etc). An attempt should be made to determine if windows were up or down and what their condition was prior to the fire. The position of the gear shift mechanism should be noted, and the ignition lock cylinder should be examined, if possible, for any signs of a key, tampering, or breaking of the lock. These components are comprised of various materials and may be damaged to varying degrees by exposure to heat; however, many may still be recovered through debris sifting and provide information relevant to the investigation.

25.9 Total Burns. "Total burns" are considered those fires that have consumed all or nearly all combustible materials. Vehicles that have burned to completion pose special problems for determining the origin and cause of the fire. If a vehicle burns to completion, there may be no witnesses to interview and possibly no fire department response. The vehicle should be identified and documented completely. An attempt should be made to determine the condition of the vehicle at the time of the fire. If components are missing, an attempt should be made to determine if they were removed prior to the fire, or removed or lost after the fire. The pre-fire condition of the engine and transmission can frequently be determined by analyzing their fluids. Fire debris from the floor and surrounding areas can be checked for the presence of ignitible liquid residues.

25.10 Special Considerations for Incendiary Vehicle Fires. Vehicles that are stolen or allegedly stolen present a special case. The likelihood that a vehicle sustained an accidental fire after being stolen is very low, but a thorough examination should still be conducted. Vehicles may be stolen for many reasons. One common motive is the theft for parts. If a thief steals a vehicle for parts, it may never be recovered. If it is recovered, it should be obvious that parts have been removed. Items of value that may be missing include wheels, most major body panels, engines and transmissions, air bags, sound system, and seats. Another common motive for vehicle theft is to commit another crime. The thief will sometimes burn the vehicle to cover any evidence such as fingerprints. This thief may not remove any item from the vehicle and may use available materials in setting the fire. Vehicles are sometimes deliberately burned and reported stolen. Such vehicles may exhibit evidence of the fraud in the form of substitution of inferior parts, such as wheels and sound systems. Sometimes a leased vehicle is deliberately burned so that the lessee can get out of the contract that he cannot afford or where the vehicle has accumulated too many miles, requiring a substantial over-mileage payment.

25.10.1 Drive train fluids (such as engine oil or transmission fluid) analysis may be performed to determine the prefire condition of the engine and possible motive for the destruction of the vehicle. Samples of debris may be analyzed for the presence of ignitible liquids. This analysis may show the method of destruction.

25.10.2 Assuming a total burn or near total burn, debris removal may reveal the presence of the vehicle's ignition lock cylinder wafer (tumblers). This is the portion of the ignition switch that holds the key. Careful debris removal and documentation may show the key still in the lock cylinder tumblers. In some cases, the arrangement of the wafer in the debris will indicate whether or not the wafer were located inside the ignition lock cylinder at the time of the fire. This information, as well as a microscopic evaluation of the wafer lands, may show if the ignition lock cylinder was forced prior to the fire. If the wafers are not found, this may indicate that the ignition lock cylinder was removed prior to the fire. The investigator may find the wafer nearly intact in another portion of the vehicle. In this case, the wafer will probably show evidence of forced removal. In order for a vehicle to be stolen without the use of the correct key, either the vehicle must be towed from the site or the vehicle must be entered, the steering wheel lock must be bypassed (broken), the transmission lever locking device must be bypassed (broken), and the ignition switch must be engaged. Failure of evidence of all of these, usually indicates that the correct key was used. Some automobiles now use a plastic headed key that contains an electronic transmitter coded to the vehicle that communicates to the vehicle engine computer. The wrong key, or no key, will typically not allow the vehicle to start.

25.11 Vehicles in Structures. Motor vehicles are often stored in structures and may be damaged by a fire that occurs within the structure. The damage may include thermal effects of the fire and crushing of the vehicle by collapse of the structure. If the vehicle is located in the area of origin, then it should be examined as a potential fire cause, as outlined in Section 25.5 through Section 25.10. Conducting a thorough examination of this vehicle may include removing it from the debris, removing the hood and roof, removing debris from the interior, and elevating the vehicle to examine its underside. Regardless of whether the vehicle is a cause or victim of the fire, a vehicle provides a substantial fuel package, which also is associated with low burning and should be carefully considered to prevent an erroneous origin conclusion in the structure.

25.11.1 Fires that start external to the vehicle may spread to the vehicle and cause the vehicle to sustain significant damage. External fire may also cause a loss of integrity of fuel system components, including fuel lines and tanks, and may cause the release of ignitible liquids.

25.11.2 Vehicles that are in structures may be there for repair or rebuilding purposes. It is important to document the state of these efforts at the time of the fire.

25.12 Recreational Vehicles. This section deals with factors related to the investigations of fires involving recreational vehicles (RV's). Information relative to the investigation of RV fires includes construction materials, systems, and failure modes. Additional information regarding RV systems and standards is available in NFPA 1192, *Standard on Recreational Vehicles*. RVs incorporate many similarities to houses and mobile homes. They are also in many cases, a motorized vehicle, containing all the intricate details of automobiles. However, because of the standards required for their lightweight construction methods they may utilize large volumes of plastics and other combustible materials. There will often be additional large fuel items like polyurethane foam in couches, mattresses, and refrigerator insulation and propane gas.

25.12.1 General. Recreational vehicles are a unique combination of both a vehicle and a residence, utilizing unique construction materials and methods. The investigator needs to have a basic familiarity of the construction methods, the systems, and associated subsystems of the unit. As with any other fire, the first step is to determine an area of origin. Recreational vehicles can be divided into six seven major areas; i.e., exterior, engine compartment, basement or storage compartment, coach, galley, lavatory, and bedroom. The size, construction, and fuel load of these compartments can vary considerably depending upon the size, style, and manufacturer of the RV.

25.12.2 Recreational Vehicle Investigation Safety. Safety should always be the investigator's first concern. Documentation can be conducted during the safety assessment. *(See Safety chapter for further information.)*

25.12.2.1 Confined Spaces. By their nature, fire investigations on RVs may often involve working in confined spaces. The investigator should be aware of confined space entry concerns (e.g., entry/egress and atmospheric issues), and appropriate precautions should be taken. Prior to entry, the investigator should ensure the space does not contain hazardous levels of explosive or toxic vapors or gases (i.e., carbon monoxide) or is not oxygen deficient. The hazards of the space should be evaluated before entering and the appropriate level of personal protective equipment should be worn. Lights and other equipment should be intrinsically safe and suitable for such environments.

25.12.2.2 Airborne Particulates. Because many RVs incorporate large quantities of fiberglass in their construction, resin used as part of the construction process generally burns away, leaving small, irritating particles of fiberglass. When dealing with issues such as refrigerator fires, some materials, such as sodium chromate, can be carcinogenic. nature. These particles, when combined with burned resin, are highly irritating to the respiratory system and The appropriate level of personal protective equipment, including respiratory protection, should be worn as indicated by the level of hazard.

25.12.2.3 Energy Sources. There may be numerous energy sources present in an RV. Electrical hazards may be from 12V DC batteries, or 120/240V AC from shore power or a generator. Fuels such as gasoline, diesel fuel, or propane, may pose a spill, fire, or explosion hazard. Pressurized containers may rupture and pneumatic or hydraulic cylinders may fail. Precautions must be taken to ensure that the hazards are disabled, minimized, or otherwise made safe to prevent personal injury, as well as the possible recurrence of a fire or explosion. The investigator should employ personal protective equipment that is appropriate for the anticipated hazards.

25.12.2.4 Stability. RV fire scenes may be unstable. Care should be exercised when accessing the interior of an RV due to potential collapse of roof, side wall sidewall, or floor. RV manufacturers often place heavy items such as air conditioners and satellite dishes on roofs. Sidewalls Side walls may appear stable but may fail just by movement of debris. Never place any portion of your body underneath an RV without proper support of the RV.

25.12.3 Recreational Vehicle (RV). A recreational vehicle is a vehicular unit primarily designed to provide temporary living quarters for recreational, camping, travel, or seasonal use that either has its own motive power or is mounted on or towed by another vehicle. The living section of most RVs is referred to as the "coach."

25.12.3.1 Motor Home. Motor homes are a vehicular unit mounted on or permanently attached to a self-propelled motor vehicle chassis or on a chassis cab or van, that is an integral part of the completed vehicle.

25.12.3.2 Fifth Wheels. "Fifth wheels" are vehicular units mounted on wheels, of such size or weight as not to require special highway permit(s). The gross trailer area is not to exceed 400 square feet in the in the set-up mode mode and it is designed to be towed by a motorized vehicle that contains a towing mechanism that is mounted above and forward of the tow vehicle's rear axle.

25.12.3.3 Camping Trailers. Camping trailers are vehicular units mounted on wheels, constructed with collapsible partial side walls that fold for towing by another vehicle and unfold at the campsite.

25.12.3.4 Travel Trailers. Travel trailers are vehicular units, mounted on wheels of such size or weight as not to require special highway permits when towed by a motorized vehicle, and a gross trailer area less than 320 ft^2.

25.12.4 Unique Systems or Components. Many RVs are equipped with a variety of components and equipment typically not found in any other vehicle.

25.12.4.1 Shore Power. 120/240V AC V AC electrical power supplied from a nearby receptacle source via a cord set.

25.12.4.2 Generator. The auxiliary power generator systems in RVs vary. Generators may be fueled by gasoline, diesel, or propane. Gasoline and diesel fueled generator engines will operate similar to the descriptions earlier in this chapter. Propane generator engines will have the propane delivered as a liquid typically through a reinforced hose. Generators may be air or liquid cooled. RV operators may also use portable generators, which are typically gasoline fueled.

25.12.4.3 Automatic Generator Starting System (AGS). A control system that automatically starts and stops engine generators when pre-set RV conditions occur, such as beginning and end of quiet time, low or high battery charge or demand, availability or loss of shore power connection, or appliance demand changes.

25.12.4.4 Electrical Converter. Because there are two different electrical systems (12V DC and 120V AC), at times there is a need to interchange their use. Some components and appliances are designed to operate at 12V DC, while at the same time, others may be designed to operate at 120V AC. An electrical converter is used to transform 120V AC input to 12V DC output. This is usually performed when the RV is connected to shore power or when the generator is in operation. The converter is also used for battery charging.

25.12.4.5 Electrical Inverter. While converters change 120V AC to 12V DC, inverters perform the opposite function. The inverter changes a 12V DC input to 120V AC output. This allows the use of regular appliances without having to operate the generator or be connected to a shore power source. The disadvantage of inverters is they can use large amounts of energy from the batteries in a short time. It should be noted that some RVs are equipped with a single unit that incorporates both the converter and inverter functions. The investigator may observe only one unit that serves both functions.

25.12.4.6 Batteries While batteries are not unique to vehicles, the quantity and arrangement in an RV may be quite different. In a motor home, there is typically a 12V DC battery or batteries for the starting and operating of the engine. The coach typically has a separate system of batteries. Depending on the manufacturer, these may be connected in a variety of ways. The manufacturer provides the separation of 12V DC systems so that the starting batteries cannot be depleted by operating the coach accessories or appliances. While these systems are separate for output, the engine-mounted generator can recharge both the coach batteries while the engine is running. Some generator systems have direct battery charging capabilities while other work through the converter. Most RVs are equipped with battery disconnects for the coach side of the 12V DC system. The batteries may not be all located in the same area of the RV. The engine starting batteries are typically located at the engine area. The coach batteries are typically located in an accessible area.

25.12.4.7 Means of Escape. The emergency exit in a RV is typically a specially constructed window to the outside of the recreational vehicle. The primary means of escape from an RV is the entry/exit door(s). Secondary or emergency means of escape are provided in the form of specially constructed hatches or windows designed for this purpose.

25.12.4.8 Liquid Holding Tank. RVs may be equipped with sewage holding tanks. Tanks for the containment of toilet waste are referred to as "black water" tanks and those for the containment of sink and shower wastes are referred to as "grey water" tanks. A fresh water tank may also be present.

25.12.5 Systems.

25.12.5.1 Electrical Systems. Recreational vehicles typically contain two different electrical wiring systems; 120V AC and 12V DC. All electrical wiring in RVs is required to comply with *NFPA 70, National Electrical Code.* Most 120V AC system components are similar to that found in residential settings. Stranded conductors are used for 12V DC and solid conductors are used for 120V AC wiring. Lighting systems, for example, may use both 12V DC and 120V AC in the same fixture. When connected to shore power, the electrical demands of the RV are met by the 120V AC input. 12V DC components are powered through the converter. When disconnected from shore power, 120V AC generator output may be the coach input power source. When not connected to any 120V AC source, 12V DC components are utilized. Solar panels may be present. These panels are typically used for battery recharging. Detailed information regarding the 12V DC systems for the chassis is generally the same as a motor vehicle.

25.12.5.1.1 Shore Power. The most common RV shore power is a 30-ampere input. This is a three-wire system. This system is designed to power up to five 120V AC circuits. Some larger RVs use a 50-ampere input. This is a four-wire system. The four wire system utilizes two hot conductors, a neutral, and a ground. Even though there are two hot conductors, RVs do not use the traditional 240V AC input. There are no 240V AC systems in RVs. Campgrounds commonly have both 30-ampere and 50 ampere receptacles available in the same housing. Many homeowners install electrical services for their RV. Adapters may also be used. For instance, an RV manufactured with a four-wire, 50 ampere system may use an 'adapter' to plug into a three-wire, 30-ampere receptacle. This may limit the branch circuits available for use in the RV.

25.12.5.2 Propane Systems. Propane is used for cooking, heating, and refrigeration. Portable propane cylinders must be constructed in accordance with U.S. Department of Transportation Specifications for LP-Gas Containers (49 CFR) or fabricated to Transport Canada (TC) requirements. A fixed propane tank must be constructed in accordance with the ASME Boiler and Pressure Vessel Code, Section VIII, "Rules for the construction of unfired pressure vessels."

25.12.5.2.1 Propane Regulators. Two-stage regulators are typically used on RVs. They are required to have a capacity exceeding the total input of all propane fueled appliances. The first stage of the regulator reduces the tank pressure of the propane to an output not to exceed 10psi. The second stage of the regulator reduces the pressure further to within the tolerances of the entire system.

25.12.5.2.2 Maximum Propane Container Capacities. NFPA 1192 limits the capacity of propane containers stored in an RV; there should be should be no more than three cylinders or two tanks. The maximum capacity of each individual cylinder is a water capacity of 105 lb (47.6 kg) or approximately 45 lb (20.4 kg) propane capacity. The maximum aggregate water capacity is 200 gallons (0.8 m^3).

25.12.5.2.3 Propane System Shielding. Many manufacturers provide thermal, impact and abrasion shielding of the container and piping systems. Measurements should to be taken of the distances from the propane container to heat producing components. The container, piping, or hoses may be shielded by a vehicle frame member or by a noncombustible baffle with an air space on both sides of the frame member or baffle. Measurements should be taken of the distances between the propane piping or hoses to heat-producing components. The piping or hose may be shielded by a vehicle frame member or by a noncombustible baffle with an air space on both sides of the frame member or baffle.

25.12.5.2.4 Stoves. There are two basic types of kitchen stoves, a cook top and a range. A cook top is an appliance mounted in a drop-in configuration in the countertop. It does not have an oven. A cook top with an attached oven is a range. Ranges are installed in a cabinet cutout. These components are fueled by propane gas. Cooking control is provided by a manifold assembly and burner control valves. Stoves are equipped with a non-adjustable propane regulator as a mandatory safety feature. Stoves may have standing pilots for the oven and spark ignition for the surface burners. Clearances are important as these devices are specifically designed for the tight spaces encountered in RVs. Residential-style appliances are not typically designed the same and are not suitable for installation in an RV.

25.12.5.2.5 Water Heaters. Water heaters are fueled by propane or electrically energized by 120V AC. 12V DC is typically required for controls. These units are normally self-contained. Propane is ignited in a burner. The heat is then passed through a flue tube inside the water tank and the hot exhaust gases are then vented to the exterior. The entire process is outside of the living quarters of the coach. Water heaters may have standing pilots or electronic igniters. Water heaters use an internal thermostat to regulate water temperature. Water heaters use a combination pressure and temperature relief valve which are designed to open and release pressure prior to reaching a dangerous level.

25.12.5.2.6 Furnaces. Forced-air and ducted furnaces operate on the principles of combustion and heat transfer. Most furnaces use 12V DC for controls and blower operation.

25.12.5.2.7 Anhydrous Ammonia Absorption Refrigerators. RV refrigerators do not normally have compressors. These refrigerators use a heat absorption system that allows anhydrous ammonia to be heated and then flow through a system of tubes. The cooling of the anhydrous ammonia provides cooling for the refrigerator interior. The anhydrous ammonia is heated by propane in a burner or a 120V AC heating element (2-way system), or additionally 12V DC heating element (3-way). Typical operation will be on electricity when available; however, the operator may select the fuel. 12V DC is typically necessary to operate the controls. Refrigerators may have standing pilots or electronic ignition. The anhydrous ammonia is combustible and may become ignited if it escapes from the tubing.

25.12.5.3 Safety Systems.

25.12.5.3.1 Smoke Alarms. The smoke detector alarm(s) should be installed as recommended by the manufacturer and labeled as being suitable for installation in recreational vehicles under the requirements of ANSI/UL 217.

25.12.5.3.2 Carbon Monoxide (CO) Alarms. The CO alarm should be installed as recommended by the manufacturer and labeled as being suitable for installation in RV's under the requirements of ANSI/UL 2034 or CSA 6.19.

25.12.5.3.3 Propane Alarms Detectors. The propane alarm detector should be installed as recommended by the manufacturer and labeled as being suitable for installation in RV's under the requirements of ANSI/UL 1484.

25.12.5.3.4 Fire Extinguishers. NFPA 1192 requires all to be manufactured install with at least a 10B:C fire extinguisher.

25.12.6 Construction.

25.12.6.1 Exterior Construction. The investigator should be familiar with the composition and construction of the RV. General construction methods use a welded steel frame to which the floor is attached. The floor system may be constructed of steel, aluminum, or wooden frames with a wood or composite sub-floor material attached. Unlike residential construction, RVs are commonly built from the inside out. All of the interior structural components are installed on the floor prior to the installation of the exterior walls. This may include some of the appliances and fixtures. The side walls are then added. These components may be wood, steel, or aluminum framing, with fiberglass, aluminum, or stainless steel exterior siding and interior wall coverings. Newer methods use a vacuum process to 'bond' all the wall materials into a thin, strong, lightweight assembly. These assemblies can include all of the necessary wiring, both 12V DC and 120V AC. The final major component is the attachment of the roof assembly. The most common style roof is a wood sheath over frame construction covered by a waterproof membrane. A one-piece fiberglass roof may be used. Manufacturers may keep on end of the RV open to facilitate the installation of large appliances or fixtures.

25.12.6.2 Interior. Due to design and weight concerns, the interior construction materials may differ from those found in a residential structure. Interiors may be constructed with fiber reinforced panels (FRP) or wood framing and exterior grade plywood or veneer. ASTM E 84 flame spread ratings of 200 or lower is required for interior finishes of walls, partitions, ceilings, exterior passage doors, cabinets, habitable areas, hallways and bath rooms, including shower/tub walls. Counter surfaces may be natural (marble or granite), synthetic (plastic laminate), or wood veneer. Additional fuel loads may exist in the form of carpeting and other synthetic materials utilized as vertical and overhead finishes. The flame spread requirements do not apply to moldings, trim, furnishings, windows, door, skylight frames, casings; interior passage doors, countertops, cabinet rails, stiles, mullions, toe kicks, or padded cabinet ends. These fuels can dramatically affect any fire incident, ultimately affecting the spread and growth of the fire.

25.12.7 Recreational Vehicle Examinations. RV dimensions, clearances, and tolerances designed into the coach are sometimes critical, depending on the system. Any examination and documentation of any evidence, including the removal and securing of items of evidence, should be performed in accordance with the guidelines specified in the Physical Evidence chapter and the ASTM standards presented in Annex A of this document. If improperly removed, the investigator may destroy key pieces of evidence and information relevant to a potential cause. *(See Section 16.3.)*

25.12.7.1 Examination of the exterior of the RV may reveal significant fire patterns. Caution should be exercised when interpreting patterns around enclosures such as the refrigerator, furnace, or water heater compartments. These compartments in the RV that are ventilated and may contain have significant fuel packages. This which may affect exterior fire patterns.

25.12.7.2 Analyzing the fire patterns can help determine whether the fire spread from inside the compartment. NFPA 1192 requires that the interior of the coach be vapor resistant to the exterior. The arrangement of the cabin, appliances, and fixtures, along with whether the windows were open or closed, may influence the fire spread characteristics and resultant fire patterns. Compartments typically used for sleeping, lounging, cooking, storage, or lavatory, may include a variety of materials and ignition scenarios.

25.12.7.3 Investigating Systems and Components Unique to Recreational Vehicles. Specialized systems unique to RVs including: Propane propane gas systems, Electrical electrical systems, heating systems, appliances, and fire safety devices systems. These systems discussed in paragraphs 25.12.5.1 through 25.12.5.3.4 should be examined and analyzed for their possible roles as ignition or abnormal heat sources, their roles as first or subsequent fuels, and their roles in fire growth and spread. These examinations and analyses should include possible failure analysis, improper installation or functioning, pre-fire damage, fuel leakage, and the functioning or non-functioning of safety and control features.

25.12.8 Ignition Sources. In most cases, the sources of ignition energy in RVs will be similar to those already described in this chapter and those associated with structural fires such as arcs, mechanical sparks, overloaded wiring, open flames, and smoking materials. Additional considerations are the added appliances described previously. Consideration should also be given to the effects of vibration, mechanical damage, rodent activity, or movement, due to being driven or towed over the road.

25.12.9 Recreational Vehicle Identification. The National Highway Traffic Safety Administration (NHTSA) requires that all RVs be identified by a 17 digit Vehicle Identification Number (VIN). Manufactures commonly apply a VIN to the coach as well as identify the make, model, and serial number of the major appliances. This information is found in the owner's documents, a label on the exterior of the coach, and a placard located in a galley cabinet. After a fire, these numbers may not be readily available. Any numbers found stamped on the chassis may be submitted to the National Insurance Crime Bureau (NICB) or the Insurance Bureau of Canada (IBC) to verify the vehicle identity.

25.12.10 Recreational Vehicle Information. There are many different sources available to the investigator to assist in origin and cause determination. Such sources include but are not limited to sales brochures, owner's manuals, and Internet based web sites such as that of the manufacturer of the unit or its components. Floor plans and options can be found on the Internet from general public sales sites.

25.12.11 Recreational Vehicles in Structures.

25.12.11.1 RVs within garages, carports, or other enclosures should be examined with consideration of the additional combustibles within the enclosure, especially when considering the origin. Heat transfer from a fire originating in the structure may damage the RV.

25.12.11.2 If applicable, the building electrical service needs to be inspected for electrical supply voltage and polarity to verify if they are consistent with the requirements of the RV electrical and associated subsystems. Information about the shore power receptacle may be important.

25.12.11.3 NFPA 1194 refers to parks or campgrounds. Similar considerations need to be made when the RV is parked at one of these properties. If applicable, all services including propane, electricity, sewer, etc.) provided by the park or campground need to be inspected and verified.

25.12.11.4 RV system operational requirements require the unit to be parked in a generally level position. The type of surface the RV was parked on should be documented. An angle gauge can be employed to measure any slope present to establish the specific angles of the scene.

25.13 Heavy Equipment. Heavy equipment includes medium- and heavy-duty trucks and mass transit vehicles (buses), and earth moving, mining, forestry, landfill, and agricultural equipment. Some of these vehicles have a significant materials-handling function in addition to the locomotive function. This equipment is usually diesel powered and may have a hydraulic transmission. Heavy equipment is subject to the same failure modes as ordinary vehicles, but also susceptible to failure due to overloading the transmission, failure of the hydraulics involved in materials handling, bearing failures, and ignition of the material being handled. Investigation of a heavy equipment fire requires knowledge of the systems involved, including the systems specific to that vehicle. Some heavy equipment may be equipped with fixed fire suppression systems.

25.13.1 Medium- and Heavy-Duty Trucks, and Buses. Most medium and heavy trucks and truck-tractors are built as incomplete vehicles commonly referred to as chassis-cab units. This is the first step for vehicles manufactured in two or more stages in accordance with Title 49 of the Code of Federal Regulations, Part 568. Subsequent stage manufacturers are typically body fabricators and installers who may complete only a portion of or the whole vehicle. Modifications and additions may be extensive and involve body, electrical, and mechanical modifications to the chassis-cab.

25.13.1.1 Typically, completed trucks and truck-tractors are custom-built units designed to meet a specific application. Truck original equipment manufacturers (OEMs) often provide choices of vehicle specifications. This availability of options and brands makes for a large number of possible feature combinations. The investigator should take care to understand how the specific vehicle was built, and by whom.

25.13.1.2 Trucks, truck-tractors, and buses are also subject to severe duty cycles, including long hours of operation, and high mileage. This makes proper maintenance and repairs, and the quality and applicability of replacement parts, important to the prevention of fires on these vehicles.

25.13.1.3 The interpretation of fire patterns is further complicated by specific ventilation effects, fuel loads in different areas of the vehicle, complex electrical systems, and the unusual progression of the fire as it is obstructed or contained by structural members in the frame and cab. Truck cabs can be built from combinations of steel, aluminum, wood, fiberglass, and plastics. The differences in material properties and the influence of venting from open windows, doors, or an opening in the cab enclosure can create fire patterns that could mislead the investigator from the true origin and cause. The extensive use of plastics and the concentration and volume of combustible materials such as hydraulic fluids, large capacity fuel filters, and engine and transmission oils can also create fire patterns that are not related to the origin.

25.13.1.4 A wide variety of truck applications, including dump, garbage, snow plow and sander, vacuum, etc., use hydraulics for their operation. A hydraulic leak can spray fluid over a wide area with possible ignition on hot exhaust components.

25.13.1.5 Because diesel vehicles often have fuel tanks with 757 L (200 gal) capacity, a fire in a truck where fire suppression is delayed, may involve all the fuel, causing severe damage. This often makes determination of the origin and cause of the fire more difficult. In cooler climates, many vehicles use programmable coolant heaters, fuel, and engine oil heaters. Cab/sleeper heaters that operate on diesel fuel are also available. These systems should be inspected for possible involvement in the fire cause.

25.13.1.6 The electrical system on these vehicles is typically more extensive and complex than other vehicles and may include high current systems, 12- or 24-volt systems, ac systems, as well as low voltage multiplexed circuits. Some engines use high amperage electric intake manifold air heaters or glow plugs. Some large gauge wiring and high capacity relays often remain energized and may fail, resulting in a fire, or fail, as a result of a fire.

25.13.1.7 Medium and heavy trucks, tractors, and buses can use either hydraulic or air brakes. Air brakes are operated with air provided by a compressor at up to 827 kPa (120 psi). Damage to an air hose (plastic tubing or rubber reinforced with steel braid) or steel tubing at the time of fire initiation, or during the fire, may significantly push and fan a fire. This may provide fire patterns that are not related to the cause of the fire. Brake and wheel fires may occur when brakes are not released and the vehicle is driven. Most electronically controlled engines can develop enough torque at low speed to overpower parking brakes. Tires may sometimes ignite while driving, while under-inflated, or if road debris is lodged between duals.

25.13.2 Mass Transit Vehicles Transit buses can be fueled by diesel, CNG, or hybrid diesel-electric. Many are equipped with catalytic converters and mufflers. Compressed air systems are similar to heavy-duty trucks. Large air-conditioning systems, 24-volt dc electric systems, a 24-volt dc 270 ampere alternator, hydraulic operated steering, wheelchair ramps and fan motors, and auxiliary coolant heaters are common. Many buses have on-board fire suppression systems. Some inner city buses use unbreakable but combustible plastic for window glass. Many buses use low combustibility or fire retardant seating. Intercity highway coaches have similar engine compartment components to transit buses but usually employ a more powerful diesel engine.

25.13.3 Earth-Moving Equipment. Wheel driven equipment is often powered from the engine mechanically. Track-driven equipment is usually powered by hydraulic motors. Almost all other moving components are hydraulically operated. Damaged hydraulic components or leaking fluid may contact a competent ignition source and result in a fire. The cause of the damage to the hydraulic system should be investigated.

25.13.4 Forestry/Logging Equipment. A fire in the equipment, if not extinguished by an on-board fire suppression system, may burn to completion. Hydraulics operate almost all of the equipment functions, and the inherent rough service conditions of forests mean that hoses can get damaged. Hydraulic fluid leaks at up to 41,368 kPa (6000 psi) can contact the turbocharger or other hot exhaust components and ignite. Accumulations of wood chips and dust may be ignited by hot components or friction.

25.13.5 Landfill Equipment. Landfill equipment has the added problem that if the equipment catches on fire the combustibles in the landfill can also ignite. The unwanted accumulation of combustible debris on landfill equipment commonly provides initial and secondary fuels. This is often compounded by less than ideal daily cleaning maintenance (housekeeping). Fire suppression systems are common in landfill equipment for this reason.

25.13.6 Agricultural Equipment. Combines, balers, and farm tractors have numerous hydraulic components with the potential for fluid leaks similar to other heavy equipment. Balers can pick up items from the field that might spark against the steel components while baling dry materials, thereby igniting the materials. No evidence of the igniting substance would be found after the fire. Worn or unlubricated bearings can overheat, igniting combustible accumulations. Disassembly of the rollers after the fire should reveal the damaged bearing. Birds and small rodents can build nests in agricultural equipment when not in use and if located near the exhaust components, the nest can be ignited by the heat of the exhaust system when the equipment is later used.

25.14 Agricultural Equipment and Implements Introduction. This section refers to the investigation of fires involving self-propelled agricultural equipment and drawn farm implements. Self-propelled agricultural equipment is any type of equipment powered by an internal combustion engine and does not have to be pushed or pulled by other equipment. Drawn farm implements must be connected to a tractor or similar equipment and pulled o
signed task. Common functions pe
clude soil cultivation, planting, fertiliz
and harvesting. Implements generally
may require connection to a tractor's po
for operation. Many implements require hy
on-board hydraulic drives or electrical power
tronic controls, and sensing equipment. Include
cussion will be the classification and description of th
types of equipment likely to be encountered, unique fe
and special hazards associated with each type.

25.14.1 Agricultural Equipment Investigation Safety. In addition to the guidance set forth in the Safety chapter, the following hazards may be encountered when performing a post-fire investigation.

25.14.1.1 Due to the nature of construction and use of agricultural equipment, an investigator may encounter particulate from burned composite materials (e.g., fiberglass), and burned or decaying vegetation. Sharp edges from re-solidified glass or metals may cause penetrating injuries or lacerations. The sheer size and bulk of agricultural equipment creates an overhead environment with potential impact hazards which can cause head or eye injury. Investigations may require working at elevations with little or no platform or support, and slippery and/or uneven surfaces posing slip or fall hazards. Ladders or other specialized equipment may be necessary for working on the upper regions of this equipment. These hazards may require the use of additional safety equipment or procedures as outlined in the Safety chapter.

25.14.1.2 Hazardous materials such as fuels, herbicides, pesticides, and other agricultural chemicals may be found in various tanks or containers located on the unit. The potentially large capacities of storage tanks (fuel, hydraulic fluid, water, chemicals) may result in large spills or pools of liquid. Chemicals may also be present in powder or granular form. Soil surrounding the involved equipment may be contaminated by these products. Exposure to these materials may result in safety concerns such as inhalation injuries, chemical burns, irritations, slips, or falls. These hazards may require the use of additional safety equipment or procedures as outlined in the Safety chapter.

25.14.1.3 The post-fire condition or location of the agricultural equipment may be unstable. Efforts similar to those outlined previously in this chapter should be employed to stabilize the equipment prior to initiating an examination, as required. Mechanical parts or components may fall, pinch, slide, shear, rupture, collapse, or release pressurized liquids or gases. Efforts should be made to identify and secure these potential hazards prior to conducting an investigation.

25.14.1.4 Agricultural equipment fires generally occur in agricultural fields, roadways, or equipment sheds. Burned equipment in fields may pose additional hazards including uneven ground, loose soil, water, mud, snow, plants, snakes and insects. Burned agricultural equipment may provide hiding places for reptiles or rodents. Agricultural equipment fires in structures may pose additional hazards such as collapse, exposure to chemicals, collapse of stacked or stored contents, or sharp or pointed tools or implements.

25.14.2 Equipment Classification and Description. For specifications of a particular make or model of equipment, refer to the manufacturer's information for that particular piece of equipment.

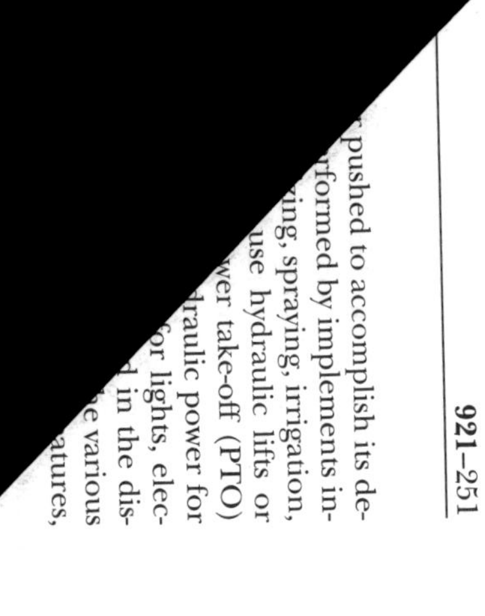

...or pulling or ...oading, plow- ...similar tasks. ...re equipped ...conventional ...inge in the ...e operator's ..., hydrostatic ...raulic trans- ...ive. Tractors

...re equipped ...ent off of the ...tractor to an ...mounted at ...

25.14.2.1.2 Hydraulic Components. Tractors may be equipped with hydraulically powered attachments. A series of hydraulic pumps, valves, and hoses, are required for operation.

25.14.2.2 Combines. Combines are used to harvest corn, soybeans, wheat and other grain crops. There are three separate functions that a combine performs during the harvesting process. First a cutting unit, generally referred to as a "header", cuts the crop to be harvested. The crop is then picked up and carried through a threshing mechanism, which separates and removes the unwanted material (residue). The remaining crop is then fed through a cleaning system and into a hopper or tank. When the hopper is full, the crop is mechanically off-loaded into a mobile transport trailer or truck. Most combines currently in use on farms today are equipped with turbocharged diesel engines. Engines are generally transverse mounted high and to the rear of the grain tank. On some older units, engines can be found to the right side of the operator's station, forward of the grain tank. Combines are generally equipped with hydrostatic drive transmissions. The front wheels on a combine are the drive wheels while the rear wheels steer. Hydraulic drive motors may be used to provide all-wheel-drive on newer equipment. *(See Figure 25.14.2.2.)*

25.14.2.2.1 Headers. There are various header configurations to match the crop to be harvested. Headers are attached to the front of the combine through a latching system that allows for changing and adjustment. Hydraulic and electrical connections are also made to the header for its operation.

25.14.2.2.2 Threshing/Crop Cleaning System. Once the crop is cut, it is carried up to a system of cylinders and plates to separate the residue from the crop. The crop then travels through a series of sieves and fan-driven air to further clean the crop. The residue is carried on a separate series of plates to the back of the combine where it may be chopped into fine pieces. A mechanical spreader may be used at the rear of the combine to scatter the residue on the ground behind the unit as it moves through the field.

25.14.2.2.3 Unloading System. After the crop is cleaned, it is fed into a hopper or grain tank. Sensors in the tank provide indication to the operator when the tank is full. An auger system moves the harvested crop through a chute and discharges it into an "attending trailer."

25.14.2.3 Forage Harvesters. Forage harvesters (also known as a silage harvester, forager or chopper) are similar equipment that harvests field crops to make silage. Silage is grass, corn, or other plants that have been chopped into small pieces, compacted, and then allowed to partially ferment to provide feed for livestock. Haylage is a similar process but uses dry hay. Forage harvesters utilize special cutting heads that cut the plants and then feeds them into a series of rotating processing drums, which

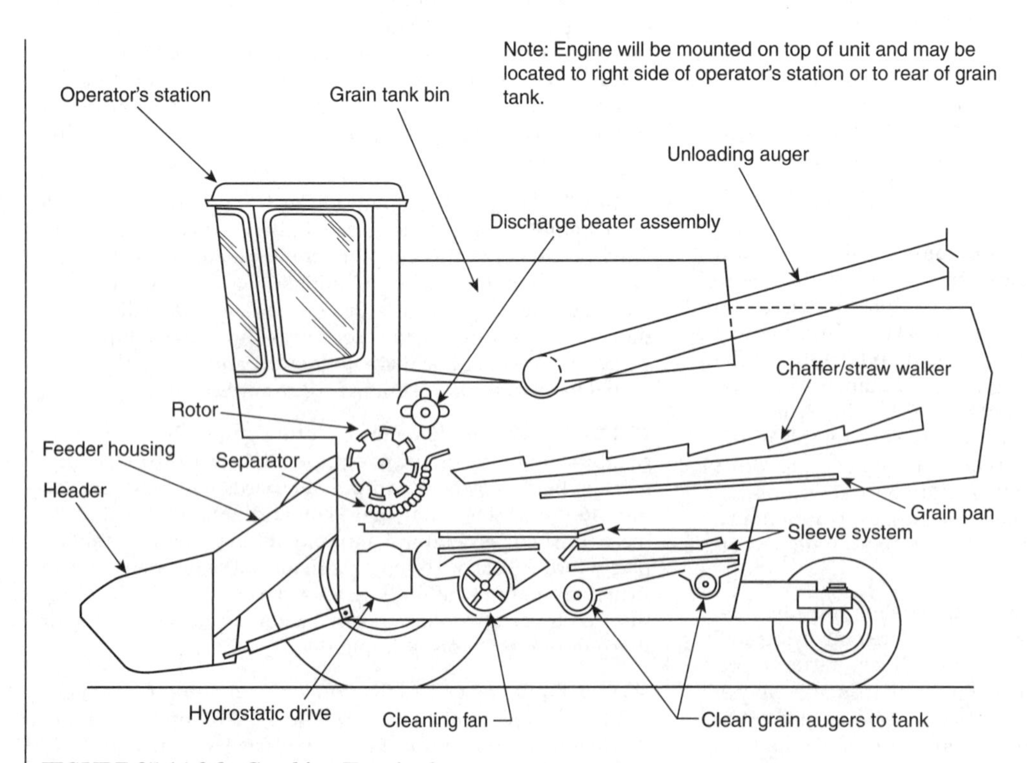

FIGURE 25.14.2.2 Combine Terminology.

utilize steel knives to chop the plant material into small pieces (silage). The silage is then fed into an accelerator device similar to a rotating fan, which forcefully pushes the silage up and out of a discharge chute and into an "attending trailer." Most forage harvesters currently in use on farms today are equipped with turbo-charged diesel engines. Engines are generally mounted in-line at the rear of the harvester. Forage harvesters are generally equipped with electronically assisted hydraulic transmissions. The front wheels on a harvester are the drive wheels while the rear wheels steer. Harvesters may be equipped with four-wheel drive.

25.14.2.4 Cotton Pickers. Cotton pickers harvest cotton using attachments called row units or picking drums. Rows of barbed spindles rotate at high speed to remove the cotton from the plant. The cotton is removed from the spindles by a counter-rotating doffer and blown upward into a large steel basket mounted at the rear of the unit. Once the basket is full, the picker is designed to dump the harvested cotton into a detached steel box with a hydraulic compactor arm called a module builder. Evolving cotton picker designs take the harvested cotton, form it into a round or square bale, wrap it, and eject the bale much like a hay baler (see below). Most cotton pickers currently in use are equipped with turbo-charged diesel engines. Engines are generally transverse mounted below and to the rear of the operator's station, near the center of the unit. Cotton pickers are generally equipped with hydrostatic drive transmissions. The front wheels on a cotton picker are the drive wheels while the rear wheels steer and provide some drive assistance. Cotton pickers are generally equipped with disc, drum brakes or sealed wet-brakes.

25.14.2.4.1 Row Units. Row units are attached to a header at the front of the cotton picker. The harvesting and separation process takes place in the row units. The individual units are attached to the header in such a way as to allow easy adjustment. Hydraulic, electrical, and water connections are made to the row units. The mechanical operations of the units are usually PTO driven. The electrical connections are for fine adjustment of the row units and operation sensors. Each row unit contains a series of vertical drums mounted with a plurality of moistened tapered spindles that pull the cotton from the bolls. As the spindles pass by rubberized "doffer" lugs, the cotton is scrubbed from the spindles. The liberated cotton is then collected at the back of the row unit where forced air is used to blow the cotton out of the unit, up a plastic chute attached to each row unit, and into the basket, which is located immediately to the rear of the operator's station, over the rear portion of the equipment. *[See Figure 25.14.2.4.1(a) and Figure 25.14.2.4.1(b).]*

FIGURE 25.14.2.4.1(a) Elevation View of Internal Components of a Row Unit.

25.14.2.4.2 Moistener System. Unique to cotton pickers is the moistening system, which consists of a water tank, a pump, and distribution system. Depending on the design, the tank may be located either inside the frame, under the basket, at the rear of the unit or immediately behind the operator's station and in front of the basket. The moistener system keeps the spindles moist during operation for more efficient picking of the cotton as it passes through the row unit.

25.14.2.4.3 Hydraulic System. Hydraulic systems operate the row units, drive wheels, braking, steering, and basket lift systems. System components include a large capacity reservoir, pump, manifold valve assembly, hoses, and couplings.

25.14.2.4.4 Lubrication System. Cotton Pickers also contain a separate lubrication system for the row units. This lubrication system consists of a non-metallic lubricant reservoir, a lubrication pump, control box, hoses and couplings. Oil-base lubricant provides an additional fuel load in the picking units in the event of a fire.

25.14.2.5 Sprayers. Sprayers may be self-propelled tractors fitted with solution tanks, booms, and a solution dispensing system specifically designed for chemical solution applications to row crops. Most sprayers have similar characteristics to farm tractors. The primary difference between the two is that sprayers are elevated for adequate ground clearance above maturing crops. Most sprayers are not equipped with PTO attachments. Sprayers may also be found as attachments to standard tractors.

25.14.2.6 Windrowers, Floaters, Spreaders, Fertilizer Applicators. Each of these pieces of equipment is designed for a specific agricultural function with physical characteristics similar to sprayers. Each piece of equipment may also be found as attachments to standard tractors.

25.14.2.7 Baling Equipment. A baler is used to compress a cut and raked crop (such as hay, straw, corn plants, or peanut plants) into round or square bales, which are bound with twine, wire, or synthetic wrap.

25.14.2.7.1 Round Balers. Round balers produce "round" or "rolled" bales of hay or silage. The crop is rolled inside the baler using a series of rubberized belts and fixed rollers. When the bale reaches a determined size, it is bound and discharged from the rear.

25.14.2.7.1.1 Baling Chamber. The baling chamber is located in the center of the unit. The chamber contains a series of rotating rubberized belts, mounted side by side, and routed around a series of fixed and floating rollers, which generate the roll or bale. Additional crop material is added to the outside of the roll. Once the roll is complete, the operator prepares the roll and it is discharged.

25.14.2.7.1.2 Gate. A hydraulic gate is located at the rear of the baler. When opened, a bale is discharged, then closed, and the formation of a new bale begins. Some models are equipped with a hydraulically operated bale ejector or ramp.

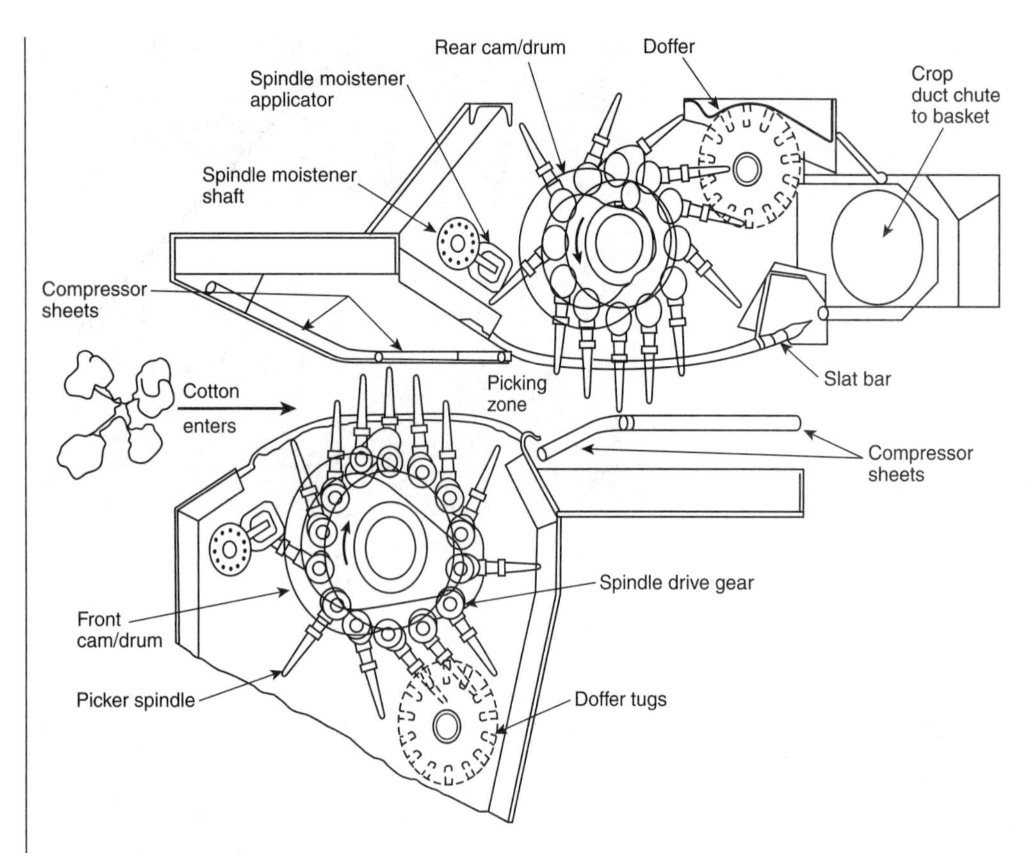

FIGURE 25.14.2.4.1(b) Schematic of a Row Unit.

25.14.2.7.2 Square Balers. A knife cuts the hay where it enters the chamber from the pickup. The mechanical plunger then rams the hay rearwards, compressing it into thin wafers or sections that form the bale. A measuring device measures the appropriate length of the bale and triggers the twine tying mechanism. As the next bale is formed, the finished bale is driven out of the rear of the baling chamber. Some square balers are equipped to hydraulically eject bales out of a chute and into a towed trailer.

25.14.2.7.3 Pickups. Pickups use a series of tines to pick up crop material that has been piled in rows and is located underneath the front of the baler. The pickup, which is driven by a chain and sprocket attached to the main drive sprocket gather the crop material and feed it into the baling chamber.

25.14.2.7.4 Chain Drive Systems. These systems have a series of chains, sprockets, and idler tension pulleys that are used to drive the pickup and the belt and roller system. These components can be found behind lift-up or swing out door panels, or behind bolted inspection panels, which provide access to fixed roller mounting points. The chain drive system is powered through a gear box on the baler by shaft attachment to the tractor's PTO.

25.14.2.7.5 Hydraulic System. Hydraulic pressure is supplied to the baler by the tractor via hydraulic line connections. On board equipment includes hydraulic lines, valves, and connectors.

25.14.2.7.6 Electrical System. Electrical circuits on the baler operate lights and sensors powered and protected by the tractor's electrical system.

25.14.3 Unique Safety Concerns.

25.14.3.1 Fluid Capacities. All fluid capacities tend to be larger on agricultural equipment. Diesel fuel capacities can be several hundred gallons to allow for uninterrupted operation over many hours. Large quantities of combustible liquids may be released prior to or during a fire, which can increase the rate and intensity of the fire. This can potentially obscure fire patterns produced in the early stages of the fire. The fuel tank construction may be non-metallic and its location may be elevated. The construction material of the liquid tank may also contribute to the overall fuel load.

25.14.3.2 Internal Equipment. Due to the construction of this type of equipment, specialized tools or assistance may be necessary in order to access internal areas. The investigator may seek assistance from persons who have experience with the specific make and model of equipment in order to complete the investigation while maintaining preservation of the evidence. Refer to the Legal Considerations chapter for detailed information regarding spoliation of evidence, notification of interested parties, and recommended procedures for destructive testing before any dismantling of the machinery.

25.14.3.3 Overhead Electrical Hazards. The normal height of agricultural equipment may exceed the height of standard-sized vehicles, such as semi trucks. Operational heights may also be greater due to dumping or other reasonably foreseeable operations. Agricultural fields crossed by or bordered by overhead high voltage electrical lines may pose an additional hazard, since these energized lines might be contacted by the agricultural equipment during normal operations. In the event that agricultural equipment has made contact with high voltage electrical lines, the investigator should take all necessary precautions to ensure that the electrical conductor (lines) have been properly de-energized, tested and grounded by qualified electric utility personnel prior to approaching the equipment for investigation. A safety perimeter of at least ten feet as required by OSHA should be established away from the affected equipment and enforced by the investigator until the electrical lines have been properly de-energized, tested and grounded by qualified electric utility personnel.

25.14.3.4 Elevated Basket. Cotton picker fires may occur in the harvested cotton collected in the basket. A typical action for the operator is to attempt to dump the cotton from the basket and move the picker away from the burning cotton. If the operator is unable to complete this action, the basket may be elevated from its normal position on the frame. The picker may be unstable because of this condition or the basket itself may need to be secured before an investigation can be initiated. In some cases, the basket may have fallen to the side of the picker, hindering the investigation of the hydraulic system. Special assistance or equipment may be necessary to gain safe access to the areas underneath the basket.

25.14.3.5 Gate Failure. On round balers, failure of the gate's hydraulic system may result in severe crushing or entrapment hazards. Care should be taken to secure the gate if open.

25.14.3.6 Hydraulic Bale Ejector. Bale ejectors may be found as an option on certain makes of square hay balers. These ejectors are hydraulically powered and controlled by electrical sensors/switches for operation. The hydraulic system is self-contained and pre-charged. The hydraulic pump is driven by the tractor's PTO. Springs may be used to assist in the ejection process. Because the system is hydraulically driven, the ejector is not a safety concern, post fire, unless the unit is placed into operation.

25.14.4 Unique Fire Cause Concerns.

25.14.4.1 Scrapping. Scrapping is the common practice of a secondary harvest of previously picked cotton fields in which scrap cotton, still attached to the boll, is harvested. During this harvest, the plant has been dead for some time and is generally very dry. In this dry condition, water is not needed to assist with spindle cleaning as it is during the primary harvest. The water pump is taken off line to protect the pump from cavitation, over-heating, and warping of the seals. During this operation, excessive amounts of crop residue can collect within the row unit causing friction with the rotating mechanical components or jammed crop residue can force these mechanical components out of alignment, resulting in metal-to-metal contact between spindles or the metal housing components of the row unit. Hot slag generated from this contact may ignite collected cotton dust or fine crop residue and be blown into the basket, igniting the harvested cotton.

25.14.4.2 Foreign Combustible Materials. The collection and build-up of foreign combustible materials may occur in various locations on agricultural equipment. These materials include combustible vegetative debris or residue, which may collect in or around exhaust system components and turbochargers. It may also collect in air filters, clogging the filter and causing higher than normal operating temperatures. Vegetative debris or residue may also collect in and around belts, chains, sprockets, rotating shafts, and other moving parts, leading to ignition by frictional heating. The presence of foreign combustible materials may become increasingly important as a fire risk with increased engine component operating temperatures necessitated by changes in emissions requirements. *(See Figure 25.14.4.2.)*

FIGURE 25.14.4.2 Collected Foreign Combustible Material on a Cotton Picker.

25.14.4.3 Maintenance. Agricultural equipment requires routine maintenance as outlined in the Equipment Operators' Manual. If the pre-determined maintenance schedule is not followed, an increase in the potential for fire may result. The type of crop to be harvested, environmental conditions at the time of harvest, and the time available to the farmer to complete the harvest all impact the farmer's ability to maintain and adhere to an acceptable and appropriate service and maintenance schedule. Interviews conducted with the equipment owner and/or operator(s) and a review of maintenance records (if available) are helpful to establishing any relevance of the involved equipment's maintenance history to the origin or cause of the fire.

25.14.5 Fuels. Many of the fuels that are found on agricultural equipment are similar to those found in automobiles and trucks. Please refer to the Fuels in Vehicles section of this chapter for detailed information on the various fuels. Agricultural equipment carries larger quantities of various ignitable liquids to: a) allow the equipment operator to cover more ground over a longer period of time between refueling stops; b) provide adequate lubrication for moving and rotating components; or c) provide adequate hydraulic fluid quantities to maintain pressure for multiple, simultaneous functions.

25.14.5.1 Diesel Fuel. Diesel fuel is the primary ignitable liquid used for agricultural equipment engines.

25.14.5.2 Biodiesel. Biodiesel is an alternative fuel, produced from, renewable resources for use in common diesel engines. Biodiesel itself contains no petroleum, but can be blended at any level with petroleum to create a biodiesel blend. Agricultural equipment may be designed to run on biodiesel fuels. Suitable blends contain a range of 2 percent–20 percent biodiesel and 80 percent to 98 percent petroleum diesel. B20 (20 percent biodiesel and 80 percent petroleum) has similar performance characteristics to No. 2 diesel fuel. B100 (100 percent biodiesel) generates slightly lower energy production than petroleum diesel (118,170 BTU/gal. vs. 129,050 BTU/gal.) and has a higher flash point than petroleum diesel (100°C–170°C vs. 60°C–80°C). (Footnote: Biodiesel Handling and Use Guidelines, Third Edition. September 2006. US Dept. of Energy) Biodiesel is lighter than water, insoluble, slightly heavier than air, and mildly volatile. If a biodiesel blend is suspected of being used, the same safety precautions should be taken as those for diesel fuel. The investigator should be aware of the possibility of self-heating when the bio diesel fuel has been in contact with rags or similar materials.

25.14.5.3 Plastics. Plastics may be found as trim, covers, or enclosures of most agricultural equipment, particularly those with enclosed cabs. Most of the rigid plastics are used for engine enclosures and wheel covers, and in the operator's station for trim panels, control panels, knobs, buttons, etc. Fuel reservoirs, water, and solution, tanks may also be constructed of plastic and can add to the over-all fuel load of the equipment.

25.14.5.4 Composite Materials. Body panels and other user interface components may be manufactured from fiberglass or other similar composite materials. Paints and resins used in the manufacture of these components may add to the over-all fuel load of the machine. The key concern is that these components may be completely destroyed, altering the significance of remaining fire patterns to sequential fire pattern analysis. Care should be taken by the investigator to recognize the presence of composite materials and consider the impact of their combustion in the post-fire evaluation of indicators of fire origin or spread.

25.14.5.5 Rubber. Rubberized components found on agricultural equipment include tires, tracks, belts, hose material, and doffer pads (cotton pickers). Of these components, tires and rubber tracks are the more significant in terms of fuel load. Agricultural tires may be very large and are constructed for heavy duty use. It is not uncommon to arrive at an agricultural equipment fire to find residue of tires burning or smoldering, several days after the intial fire. Smoke production from a burning or smoldering tire may interfere with the investigation and may require additional suppression efforts before proceeding.

25.14.6 Ignition Sources. As with other motorized equipment, many of the same ignition factors and scenarios exist with agricultural equipment.

25.14.6.1 Electrical Sources. Agricultural equipment may be constructed with a single or multiple battery system. Additional lighting or accessories may demand larger charging or reserve electrical capacities. Refer to the Electrical Sources and Motor Vehicle Electrical System sections of this chapter.

25.14.6.2 Hot Surfaces. Exhaust manifolds, turbochargers, and other related components are generally considered the primary concern for hot surface ignition of combustibles or ignitable liquids in agricultural equipment. Much of what is written in the hot surfaces section of this chapter will also apply to agricultural equipment. Some differences exist such as the configuration of exhaust system components which may lead to the build-up of crop residues. Failure to maintain manufactures required shielding or clearances between hot surfaces and combustible materials may result in ignition. Changes in emissions technology will create other hot surfaces with additional clearances requirements to combustibles.

25.14.6.3 Mechanical Sparks. Mechanical sparks can be created by foreign metals picked up and passed through row units on cotton pickers. Forced mis-alignment from jammed crop residue or from equipment failure that allows spindles to have direct contact with compression plates or other steel components can also generate mechanical sparks. Spindles, which are made of hardened steel with a chrome/alloy coating, can break and be passed through the unit. Mechanical spark events are of a short duration and are a minimally-viable source of ignition. The heat energy of mechanical spark particles is generally insufficient to ignite ordinary combustible materials except in rare circumstances. It may be possible to cause smoldering combustion in fine accumulations of dust. Mechanical sparks may ignite cotton dust or fine crop residue at the base or rear of the row unit. The cotton dust may smolder or flame as a result of forced air injected through the compartment. Harvested cotton passing through the area may then be ignited and blown up into the basket where the cotton continues to smolder or burn.

25.14.6.4 Friction. Friction occurs with the application of brakes or the normal operation of a slip clutch. Friction heating may occur as a result of improper application of a component function, poor maintenance, inadequate or improper repairs, or part failure. The heat generated may be quite substantial, depending on the type of materials involved, speed of the action between the materials, and the ability of the generated heat to dissipate. Sufficient friction heat may be generated to ignite nearby combustibles, which could include tires, rubberized belts, hoses, lubricating grease, or vegetative debris or crop residue.

25.14.6.5 Brakes. Agricultural equipment may be equipped with disc or drum brakes. Some equipment may be equipped with park brakes. Friction heat generated through normal braking operations is easily handled by the design of the system and the types of materials used. When brakes are misapplied, excessive friction heat occurs. Misapplication of brakes can occur in situations where the equipment operator fails to release the park brake before moving the machine, repeatedly uses brakes for steering, or improper replacement/adjustment of brake components.

25.14.6.6 Slip Clutch. A slip clutch is a safety device which utilizes a friction plate or disc in a coupling configuration with a rotating drive shaft. Excessive forces on the drive system, such as that caused by "plugging" a machine with too much crop material will cause the clutch to "slip" thereby preventing damage to drive components. A slip clutch must be torqued to proper specifications or the clutch may slip too frequently or not at all. Friction heat from a "slipping" clutch may be a competent ignition source in the presence of dry crop residue.

25.14.6.7 Rotating Shafts. Rotating shafts can be a source of friction heat, particularly in situations where the equipment has not been regularly cleaned of vegetative debris or residue. When foreign debris becomes packed into areas around these shafts, the normal operational rotation of the shaft can generate sufficient friction heat to ignite the foreign debris.

25.14.6.8 Rotating Bearings. Rotating bearings require regular lubrication via a grease fitting. Other bearings are sealed and are permanently lubricated. Rotating bearings may fail from a lack of lubrication or excessive lubrication, which results in failure of a grease seal. Sealed bearings can fail because of insufficient lubrication during manufacture or from physical damage to the bearing from outside force such as contact with a foreign object. Bearings also fail for other reasons: shaft or roller misalignment, improper installation, and broken mounting bolts. Loose or frozen bearings are both indicators of bearing failure. Particular interest should be given to developing a maintenance and use history for the involved equipment if bearing failure or excessive friction is a concern. *[See Figure 25.14.6.8(a) and Figure 25.14.6.8(b).]*

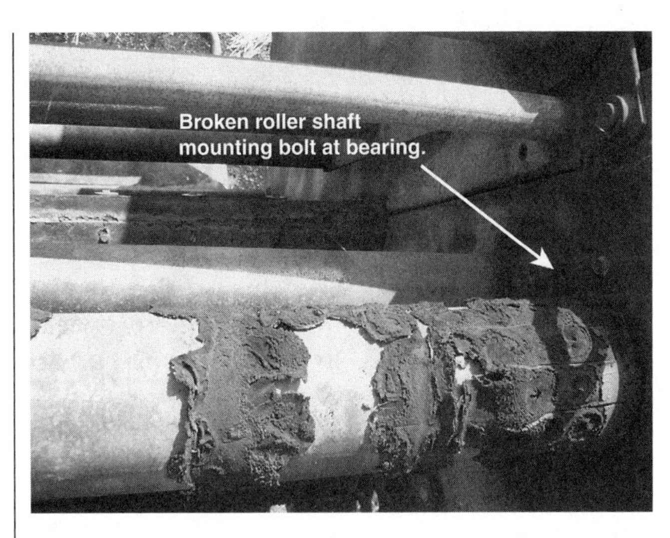

FIGURE 25.14.6.8(a) Failed Bearing Mounting for a Rotating Shaft on a Round Hay Baler.

FIGURE 25.14.6.8(b) Recovered Head of Shaft Mounting Bolt for Roller Shown in Figure 25.14.6.8 (a).

25.14.6.9 Electronics and Aftermarket Equipment. Much like automobiles and trucks, farm machinery now incorporates a number of technologically advanced electronics in the operator's station. In addition, farmers and operators may also incorporate any number of after-market electronics into the operator's station for either business or pleasure. Agricultural equipment such as combines and cotton pickers now utilize a number of technologies to assist with harvesting. GPS integrated computer systems are utilized to assist with auto-guidance accuracy, yield and moisture mapping, spraying, prescription planting, and maximizing productivity for optimal crop yields. Data loggers and/or monitors are fast becoming common equipment. In certain types of equipment, electronic, joystick controls are used in conjunction with computer aided guidance systems for precise handling. Such electronic components utilize low-current circuits protected by fuses or circuit breakers. Most cab-equipped units will also have as options an audio system or two-way communications.

25.15 Hybrid Vehicles. The investigator is cautioned to obtain the information specific to the hybrid vehicle prior to the investigation.

25.15.1 Hybrid Vehicle Investigation Safety. As a safety precaution, the investigator should approach the hybrid vehicle as though the high voltage system is energized. Before inspecting a hybrid vehicle the investigator should be familiar with the high voltage system. Most hybrid vehicles will have a manual disconnect means to isolate high voltage to the battery pack. Investigators should ensure that the disconnect is in the isolation position before beginning any physical inspection. As an additional precaution, a voltmeter should be used to check whether high voltage is present on any suspect wiring or component.

WARNING: *Because the high voltage battery potential could range up to 600 volts, opening the battery pack could be extremely hazardous and should not be attempted by untrained personnel.*

25.15.2 Hybrid Vehicle Technology. Hybrid vehicles contain as means of propulsion both an internal combustion engine and an electric drive system. The electrical system of a hybrid vehicle will consist of a high voltage system used for traction and a regular automotive 12-volt system used for all other electrical loads. The high voltage system (up to 600 volts) will consist of a high voltage battery, a high voltage traction motor, a generator for charging the battery from the internal combustion engine, an electronic controller, and a converter for stepping voltage between the high and low voltage systems. Hybrid vehicles also contain a standard 12-volt automotive battery. Hybrid vehicles can have an integrated alternator and starter motor as part of the drive train. As hybrid technology evolves, other automotive loads, such as power steering and air conditioning, may become powered from the high voltage bus. The high voltage system does not use the vehicle chassis for ground return and so there will be two conductors to every high voltage load. It is standard to reserve the color orange for all high voltage wiring. The high voltage system will most likely be protected by an automatic disconnecting means that may be activated by any number of events including KEY OFF, detection of a ground path, detection of an open high voltage connector, sensing of a crash, etc. Most high voltage systems will also contain a circuit protection device as part of the battery pack to protect against a direct short circuit.

25.15.3 Investigation of Hybrid Vehicle Fires. The investigation should proceed as usual for vehicles except that the high voltage system should be considered as a potential ignition source. Investigators should obtain vehicle-specific information regarding the type and location of all high voltage components, the manufacturer's design to reduce the risk of unintended electrical energy release from the high voltage system, and whether that system has been compromised. The investigator should determine if the high voltage electrical system is intact and whether the circuit is open or closed. It should be determined if the circuit protection device is on, off, tripped, or damaged.

25.16 Towing Considerations. Pre- or post-crash towing or vehicle transport can result in mechanical damage visible at the time of the fire investigation. Sometimes this damage involves the fuel system of the vehicle. Such fuel system damage might include tank puncture or grind-through, hose clamp loss, or fuel line separation. To avoid the potential loss of evidence, the vehicle can be wrapped prior to removal from the scene.

25.16.1 The investigator should identify and document the nature of post-incident damage. Damage to metallic parts that exposes fresh metal indicates that the damage was likely inflicted after the fire was extinguished. Damage to the area surrounding the damage of interest may also indicate the nature of the damage and the time at which it occurred.

25.16.2 Pre- or post-crash towing or vehicle transport can result in fire initiation. Non-damaged vehicles, such as a semi-trailer, may initiate a fire due to cargo, a malfunctioning power unit, or an electrical condition. Frictional heat aspects, such as from brake drag due to residual air or hydraulic system pressure, or crash-applied (parking) brakes, can be misidentified as an ignition source. Be aware that towing, moving, or lifting a vehicle can be the source of damage to the exterior of the vehicle. It can also lead to fuel system component displacement as well as vehicle content relocation of concern during a fire investigation. Any such condition noted should be fully documented with photographs and notes.

25.17 Hydrogen-Fueled Vehicles. The investigator is cautioned to obtain information specific to the hydrogen-fueled vehicle prior to the investigation.

25.17.1 Hydrogen-Fueled Vehicle Investigation Safety. The investigator should be aware of the potential hazards associated with hydrogen vehicles, such as the extremely wide flammability range (4 percent–75 percent), ease of ignition, and high-voltage battery potential, and proceed with caution. The hydrogen systems are often maintained at high pressures. Immediately after a fire or crash, the hydrogen tank or downstream fuel system may be venting fuel. The escaping hydrogen from high pressure systems may auto-ignite. If the tank or fuel system is damaged due to a crash or fire it can result in a hazardous condition, and a hydrogen release may occur at a later time. An investigation should not commence until all venting of the hydrogen is completed or the fuel system is stabilized. Specific vehicle models may have an electronic or mechanical means for safely venting the hydrogen before commencing a vehicle fire investigation. Hydrogen has a very low ignition energy of 0.018 mJ and can be ignited by an energy source such as a static electric discharge *(See Table 25.3.2)*. Hydrogen burns with a pale blue, almost invisible flame that can be difficult to detect. It diffuses rapidly and is lighter than air, but hydrogen has a large range of flammability that may create a safety concern where hydrogen may have collected. Ignition of hydrogen may result in an explosion under some conditions and leaks in confined spaces may be hazardous. Hydrogen vehicles can also be battery hybrids, and so all of the precautions for high voltage electrical safety from the Hybrid Vehicle section apply.

25.17.2 Hydrogen-Fueled Vehicle Description. Hydrogen-fueled vehicles may be powered by internal combustion engines or by a fuel cell powering an electric drive.

25.17.2.1 A fuel cell powered vehicle utilizes electrical power to drive the vehicle and accessories such as the air conditioning compressor and the power steering pump. The fuel cell and major electrical components are cooled by a system similar to those used to cool internal combustion engines. The surface temperatures of under-hood components are typically lower than those associated with internal combustion engine vehicles.

25.17.2.2 The manufacturer's manuals should be consulted for fuel system configuration details. Hydrogen vehicles, parking garages, and fueling stations may have hydrogen sensors designed to set off an alarm when leaking hydrogen is detected. Some vehicle manufacturers have proposed using reformers to supply the source of hydrogen. A reformer uses chemical processes to free hydrogen from a standard fuel such as gasoline. Vehicles of this type will most likely have a gasoline fuel system as well. Hydrogen vehicles should have standardized markings as recommended by SAE J2578, with a diamond label consisting of white letters on a blue background, which caption: "CHG" or "LH2," referring to compressed or liquid hydrogen respectively *(see Figure 25.17.2.2)*.

FIGURE 25.17.2.2 Hydrogen Labels.

25.17.2.3 Vehicles fueled with hydrogen may store the hydrogen as a high-pressure compressed gas (34.5 MPa – 69.0 MPa [5,000 psi – 10,000 psi]), in a moderate pressure tank (up to 10.3 MPa [1500 psi]) as a hydride, or in the form of a cryogenic liquid at low pressure. These storage devices and the downstream fuel system will be protected by one or more pressure relief devices (PRDs) or rupture disks. The devices may be actuated by either high pressure or high temperature resulting from a vehicle fire. There will also be one or more pressure regulators to reduce the pressure of the hydrogen provided by the storage vessel pressure to the fuel cell or the internal combustion engine. Additionally, there will be a fill line, one or more electrically actuated normally closed shut-off valves, and vent lines for the PRDs and the fuel cell exhaust.

Chapter 26 Wildfire Investigations

26.1* Introduction. Wildfire investigation involves specialized techniques, practices, equipment, and terminology. While the basic principles of fire science and dynamics are the same in a wildfire, the fire development and spread is influenced by different factors such as wildland fuels, fire weather, topography, and unconfined burning.

26.1.1 The purpose of this chapter is to identify and explain those aspects unique to wildfire investigation. This chapter is intended as a basic introduction and the user is urged to consult the reference material listed in Annex A and Annex B. As with other types of fire investigation covered in this guide, specialized personnel may be needed to provide technical assistance.

26.2 Wildfire Fuels. Keen observation of variations in wildfire fuels is essential to accurately analyze fire behavior.

26.2.1 Fuel Condition Analysis. In a wildland, great differences exist in the character of flammable materials. Deep duff, newly fallen dead leaves, clumps of grass, litters of dry twigs and branches, downed logs, low shrubs, green tree branches, hanging moss, snags, and many other types of material may be present. Each of these materials has distinctive burning characteristics. The flammability of a particular fuel matrix is governed by the burning characteristics of individual materials and by the combined effects of the various types of materials present.

26.2.1.1 Before fuel condition can be analyzed, the physical characteristics of combustible wildland materials must be classified. Such a classification permits the identification of the fuel factors that influence flammability. After the fuel has been classified properly, topographic and weather factors must be considered before the rate of spread and the general behavior of fires in that fuel can be determined.

26.2.1.2* Forest fuels are varied and complex therefore it is necessary to develop a systematic approach to fuel condition analysis. First, the fuel matrix is subdivided into three broad vertically arranged categories: ground, surface, and aerial fuels. Wildland fire investigators must concern themselves with how fuels are stratified, their physical characteristics, and the fuel quantity in the general origin area to determine the fire behavior context. Within these broad categories, fuels are broken down into four major fuel groups. These groups are described as grass, shrub, timber litter, and logging debris. Each group is further divided into fuel-type models that are an identifiable association of fuel elements for distinctive species, fuel form, size, and arrangement. These characteristics determine a predictable rate of spread or resistance to control under specified weather conditions. Finally, fuels are also classified according to their size as it relates to their fuel moisture retention characteristics. Dead fuels are grouped according to 1, 10, 100 and 1000 hour time lag fuels and living fuels are grouped as herbaceous (either annual or perennial) or woody.

26.2.2 Ground Fuels. Ground fuels include all flammable materials located between the mineral soil layer and the ground surface. These fuels typically include twig, leaf and needle litter, and decomposing vegetation such as duff, peat moss, buried limbs and roots. Buried limbs and roots can burn along their entire length and ignite a surface fire in a different location. When sufficiently dry, ground fuels can be a very receptive fuel bed for a wide variety of ignition sources due to their high surface area-to-volume ratio. They can also smolder for long periods of time before transitioning to open flame. Ground fuels by themselves are not typically associated with rapid fire progression; however they can contribute to significant long term fire residency, depending upon the depth of the materials.

26.2.2.1 Duff. Duff seldom has a major influence on the spread rate of fire because it is typically moist and tightly compressed so that little of its surface is freely exposed to air, and its rate of combustion is slow. In forest fires, most of the duff can be consumed down to mineral soil. Occasionally, duff contributes to the rate of spread by furnishing a path for the fire to creep along between patches of more flammable material.

26.2.2.2 Roots. Roots are not an important factor in rate of fire spread, as the greatly restricted air supply prevents rapid combustion. However, fires can burn slowly in roots. Some fires have escaped control because a root provided an avenue for the fire to cross the control line. Large roots from dead and partially decayed vegetation will more readily spread fire than the roots of live vegetation.

26.2.3 Surface Fuels. Surface fuels are those flammable materials located from the surface of the ground to approximately 2 m (6 ft) above the surface. Surface fuels include grasses, leaves, twigs, needles, field crops, slash and downed limbs. Surface fuels in the one-hour time lag fuel moisture category are the most common materials first ignited. Surface fuels contribute to rapid fire propagation and the rate of fire spread. They also serve as the primary fuel ladder to aerial fuels.

26.2.3.1 Fine Dead Wood. Fine dead wood consists of twigs, small limbs, bark particles, and rotting material. Normally, the fine dead wood classification is confined to material with a diameter of less than 25.4 mm (1 in.). These fuels are included in the 1 and 10 hour time lag fuel category. These fine dead surface fuels are among the most important of all materials influencing the rate of fire spread and general fire behavior in forest areas. Fine dead wood ignites easily and often provides the main avenue for carrying fire from one area to another. It is the kindling material for larger, heavier fuels.

26.2.3.1.1 In areas where a great volume of fine dead wood exists, a fire can develop rapidly. The greatest volume of fine dead wood is usually found in areas containing logging slash. Under dry conditions, the strong convection currents created by the intense heat transport burning embers and carry them out ahead, causing spot fires beyond the main fire front.

26.2.3.1.2 Granulated dry rotten wood, while not an especially important factor in the rate of fire spread, is a highly ignitable fuel. Embers from the main fire often cause spot fires in rotten wood lying on the ground or in hollow places on old logs or stumps.

26.2.3.2 Dead Leaves and Coniferous Litter. As leaves and coniferous litter decay on the ground, they gradually become part of the duff layer. Before this decay takes place, however, leaves and coniferous litter are a highly flammable material and should be considered surface fuels not ground fuels. In many forests, lower level surface fuels may be composed primarily of needles dropped from coniferous trees. Ponderosa pine needles, for example, are extremely flammable because their large size and shape create a jumbled arrangement, allowing free circulation of air. Smaller needles, like those of Douglas fir, generally burn less intensely, as they are more tightly compacted.

26.2.3.2.1 Needles that are still attached to dead branches are especially flammable because they are exposed freely to air and are not typically in direct contact with the more moist material on the ground. Needles remaining on fallen limbs form highly combustible kindling for larger material. For this reason, logging slash containing dry needles is dangerous fuel.

26.2.3.3 Grass. Grass, weeds, and other small annual or perennial plants are important surface fuels that influence rate of fire spread. The key factor in these fuels is the degree of curing. Succulent green grass acts as a fire barrier. During the course of a normal fire season, however, grass gradually becomes drier and more flammable as the plant cures or as the stems and leaves die due to lack of moisture. At this time, the grass cover becomes easily ignitible. Cured grass, if present in a large and uniform volume, becomes the most flammable, fastest spreading surface fuel. Grass and other small plants occur on the floor of almost all forests. Fire investigators need to determine the volume and continuity of the grass cover. In dense forests where little sunlight reaches the ground, very little grass is found. In more open forests, such as in mature stands of pine, there may be a large amount of grass fuel. If there is a more-or-less continuous cover of dry grass on the forest floor, the spread rate of a fire will be governed largely by that cover, rather than by the heavier fuels normally associated with a forest. Fires in dry grass often have high rates of spread.

26.2.3.4 Downed Logs, Stumps, and Large Limbs. Heavy fuels, such as downed logs, stumps, and large limbs comprise the 100 and 1000 hour time-lag fuel category and therefore require long periods of hot, dry weather before they become highly flammable. When such material reaches a dry state, however, high intensity fires may develop. The most dangerous heavy fuels are those containing stringers of dry wood, or many large checks and cracks. Smooth-surfaced material is less flammable, as it dries out more slowly, has little surface exposed to air, and contains less attached kindling fuel.

26.2.3.4.1 Sustained high intensity fires may develop in piles of downed logs and large limbs or in crisscrossed windfalls, as the various fuel components radiate heat to each other. Isolated individual limbs and logs will not burn very intensely unless the fire is supported by large accumulations of fine dead wood.

26.2.3.5 Low Brush and Reproduction. Low brush, tree seedlings, and small saplings are classified as surface fuels. This understory vegetation may either accelerate or slow down the spread rate of a fire. During the early part of a fire season, the shade normally provided by understory vegetation prevents other surface fuels from drying out rapidly. As the season progresses, however, continued high air temperature and low relative humidity dry out both the fuel lying on the ground and the understory vegetation. When this happens, most of the low vegetation, particularly small coniferous trees, become contiguous fuel that sustains fire spread.

26.2.3.5.1 The understory vegetation in a forest often provides a link between surface fuels and aerial fuels. The crowns of small trees may catch fire and, in turn, spread the fire to aerial fuels in the forest canopy. Either thickets of tree reproduction or dead brush may provide the first means for a surface fire to flare up and spread into the crowns of the overstory trees.

26.2.4 Aerial Fuels. Aerial fuels are those flammable materials located from approximately 2 m (6 ft) above the surface to the crowns of the canopy. These fuels include tree branches, leaves, needles, snags, moss, and tall brush. These fuels are only infrequently the materials first ignited and typically require significant amounts of heat from surface fuels. Combining steep slopes and/or higher wind speeds can easily transition the fire to a running crown fire. Aerial fuels can contribute to rapid fire spread, primarily through the generation of aerial firebrands.

26.2.4.1 Tree Branches and Crowns. The live needles of coniferous trees are a highly flammable fuel. Their arrangements on the tree branches allow free circulation of air. In addition, the upper branches of trees are more freely exposed to wind and sun than many surface fuels. These factors, plus the volatile oils and resins in coniferous needles, make tree branches and crowns important components in aerial fuels.

26.2.4.1.1 Tree branches and crowns are fuels that can ignite quickly with decreased relative humidity. Crown fires seldom occur when relative humidity is high. However, coniferous needles dry out quickly when exposed to hot, dry air. The dryness of needles is influenced by the transpiration process in a tree. When the ground is moist, trees release a large amount of moisture into the air through the leaves. As the ground becomes drier, the transpiration process slows, and, as a result, leaves and branches become drier and more flammable.

26.2.4.1.2 Dead branches on trees are an important aerial fuel. Concentrations of dead branches, such as those found in insect or disease killed stands, may enable fire to spread from tree to tree. Concentrations of dead branches on the lower trunks of trees may provide an additional avenue for fires to spread from surface fuels to aerial fuels. The most flammable dead branches are those still containing needles.

26.2.4.2 Tree Moss. Moss hanging on trees is the lightest and the most apt to ignite of all aerial fuels. Moss is important principally because it provides a means of spreading fires from surface fuels to aerial fuels or from one aerial component to another. Like other light fuels, moss reacts quickly to changes in relative humidity. During dry weather, crown fires may develop easily in heavily moss-covered stands.

26.2.4.3 High Brush. Crowns of high brush, above 2 m (6 ft.), are classified as aerial fuels because they are separated distinctly by distance from ground and surface fuels. In many forest regions, heavy stands of brush may develop in old burns, and they often form the principal vegetative cover in such areas. Crown fires in brush fuels ordinarily do not occur unless heavy surface fuels are present to develop the required heat. In some brush stands, however, a high proportion of dead stems may create a sufficient volume of fine dead aerial fuels to permit very hot and fast spreading crown fires. Key factors in evaluating the behavior of fires in high brush are volume, arrangement, the general condition of surface fuels, and the presence of fine dead aerial fuels.

26.2.5 Species. The species of vegetation involved can determine the rate of spread and intensity of the fire. Each type of vegetation has different characteristics, that is, size, moisture content, shape, and density.

26.2.6 Fuel Size. A major factor governing both the ignitibility and burning rate of a fuel is its diameter. The smaller the fuel element (having a larger surface area-to-mass ratio), the easier it will be to ignite and the faster it will be consumed. These smaller fuels are classified as fine fuels and are comprised of such items as seedlings and small trees, twigs, dry grass, brush, dry field crops, and pine needles and cones. Larger-diameter fuels, classified as heavy fuels, are much more difficult to ignite, and burn at slower rates than that of light fuels. Most often, it requires burning light fuels to serve as kindling to ignite the heavier fuels. Some examples of heavy fuel classifications are large-diameter trees and brush, large limbs, logs, and stumps.

26.2.7 Fuel Moisture Content. The amount of moisture present in the fuel plays a major role in determining the ignitibility and the rate of fire spread. As the vegetation (fuel) dries out, it becomes more readily ignitible and will burn with greater intensity. Green vegetation, or vegetation having high moisture content is more difficult to ignite and will burn more slowly due to the moisture within the vegetation requiring more heat to evaporate that moisture. Once the moisture is evaporated, the temperature will continue to rise to the fuel's ignition temperature. The moisture content of the fuel will vary, depending on the type and condition of the vegetation, solar exposure, weather, and geographic location. Fuel moisture content in dead fuels is rated by four broad time lag categories. This is an indication of the rate a fuel gains or loses moisture due to changes in its environment, or the time necessary for a fuel particle to gain or lose approximately 63 percent of the difference between its initial moisture content and its equilibrium moisture content. Dead fuels are grouped into four classes based on their size: 1-hour fuels, less than ¼ in. (0.64 cm) in diameter; 10-hour, ¼ to 1 in. (0.64 to 2.5 cm) in diameter; 100-hour, 1 to 3 inches in diameter; and 1000-hour, 3 to 8 in. (7.6 to 20.3 cm) in diameter.

26.2.8 Oil Content. The presence of oil within vegetation typically increases the fuel's ease of ignition, fire intensity, and spread factors.

26.3 Weather. Weather plays a substantial role in the behavior of wildfires. Weather elements can be described as the state of the atmosphere with respect to atmospheric stability, temperature, relative humidity, wind velocity, cloud cover, and precipitation.

26.3.1 Weather History. An important item to consider during an investigation is the weather history. Weather history is a description of atmospheric conditions over the preceding few days or several weeks. Weather elements should be analyzed to determine what influences they may have had on the fire's ignition and burning characteristics.

26.3.2 Temperature. The temperature of the ambient air directly influences the temperature of the fuel. The sun is one factor that affects temperature. As the radiant solar energy of the sun heats the ground and vegetation, the fuel becomes more susceptible to ignition. Temperature differences of more than 10°C (18°F) can occur between shaded areas and areas exposed to direct sunlight. Another factor influencing temperature is altitude. The lower air pressure at high altitudes allows the air to expand and cool.

26.3.3 Relative Humidity. Humidity is the measure of water vapor in the air. Humidity is usually expressed as relative humidity, which is the ratio of the amount of moisture in the air to the amount that a known volume of air at that particular temperature can hold, expressed as a percentage. The moisture in the air directly affects the amount of fuel moisture and vice versa. Dry air will draw moisture out of the vegetation, making it more susceptible to fire. Fine fuels are more responsive to relative humidity than are large fuels. If the air has a high relative humidity, the vegetation will take in some of that moisture, making it less susceptible to ignition, or the moisture may slow the rate of spread of a fire in progress. Warm air can hold more water than can cool air. This is illustrated by early morning dew. As the night air cools, the air loses its ability to hold moisture. The moisture condenses at the air's dew point and forms the dew.

26.3.4 Wind Influence. Wind greatly affects the speed at which a fire will spread. Wind pushes the flame ahead, resulting in a preheating of the fuel. The wind also assists in drying out vegetation, increasing the ease of ignition. Wind further results in the transport of airborne firebrands that have been carried aloft by the heated convective air column. The accompanying fire wind can blow embers and hot sparks ahead of the main fire to ignite secondary fires in areas of unburned fuel. The different types of wind influencing fire behavior are classified as meteorological, diurnal, foehn, and fire winds.

26.3.4.1 Meteorological Winds. Meteorological winds are caused by atmospheric pressure differentials in upper level air masses that generate regional weather patterns. The earth's rotation, oceans, and topographic features create these major air movements to form wind and pressure belts.

26.3.4.2 Diurnal Winds. Diurnal winds are formed by solar heating and nighttime cooling. As air warms during daytime, the rising air creates upslope and up valley winds. When air cools after sunset, this denser, heavier air sinks and causes downslope and down valley winds.

26.3.4.3 Foehn Winds. Foehn winds are a dry, downslope wind that is formed as a result of air flow between pressure gradients. As the air flows downhill, it is accelerated by gravity and is heated through the process of adiabatic compression. Foehn winds can raise temperatures by as much as 30°C (54°F) and drop relative humidity significantly in just a matter of hours. These winds are also frequently channeled through mountain passes and drainages which further increases their velocity. It is not uncommon to see wind speeds in excess of 70 mph associated with this phenomenon. In the western United States, the most well-known Foehn winds are the Santa Ana in Southern California and the Chinook winds associated with the Rocky Mountains.

26.3.4.4 Fire Winds. Fire winds are caused by the fire itself. These winds are the result of the rising expanding fire plume. Fire winds influence the spread of the fire.

26.4 Topography. Topography relates to the form of natural or man-made earth surfaces. Topography affects the intensity and spread of a fire. Winds, as well, are greatly affected by the topography of the land.

26.4.1 Slope. The slope of an area is determined by measuring the rise over run, or the change in elevation over a given distance. Zones on a given slope are often described as top third, middle third, and bottom third. Slope allows the fuel on the uphill side to be preheated more rapidly than if it were on level ground as the distance between the flame and the fuel is decreased. This allows the fire to burn with more intensity and speed. Wind currents, which prevalently move uphill during the day, can accelerate the fire uphill. Slope may also result in burning debris rolling down the hill, starting spot fires below the primary fire and across control lines.

26.4.2 Aspect. The aspect of the slope is the compass direction in which the slope faces. This is an important consideration because solar heating on the fuel and ground surface raises the potential for ignition and increases the rate of spread. Slopes facing the sun are typically drier, and they may have a more combustible fuel type or character of vegetation resulting in a greater ease of ignition and faster spread. Sun angle and the amount of radiant energy a particular area receives daily or over a season has a direct affect on forest fuels.

26.5 Fire Shape. The parts of a wildland fire are generally described in relation to the type of fire spread occurring at that portion of the perimeter as illustrated in Figure 26.5.

26.5.1 Fire Head. The portion of a fire that is moving most rapidly is called the fire head. The direction the local wind is blowing while the fire is burning primarily determines the route of the head's advance, subject to influences of slope and other topographic features. Large fires burning in more than one drainage or fuel type can develop additional heads. The

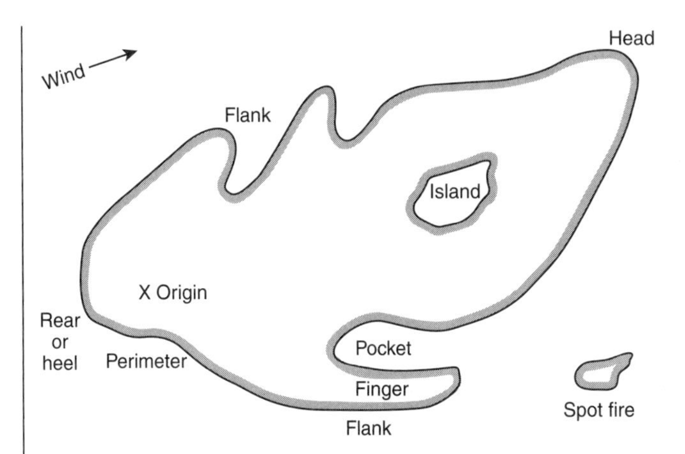

FIGURE 26.5 Anatomy of Fire Showing Fire Head and Heel (Rear).

head is generally the area of most rapid fire spread and greatest fire intensity. Fire spread at the head of the fire is referred to as advancing fire.

26.5.2 Fire Flanks. The fire flanks are located on either side of the head. These are the parts of a fire's perimeter that are roughly parallel to the main direction of fire spread. Fire progression on the flanks of the fire is characterized by less intense fire behavior than at the head of the fire. Fire spread on the flanks of the fire is referred to as lateral fire.

26.5.3 Fire Heel. The fire heel is located at the opposite end of the fire from the head. The fire at the heel is less intense and is easier to control. Generally, the fire at the heel will be backing or burning slowly against the wind or downslope. Fire spread at the heel of the fire is referred to as backing fire.

26.5.4 Factors Affecting Fire Spread. There are numerous factors to be considered when assessing the shape of a wildland fire in order to help locate the general origin area. The major factors are the wind speed, wind direction and the slope. These factors relate directly to the resulting speed and direction of fire spread and create the largest fire effects resulting in the easiest indicators to read.

26.5.4.1 Lateral Confinement. When wildfires are confined by landforms such as gullies, ravines, or narrow valleys, convective heating by confined gases and radiation feedback from flames and burning vegetation increases the heat release rate of the burning fuels. Rapid fire spread is also enhanced by the acceleration and channeling of wind through these topographical features. These factors may result in a more rapid combustion and spread than that of an unconfined vegetation fire.

26.5.4.2 Fuel Influence. Following wind and slope, fuel type and characteristics provide the third greatest influence on the rate of spread and fire intensity.

26.5.4.3 Suppression. Fire suppression is the combination of all activities that lead to the extinguishment of the fire. Suppression includes everything from the initial stage of fire discovery to the final stage of completely extinguishing the fire. Protection by fire crews of potential areas of origin is of extreme importance in establishing the fire origin and cause. It may be useful for the investigator to review and analyze the fire suppression tactics and their effects on spread to assist in identifying the fire origin.

26.5.4.3.1 Fire Breaks. Fire breaks, fire lines, or control lines are any natural or man-made barriers used to slow, stop, or reroute the direction of the fire spread by separating the fuel from the fire. Natural fire break examples are bodies of water, cliffs, and areas lacking vegetation or areas where the fuel moisture is higher than that of surrounding area. Examples of man-made breaks include roads, fire lines, pre-burned areas, and barriers of water, retardant, or foam.

26.5.4.3.2 Air Drops. An air drop is the aerial application of water or retardant mixture directly onto the fire, onto the threatened area, or along a strategic position ahead of the fire to stop or slow the spread of fire. Air drops may alter fire indicators in or near drop zones.

26.5.4.3.3 Firing Out. Firing out is the process of burning the fuel between a fire break and the approaching fire to extend the width of the fire barrier. These fires are normally started at a fire control line (normally on the downwind side of the fire) and are burned back toward the leading edge of the fire. Several different sources of ignition are used, including drip torches, fusees, matches, and helitorches.

26.5.5 Other Natural Mechanisms of Fire Spread. The direction and rate of fire spread may be altered by natural or self-induced means.

26.5.5.1 Embers and Firebrands. Embers and firebrands can be lofted by the convective column and fall out or be blown by wind into unburned fuel great distances from the original fire. These embers often start new fires outside the perimeter of the main fire. These new fire origins are referred to as spot fires. Care should be taken to distinguish spot fires from possible independent fires unassociated with the main fire.

26.5.5.2 Fire Storms. A fire storm is an intense and violent fire created by strong convective winds produced by a large plume associated with atmospheric instability. The indrafts created by the fire's convection column may be strong enough to uproot vegetation and propel small rocks. One characteristic of a fire storm is the tornadic fire whirls, which accompany the powerful indrafts.

26.5.5.3 Animals. Animals and birds can spread fire from flaming fur or feathers. Animals have been set afire accidentally and deliberately to start a wildfire. Burned animals that are fire victims can start fires in unburned areas during their flight. A bird's feathers or an animal's fur can be ignited by contact with power lines and can start a wildfire when the electrocuted body falls to the ground.

26.6 Indicators. The indication of the direction of a fire's spread is imprinted on partially burned fuels and noncombustible objects. These visual fire effects may include differential damage, char patterns, discoloration, carbon staining, shape, location, and condition of residual unburned fuel. Analysis of the directional pattern shown by multiple indicators in a specific area will identify the path of fire spread through this site. By applying a systematic approach to backtrack the spread of the fire, the investigator can retrace the path of the fire to the point of origin.

26.6.1 Wildfire V-Shaped Patterns. Wildfire V-shaped patterns, as shown in Figure 26.6.1, are horizontal ground surface burn patterns generated by the fire spread. When viewed from above, they are generally shaped like the letter "V." These are not to be confused with the traditional plume-generated vertical V patterns associated with structure fires. These V-shaped patterns are affected by wind direction, and/or the slope on which the fuel is located. As the fire spreads in the direction of the wind or up a slope, the widening legs of the V are created. The width of the pattern increases as the fire advances from the area of ignition. The origin of the heat source that created

FIGURE 26.6.1 V-Shaped Pattern.

the pattern often is found at or near the base or most narrow point of the pattern. Therefore, the analysis of these horizontal V-shaped patterns can be useful in identifying a general location of the fire origin.

26.6.2 Degree of Damage. The degree of damage to a fuel is an indication of the fire's intensity, duration, and direction. Leaves, branches, and limbs will show greater damage on the side from which the fire approached. As shown in Figure 26.6.2, this is one of the useful indicators in determining the direction of advancing fire spread. Also, items lying on and protecting fuels leave a pattern that can assist in locating the origin. Vegetation on the side of an object exposed to an oncoming fire front will be burned away, while the basal stems of vegetation adjacent to the reverse (shielded) side will remain only partially burned. Also, items lying on and protecting fuels leave a similar pattern on residual vegetation. This indicator is closely related to the indicator discussed in Exposed and Protected Fuels.

FIGURE 26.6.2 Degree of Damage.

26.6.3 Grass Stems. The charred remains of grass stems left in the fire's wake will have different appearances depending upon the direction of the fire's travel. In advancing fire areas, the flames will attack the stem from the top and burn them to ground level, completely consuming all but the very base of the stem. Advancing areas are typically characterized by an absence of residual stems. Grass that grows in clumps may not be entirely consumed, showing protection on the side opposite the direction the fire came from. When this occurs in advancing areas, the residual basal stalks in the clump may show an angle of char that is steeper than the slope and exhibit cupping on the tips, with the low side of the cup on the side facing the direction the fire came from. In areas of backing fire spread, and occasionally in the lateral areas, the flames will first attack the stalk at the base, toppling the remainder of the stalk into the burned area as shown in Figure 26.6.3. The remaining grass heads will point generally in the direction the fire came from.

26.6.4 Angle of Char. Angle of char indicators are divided into two groups based on the types of fuels they occur in. Angle of char can occur in pole type fuels (tree trunks, utility

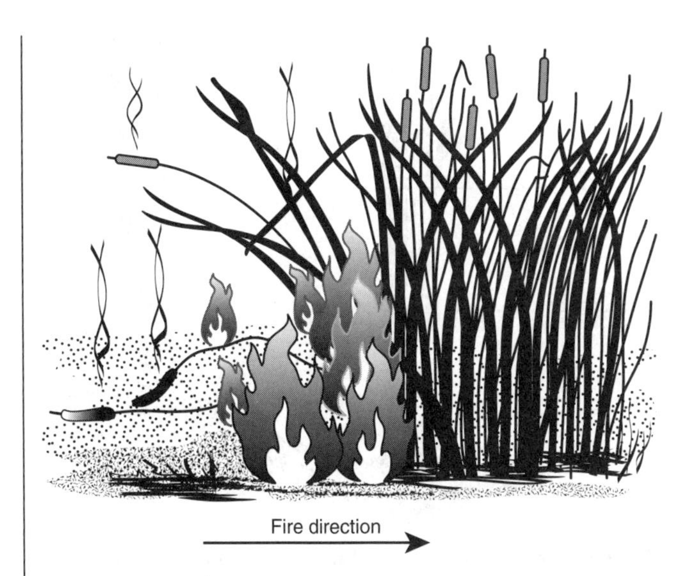

FIGURE 26.6.3 An Example of Grass Stems Indicating the Direction of Backing Fire Movement (left to right).

poles, fence posts, etc.) or in the foliage crowns of brush or timber type fuels.

26.6.4.1 Angle of Char, Pole Type Fuels. Standing, pole type fuels are burned at an angle that corresponds to the flame angle and height associated with the area of fire progression. Reliability is generally greater on individual specimens in open canopy settings. On pole-type vertical fuels, an eddy vortex creates flame-wrap on the side opposing the oncoming fire, leaving a characteristic angle of char. On fires backing against the wind or down slope, the char angle will be parallel to the slope angle, see Figure 26.6.4.1(a). Accumulation of debris may cause char up the side of the tree above the debris, but it will have little effect on the char pattern around the rest of the tree. A fire advancing with the wind or upslope will exhibit a char pattern that is steeper than the slope, see Figure 26.6.4.1(b).

26.6.4.2 Angle of Char, Foliage Crown. On foliage crowns, the flaming front will consume or char fuels at an angle that is consistent with the fire's direction of travel. Backing fire will leave angle of char patterns parallel to the slope. Advancing fire will leave angle of char steeper than the slope due to the flame front entering low on the exposed side and exiting high on the back side see Figure 26.6.4.2(a). Height of char angle is often correlated to fire intensity. This pattern is best viewed from the side of the object. Figure 26.6.4.2(b) shows the typical effect on the crown of trees or brush as a fire starts at point "A" and moves out, slowly building up heat and speed. At the point of origin (point A), the fire is still relatively cool as surface fuels are burned, but the tree's crown is left mostly intact. Farther from the point of origin, the fire has become more intense, and more crown is burned. All the crowns may be burned as the fire intensifies.

26.6.5 White Ash Deposit. White ash can be the byproduct of combustion. More white ash will be created on the sides of objects exposed to greater amounts of heat and flame. Ash is often dispersed downwind and deposited on the windward sides of objects. Ash can also be used to reconstruct probable fuel volumes. Fuels facing the advancing fire will appear lighter on the side facing the oncoming fire and darker on the side opposite the direction the fire came from. Ash indicators

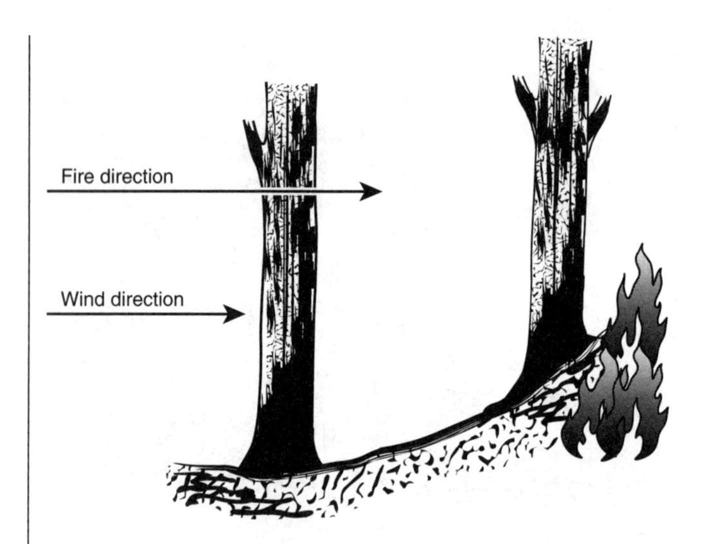

FIGURE 26.6.4.1(a) A Fire Burning Uphill or with the Wind, Creating Char Patterns That Slope Greater Than the Ground Slope.

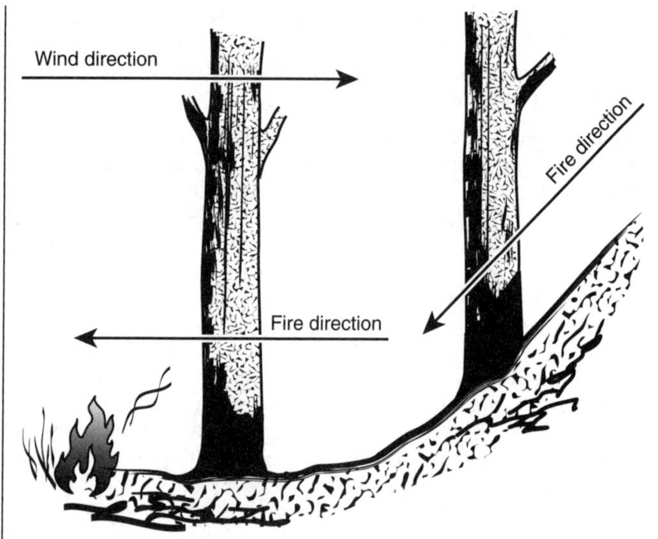

FIGURE 26.6.4.1(b) A Fire Burning Downslope or Against the Wind, Creating Char Patterns That Are Even or Parallel to the Ground Slope.

FIGURE 26.6.4.2(a) Example of Char Patterns Created by the Way a Fire Moves Through Trees and Brush.

can begin to quickly degrade and lose reliability after only a few hours or when exposed to moisture or high winds. White ash deposits on tree boles will be on the side facing the oncoming advancing fire. By comparing and contrasting the two opposing sides, the investigator can distinguish the side facing the oncoming fire as it has more white ash present. White ash can also reveal the direction of fire travel in grass fuels. White ash can remain on the exposed sides of grass stems and clumps. When looking in the direction the advancing fire spread, the burned area will appear lighter. When viewed looking back towards the area the fire came from, the burned area will appear darker.

26.6.6 Cupping. Cupping is a concave or cup-shaped char pattern on grass stem ends, small stumps and the terminal ends of brush and tree limbs. Limbs and twigs on the side facing the oncoming fire will have their tips burned off by the approaching flames leaving a rounded or blunt end. On the opposing side, twigs and limbs will be exposed to flames from underneath, along the base to the terminal end, creating a tapered point. Therefore, in advancing areas of the fire, twigs and limbs on the side opposite the direction the fire came from will show a sharply pointed or tapered end. Limbs of the brush or tree on the side facing the oncoming fire will usually be blunt or rounded off. Stumps, terminal ends of upright twigs and the remains of grass stems can also exhibit a tapered point, with the sharp end on the non-exposed side, as shown in Figure 26.6.6. The low side of the cup will face the oncoming fire. This indicator is usually not associated with backing areas of the fire, except in areas of steep slopes or under high wind conditions. Partially charred branch tips may sometimes be found on the ground on the oncoming fire side of brush and small trees, where they have fallen after being burned off. Large diameter stumps and limbs should not be considered when using this indicator due to their longer term fire residency.

26.6.7 Die-Out Pattern. As a fire enters different fuel types, areas where there is increased fuel moisture or other locations where conditions cause a decrease in rate of spread and intensity, progress may slow or the fire may self extinguish. These areas will exhibit fingers and islands of unburned or partially burned fuels. This pattern is most often associated with the lateral and backing areas of the fire, however these areas should not be assumed to be the origin of the fire. These areas may be useful as macro-scale indicators to establish general fire progression.

26.6.8 Exposed and Protected Fuels. A non-combustible object or the fuel itself shields the unexposed side of a fuel from heat damage. Fuels will be unburned or exhibit less damage on the side shielded from the advancing fire. Look for charring, staining, white ash and clean burn lines on exposed sides

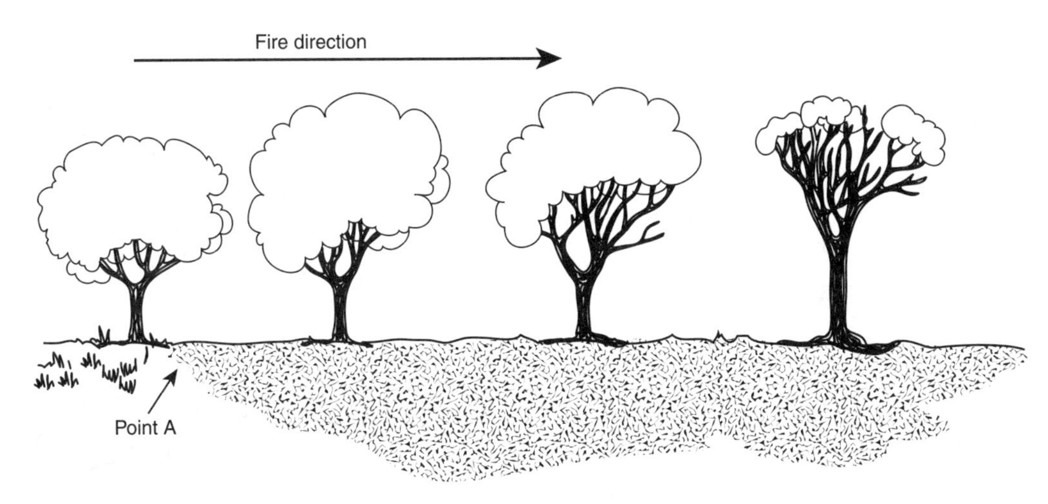

FIGURE 26.6.4.2(b) Progressive Crown Burning from the Point of Origin (Point A)

FIGURE 26.6.6 Cupping.

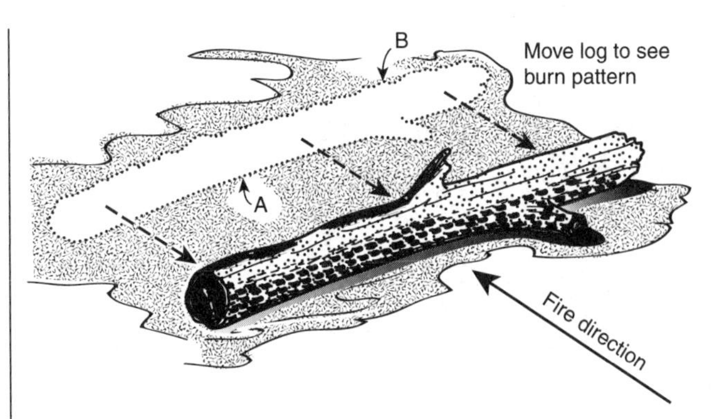

FIGURE 26.6.8 Clean Burn Line on the Front Side (Point A) and a Ragged Burn Line (Point B) on the Other Side, Showing That the Fire Moved from Point A to Point B.

of fuels and non-combustible objects. Compare and contrast to the opposing sides of objects. Lift or remove objects to detect the exposed and protected fuels. Objects resting on top of ground and surface fuels will protect the fuels on the unexposed side. Surface fuels on the exposed side will exhibit a clean burn line see point "A" in Figure 26.6.8. Surface fuels on the protected side will appear ragged and uneven, see point "B" in Figure 26.6.8.

26.6.9 Staining and Sooting. Staining is caused by hot gases, resins and oils condensing on the surface of objects. This occurs most commonly with non-combustible objects such as metal cans, glass bottles, or rocks. Stains will appear on the side of the object exposed to the flames as shown in Figure 26.6.9(a). These yellow to dark brown stains will often feel tacky to the touch and may be covered with a thin layer of white ash. Closely related to staining is sooting. Carbon soot is caused by incomplete combustion and the natural fatty oil content in some vegetation. Carbon soot is typically more heavily deposited on the side facing the approaching fire. Soot will be deposited on the side of fence wires facing toward the origin and can be detected by rubbing your fingers along the wire. On larger objects, soot deposits can also be noticed by rubbing your hand across the surface. In many cases there will be other indicators, such as protected fuel or staining. When checking a wire fence for soot, check the lower wires as they will show more evidence of soot than higher wires as shown in Figure 26.6.9(b).

26.6.10 Depth of Char. Char on limbs, trunks, and finished lumber products exhibit a fissured or scale-like appearance. Wood materials lose mass and shrink as they burn, forming a scale-like surface. Compare and contrast the amount of charring on all sides of the object. The side with the deepest charring will typically be on the side facing the oncoming fire. Figure 26.6.10 shows that the char on the fence posts is deeper on the exposed side, as indicated by the arrow. This means that the fire moved from left to right.

26.6.11 Spalling. Spalling will appear as shallow, light-colored craters or chips in the surface of rocks, as shown in Figure 26.6.11, within the fire area. They will usually be

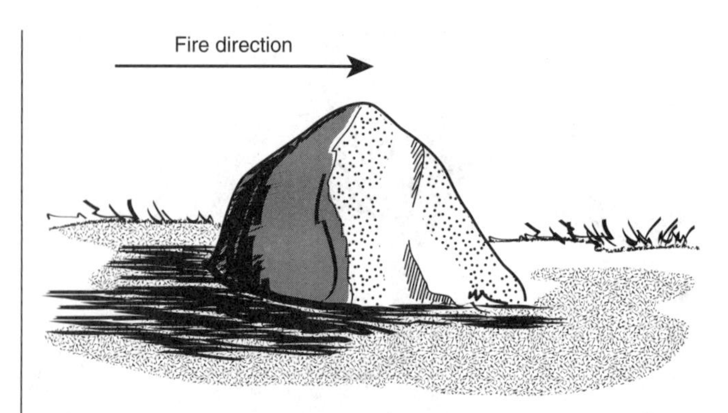

FIGURE 26.6.9(a) Staining (Shaded Area) of Noncombustible Objects by Vaporized Fuels and Minute Particles Carried by the Fire.

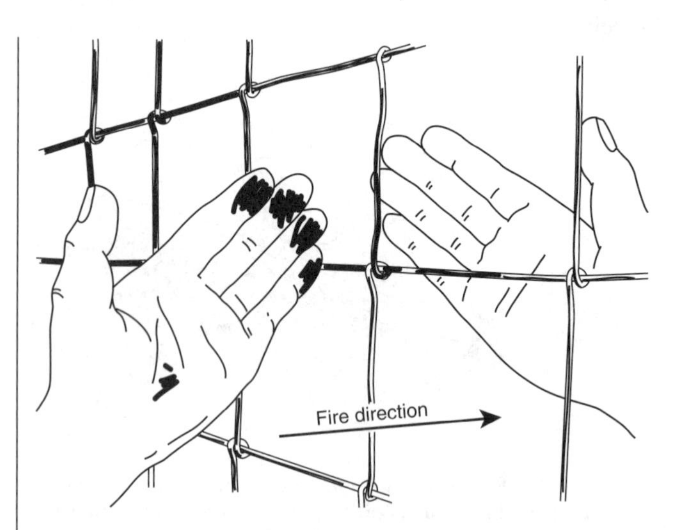

FIGURE 26.6.9(b) Soot Deposited on the Side of Fences Facing the Approaching Fire. The soot can be noticed by rubbing a hand along the wire.

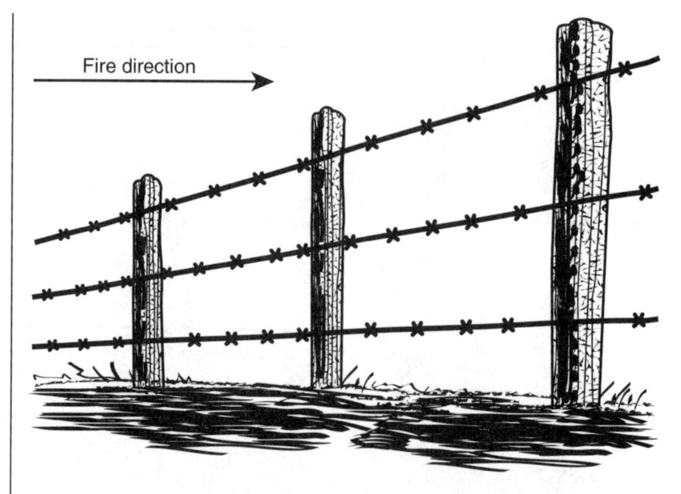

FIGURE 26.6.10 Greater Depth of Char on Side of Fencepost, Indicating the Fire Moved from Left to Right.

FIGURE 26.6.11 Spalling.

accompanied by slabs or flakes exfoliated from the surface of the rock. Spalling is caused by a breakdown in the tensile strength of the rock's surface that has been exposed to heat. Spalling is generally associated with advancing fire areas and will appear on the side of the rock exposed to the flames.

26.6.12 Foliage Freeze. When leaves and small stems are heated, especially in the advancing areas of the fire, they tend to become soft and pliable and are easily bent in the direction of the prevailing wind or drafts created by the fire. They often remain pointed in this direction (freeze) as they cool following the passage of the flame front. While this indicator is almost always an accurate reflection of wind direction at that precise point, it may not always coincide with fire direction. Validate freezing indicators with other indicator categories nearby to confirm the fire's direction. *(See Figure 26.6.12.)*

26.6.13 Curling. Curling occurs when green leaves curl inward toward the heat source. They fold in the direction the fire is coming from. This usually occurs with slower moving, lighter burns associated with backing and lateral fire movement. *(See Figure 26.6.13.)*

26.7 Origin Investigation. The first objective of a wildfire investigation is to identify the area of origin. Considering the factors of wind, topography, and fuels, the origin is normally located close to the heel or rear of the fire. *(See Figure 26.7.)*

26.7.1 Initial Area of Investigation. The initial area of investigation can be determined from information from first-arriving fire fighters and eyewitnesses. The fire fighters and witnesses can verify the location and size of the fire during its early involvement. These accounts will assist the investigator in narrowing down the search area.

26.7.1.1 Observations of Reporting Parties. The reporting party or parties may have seen the fire at an early stage and can provide valuable information that may assist in determining origin and cause. The observations of the reporting parties are important because origin investigation can be narrowed to the area burned at the time of the report. They may also be able to verify or supplement some of the information obtained from other witnesses and attack crews.

26.7.1.2 Observations of Initial Attack Crew. The initial attack crew can play a vital role in the investigation by making obser-

FIGURE 26.6.12 Foliage Freeze.

FIGURE 26.6.13 Curling.

vations while en route and upon arrival on the scene. Crew members may be able to provide valuable information pertaining to the investigative area, the identification of people or vehicles leaving the area, the weather, the condition of locks and gates, and damage or abnormal conditions. If any potential evidence is found, it should be marked and protected. Questions to ask first-arriving crews include the location of the fire upon arrival, the direction the fire was spreading, general fire behavior, descriptions of persons or vehicles in or leaving the area, and weather conditions.

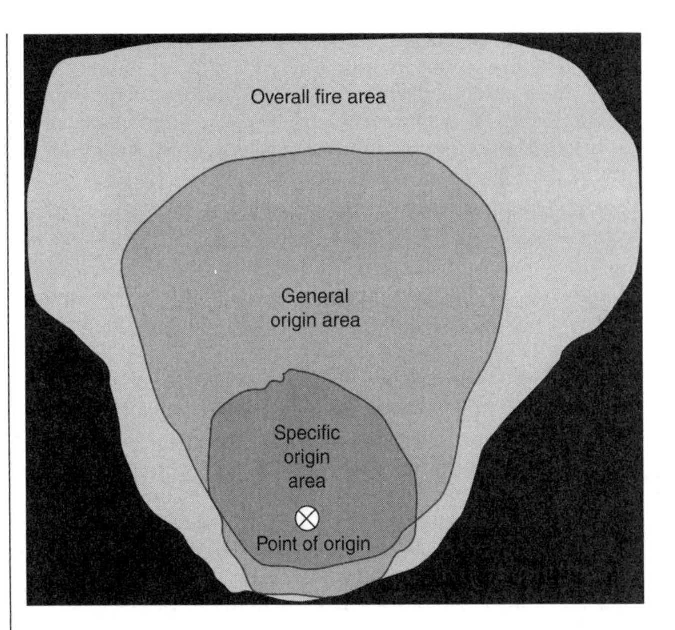

FIGURE 26.7 Anatomy of Origin Area.

26.7.1.3 Observations of Airborne Personnel. Airborne personnel can make the same types of observations as ground crews but could have a valuable different perspective of the potential area of origin, direction of fire spread, and people or vehicles leaving the area. The information should be given to ground personnel promptly and as accurately as possible. Photographs of the fire area taken by airborne personnel often prove invaluable in showing areas burnt, the direction of fire travel, and intensity at a given location and point in time.

26.7.1.4 Observations of Other Witnesses. Witnesses can often provide vital information in the investigation of a wildfire. Witnesses can provide information about vehicles or people that may have been in the area. They are often familiar with the area and can give information as to the area of origin and a possible cause. They can also provide information on the condition of the fire, such as smoke conditions, intensity, rate of spread, and weather.

26.7.1.5 Satellite Imaging or Remote Sensing. Satellite or imaging tools that are used primarily to establish fire suppression tactics can also be utilized to assist in the establishment of the area of origin, based on the direction of fire spread and data indicating the fire location when it was first detected. Pre-fire imagery can provide information regarding pre-fire fuel conditions and activity in the area.

26.7.2 General Origin Area. The area of the fire that the investigator can narrow down based on macro-scale indicators, witness statements, and analysis of fire behavior. It may be a limited area on a small fire, or several acres on a large fire.

26.7.2.1 Specific Origin Area. The smaller area, within the general origin area, where the fire's direction of spread was first influenced by wind, fuel, or slope. Generally this area is characterized by subtle and micro-scale fire direction indicators. It will usually be no smaller than about 5 ft × 5 ft, and may be substantially larger, depending on fire spread indicators and other factors. It is typically characterized by less intense burning.

26.7.2.2 Point of Origin. Contained within the specific origin area will be the precise location where the ignition source came into contact with the material first ignited and sustained combustion occurred. Physical evidence of the actual ignition source is likely to be located at or within close proximity to the point of origin.

26.7.3 General Origin Investigation Techniques. Once the location of the general origin has been determined, the cause of the fire's ignition must still be determined. The investigator should conduct the examination of this area with minimum disturbance, seeking to expose evidence of fire spread from the origin, rather than destroying or removing it during the investigation. A photographic record should be continued throughout the investigative process.

26.7.3.1 Protection of General Origin. The integrity of the fire scene needs to be preserved. Foot and vehicle traffic moving through the area must be kept to a minimum or eliminated. Firefighters should enter this area only if necessary for fire operations. Evidence should not be handled or removed without documentation. The area should be cordoned off by the use of flags, tape or posting personnel at the scene to restrict access.

26.7.3.2 Identifying Evidence. During the investigation, metal flagging stakes can be used to mark items that could be considered potential evidence. Labeled flags can also be used to mark the position of an item of evidence that has been removed from the scene. Care must be continually exercised not to destroy evidence, such as tracks left by individuals or vehicles in the suspected area of origin. The surrounding area should be examined for other evidence.

26.7.3.3 General Principles of Burn Pattern Interpretation. When analyzing a fire's progression using burn patterns, the investigator should keep in mind the following general principles. The interpretation should be based on the majority of the indicators within an indicator category, the totality of the indicators, and fire behavior principles. A single indicator may only be accurate within a 180° arc. Indicators will usually become less pronounced near the origin. The indicators should be documented as they are observed. Work from the area of most intense burning, to the area of least intense burning, following the fire's advancing spread back to the origin. Preliminary fire shapes may be primarily dependent on effective wind speeds, the midflame windspeed adjusted for the effect of slope on fire spread. Direction of spread will be influenced by obstacles.

26.7.3.4 Walk the Exterior Perimeter of the General Origin. Due to the influence of light, shadows and terrain it is recommended that the investigator walk the perimeter of the general origin area at least twice: once in a clockwise direction and once in a counterclockwise direction. Look at the unburned area as well as the burned area. Examine and mark directional burn pattern indicators at the perimeter of the general origin with colored flags or other appropriate markers as they are located. If relevant physical evidence is located, it is standard practice to protect and mark it with white flags.

26.7.3.5 Identify Advancing Fire. Identify the initial run, the rapid advance at the head of the fire that the fire made. Frequently this will be the area which shows the cleanest burn, and may be characterized by a classic "V" or "U"-shaped pattern. It is often bounded on both flanks by lateral fire spread indicators showing less complete consumption of fuels.

26.7.3.6 Enter the General Origin Area. Photograph the general origin area prior to entering. Once the initial run has been identified, the general origin area should be entered from the head or advancing side of the run; this is the side farthest away from the suspected point of origin. The reason to enter from the advancing side is that the burn pattern indicators are more obvious in this area and the investigator is less likely to disturb the specific origin area than if the general origin area is entered from the heel or backing side of the fire. There is one exception to this rule: if the general origin is on a very steep slope where material and soil may be dislodged by the investigator and roll down hill and disturb the specific origin, then the investigator may be forced to enter from the backing or heel side of the general origin area and work up hill.

26.7.3.7 Working the General Origin Area. A suggested method for working the general origin area is as follows: Enter from the advancing area. Work across the run until the lateral transition zone is reached. Move several feet closer towards the origin and re-cross the advancing run to the opposing lateral transition zone. Repeat above steps until specific origin area is reached Figure 26.7.3.7. Document each indicator located with a visible marker. Color-coded surveyors' flags have been found to be the most visible and easiest markers to use. Standard recommended colors are: red for advancing fire indicator, yellow for lateral fire indicator, blue for backing fire indicator, and white for evidence.

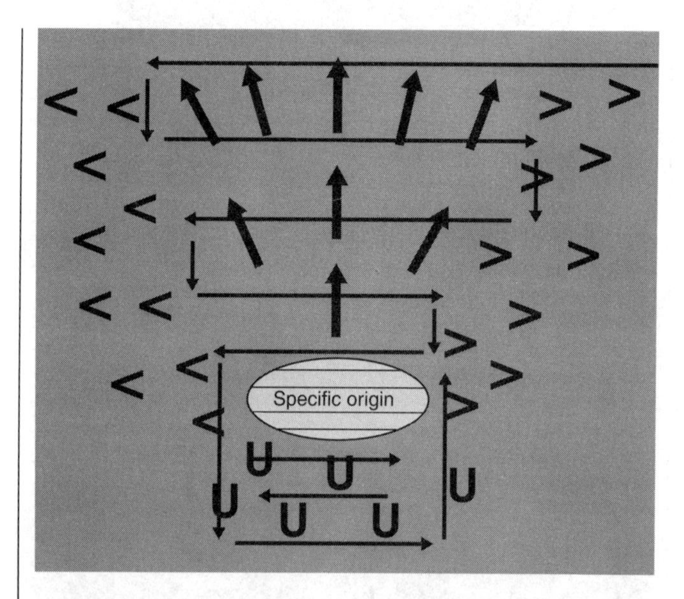

FIGURE 26.7.3.7 Working the General Origin Area.

26.7.4 Specific Origin Investigation Techniques. The following are suggested techniques for investigating the specific origin.

26.7.4.1 Walk Specific Origin Perimeter. Just as with the general origin area, walk the perimeter of the specific origin at least twice, once clockwise and once counter clockwise before entering. During this walk carefully examine the specific origin area, note and mark any items of potential evidence, and continue to note and mark burn pattern indicators. Consider using a spotting scope or binoculars to first visually search this area prior to entering it.

26.7.4.2 Establish Grid Lanes. Use colored twine and four stakes to establish a grid lane. Each lane should be 12 in. to 18 in. in width and be oriented perpendicular to the first fire run. The grid lane should extend from the lateral indicators on one flank to the lateral indicators on the opposite flank on the advancing side of the fire. Number and photograph each lane prior to searching. Measure each lane end to reference point(s), if needed.

26.7.4.3 Search Each Lane. Search each lane visually and then visually with magnification. Once the surface layer has been examined; remove lightweight debris and ash by brushing or blowing. Many investigators find it useful to use a ruler or straight edge to help focus their search pattern. Continue locating and marking indicators with flags as each lane is searched. After the visual search employ a strong magnet to search for ferrous metals. Additionally, a metal detector should be used to search for nonferrous metals. In some cases it may be necessary to screen the remaining debris for evidence. Continue this process, one lane at a time, until the point of origin is reached and/or an ignition source is located. After any evidence has been documented and secured, continue searching past the point of origin or evidence until clear backing fire indicators are encountered.

26.7.5 Search Equipment. Different tools are used by investigators in search of the origin and cause of wildfires.

26.7.5.1 Magnifying Glass. A magnifying glass or reading glasses allows the investigator to see evidence that may not be visible without magnification. It also enhances small details that would otherwise go unnoticed.

26.7.5.2 Magnet. A pull-release type magnet with a stainless steel bottom, with a rating of at least 1.36 kg (3 lbs), is used to locate ferrous metal fragments or particles. Moving a magnet over the burned area will cause such materials to be attracted to the magnet. When an item is located, the pull-release can be operated and the item will fall into an appropriate evidence container.

26.7.5.3 Straight Edge. A straight edge can be used to segment the origin area. By reducing the search area, the investigator will find it easier to focus on small objects. This is very helpful when using a magnifying glass.

26.7.5.4 Probe. A probe is particularly useful in uncovering small pieces of evidence from the surrounding vegetation, for example, removing grass stems from the underlying matchbook.

26.7.5.5 Comb. A wide-gap comb can be used to separate evidence from debris. It also works well for picking up evidence without damaging it. By using the comb like a scoop, the investigator can pick up small pieces of evidence while letting burnt grass and other debris sift out through the teeth. Hair picks work exceptionally well for this task.

26.7.5.6 Hand-Held Lights. Hand-held lights assist in locating items in lowlight areas. They also eliminate shadows.

26.7.5.7 Air Blower. A small air blower (sold for camera cleaning) is useful to separate light ash from items of interest. Air expelled through pursed lips provides similar results, but is not as controlled.

26.7.5.8 Metal Detector. Metal detectors are used to locate ferrous and nonferrous metal objects that may be of evidentiary value.

26.7.5.9 Sifting Screen. Sifting screens of various sizes assist in the separation of a suspected item of evidence from the surrounding dirt and vegetation.

26.7.5.10 Global Positioning Satellite (GPS) Recorder. A GPS may be utilized to obtain the accurate longitudinal and latitudinal position of the fire origin. The position can be cross-referenced against site survey information, lightning strike data, aerial photography, or satellite imagery.

26.8 Fire Cause Determination. The objective of every origin and cause investigation is to establish the cause of the fire and to confirm this finding by identifying and, if possible, recovering the heat source or ignition device *(see Fire Cause Determination chapter)*. If the fire was ignited intentionally, the ignition source may have been discarded nearby or removed from the scene.

26.8.1 Natural Fire Cause. Wildfires are not always started by the actions of people. Many are ignited by natural causes such as lightning and some occasionally by volcanoes.

26.8.1.1 Lightning. Lightning is a well-recognized cause of wildfires, particularly in forested areas, with lightning striking trees, power lines, and rock outcrops. The action of the lightning strike in a tree can splinter the trunk and can form glassy clumps, called fulgurites, which are formed in the soil by melting sand in the root area. Lightning may simply strike the ground, igniting nearby fuels. When lightning is suspected, a GPS may be used to gather data, which then can be provided to a lightning detection service for confirmation of that activity. Lightning-caused fires may smolder undetected for several weeks after a lightning strike before conditions change and the fire transitions to an active wildfire.

26.8.2 Human Fire Cause. Human-caused fires are a result of human action or omission and are classified as accidental or incendiary. Accidental fires involve all those for which the proven cause does not involve an intentional human act to ignite or spread fire into an area where the fire should not be. The incendiary fire is one intentionally ignited under circumstances in which the person knows that the fire should not be ignited.

26.8.2.1 Campfire. A circle of rocks, a pit with a large amount of ash, or a pile of wood is a good indicator of a campfire. Even camp areas that have burned completely leave evidence of their prior existence. Discarded food containers, metal tent stakes, or metal grommets from a tent may be found, indicating the possibility of a campfire.

26.8.2.2* Smoking. Discarded ignited smoking materials, such as cigarettes, cigars, pipe tobacco, and matches, can start wildfires. Evidence of these ignitions may survive at or near the point of origin. The ash and filter of a cigarette butt or a burned match may be identifiable at the point of origin.

26.8.2.2.1 Smoldering smoking materials require receptive fuels for ignition. Fuel factors such as type, size, moisture content, temperature and arrangement affect the potential for ignition. The fuel is typically fine or powdery, such as litter or punky wood, which is conducive to ignition of smoldering combustion.

26.8.2.2.2 Ignition of wildland fuels by smoldering smoking materials is very sensitive to environmental conditions at the point of ignition. These factors include relative humidity, temperature, wind speed, and wind direction. There can be significant differences between microclimatic conditions at the point of origin and readings from distant weather stations. Typically relative humidity is below 25 percent for ignition to occur but the complexity of interactions of conditions leading to ignition, especially microclimatic variations including wind and impinging solar radiation, preclude that percentage as being absolute.

26.8.2.2.3 The position of the smoldering smoking material on the fuel is a significant factor in ignition. The area of contact between the ignition source and the fuel and the duration of that contact influences ignition of the fuel. The orientation of the ignition source on the fuel, such as whether the glowing tip of a cigarette is aligned downward or upward which would influence preheating of the adjacent fuel, is another factor for ignition.

26.8.2.3* Debris Burning. Fires occur at dumpsites, timber harvesting operations, and land clearing operations as well as at residences from garbage and other debris set on fire. These fires can spread to the neighboring vegetation. Burn barrels or incinerators may be a consideration as a fire cause. In windy conditions, hot ash and debris can blow away from the debris burn and start a fire some distance away. Large woody debris piles, especially if mixed with soil, have been known to hold long term thermal residency for many months, including over winter, before escaping into adjacent wildland. Witnesses are often useful in determining whether or not debris burning was the cause of the fire. A prescribed fire is a fire resulting from intentional ignition by a person or a naturally caused fire that is allowed to continue to burn according to approved plans to achieve resource-management objectives.

26.8.2.4 Incendiary. These fires are sometimes set in more than one location, and in areas that are frequently traveled. A time-delay ignition device may have been used. Items to look for include matches, fuses, cigarettes, rope, rubber bands, tape, candles, and wire. For further information on incendiary fires, refer to the Incendiary Fire chapter.

26.8.2.5 Equipment Use. Vehicles and power machinery can cause wildfires in countless ways, from operating failures, overheated equipment, exhaust particles, ignition of fuel leaks and spills, and friction. Any power or motorized equipment that uses electricity or flammable products in its operation or that creates ignition temperatures in its operation is capable of starting a fire when being operated in or adjacent to combustible vegetation. This includes vehicles, powered portable and mobile machinery, harvesting and construction equipment, chain saws, grinding tools, and cutting torches. Defective or failed parts add to the fire potential through friction such as heating of bearing-worn brakes, "frozen" shafts, or abrasion.

26.8.2.6 Railroad. Wildfires are sometimes started along railroads. Occasionally, fire intended to clear a right-of-way will escape. Diesel and diesel-electric powered locomotives can cause trackside fires from exhaust particles, ignition of external build-ups of lubricating oil, and exhaust and fuel line failure, while rolling stock can start fires from hot brake metal and overheated wheel bearings (hot box). Fires can also result from derailments, cutting or grinding on rails, or warning flares.

26.8.2.7 Fire Play. Fire play-caused wildfires are those started by children 12 years of age or younger. These fires are often motivated by normal curiosity and the use of fire in experimental or play fashion. Matches or lighters are the most frequent ignition source. These cases often involve multiple children. These fires most commonly occur around residences, schools, playgrounds, campsites, wooded areas and other areas frequented by children. Determine if the actions are due to normal curiosity or a pathological behavior. Consider referral to juvenile authorities and/or a juvenile firesetter intervention program.

26.8.2.8 Fireworks. Fireworks provide means of ignition through sparks and flaming debris. Sparklers are a smaller hazard, but may ignite dry grass or other fuels. Most sparklers include a metal (wire) or wood core that may be found at or near the point of fire origin. The remains of fireworks or their packaging may be found near the area of origin. Some fireworks have the potential to create small indentations in the ground due to their explosive force.

26.8.2.9 Utilities. Public and private utilities often are present through wildfire areas and therefore provide a potential ignition source.

26.8.2.9.1 Electricity. Overhead power lines may cause wildfires when trees contact a conductor and ignite the branch or foliage involved. This contact may leave unique fire damage on the portion of the tree that made contact and create a pit or flash mark on the power conductor. After ignition, burning portions of the tree may fall to the ground and ignite surface fuels. In addition to tree contact, conductors may be blown against each other (phase to phase) during a windstorm, creating a hot metal globule that falls to the ground. Conductors and transformers may fail, starting the pole or other equipment on fire, and may drop flaming or hot material onto the ground. Underground conductors can be damaged by heavy equipment or digging operations, resulting in fire. Electric fences are likewise a source of energy resulting in ignition of combustible materials.

26.8.2.9.2 Oil and Gas Drilling. Oil and gas drilling activities in wildfire areas can cause a fire. Most of the hazardous activities associated with fire take place during the drilling operations. A number of these hazards have been discussed in the smoking, equipment use, and electric paragraphs. A well blowout and subsequent ignition may cause a wildfire to ensue. Depending on the minerals to be extracted, gas-fired equipment such as separators may be present at the well site after the drilling process has been completed. Likewise, the proximity of pipelines carrying gas or liquid fuel may provide sources of the initial fuel or may be a contributing factor to wildfire spread.

26.8.2.10 Miscellaneous. Wildfire causes that cannot be properly classified under other cause groupings are usually classified as miscellaneous. This is not an all inclusive list but is presented as examples of other wildland fire causes.

26.8.2.10.1 Spontaneous Heating. Some types of fuel can ignite spontaneously from internal heating caused by biological and chemical action. This action is most likely to occur on warm, humid days in decomposing piles of organic material such as hay, grains, feeds, manure, sawdust, wood chip piles, and harvested piled peat moss.

26.8.2.10.2 Sunlight Refraction and Reflection. The sun's rays can be focused to a point of intense heat if concentrated or focused by either a transparent object that is spherical or cylindrical in its cross section or by a concave, highly reflective surface. This refraction or reflection process bends light rays, similar to that which occurs with a magnifying glass or mirror. Fires started by these items are very rare occurrences; however, objects possessing these characteristics recovered from the specific origin area may need to be carefully examined for purposes of exclusion or inclusion.

26.8.2.10.3 Firearms. The use of black powder firearms, and modern firearms discharging tracer, incendiary and steel core ammunition can cause wildfires. Black powder caused fires are frequently the result of burning patch material rather than wildland fuels ignited directly by the burning black powder. The burning chemical compounds contained within the projectile of tracer and incendiary ammunition can ignite wildland fuels. Ammunition with steel cores, such as armor piercing ammunition and certain brands of 7.62 x 39 mm.

ammunition, can cause fires when the steel core strikes a rock or other material hard enough to cause sparks.

26.9 Evidence. The protection, preservation, collection, and documentation of evidence is similar for wildfire, structure, and vehicle fire investigations. Refer to the Physical Evidence chapter for information and guidance.

26.10 Special Safety Considerations. Safety considerations vary from those present in structural fire investigation; however, the basic principles are universal. Refer to the Safety chapter for further information.

26.10.1 Hazards. Safety is a major concern during all fires. Investigating wildfires carries its own specific hazards that must be considered while on the fire scene. The investigator should log in with the Incident Commander and should keep firefighting personnel aware of his or her location. The investigator must be aware of areas where the fire may still be burning, or where fire has been extinguished but could rekindle. The investigator should be mindful of the safety requirements of LCES (Lookout, Communication, Escape route, Safety area). An avenue of escape must always be present and continually evaluated as fire conditions change. A rekindle and availability of fuels, along with a change in the wind direction, may create hazards as extreme as a crowning fire, which may block the previously planned escape path.

26.10.1.1 Other associated hazards are falling debris from fire weakened trees and limbs. Leaning against or putting weight on trees could cause them to topple or break. Even light winds may cause branches to break and fall to the ground. Additional hazards may be present in sloped terrain. Logs or rocks can be loosened from their lodged location as a result of the fire, from suppression operations, or from the actions of an investigator. When the root structure is destroyed by fire, the soil may begin to lose its stability, which in turn can cause slides that may injure the investigator or destroy evidence.

26.10.1.2 The weather can cause or contribute to hazards. Rain can create slippery footing. Lightning may be a concern as well. During lightning conditions, the investigator should not stand under trees, but rather move to open space, staying low without lying down.

26.10.1.3 Underground burning in a smoldering stage can erupt into flaming combustion if the fire burns to the surface or if the top layer of soil is disturbed, exposing the heated fuel to air. This phenomenon most commonly occurs where there is heavy peat content in the soil, or in areas where large sawdust piles were left from logging or sawmill operations.

26.10.2 Personal Protective Equipment. Protective clothing and safety equipment will vary depending on the circumstances. If investigating while the fire is still in progress in the investigator's immediate vicinity, the investigator will need to comply with applicable federal, state, provincial, or local requirements for personal protective equipment applicable to wildfire fighting.

26.10.2.1 Fire investigators in such situations should conform to NFPA 1977, *Standard on Protective Clothing and Equipment for Wildland Fire Fighting,* and appropriate sections of NFPA 1500, *Standard on Fire Department Occupational Safety and Health Program.* The types of equipment addressed in NFPA 1977 are head protection, hood, body protection, hand protection, foot protection, and eye protection. In addition, respiratory protection is necessary and will be dictated by the type of hazardous atmosphere encountered. This type of safety equipment is perhaps the most important but the most often overlooked.

26.10.2.2 When investigating at a scene where the fire has been completely extinguished, the investigator will still need to comply with any regulatory standards. Even in the absence of mandatory regulations, recognized industry standards ought to be followed for the investigator's own protection. Refer to the Safety chapter for further information.

26.11 Sources of Information. Information sources on wildfires can be found in Annex B and Section B.11.

Chapter 27 Management of Complex Investigations

27.1 Scope. This chapter addresses those issues that are unique to managing investigations that are complex due to size, scope, or duration. Complex investigations generally include multiple simultaneous investigations and involve a significant number of interested parties. A complex investigation may arise from a fire or explosion incident that involves circumstances such as fatalities or injuries, fires in high-rise buildings, large complexes or multiple buildings, or fires and explosions in industrial plants or commercial properties, but may not always be large in size or magnitude. The methodology of this chapter may be applied to other investigations not considered complex.

27.1.1 Governmental Inquiry.

27.1.1.1 Public sector agencies that have jurisdiction for investigating an incident are governed by their jurisdictional legislative and statutory authority, laws, regulations, rules and policies; therefore, the methodology of this chapter may not pertain in its entirety to their investigations.

27.1.1.2 It is understood that the governmental entities having public safety authority and responsibility normally conduct an initial investigation. The officials who conduct the governmental investigation should attempt to follow industry standards and guidelines and preserve (or leave in place) the physical evidence in order that interested parties may have the opportunity conduct a complex investigation as set forth in this chapter. Government investigators may find it necessary to alter evidence during the various missions of the agencies. Evidence should not be altered, damaged, or destroyed unless doing so is unavoidable and furthers the purpose of the various missions of the agency. If evidence must be altered in some way, public authorities should consider consulting with other interested parties so as not to impair the integrity of the public investigation. The governmental investigators should document their investigation. At the point that there are no longer significant public safety issues or the investigators have determined that their inquiry is not a criminal investigation, the governmental investigators may choose to participate in the complex investigation process or terminate their inquiry.

27.1.2 Intent. This chapter is intended to provide a framework for the management of complex investigations. The use of the term investigation in the singular in this chapter refers to the multiple simultaneous investigations conducted by the interested parties. The degree to which the guidelines in this chapter apply will vary depending on the specific incident.

27.1.3 Purpose. The purpose of this chapter is to provide guidance for the management and coordination of investigative activities among multiple interested parties, which affords an opportunity for all to investigate the incident, to protect their respective interests, and to allow for cost-effective and expeditious investigations. It is not the purpose of this chapter to instruct interested parties how to investigate the incident. The organization of investigative teams, functions, and activities are provided in Chapter 14. The scene examination should be conducted according to the principles recommended in this guide.

27.1.4 Interested Parties. As used in this document, "interested parties" means any person, entity, or organization, including their representatives, with statutory obligations or whose legal rights or interests may be affected by the investigation of a specific incident. Interested parties may include persons or groups conducting fire or explosion investigations on behalf of public safety agencies (such as fire departments, law enforcement, fire marshals, code enforcement agencies, or criminal prosecutors), property owners, or existing or potential parties to civil or criminal litigation and their attorneys (such as insurance interests, fire victims, and criminal defendants or potential criminal defendants).

27.1.5 Chapter Definitions.

27.1.5.1 Interested Party. Any person, entity, or organization, including their representatives, with statutory obligations or whose legal rights or interests may be affected by the investigation of a specific incident.

27.1.5.2 Investigation Site. For the purpose of this chapter, the terms "site" and "scene" will be jointly referred to as the "investigation site," unless the particular context requires the use of one or the other word.

27.1.5.3 Investigative Team. A group of individuals working on behalf of an interested party to conduct an investigation into the incident.

27.1.5.4 Protocol. A description of the specific procedures and methodologies by which a task or tasks are to be accomplished.

27.1.5.5 Scene. The specific physical location of the incident within the site. The area or areas (structure, vehicle, boat, piece of equipment, etc.) designated as relevant to the investigation of the incident because it may contain physical damage or debris, evidence, victims, or incident-related hazards. This is the area to be processed during the scene examination.

27.1.5.6 Site. The general physical location of the incident, including the scene and the surrounding area deemed significant to the process of the investigation and support areas.

27.1.5.7 Understanding or Agreement. A written or oral consensus between the interested parties concerning the management of the investigations.

27.1.5.8 Work Plans. An outline of the tasks to be completed as part of the investigation including the order or timeline for completion. See Chapter 14.

27.2 Basic Information and Documents. Copies of blueprints, building plans, site plans, schematics, manuals, and other documents needed for the management of the investigation site should be obtained from the property owner, contractor, architect, or the public building department to orient investigators to the site, place buildings or items in relation to each other, or assist in understanding building systems.

27.3 Communications Among Interested Parties.

27.3.1 Notice to Interested Parties. Notification should be given to all known, interested parties in an expeditious manner to allow them the opportunity to examine the scene as early as possible and minimize claims of spoliation. See 11.3.5; see also ASTM E 860, *Standard Practice for Examining and Testing Items That Are or May Become Involved in Product Liability Litigation.* Initial notice should be provided by the entity in control of the scene to any known interested parties.

27.3.1.1 Entity in Control. The entity having control of the site should provide notice to all known interested parties if it intends to conduct an investigation at the site of the incident. If the entity controlling the site does not intend to conduct an investigation at the site, or if it fails to initiate notification, any interested party intending to conduct such an investigation may initiate the notification process.

27.3.1.2 All Interested Parties. All interested parties should notify the entity controlling the investigation site in a timely manner of other interested parties that should be included in the investigation. Responsibility for and timing of the notification will vary according to such factors as the jurisdiction, whether the interested party giving notice is public or private, whether criminal conduct is implicated, and applicable laws and regulations. See 11.3.5.1.

27.3.1.3 Roster of Interested Parties. All interested parties, particularly the private sector parties, must take steps to ensure that timely notification is given to all other interested parties. A roster identifying all interested parties should be created and shared. All notices to interested parties should also be shared.

27.3.1.4 Notification of Changes. In circumstances when control of or access to the scene changes or the scene or evidence could be altered, notification should be given to known interested parties. Such circumstances may include:

(1) When the public sector has control of the scene, determines that a crime has not been committed, and allows the private sector to participate
(2) When the public sector has transferred control of the scene
(3) When an investigation by a private sector party has started and it has been determined that there are additional interested parties

27.3.1.5 Making Notification. Notification can be made by telephone, letter, fax, or e-mail. Oral notification should be confirmed in writing. Identification of interested parties can sometimes be difficult. Methods by which interested parties can be identified include examination of product labels, documents such as contracts or service records, receipts, etc. Contact information can be obtained from numerous sources, such as an Internet search. Communication can be established with the interested party's legal, risk management, or other similar department, to identify to whom notification should be directed.

27.3.1.6 Content of Notification. Notification to interested parties should include the date of the incident; the nature of the incident; the incident location; the nature and extent of loss; damage, death, or injury to the extent known the notified party's potential connection to the incident; next action date; circumstances affecting the scene (such as pending demolition orders or environmental conditions); a request to reply by a certain date; contact information to whom the notified person is to reply; and the identity of the individual or entity controlling the scene. Suggestions for the content of this notice are not mandatory, but may assist in efficient site management and avoid delays. The notification should include a roster of all parties to whom notice has been provided, which should include name, address, and telephone number. The roster should be updated with any additional, later-identified parties on a regular basis.

27.3.1.7 Subsequent Notifications. As additional interested parties are identified, they should be notified of the investigation. Subsequent notifications may be made by any interested party; however, it is incumbent upon that party to provide a copy of the roster to the newly notified parties and a copy of the new notification to the prior notified parties. Every interested party, especially the entity controlling the investigation site, should receive a copy of the subsequent notification.

27.3.2 Meetings.

27.3.2.1 Preliminary Meeting. The entity controlling the investigation site should call a preliminary meeting of the interested parties before beginning the investigation to determine the needs for protocols and work plans, how the scene will be processed, access to investigation site and any restrictions, safety and environmental considerations, etc. During this preliminary meeting interested parties should develop an understanding or agreement regarding how the investigation will progress.

27.3.2.2 Meetings as the Investigation Progresses. Meetings should be held to ensure everyone is signed in, reminded of safety considerations, and advised of changes in conditions, work plans, and future scheduling, and to make introductions of new participants. It is suggested that short meetings be held daily and more thorough meetings be held weekly.

27.3.3 Web Site. The development of a secure web site can be a useful tool in the investigation management. If used, it should be made part of the overall agreement between the interested parties. Web sites are an efficient means of communicating information and can be constructed in a manner that limits access to the interested parties or may limit access to only certain areas of the web site to certain interested parties. The content of the web site may include: photos, videos, contact list/roster, proprietary and nonproprietary information that has been exchanged, protocols, understandings or agreements, work schedule of events, incident action plan (IAP), evidence logs, copies of reports from the public sector, public documents, sign-in sheets for each investigative day, safety instructions, and anything else that the interested parties want to share. See also 27.4.6.2.

27.3.4 Additional Dissemination of Information. An on-site bulletin board or e-mail messages can be used to disseminate information to the investigative teams, to post daily activity reports, and to facilitate the exchange of information. Another method to disseminate information is an incident action plan (IAP) from an incident command system. See 27.5.1.

27.4 Understandings and Agreements.

27.4.1 Purposes. Due to the complexity of the investigation and to ensure that all known interested parties are afforded an opportunity to investigate the incident and protect their respective interests, understandings or agreements should be developed as early as possible. Such understandings or agreements will assist in the coordination of the investigation in an efficient and systematic way, while minimizing disputes. Items on which the parties may wish to have a common understanding or agreement include the following:

(1) Safety/environmental hazards
(2) Control and access to the site
(3) Cost sharing
(4) Scheduling
(5) Communication
(6) Logistics
(7) Protocols
(8) Tagging, removal, custody and storage of evidence
(9) Documentation of the scene/investigation
(10) Web site
(11) Interviewing of witnesses
(12) Sharing of proprietary/nonproprietary information
(13) Evidence examination and testing

27.4.2 Scheduling. Effort should be made to accommodate the schedules of the greatest number of interested parties and their representatives in order to allow them to participate in the investigation. It may not always be feasible to accommodate all interested parties.

27.4.3 Cost Sharing. Various aspects of the investigation lend themselves to the sharing of costs. These items may include evidence storage, heavy equipment, debris removal, personal comfort items, security, first aid, and specialized tests or examinations of evidence. The interested parties should identify those aspects of the investigation for which the costs will be shared. A written agreement should be prepared and signed by all interested parties, setting forth the costs to be shared. Interested parties unwilling to share in costs may not receive the benefits of joining in the cost-shared activities.

27.4.4 Nondisclosure Agreements. In order to protect confidential, trade secret, and proprietary information, an interested party may require a nondisclosure or confidentiality agreement. These agreements facilitate the sharing of information among interested parties by limiting who may have access to the information.

27.4.5 Protocols. Protocols can apply to many aspects of the investigation, including how the scene examination should be conducted and how evidence will be collected, examined, tested, and stored. Interested parties should develop a mutually agreed upon protocol. Any objections to the protocol should be set forth in writing whenever practical, specifying the portion(s) objected to, the reasoning behind the objection, and a recommended alternative procedure.

27.4.6 Information Sharing. Information sharing breaks down the barriers to obtaining information by allowing interested parties to exchange factual data. For example, one party may have obtained the public fire report. Sharing this information may help expedite the investigation. Sharing information that does not compromise a party's investigation but will facilitate a swift and cost-effective completion of the investigation is encouraged. Information sharing may also assist in the identification and collection of fire scene evidence.

27.4.6.1 Timeliness. Sharing information between the public and private sector in a timely manner is important to all investigators to assist in the preservation of evidence and the accurate determination of origin, cause, and responsibility.

27.4.6.2 Public Records of Communications. All information provided to public sector investigators and communications with the public sector investigators, including e-mails, may become a matter of public record under open records law requests.

27.4.7 Interviews. Interviews can be conducted while other activities are being performed. Interested parties should try to secure interviews with witnesses, such as emergency response personnel, passersby, neighbors, property owner(s), employee(s), tenant(s), or other people who may have information.

However, there may be legal constraints in interviewing particular witnesses. Whether to conduct joint interviews of particular witnesses may be included in the understandings or agreements. For witness interviews generally, see Section 13.4.

27.4.8 Amendments to Agreements. As the investigation progresses, it may be necessary to amend the understandings or agreements. If the original understanding or agreement was in writing, any amendments should also be in writing. All interested parties must be notified of the suggested change and have an opportunity to agree, disagree, or provide input regarding the amendment. Amendments can be made to the agreement during the scene investigation by agreement of the parties present to the detriment of parties who choose not to attend, provided that the amendment is within the original scope of the agreement. Interested parties may not object to site amendments if the party chose not to attend.

27.4.9 Disagreements. Disagreements may arise concerning any number of issues relating to the investigation. Good faith efforts should be made to resolve disagreements among the interested parties as expeditiously as possible to allow the investigation to move forward with minimal delay and expense. If the dispute is not resolved it may be necessary to seek a ruling from a court to address the issue, including a temporary injunction or restraining order pending the court's decision.

27.5 Management of the Investigation. Proper investigation site management will limit the potential for safety-related incidents, facilitate investigative activities, address potential concerns about evidence spoliation, and assist in many other aspects of the investigation. The implementation of an effective incident management system prior to the start of the investigative process will benefit all interested parties.

27.5.1 Organizational Models. The management of the investigation involves many interested parties performing joint and independent tasks simultaneously. A variety of organizational models can be used to manage the investigation so that the various interested parties participating in the investigation can work in a coordinated manner.

27.5.2 Control of the Site and Scene. Which entity or agency controls the investigation site may be governed by statutes, jurisdiction, contract, ownership, or court order, and may change as time passes. For example, the initial control of the investigation site will most likely be a public authority, such as a fire department. Control may then transfer to another agency, such as a law enforcement or regulatory agency, the building owner or insurer, or a court.

27.5.2.1 Securing the Site and Scene. One of the first tasks to be completed is the establishment of investigation site security. Entrance should be limited to those individuals necessary to provide for safety; prevent the removal, destruction, or loss of items of evidentiary value; and prevent undue change or additional damage to the scene. It may be necessary to hire private security personnel or to install barriers to obtain the level of security needed. Security should be maintained continuously until the investigation site activities are complete. First responders to the scene should establish and maintain initial control and security of the investigation site. As interested parties are notified and arrive at the investigation site, one of the first duties is to ensure that security at the site has been established.

27.5.2.2 Delegation of Control. The entity controlling the investigation site determines who has control of the management of the investigation and may delegate control of the management of the investigation to its representative or to others.

27.5.2.3 Transfer of Control. Transfer of control between interested parties should be planned in advance. For example, a public investigatory agency should tell the owner or insurer when they will be transferring control so the receiving party can make arrangements for safety and security or other immediate needs. Upon transfer of control, the entity relinquishing control should provide information or documentation relating to alterations to the site, alterations to the scene, evidence collected, contamination of the site or scene, and other information that could affect other interested parties' investigations.

27.5.2.4 Site and Scene Access.

27.5.2.4.1 Control of the Site. The decision regarding investigation site access rests with the entity having control of the investigation site. In some cases, the controlling party may need to accept input from the site owner or tenant to protect proprietary or trade secret information that structure damage may now reveal, post-fire. A representative of each interested party should have access to the scene as soon as possible. This is important for organizational planning. Coordination of access during the scene examinations, reconstruction, and evidence removal may require planning so the parties can be fairly represented. If the size of the group is an issue, access may be limited by allowing only one person per interested party to attend.

27.5.2.4.2 Establishing Procedures for Access. Procedures for obtaining access to the site and scene should be established. The procedures should set forth who will be admitted to the site and scene, under what circumstances, and for what purposes. The controlling party may establish different areas or zones with different levels of security and access within the site and scene. These procedures should be communicated to any security providers as well as all interested parties. Clear markings, barriers, or personnel should be in place to prevent accidental incursions into limited access areas.

27.5.2.4.3 Monitoring Entry to the Site. Entry to the investigation site should be closely monitored. Procedures should be adopted to record the identity of each person who enters and leaves the investigation site and the time. Each person who enters the investigation site should identify their name, telephone number, fax number, e-mail address, company affiliation, and interested party who they represent. In general, any person refusing to identify this information may be reasonably barred entry from the site and the investigation can continue in that person's absence. A roster of all individuals who enter the investigation site should be made available to all interested parties. Not only does this keep a record of who was at the investigation site, but allows for easy communication with those individuals should a change in the work plan or schedule occur. Personal protective equipment requirements may be necessary.

27.5.2.4.4 Access Control. Additional access control may be required for the scene due to safety and other considerations. Methods of identification, monitoring, and the keeping of records of persons present in the scene and their location should be adopted. Entry permits, accountability systems, or log books detailing entry may be sufficient; however new technologies such as bar coded credentials may be more efficient to record entry, or GPS may be used to track an individual's location within the scene.

27.5.2.4.5 Escorts. For safety and security reasons and to maintain the integrity of the scene, escorts may be used to allow interested parties controlled access. Escorted parties must follow the escort's reasonable directives, but may express objections and request re-entry regarding the resolution of those objections. Unresolved objections should be recorded and handled as with disagreements, see 27.4.9.

27.5.2.4.6 Public Sector Concerns. Public sector organizations should consider the needs of other interested parties. They should provide the other interested parties with briefings and investigation status information as well as limited and controlled access to the scene and any evidence collected if it does not compromise their investigation or violate agency regulations or laws. In most cases, other than criminal investigations, public authorities should provide walk-throughs for observations and photographing or make their photographs available to other interested parties during the investigation. Alternatively, the public authorities may conduct a joint investigation with private parties.

27.5.2.4.7 Occupant Access and Control. Allowing owners, tenants, and occupants limited access to the investigation site is determined by the entity controlling the investigation site. If possible, the investigation should be conducted in a manner that allows the occupants limited access to the investigation site. Allowing the tenants/occupants access is important, but the investigation site should still be secured. This effort may help to reduce the interruption to business and the impact of the fire. Occupant access to the investigation site should be allowed only after the safety of the area has been established.

27.5.2.4.8 Decontamination In and Out. The chance of contamination can be reduced by establishing decontamination procedures and facilities on site. For example, the entity controlling the investigation site may require all parties entering to follow a protocol for washing footgear and tools. Areas and equipment for decontamination upon exiting may also be provided to reduce hazards associated with fire scene debris.

27.5.2.5 Site-Specific Restrictions or Requirements. The entity controlling the investigation site should inform interested parties of any restrictions and requirements that will be imposed upon them, such as what items are prohibited from the site, whether confidentiality agreements or waivers must be signed, and what personal protective equipment may be necessary.

27.5.2.6 Scene Integrity. Regardless of who has control of the scene, in order to maintain scene integrity, contaminants, such as smoking materials, ignition sources, food and beverage containers, and wrappers, should not be brought into the scene. Supplies necessary for the scene examination should not be disposed of in the scene.

27.5.2.7 Release of Information. When multiple public agencies are involved in an investigation, a system to coordinate public dissemination of information should be established.

27.6 Evidence.

27.6.1 Evidence Control. Evidence should be handled and secured as discussed in Chapter 16. An agreed-upon evidence protocol should be in place before the scene is processed. This protocol should include the means of identifying, documenting, marking, preserving, transporting, and storing evidence.

27.6.1.1 Evidence Custodian. As part of the understanding and agreements, one or more evidence custodians should be designated to manage various aspects of the evidence collection process, including a master evidence log, bagging and tagging, preservation, and storage of evidence.

27.6.1.2 Interested Party Responsibility. An interested party is responsible for identifying the evidence it wants collected and providing sufficient information to the evidence custodian to ensure the collection of the identified evidence and the preservation of the scene for further investigative activities.

27.6.2 Evidence Removal from the Scene. All known interested parties should have the opportunity to document evidence items in place prior to the items being disturbed or removed from the scene. Evidence should be removed according to agreed-upon evidence protocols.

27.6.3 Evidence Storage. Storage of evidence requires a secure location with a log-in sheet for persons accessing the area. All evidence collected should be available to all interested parties involved in the investigation. The interested parties should agree to procedures for accessing stored evidence, such as giving notice to all interested parties and requiring the presence of the evidence custodian when access to the evidence is given.

27.6.4 Evidence Inspections.

27.6.4.1 Nondestructive Inspections. Inspections may be conducted so that interested parties can view, make notes, photograph, measure, and make other visual observations of the evidence in a nondestructive manner. All interested parties need not be present for such nondestructive inspections.

27.6.4.2 Destructive Inspections. Inspections may occur where it could be reasonably anticipated that evidence will be altered or destroyed. Before such inspections occur, all interested parties should receive notice and a protocol should be agreed upon for the inspections.

27.6.4.3 Testing of Evidence. Prior to any testing of evidence, all interested parties should be notified, protocols should be developed, and testing facilities should be agreed upon. Each of the interested parties may decide to have its own expert view or participate in the testing.

27.7 Logistics. Providing logistical support will facilitate the investigation, protect the health and well-being of the investigators, protect the investigation site and the evidence, and reduce conflicts. The cost of providing logistical support should be addressed in the agreements or understandings among the parties.

27.7.1 Transportation. Travel to and within the investigation site by a large number of people may be difficult because of its location, size, and other factors. Arrangements for the transportation of investigators and their equipment to the investigation site or for parking on the site should be addressed as early as possible during the investigation by representatives of all the interested parties. Shuttle service from off-site parking areas, designated parking facilities, parking permits, and passenger/equipment drop-off points can be utilized to accommodate the needs of all interested parties.

27.7.2 Equipment. The use of equipment, such as construction vehicles, lifting apparatus, radiograph equipment, and modes of transportation may be necessary. The type and specific use of equipment should be part of the agreements and understandings among the interested parties.

27.7.3 Investigation Site Security. Investigation site security may include physically erected barriers around the site and scene, the use of public or private guards, and the issuance of identification badges. The location or nature of the activity conducted at the investigation site might require background investigations, security clearances, and specialized safety training for all persons entering.

27.7.4 Decontamination. Decontamination stations should be established at appropriate points, such as scene entrances and exits and food or beverages areas.

27.7.5 Environmental. Environmental concerns, such as spills, air quality, disposal, and regulations should be addressed. The entity controlling the investigation site may need to arrange for environmental monitoring, control, and clean-up services. Additional information may be obtained from the appropriate governmental agency regulating environmental concerns, such as the Environmental Protection Agency (EPA), the United States Coast Guard, the state and local environmental authorities.

27.7.6 Communications. A large incident area may create communication problems for personnel working within the investigation site. A command post using temporary trailers or other facilities may be needed. Communication may be provided through either mobile or fixed means, such as portable radios, temporary hardwired phone systems, loudspeaker systems, or cellular telephones. Depending on the size and scope of the incident, multiple methods of communication may be required between the entity controlling the investigation site, interested parties, teams entering the scene, and others. Additionally, a preset signal, such as an alarm or air horn, can be used to notify persons on site to immediately exit the scene and report to the safe meeting area.

27.7.7 Sanitary and Comfort Needs. Provisions should be made for sanitary facilities and drinking water. An uncontaminated area should be available for eating, resting, and meeting. Weather conditions may require that provisions be made for a shaded resting place or a warm-up area.

27.7.8 Trash Disposal and Removal. Containers for the disposal of expendables and ordinary trash should be provided so trash is not introduced into the investigation site. Trash receptacles should be removed from the site as necessary.

27.7.9 Snow and Ice Removal. Snow may be present at the commencement of the investigation or may fall on the investigation site during the investigation. Suppression activities may also result in ice covering the investigation site, which presents a safety hazard and an impediment to the investigation. If the investigation cannot wait for the snow or ice to melt by natural means, structures can be tented or tarped and heaters can be used to melt the snow or ice. If heaters are to be used, extreme care should be taken not to introduce fuels or products of combustion that might contaminate the scene. Care must also be taken to avoid the introduction of melting agents into the scene.

27.7.10 Lighting. Temporary lighting may be required. It may be better to install temporary lighting than to use portable lights and flashlights for extended periods of time. The need to install temporary lighting will be a function of current lighting conditions, the estimated time the investigation will take, and the availability of other lighting and electrical power. Whenever practical, photographs should be taken prior to the installation of temporary lighting to allow better documentation of the undisturbed scene.

27.7.11 Evidence Storage. Provisions should be made to store evidence on site, off site, or both. The storage facility should be appropriate for the type of material being stored. Temporary on-site storage may occur, with subsequent evidence relocation to a more permanent location.

27.8 Site and Scene Safety. Safety is the responsibility of all interested parties, but in some situations it may be necessary to designate a safety officer who will be responsible for monitoring conditions at the investigation site to ensure the safety of all interested parties. The understandings and agreements should address safety issues for the interested parties or anyone else who may enter the investigation site, such as tenants, contractors, or other logistical support personnel. A site safety assessment will need to be conducted prior to any investigative activities or removal of debris. Safety concerns may be of such a magnitude that the retention of a company specializing in the evaluation of site hazards, the implementation of safety plans, and the providing of safety officers should be considered. *(See Chapter 12.)* Site personnel must follow the safety officer's directives, but may express objections and request reentry. A preliminary safety briefing specific to the particular site and scene may be required for anyone entering the investigation site. A record of those having attended such a safety briefing should be maintained. Unresolved objections should be recorded and handled as with disagreements, see 27.4.9.

27.8.1 Continual Safety Monitoring. It may be necessary to continue to monitor the air quality, environmental conditions, structural stability, and other hazards during the investigation. For example, a building's ventilation system should be evaluated for possible use in improving air quality. If the air or the environment cannot be rendered safe, the interested parties may be required to wear personal protective equipment.

Chapter 28 Marine Fire Investigations

28.1* Introduction. This chapter deals with factors related to the investigations of fires involving recreational boats generally defined as less than 19 m (65 ft) in length. Included in this discussion are motor boats, sailing boats, yachts, and other watercraft. Information relative to the investigation of boat fires includes construction materials, systems, and failure modes. Additional information regarding boat systems and standards is available in NFPA 302, *Fire Protection Standard for Pleasure and Commercial Motor Craft.* For investigation of boat fires outside U.S. boundaries, additional information may be obtained in local and regional standards and codes. The use of this chapter in the investigation of marine fires within vessels exceeding this length may still be of value. *(See Figure 28.1(a) and Figure 28.1(b) for common terminology.)*

28.2 Powerboat and Sailboat Terminology. The following are terms that apply to Figure 28.1(a) and Figure 28.1(b):

(1) *Accommodation space.* Space designed for living purposes.
(2) *Adrift.* Loose, not on moorings or towline.
(3) *Afloat.* (1) In a floating position or condition; (2) On a boat or ship away from the shore; at sea. Borne on or as if on the water.
(4) *Aft.* Toward the rear of the vessel.
(5) *Aground.* Touching or fast to the bottom.
(6) *Below.* Beneath the deck.
(7) *Boat.* Any vessel manufactured or used primarily for noncommercial use; leased, rented, or chartered to another for the latter's noncommercial use; or operated as an un-inspected passenger vessel subject to the requirements of 46 CFR Chapter 1, subchapter C.
(8) *Bulkhead.* A vertical partition separating compartments.
(9) *Cabin.* A compartment for passengers or crew.
(10) *Capsize.* To turn over.
(11) *Deck.* A permanent covering over a compartment, hull or any part thereof.
(12) *Dock.* A protected water area in which vessels are moored. The term is often used to denote a pier or a wharf.
(13) *Dorade vent.* Designed deck box ventilation to keep water out with a baffle while letting air in below decks.

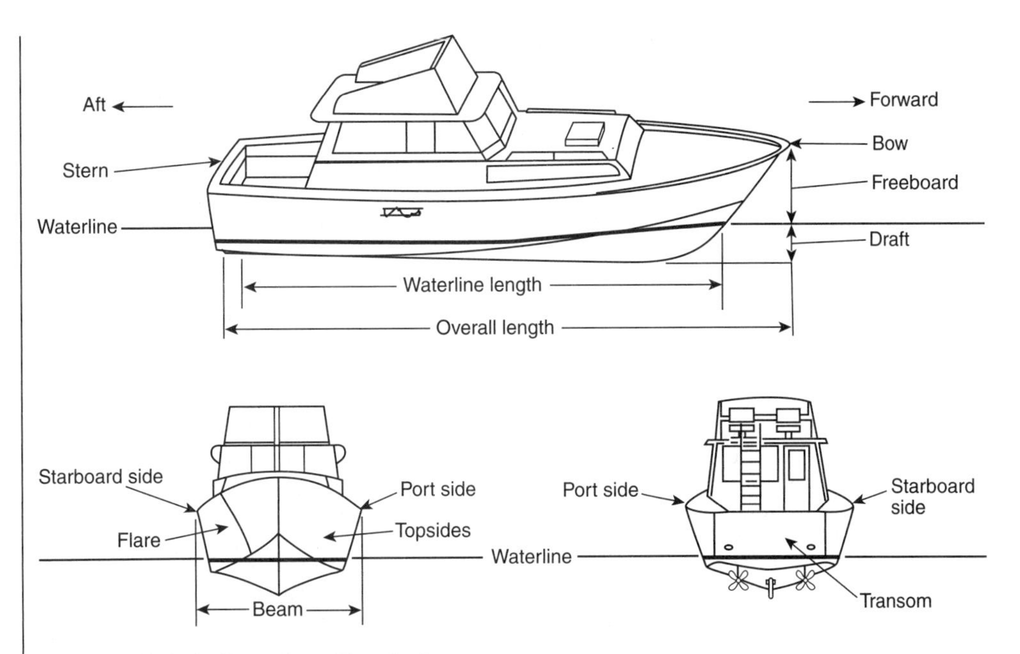

FIGURE 28.1(a) Powerboat Terminology.

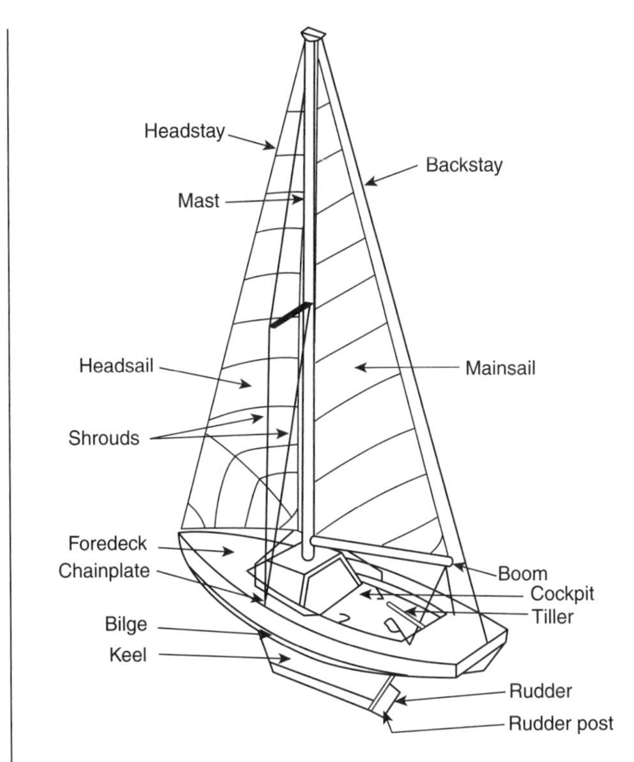

FIGURE 28.1(b) Sailboat Terminology.

(14) *Fender.* A cushion, placed between boats, or between a boat and a pier, to prevent damage.
(15) *Forward.* Toward the bow of the boat.
(16) *Freeboard.* The vertical distance between the water line and the gunwale.
(17) *Galley.* The kitchen area of a boat.
(18) *Gear.* A general term for ropes, blocks, tackle, and other equipment.
(19) *Gunwale.* The upper edge or surface of a boat's side.
(20) *Hatch.* An opening in a boat's deck fitted with a watertight cover.
(21) *Hold.* A compartment below deck in a large vessel, used solely for carrying cargo.
(22) *Hull.* The structural body of a boat, not including superstructure, mast or rigging.
(23) *Inboard.* (1) More toward the center of a boat, inside; (2) An engine fitted inside a boat.
(24) *Inboard/Out-Drive (I/O).* A propulsion system consisting of an engine fitted inside a boat with a stern drive, similar to the lower unit of an outboard motor attached to the transom.
(25) *Outboard.* (1) Toward or beyond the boat's sides. (2) A detachable engine mounted on a boat's stern.
(26) *Overboard.* Over the side or out of the boat.
(27) *Port.* The left side of a boat when looking forward.
(28) *Rub Rail.* The rubberized, plastic, or metal bumper that extends along both sides of the vessel, usually immediately below the gunwales.
(29) *Shore Power.* Electrical power supplied from shore via a cord set.
(30) *Sole.* (1) Cabin or salon floor; (2) Timber extensions on the bottom of the rudder. (3) The molded fiberglass deck of a cockpit.
(31) *Starboard.* The right side of a boat when looking forward.
(32) *Superstructure.* The cabins and other structures above deck.
(33) *Topside.* The sides of a vessel between the waterline and the deck; sometimes referring to onto or above the deck.
(34) *Transom.* The stern cross-section of a square sterned boat.
(35) *Underway.* Vessel in motion, i.e., when not moored, at anchor, or aground.
(36) *Vessel.* Includes every description of watercraft, other than seaplane on the water, used or capable of being used as a means of transportation on the water.
(37) *Waterline.* A line painted on a hull which shows the point to which a boat sinks when it is properly trimmed.

28.3 Boat Investigation Safety.

28.3.1 As with land-based fire investigations, safety should always be the investigator's first concern. Scene documentation can also be conducted during safety assessment. See Chapter 12 for further information.

28.3.2 Boats on land should initially be inspected to determine if they are stable before boarding. If not, the boat should be stabilized prior to boarding. Boat shore power connections and the battery supply circuit should be de-energized if they present a safety hazard. The batteries and direct current systems should be inspected with caution. The investigator should employ personal protective equipment that is appropriate for the anticipated hazards.

28.3.3 Boats afloat should be approached with extreme caution, as the boat may have water within the hull. Any volume of water within the hull may create an unstable condition within the boat that could cause it to capsize or sink. The water should be removed from the boat (this is called dewatering). Then the boat can be removed from the water and stabilized on land, if necessary, prior to continuing the investigation. The investigator should consider wearing a personal floatation device while working on or near the water.

28.3.4 Underwater inspections of submerged boats present a potential entrapment hazard for a diver.

28.3.5 Specific Safety Concerns.

28.3.5.1 Confined Spaces. By their nature, fire investigations on boats may involve working in confined spaces. The investigator should be aware of confined space entry concerns (e.g. entry/egress and atmospheric issues), and appropriate precautions should be taken. Prior to entry, the investigator should ensure the space does not contain hazardous levels of explosive or toxic vapors or gases (i.e., carbon monoxide) or is not oxygen deficient. The hazards of the space should be evaluated before entering, and the appropriate level of personal protective equipment should be worn. Lights and other equipment should be intrinsically safe and suitable for such use.

28.3.5.1.1 Prior to entering spaces covered by automatic fire extinguishing systems, the investigator should ensure that the system is disabled or inoperative. Should the system operate while the investigator is inside the compartment, the atmosphere may become toxic or oxygen deficient.

28.3.5.2 Airborne Particulates. In a fiberglass reinforced plastic (FRP) boat fire, the resin used as part of the construction process generally burns away, leaving small, irritating particles of fiberglass. These particles, when combined with burned resin, are highly irritating to the respiratory system and the appropriate level of personal protective equipment, including respiratory protection, should be worn as indicated by the level of hazard.

28.3.5.3 Identify and Assess Energy Sources. There are numerous energy sources onboard a boat; precautions must be taken to ensure they are disabled to prevent personal injury, as well as possible recurrence of a fire.

28.3.5.3.1 Batteries. After proper documentation and photography, battery cables, should be disconnected at each battery, as there may be more than one battery and multiple locations. Prior to disconnecting the battery cables, the investigator should ensure the atmosphere is properly ventilated and free of explosive vapors. During disconnection, arcs from battery terminals can cause a fire or explosion.

28.3.5.3.2 Inverters. Inverters that convert dc to 120/240-volt ac power are often found on boats and, after proper documentation and photography, should be disabled by disconnecting the dc input power.

28.3.5.3.3 Shore Power. After proper documentation, shore power sources (cord sets) should be de-energized and disconnected. This should be done prior to conducting an investigation.

28.3.5.4 Fuel Leaks. Engine fuels, heating fuels, and cooking fuels, including propane from onboard LP gas systems can leak due to fire and explosion damage or during investigative activity. Such leaks can produce additional fire hazards. Investigators should be alert to the presence of such leaks and be prepared to prevent or mitigate these hazards.

28.3.5.5 Sewage Holding Tank. Boats that contain sewage holding tanks may accumulate methane gas. Proper venting and explosion prevention techniques must be employed during the fire scene investigation. Sewage holding tanks may be damaged, and if the contents escape, a biological hazard may be encountered. Care should be taken when working in areas containing sewage.

28.3.5.6 Hydrogen Gas. Hydrogen gas may be present in any battery compartment. Care should be taken to avoid static discharges, short circuiting, and arcing during the removal of battery cables and batteries.

28.3.5.7 Other Hydrocarbon Contaminants. The various fluids found onboard boats, as well as the residue left from combustion, contain hydrocarbons that are generally environmental contaminants. Additionally, the presence of these contaminants poses a fire hazard and also presents a danger for slips and falls.

28.3.5.8 Stability. One of the most dangerous situations an investigator may encounter when examining a boat afloat following a fire is lack of stability. Water entering bilges and lower sections of the boat from fire-fighting activities can cause instability. The investigator should ensure the boat is stable and safe to work in prior to entry for investigation purposes. The boat should be de-watered prior to entry for the investigation. If the boat lists to one side, or turns over, the investigator may become trapped.

28.3.5.9 Damage to the Structure of the Boat. The boat may have sustained sufficient damage due to the fire, causing it to become unsafe. The structural integrity may be compromised and the weight of the investigator may cause collapse of decking, further resulting in instability. Fires involving the interior can cause structural damage, weakening of the decking, or weakening of structural supports, similar to damage of land-based structures.

28.3.5.10 Wharves, Docks, and Jetties. Because wharves, docks, and jetties are close to water, seaweed infestation, slime, and slippery surfaces become a hazard to the investigator. The investigator should take care when entering such areas and accessing boats subject to fire damage. Care should also be taken to ensure the structural stability of wharves, docks, and jetties that are attached or in contact with a vessel that has been subjected to fire. Listing vessels may cause these structures to be weakened and/or collapse if they are in contact with the structure, or still tied to these structures.

28.3.5.11 Submerged Boat. Boats under the water or submerged should be inspected and photographically documented prior to recovery, if possible. When it becomes necessary to conduct an investigation of a boat that has sunk, or is partially submerged, use qualified personnel to carry out the underwater observation and scene search. Recovery efforts should be managed in order to reduce disruption and alteration of the boat and its appurtenances prior to the commencement of topside or land-based investigation activities. Underwater operations may be monitored from the surface. Only qualified divers should conduct underwater investigations.

28.3.5.12 Visual Distress Signals and Pyrotechnics. Many boats carry distress-signaling devices. These devices can consist of pyrotechnics that can be harmful to the investigator if accidental activation occurs. The investigator should take care to discover and secure any such devices.

28.3.6 Openings. Boats are fitted with all types of openings, some for access to below-deck spaces, and others for access to storage. Some openings are evident, others are not; some are covered by hatches. After a fire, damage may have occurred to hatches and other types of covers concealing openings in the deck. These deck areas may be covered by water and openings may be concealed, creating a hazard for the unwary investigator.

28.4 System Identification and Function.

28.4.1 Fuel Systems: Propulsion and Auxiliary.

28.4.1.1 Vacuum/Low Pressure Carbureted. Carbureted inboard and inboard/out-drive (I/O) engine systems onboard boats differ from automotive engines in that the carburetors do not allow for the escape of fuel from the venturi in the event of flooding. This is accomplished, in part, by a change in the gasket set for the carburetor when in marine use. Carburetors in marine use are required to meet 33 CFR 183.526, "Carburetors." Carburetors are required to be fitted with backfire flame arresters.

28.4.1.2* High Pressure/Marine Fuel Injection Systems (Including Return Systems). Fuel injection systems on inboard and I/O engines in boats generally include a throttle body, plenum and fuel rail assembly, knock sensor, and engine control module. The design of the system is unique to each marine engine manufacturer. In some instances, the fuel injection system for a particular model engine may have three or more different versions. It is important that the investigator record the engine serial number and contact the manufacturer to establish which system design is applicable. The fuel storage and delivery system onboard boats are low-pressure systems and the tank is vented.

28.4.1.3* Diesel. Diesel engines installed in boats utilize manufacturer specific fuel injection systems. Some of these engines require a 24-volt starting system, while the boat operates on a 12-volt system. The combustion air requirements and ambient engine room temperature require large exchanges of air in the compartment. The fuel system requirements are similar to gasoline.

28.4.2 Fuel Systems: Cooking and Heating.

28.4.2.1* Liquefied Petroleum Gases. Liquefied petroleum gases (LPG) can be used as a cooking or heating fuel. The LPG cylinder must be within an enclosure that is vented from the bottom to the outside of the boat. If LPG-fueled cooking devices are installed then all other devices must be ignition protected. Cooking devices with integral LPG containers of less than 450 g (16 oz) are covered under ABYC A-30, *Cooking Appliances with Integral LPG Cylinders*. LPG can be used as a cooking or heating fuel when the appliance meets the requirements of ABYC A-26, *LPG and CNG Fueled Appliances*.

28.4.2.2* Compressed Natural Gas. Compressed natural gas (CNG) has been used as a cooking fuel onboard some boats since the 1980s; however, its popularity has declined in recent years. The CNG cylinder may be located in accommodation spaces if the cylinder is less than 2.8 m^3 (100 ft^3). CNG can be used as a cooking or heating fuel when the appliance meets the requirement of ABYC A-26, *LPG and CNG Fueled Appliances*.

28.4.2.3* Alcohol. Alcohol may be utilized as a fuel in galley ranges. The fuel tank may be an integral component of the stove or a separate container. The fuel can be pressurized by the use of a hand pump.

28.4.2.4 Solid Fuels. Solid fuels, including wood and charcoal, are used in wood burning appliances as covered in NFPA 302, *Fire Protection Standard for Pleasure and Commercial Motor Craft*.

28.4.2.5 Diesel. Diesel fuel is sometimes used for cooking and heating appliances when in accordance with NFPA 302, *Fire Protection Standard for Pleasure and Commercial Motor Craft*, and ABYC A-3, *Galley Stoves*, and ABYC A-7, *Boat Heating Systems*.

28.4.3 Turbochargers/Super Chargers. Diesel and gasoline inboard and I/O powered boats can be fitted with turbochargers/super chargers. *(See Figure 28.4.3)* These units often use engine lubrication oil for lubrication and cooling purposes. The units are fitted with heat blankets, water jackets, or a combination of both as their operating temperatures exceed the ignition temperature of surrounding engine room materials. Turbochargers are powered by hot exhaust gases and send pressurized air into the combustion intake. Super chargers, powered by an engine drive belt, inject compressed air into the combustion intake.

28.4.4 Exhaust System.

28.4.4.1 Dry exhaust systems generally exhaust propulsion or auxiliary combustion gases vertically through a pipe that is covered with a heat-insulating blanket. This is not commonly found in recreational boats.

28.4.4.2 Wet exhaust systems utilize seawater (fresh or salt) that is injected into the exhaust elbow of the engine and travels with the exhaust until it exits the boat. The muffler (if provided) and the hoses of the wet system are specifically designed for the expected temperature and environmental conditions in accordance with ABYC P-1, *Installation of Exhaust Systems*, and NFPA 302, *Fire Protection Standard for Pleasure and Commercial Motor Craft*.

28.4.4.3 De-watered exhaust systems remove the water from the exhaust stream at the muffler and usually route it to the transom while the exhaust gases are routed through the bottom of the boat, in accordance with ABYC P-1, *Installation of Exhaust Systems*.

28.4.5 Electrical Systems.

28.4.5.1 Alternating current (ac) on boats is provided by shore power, generators, or inverters. When alternating current is supplied via a "shore power" receptacle it should meet NFPA 303, *Fire Protection Standard for Marinas and Boatyards*, and *NFPA 70, National Electrical Code*, Article 555. Alternating current systems onboard boats should meet 33 CFR 183.435, "Conductors in Circuits of 50 Volts or More." Additional information on ac systems can be found in Chapter 8 of NFPA 302, *Fire Protection Standard for Pleasure and Commercial Motor Craft*, and ABYC E-11, *AC and DC Electrical Systems*, and 33 CFR 183.401–183.460, "Electrical Systems."

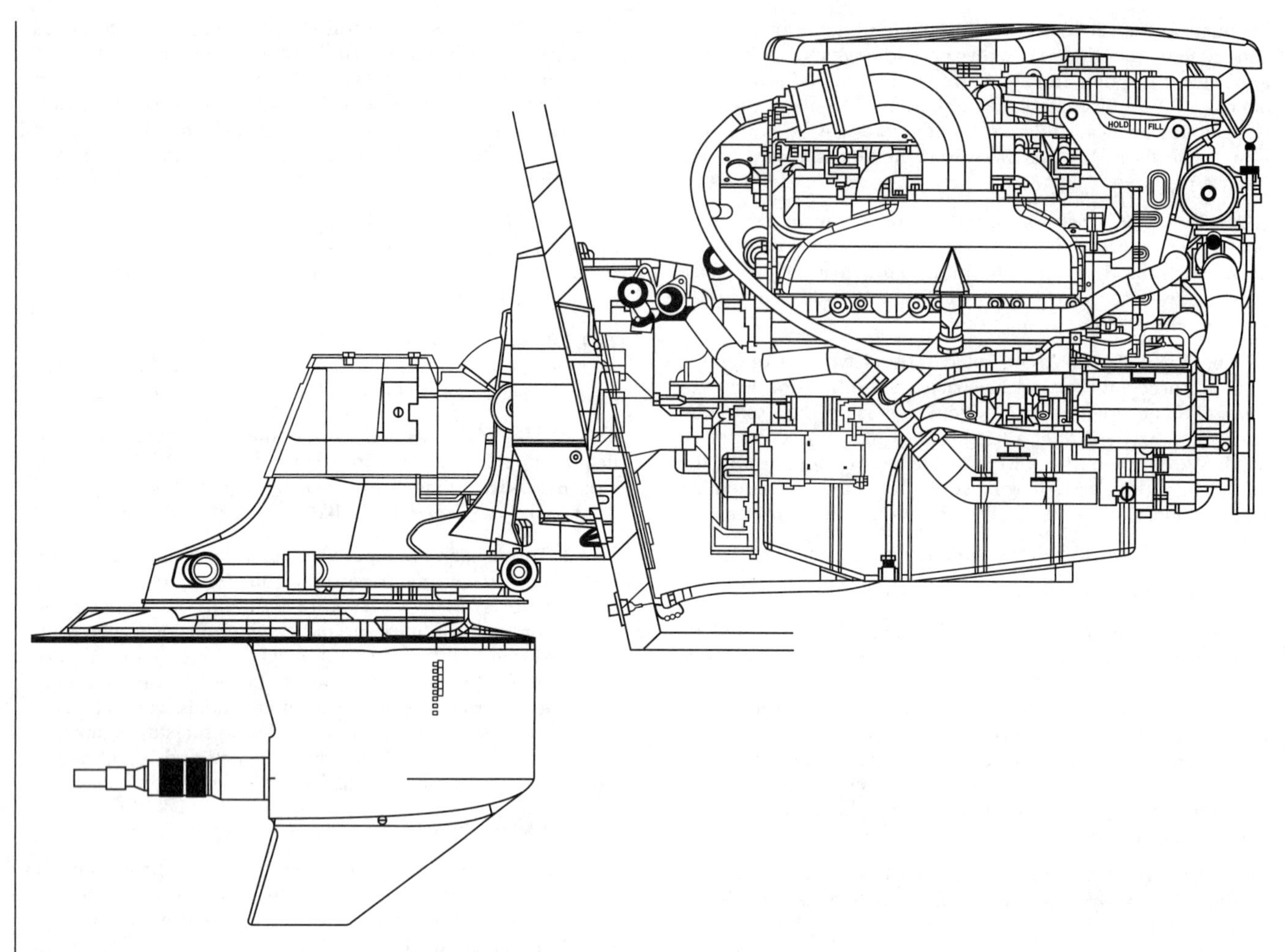

FIGURE 28.4.3 Inboard/Out-Drive Profile.

28.4.5.2 Direct current (dc) systems onboard boats are typically supplied by batteries and provide electrical power for lighting, communications, and navigational equipment. It should meet 33 CFR 183.430, "Conductors in Circuits of 50 Volts or Less," and 33 CFR 183.440, "Secondary Circuits of Ignition Systems." Additional information on dc systems can be found in NFPA 302, *Fire Protection Standard for Pleasure and Commercial Motor Craft,* ABYC E-11, *AC and DC Electrical Systems,* and 33 CFR 183.401–183.460, "Electrical systems."

28.4.6 Engine Cooling Systems.

28.4.6.1 Liquid coolant systems on inboard and I/O boats are categorized into two groupings: seawater cooled and closed cooled. Seawater-cooled engines pick up seawater via a water pump and circulate it through the engine and discharge it through the exhaust system. Closed-cooled systems circulate the seawater through a heat exchanger and then into the exhaust system. The heat exchanger coolant is usually a 50/50 mixture of propylene glycol and water. Inboard, I/O, and auxiliary engine applications in boats are not air cooled.

28.4.7 Ventilation.

28.4.7.1 Permanently installed fuel tanks are vented from the top via a hose routed overboard and are equipped with a flame arrester. The ventilation of fuel tanks is mandated under 33 CFR 183.520, "Fuel tank vent systems." Additional information can be found in 5.6.10–5.6.12 of NFPA 302, *Fire Protection Standard for Pleasure and Commercial Motor Craft,* and ABYC H-24.13, *Gasoline Fuel Systems.*

28.4.7.2* Gasoline fuel tank compartments should be naturally ventilated (not forced) to the atmosphere in accordance with 33 CFR 183.601, "Ventilation—applicability," and 33 CFR 183.620, "Natural ventilation systems." For requirements regarding ventilation of diesel tank compartments refer to ABYC H-32, *Ventilation of Boats Using Diesel Fuel.*

28.4.7.3* Gasoline engine and machinery compartments are required to be power ventilated in accordance with 33 CFR 183.610, "Powered ventilation system."

28.4.7.4 Accommodation spaces in boats may be fitted with hatches and port lights (portholes), which can provide natural ventilation. Some boats are fitted with dorade vents or other ventilation cowls on the cabin top that provide ventilation at all times. The presence of these devices may result in unusual fire ventilation patterns that should be considered during the investigation.

28.4.7.5 The bilges in the engine and fuel compartments are required to be ventilated as specified in this chapter. The bilge in the accommodation areas is not required to be ventilated.

28.4.8 Transmissions.

28.4.8.1 Mechanical gear transmissions are generally not found on inboard engines. Outboard and I/O propulsion engines have a mechanical transmission located in the gear case.

28.4.8.2 Hydraulic-geared transmissions are found on inboard and some I/O engines (performance boats). These units have their own oil and cooler. The lube oil is generally SAE 90 weight.

28.4.9 Accessories.

28.4.9.1 Air-conditioning compressors are integrated in heat exchanger units that utilize seawater as a medium to transfer the heat to and from the coils. These units are generally located within the accommodation space, but may be located in the engine room.

28.4.9.2 Power steering is accomplished by a pump that is belt driven and located on the propulsion engine.

28.4.9.3 Refrigeration compressors are often located within the appliance and are usually powered by ac and/or dc current. In the event that an LPG refrigerator is in use in an accommodation space, it is required to meet UL 1500, *Standard for Safety Ignition Protection Test for Marine Products.*

28.4.9.4 Electrical power generation is to be provided by permanently mounted systems utilizing the propulsion engine fuel type (gasoline or diesel) with its own fuel delivery system that meets all applicable requirements of the fuel type. Units are usually mounted in the engine room and, if gasoline powered, should be ignition protected as required by 33 CFR 183.410, "Ignition protection." Other sources of electrical power supply that should be considered in the investigation are battery chargers, inverters, portable generators, etc.

28.4.9.5 Hydraulic Systems.

28.4.9.5.1 Trim tabs are powered via a pump motor that is generally ignition protected and that is located adjacent to a small hydraulic tank mounted near the transom.

28.4.9.5.2* Hydraulic thruster systems are available, but are not generally used in boats. Typically, thrusters are electrically driven motorized propellers located in a tunnel forward in the boat. These systems are generally 12 or 24 volts dc and utilize a 200 amp fuse installed near the thruster.

28.4.9.5.3 Hydraulic steering systems use either mechanical or electric pumps and piping, either rigid or flexible, in a closed system to assist in steering the boat. This closed system contains hydraulic fluid along with electrical connections that may pose an ignition source for escaping fluids.

28.5 Exterior.

28.5.1 Hull Construction. The investigator should be familiar with the composition of the boat. General building materials for boats are wood, steel, aluminum, ferrocement, and fiberglass-reinforced plastic (FRP). Wood and FRP construction materials should be considered when identifying fuel loads. Some FRP boats utilize core material such as balsa or high-density foam, which during a fire acts as a thermal conduction insulator.

28.5.2 Superstructure Construction Material. Materials used in construction of cabins and other structures above the main deck are generally the same as used for the hull.

28.5.3 Deck. Decks are generally the same as hull construction materials; however, they may be inlayed with wood, which adds to the fuel load.

28.5.4 Exterior Accessories. Accessories are items or equipment located on the exterior of a boat, such as communications equipment, antennas, navigational aids, search lights, navigation lights, outriggers, handrails, lines, fenders, personal floatation devices (PFDs), life rafts, seat cushions, masts, booms, and sails (sheets). Some or all of these items or equipment may be combustible.

28.6 Interior.

28.6.1 Construction Materials. Due to weight concerns, the interior construction of a boat may utilize different materials than those found in a normal dwelling. Interiors are generally found to be constructed from FRP material and/or conventional building materials, such as solid stock woods and exterior grade plywood and veneers. Noncombustible materials such as steel or aluminum may be used for structural components such as bulkheads.

28.6.2 Finishes. Interior finishes, including wood paneling and carpeted floor (sole), are generally similar to those found in residences and recreational vehicles. Abnormal fire loading may exist in the form of carpeting and other synthetic materials utilized as vertical and overhead finishes. Organic oils (linseed or tung oil) and varnish, paints, etc., that are used on interior finishes may contribute to ease of ignition and fire spread. Counter surfaces may be synthetic (plastic laminate) or wood veneer.

28.6.2.1 Accommodation Furnishings. Accommodation furnishings are those items within a boat that are located in areas that are designed to be occupied by people. Fabric-covered foams utilized in beds and other types of furniture are commonly found, and are similar to those in residences and some motor vehicles, including recreational vehicles.

28.6.2.2 Interior Accessories. Interior accessories may include candles, oil lamps, and other open flame devices. Small appliances that are normally found in residences, including televisions, radios, stereos, and fans are used in boats. Such interior accessories can be a source of ignition.

28.6.2.3 Engine/Machinery Compartments. Engine and machinery compartments on boats may contain, in addition to propulsion engines, batteries, auxiliary generator sets inverters (for the production of 120/240-volt ac power onboard the boat), storage tanks (fuel, water, sewage, hydraulic fluid), bilge pumps, ac and dc electrical system wiring, and in some cases ac and dc electrical panels. Engine compartments may be equipped with fixed automatic fire-extinguishing system equipment (either "engineered" or "pre-engineered"). Prior to entering spaces covered by automatic fire-extinguishing systems, the investigator should ensure that the system is disabled or inoperative. Should the system operate while the investigator is inside the compartment, the atmosphere may become toxic, or render the area uninhabitable due to a lack of oxygen from the discharge of the extinguishing agent.

28.6.2.4 Flammable/Explosive Vapor Detectors. These detectors are usually located in the engine compartment(s) and are designed to detect fugitive ignitible vapors (*see Figure 28.6.2.4*). Detectors may also be found in the lower portion of accommodation spaces. Locating and examining these detectors and their operating circuits may provide data indicating whether the device was functional at the time of the fire.

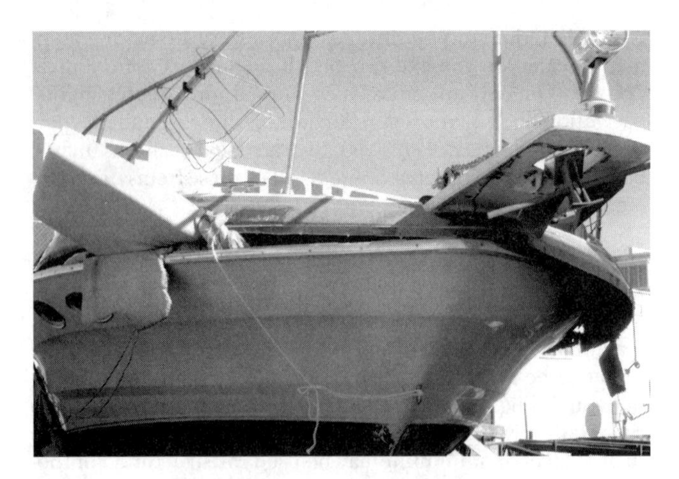

FIGURE 28.6.2.4 Interior Vapor Explosion.

28.6.2.5 Storage and Holds. Storage and holds are often found in lazarettes, below deck, or under bunks in accommodation spaces. These are provided for the storage of miscellaneous gear and supplies and may contain many different types of combustible materials that should be taken into consideration when conducting a fire investigation. Improper storage of rags and materials containing remnants of organic oils used in maintaining interior finishes may be susceptible to spontaneous heating.

28.6.2.6 Fuel Tanks. Fuel tanks on boats are generally constructed of steel, aluminum, cross-linked polyethylene, or fiberglass and fiberglass with fire-retardant resin. These tanks are subject to rigid construction and testing standards pursuant to 33 CFR 183, Subpart J. Fuel fittings and hoses are required to be approved and display a U.S. Coast Guard designation. The investigator may encounter fittings and hoses made of neoprene, synthetic rubber, multilayer flexible laminates, metal, or nylon, all of which may or may not be approved. They may be a source of leakage due to deterioration caused by vibration, age, heat, mechanical damage, etc. or, from corrosion due to water (particularly salt water). The investigator should ensure that the fittings and hoses are suitable for the particular use.

28.7 Propulsion Systems.

28.7.1 Electric Systems. Although generally not used for the main source of propulsion, many boats are equipped with electric trolling motors powered by batteries. These systems include the motor, one or more batteries, and the electrical conductors used to supply the energy from the batteries to the motor.

28.7.2 Fuels for Boats with Motorized Propulsion Systems.

28.7.2.1 Fuel Systems. Fuel systems encountered in boats include those for the engine, appliances, and electrical generators.

28.7.2.1.1 Engines. Three general categories of commonly encountered engines and their attendant fuel systems are outboard engines with self-contained or portable fuel tanks; inboard gasoline, and inboard diesel engines.

28.7.2.1.1.1 Outboard Engines (Outboard Motors). Outboard engines include two-cycle and four-cycle gasoline engines, with either carbureted or fuel-injected systems. Two-cycle engines use fuel and oil mixtures, with the oil either premixed with the gasoline, or supplied in a separate reservoir and automatically mixed with the fuel. The basic principles of four-cycle systems are similar to automobile engines as discussed in Chapter 25. With outboard engines the fuel is delivered by means of a low-pressure fuel delivery system that includes a pump. The high pressure needed for fuel-injected systems is generated by an engine-mounted fuel pump. *(See Figure 28.7.2.1.1.1.)*

FIGURE 28.7.2.1.1.1 Outboard Fuel Filter Fire.

28.7.2.1.1.2 Inboard Gasoline Engines. Inboard gasoline engines are generally four-cycle engines. The fuel system includes built in fuel tanks, fuel lines, filters, and fuel pumps for delivering fuel to the engine. The tanks are required to be vented. When multiple fuel tanks are used, they are often connected with an equalizing line, commonly known as a manifold, to ensure the fuel weight remains evenly distributed. The plate around the fill port is required to be connected to the boat ground, and proper fueling procedure requires that the supply nozzle be in contact with the deck plate before the dock pump is started to prevent static discharge during refueling. The purpose of the fill plate is to prevent incidental fuel spills from draining into the hull. All fittings to the fuel tank are required to be located on the top of the tank, and fuel lines must be routed above tank level to prevent accidental draining of the tank from an open valve or leaking line. Boats built after 1977 are required to be equipped with an anti-siphon valve installed at the tank if the top of the engine is below the level of the top of the fuel tank (see 7.5.3.5 of NFPA 302, *Fire Protection Standard for Pleasure and Commercial Motor Craft*). Fuel hose requirements are specified in NFPA 302 and other applicable references such as 33 CFR 183.555, "Fuel systems."

28.7.2.1.1.3 Diesel Engines. The basic operating principles of inboard diesel engines are similar to automotive diesel systems, as discussed in Chapter 25. The high-pressure fuel injection systems for diesel engines typically include a low-pressure return line to the fuel tank.

28.7.2.1.1.4 Propulsion System Fluids. All propulsion systems contain lubricating fluids that may be sampled and analyzed after a fire to determine prefire conditions.

28.7.2.2 Appliance Fuel Systems. Liquefied petroleum gas (LPG) is the most commonly used fuel for onboard cooking, heating, and refrigeration systems. Considerations are similar to those for automotive systems, as discussed in Chapter 26. An added consideration for boats is the risk that vapors from leaks may accumulate below deck or in the hull.

28.7.2.3 Electric Generators. Electric generators may be used to supply power for a variety of appliances on a boat. Generators are typically powered using a separate fuel line from the boat's fuel tank.

28.7.3 Other Fuel Systems Used for Propulsion.

28.7.3.1 Most boats are powered by conventional fuels such as diesel and gasoline; however, there are other fuels (e.g., wood, charcoal, coal, and paraffin) that are still being used in steam-powered propulsion systems.

28.8 Ignition Sources. The sources of ignition energy in boat fires are often the same as those associated with structural and vehicle fires (e.g., arcs, overloaded wiring, open flames, and smoking materials). There are some unique sources that should be considered, such as the hot surfaces of turbochargers and manifolds. Because some of these ignition sources may be difficult to identify following a fire, the descriptions in 7.4.1 through 7.4.5 are provided to assist in their recognition.

28.8.1 Open Flames. A common source of an open flame in a carbureted engine is a backfire through an unprotected carburetor. *(See Figure 28.8.1.)* In order to prevent these flames from causing a fire, all inboard and I/O boats are requried to have an approved backfire flame arrestor attached to the air intake with a flame-tight connection. Backfire flame arrestors must be approved for marine use by the United States Coast Guard or comply with Society of Automotive Engineers Standard SAE J-1928. A frequent cause of engine compartment fires is the failure of the owner to keep the arrestor clean and free of oil or gasoline residues. In boats, operating burners and ovens are open-flame ignition sources. Since 1977, appliances with pilot flames have been prohibited on gasoline fueled boats pursuant to U.S. Federal law *(see 33CFR183.410, "Ignition protection").*

28.8.2 Electrical Sources. When the engine is not running, the sources of electrical power in a boat are the batteries, inverters, or generators. *(See Figure 28.8.2.)* Without a battery disconnect switch, a number of components remain electrically connected to the battery, even though the ignition switch is off and the engine is off. These circuits and components, such as an alternator, ignition switch, or inverter and dc power system, can fail when the engine is off. For instance, the bilge pump is normally installed to remain energized at all times. The investigator should determine if the boat's engine(s) was running at the time of the fire. A boat with its engine running has more potential sources of ignition (e.g., ignition wires, alternators, electrical fuel pumps) than when not running. Although marine-rated replacement parts are required by U. S. Federal law, non-marine parts that are not ignition protected may be present. An example is replacing a marine-approved alternator with an automotive alternator. Protection of electrical circuits in boats is provided by fuses, circuit breakers, or fusible links. As with structures or land-based vehicles, any of these safety devices can be altered, bypassed, or fail. The installation of additional equipment can affect the way a safety device will operate. Unlike the electrical systems of most structures, a boat's electrical systems often use both alternating current (ac) and direct current (dc) systems. Those metallic components located beneath the water line are electrically connected or bonded to the negative side of the dc system to provide lightning protection and to prevent galvanic corrosion *(see Chapter 11 of* NFPA 302, *Fire Protection Standard for Pleasure and Commercial Motor Craft)*. The negative side of the battery is generally connected to the engine block. The positive side of the battery supplies current to fuses, fuse blocks, or a circuit breaker panel and to all of the direct current electrical equipment. With few exceptions, all circuits are required to consist of two conductors: positive and negative. No metallic portion of the boat may be used for the negative return as per NFPA 302, *Fire Protection Standard for Pleasure and Commercial Motor Craft.* Any time an energized positive conductor, terminal end, or component comes in contact with a grounded conductor or surface, a completed circuit will occur and may provide a source of ignition. Boats may have onboard batteries and wiring similar to those utilized in motor vehicles, as well as alternating current (ac) conductors and fixtures similar to a structure. They may also be equipped with a converter that changes the ac (household) current into dc power for battery charging, lighting, and other uses. Boats may be equipped with an onboard generator or a dc to ac inverter to provide ac current when shore power is not available *(see Chapter 10 of* NFPA 302, *Fire Protection Standard for Pleasure and Commercial Motor Craft)*.

FIGURE 28.8.1 Inboard Carburetor Fire.

FIGURE 28.8.2 Power Source Generator.

28.8.2.1 Overloaded Wiring. Faults in wiring can raise the conductor temperature to the point of deteriorating, melting, or igniting the insulation, particularly in bundled cables such as the wiring harnesses or the accessory wiring routed within the boat, through bulkheads, and in cable trays where the heat generated is not readily dissipated. This can occur without activating the overcurrent circuit protection devices. The addition of accessories, such as radios, GPS systems, and radar may contribute to the overloading of the original factory wiring. The prefire history of aftermarket additions and any prior electrical malfunction should be evaluated in the fire analysis.

28.8.2.2 Electrical Short Circuiting and Arcs. Electrical arcing can result when a conductor's insulation becomes worn, brittle, cracked, or otherwise damaged, allowing it to contact a grounded conductor or surface. Insulation failure may occur as a result of chafing, crushing, or cutting of wires. Battery and starter cables are not provided with overcurrent protection and are designed to carry high currents. capable of igniting materials such as engine oil accumulations, some plastic materials, and electrical wiring insulation.

28.8.2.3 Electrical Connections. In addition to the electrical failures associated with a structure or vehicle electrical system, corrosion-induced failures may occur from the boat being exposed to water and salt. This corrosion may result in high-resistance heating, providing energy sufficient for ignition of adjacent common combustibles. The electrical connections on a boat should be protected in a similar manner as those comprising the electrical systems for motor vehicles. A detailed discussion of poor connections is located in Chapter 8.

28.8.2.4 Lightning. Lightning can cause a boat fire. Due to the size of the lightning charge, an unanticipated current path can occur in any part of the boat from a direct strike. This unanticipated current path can cause the conducting material to ignite, or can ignite nearby combustible material. Lightning strikes, either directly to the boat or indirectly near the boat, can damage electrical system components, creating electrical failures and potentially a fire. An expanded discussion on lightning can be read in Chapter 8.

28.8.2.5 Static Electricity and Incendive Arcs. Static electricity and incendive arcs can be an important consideration as a source of ignition in a fire or explosion on a boat. Gasoline vapors are often present during the refueling operation, creating a readily ignitible fuel for this type of ignition source. These vapors may typically collect in below-deck compartments (e.g. engine compartments, cabins, etc.) Fuel system leaks can result in the fuel collecting in the bilge of the boat with the vapors migrating to areas where static electricity may occur from normal activity. A detailed discussion regarding the investigation of static electric ignitions and incendive arcs can be found in the static electricity section of Chapter 8.

28.8.3 Hot Surfaces.

28.8.3.1 Manifolds. Exhaust manifolds and manifold components can generate high temperatures sufficient to ignite diesel spray and vaporized gasoline. Engine oil and transmission fluid coming in contact with a hot manifold can ignite. These fluids may ignite after the engine is shut off due to the loss of cooling water flowing through the engine. When the engine is shut off, the water flow ceases, and manifold temperatures may rise to a level sufficient to ignite atomized fluids or fuel vapors.

28.8.3.2 Exhaust Systems. Commonly, the exhaust systems on boats are different than the exhaust systems found on motor vehicles. One difference may be that the exhaust piping and hoses on a boat are water-cooled. Marine engines can be cooled by water pumped from outside of the boat that passes through the engine and is used to cool the exhaust system. The water is discharged through the exhaust system overboard with the exhaust gasses. The exhaust system on a boat can produce temperatures sufficient to ignite combustibles present in the engine compartment if certain conditions exist, such as a reduced amount of or no cooling water, heavy loading of the engine, rough water conditions, and inadequate spacing from the hot surfaces of the exhaust system to combustibles.

28.8.3.3 Cooking Surfaces. Boat cooking surfaces, if present, usually include a range or oven. The heat is provided by either electricity or by burning propane, compressed natural gas, or alcohol. The hot surface of the electrical heating element or the open flame from the normally operating gas range, oven, or cooking surface can ignite common combustibles found on a boat in the same manner as would occur in a structure. *(See Figure 28.8.3.3.)* The same analysis used in a structure should be used in a boat when a cooking surface is considered as a source of ignition. A further discussion on appliances can be found in Chapter 16.

FIGURE 28.8.3.3 Cooking Surface Fire.

28.8.3.4 Heating Systems. Some boats utilize fuel gas or electric space heaters. Combustibles in too close a proximity to these heat-producing devices may result in ignition. Some boats utilize reverse cycle heating/cooling units that operate on 120 volts ac and may provide an electrical ignition source.

28.8.4 Mechanical.

28.8.4.1 Bearing Failures. The failure of the main bearing in the engine does not usually result in a fire; however, a failure of the bearing in a pulley, motor, alternator, or pump may result in a fire when combustible materials are in contact or in close proximity to the failed bearing. The damage to the bearing after the fire will exhibit physical damage to the race or bearings. These failures of the bearing may result in the release or expelling of hot metal from the bearing.

28.8.4.2 Friction. The drive belts used on some marine engines are the same as those used in motor vehicles. Belts can be ignited if the component pulley fails or locks up and the engine continues to run. The slipping of the belt across the pulley of the failed component can produce temperatures sufficient to ignite the belt, thereby resulting in a fire.

28.8.5 Smoking Materials. In some circumstances, upholstery, fabrics, and materials may be ignited by a cigarette. For example, ignition may occur if a lighted cigarette is insulated (e.g., falling between the back and seat cushions, or into paper, tissues, or other debris). This scenario could occur while the boat is being trailered. Urethane foam seating, once ignited, burns readily and adds substantially to the intensity of a boat fire.

28.9 Documenting Boat Fire Scenes. In general, the requirements for recording and documenting a boat fire incident are similar to those for structures and vehicles. Whenever possible, the boat should be examined in place at the scene. In many cases, however, the investigator may not have the opportunity to view the boat in place. For many reasons, the boat may have been moved before the investigator reaches the scene. Frequently, part of the documentation takes place at a salvage yard, repair facility, marina, boatyard, or warehouse.

28.9.1 On Land. Initially, the investigator should determine if the boat was damaged at its present location or relocated postfire. If it is determined that the fire occurred at another location, that area should be examined if possible. It should be determined if shore power was connected to the boat at the time of the fire. The means of connection to external power should be identified and recorded. Documentation should include other potential ignition sources and fuels in the immediate vicinity of the boat at the time of the fire. It should be determined if the boat was on a trailer or properly shore stored at the time of the fire. An improperly stored boat may be evidence of abandonment, neglect, or theft.

28.9.2 In Water. As on land, the investigator should determine if the boat that is in the water was damaged at its present location or relocated postfire. If it is determined that the fire occurred at another location, that area should be examined if possible. It should be determined if shore power was connected to the boat at the time of the fire. The means of connection to external power should be identified and recorded. Documentation should include other potential ignition sources and fuels in the immediate vicinity of the boat at the time of the fire. It should be determined if the boat was moored, anchored, or underway. It may be necessary, when feasible, to conduct an underwater examination at the fire location for potential evidence.

28.9.2.1 Moored. The location where the boat was moored should be examined for the presence of potential ignition sources and fuels that are not normally considered part of the boat. The examination should include the type or style of structure (slip, dock, seawall, etc.) to which the boat was moored. *(See Figure 28.9.2.1.)* The investigator should consider and document the construction materials and the type of structure because those materials may affect the fire intensity and spread. Furthermore, the investigator should consider the possibility of the fire having spread from another boat or other outside source (exposure fire).

28.9.2.2 Anchored and Underway. The investigator should determine if the fire occurred while the boat was anchored, adrift, or underway, as each may provide different or unique ignition sources. As an example, while at anchor, generators, cooking appliances, etc. may be in operation that would not generally be operating while underway.

28.9.2.3 Underwater. Underwater examination should only be conducted by qualified personnel. When a boat is submerged as the result of a fire, an underwater examination of the boat and the surrounding area should be conducted and documented, if possible. The main purpose of this type of examination is to document the position and condition of the boat prior to attempts to raise the boat. Raising may result in the displacement or loss of evidence. *(See Figure 28.9.2.3.)* Damage to the remaining structural integrity of the boat may occur during raising operations.

FIGURE 28.9.2.1 Generic Boat Fire.

28.9.3 Boat Identification. Boats, like vehicles, possess unique identifiers, which may vary, depending on the manufacturer and local regulations.

28.9.3.1 Hull Identification Number (HIN). Every boat produced or imported into the United States is required to have a HIN, marked in a primary and a secondary location. The primary location varies, depending on the year of manufacture, but is typically located on the right rear (starboard transom) below the boat's rub rail. The HIN is required to be permanently affixed to the hull of the boat by means of stamping, engraving, or the attachment of a plastic or metal plate. *(See Figure 28.9.3.1.)* As a result of a fire the primary HIN may no longer be available and therefore the information may be obtained from the secondary (duplicate) HIN. The location of

FIGURE 28.9.2.3 Sunken Boat Recovery.

FIGURE 28.9.3.1 Hull Identification Number (HIN).

the secondary HIN must be obtained through the manufacturer, as its location is not standardized. In the case of inboards or I/Os, the HIN may be determined by contacting the manufacture of the boat and providing them with the engine identification numbers. The manufacturer ID code is the first three letters of the HIN. Homemade boats are likewise required to have HIN numbers that are assigned by the state through the registration process, as required in 33 CFR 181.23, "Hull identification numbers required."

28.9.3.2 Registration Numbers. Boat registration configuration and locations are mandated under 33 CFR 173, Subpart B. Generally, the registration number will appear near the bow of the boat. Ownership information can be obtained through the local regulating authority based on that registration information.

28.9.3.3 U. S. Coast Guard Documentation Numbers. Some boats may not display registration numbers. However, they may have been assigned U.S. Coast Guard documentation numbers that are required to be permanently marked on the inside of the boat. The documentation papers are required to be kept with that boat.

28.9.3.4 Boat Name and Hailing Port. If the boat is properly documented, the hailing port and name will follow documentation requirements. In years past, the hailing port was the location where the boat was normally moored. Currently, the hailing port can be a location chosen by the boat owner. When state-registered, there is no requirement for a boat name or hailing port.

28.9.3.5 Boat History. The general history of the boat, for example, age, alterations, previous owners, repairs, and maintenance records, may provide valuable information to the investigator.

28.9.3.6 Fire Scene History. It is recommended that an attempt be made to develop a scenario of the events leading up to the fire, as well as the progression of the fire itself. The operator of the boat, passengers, bystanders, the fire department, and police personnel should be interviewed. This information may be used to assist in the investigation.

28.9.3.6.1 Actions Before the Fire. Information regarding the operation of the boat immediately prior to the fire may be obtained from the operator or owner to determine the following:

(1) When the boat was last operated and for how long
(2) The total number of hours on the engine
(3) If the boat systems were operating normally (electrical malfunctions, engine stalling, abnormal handling such as steerage, etc.)
(4) When the boat was last serviced e.g.: oil change, repairs
(5) When the boat was last fueled, including the amount and type of fuel
(6) When and where the boat was last moored
(7) When the last time the boat was observed prior to the fire
(8) What auxiliary equipment was reported present on the boat (e.g., single side band [SSB] radio, CB, GPS, VHF radio, depth finder, fish finders, radar)
(9) Personal items that were present on the boat (e.g., clothing, tools, recreational items, fishing gear, skis)

28.9.3.6.2 Actions During the Fire. If the boat was occupied or being operated at the time of the fire, the following information may assist in the investigation:

(1) How long the boat had been in operation and at what speeds
(2) The route
(3) Water and weather conditions
(4) When and where the odor, smoke, or flame was first noticed
(5) How the boat was reacting at the time the fire was first noticed (e.g., electrical malfunctions, erratic gauge indications, motor racing, etc.)
(6) Actions taken by the operator when the fire was first noticed (e.g., anchoring, actions if any were taken to extinguish the fire via portable or automatic systems, power disconnection, if and when public officials were notified)
(7) The length of time the fire burned prior to help being summoned as well as when help arrived
(8) The estimated time the fire burned until extinguished and by what means
(9) Other observations

28.9.3.6.3 Actions After the Fire. Information regarding actions or events after the fire may assist the investigator. Those may include the following:

(1) Information regarding salvage operations if applicable (raising, towing, etc.)
(2) Environmental Protection Agency (EPA) and similar agency actions, if any
(3) Actions and location of occupants and fire victims
(4) Actions taken by other public officials
(5) Previous investigations

28.9.4 Boat Particulars. Once the boat has been positively identified as being the subject of the investigation, the mechanical functions of that particular boat, its material composition, and its fire reactivity should be reviewed. The investigator may find it helpful to examine an exemplar boat, if one is available, as well as the appropriate sales literature and service manuals. Recall information regarding fire causes in boats of the manufacturer, model, and year may be obtained from the United States Coast Guard at www.uscgboating.org/recalls and from the Consumer Product Safety web site at www.cpsc.gov.

28.10 Boat Examination.

28.10.1 General. As with structure fires, the first step is to determine an area of origin. Most boats can be divided into topside/cockpit and three major compartments. *(See Figure 28.10.1.)* The compartments are the engine/fuel compartment (sometimes the fuel is in a separate compartment), the accommodation compartment or cabin, and the bilge. The size, construction, and fuel load of these compartments can vary considerably.

FIGURE 28.10.1 Marina Fire.

28.10.1.1 Examination.

28.10.1.1.1 Examination of the exterior of the boat may reveal significant fire patterns. Caution should be exercised when interpreting patterns around vents for the engine and fuel compartments, as the power blowers may have been operating during the fire, thus affecting the pattern.

28.10.1.1.2 An accommodation compartment fire will frequently cause failure at the cabin top from the inside out and may leave the other boat areas largely intact. Analyzing the fire patterns will determine whether the fire spread from inside the compartment. Cabins are relatively airtight when ventilation openings are closed and some fires may self extinguish. Cabins may have galley equipment that could ignite other combustibles. The spatial arrangements of the cabin, due to its shape, may influence the fire spread characteristics.

28.10.1.1.3 Accommodation compartments are those typically used for sleeping, lounging, cooking, storage, head, etc. These multiple configurations result in varied potential ignition sources and fuel loads. Since many of these compartments have limited vertical space, sloping hulls, and offsets, normal structural patterns relied upon by the investigator may not be present.

28.10.1.1.4 Engine and fuel compartment fires typically are fuel vapor–related and this may consume the compartment, resulting in fire spread to the cabin and other portions of the boat. Attention should be given to the carburetor or fuel injection systems on the engine, as well as the ignition system components, fuel delivery systems, and fuel tanks. The exhaust system should be inspected for evidence of heat failure (often due to water starvation), which may result in combustion of nearby boat components.

28.10.1.1.5 Bilge areas are relatively air stagnant, and water that accumulates within the bilge is frequently pumped overboard via a bilge pump that incorporates a float switch (normally ignition protected). The bilge within each compartment generally is segregated from other compartments, allowing heavier-than-air fuel gases to accumulate within the bilge. Gasoline, diesel, or oil may accumulate in the bilge and will float on top of water in the bilge. Due to the difficulty of cleaning bilges, remnants of these materials may be present at the time of the fire. Once the compartment(s) of origin has been established, a detailed inspection should be made.

28.10.2 Examination of Boat Systems. The individual systems within the boat should be examined to determine their role, if any, in the cause of the fire using the system identifications and functions described in Section 28.4.

28.10.2.1 The investigator should inspect the fuel tank for edge or bottom failure or corrosion penetrations that may allow the release of fugitive fuel. The conditions of the fuel fill and vent hoses should be inspected for deterioration, chafing, or other damage. The static ground from the fuel tank to the deck fuel fill should be located and its electrical continuity verified.

28.10.2.1.1 Fuel tanks, when exposed to heat, generally exhibit a line of demarcation that represents the fuel level at the time the fire was extinguished. Plastic fuel tanks may still be intact, indicating the fuel level at the time the fire was extinguished.

28.10.2.2 Switches, Handles, and Levers. During the inspection of the boat interior, the position of switches should be noted to determine their position. *(See Figure 28.10.2.2.)* An attempt should be made to determine if port lights and hatches were open or closed prior to the fire. The position of the battery switches, generator, shore power transfer switch, etc., should be noted. The ignition switches should be examined, if possible, for any signs of a key, tampering, or breaking of the lock. Most of these elements are made of materials that may be easily consumed in a fire; however, there may be enough residue left to assist in the investigation.

28.11 Boats in Structures.

28.11.1 Boats within garages and carports should be examined with consideration of the combustibles within the garage or carport, especially when considering the origin. Heat transfer from a fire originating in the structure may ignite the boat.

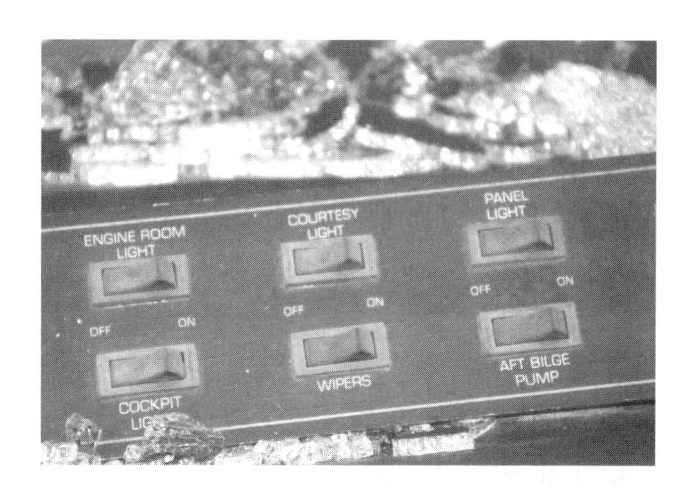

FIGURE 28.10.2.2 Switch Positions.

28.11.2 The storage of boats in buildings (including dry rack) is subject to the standards in Section 5.2 of NFPA 303, *Fire Protection Standard for Marinas and Boatyards.*

28.11.3 During origin and cause determination, boats moored afloat in covered docks should be examined with consideration of the reflected radiant energy from the hot layer developed under the roof and the role of other relevant compartment fire dynamics. These docks are usually floating, and the electrical service to and on the dock should likewise be examined when evaluating the cause.

28.12 Legal Considerations. Unique circumstances encountered in a boat fire may result in the fire investigation being impacted by admiralty law. Admiralty law is a specific area of law that will require specialized legal assistance. Other legal considerations during a boat fire investigation are similar to those involving other investigations and may include the right to investigate, the right to enter, and spoliation and preservation of evidence. A full discussion on legal considerations may be found in Chapter 11.

Annex A Explanatory Material

Annex A is not a part of the recommendations of this NFPA document but is included for informational purposes only. This annex contains explanatory material, numbered to correspond with the applicable text paragraphs.

A.1.4 A complete list of conversion factors can be found in the *NFPA Fire Protection Handbook* or in ASTM E 380, *Standard for Metric Practice.*

A.3.2.1 Approved. The National Fire Protection Association does not approve, inspect, or certify any installations, procedures, equipment, or materials; nor does it approve or evaluate testing laboratories. In determining the acceptability of installations, procedures, equipment, or materials, the authority having jurisdiction may base acceptance on compliance with NFPA or other appropriate standards. In the absence of such standards, said authority may require evidence of proper installation, procedure, or use. The authority having jurisdiction may also refer to the listings or labeling practices of an organization that is concerned with product evaluations and is thus in a position to determine compliance with appropriate standards for the current production of listed items.

A.3.2.2 Code. The decision to designate a standard as a "code" is based on such factors as the size and scope of the document, its intended use and form of adoption, and whether it contains substantial enforcement and administrative provisions.

A.3.2.3 Guide. There are other standards-making bodies that define the word *guide* differently.

A.3.2.4 Recommended Practice. There are other standards-making bodies that define the term *recommended practice* differently.

A.3.2.5 Standard. There are other standards-making bodies that define the word *standard* differently.

A.3.3.1 Absolute Temperature. The Celsius and Fahrenheit scales are relative scales that use "degrees." The Kelvin and Rankine scales are absolute scales that do not utilize the term *degree.* Absolute zero is the lowest possible temperature, with 0 K being equal to −273°C, and 0 R equal to −460°F; 273 K corresponds to 0°C, and 460 R corresponds to 0°F.

A.3.3.11 Arson. Precise legal definitions vary among jurisdictions, wherein it is defined by statutes and judicial decisions.

A.3.3.28 Combustible. A combustible material is capable of burning, generally in air under normal conditions of ambient temperature and pressure, unless otherwise specified; combustion can occur in cases where an oxidizer other than the oxygen in air is present (e.g., chlorine, fluorine, or chemicals containing oxygen in their structure).

A.3.3.29 Combustible Gas Indicator. Some units can indicate the percentage of the lower explosive limit of the air–gas mixture.

A.3.3.61 Fire Dynamics. See sources in Section B.1.

A.3.3.67 Fire Science. See sources in Section B.2.

A.3.3.90 Heat. Heat is measured in joules, calories, or Btus. Heat is not measured in °F or °C. The joule is the preferred unit. A joule/second is a watt. *(See 3.3.184.)*

A.3.3.93 Heat of Ignition. Heat energy comes in various forms and usually from a specific object or source. Therefore, the heat of ignition is divided into two parts: (a) equipment involved in ignition and (b) form of heat of ignition.

A.3.3.94 Heat Release Rate (HRR). The heat release rate of a fuel is related to its chemistry, physical form, and availability of oxidant and is ordinarily expressed as Btu/sec or kilowatts (kW).

A.3.3.101 Ignition Temperature. Reported values are obtained under specific test conditions and might not reflect a measurement at the substance's surface. Ignition by application of a pilot flame above the heated surface is referred to as pilot ignition temperature. Ignition without a pilot energy source has been referred to as autoignition temperature, self-ignition temperature, or spontaneous ignition temperature. The ignition temperature determined in a standard test is normally lower than the ignition temperature in an actual fire scenario.

A.3.3.116 Noncombustible Material. Materials that are reported as passing ASTM E 136, *Standard Test Method for Behavior of Materials in a Vertical Tube Furnace at 750 Degrees C,* shall be considered noncombustible materials.

A.3.3.122 Overload. A fault, such as a short circuit or ground fault, is not an overload. *(See also 3.3.120.)*

A.3.3.125 Plastic. Plastics are usually made from resins, polymers, cellulose derivatives, caseins, and proteins. The principal types are thermosetting and thermoplastic.

A.3.3.163 Spontaneous Heating. The process results from oxidation, often aided by bacterial action where agricultural products are involved.

A.3.3.167 Temperature. The lowest possible temperature is absolute zero on the Kelvin temperature scale (−273° on the Celsius scale). At absolute zero it is impossible for a body to release any energy.

A.3.3.169 Thermal Expansion. The amount of this increase per degree temperature, called the coefficient of thermal expansion, is different for different substances.

A.4.1 For additional information, see the following publications:

Cooke, R. A., and R. H. Ide. 1985. *Principles of Fire Investigation.* Gloucestershire, UK: Institution of Fire Engineers, pp. 135–137.

DeHaan, J. D. 1997. *Kirk's Fire Investigation.* 4th ed. Upper Saddle River, NJ: Brady/Prentice-Hall, Inc.

Kennedy, J., and P. Kennedy. 1985. *Fires and Explosions — Determining Cause and Origin.* Chicago, IL: Investigations Institute.

A.4.3.4 Additional guidance can be found in ASTM E 678, *Standard Practice for Evaluation of Technical Data.*

A.4.3.5 The inductive method is the reasoning process by which a person starts from a particular experience and proceeds to generalizations. The person may start with, "All apples I have eaten were sweet." From this he or she induces that apples are sweet. But the next apple may not be sweet. The inductive method leads to probabilities, not certainties. It is the basis of the common sense upon which persons act. It is also used in scientific discovery. Scientists use both induction and deduction. In deduction, scientists begin with generalizations. They deduce particular assertions from them. They might test their assertions by experiments, then confirm, revise, or reject their original generalizations. Using only deduction, people would ignore experience. Using only induction, they would ignore relationships among facts. By combining these methods, science unifies theory and practice.

A.4.3.6 This discussion is meant to specifically allow for logic-based "thought experiments." In such an experiment, one sets up a premise and tests it against the data. An example of a thought experiment is, "If the door was closed during the fire, then there should be mirror-image patterns on the matching surfaces of the hinges." It is not necessary to burn the door in the open position and in the closed position and compare results. The finding of mirror-image patterns, combined with the investigator's knowledge, allows for conclusions to be made about many aspects of the fire without the need for a physical experiment. Deductive logic in the design and implementation of the thought experiment, however, is still a requirement for valid hypothesis testing.

The deductive method is the process of reasoning from which we draw conclusions by logical inference from given premises. If we begin by accepting the propositions that "all Greeks have beards" and that "Zeno is a Greek," we may validly conclude, "Zeno has a beard." We refer to the conclusions of deductive reasoning as valid, rather than true, because we must distinguish clearly between that which follows logically from other statements and that which is the case. Starting premises may be articles of faith or assumptions. Before we can consider the conclusions drawn from these premises as valid, we must show that they are consistent with each other and with the original premise. Mathematics and logic are examples of disciplines that make extensive use of the deductive method. The scientific method requires a combination of induction and deduction.

A.4.3.6.1 Vaughn, Lewis, *The Power of Critical Thinking, Effective Reasoning About Extraordinary Claims,* 2nd ed. Oxford University Press, New York/Oxford, 2008; Damer, Edward, T. *Attacking Faulty Reasoning. A Practical Guide to Fallacy-Free Thinking,* 4th ed., Wadsworth Thomson Learning, Belmont, CA, 2001.

Kahane, Howard and Cavender, Nancy, *Logic and Contemporary Rhetoric, The Use of Reason in Everyday Life,* 9th ed. Wadsworth-Thomson Learning, Belmont, CA, 2002.

A.4.3.9 For a discussion of concrete examples of confirmation bias and its potential for causing erroneous interpretations of data, see Wason, P. C.(1960)

A.5.1.1 Because materials in Chapter 5 are tutorial in nature, the references in this chapter are primarily textbooks and handbooks. Only where material has not yet reached the textbook/handbook literature or where the material is very specific to fire investigation (not of sufficient interest to the general fire science community) are primary references provided.

The following are textbooks in fire science that include coverage of most of the topics in Chapter 5:

Drysdale, D. (1999), *An Introduction to Fire Dynamics,* Second Edition, John Wiley and Sons, Chichester, England.

Friedman, R. (1998), *Principles of Fire Protection Chemistry and Physics,* Third Edition, National Fire Protection Association, Quincy, MA.

Quintiere, J. (1998), *Principles of Fire Behavior,* Delmar Publishers, Albany, NY.

Karlsson, B., Quintiere, J. (2000), *Enclosure Fire Dynamics,* CRC Press, Boca Raton, FL.

Quintiere, J. (2006), *Fundamentals of Fire Phenomena,* John Wiley and Sons, Chichester, England.

A.5.2 The following is a list of references about fire chemistry:

Drysdale, D. (2003), "Chemistry and Physics of Fire," *NFPA Fire Protection Handbook,* 19th ed., Section 2.3.

Friedman, R. (1998), *Principles of Fire Protection Chemistry and Physics,* Third Edition, National Fire Protection Association, Quincy, MA.

Grand, A., Wilkie, C. (2000), *Fire Retardancy of Polymeric Materials,* Marcel Dekker, New York.

Fire, F. (1991), *Combustibility of Plastics,* Van Nostrand Reinhold, New York.

Troitzsch, J. (2004), *Plastics Flammability Handbook,* Hanser Publishers, Munich.

Cullis, C., Hirschler, M. (1981), *The Combustion of Organic Polymers,* Clarendon Press, Oxford.

Aseeva, R., Zaikov, G. (1981), *Combustion of Polymer Materials,* Hanser Publishers, Munich.

Beyler, C. and Hirshcler M. (2002), "Thermal Decomposition of Polymers," *SFPE Handbook of Fire Protection Engineering,* Ed. P. DiNenno, National Fire Protection Association, Quincy, MA.

Beyler, C. (2002), "Flammability Limits of Premixed and Diffusion Flames," *SFPE Handbook of Fire Protection Engineering,* Ed. DiNenno, National Fire Protection Association, Quincy, MA.

Simmons, R. (1995), "Fire Chemistry," *Combustion Fundamentals of Fire,* Ed. G. Cox, Academic Press, London.

A.5.2.3.4 For more information on diffusion flames, consult the following references:

Beyler, C. (2002), "Flammability Limits of Premixed and Diffusion Flames," *SFPE Handbook of Fire Protection Engineering,* Ed. P. DiNenno, National Fire Protection Association, Quincy, MA.

Babrauskas, V., (2003), Chapter 4 – Gases, *Ignition Handbook,* Fire Science Publishers, Issaquah, WA.

A.5.3 The following is a list of references on products of combustion:

Mulholland, G. (2002), "Smoke Production and Properties," *SFPE Handbook of Fire Protection Engineering,* Ed. P. DiNenno, National Fire Protection Association, Quincy, MA.

Tewarson, A. (2002), "Generation of Heat and Chemical Compounds in Fires," *SFPE Handbook of Fire Protection Engineering,* Ed. P. DiNenno, National Fire Protection Association, Quincy, MA.

A.5.4 For further information of fluid flows, see the following:

Zukoski, E (1995), "Properties of Fire Plumes," *Combustion Fundamentals of Fire,* Ed. G. Cox, Academic Press, London.

Kandola, B.(2002), "Introduction to Mechanics of Fluids," *SFPE Handbook of Fire Protection Engineering,* Ed. P. DiNenno, National Fire Protection Association, Quincy, MA.

Heskestad, G. (2002), "Fire Plumes, Flame Height, and Air Entrainment," *SFPE Handbook of Fire Protection Engineering,* Ed. P. DiNenno, National Fire Protection Association, Quincy, MA.

Alpert, R. (2002), "Ceiling Jet Flows," *SFPE Handbook of Fire Protection Engineering,* Ed. P. DiNenno, National Fire Protection Association, Quincy, MA.

Emmons, H. (2002), "Vent Flows," *SFPE Handbook of Fire Protection Engineering,* Ed. P. DiNenno, National Fire Protection Association, Quincy, MA.

A.5.5 For further information on heat transfer, see the following:

Rockett, J., Milke, J. (2002), "Conduction of Heat in Solids," *SFPE Handbook of Fire Protection Engineering,* Ed. P. DiNenno, National Fire Protection Association, Quincy, MA.

Atreya, A. (2002), "Convection Heat Transfer," *SFPE Handbook of Fire Protection Engineering,* Ed. P. DiNenno, National Fire Protection Association, Quincy, MA.

Tien, C., Lee, K., Stretton, A. (2002), "Radiation Heat Transfer," *SFPE Handbook of Fire Protection Engineering,* Ed. P. DiNenno, National Fire Protection Association, Quincy, MA.

A.5.5.2.2 For further information on flame spread properties, see the following:

Quintiere, J., Harkleroad, M. (1985), "New Concepts for Measuring Flame Spread Properties," *Fire Safety Science and Engineering,* STP 882, American Society for Testing and Materials, Philadelphia, PA.

A.5.6 For further information on fuel load, fuel packages, and properties of flame, see the following:

Babrauskas, V., Grayson, S. (1992), eds., *Heat Release in Fires,* Elsevier Applied Science, London.

Kransny, J., Parker, W., Babrauskas, V. (2001), *Fire Behavior of Upholstered Furniture and Mattresses,* Noyes Publications, Park Ridge, NJ.

Zukoski, E (1995), "Properties of Fire Plumes," *Combustion Fundamentals of Fire,* Ed. G. Cox, Academic Press, London.

Heskestad, G. (2002), "Fire Plumes, Flame Height, and Air Entrainment," *SFPE Handbook of Fire Protection Engineering,* Ed. P. DiNenno, National Fire Protection Association, Quincy, MA.

Babrauskas, V. (2002), "Heat Release Rates," *SFPE Handbook of Fire Protection Engineering,* Ed. P. DiNenno, National Fire Protection Association, Quincy, MA.

Gottuk, D., White, D. (2002), "Liquid Fuel Fires," *SFPE Handbook of Fire Protection Engineering,* Ed. P. DiNenno, National Fire Protection Association, Quincy, MA.

Lattimer, B. (2002), "Heat Fluxes from Fires to Surfaces," *SFPE Handbook of Fire Protection Engineering,* Ed. P. DiNenno, National Fire Protection Association, Quincy, MA.

Beyler, C. (2002), "Fire Hazard Calculations for Large, Open Hydrocarbon Fires," *SFPE Handbook of Fire Protection Engineering,* Ed. P. DiNenno, National Fire Protection Association, Quincy, MA.

A.5.6.4.6 For additional information, see the following references:

(1) Beyler, C. L., "Plumes and Ceiling Jets," *Fire Safety Journal,* 1986.
(2) Hasemi, Y., and Tokunaga, T., "Some Experimental Aspects of Turbulent Diffusion Flames and Buoyant Plumes from Fire Sources Against a Wall and in a Corner of Walls," *Combustion Science and Technology,* 40, pp. 1–17, 1984.
(3) Mizuno, T., and Kawagoe, K., "Burning Rate of Upholstered Chairs in the Center, Alongside a Wall and in a Corner of a Compartment," *Fire Safety Science,* Proceedings of the First International Symposium, pp. 849–857, 1985.
(4) Zukoski, E. E., "Properties of Fire Plumes," *Combustion Fundamentals of Fire,* Cox, Ed. G., Academic Press, London, 1995.
(5) Karlsson, B., and Quintiere, J. G., "Fire Plumes and Flame Height," *Enclosure Fire Dynamics,* CRC Press, New York, 2000.
(6) Mowrer, F. W., and Williamson, R. B., "Estimating Room Temperatures from Fires Along Walls and in Corners," *Fire Technology,* Vol. 23, No. 2, pp. 133–145, 1987.
(7) Back, J., Beyler, C., DiNenno, P., "Wall Incident Heat Flux Distributions Resulting from Adjacent Fire," Proceedings of the Fourth International Symposium on Fire Safety Science, International Association of Fire Safety Science, 1994.
(8) Heskestad, G., "Fire Plumes, Flame Height, and Air Entrainment", Section 2, Chapter 1, *SFPE Handbook of Fire Protection Engineering,* 3rd ed., Ed. DiNenno, P.J., Society of Fire Protection Engineers, National Fire Protection Association, Quincy MA, 2002.

A.5.6.4.7 For further information on flame lengths under ceilings, see Babrauskas, V., Flame Lengths Under Ceilings, *Fire and Materials* 4, 119–126 (1980).

A.5.6.6.2 Figure 5.6.6.2(c) is adopted from Beyler, C. (2002), *SFPE Handbook of Fire Protection Engineering,* Ed. P. DiNenno et al, National Fire Protection Association, Quincy, MA.

A.5.7 Extensive treatment of the wide variety of means of ignition is available, as seen in the following:

Babrauskas, V. (2003), *Ignition Handbook,* Fire Science Publishers, Issaquah, WA.

Bowes, P. (1984), *Self-heating: evaluating and controlling the hazards,* Building Research Establishment, Her Majesty's Stationery Office.

Beyler, C. (2002), "Flammability Limits of Premixed and Diffusion Flames," *SFPE Handbook of Fire Protection Engineering,* Ed. P. DiNenno, National Fire Protection Association, Quincy, MA.

Gray, B. (2002), "Spontaneous Combustion and Self-Heating," *SFPE Handbook of Fire Protection Engineering,* Ed. P. DiNenno, National Fire Protection Association, Quincy, MA.

Kanury, A. (2002), "Ignition of Liquid Fuels," *SFPE Handbook of Fire Protection Engineering,* Ed. P. DiNenno, National Fire Protection Association, Quincy, MA.

Kanury, A. (2002), "Flaming Ignition of Solid Fuels," *SFPE Handbook of Fire Protection Engineering,* Ed. P. DiNenno, National Fire Protection Association, Quincy, MA.

A.5.7.4.1.3.9 For further information, see the following:

Babrauskas, V. (2003), *Ignition Handbook,* Fire Science Publishers, Issaquah, WA.

Babrauskas, V., Gray, B. F., and Janssens, M. L., Prudent Practices for the Design and Installation of Heat-Producing Devices near Wood Materials, *Fire and Materials,* 31, pp. 125–135.

Beyler, C., Gratkowski, M., Sikorsky, J., ISFI 2006.

Babrauskas, V., "Ignition of Wood, A Review of the State of the Art," Journal of Fire Protection Engineering, 12(3), 2002, pp. 163–189.

Cuzzillo, B., Pagni, P., "The Myth of Pyrophoric Carbon," *Fire Safety Science, Proceedings of the Sixth International Symposium,* International Association for Fire Safety Science, 2003.

A.5.7.4.1.3.10 For further information, see the following:

Wolters, F., Pagni, P., Frost, T., Cuzzillo, B., "Size Constraints on Self Ignition of Charcoal Briquets," *Fire Safety Science, Proceedings of the Sixth International Symposium,* International Association for Fire Safety Science, 2003.

A.5.7.4.1.3.11 For further information, see the following:

Babrauskas, V., *Ignition Handbook,* Fire Science Publishers/ Society of Fire Protection Engineers, Issaquah, WA (2003)

A.5.7.4.1.4 For further information, see the following:

Fire Protection Handbook, 19th edition, National Fire Protection Association, Quincy, MA, 2003.

A.5.7.4.1.7.2 For further information, see the following:

Babrauskas, V., and Krasny, J. F., "Upholstered Furniture Transition from Smoldering to Flaming," *Journal of Forensic Sciences* 42, 1029–1031 (1997).

A.5.8 For further information, see the following:

Quintiere, J. (2002), " Surface Flame Spread," *SFPE Handbook of Fire Protection Engineering,* Ed. P. DiNenno, National Fire Protection Association, Quincy, MA.

Gottuk, D., White, D. (2002), "Liquid Fuel Fires," *SFPE Handbook of Fire Protection Engineering,* Ed. P. DiNenno, National Fire Protection Association, Quincy, MA.

Fernandex-Pello, C. (1995), "The Solid Phase," *Combustion Fundamentals of Fire,* Ed. G. Cox, Academic Press, London.

A.5.8.1.3.1 For further information, see the following:

Drysdale, D. Fire Dynamics, *ISFI 2006 Proceedings, International Symposium on Fire investigation Science and Technology,* National Association of Fire Investigators, Sarasota, FL, 2006.

Drysdale D. "Learning from experience - Fire Investigation in Great Britain," Fire Engineers and Fire Prevention magazines, Institution of Fire Engineers, November 2000.

Atkinson, G., Drysdale, D. D., Wu, Y., "Fire driven flow in an inclined trench." *Fire Safety Journal.* 25, pp. 141–158, (1995).

P. Woodburn and D. D. Drysdale, Fires in inclined trenches, Part I: The dependence of the critical angle on the trench and burner geometry, *Fire Safety Journal* 31, pp. 143–164 (1998).

P. Woodburn and D. D. Drysdale, "Fires in inclined trenches, Part II: Time dependent flow."

A.5.8.3.4 For further information, see the following:

Atreya, A. (1984), Fire Growth on Horizontal Surfaces of Wood," *Combustion Science and Technology,* 39, pp. 163–194.

A.5.8.3.5 For further information, see the following:

Orloff, L., deRis, J., Markstein, G. (1974), "Upward Turbulent Fire Spread and Burning of Fuel Surface," *Fifteenth Symposium (International) on Combustion,* The Combustion Institute, Pittsburgh, PA, pp. 183–192.

Wu, P., Orloff, L., Tewarson, A. (1996), "Assessment of Material Flammability with the FG propagation model and Laboratory Test Methods," *13th Joint Panel Meeting of the UJNR Panel on Fire Research and Safety,* Gaithersburg, MD.

McGrattan, K., Hamins, A. and Stroup, D. (1998). Sprinkler, Smoke & Heat Vent, Draft Curtain Interaction — Large Scale Experiments and Model Development. Technical Report NISTIR 6196-1, National Institute of Standards and Technology, Gaithersburg, MD.

Saito, K., Quintiere, J. G., Williams, F. A. (1986), "Upward Turbulent Flame Spread," *Fire Safety Science. Proceedings. 1st International Symposium,* International Association for Fire Safety Science, Hemisphere Publishing Corp., New York, C. E. Grant, P. J. Pagni, eds., pp. 75–86.

Grant, G., Drysdale, D. (1995), "Numerical Modeling of Early Flame Spread in Warehouse Fires, *Fire Safety Journal,* 24(3), pp. 247–278.

A.5.9 For further information, see the following:

Thomas, P. (1995), "The Growth of Fire-Ignition to Full Involvement," *Combustion Fundamentals of Fire,* Ed. G. Cox, Academic Press, London.

A.5.10.4.5 For further information, see the following:

Babrauskas, V., "Upholstered Furniture Room Fires—Measurements, Comparison with Furniture Calorimeter Data, and Flashover Predictions," *Journal Fire Sciences* 2, 5–19 (1984).

Babrauskas, V., Peacock, R. D., and Reneke, P. A., "Defining Flashover for Fire Hazard Calculations. Part II." *Fire Safety Journal* 38, 613–622 (2003).

A.5.10.6.2 Mowrer, F., Williamson, B. (1987), "Estimating Room Temperatures from Fires along Walls and in Corners," *Fire Technology,* 23(2), pp. 133–145.

A.6.2.2.2 For more information, consult the following publications:

Babrauskas, V. "Temperatures in Flames and Fires," Fire Science and Technology, Inc., Issaquah, WA, 2006.

McCaffrey, B. J., *Purely Buoyant Diffusion Flames: Some Experimental Results* (NBSIR 79 1910). [U.S.] Natl. Bur. Stand., Gaithersburg, MD (1979).

Audoin, L., Kolb., G., Torero, J. L., and Most., J. M., "Average Centerline Temperatures of a Buoyant Pool Fire Obtained by Image Processing of Video Recordings," *Fire Safety Journal* 24, 107–130 (1995).

Cox, G., and Chitty, R., "A Study of the Deterministic Properties of Unbounded Fire Plumes," *Combustion and Flame* 39, 191–209 (1980).

Smith, D. A., and Cox, G., "Major Chemical Species in Turbulent Diffusion Flames," *Combustion and Flame* 91, 226–238 (1992).

Quintiere, J. G., *Principles of Fire Behavior,* Delmar Publishers, Albany, NY, 1998, pp. 48–49.

A.6.2.4.4 For more information, see the following references:

Babrauskas, V., "Wood Char Depth: Interpretation in Fire Investigations," *Proceedings of ISFI 2004, International Symposium on Fire Investigation,* Fire Service College, Morton in Marsh, England, (June 2004).

Babrauskas, V., *Charring Rate of Wood as a Tool for Fire Investigations,* Interflam 2004, Interscience Communications, London (July 2004).

A.6.2.5 See Buenger, B. 2003, *The Impact of Wildland and Prescribed Fire on Archaeological Resources,* thesis, University of Kansas.

A.6.2.5.1.1 For more information see the following references:

Sanderson, J. L, "Tests Add Further Doubt to Concrete Spalling Theories", FireFindings, Fall. 1995 volume 3.

Khoury, G. A., "Effect of Fire on Concrete and Concrete Structures", Progress in Structural Engineering Materials 2:229–447 (Cite in Forensic Fire Scene Reconstruction).

Canfield, D., "Causes of Spalling of Concrete at Elevated Temperatures", Fire and Arson Investigator 34, June 1984.

Schroeder, R. A, (Post Fire Analysis Of Construction Materials", Dissertation, Graduate Division, University of California, Berkeley, Spring 1999.

Midkiff, C. R., "Spalling of Concrete as an Indicator of Arson", Fire and Arson Investigator, 41, December 1990.

Beland, B., "Spalling of Concrete", Fire and Arson Investigator, 44 September 1993.

A.6.2.5.1.2 For more information see the following references:

Sanderson, J. L, "Tests Add Further Doubt to Concrete Spalling Theories," *Fire Findings*, Fall 1995, volume 3.

Smith, F. P., "Concrete Spalling: Controlled Fire Tests and Review," *Journal of Forensic Science* 31, 1991.

Midkiff, C. R., "Spalling of Concrete as an Indicator of Arson," *Fire and Arson Investigator* 41, December 1990.

A.6.2.5.1.3 See Khoury, G. A., "Effect of Fire on Concrete and Concrete Structures," *Progress in Structural Engineering Materials* 2, pp. 229–447.

For further information see the following references:

Smith, F. P., "Concrete Spalling: Controlled Fire Tests and Review," *Journal of Forensic Science* 31, 1991.

Schroeder, R. and Williamson, R. "Application of Materials Science to Fire Investigation," *Fire and Materials*, 2001, Interscience, London, UK.

See Bostrom, L., "Methodology for Measurement of Spalling of Concrete," *Fire and Materials*, 2005 Conference, Interscience, London, UK.

See Smith, F. P., "Concrete Spalling: Controlled Fire Tests and Review," *Journal of Forensic Science* 31, 1991.

A.6.2.6 For more information see the following references:

NFPA Fire Protection Handbook, 19th edition, A. E. Cote, ed., National Fire Protection Association, Quincy, MA, 2003.

DeHann, J. D. *Kirk's Fire Investigation*, 5th edition, Prentice Hall, Upper Saddle River, NJ, 2002.

Gray, H. B., et al. *Braving the Elements.* University Science Books, Sausalito, CA 1995.

A.6.2.6.4 For more information on the metallurgical determination of melted steel, see Lentini, J. J., D. M. Smith, and R. W. Henderson. "Baseline Characteristics of Residential Structures Which Have Burned to Completion: The Oakland Experience," *Fire Technology*, 28(3), 1992.

A.6.2.7 For more information see the following references:

Cooke, R. A. *Principles of Fire Investigation*, Gloucestershire, UK: The Institution of Fire Engineers, 1985, pp. 386–387.

Aston University. *Assessment of Fire Damaged Concrete Using Color Analysis.* Interflam '96, March 26–28. London: Interscience Communications Ltd.

A.6.2.8.6 For more information see the following references:

Brown, LeMay, Bursten, *Chemistry: The Central Science*, Sixth Ed, Prentice Hall, Englewood Cliffs, New Jersey, 1994.

Umland, J. B., and J. M. Bellama. *General Chemistry*, Second Edition, West Publishing Company, Saint Paul, MN, 1996.

A.6.2.8.6.1 For more information see the following references:

Beland, B., Roy, C., Tremblay, M. "Copper-Aluminum Interaction in Fire Environments," *Fire Technology*, Volume 19, No. 1, pp. 22–30, February 1983.

Beland, B. "Behaviour of Electrical Contacts Under Fire Conditions," *Fire and Arson Investigator*, Vol. 38, No. 1, September 1987.

Fire Findings. "Alloying Helps Explain How Some Materials Melt at Temps Lower Than Expected," *Fire Findings*, Vol. 4, No. 4, pp. 1–3, Fall 1996.

A.6.2.8.6.2 See A.6.2.8.6.1.

A.6.2.8.6.3 See A.6.2.8.6.1.

A.6.2.9 For more information see the following references:

Specification for the Design, Fabrication, and Erection of Structural Steel for Buildings. New York: American Institute of Steel Construction, 1978.

Manual of Steel Construction: Load and Resistance Factor Design. Third Ed. Chicago: American Institute of Steel Construction, 2001.

Dill, F. H. "Structural Steel after a Fire," *Proceedings of the 1960 National Engineering Conference.* Chicago: American Institute of Steel Construction.

Lie, T. T. *Fire and Buildings.* London: Applied Science, 1972.

Malhotra, H. L. *Design of Fire-resistant Structures.* New York: Chapman and Hall, 1982.

Hamarthy, T. Z. ASTM STP 422, *Standard Test Method for Specific Optical Density of Smoke Generated Solids.* Philadelphia: American Society for Testing and Materials, 1967.

Kirby, B. R. and Preston, R. R. *High Temperature Properties of Hot-Rolled Structural Steels for Use in Fire Engineering Design Studies.* London: British Steel Corp., 1988.

Larsson, T. and Pettersson, O. *Buckling of Fire-Exposed Steel Columns, Partially Restrained with Respect to Longitudinal Expansion.* Lund, Sweden: Division of Structural Mechanics and Concrete Construction, Lund Institute of Technology, 1974.

Thomas, G. "Thermal Properties of Gypsum Plasterboard at High Temperatures." Wellington, New Zealand: Victoria University of Wellington. *Fire and Materials* 26(1), 37–45, January/February 2002.

A.6.2.10 For more information see the following references:

SFPE Handbook of Fire Protection Engineering. 3rd Ed, P. J. DiNenno, ed. Bethesda, MD: Society of Fire Protection Engineers, 2002.

Kramps, N. "Damage Control — Limiting Secondary Losses After Fire." *Fire Prevention*, Dec. 1987, pp. 26–29.

A.6.2.10.3 For more information on research into Acoustic Soot Agglomeration on smoke alarms see the following references:

Worrell, C. L., Roby, R. J., Streit L., and Torero, J. L., "Enhanced Deposition, Acoustic Agglomeration, and Chladni Figures in Smoke Detectors," Fire Technology, 37, 343-362, (USA: Kluwer Academic Publishing, 2001).

Worrell, C.L., Lynch, J.A., Jommas, G., Roby, R.J., Streit, L., and Torero, J.L., "Enhanced Soot Deposition, Acoustic Agglomeration, and Chladni Figures in Smoke Detectors," Fire Technology, 39, 309-346, 2003.

Kennedy, Patrick M., Kennedy, Kathryn C., and Gorbett, Gregory E., "A Fire Analysis Tool – Revisited: Acoustic Soot Agglomeration in Residential Smoke Alarms," InterFlam 2004 Proceedings, InterScience Communications, London, 2004.

Phelan, Patrick. An Investigation of Enhanced Soot Deposition on Smoke Alarm Horns. M. S. Thesis, Worcester Polytechnic Institute. 2004.

Mealy, C.L., and Gottuk, D.T., "Full-Scale Validation Tests of a Forensic Methodology to Determine Smoke Alarm Response," ISFI 2008 Proceedings, International Symposium on Fire Investigation Science and Technology, NAFI, Sarasota, FL, 2008.

The photos in Figure 6.2.10.3(a) through Figure 6.2.10.3(d) are from laboratory research tests reported in Kennedy, P.M., Kennedy, K.C., and Gorbett, G.E., "A Fire Analysis Tool – Revisited: Acoustic Soot Agglomeration in Residential Smoke Alarms," InterFlam 2004 Proceedings, InterScience Communications, London, 2004.

A.6.2.12 For more information see the following references:

Chu Nguong, Ngu. "Calcination of Gypsum Plasterboard under Fire Exposure," *Fire Engineering Research Report* 04/6, May 2004.

Lawson, J. *An Evaluation of Fire Properties of Generic Gypsum Board Products,* (NBSIR 77-1265). Center for Fire Research, NIST, Washington DC, 1977.

McGraw, J. and Mower, F. *Flammability of Painted Gypsum Wallboard Subjected to Fire Heat Fluxes,* Interflam 99 Proceedings, Interscience Communications Ltd. London, 1999.

Mower, F. *Calcination of Gypsum Wallboard in Fire.* Presentation at NFPA World Fire Safety Congress, Anaheim, CA May 2001.

Posey, J. E. and Posey, E. P. "Using Calcination of Gypsum Wallboard to Reveal Burn Patterns," *Fire and Arson Investigator* 33, March 1983, pp. 17–19.

Schroeder, R. and Williamson, R. "Application of Materials Science to Fire Investigation," *Fire and Materials* 2001, Interscience, London, UK.

A.6.2.12.1.3 Kennedy, P., Kennedy, K., Hopkins, R. *Depth of Calcination Measurement in Fire Origin Analysis,* Proceedings of the Fire and Materials Conference 2003, Interscience Communications, London, 2003.

A.6.2.13 For more information see the following references:

Babrauskas, V. *Glass Breakage in Fires.* Issaquah, WA: Fire Science and Technology Inc., 2005.

Keski-Rahkonen, O. "Breaking of Window Glass Close to Fire," *Fire and Materials,* 12(2), June 1988, pp. 61–69.

Lentini, J. "Behavior of Glass at Elevated Temperature," *Journal of Forensic Science,* Vol. 37, No. 5, Sept 1992, 1358–1362.

Lentini, J. J., Smith, D. M., Henderson, R. W. "Baseline Characteristics of Residential Structures Which Have Burned to Completion: The Oakland Experience," *Fire Technology,* NFPA, 1992: pp. 195–214.

Mowrer, F. P., *Window Breakage Induced By Exterior Fires.* NIST-GCR-98-751, Washington, DC: U.S. Dept. of Commerce, June, 1998.

Skelly, M. J., Roby, R. J., Beyler, C. L. "Experimental Investigation of Glass Breakage in Compartment Fires," *Journal of Fire Protection Engineering,* 3(1), March 1991: pp. 25–34.

A.6.2.14 For more information see the following references:

American Society for Metals, *Glossary of Metallurgical Terms and Engineering Tables.* Material Park, OH: ASM International, 1985.

Lentini, J., et al., "Baseline Characteristics of Residential Structures Which Have Burned to Completion: The Oakland Experience." *Fire Technology* 28, August 1992 pp. 195–214.

Tobin, W. and Monson, K. "Collapsed Springs in Arson Investigations: A Critical Metallurgical Evaluation," *Fire Technology* 25, November 1989, pp. 317–335.

A.6.2.17 For more information, see the following references:

Bohnert, M., Rost, T. and Pollack, S. "The Degree of Destruction of Human Bodies in Relation to the Duration of the Fire," *Forensic Science International* 95, pp. 11–21, 1998.

Pope, E. J., O. C. Smith, and T. G. Huff. "Exploding Skulls and Other Myths About How the Human Body Burns," *Fire and Arson Investigator,* vol 54, #4, April 2004.

Pope, E. J., O. C. Smith. "Identification of Traumatic Injury in Burned Cranial Bone: An Experimental Approach," *Journal of Forensic Science,* May 2004, vol 49, #3.

Smith, O. C. and Pope, E. J. *Burning Extremities: Patterns of Arms, Legs, and Preexisting Trauma.* Presentation Lecture, American Academy of Forensic Sciences (AAFS) 55th Annual Meeting, Chicago, February 17–22, 2003.

SFPE. *Engineering Guide to Predicting 1st and 2nd Degree Skin Burns,* Bethesda, MD: SFPE Task Group on Engineering Practices, Society of Fire Protection Engineers, 2000.

A.6.3.2.1 For more information, see Beyler, C., "Fire Plumes and Ceiling Jets", *Fire Safety Journal,* 11, 1986, pp. 53–75 and Lattimer, B. Y., "Heat Fluxes from Fires to Surfaces." *The SFPE Handbook of Fire Protection Engineering,* 3rd ed., P. J. DiNenno, ed. NFPA, Quincy, MA, 2002.

A.6.3.2.2.3 For more information, see sources listed in Section B.3.

A.6.3.3.2.2 For more information on floor covering critical radiant flux, see NFPA 253, *Standard Method of Test for Critical Radiant Flux of Floor Covering Systems Using a Radiant Heat Energy Source,* Chapter 5, as well as the NFPA *Fire Protection Handbook,* 19th edition, Section 12, Chapter 3.

Also see Fang, J. B., and J. N. Breese, *Fire Development in Basement Rooms.* Gaithersburg, MD: NIST, 1980.

A.7.1 For more information see the following publications:

Brannigan, F. L., *Building Construction for the Fire Service,* 3rd Ed., *NFPA Journal,* 1992.

NFPA 220, *Standard on Types of Building Construction,* 2006 edition.

A.8.1 For more information see the following publications:

NFPA 70, National Electrical Code®, 2008 edition.

NFPA 77, *Recommended Practice on Static Electricity,* 2007 edition.

National Electrical Code Handbook, Quincy, MA: NFPA, 2008.

Beland, B. "Considerations on Arcing as a Fire Cause," *Fire Technology* 18, No. 2 (1982): 188–202.

Beland, B. "Electrical Damages — Cause or Consequence?" *Journal of Forensic Science* 29 (3) (July 1984): 747–761.

Beland, B. "Examination of Electrical Conductors Following Fire," *Fire Technology* 16, No. 4 (1980): 252–258.

Beland, B., C. Roy, and M. Tremblay. "Copper-Aluminum Interactions in Fire Environments," *Fire Technology* 19, No. 1 (February 1983): 22–30.

Cahill, P. L., and J. H. Dailey. U.S. Department of Transportation. *Aircraft Electrical Wet-Wire Arc Tracking,* DOT/FAA/CT-88/4. Washington, DC: Federal Aviation Administration, August 1988.

Campbell, F. J. *Flashover Failures from Wet-Wire Arcing and Tracking.* NRL Memorandum Report 5508. Washington, DC: Naval Research Laboratory, December 17, 1984.

Campbell, J. A., National Advisory Committee for Aeronautics. *Appraisal of the Hazards of Friction-Spark Ignition of Aircraft Crash Fires,* Technical Note 4024. Cleveland, OH: Lewis Flight Propulsion Laboratory, May 1957.

Ettling, B. "Ignitibility of PVC Electrical Insulation by Arcing," *IAAI Oregon Chapter Newsletter* 1, No. 4 (March 1997).

Ettling, B. "The Overdriven Staple as a Fire Cause," *Fire and Arson Investigator,* March 1994.

Ettling, B. "Arc Marks and Gouges in Wires and Heating at Gouges," *Fire Technology* 17, No. 1 (1981): 61–68.

Ettling, B. "Glowing Connections," *Fire Technology* 18, No. 4 (1982): 344–349.

Bustin, W. M., and W. G. Duket. *Electrostatic Hazards in Petroleum Industry.* Research Studio Press, July 1983.

Mil-Std–202F, *Test Method for Electronic and Electrical Components.* "Method 301, Dielectric Withstand Voltage." U.S. Department of Defense Military Test Standard, 1998.

Mil-Std–202F, *Test Method for Electronic and Electrical Components.* "Method 302, Insulation Resistance."

API 2214, *Spark Ignition Properties of Hand Tools,* American Petroleum Institute, 1989.

API 2216, *Ignition Risk of Hydrocarbon Vapors by Hot Surfaces in the Open Air,* 2nd edition, American Petroleum Institute, 1991.

API RP 2003, *Protection Against Ignitions Arising Out of Static, Lightning, and Stray Currents,* American Petroleum Institute, 1991.

A.8.3.1 For more information, see *NFPA 70, National Electrical Code®*, 2008 edition, or the NFPA *National Electrical Code Handbook,* 2008 edition.

A.8.7.4 See Figure 8.10.6.2.2 for a photograph of melted solidified aluminum conductor.

A.8.9.4.5 Additional information on arc tracking is found in the following publications:

Campbell, F. J. *Flashover Failures from Wet-Wire Arcing and Tracking,* 1984, and in Cahill, P. L., and J. H. Dailey, *Aircraft Electrical Wet-Wire Arc Tracking,* Washington, DC: FAA, 1988.

Babrauskas, V. "Mechanisms and Modes for Ignition of Low-voltage, PVC- insulated Electrotechnical Products," *Fire and Materials* 30 (2006): pp. 150–174.

Babrauskas, V., *Ignition Handbook,* Fire Science Publishers/ Society of Fire Protection Engineers, Issaquah WA, 2003.

A.8.10.2 For more information, see Beland, B. "Considerations on Arcing as a Fire Cause," and Beland, B. "Electrical Damages — Cause or Consequence?"

A.8.10.3 For more information, see Beland, "Considerations on Arcing as a Fire Cause," and Beland, B. "Electrical Damages — Cause or Consequence?"

A.8.10.3.4 For more information, see *IEEE Transactions,* IEEE Industrial and Commercial Power System Conference, Detroit, MI, 1971.

A.8.10.4 For more information, see Ettling, B. "Glowing Connections," *Fire Technology* 18(4) (1982): 344–349.

A.8.10.5 For more information, see Beland, B. "Examination of Electrical Conductors Following Fire," *Fire Technology* 16(4) (1980): 252–258.

A.8.10.6.3 For more information, see Beland, B. et al. "Copper-Aluminum Interactions in Fire Environments," *Fire Technology* 19(1) (February 1983): 22–30.

A.8.10.6.4 For more information, see Ettling, B. "Arc Marks and Gouges in Wires and Heating at Gouges," *Fire Technology* 17(1) (1981): 61–68.

A.8.11.8 For more information, see Ettling, B., "The Overdriven Staple as a Fire Cause," *Fire Arson Investigator,* March 1994, and Ettling, B. "Ignitibility of PVC Electrical Insulation by Arcing," *IAAI Oregon Chapter Newsletter,* 1997.

A.8.12.2.2 For more information on static in ignitible liquids, see API RP 2003, *Protection Against Ignitions Arising Out of Static, Lightning, and Stray Currents,* 1991.

A.8.12.2.4 For more information on switch loading, see 5.6.10, 1996 edition of NFPA 30, *Flammable and Combustible Liquids Code,* Annex B of NFPA 385, *Standard for Tank Vehicles for Flammable and Combustible Liquids,* 2007 edition; API RP 1004, *Bottom Loading and Vapor Recovery for MC-306 Tank Motor Vehicles,* 2003; and API RP 2013, *Cleaning Mobile Tanks in Flammable or Combustible Liquid Service,* 1991.

A.8.12.3 For more information, see NFPA 77, *Recommended Practice on Static Electricity,* 2007 edition.

A.8.12.4 For more information, see NFPA 77, *Recommended Practice on Static Electricity,* 2007 edition.

A.8.12.8 For additional information, see NFPA *Fire Protection Handbook,* Section 3, Chapter 3, 18th edition.

A.9.1 For more information see the following publications:

Liquefied Petroleum Gases Handbook, 2nd (1989) edition.

National Fuel Gas Code Handbook, 4th (1999) edition.

American Gas Association. *Gas Engineers Handbook.* New York: Industrial Press, 1965.

O'Loughlin, J. R., and C. F. Yokomoto. "Computation of One-Dimensional Spread of Leaking Flammable Gas." *Fire Technology* 25, No. 4 (November 1989): 308–316.

Rabinkov, V. A. "The Distribution of Flammable Gas Concentrations in Rooms." *Fire Safety Journal* 13 (1988): 211–217.

NFPA 54, *National Fuel Gas Code,* 2006 edition.

NFPA 58, *Liquefied Petroleum Gas Code,* 2008 edition.

Gas odorization technology:

The Gas Odorization Manual, American Gas Association, 1996. This is an updated version of the section on gas odorization in the *Gas Engineers Handbook.*

Wilson, G. G., and A. A. Attari, eds. *Odorization III,* Institute of Gas Technology, Chicago, IL (1993).

Jacobus, O. J., and J. S. Roberts. "What Constitutes Adequate Odorization of Fuel Gases?" *IGT Odorization Symposium* 1995, IGT, 1995.

Andreen, B. H., and R. L. Kroencke. "Stability of Mercaptans Under Gas Distribution System Conditions," *American Gas Journal* 48 May 1965.

Andreen, B. H., and R. L. Kroencke. "Stability of Mercaptans Under Gas Distribution System Conditions," *Proceedings of the American Gas Association,* 136, 1964.

Campbell, I. D., N. A. Chambers, and O. J. Jacobus. "The Chemical Oxidation of Ethyl Mercaptan in Steel Vessels," *IGT Odorization Symposium* 1994, IGT, 1994.

Campbell, I. D. "Factors Affecting Odorant Depletion in LPG," *Proceedings Symposium on LP-Gas Odorization,* 28, 1989.

Campbell, I. D. "Odorant Depletion in Portable Cylinders," *Proceedings of the Symposium on LP-Gas Odorization,* 1990.

Holmes, S. A., P. B. Van Benthuysen, D. C. Lancaster, and M. A. Tiller. "Laboratory and Field Experience with Ethyl Mercaptan Odorant in Propane," *Proceedings of the Symposium on LP-Gas Odorization,* 1990.

Kopidlansky, R. "Odorant Conditioning of New Distribution Systems," *IGT Odorization Symposium* IGT, 1995.

Sanders, R. "OSHA and DOT Odorization Regulations," *IGT Odorization Symposium,* IGT, 1995.

Johnson, J. L. "1965 Report on Project PB-48, Stability of Odorant Compounds," *Proceedings of the 1966 AGA Distribution Conference,* American Gas Association, 1966.

Johnson, S. J. "Ethyl Mercaptan Odorant Stability in Stored Liquid Propane," *Proceedings of the Symposium on LP-Gas Odorization,* 1989.

McHenry, W. B., and H. M. Faulconer. "Summary Report on Odorant Investigations," *Proceedings of the Symposium on LP-Gas Odorization,* 1, 1990.

Roberts, J. S., and D. W. Kelly, "Selection and Handling of Natural Gas Odorants." In Wilson and Attari, eds. *Odorization III,* 29, IGT, 1993. *Proceedings of the Symposium of LP-Gas Odorization,* 1989 and 1990. Copies may be available from the National Propane Gas Association or the Gas Processors Association.

Sampling and testing LP-Gas by stain tubes:

ASTM D 5305, *Standard Test for Determination of Ethyl Mercaptan in LP-Gas Vapor,* 1992.

ASTM D 1265, *Standard Method for Sampling Liquefied Petroleum (LP) Gases*, 1982.

Soil adsorption of odorants:

Nomura, K., and R. Organ. "Reaction of LP-Gas Odorants and Soils," *Proceedings of the Symposium on LP-Gas Odorization*, 76, 1990.

Parlman, R. M., and R. P. Williams. "Penetrabilities of Gas Odorant Compounds in Natural Soils," *Proceedings of the AGA Distribution Conference*, May 6, 1979.

Sullivan, F. "New Gas Odorants." In G. G. Wilson and A. A. Attari, eds., *Odorization III* 209, IGT, 1993.

Tarman, P. B., and H. R. Linden. "Soil Adsorption of Odorant Compounds," *IGT Research Bulletin* #33 July 1, 1961.

Williams, R. P. "Soil Penetrabilities of Natural Gas Odorants," *Pipeline and Gas Journal*, March 1976.

A.9.2 Further information on the properties of fuel gas can be found in the NFPA *Liquefied Petroleum Gases Handbook*, 1995, and the NFPA *National Fuel Gas Code Handbook*, 2003.

A.9.3.1 For more information, see Title 49, Code of Federal Regulations, Part 178, "Specifications for Packagings."

A.9.4.1.1 For more information, see ASME *Boiler and Pressure Vessel Code*. New York, NY: American Society of Mechanical Engineers, 1995.

A.9.4.1.2 For more information, see ASME *Boiler and Pressure Vessel Code*. New York, NY: American Society of Mechanical Engineers, 1995.

A.9.7.1.3 For more information, see U.S. Consumer Products Safety Commission Working Group on Gas Voluntary Standards, *Position Paper on Gas Water Heaters to Prevent Ignition of Flammable Vapors*.

A.9.8 For more information, see American Gas Association, *Gas Engineers Handbook*. New York, NY: Industrial Press, 1965.

A.9.9.4.8 For more information, see Kennedy, P., and J. Kennedy, *Explosion Investigation and Analysis*. Chicago, IL: Investigations Institute, 1990.

A.9.9.5.3 For more information, see NFPA *Liquefied Petroleum Gases Handbook*, and the NFPA *National Fuel Gas Code Handbook*.

A.9.9.9 For more information, see American Gas Association, *Gas Engineers Handbook*. New York, NY: Industrial Press, 1965.

A.9.9.9.2 See sources listed in Section B.6.

A.10.1 For more information, see the following publications:

Bryan, J. L. *An Examination and Analysis of the Dynamics of Human Behavior in MGM Grand Hotel Fire*. Quincy, MA: National Fire Protection Association, 1983.

"Kentucky State Police Investigative Report to the Governor, Beverly Hills Supper Club Fire." Frankfurt: Kentucky State Police, 1977.

SFPE. "Behavioral Response to Fire and Smoke." In *SFPE Handbook of Fire Protection Engineering*, ed. P. DiNenno. Quincy, MA: National Fire Protection Association, 2002.

Bryan, J. L. "A Study of the Survivors' Reports on the Panic in the Fire at Arundel Park Hall, Brooklyn, Maryland, on January 29, 1956." Fire Protection Curriculum, University of Maryland, College Park, MD, 1957.

Latine, R., and J. M. Darley. *Journal of Personality Social Psychology* 10(215), 1968.

A.10.3.2.1 See Latine, R. and J. M. Darley, *Journal of Personality Social Psychology* 10 (215), 1968.

A.10.3.2.2 See Bryan, "A Study of the Survivor's Reports on the Panic in the Fire at Arundel Park Hall, Brooklyn, Maryland, on January 29, 1956." Fire Protection Curriculum, University of Maryland, College Park, MD, 1957.

A.10.3.3.4 For further information, see J. L. Bryan. NFPA *Fire Protection Handbook*, 19th ed., Section 8, Ch. 1, Quincy, MA: National Fire Protection Association, 2003.

A.11.1 While many of the basic legal rules and concepts are similar in Canada, important differences also exist. An investigator should be cautious about applying the legal rules outlined in this chapter to investigations governed by Canadian law. For an explanation of the relevant Canadian legal, procedural, and evidentiary rules addressed in this chapter, see T. D. Hewitt, *Fire Loss Litigation in Canada: A Practical Guide*. Toronto: Carswell, current edition.

For more information, see Black, H. C., *Black's Law Dictionary*, 8th edition, West Pub. Co., 2004.

A.11.5.1 Rules 701, 702 and 703 of the *Federal Rules of Civil Procedure* are printed below:

Rule 701. Opinion Testimony by Lay Witnesses

If the witness is not testifying as an expert, the witness' testimony in the form of opinions or inferences is limited to those opinions or inferences which are (a) rationally based on the perception of the witness, (b) helpful to a clear understanding of the witness' testimony or the determination of a fact in issue, and not based on scientific, technical or other specialized knowledge.

Rule 702. Testimony by Experts

If scientific, technical, or other specialized knowledge will assist the trier of fact to understand the evidence or to determine a fact in issue, a witness qualified as an expert by knowledge, skill, experience, training, or education, may testify thereto in the form of an opinion or otherwise, provided that (1) the testimony is sufficiently based upon reliable facts or data, (2) the testimony is the product of reliable principles and methods, and (3) the witness has applied the principles and methods reliably to the facts of the case.

Rule 703. Bases of Opinion Testimony by Experts

The facts or data in the particular case upon which an expert bases an opinion or inference may be those perceived by or made known to the expert at or before the hearing. If of a type reasonably relied upon by experts in the particular field in forming opinions or inferences upon the subject, the facts or data need not be admissible in evidence in order for the opinion or inference to be admitted. If the facts or data are otherwise inadmissible, they shall not be disclosed to the jury by the proponent of the opinion or inference unless their probative value substantially outweighs their prejudicial effect. *[Federal Rules of Civil Procedure: 701–703]*

A.12.1 For additional information concerning safety requirements or training, see appropriate local, state, or federal occupational safety and health regulations and Munday, J. W., *Safety at Scenes of Fire and Related Incidents*. London: Fire Protection Association, 1994.

Donahue, M. "Safety and Health Guidelines for Fire and Explosion Investigators," Stillwater, OK: Fire Protection Publications, International Fire Service Training Association, Oklahoma State University, 2002.

A.12.1.1 NTIS Publication number PB-94-195047, May 1994.

A.12.1.1.1 Examples of such studies are described in A.12.1.1.1(A) through (C).

(A) In 1998, the Phoenix (Arizona) Fire Department conducted a comprehensive air monitoring study designed to characterize fire fighter exposures during overhaul operations. The study concluded that numerous toxic air contaminants were present during fire overhaul that exceeded occupational permissible exposure limits (PELs). The researchers found that without the use of respiratory protection, fire fighters were exposed to these irritants, chemical asphyxiants, and carcinogens.

(B) A National Fire Protection Association study tallied an average of 4,585 injuries per year related to smoke inhalation, gas inhalation, and respiratory distress in fire fighters between 1996 and 2006.

(C) In 2006, researchers from the University of Cincinnati, College of Medicine compiled data from over 32 studies that investigated the propensity of fire fighters to develop cancer and found that fire fighters had probable cancer risks for multiple myeloma, non-Hodgkin's lymphoma, and prostate and testicular cancers.

A.12.3.3.2 For additional information, see the following:

"Who sets the rules for electrical testing and safety?" originally published by Fluke Corporation and cited from the Cole-Parmer Technical Library (http://www.coleparmer.com/techinfo/pring.asp?htmlfile=fluke_electricalsafety.htm&=293 accessed December 9, 2006.

Fire Fighter Fatality Investigation Report F99-28 — CDC/NIOSH, as accessed at http://www.cdc.gov/niosh/fire/reports/face9928.html on December 9, 2006.

NJ FACE Investigation Report 04-NJ-059, dated October 4, 2005.

A.12.4.1 Figure A.12.4.1 is a hazard and risk assessment sample form.

NAFI Sample Hazard and Risk Assessment

NATIONAL ASSOCIATION OF FIRE INVESTIGATORS NAFI

Location: Date: Assessment Conducted By:

Task: General Scene Safety Survey

Type of Hazard	Risk	Control Methodology		
A. Physical Hazards	H M L	Engineering	Administrative	PPE
B. Structural Hazards				
C. Electrical Hazards				
D. Chemical Hazards				
E. Biological Hazards				
F. Mechanical Hazards				

FIGURE A.12.4.1 Sample Hazard and Risk Assessment.

A.12.4.2.1 OSHA Model Plans and Programs for the OSHA Bloodborne Pathogens and Hazard Communications Standards. OSHA Publication 3186-06N, (2003). Also available as a 521 KB PDF, 29 pages. Provides a model hazardous communication program with an easy-to-use format to tailor to the specific requirements of your establishment.

A.12.6.1.3 For specific guidance in decontamination of turnout gear, NFPA 1851, *Standard on Selection, Care, and Maintenance of Protective Ensembles for Structural Fire Fighting and Proximity Fire Fighting*, should be consulted.

For incidents where hazardous materials are confirmed to be present, decontamination procedures may be obtained from NFPA Supplement 10, *Guidelines for Decontamination of Fire Fighters and Their Equipment Following Hazardous Materials Incidents*.

A.12.6.2 National Standards. In addition to occupational laws and regulations there are standards that are promulgated in relation to the various components of firefighting PPE. This gear is listed in NFPA 1971, *Standard on Protective Ensembles for Structural Fire Fighting and Proximity Fire Fighting*. This standard provides minimum requirements for the following components:

(1) Turnout coat
(2) Turnout pants
(3) Helmet with eye shield
(4) Fire fighting-type gloves
(5) Fire fighting-type boots

In addition to the NFPA standards that may be adopted, there are mandatory regulations as to the use of PPE. These mandatory requirements come in the form of OSHA standards; OSHA 29, CFR 1910.132 to 1910.139 covers mandates for PPE. The scene itself will dictate which level of protection will be required.

For additional information concerning the use of PPE, see 29 CFR 1910.134 Subpart I, Personal Protective Equipment, 29 CFR 1926 Subpart E, Personal Protective Equipment, and NFPA 1500, *Standard on Fire Department Occupational Safety and Health Program*; *Respiratory Protection for Fire and Emergency Services*, 1st Edition, IFSTA; NFPA 472, *Standard for Competence of Responders to Hazardous Materials/Weapons of Mass Destruction Incidents*; NFPA 1404, *Standard for Fire Service Respiratory Protection Training*; NFPA 1851, *Standard on Selection, Care, and Maintenance of Protective Ensembles for Structural Fire Fighting and Proximity Fire Fighting*; NFPA 1852, *Standard on Selection, Care, and Maintenance of Open-Circuit Self-Contained Breathing Apparatus (SCBA)*; NFPA 1981, *Standard on Open-Circuit Self-Contained Breathing Apparatus (SCBA) for Emergency Services*; NFPA 1992, *Standard on Liquid Splash-Protective Ensembles and Clothing for Hazardous Materials Emergencies*; NFPA 1994, *Standard on Protective Ensembles for First Responders to CBRN Terrorism Incidents*; and *Safety and Health Guidelines for the Fire and Explosion Investigator*, Donahue, IFSTA.

A.13.6.1.3 NFPA–sponsored fire investigation training programs include The National Fire, Arson, and Explosion Investigation Training Program; The National Advanced Fire, Arson, and Explosion Investigation Program; The National Vehicle Fire Investigation Training Program; The National Seminar of Fire Analysis Litigation; NFPA two-day seminars on NFPA 921; the InterFlam International Fire Engineering Conference; and the International Symposium on Fire Investigation Science and Technology (ISFI).

A.14.1 U.S. Bureau of Mines, *Fire and Explosion Manual for Aircraft Accident Investigations*, AD-771191, August 1973.

U.S. Bureau of Mines, *Investigation of Fire and Explosion Accidents in the Chemical Mining and Fuel-Related Industries — A Manual*, Report 680, Kuchta, 1985.

Smith, D. W. "Firefighter's Role and Responsibility in Preserving the Fire Scene and Physical Evidence," *The Times*, NFPA Fire Service Section (3), September 1995, p. 6.

A.15.1 For relevant forms that can be used to record the photographs taken and to sketch the scene, see NFPA 906, *Guide for Fire Incident Field Notes*, Form 906.8 (Photograph) and Form 906.9 (Sketch). NFPA 170, *Standard for Fire Safety and Emergency Symbols*, 2006 edition, provides symbols useful in diagramming a fire or explosion scene. Helpful information can also be found in "Formats for Fire Hazard Inspecting, Surveying, and Mapping," in the NFPA *Fire Protection Handbook*, 19th edition, Quincy, MA: National Fire Protection Association, 2003.

A.15.3.2 Figure A.15.3.2(a) through Figure A.15.3.2(i) provide sample forms that can be used for data collection.

A.15.4.4(D) Many references, such as NFPA 170, *Standard for Fire Safety and Emergency Symbols*, 2006 edition, are available that can be used for assistance.

A.15.4.6.4 Further information can be obtained by contacting specification organizations such as the Construction Specifications Institute (CSI), 601 Madison Street, Alexandria, VA 22314-1791.

A.15.5 For further information see Smith, Dennis W., "Must Fire Investigators Prepare a Written Investigation Report?" *NFPA Fire Journal*, May/June 2005, p. 65.

A.16.1 For relevant forms that can be used to document physical evidence, see Figure A.15.3.2(a), Figure A.15.3.2(b), Figure A.15.3.2(c), and Figure A.15.3.2(e).

ASTM E 860, *Standard Practice for Examining and Testing Items That Are or May Become Involved in Product Liability Litigation*, 1997.

ASTM E 1188, *Standard Practice for Collection and Preservation of Information and Physical Items by a Technical Investigator*, 1995.

ASTM E 1459, *Standard Guide for Physical Evidence Labeling and Related Documentation*, 1992.

IAAI, *A Pocket Guide to Accelerant Evidence Collection*. Massachusetts Chapter, 1992.

IAAI Forensic Science Committee. "Position on Comparison Samples." *Fire and Arson Investigator* 41 (2) December 1990.

A.16.2.2 For further information see ASTM E 860, *Standard Practice for Examining and Testing Items that are or May Become Involved in Litigation*, 1997; ASTM E 1188 *Standard Practice for Collection and Preservation of Information and Physical Items by a Technical Investigator*, 1995; ASTM E 1492, *Standard Practice for Receiving, Documenting, Storing, and Retrieving Evidence in a Forensic Science Laboratory*, 1992; and ASTM E 1459, *Standard Guide for Physical Evidence Labeling and Related Documentation*, 1992.

A.16.3 For more information, see Smith, D.W. "Firefighter's Role and Responsibility in Preserving the Fire Scene and Physical Evidence," *The Times, NFPA Fire Service* 3, September 1995, p. 6.

A.16.4.2 For further information, see Massachusetts Chapter of International Association of Arson Investigators, *A Pocket Guide to Accelerant Evidence Collection*, 2007.

A.16.5.1.2 For further information see ASTM E 860, *Standard Practice for Examining and Testing Items that are or May Become Involved in Litigation*, 1997; ASTM E 1188, *Standard Practice for Collection and Preservation of Information and Physical Items by a Technical Investigator*, 1995; ASTM E 1492, *Standard Practice for Receiving, Documenting, Sorting, and Retrieving Evidence in a Forensic Science Laboratory*, 1992; and ASTM E 1459, *Standard Guide for Physical Evidence Labeling and Related Documentation*, 1992.

FIRE INCIDENT FIELD NOTES

Agency: ______________________ File No: ______________

TYPE OF OCCUPANCY

Location/ Address						
Property Description	Structure	Residential	Commercial	Vehicle	Wildland	Other
Other Relevant Info						

WEATHER CONDITIONS

Indicate Relevant Weather Information					
	Visibility	Rel. humidity	GPS	Elevation	Lightning
	Temperature	Wind direction	Wind speed	Precipitation	

OWNER

Name			DOB
d/b/a (if applicable)			
Address			
Telephone	Home	Business	Cellular

OCCUPANT

Name			DOB
d/b/a (if applicable)			
Permanent Address			
Temporary Address			
Telephone	Home	Business	Cellular

DISCOVERED BY

Incident Discovered by	Name		DOB
Address			
Telephone	Home	Business	Cellular

 NFPA 921 (p. 1 of 2)

FIGURE A.15.3.2(a) Sample Form for Collecting Fire Incident Field Data.

FIRE INCIDENT FIELD NOTES (Continued)

File No: ____________________

REPORTED BY

Incident Reported by	Name		DOB
Address			
Telephone	Home	Business	Cellular

INVESTIGATION INITIATION

Request Date and Time	Date of request	Time of request
Investigation Requested by	Agency name	Contact person/Telephone no.
Request Received by	Agency name	Contact person/Telephone no.

SCENE INFORMATION

Arrival Information	Date		Time		Comments	
Scene Secured	Yes	No	Securing agency		Manner of security	
Authority to Enter	Contemporaneous to exigency		Consent		Warrant	
			Written	Verbal	Admin.	Crim.	Other
Departure Information	Date		Time		Comments	

OTHER AGENCIES INVOLVED

	Dept. or Agency Name	Incident No.	Contact Person/Phone
Primary Fire Department			
Secondary Fire Department(s)			
Law Enforcement			
Private Investigators			

ADDITIONAL REMARKS

____

____

____

____

____

 NFPA 921 (p. 2 of 2)

FIGURE A.15.3.2(a) *Continued*

CASUALTY FIELD NOTES

Agency: ______ Incident date: ______ Case number: ______

DESCRIPTION

Name: ______ DOB: ______ Sex/Race: ______

Address: ______ Phone: ______

Other identifiers: ______

Description of clothing and jewelry: ______

Occupation: ______ Place of employment: ______

Marital status: ______

Victim's doctor: ______ Victim's dentist: ______

Smoker: ❑ Yes ❑ No ❑ Unknown

CASUALTY TREATMENT

Treated at scene: ❑ Yes ❑ No By: ______

Transported to: ______ Remarks: ______

SEVERITY OF INJURY

❑ Minor ❑ Moderate ❑ Severe ❑ Fatal

Describe injury: ______

NEXT OF KIN

Name: ______ Address: ______ Phone: ______

Relationship: ______ Notified on ___/___/___ By: ______

FATALITY INFORMATION

Where was victim initially found: ______

Who located victim: ______

Body position when initially found: ______

Victim's appearance: ______

Body removed by: ______ To: ______

Photographed in place: ❑ Yes ❑ No Significant blood present under/near victim: ❑ Yes ❑ No

MEDICAL EXAMINER/CORONER

Agency: ______

Date of examination: ___/___/___ Location: ______

Autopsy requested: ❑ Yes ❑ No Autopsy completed: ❑ Yes ❑ No Copy attached: ❑ Yes ❑ No

Full body x-rays: ❑ Yes ❑ No Other x-rays: ______

Identification made from: ❑ Physical appearance ❑ Dental records ❑ Fingerprints ❑ Prior injury comparison
❑ Other: ______

Condition of trachea: ______

Evidence of pre-fire injury: ❑ Yes ❑ No Type/location: ______

Blood samples taken: ❑ Yes ❑ No Other specimens collected: ______

CO level: ______ Blood alcohol: ______ Other: ______

Cause of death: ______

COMPLETE BODY DIAGRAM ON REVERSE

 NFPA 921 (p. 1 of 2)

FIGURE A.15.3.2(b) Sample Form for Collecting Casualty Field Data.

CASUALTY FIELD NOTES (Continued)

REMARKS

____

____

____

____

____

____

BODY DIAGRAM

Indicate parts of body injured: ❑ None ❑ Blisters (red marker) ❑ Burns (black marker)

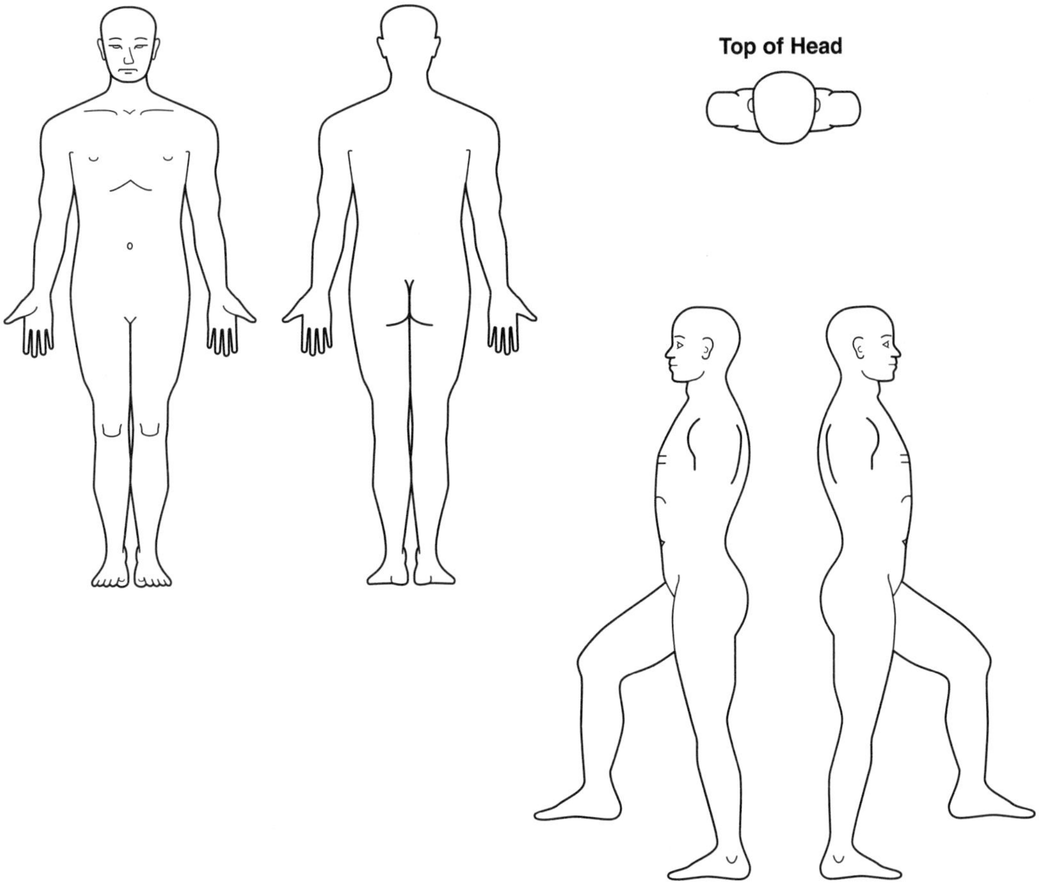

Top of Head

Fire Investigation Data Sheet/Attachment: ______________________ Initials: ________
Body Diagram

 NFPA 921 (p. 2 of 2)

FIGURE A.15.3.2(b) *Continued*

WILDFIRE NOTES

Agency: ____________________ File number: ____________________

PROPERTY DESCRIPTION

Fire damage: ❑ Less than acre ______ No. acres	Other properties involved:
Security: ❑ Open ❑ Fenced ❑ Locked gate	Comments:

FIRE SPREAD FACTORS

Type fire: ❑ Ground ❑ Crown	Factors: ❑ Wind ❑ Terrain	Comments:

AREA OF ORIGIN

PEOPLE IN AREA

At time of fire: ❑ Yes ❑ No ❑ Undetermined	Comments:

IGNITION SEQUENCE

Heat of ignition:

Material ignited:

Ignition factor:

If equipment involved: Make: Model: Serial no.:

Comments:

 NFPA 921

FIGURE A.15.3.2(c) Sample Form for Collecting Wildfire Data.

EVIDENCE FORM

Case #: ____________

Date of incident: ___/___/___ Storage location: ____________

Item No.	Description	Location		
____	____________	____________	Destroyed	Released
____	____________	____________	Destroyed	Released
____	____________	____________	Destroyed	Released
____	____________	____________	Destroyed	Released
____	____________	____________	Destroyed	Released
____	____________	____________	Destroyed	Released
____	____________	____________	Destroyed	Released
____	____________	____________	Destroyed	Released
____	____________	____________	Destroyed	Released
____	____________	____________	Destroyed	Released

How was evidence received? Date received: ___/___/___ Date stored: ___/___/___

❑ Removed from scene by investigator.

❑ Received by investigator from: ____________
Name, Company, or Dept.

Received via: ❑ UPS ❑ FedEx ❑ Airborne ❑ U.S. Mail ❑ In person ❑ Freight ____________
Name of Company

❑ Other: ____________
Describe

____________ Received by

____________ Case Investigator

LOCATION EVIDENCE REMOVED

____________ Owner

____________ State Zip Phone

____________ Company

____________ Address 2

____________ Address 1

____________ City

____________ City

____________ State Zip Phone

 NFPA 921 (p. 1 of 2)

FIGURE A.15.3.2(d) Sample Form for Collecting Evidence.

EVIDENCE FORM (Continued)

INTERNAL EXAMINATION

Investigator	Date Pulled	Date Examined	Date Returned

EVIDENCE DESTRUCTION

Authorized by — Date

Investigator's Authorization — Date

Destroyed by — Date

EVIDENCE RELEASE

Signature of Person Receiving Evidence

Person Receiving Evidence (Please Print) — Date

Company Name

Address

City — State — Zip Code

Authorized by — Date

Investigator's Authorization — Date

Released via

REMARKS

EXAMINATION BY OTHERS

Name — Date of Examination

Company

Address

City — State — Zip — Phone

Authorized by

Investigator's Authorization — Date

Name — Date of Examination

Company

Address

City — State — Zip — Phone

Authorized by

Investigator's Authorization — Date

Name — Date of Examination

Company

Address

City — State — Zip — Phone

Authorized by

Investigator's Authorization — Date

NFPA 921 (p. 2 of 2)

FIGURE A.15.3.2(d) ***Continued***

VEHICLE INSPECTION FIELD NOTES

Job # ________ File # ________ Date of Occurrence ________
Insured ________ Date of Assignment ________
Address (City, State) ________ Date of Receipt ________
Loss Location ________ Date of Inspection ________
________ Insp Location ________
Stolen? ❑ Yes ❑ No Recovered by ________ Time of Inspection ________
Police Report ________ Fire Report ________
# of Keys ________ Alarm System? ❑ Yes ❑ No Alarm Type ________
Hidden Keys? ❑ Yes ❑ No Location ________

VEHICLE

Make ________ Model ________ Year ________
VIN ________ Odometer ________

EXTERIOR

Tires	Tire Type	Wheel Type	Tire Tread Depth	Lugs	Missing
LF					
LR					
RR					
RF					
SP					

Doors	Glass Y/N	Window UP/DOWN	Locked Y/N	Open/Closed	Prior Damage
LF					
LR					
RR					
RF					

Body Panels	Construction	Condition	Prior Damage
F Bumper			
Grill			
LF Fender			
LR Quarter			
R Bumper			
RR Quarter			
RF Fender			
Hood			
Roof			
Trunk			

UNDER HOOD	Intact	Missing	Parts Missing	Condition
Engine				
Battery				
Belts & Hoses				
Wiring				
Accessories				

FLUIDS	Level	Condition	Sample Taken
Oil			
Transmission			
Radiator			
Pwr Steer			
Brake			
Clutch			

ATS 851B, 8/97 NFPA 921 (p. 1 of 2)

FIGURE A.15.3.2(e) Sample Form for Vehicle Inspection. *(Source: Applied Technical Services, Inc.)*

VEHICLE INSPECTION FIELD NOTES (Continued)

Job # ____________

INTERIOR	Intact	Missing	Parts Missing	Condition
Dash Pod				
Glove Box				
Strg Column				
Ignition				
Front Seat				
Rear Seat				
Rear Deck				
			Make/Model	
Stereo				
Speakers				
Accessories				
FLOOR			**Sample Taken**	
LF				
LR				
RR				
RL				

PERSONAL EFFECTS IN THE INTERIOR

TRUNK OR CARGO AREA

AFTERMARKET ITEMS NOT PREVIOUSLY DESCRIBED

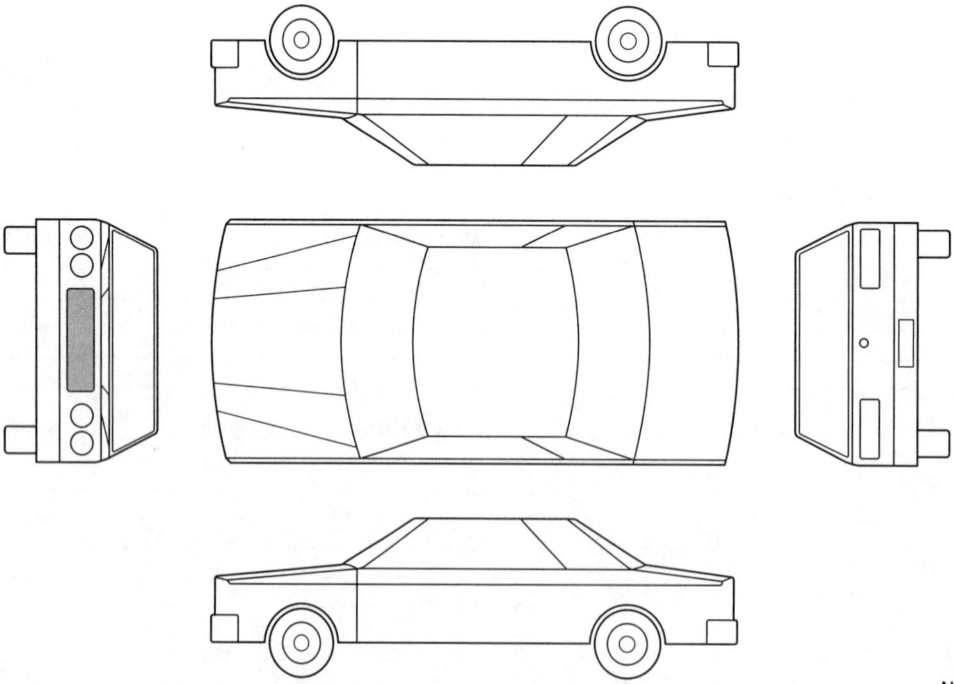

ATS 851B, 8/97

NFPA 921 (p. 2 of 2)

FIGURE A.15.3.2(e) *Continued*

PHOTOGRAPH LOG

Roll #: ______________________ Exposures: ______________________

Case #: Date:

Camera make/type: Film type: Film speed: ASA:

Number	Description	Location
1)		
2)		
3)		
4)		
5)		
6)		
7)		
8)		
9)		
10)		
11)		
12)		
13)		
14)		
15)		
16)		
17)		
18)		
19)		
20)		
21)		
22)		
23)		
24)		
25)		
26)		
27)		
28)		
29)		
30)		
31)		
32)		
33)		
34)		
35)		
36)		

Photos taken by: ______________________ Initials: __________

 NFPA 921

FIGURE A.15.3.2(f) Sample Form for Photograph Log.

ELECTRICAL PANEL DOCUMENTATION

Fire location:	Date:	Case #:
Panel location:	Main size:	Fuses: ☐
		Circuit breakers: ☐

LEFT BANK

#	Rating Amps	Labeled Circuit	Status
1	—		—
3	—		—
5	—		—
7	—		—
9	—		—
11	—		—
13	—		—
15	—		—
17	—		—
19	—		—
21	—		—
23	—		—
25	—		—
27	—		—
29	—		—

Notes:

RIGHT BANK

#	Rating Amps	Labeled Circuit	Status
2	—		—
4	—		—
6	—		—
8	—		—
10	—		—
12	—		—
14	—		—
16	—		—
18	—		—
20	—		—
22	—		—
24	—		—
26	—		—
28	—		—
30	—		—

Notes:

Documented by: ____________________

 NFPA 921

FIGURE A.15.3.2(g) Sample Form for Electrical Panel Data.

STRUCTURE FIRE

Agency: ____________________ Case number: ____________________

TYPE OF OCCUPANCY

Residential	Single family	Multifamily	Commercial	Governmental
Church	School	Other:		

Estimated age:	Height (stories):	Length:	Width:

PROPERTY STATUS

Occupied at time of fire? ❑ Yes ❑ No	Unoccupied at time of fire? ❑ Yes ❑ No	Vacant at time of fire? ❑ Yes ❑ No
Name of person last in structure prior to fire:	Time and date in structure:	Exited via which door/egress:
Remarks:		

BUILDING CONSTRUCTION

Foundation Type	Basement		Crawl space		Slab		Other:
Material	Masonry		Concrete		Stone		Other:
Exterior Covering	Wood	Brick/Stone	Vinyl	Asphalt	Metal	Concrete	Other:
Roof	Asphalt	Wood		Tile	Metal		Other:
Type of Construction	Wood frame	Balloon	Heavy timber	Ordinary	Fire resistive	Non-combustible	Other:

ALARM/PROTECTION/SECURITY

Sprinklers ❑ Yes ❑ No	Standpipes ❑ Yes ❑ No	Security camera(s) ❑ Yes ❑ No
Smoke detectors ❑ Yes ❑ No	Hardwired ❑ Yes ❑ No	Battery ❑ Yes ❑ No
Were batteries in place? ❑ Yes ❑ No	Location(s):	

Hidden keys ❑ Yes ❑ No where:	Security bars: Windows? ❑ Yes ❑ No Doors? ❑ Yes ❑ No

Remarks:

 NFPA 921 (p. 1 of 2)

FIGURE A.15.3.2(h) Sample Form for Structure Fire Data.

STRUCTURE FIRE (Continued)

CONDITION OF DOORS/WINDOWS

Doors	Locked	Unlocked but closed	Open	
	Forced entry? ❑ Yes ❑ No	Who forced if known?		
Windows	Secure	Unlocked but closed	Open	Broken
	Broken by first responders? ❑ Yes ❑ No	Remarks:		

FIRE DEPARTMENT OBSERVATIONS

Name of first on scene:	Department:
General observations:	
Obstacles to extinguishment?	First-In Report attached? ❑ Yes ❑ No

UTILITIES

Electric	❑ On ❑ Off ❑ None	❑ Overhead ❑ Underground	
	Company:	Contact:	Telephone:
Gas/Fuel	❑ On ❑ Off ❑ None	❑ Natural ❑ LP ❑ Oil	
	Company:	Contact:	Telephone:
Water	Company:	Contact:	Telephone:
Telephone	Company:	Contact:	Telephone:
Other	Company:	Contact:	Telephone:

COMMENTS:

 NFPA 921 (p. 2 of 2)

FIGURE A.15.3.2(h) *Continued*

COMPARTMENT FIRE MODELING

Room Number ________ Use ________________________________

Size (use diagrams if possible) | Wall/floor/ceiling

Construction

Length ____________________ | ____________________

Width ____________________ | ____________________

Height ____________________ | ____________________

Lining materials that represent over 10% of room lining
(Include thickness, density, and other material characteristics if known.)

Wall Material	Percentage of Area Involved

Ceiling Material	

Floor or Floor Covering Material	

Doors, Windows, and Other Openings [Enter all heights as distance above floor. If door sill is at floor, enter zero (0).]

Openings	To Top	To Sill	Width	Changes During Fire (How?)[1]

[1] For example: "Window broke at 10:33" or "Door was closed until opened by escaping occupant, then left open — Exit Time 10:30."

 NFPA 921 (p. 1 of 2)

FIGURE A.15.3.2(i) Sample Form for Compartment Fire Modeling Fire Data.

COMPARTMENT FIRE MODELING (Continued)

Heating, Ventilation, and Air Conditioning (HVAC). Include air flows from HVAC systems. Give rates and positions of supply and return or exhaust in this room. Also sizes and types of ducts/diffusers.

____

____

____

Tightness of Walls, Closed Windows, Door Fits, etc. (Unless fit is very loose, classify as tight, average, or loose. If fit is very loose, try to get size, number, and location of cracks, etc.)

Doors ______________________________

Windows ______________________________

Inside Walls ______________________________

Exterior Walls ______________________________

Fire History (List all significant events involving progress of the fire.)

Time (hard or soft)	Event
e.g., 1:10 a.m.	sofa involved, flames 3 feet high
1:17 a.m.	room flashover
1:19 a.m.	large fire plume into hallway
1:23 a.m.	smoke out of third floor window

Initial Fuel Item(s) Description

Description	Size	Material
e.g., sofa	full	polyurethane, with cotton upholstery

Suspected Igniter (List igniter if known with qualification on confidence.)

Igniter: e.g., cigarette

Confidence: probable

 NFPA 921 (p. 2 of 2)

FIGURE A.15.3.2(i) ***Continued***

A.16.5.4.6 For more information, see IAAI Forensic Science Committee, "Position on Comparison Samples," *Fire and Arson Investigator.* Vol. 41, No. 2 (December 1990).

A.16.5.4.7 See IAAI Forensic Science Committee, "Position on the Use of Accelerant Detection Canines," Fire and Arson Investigator, 45, No 1 (September 1994).

A.16.5.4.7.2 For more information, see Kurz et al., "Evaluation of Canines for Accelerant Detection at Fire Scenes"; DeHaan, "Canine Accelerant Detection Teams: Validation and Certification"; and Tindall and Lothridge, "An Evaluation of 42 Accelerant Detection Canine Teams."

A.16.5.4.7.4 See Lentini, J.J., J.A. Dolan, and C. Cherry. "The Petroleum-Laced Background." *Journal of Forensic Science* 45 (5) (2000): 968–989.

A.17.4.2 For more information on heat and flame vector analysis, see the following references:

Shanley, J. H. Jr.; Alletto, W. C.; Corry, R.; Herndon, J.; Kennedy, P. M.; and Ward, J.; "Federal Emergency Management USFA Fire Burn Pattern Tests." Report of the United States Fire Administration Program for the Study of Fire Patterns, FA 178; p. 221 July 1997.

Kennedy, J. and P. Kennedy. *Fires and Explosions — Determining Cause and Origin.* Chicago: Investigations Institute, 1985, pp. 442–444.

Shanley, J., and P. Kennedy. "Program for Study of Fire Patterns," Annual Conference on Fire Research (Gaithersburg, Maryland: National Institute of Standards and Technology (NIST), Building and Fire Research Laboratory (BFRL), 1992).

Kennedy, P. and J. Shanley, "Report on the USFA Program for the Study of Fire Pattern," Interflam '96 Proceedings (London: Interscience Communications, 1996), pp. 971–975.

Kennedy, P. "NFPA 921, 'Heat and Flame Vector Analysis,' and Current Fire Pattern Analysis Research," Interflam '99 Proceedings. London: Interscience Communications, 1999, pp. 1311–1316.

Icove, D. and J. DeHaan, *Forensic Fire Scene Reconstruction.* Upper Saddle River, NJ: Pearson/Prentice Hall, 2004, pp. 102–105.

DeHaan, J. *Kirk's Fire Investigation*, 6th ed., Upper Saddle River, NJ: Pearson/Prentice Hall, 2007, p. 238.

A.17.4.5.1 This procedure is more likely to be successful in locations where circuits are arranged in a radial system, and "upstream" and "downstream" orientations are readily discernable. In some locations, such as the United Kingdom, circuits are generally arranged in a "ring" system. In ring systems arc mapping is still useful in testing an origin hypothesis, but some sequential data may not be evident. "Upstream" and "downstream" may also be difficult to discern in cases where the conductors are back-fed through a pre-fire wiring error, or through uninterruptible power supply (UPS) systems employing batteries.

A.18.3.4 See ASTM G145-08, *Standard Guide for Studying Fire Incidents in Oxygen Systems,* and NFPA 430, *Code for the Storage of Liquid and Solid Oxidizers.*

A.18.6.4.1 Example reference sources include the following:

(1) FIREDOC, an online searchable fire library maintained by the NIST Building and Fire Research Lab (http://fris2.nist.gov/starweb/FireDoc/servlet.starweb?path=FireDoc/firedoc.web)
(2) The *SFPE Handbook of Fire Protection Engineering,* fourth edition, DiNenno, P.J. (ed.), 2008 (available from the Society of Fire Protection Engineers and NFPA).
(3) The International Association of Fire Safety Science (http://iafss.org/) websites has all symposia and the Fire Research Notes searchable on line. The Fire Research Notes are the comprehensive record of research conducted by the UK Fire Research Station (FRS) over the period from 1952 to 1978 on all aspects of fire and explosion safety.

A.18.6.5 For more information, see the following: Smith, Dennis W., "The Pitfalls, Perils and Reasoning Fallacies of Determining the Fire Cause in the Absence of Proof: The Negative Corpus Methodology," ISFI Proceedings 2006, International Symposium on Fire Investigation Science and Technology, National Association of Fire Investigators, Sarasota, FL, 2006, pp. 313-325; and, The National Fire Investigator, Spring 2007, NAFI, p.4-11.

A.18.6.5.2 For more information, see the following: Smith, Dennis W., "The Pitfalls, Perils and Reasoning Fallacies of Determining the Fire Cause in the Absence of Proof: The Negative Corpus Methodology," ISFI Proceedings 2006, International Symposium on Fire Investigation Science and Technology, National Association of Fire Investigators, Sarasota, FL, 2006, pp. 313-325; and, The National Fire Investigator, Spring 2007, NAFI, p.4-11.

A.19.1 Fore more information, see the following:

ANSI Z535.4, *Product Safety Signs and Labels,* 1998.

ANSI Z535.5, *Accident Prevention Tags,* 1998.

UL 969, *Standard for Marking and Labeling Systems,* 1995.

FMC Corporation, *FMC Product Safety Sign and Label System Manual.*

A.20.1 For more information, see the following:

Kimamoto, H., and E. J. Henley. *Probabilistic Risk Assessment and Management for Engineers and Scientists.* IEEE Press, 1996.

Fire test methods:

NFPA 253, *Standard Method of Test for Critical Radiant Flux of Floor Covering Systems Using a Radiant Heat Energy Source,* 2006 edition.

ASTM D 56, *Standard Test Method for Flash Point by Tag Closed Tester,* 2002.

ASTM D 92, *Standard Test Method for Flash and Fire Points by Cleveland Open Cup,* 2002.

ASTM D 93, *Standard Test Method for Flash Point by Pensky-Martens Closed Cup Tester,* 2002.

ASTM D 1230, *Standard Test Method for Flammability of Apparel Textiles,* 2001.

ASTM D 1310, *Standard Test Method for Flash Point and Fire Point of Liquids by Tag Open-Cup Apparatus,* 2001.

ASTM D 1929, *Standard Test Method for Determining Ignition Temperature of Plastics,* 2001.

ASTM D 2859, *Standard Test Method for Flammability of Finished Textile Floor Covering Materials,* 1993.

ASTM D 3065, *Standard Test Methods for Flammability of Aerosol Products,* 2001.

ASTM D 3828, *Standard Test Methods for Flash Point by Small Scale Closed Tester,* 2002.

ASTM D 4809, *Standard Test Method for Heat of Combustion of Liquid Hydrocarbon Fuels by Bomb Calorimeter (Precision Method),* 2000.

ASTM D 5305, *Standard Test Method for Determination of Ethyl Mercaptan in LP-Gas Vapor,* 1997.

ASTM E 84, *Standard Test Method for Surface Burning Characteristics of Building Materials*, 2003.

ASTM E 108, *Standard Test Method for Fire Tests of Roof Coverings*, 2000.

ASTM E 119, *Standard Methods of Tests of Fire Endurance of Building Construction and Materials*, 2000.

ASTM E 603, *Standard Guide for Room Fire Experiments*, 2001.

ASTM E 648, *Standard Test Method for Critical Radiant Flux of Floor-Covering Systems Using a Radiant Heat Energy Source*, 2000.

ASTM E 659, *Standard Test Method for Autoignition Temperature of Liquid Chemicals*, 2000.

ASTM E 681, *Standard Test Method for Concentration Limits of Flammability of Chemicals*, 2001.

ASTM E 800, *Standard Guide for Measurement of Gases Present or Generated During Fires*, 2001.

ASTM E 906, *Standard Test Method for Heat and Visible Smoke Release Rates for Materials and Products*, 1999.

ASTM E 1226, *Test Method for Pressure and Rate of Pressure Rise for Combustible Dusts*, 2000.

ASTM E 1352, *Standard Test Method for Cigarette Ignition Resistance of Mock-up Upholstered Furniture Assemblies*, 2002.

ASTM E 1353, *Standard Test Methods for Cigarette Ignition Resistance of Components of Upholstered Furniture*, 2002.

ASTM E 1354, *Standard Test Method for Heat and Visible Smoke Release Rates for Materials and Products Using an Oxygen Consumption Calorimeter*, 2003.

ANSI/UL 263, *Standard for Safety Fire Tests of Building Construction and Materials*, 2003.

A.20.3.1 For more information on fault trees, see NFPA *Fire Protection Handbook*, 18th ed., pp. 37–41; NFPA 550, *Guide to the Fire Safety Concepts Tree*, 2007 edition; and Kimamoto and Henley, *Probabilistic Risk Assessment and Management for Engineers and Scientists*, Ch. 4.

A.20.4.1.6 For an example of a V&V document relating to a fire model, see *Engineering Guide—Evaluation of the Computer Fire Model DETACT-QS*, Society of Fire Protection Engineering, Bethesda, MD, 2002.

A.20.4.7 For additional information on egress analysis, see Section B.7.

A.20.4.8 More extensive discussions of fire dynamics analyses and models are included in the NFPA *Fire Protection Handbook*, and *The SFPE Handbook of Fire Protection Engineering*.

A.20.5.2 Figure A.15.3.2(i) gives examples of forms that can be used in data collection for compartment fire modeling.

A.21.1 The references in Section B.8 are of value when considering additional technical information on explosions involving structures or vessels. For more information see the following:

ASME Boiler and Pressure Vessel Code, Section VIII, 2007.

Baker, W. E. *Explosion Hazards and Evaluations*. Amsterdam–New York: Elsevier Publishers, 1983.

Baker, W. E., and M. J. Tang. *Gas, Dust, and Hybrid Explosions*. Amsterdam–New York: Elsevier Publishers, 1991.

Bartknecht, W. *Dust Explosions*. Berlin–New York: Springer-Verlag, 1989.

Bartknecht, W. *Explosions — Course Prevention Protection*. New York: Springer-Verlag, 1980.

Best, R., and W. L. Walls. "Hot Water Heater BLEVE in School Kills Seven." *Fire Journal* 76, No. 5 (September 1982): pp. 20–24 and 104–105.

Bodurtha, F. T. *Industrial Explosion Prevention and Protection*. New York: McGraw Hill, 1980.

Cashdollar, K., and M. Hertzberg. *Explosibility and Ignitibility of Plastic Abrasive Media*. Internal Report No. 4657, June 1987.

Center for Chemical Process Safety. *Guidelines for Evaluating the Characteristics of Vapor Cloud Explosions, Flash Fires and BLEVEs*. Washington, DC: American Institute of Chemical Engineers, 1994.

Chemical Propulsion Information Agency. *Hazards of Chemical Reactants and Propellants*, CPIA 394, September 1984.

Eckhoff, R. K. *Dust Explosions in the Process Industries*. Butterworth–Heineman, 1997.

Fleischmann, C. M., P. J. Pagni, and R. B. Williamson, "Exploratory Backdraft Experiments," *Fire Technology* 29(4), 1993.

Fleischmann, C. M., P. J. Pagni,, and R. B. Williamson. "Quantitative Backdraft Experiments," *4th International Symposium on Fire Safety Science*, International Association for Fire Safety Science, 1994.

Gottuk, D. T., M. J. Peatross, J. P. Farley, and F. W. Williams, "The Development and Mitigation of Backdraft: A Real-Scale Shipboard Study," *Fire Safety Journal*, 33(4), November 1999.

Gottuk, D. T., F. W. Williams, and J. P. Farley. "The Development and Mitigation of Backdrafts: A Full-Scale Experimental Study," *Fire Safety Science—Proceedings of the Fifth International Symposium*, International Association of Fire Safety Science, 1997.

Gugan, K. *Unconfined Vapor Cloud Explosions*. Houston, TX: Gulf Publishing, 1978.

Harris, R. J. *The Investigation and Control of Gas Explosions in Buildings and Heating Plants*. London and New York: E&FN Spoon, Ltd., 1983.

Hertzberg, M., and K. Cashdollar. "Domains of Flammability and Thermal Ignitibility for Pulverized Coals and Other Dusts; Particle Size Dependence and Microscopic Residue Analyses." *19th Symposium on Combustion*. Combustion Institute, 1982, pp. 1169–80.

Institute of Makers of Explosives. *Glossary of Commercial Explosives Industry Terms*. Washington, DC: Safety Library Publications, No. 12, 1985.

Kennedy, P., and J. Kennedy. *Explosion Investigation and Analysis*. Chicago, IL: Investigations Institute, 1990.

National Fire Academy, Open Learning Fire Service Program. *Fire Dynamics Course Guide/Reader, Unit 4 — Explosions*. Emmitsburg, MD: National Emergency Training Center.

Nettleton, M. A. *Gaseous Detonations; Their Nature, Causes and Control*. New York: Routledge, Chapman and Hall, 1987.

NFPA 68, *Standard on Explosion Protection by Deflagration Venting*, 2008 edition.

NFPA 69, *Standard on Explosion Prevention Systems*, 2008 edition.

NFPA 654, *Standard for the Prevention of Fire and Dust Explosions from the Manufacturing, Processing, and Handling of Combustible Particulate Solids*, 2006 edition.

Prugh, R. W. "Quantitative Evaluation of 'BLEVE' Hazards," *Journal of Fire Protection Engineering* 3(1) (March 1981).

Stull, D. R. "Fundamentals of Fire and Explosion,"AICHE Monograph Series 73(10), 1977.

Title 49, Code of Federal Regulations, Part 178, "Specification for Packagings," U.S. Government Printing Office, Washington, DC.

U.S. Army Technical Manual, TM 5-1300, Revision 1, "Structures to Resist the Effects of Accidental Explosions," Washington, DC: U.S. Government Printing Office.

USBM. *Fire and Explosion Manual for Aircraft Accident Investigations.* Washington, DC: U.S. Bureau of Mines, 1973.

USBM. *Investigation of Fire and Explosion Accidents in the Chemical Mining and Fuel-Related Industries — A Manual.* Washington, DC: U.S. Bureau of Mines, 1985.

USBM. *Pressure Development in Laboratory Dust Explosions.* Washington, DC: U.S. Bureau of Mines.

USBM. *Preventing Ignition of Dust Dispersion by Inerting,* Report of Investigations 6543. Washington, DC: U.S. Bureau of Mines, 1964, p. 12.

Walls, W. J. "Just What Is a BLEVE?" *Fire Journal* 72(6) (November 1978): 46–47.

Yallop, J. *Explosion Investigation.* Harrogate, UK: Forensic Science Society Press, 1980.

Zabetakis, M. G. *Flammability Characteristics of Combustible Gases and Vapors,* Bulletin 627, 1965.

Zalosh, R. "Explosion Protection," *The SFPE Handbook of Fire Protection Engineering.* Quincy, MA: Society of Fire Protection Engineers, 1995, Section 2, Chapter 5.

A.21.2 For more information, see the NFPA *Fire Protection Handbook,* 19th ed., Section 1, Chapter 6.

A.21.2.2 For more information on BLEVEs, see the following publications: Walls, W. J. "Just What Is a BLEVE?" *Fire Journal* 72(6) (November 1978): 46–47 and Best, R. and W. L. Walls, "Hot Water Heater BLEVE in School Kills Seven," *Fire Journal* 76(5) (September 1982): 20–24 and 104–105.

A.21.2.3 For more information, see the NFPA *Fire Protection Handbook,* 19th ed., Section 1, Chapter 6.

A.21.3.2 For more information, see Kennedy, P., and J. Kennedy. *Explosion Investigation and Analysis.* Chicago, IL: Investigations Institute, 1990.

A.21.4.3.2 For additional information on Fireballs See: Lees' Loss Prevention in the Process Industries, 16.15 Fireballs, Babrauskas, Vytenis, Ignition Handbook, p.524-527.

A.21.5.3 For more information, see the following publications:

Bartknecht, W. Explosions — Course Prevention Protection. New York, NY: Springer-Verlag, 1980.

NFPA 68, *Standard on Explosion Protection by Deflagration Venting,* 2008 edition.

Zalosh, R. "Explosion Protection," The SFPE Handbook of Fire Protection Engineering, Quincy, MA NFPA, 1995.

A.21.5.4 For more information, see NFPA 68, *Standard on Explosion Protection by Deflagration Venting,* 2007 edition.

A.21.7.3 Additional information may be found in Yallop, J. *Explosion Investigation.* Harrogate, UK: Forensic Science Society Press, 1980.

A.21.8.1 For more information, see NFPA 77, *Recommended Practice on Static Electricity,* 2007 edition.

A.21.8.2.1 The following references provide additional information on flame speeds:

Harris, R. J. *The Investigation and Control of Gas Explosions in Buildings and Heating Plants.* London and New York: E & FN Spoon, Ltd., 1983.

NFPA 68, *Standard on Explosion Protection by Deflagration Venting,* 2007 edition.

A.21.8.2.1.4 Laminar burning velocity has sometimes been referred to in the historical literature as the fundamental flame speed, or the laminar flame speed. This document specifically refers to these properties in the terms as discussed in sections 21.8.2.1.4 through 21.8.2.1.6.

A.21.8.2.2 For more information, see the following publications:

Kennedy, J., and P. Kennedy. *Fires and Explosions — Determining Cause and Origin.* Chicago, IL: Investigations Institute, 1985.

O'Loughlin, J. R., and C. F. Yokomoto, "Computation of One-Dimensional Spread of Leaking Flammable Gas." *Fire Technology* 25(4) (November 1989): 308–16.

Rabinkov, V. A. "The Distribution of Flammable Gas Concentrations in Rooms," *Fire Safety Journal* 13 (1988): 211–17.

A.21.8.2.2.5 Harris (1983) pp. 18-25.

A.21.8.2.2.6 Kennedy, P., and J. Kennedy. Explosion Investigation and Analysis. Chicago, IL: Investigations Institute, 1990. pp. 74-81.

A.21.8.4 For more information, see Harris, R. J. *The Investigation and Control of Gas Explosions in Buildings and Heating Plants.* London and New York: E & FN Spoon, Ltd., 1983.

A.21.9.1.1 NFPA 68, *Standard on Explosion Protection by Deflagration Venting,* provides a more complete introduction to fundamentals of dust explosions.

A.21.9.2 For more information, see the following publications:

Hertzberg, M. and K. Cashdollar, "Domains of Flammability and Thermal Ignitibility for Pulverized Coals and Other Dusts: Particle Size Dependence and Microscopic Residue Analyses," *19th Symposium on Combustion.* Pittsburgh, PA: Combustion Institute, 1982, pp. 1169–80.

Hertzberg, M., and K. Cashdollar. *Explosibility and Ignitibility of Plastic Abrasive Media.* Washington, DC: U.S. Bureau of Mines, Internal Report No. 4657, June 1987.

A.21.9.3 For more information, see USBM, *Pressure Development in Laboratory Dust Explosions.* Washington, DC: U.S. Bureau of Mines.

A.21.9.5 For more information, see USBM, *Preventing Ignition of Dust Dispersion by Inerting,* Washington, DC: U.S. Bureau of Mines, Report of Investigations 6543. 1964, p, 12.

A.21.10 For more information, see the following:

Gottuk, D. T., M. J. Peatross, J. P. Farley, and F. W. Williams, "The Development and Mitigation of Backdraft: A Real-Scale Shipboard Study," *Fire Safety Journal,* 33 (4) (November 1999): 261–282.

Gottuk, D. T., F. W. Williams, and J.P. Farley. "The Development and Mitigation of Backdrafts: A Full-scale Experimental Study," *Fire Safety Science — Proceedings of the Fifth International Symposium,* International Association of Fire Safety Science, 1997, pp. 935–946.

Fleischmann, C. M., P. J. Pagni, and R. B. Williamson, "Exploratory Backdraft Experiments," *Fire Technology* 29(4) (1993), 298–316.

Fleischmann, C. M., P. J. Pagni, and R. B. Williamson, "Quantitative Backdraft Experiments," *4th International Symposium on Fire Safety Science,* International Association for Fire Safety Science, 1994, pp. 337–348.

A.21.12 For more information, see Institute of Makers of Explosives, *Glossary of Commercial Explosives Industry Terms.* Washington, DC: Safety Library Publications No. 12, 1985.

A.21.18.4 For condensed phase explosives, a methodology such as that presented in U.S. *Army Technical Manual,* TM5-1300, "Structures to Resist the Effects of Accidental Explosions" (U.S. GPO, Washington, DC), can be used to estimate the amount and configuration of explosives necessary to cause the damage. Far-field effects are associated with the explosion damage that results from the air blast fronts and flying fragments. ATNT energy equivalent of the exploding system can be deduced from projectile weight, distance, and resulting damage. More details can be found in National Fire Academy, *Fire Dynamics Course Guide/Reader, Unit 4 — Explosions.* Emmitsburg, MD: National Emergency Training Center.

A.22.1 For more information, see the following:

Carman, S. *High Temperature Accelerants, A Study of HTA Fires Reported in the United States and Canada Between January 1981 and August 1991,* Sacramento, CA: U.S. Bureau of Alcohol, Tobacco and Firearms, October 1994.

DeHaan, J. "Canine Accelerant Detection Teams: Validation and Certification," *CAC News,* California Association of Criminalists, July 1994.

Douglas, J. E., A. W. Burgess, A. G. Burgess, and R. K. Ressler. *Crime Classification Manual.* New York: Lexington Books, 1992.

Icove, D. J. *Incendiary Fire Analysis and Investigation.* Open Learning Fire Science Program Course. Lexington, MA: Ginn Custom Publishing, 1983.

Icove, D. J., V. B. Wherry, and J. D. Schroeder. *Combatting Arson for Profit.* Columbus, OH: Battelle Press, 1980.

Kurz, M., et al. "Evaluation of Canines for Accelerant Detection at Fire Scenes," *Journal of Forensic Sciences* 39(6) (November 1994): 1528–36.

Sapp, A. D., T. G. Huff, G. P. Gary, and D. J. Icove. "A Motive-Based Offender Analysis of Serial Arsonist," Department of Justice/Federal Bureau of Investigation/Federal Emergency Management Agency.

Tindall, R., and K. Lothridge. "An Evaluation of 42 Accelerant Detection Canine Teams." *Journal of Forensic Sciences* 40(4) (July 1995): 561–64.

Douglas, J. E., A. W. Burgess, A. G. Burgess, and R. K. Kessler. *Crime Classification Manual.* New York, NY: Lexington Books, 1997.

Restatement of the Law, Second Torts, 402A, American Law Institute, Philadelphia, PA.

Model Penal Code, Section 220.1(1), American Law Institute, Philadelphia, PA.

A.22.2.4 See Carman, S. *High Temperature Accelerants, A Study of HTA Fires Reported in the United States and Canada Between January 1981 and August 1991.* Sacremento, CA: U.S. Bureau of Alcohol, Tobacco and Firearms, October 1994.

A.22.4.2.1 For additional information, see Icove, D. J, *Incendiary Fire Analysis and Investigation.* Open Learning Fire Service Program Course. Lexington, MA: Ginn Custom Publishing, 1983.

A.22.4.9 See Douglas et al., *Crime Classification Manual,* New York, NY: Lexington Books, 1992, p. 165.

A.22.4.9.3 See Sapp et al., *A Motive-Based Offender Analysis of Serial Arsonist,* Department of Justice pp. 4 and 71.

The following are the numerical classifications for firesetting used by the *Crime Classification Manual.*

(1) **Vandalism** (200)
  (a) Willful and malicious mischief (201)
  (b) Peer acceptance/group pressure (202)
  (c) and other — not classified further (209)
(2) **Excitement** (210)
  (a) Thrill seeking (211)
  (b) Attention seeking (212)
  (c) Recognition (213)
  (d) Sexual gratification or perversion (214)
(3) **Revenge** (220)
  (a) Personal retaliation (221)
  (b) Societal retaliation (222)
  (c) Institutional retaliation (223)
  (d) Group retaliation (224)
  (e) Intimidation (225)
  (f) Other (229)
(4) **Crime concealment** (230)
  (a) Murder (231)
  (b) Suicide (232)
  (c) Breaking and entering or burglary (233)
  (d) Embezzlement (234)
  (e) Larceny (235)
  (f) Destroying records or documents (236) and other (239)
(5) **Profit** (240)
  (a) Fraud fires (241)
    i. Fraud to collect insurance (241.01)
    ii. Fraud to liquidate property (241.02)
    iii. Fraud to dissolve a business (241.03)
    iv. Fraud to conceal a loss or liquidate inventory (241.04)
  (b) Employment (242)
  (c) Parcel/property clearance (243)
  (d) Competition (244)
  (e) Other (249)
(6) **Extremist** (250)
  (a) Terrorism (251)
  (b) Riot/civil disturbance (253)

The numerical classifications, which appear in the parentheses behind the classifications and the subclassifications, correspond with the classifications of the *Crime Classification Manual.*

A.22.4.9.3.3 See DeHaan, *Kirk's Fire Investigation,* p. 400.

A.22.4.9.3.4 See Sapp et al., *A Motive-Based Offender Analysis of Serial Arsonist,* pp. 7–8, 28, 31, 38, 41. Also see DeHaan, *Kirk's Fire Investigation,* page 402.

A.22.4.9.3.5 See DeHaan, *Kirk's Fire Investigation,* p. 402.

A.22.4.9.3.6 See DeHaan, *Kirk's Fire Investigation,* pp. 396–398.

A.22.4.9.3.7 See DeHaan, *Kirk's Fire Investigation,* p. 403.

A.23.2 Birky, M., B. Halpin, Y. Caplan, R. Fisher, J. McAllister, and A. Dixon. Fire Fatality Study, Fire and Materials 3, 1979, 211-217.

Lundquist, P., L. Rammer, and B. Sorbo. Role of Hydrogen Cyanide and Carbon Monoxide in Fire Casualties: A Prospective Study, Forensic Sci. Int. 43, 1989, 9-14.

Purser, D.A. Interactions among carbon monoxide, hydrogen cyanide, low oxygen hypoxia, carbon dioxide and inhaled irritant gases, In Carbon Monoxide Toxicity, 1st ed. CRC Press Ltd., Boca Raton, Florida, 2000, pp. 157-191.

Purser, D.A. and W.D. Woolley. Biological effects of combustion atmospheres, J. Fire Sci. 1, 1982, 118-144.

Purser, D.A. Behavioral impairment in smoke environments. Toxicology. 115, 1996, 25-40.

A.23.2.1 DeHaan, J. "The Dynamics of Flash Fires Involving Flammable Hydrocarbon Liquids." American Journal of Forensic Medicine and Pathology, 7(1), 1996, 24-31.

Gann, R.J., V. Babrauskas, and R.D. Peacock. "Fire Conditions for Smoke Toxicity Measurements." Fire and Materials, 18(3), May/June 1994.

Gottuk, D.T., and B.Y. Lattimer. "Effect of Combustion Conditions on Species Production," Section 2/Chapter 5, In The SFPE Handbook of Fire Protection Engineering, 3rd edition. Edited by P.J. DiNenno et al. Quincy, MA: NFPA, 2002.

Hall, J.R., and B. Harwood. "Smoke or Burns- Which Is Deadlier?" NFPA Journal, January/February 1995.

Hirschler, M.M., S.M. Debanne, J.B. Larsen, and G.L. Nelson, eds. Carbon Monoxide and Human Lethaility: Fire and Non-Fire Studies. New York: Elsevier Applied Science, 1993.

Nelson, G.L. "Carbon Monoxide and Fire Toxicity: A Review and Analysis of Recent Work," Fire Tech., 34(1), 1998.

Penney, D.G., Carbon Monoxide Poisoning, CRC Press, Taylor & Francis, Boca Raton, 2008.

Purser, D.A. "Toxicity Assessment of Combustion Products," Section 2, Chapter 6, In The SFPE Handbook of Fire Protection Engineering, 3rd edition. Edited by P.J. DiNenno et al. Quincy, MA: NFPA, 2002.

Tewarson, A. "Generation of Heat and Chemical Compounds," Section 3/Chapter 4, In The SFPE Handbook of Fire Protection Engineering, 3rd edition. Edited by P.J. DiNenno et al. Quincy, MA: NFPA, 2002.

World Health Organization, Environmental Health Criteria 213: Carbon Monoxide, 2nd Edition, 1999.

A.23.2.1.1 World Health Organization, Environmental Health Criteria 213: Carbon Monoxide, 2nd Edition, 1999.

A.23.2.1.3 Kunsman, G. and B. Levine. Carbon Monoxide/ Cyanide. In Principles of Forensic Toxicology, 2nd ed., AACC Press, Washington, D.C., 2006, 359-372.

A.23.2.1.4 Carney, A., E. Anderson. The System Approach to Brain Blood Flow. Advances in Neurology, 30, 1981, 1-30.

Radford, E., T. Drizd. Blood Carbon Monoxide Levels in Persons 3-74 Years of Age: United States, 1976-80. National Center for Health Statistics: Advanced Data, United States Department of Health and Human Services, Number 76, March 17, 1982.

Raud, J., M. Mathieu-Nolf, N. Hampson, S. Thom. Carbon Monoxide poisoning- a public health perspective. Toxicology, 145, 2000, 1-14.

Stewart, R.D. The effect of carbon monoxide on humans. Ann. Rev. Pharm. 15, 1975,409-422.

Stewart, R.D., E.D. Baretta, L.R. Platte et al. Carboxyhemoglobin levels in American blood donors. J.Am.Med.Asso.229, 1974, 1187-1195.

World Health Organization, Environmental Health Criteria 213: Carbon Monoxide, 2nd Edition, 1999.

Zwart, A., E.J. Van Kampen. Dyshaemoglobin, especially carboxyhaemoglobin, levels in hospitalized patients, Clin. Chem., 31, 1985, 945.

A.23.2.1.5 Bruce, M.and E. Bruce. Analysis of factors that influence rates of carbon monoxide uptake, distribution, and washout from blood and extravascular tissues using a multicompartment model, J Appl Physiol 100, 2006, 1171-1180.

Penney, D.G., Carbon Monoxide Poisoning, CRC Press, Taylor & Francis, Boca Raton, 2008.

Peterson, J. and R. Stewart. Absorption and Elimination of Carbon Monoxide by Inactive Young Men. Arch Environ Health. 21, 1970, 165-171.

Shimazu, T. Half-life of Blood Carboxyhemoglobin. Chest. 119, 2001, 661-663.

Stewart, R.D. The effect of carbon monoxide on humans. Ann. Rev. Pharm. 15, 1975,409-422.

Weaver, L., S. Howe, R. Hopkins, K. Chan. Carboxyhemoglobin Half-life in Carbon Monoxde-Poisoned Patients Treated With 100% Oxygen at Atmospheric Pressure. Chest. 117, 2000, 801-808.

World Health Organization, Environmental Health Criteria 213: Carbon Monoxide, 2nd Edition, 1999.

A.23.2.1.6 Penney, D.G., Carbon Monoxide Poisoning, CRC Press, Taylor & Francis, Boca Raton, 2008.

A.23.2.1.7 Birky, M., B. Halpin, Y. Caplan, R. Fisher, J. McAllister, and A. Dixon. Fire Fatality Study, Fire and Materials 3, 1979, 211-217.

A.23.2.2 Altman, R.E., The Microdetermination of Cyanide in Fire Fatalities. University of Maryland, Baltimore. Doctor of Philosophy, 1976.

Ballantyne, B. and H. Salem. Experimental, Clinical, Occupational Toxicology, and Forensic Aspects of Hydrogen Cyanide with Particular Reference to Vapor Exposure. In Inhalation Toxicology, 2nd ed. CRC, Taylor and Francis, Boca Raton, FL, 2006, pp. 717-802.

Baud, F.J., P. Barriot, V. Toffis, B. Riou, E. Vicaut, Y. Lecarpentier, R. Bourdon, A. Astier, and C. Bismuth. Elevated blood cyanide concentrations in victims of smoke inhalation. New Engl. J. Med. 325, 1991, 1761-1766.

Canfield, D., A. Chaturvedi, and K. Dubowski. Interpretation of Carboxyhemoglobin and Cyanide Concentrations in Relation to Aviation Accidents. Civil Aerospace Medical Institute, Federal Aviation Administration, Final Report, May 2005.

Chaturvedi, A., B. Endecott, R. Ritter, D. Sanders. Variations in Time-to-Incapacitation and Blood Cyanide Values for Rats Exposed to Two Hydrogen Cyanide Gas Concentrations. Office of Aviation Medicine, Washington, D.C., 1993.

DeRosa, M. and C. Litton, "Hydrogen Cyanide and Smoke Particle Charatersitics During Combustion of Polyurethane Foams and Other Nitrogen-Containing Materials", U.S. Bureau of Mines, 1991.

Esposito, F.M.. Blood and air concentrations of benzene, carbon monoxide and hydrogen cyanide following inhalation of these gases or thermal decomposition products of polymers releasing these toxicants, Ph.D. dissertation submitted to Graduate School of Pubic Health, University of Pittsburgh, September 1987.

Esposito, F. and Y. Alarie. Inhalation Toxicity of Carbon Monoxide and Hydrogen Cyanide Gases Released During the Thermal Decomposition of Polymers. J. Fire Sci. 6, 1988,195-242.

Ferrari, L., M. Arado, L. Giannuzzi, G. Mastrantonio, and M. Guatelli. Hydrogen Cyanide and carbon monoxide in blood of convicted dead in a polyurethane combustion: a proposition for the data analysis, Forensic Sci. Int. 121, 2001,140-143.

Levin, B., M. Paabo, M.L. Fultz, and C. Bailey. Generation of Hydrogen Cyanide from Flexible Polyurethane Foam Decomposed under Different Combustion Conditions. Fire and Materials 9, 1985, 125-134.

Okae, M., K. Yamamoto, Y. Yamamoto, and Y. Fukui. Cyanide, Carboxyhemoglobin and Blood Acid-Base State in Animals Exposed to Combustion Products of Various Combinations of Acrylic Fiber and Gauze. Forensic Sci. Int. 42. 1989, 33-41.

Purser, D.A., P. Grimshaw and K.R. Berrill. Intoxication by cyanide in fires: a study in monkeys using polyacrylonitrile. Arch. Environ. Hlth. 39, 1984, 394-400.

Simeonova, F. and L. Fishbein. Hydrogen Cyanide and Cyanides: Human Health Aspects. Concise International Chemical Assessment Document 61, World Health Organization, first draft, Geneva 2004.

United States Department of Health and Human Services, A Toxicological Profile for Cyanide, Agency for Toxic Substances and Disease Registry, September 2004.

A.23.2.2.1 Birky, M., M. Paabo, J. Brown. Correlation of Autopsy Data and Materials Involved in the Tennessee Jail Fire. Fire Safety Journal. 2, 1979/80, 17-22.

Gill, J., L. Goldfeder, M. Stajic. The Happy Land Homicides: 87 Deaths Due to Smoke Inhalation. J. Forensic Sci. 48, 2003, 161-163.

Levin, B.C., P.R. Rechani, J.L. Gurman, F. Landron, H.M. Clark, M.F. Yoklavich, J.R. Rodriguez, L. Droz, F. Mattos de Cabrera, and S. Kaye. Analysis of carboxyhemoglobin and cyanide in blood from victims of the Dupont Plaza Hotel fire in Puerto Rico. J. Forensic Sci. 35, 1990, 151-168.

Nelson, G.. Carbon Monoxide and Fire Toxicity: A Review and Analysis of Recent Work. Fire Tech. 34, 1998, 39-58.

A.23.2.2.4 Anderson, C.J., and W.A. Harland. Fire deaths in the Glasgow area: III , The role of hydrogen cyanide Med. Sci. Law 22, 1982, 35-39

Ballantyne, B., J. Bright, and P. Williams. An experimental assessment of decreases in measurable cyanide levels in biological fluids. J. Forensic Sci. Soc. 13, 1973, 111-117.

Ballantyne, B., J. Bright, and P. Williams. The post-mortem rate of transformation of cyanide. Forensic Sci. 3, 1974, 71-76.

Ballantyne, B. Changes in blood cyanide as a function of storage time and temperature. J. Forensic Sci. Soc.16, 1976, 305-310.

Ballantyne, B. In vitro production of cyanide in normal human blood and the influence of thiocyanate and storage temperature. Clin. Toxicol. 11, 1977, 173-193.

Ballantyne, B. Artifacts in the definition of toxicity by cyanides and cyanogens. Fund. Appl. Toxicol 3, 1983, 400-408.

Ballantyne, B. and T. Marrs. Clinical and Experimental Toxicology of Cyanides. Wright, Bristol, U.K., 1987.

Bright, J.E., R.H. Inns, N.J. Tuckwell, and T.C.Marrs. The effect of storage upon cyanide in blood samples. Hum. Exp. Toxicol. 9, 1990, 125-129.

Chikasue, F., M. Yashiki, T. Kojima, T. Miyazaki. Cyanide distribution in five fatal cyanide poisonings and the effect of storage conditions on cyanide concentrations in tissues. Forensic Sci. Int. 38, 1988, 173-183.

Clark, C.J., D. Campbell, and W.H. Reid. Blood carboxyhaemoglobin and cyanide levels in fire survivors". Lancet 20, 1981, 1332-1336.

Curry, A.S.. Cyanide poisoning. Acta Pharmacolgica et Toxicologica cta. 20, 1963, 291-294.

Curry, A.S., D.E. Price, and E.R.Rutter. The production of cyanide in post mortem material. Acta Pharmacolgica et Toxicologica 25, 1967, 339-344.

Egekeze, J.O., and F.W. Oehme. Direct potentiometric method for the determination of cyanide in biological materials. J. Anal. Toxicol. 3, 1979, 119-124.

Karhunen, P.J., I. Lukkari, and E. Vuori. High cyanide level in a homicide victim burned after death: Evidence of post-mortem diffusion. Forensic Sci. Int. 49, 1991, 179-183.

Lokan, R.J., R.A. James, and R.B. Dymock. Apparent post-mortem production of high levels of cyanide in blood. J. Forensic Sci. Soc. 27, 1987, 253-259

McAllister, J.L., Roby, R., Levine, B., Purser, D., "Stability of Cyanide in Cadavers and in Postmortem Stored Tissue Specimens, a Review", Journal of Analytical Toxicology, Volume 32, Number 8, p. 612-620, October, 2008.

Moriya, F. and Y. Hashimoto. Potential for error when assessing blood cyanide concentrations in fire victims. J. Forensic Sci. 46, 2001, 1421-1425.

Narayanaswami, K., S. Jand, and R.S. Kotangale. Changes in cyanide as a function of storage time in autopsy material by UV Spectrometry. Ind. J. Med. Res. 73, 1981, 291-295.

Seto, Y. Stability and spontaneous production of blood cyanide during heating. J. Forensic Sci. 41, 1996, 465-468.

Vesey, C.J. and J. Wilson.. Red cell cyanide. J. Pharm. Pharmacol. 30, 1978, 20-26.

A.23.2.2.5 Purser, D.A. "Toxicity Assessment of Combustion Products," Section 2, Chapter 6, In The SFPE Handbook of Fire Protection Engineering, 3rd edition. Edited by P.J. DiNenno et al. Quincy, MA: NFPA, 2002.

A.23.2.2.6 Kimmerle, M.G. Aspects and Methodology for the Evaluation of Toxicological Parameters During Fire Exposure. Combustion Toxicol. 1, 1974, 4-51.

Purser, D.A. "Toxicity Assessment of Combustion Products," Section 2, Chapter 6, In The SFPE Handbook of Fire Protection Engineering, 3rd edition. Edited by P.J. DiNenno et al. Quincy, MA: NFPA, 2002.

A.23.2.4.1 Purser, D.A. "Toxicity Assessment of Combustion Products," Section 2, Chapter 6, In The SFPE Handbook of Fire Protection Engineering, 3rd edition. Edited by P.J. DiNenno et al. Quincy, MA: NFPA, 2002.

A.23.2.5 Purser, D.A. "Toxicity Assessment of Combustion Products," Section 2, Chapter 6, In The SFPE Handbook of Fire Protection Engineering, 3rd edition. Edited by P.J. DiNenno et al. Quincy, MA: NFPA, 2002.

A.23.2.5.1 Bull, J. P., and J. C. Lawrence. "Thermal Conditions to Produce Skin Burns," Fire and Materials 3(2) (June 1979): 100–05.

Derksen, W. L., T.I. Monohan, and G.P. deLhery, "The Temperature Associated with Radiant Energy Skin Burns," Temperature — Its Measurement and Control in Science and Industry, 3(3), pp. 171–75, 1963.

Society of Fire Protection Engineers, "Engineering Guide: Predicting 1st and 2nd Degree Skin Burns from Thermal Radiation," Society of Fire Protection Engineers, 2000.

Stoll, A., and L.C. Green, "Relationship between Pain and Tissue Damage due to Thermal Radiation," Journal of Applied Physiology 14, pp. 373–383, 1959.

Stoll, A. M., and M.A. Chianta, "Method and rating system for evaluation of thermal protection," Aerospace Medicine 40, pp. 1232–1238, 1969.

A.23.2.6 Moritz, A.R., F.C. Henriques, F.R. Dutra, and J.R. Weisiger, Studies of thermal injury. IV. Exploration of casualty-producing attributes of conflagrations. The local and systemic effects of generalized cutaneous exposure to excessive circumambient (air) and circumradiant heat of varying duration and intensity, Arch. Pathol., 43, 1947, 466.

Zikria, BA, W.Q. Sturner, N.K. Astarjian, C.L. Fox, Jr, J.M. Ferrer, Jr. Respiratory tract damage in burns: pathophysiology and therapy. Ann N Y Acad Sci. 150(3), 1968, 618–626.

A.23.3 DeHaan, J. D., S.J. Campbell, and S. Nurbaksh, "Combustion of Animal Fat and Its Implications for the Consumption of Human Bodies in Fires," Science and Justice 39 (1), 1999, 27–38.

DeHaan, J.D. and S. Nurbaksh "Sustained Combustion of an Animal Carcass and Its Implications for the Consumption of Human Bodies in Fires," J Forensic Sci, 46(5), 2001, 1076-1081.

DeHaan, J.D., D.J. Brien, R. Large. Volatile Compounds from the Combustion of Human and Animal Tissue. Science and Justice 44 (4):223-236, 2004.

DeHaan, J.D. and E. Pope. Combustion Properties of Human and Large Animal Remains. InterFlam 2007 Conference, 11th Proceeding, Volume 2, September 3-5, 2007, London, England.

Pope, Pope, O.C. Smith, T. Huff, Exploding skulls and other myths about how the human body burns, Fire and Arson Investigator 55(4). 2004, 23-28.

Pope, E., The Effects of Fire on Human Remains: Characteristic of Taphonomy and Trauma. University of Arkansas. Doctor of Philosophy, 2007.

A.23.5.8.2 Beasley, V., Veterinary Toxicology. Online reference for veterinary toxicology. www.ivis.org/advances/Beasley/cpt18a/IVIS.pdf. International Veterinary Information Service (IVIS), Ithaca, NY. (2004).

A.23.6.1 Posters are sometimes used to remind persons of the law requiring the reporting of burn injuries. An example is shown in Figure A.23.6.1.

(Please Post)

REPORT ALL BURNS IMMEDIATELY!

It's the law!*

REPORT ALL BURN INJURIES IMMEDIATELY: 1-800-345-5811
NYS DEPARTMENT OF STATE OFFICE OF FIRE PREVENTION AND CONTROL
Burn Injury Report (File within 72 hours)
Print or Type
FOR OFPC USE ONLY
CONTROL #
1) VICTIM'S NAME (Last, First, M.I.)
2) DATE OF BIRTH
3) VICTIM'S ADDRESS (NUMBER, STREET)
STATE ZIP CODE
CITY, TOWN, POST OFFICE
4) ADDRESS WHERE BURN OCCURRED (Number, Street)
STATE ZIP CODE
CITY, TOWN, POST OFFICE
6) TIME OF INJURY HRS.
7) PERCENT BURNED %
8) DEGREE(S) OF BURN(S) ☐ 1st ☐ 2nd ☐ 3rd
6 ❑ LEG
7 ❑ FOOT
❑ ARM
HAND
10) INJURY SEVERITY (Place appropriate number in box)
1 MODERATE (treated and released)
2 SERIOUS (hospitalized)
3 LIFE THREATENING (deaths imminent and/or probable)
4 DEAD ON ARRIVAL
appropriate number in box)
istic, corrosive or irritating substance
furnace, iron, steam exhaust pipe, etc.

Call the NYS Office of Fire Prevention and Control
24-hour toll-free hotline:
1-800-345-5811

- 2nd and 3rd degree burns 5 PERCENT OR MORE S.B.A.
- Burns to the UPPER RESPIRATORY TRACT
- LARYNGEAL EDEMA due to inhalation of super-heated air
- Any burn that may result in death

***Section 265.28 of the New York State Penal Law**

FIGURE A.23.6.1 Example of a Poster Reminding Persons of the Law Requiring the Reporting of Burn Injuries.

A.23.8.1.1 Alternate descriptions of degrees of burn to skin are superficial, partial, and full-thickness burns.

A.23.9 DiMaio, V. and D. DiMaio, Forensic Pathology, Second Edition, CRC Press, Boca Raton, 2001.

Levine, B. Principles of Forensic Toxicology, 2nd ed., AACC Press, Washington, D.C., 2006.

A.23.9.2.2 Ambach, E., W. Paarson, K. Zehethofer, and R. Scheithauer. Sixth International Symposium on Human Identification. Institute for Forensic Medicine, University of Innsbruck, Austria, 1995.

DiZinno, J., et al. "The Waco, Texa Incident: The Use of DNA Analysis to Identify Human Remains." Fifth International Symposium on Human Identification, 1994.

A.23.9.8 Spitz, W. U. Spitz and Fisher's Medicolegal Investigation of Death. 3rd ed. Springfield, IL: Charles C. Thomas, 1993.

A.23.9.9 Pope, E., The Effects of Fire on Human Remains: Characteristic of Taphonomy and Trauma. University of Arkansas. Doctor of Philosophy, 2007.

A.23.10.6 Copper, L.Y., Compartment Fire-Generated Environment and Smoke Filling., Section 3/Chapter 10, In The SFPE Handbook of Fire Protection Engineering, 3rd edition. Edited by P.J. DiNenno et al. Quincy, MA: NFPA, 2002.

A.23.10.8 Copper, L.Y., Compartment Fire-Generated Environment and Smoke Filling., Section 3/Chapter 10, In The SFPE Handbook of Fire Protection Engineering, 3rd edition. Edited by P.J. DiNenno et al. Quincy, MA: NFPA, 2002.

DeRosa, M and C. Litton, "Hydrogen Cyanide and Smoke Particle Characteristics, During Combustion of Polyurethane Foams and Other Nitrogen-Containing Materials", U.S. Bureau of Mines, 1991.

Gann, R., J. Averill, E. Johnsson, M. Nyden, and R. Peacock. Smoke Component Yields from Room-scale Fire Tests. National Institute of Standards and Technology, Gaithersburg, MD, 2003.

Gottuk, D. Generation of Carbon Monoxide in Compartment Fires. Virginia Polytechnic Institute, Blacksburg, VA, 1992.

Lattimer, B.Y., U. Vandsburger, and R.J. Roby. Carbon Monoxide Levels in Structure Fires: Effects of Wood in the Upper Layer of a Post-Flashover Compartment Fire. Fire Tech. 34, 1998, 325-355.

Levin, B., M. Paabo, M.L. Fultz, and C. Bailey. Generation of Hydrogen Cyanide from Flexible Polyurethane Foam Decomposed under Different Combustion Conditions. Fire and Materials 9, 1985, 125-134.

Purser, D.A., J.A. Rowley, P.J. Fardell, and M. Bensilum. Fully enclosed design fires for hazard assessment in relation to yields of carbon monoxide and hydrogen cyanide. In The Proceedings of the Interflame '99 Conference. Interscience Communications Ltd., London, United Kingdom, 1999, pp. 1163-1169.

Purser, D.A. "Toxicity Assessment of Combustion Products," Section 2, Chapter 6, In The SFPE Handbook of Fire Protection Engineering, 3rd edition. Edited by P.J. DiNenno et al. Quincy, MA: NFPA, 2002.

A.23.10.8.2 Coburn, R.F., R.E. Forster, and P.B. Kane. Considerations of the Physiological Variables That Determine the Blood Carboxyhemoglobin Concentration in Man. J. Clin. Invest. 44, 1965, 1899-1910.

Handbook of respiration, Saunders, Philadelphia (1955)

Peterson, J. and R. Stewart. Predicting the Carboxyhemoglobin Levels Resulting from Carbon Monoxide Exposures. J. Appl. Physiol. 39, 1975, 633-638.

A.24.1 For more information, see the following: U.S. Consumer Products Safety Commission Working Group on Gas Voluntary Standards. *Position Paper on Gas Water Heaters to Prevent Ignition of Flammable Vapors.*

A.24.5.1.2 For additional information on steel and the effect of fire on it, see Lentini et al., "Baseline Characteristics of Residential Structures Which Have Burned to Completion: The Oakland Experience."

For additional information on the alloying of metals during a fire, see Beland B., "Electrical Damages— Cause or Consequence?", and Beland B. et al., "Copper-Aluminum Interactions in Fire Environments."

A.24.5.8.2 For more information, see Annex C, NEMA document, National Electrical Manufacturers Association, 1300 North 17th Street, Suite 1847, Rosslyn, VA 22209, September 10, 2000.

A.25.1 For additional information, see the following:

NFPA 385, *Standard for Tank Vehicles for Flammable and Combustible Liquids*, 2007 edition.

API RP 1004, *Bottom Loading and Vapor Recovery for MC-306 Tank Motor Vehicles*, 1988.

API RP 2013, *Cleaning Mobile Tanks in Flammable or Combustible Liquid Service*, 1991.

Green, T. SAE Paper 980561, "Automotive Fuel Line Siphoning."

Cole, L. *The Investigation of Motor Vehicle Fires: A Guide for Law Enforcement, Fire Department and Insurance Personnel*, 3rd ed. Lee Books, 1992.

Severy, D. M., D. M. Blaisdell, and J. F. Kerkhoff. "Automobile Collision Fires," SAE 741180, 1974.

A.25.3.1.1 For additional data regarding hot surface ignition see the following publication: Ebersole, R., Matusz, L., Modi, M. and Orlando, R., Hot Surface Ignition of Gasoline-Ethanol Fuel Mixtures, SAE publication 2009-01-0016.

A.25.4.2.5 For additional information, see the following:

Beyler, C. L., and Gratkowski, M. "Electrical Fires Caused By Arc Tracking In Low-Voltage (12-14V) Systems," U.S. Vehicle Fire Trends and Patterns, August 2005.

Stimitz, J. S., and R. V. Wagner. "*Two Test Methodologies for Evaluating the Resistance to Electrical Arcing Properties of Polymeric Materials for Use in Automotive 42 V Applications*" SAE 2004 World Congress & Exhibition, March 2004, Detroit, MI.

A.25.4.2.6 For more information, see Severy et al., "Automobile Collision Fires."

A.25.4.3.1 For more information, see Colwell, J. D. and Biswas, K. (2009) "Steady-State and Transient Motor Vehicle Exhaust System Temperatures", SAE Paper 2009-01-0013.

A.25.5.1.2 Green, T. SAE 980 561, "Automotive Fuel Line Siphoning."

A.26.1 For more information, see the following:

Babrauskas, V. Ignition Handbook, Fire Science Publishers/ Society of Fire Protection Engineers, 2003.

Ford, Richard.T. Investigation of Wildfires. Bend, OR: Maverick Publication, 1995.

NWCG, Glossary of Wildland Fire Terminology, National Wildfire Coordinating Group, 2007.

NWCG,Wildland Fire Origin and Cause Determination FI-210 Student Workbook, National Wildfire Coordinating Group, 2005.

NWCG Wildfire Origin & Cause Determination Handbook, National Wildfire Coordinating Group Handbook 1, 2005.

NWCG Intermediate Wildland Fire Behavior S-290 Student Workbook & CD-Rom, National Wildfire Coordinating Group, 2007.

NWCG Fireline Handbook, National Wildfire Coordinating Group Handbook 3, March 2004.

NWCG Fireline Handbook Appendix B Fire Behavior, National Wildfire Coordinating Group, April 2006.

Schroder, Mark J.; Charles C. Buck, Fire Weather, USDA Agriculture Handbook 360, May 1970.

A.26.2.1.2 Anderson, H. E., "Aids to Determining Fuel Models for Estimating Fire Behavior", Intermountain Forest and Range Experiment Station, General Technical Report INT-122, April 1982. Scott, Joe H; Burgan, Robert E., Standard Fire Behavior Fuel Models; A Comprehensive Set For Use With Rothermel's Surface Fire Spread Model. General Technical Report RMRS-GTR-153, 2005, USDA Forest Service, Rocky Mountain Research Station.

A.26.8.2.2 Countryman, C; "Ignition of Grass Fuels by Cigarettes", US Forest Service Fire Management Notes, Volume 44, No. 3, 1983.

A.26.8.2.3 Steensland, P., "Long Term Thermal Residency in Woody Debris Piles", International Association of Arson Investigators, Denver, 2008.

A.28.1 For Nautical Terms see Chapman's *Piloting Seamanship and Small Boat Handling.*

A.28.4.1.2 See American Boat and Yacht Council (ABYC) H-24.14, *Gasoline Fuel Systems,* 2005.

A.28.4.1.3 See ABYC H33.14.

A.28.4.2.1 See NFPA 302, 2004 edition, and ABYC A-1; A-3, *Galley Stoves,* 2007; and A-26, *LPG and CNG Fueled Appliances,* 2006.

A.28.4.2.2 See ABYCA A-22 and NFPA 302, *Fire Protection Standard for Pleasure and Commercial Motor Craft,* 2004 edition.

A.28.4.2.3 See ABYC A-3, *Galley Stoves,* 2007; and NFPA 302, *Fire Protection Standard for Pleasure and Commercial Motor Craft,* 2004 edition.

A.28.4.7.2 See NFPA 302, *Fire Protection Standard for Pleasure and Commercial Motor Craft,* 2004 edition; and ABYC H-2.

A.28.4.7.3 For additional information refer to NFPA 302 and ABYC H-2. For requirements regarding power ventilation of diesel applications refer to ABYC H-32, *Ventilation of Boats Using Diesel Fuel,* 2007.

A.28.4.9.5.2 For additional information refer to ABYC H-30, *Standard and Technical Reports,* 2001.

Annex B Bibliography

This annex is not a part of the recommendations of this NFPA document but is included for informational purposes only.

B.1 Fire Dynamics. For sources on fire dynamics, see the following:

Drysdale, D. *An Introduction to Fire Dynamics.* 2nd edition, ch. 1, p. 1, middle of third paragraph. New York, NY: John Wiley and Sons, 1999.

Friedman, R. *Principles of Fire Protection Chemistry and Physics,* 3rd edition, ch. 6, p. 75. Quincy, MA: National Fire Protection Association, 1998.

B.2 Fire Science. For more information on fire science, see the following:

NFPA *Fire Protection Handbook,* 18th ed., Section 1, 1997.

Quintere, J. G. *Principles of Fire Behavior.* Delmar Publishers, 1997.

Drysdale, D. *An Introduction to Fire Dynamics,* 2nd ed., 1999.

Friedman, R. *Principles of Fire Protection Chemistry and Physics,* 3rd ed., 1998.

B.3 Effects of Room Ventilation on Pattern Magnitude and Location. See the following sources:

Kennedy, P., and J. Shanley. "Report on the United States Fire Administration (USFA) Program for the Study of Fire Patterns," *Proceedings, InterFlam 96,* Interscience Ltd., London, 1966.

Shanley, J. *USFA Fire Burn Pattern Tests,* United States Fire Administration, Emmitsburg, MD, FA 178, 1997.

B.4 Calcination. See the following sources:

Kennedy, P., and J. Shanley. "Report on the United States Fire Administration (USFA) Program for the Study of Fire Patterns," *Proceedings, InterFlam 96,* Interscience Ltd., London, 1966.

Lawson, J. R. "An Evaluation of Fire Properties of Generic Gypsum Board Products," NBSIR 77-1265, U. S. Department of Commerce, National Bureau of Standards, August, 1977.

Shanley, J. *USFA Fire Burn Pattern Tests,* United States Fire Administration, Emmitsburg, MD, FA 178, July 1997.

B.5 Pattern Geometry. See the following sources:

Patorti, A. D. Jr. (NIST), "Full Scale Room Burn Pattern Study," *NIJ Report 601-97, Publication #169 281,* National Institute of Justice, National Criminal Justice Reference Service, Washington DC, 1997.

Stanley, J. H. Jr., Alletto, W. C., Corry, R., Herndon, J., Kennedy, P. M., Ward, J. "Federal Emergency Management USFA Fire Burn Pattern Tests," *Report of the United States Fire Administration Program for the Study of Fire Patterns.* FA 178, p. 221, July 1997.

B.6 Odorant Removal from Gas.

B.6.1 For a current discussion of gas odorization technology see the following:

The Gas Odorization Manual, American Gas Association, 1996. This is an updated version of the section on gas odorization in the *Gas Engineers Handbook.*

B.6.2 For a deeper understanding of gas odorization issues, see the following:

Wilson, G. G., and A. A. Attari, eds. *Odorization III,* Institute of Gas Technology, Chicago, IL (1993).

B.6.3 Also see the following:

Jacobus, O. J., and J. S. Roberts. "What Constitutes Adequate Odorization of Fuel Gases?" *IGT Odorization Symposium 1995,* IGT, 1995.

Andreen, B. H., and R. L. Kroencke. "Stability of Mercaptans Under Gas Distribution System Conditions," *American Gas Journal* 48, May 1965.

Andreen, B. H., and R. L. Kroencke. "Stability of Mercaptans Under Gas Distribution System Conditions," *Proceedings American Gas Association* 136, 1964.

Campbell, I. D., N. A. Chambers, and O. J. Jacobus. "The Chemical Oxidation of Ethyl Mercaptan in Steel Vessels," *IGT Odorization Symposium 1994,* IGT, 1994.

Campbell, I. D. "Factors Affecting Odorant Depletion in LPG," *Proceedings Symposium on LP-Gas Odorization,* 28, 1989.

Campbell, I. D. "Odorant Depletion in Portable Cylinders," *Proceedings of the Symposium on LP-Gas Odorization,* 1990.

Holmes, S. A., P. B. Van Benthuysen, D. C. Lancaster, and M. A. Tiller. "Laboratory and Field Experience with Ethyl Mercaptan Odorant in Propane," *Proceedings of the Symposium on LP-Gas Odorization,* 1990.

Kopidlansky, R. "Odorant Conditioning of New Distribution Systems," *IGT Odorization Symposium 1995,* IGT, 1995.

Sanders, R."OSHA and DOT Odorization Regulations," *IGT Odorization Symposium 1995,* IGT, 1995.

Johnson, J. L. "1965 Report on Project PB-48, Stability of Odorant Compounds," *Proceedings 1966 AGA Distribution Conference,* American Gas Association, 1966.

Johnson, S. J. "Ethyl Mercaptan Odorant Stability in Stored Liquid Propane," *Proceedings of the Symposium on LP-Gas Odorization,* 1989.

McHenry, W. B., and H. M. Faulconer. "Summary Report on Odorant Investigations," *Proceedings of the Symposium on LP-Gas Odorization,* 1, 1990.

Roberts, J. S., and D. W. Kelly, "Selection and Handling of Natural Gas Odorants." In G. G. Wilson and A. A. Attari, eds., *Odorization III,* 29, IGT, 1993.

B.6.4 For testing LP-Gas by stain tubes and for sampling, see the following:

ASTM D 5305 *Standard Test for Determination of Ethyl Mercaptan in LP-Gas Vapor,* 1992.

ASTM D 1265 *Standard Method for Sampling Liquefied Petroleum (LP) Gases,* 1982.

B.6.5 For a current understanding of most LP odorization issues, see the following:

Proceedings of the Symposium of LP-Gas Odorization, 1989 and 1990. Copies may be available from the National Propane Gas Association or the Gas Processors Association.

B.6.6 For discussions of soil adsorption of odorants, see the following:

Nomura, K., and R. Organ. "Reaction of LP-Gas Odorants and Soils." *Proceedings of the Symposium on LP-Gas Odorization,* 76, 1990.

Parlman, R. M., and R. P. Williams. "Penetrabilities of Gas Odorant Compounds in Natural Soils." *Proceedings of the AGA Distribution Conference,* May 6, 1979.

Sullivan, F. "New Gas Odorants." In Wilson and Attari, eds., *Odorization III.* 209, IGT, 1993.

Tarman, P. B., and H. R. Linden. "Soil Adsorption of Odorant Compounds," *IGT Research Bulletin* #33, 7/1/61.

Williams, R. P. "Soil Penetrabilities of Natural Gas Odorants." *Pipeline and Gas Journal,* March 1976.

B.7 Egress Analysis. See the following sources:

Bryan, J. L. *An Examination and Analysis of the Dynamics of Human Behavior in MGM Grand Hotel Fire.* Quincy, MA: NFPA, 1983.

"Kentucky State Police Investigative Report to the Governor, Beverly Hills Supper Club Fire." Kentucky State Police, Frankfurt, 1977.

P. DiNenno, Ed. *SFPE Handbook of Fire Protection Engineering,* Section 3, Ch. 1, "Behavioral Response to Fire and Smoke," 2nd edition, Quincy, MA: NFPA, 1995.

B.8 Explosions. For more information on explosions involving structures or vessels, see the following:

Baker, W. E. *Explosion Hazards and Evaluations.* New York, NY: Elsevier Publishers, 1983.

Baker, W. E. and M. J. Tang, *Gas, Dust, and Hybrid Explosions.* New York, NY: Elsevier Publishers, 1991.

Bartknecht, W.*Dust Explosions.* Berlin–New York: Springer-Verlag, 1989.

Bodurtha, F. T. *Industrial Explosion Prevention and Protection.* New York, NY: McGrawHill, 1980.

Center for Chemical Process Safety. *Guidelines for Evaluating the Characteristics of Vapor Cloud Explosions, Flash Fires and BLEVEs.* Washington, DC: American Institute of Chemical Engineers, 1994.

Chemical Propulsion Information Agency, *Hazards of Chemical Reactants and Propellants.* CPIA 394, September 1984.

Eckhoff, R. K. *Dust Explosions in the Process Industries.* New York, NY: Butterworth-Heinemann, 1997.

Gugan, K. *Unconfined Vapor Cloud Explosions.* Houston, TX: Gulf Publishing, 1978.

Harris, R. J. *The Investigation and Control of Gas Explosions in Buildings and Heating Plants.* London and New York: E & FN Spoon, Ltd., 1983.

Nettleton, M. A. *Gaseous Detonations; Their Nature, Causes and Control.* New York, NY: Routledge, Chapman and Hall, 1987.

NFPA 68, *Standard on Explosion Protection by Deflagration Venting,* 2007 edition.

NFPA 69, *Standard on Explosion Prevention Systems,* 2008 edition.

Prugh, R. W. "Quantitative Evaluation of 'BLEVE' Hazards," Journal of Fire Protection Engineering 3(1), March 1981.

Stull, D. R. *Fundamentals of Fire and Explosion.* AICHE Monograph Series No. 10, Vol. 73. New York, NY: American Institute of Chemical Engineers, 1977.

U. S. Army,"Structures to Resist the Effects of Accidental Explosions."

USBM, *Fire and Explosion Manual for Aircraft Accident Investigations.* Washington, DC: U.S. Bureau of Mines, 1973.

USBM, *Investigation of Fire and Explosion Accidents in the Chemical Mining and Fuel-Related Industries — A Manual.* Washington, DC: U.S. Bureau of Mines, 1985.

B.9 Victim Identification. For information on victim identification, see the following:

Ambach, E., W. Paarson, K. Zehethofer, and R. Scheithauer. *Sixth International Symposium on Human Identification.* Institute for Forensic Medicine, University of Innsbruck, Austria, 1995.

DiZinno, J., et al. "The Waco, Texas Incident: The Use of DNA Analysis to Identify Human Remains." *Fifth International Symposium on Human Identification,* 1994.

B.10 Carbon Monoxide. For information on carbon monoxide, see the following:

Gottuk, D. T., and R. J. Roby. "Effect of Combustion Conditions on Species Production," Section 2/Chapter 7, *SFPE Handbook of Fire Protection Engineering,* 2nd edition. Edited by P. J. DiNenno et al. Quincy, MA: NFPA, 1995.

Tewarson, A."Generation of Heat and Chemical Compounds," Section 3, Ch. 4, *SFPE Handbook of Fire Protection Engineering*, 2nd ed. Ed. P. J. DiNenno et al. Quincy, MA: NFPA, 1995.

Purser, D. A."Toxicity Assessment of Combustion Products," Section 2, Ch. 8, *SFPE Handbook of Fire Protection Engineering*, 2nd ed. Ed. P. J. DiNenno et al. Quincy, MA: NFPA, 1995.

Nelson, G. L."Carbon Monoxide and Fire Toxicity: A Review and Analysis of Recent Work," *Fire Technology*, 34(1), 1998.

Hirschler, M. M., S. M. Debanne, J. B. Larsen, and G. L. Nelson, eds. *Carbon Monoxide and Human Lethality: Fire and Non-Fire Studies*. New York: Elsevier Applied Science, 1993.

Hall, J. R., and B. Harwood. "Smoke or Burns — Which Is Deadlier?" *NFPA Journal*, January/February 1995.

Gann, R. J., V. Babrauskas, and R. D. Peacock. "Fire Conditions for Smoke Toxicity Measurements." *Fire and Materials*, 18(3), May/June 1994.

DeHaan, J. "The Dynamics of Flash Fires Involving Flammable Hydrocarbon Liquids." *American Journal of Forensic Medicine and Pathology*, 7(1), 1996, 24–31.

DeHaan, J. D., Campbell, S. J., and Nurbaksh, S. "Combustion of Animal Fat and Its Implications for the Consumption of Human Bodies in Fires," *Science and Justice*, 39 (1), 1999, 27–38.

B.11 Wildfires. For sources on wildfires, see the following:

Ford, R. T. *Investigation of Wildfires*. Bend, OR: Maverick Publications, 1995.

International Fire Service Training Association. *Fundamentals of Wildland Firefighting*. 3rd edition. Stillwater, OK: IFSTA, 1998.

Teie, W. C. *Firefighters Handbook on Wildland Firefighting Strategy, Tactics & Safety*. Rescue, CA: Deer Valley Press, 3rd edition, 2005.

Uman, M. A. All About Lightning. New York, NY, Dover Publications, 1986.

Annex C Informational References

C.1 Referenced Publications. (Reserved)

C.1.1 Other Publications. Bolstad-Johnson, Dawn M., et. al, Characterization of Firefighter Exposures During Fire Overhaul, Phoenix Fire Department / University of Arizona Prevention Center / Arizona State University, 1998.

Grant, C., "Respiratory Exposure Study for Fire Fighters And Other Emergency Responders", The Fire Protection Research Foundation, National Fire Protection Association, Quincy, MA, December 2007.

LeMasters, G., Genaidy, A., Succop, P., et al, "Cancer Risk Among Firefighters: A Review and Meta-Analysis of 32 Studies", Journal of Occupational and Environmental Medicine, Volume 48, Number 11, November 2006, pp. 1189-1202.

Kinnes, G., Hine, G., Health Hazard Evaluation Report 96-0171-2692, Bureau of Alcohol, Tobacco, and Firearms, Washington, D.C, May 1998.

Snyder, E., Health Hazard Evaluation Report 2004-0368-3030, Bureau of Alcohol, Tobacco, Firearms and Explosives, Austin, TX, January 2007.

C.2 Informational References. The following documents or portions thereof are listed here as informational resources only. They are not directly referenced in this guide.

ASTM S1-10, *IEEE Standard for Use of International System of Units (SI). The Modern Metric System*, 1991.

Putorti, A. J. Jr. (NIST), "Full Scale Room Burn Pattern Study," NIJ Report 601-97, Publication #169 281, National Institute of Justice, National Criminal Justice Reference Service, Washington, DC, 1997.

Smith, H. W., *Strategies of Social Research, The Methodological Imagination*, Englewood Cliffs, NJ: Prentice Hall, 1975, pp. 58, 61.

NFPA 550, *Guide to the Fire Safety Concepts Tree*, 2007 edition.

NFPA 1033, *Standard for Professional Qualifications for Fire Investigator*, 2009 edition.

National Association of Fire Investigators, "The National Fire Investigator."

Kennedy, P., and J. Shanley. FA 178, "USFA Fire Burn Pattern Tests — Program for the Study of Fire Patterns," Emmitsburg, MD: U.S. Fire Administration, July 1997.

Ackland, P. J., and A. E. Willey. *Fire Investigation in Commercial Kitchen Systems*, Summerland, B.C.: Philip Ackland Holdings Ltd., 2005.

Anon. "Safety Data for Linseed Oil."

Applied Technical Services, Inc.

Babrauskas, V. "Estimating Room Flashover Potential," *Fire Technology* 16, No. 2 (May 1980): 94–103.

Babrauskas, V., "Pyrophoric Carbon. . . The Jury Is Still Out," *Fire and Arson Investigator*, Vol. 51, No. 2, 12–14 Jan. 2001.

Cuzzillo, B. R., P. J. Pagni, R. B. Williamson, and R. A. Schroeder, "The Verdict Is In: Pyrophoric Carbon Is Out," *Fire and Arson Investigator* 53(1), October 2002, pp. 19–21.

Donahue, M. *Safety and Health Guidelines for Fire and Explosion Investigators*. Stillwater, OK: Fire Protection Publications, 2002.

Ohlemiller, T. J. "Smoldering Combustion," *The SFPE Handbook of Fire Protection Engineering*. Quincy, MA: Society of Fire Protection Engineers, 1995, Section 2, Chapter 11.

Bennett, C. O., and J. E. Myers. *Momentum, Heat, and Mass Transfer*, 2nd edition, New York: McGraw-Hill Book Co., 1974.

Fang, J. B., and J. N. Breese. "Fire Development in Residential Basement Rooms." National Bureau of Standards, NBSIR 80-2120, Gaithersburg, MD, 1980.

Guide to Plastics. New York: McGraw-Hill, 1989.

Tan, S. H. "Flare System Design Simplified," *Hydrocarbon Processing*, 1967, pp. 172–176.

Flick, E. W. *Industrial Solvents Handbook*, 5th edition. Park Ridge, NJ: Noyes Data Corp., 1998.

C.3 References for Extracts in Informational Sections. (Reserved)

Index

-E-

-F-

-G-

-H-

-I-

-N-

-O-

-P-

-R-

-S-

Cou/W 1 2 3 4 5 6 14 13 12 11

Sequence of Events Leading to Issuance of an NFPA Committee Document

Step 1: Call for Proposals

- Proposed new Document or new edition of an existing Document is entered into one of two yearly revision cycles, and a Call for Proposals is published.

Step 2: Report on Proposals (ROP)

- Committee meets to act on Proposals, to develop its own Proposals, and to prepare its Report.
- Committee votes by written ballot on Proposals. If two-thirds approve, Report goes forward. Lacking two-thirds approval, Report returns to Committee.
- Report on Proposals (ROP) is published for public review and comment.

Step 3: Report on Comments (ROC)

- Committee meets to act on Public Comments to develop its own Comments, and to prepare its report.
- Committee votes by written ballot on Comments. If two-thirds approve, Report goes forward. Lacking two-thirds approval, Report returns to Committee.
- Report on Comments (ROC) is published for public review.

Step 4: Technical Report Session

- "*Notices of intent to make a motion*" are filed, are reviewed, and valid motions are certified for presentation at the Technical Report Session. ("Consent Documents" that have no certified motions bypass the Technical Report Session and proceed to the Standards Council for issuance.)
- NFPA membership meets each June at the Annual Meeting Technical Report Session and acts on Technical Committee Reports (ROP and ROC) for Documents with "certified amending motions."
- Committee(s) vote on any amendments to Report approved at NFPA Annual Membership Meeting.

Step 5: Standards Council Issuance

- Notification of intent to file an appeal to the Standards Council on Association action must be filed within 20 days of the NFPA Annual Membership Meeting.
- Standards Council decides, based on all evidence, whether or not to issue Document or to take other action, including hearing any appeals.

Committee Membership Classifications

The following classifications apply to Technical Committee members and represent their principal interest in the activity of the committee.

M *Manufacturer:* A representative of a maker or marketer of a product, assembly, or system, or portion thereof, that is affected by the standard.

U *User:* A representative of an entity that is subject to the provisions of the standard or that voluntarily uses the standard.

I/M *Installer/Maintainer:* A representative of an entity that is in the business of installing or maintaining a product, assembly, or system affected by the standard.

L *Labor:* A labor representative or employee concerned with safety in the workplace.

R/T *Applied Research/Testing Laboratory:* A representative of an independent testing laboratory or independent applied research organization that promulgates and/or enforces standards.

E *Enforcing Authority:* A representative of an agency or an organization that promulgates and/or enforces standards.

I *Insurance:* A representative of an insurance company, broker, agent, bureau, or inspection agency.

C *Consumer:* A person who is, or represents, the ultimate purchaser of a product, system, or service affected by the standard, but who is not included in the *User* classification.

SE *Special Expert:* A person not representing any of the previous classifications, but who has a special expertise in the scope of the standard or portion thereof.

NOTES:

1. "Standard" connotes code, standard, recommended practice, or guide.
2. A representative includes an employee.
3. While these classifications will be used by the Standards Council to achieve a balance for Technical Committees, the Standards Council may determine that new classifications of members or unique interests need representation in order to foster the best possible committee deliberations on any project. In this connection, the Standards Council may make appointments as it deems appropriate in the public interest, such as the classification of "Utilities" in the National Electrical Code Committee.
4. Representatives of subsidiaries of any group are generally considered to have the same classification as the parent organization.

NFPA Document Proposal Form

NOTE: All Proposals must be received by 5:00 pm EST/EDST on the published Proposal Closing Date.

For further information on the standards-making process, please contact the Codes and Standards Administration at 617-984-7249 or visit www.nfpa.org/codes.

For technical assistance, please call NFPA at 1-800-344-3555.

FOR OFFICE USE ONLY

Log #:

Date Rec'd:

Please indicate in which format you wish to receive your ROP/ROC ☐ **electronic** ☐ **paper** ☒ **download**
(Note: If choosing the download option, you must view the ROP/ROC from our website; no copy will be sent to you.)

Date April 1, 200X **Name** John J. Doe **Tel. No.** 716-555-1234

Company Air Canada Pilot's Association **Email**

Street Address 123 Summer Street Lane **City** Lewiston **State** NY **Zip** 14092

******If you wish to receive a hard copy, a street address MUST be provided. Deliveries cannot be made to PO boxes.***

Please indicate organization represented (if any)

1. (a) NFPA Document Title National Fuel Gas Code **NFPA No. & Year** 54, 200X Edition

(b) Section/Paragraph 3.3

2. Proposal Recommends (check one): ☐ new text ☒ revised text ☐ deleted text

3. Proposal (include proposed new or revised wording, or identification of wording to be deleted): [Note: Proposed text should be in legislative format; i.e., use underscore to denote wording to be inserted (<u>inserted wording</u>) and strike-through to denote wording to be deleted (~~deleted wording~~).]

Revise definition of effective ground-fault current path to read:

3.3.78 Effective Ground-Fault Current Path. An intentionally constructed, permanent, low impedance electrically conductive path designed and intended to carry ~~underground~~ <u>electric</u> fault <u>current</u> ~~conditions~~ from the point of a ground fault on a wiring system to the electrical supply source.

4. Statement of Problem and Substantiation for Proposal: (Note: State the problem that would be resolved by your recommendation; give the specific reason for your Proposal, including copies of tests, research papers, fire experience, etc. If more than 200 words, it may be abstracted for publication.)

Change uses proper electrical terms.

5. Copyright Assignment

(a) ☐ **I am the author of the text or other material (such as illustrations, graphs) proposed in the Proposal.**

(b) ☒ **Some or all of the text or other material proposed in this Proposal was not authored by me. Its source is as follows:** (please identify which material and provide complete information on its source)

ABC Co.

I hereby grant and assign to the NFPA all and full rights in copyright in this Proposal and understand that I acquire no rights in any publication of NFPA in which this Proposal in this or another similar or analogous form is used. Except to the extent that I do not have authority to make an assignment in materials that I have identified in (b) above, I hereby warrant that I am the author of this Proposal and that I have full power and authority to enter into this assignment.

Signature (Required)

PLEASE USE SEPARATE FORM FOR EACH PROPOSAL

Mail to: Secretary, Standards Council · National Fire Protection Association
1 Batterymarch Park · Quincy, MA 02169-7471 OR
Fax to: (617) 770-3500 OR Email to: proposals_comments@nfpa.org

NFPA Document Proposal Form

NOTE: All Proposals must be received by 5:00 pm EST/EDST on the published Proposal Closing Date.

For further information on the standards-making process, please contact the Codes and Standards Administration at 617-984-7249 or visit www.nfpa.org/codes.

For technical assistance, please call NFPA at 1-800-344-3555.

FOR OFFICE USE ONLY
Log #:
Date Rec'd:

Please indicate in which format you wish to receive your ROP/ROC ☐ **electronic** ☐ **paper** ☐ **download**
(Note: If choosing the download option, you must view the ROP/ROC from our website; no copy will be sent to you.)

Date ______ **Name** ______ **Tel. No.** ______

Company ______ **Email** ______

Street Address ______ **City** ______ **State** ______ **Zip** ______

******If you wish to receive a hard copy, a street address MUST be provided. Deliveries cannot be made to PO boxes.***

Please indicate organization represented (if any) ______

1. (a) NFPA Document Title ______ **NFPA No. & Year** ______

(b) Section/Paragraph ______

2. Proposal Recommends (check one): ☐ new text ☐ revised text ☐ deleted text

3. Proposal (include proposed new or revised wording, or identification of wording to be deleted): [Note: Proposed text should be in legislative format; i.e., use underscore to denote wording to be inserted (inserted wording) and strike-through to denote wording to be deleted (~~deleted wording~~).]

4. Statement of Problem and Substantiation for Proposal: (Note: State the problem that would be resolved by your recommendation; give the specific reason for your Proposal, including copies of tests, research papers, fire experience, etc. If more than 200 words, it may be abstracted for publication.)

5. Copyright Assignment

(a) ☐ **I am the author of the text or other material (such as illustrations, graphs) proposed in the Proposal.**

(b) ☐ **Some or all of the text or other material proposed in this Proposal was not authored by me. Its source is as follows:** (please identify which material and provide complete information on its source)

I hereby grant and assign to the NFPA all and full rights in copyright in this Proposal and understand that I acquire no rights in any publication of NFPA in which this Proposal in this or another similar or analogous form is used. Except to the extent that I do not have authority to make an assignment in materials that I have identified in (b) above, I hereby warrant that I am the author of this Proposal and that I have full power and authority to enter into this assignment.

Signature (Required) ______

PLEASE USE SEPARATE FORM FOR EACH PROPOSAL

Mail to: Secretary, Standards Council · National Fire Protection Association
1 Batterymarch Park · Quincy, MA 02169-7471 OR
Fax to: (617) 770-3500 OR Email to: proposals_comments@nfpa.org

6/09-C